Contemporary Ecology Research in China

Wenhua Li
Editor-in-Chief

# Contemporary Ecology Research in China

Higher Education Press

Springer

*Editor-in-Chief*
Wenhua Li
Institute of Geographic Sciences and Natural
Resources Research
Chinese Academy of Sciences
Beijing
China

Sponsors: State Key Laboratory of Urban and Regional Ecology
Key Laboratory of Ecosystem Network Observation and Modeling,
Chinese Academy of Sciences

ISBN 978-3-662-48374-9 ISBN 978-3-662-48376-3 (eBook)
DOI 10.1007/978-3-662-48376-3

Jointly published with Higher Education Press, Beijing, China
ISBN: 978-7-04-043706-5 Higher Education Press, Beijing, China

Library of Congress Control Number: 2015949402

Springer Heidelberg New York Dordrecht London


Printed on acid-free paper

Springer-Verlag GmbH Berlin Heidelberg is part of Springer Science+Business Media
(www.springer.com)

# Preface

China lies in the east of Asia with a vast expanse and extremely complex geographical and natural conditions. In the territory of China, highly diversified ecosystems with different plant species combinations had been formed and nurtured. These provide us an unusual natural laboratory to discover the rule of nature and find out the way to harmonize the relationship between nature and human beings. Thanks to our ancestors, during over 5000 years' history, the Chinese created a brilliant civilization in which the holistic thought and harmonic relationship between human and nature is the core of national philosophy. On the other hand, China's environment is characterized by its vulnerability and fragility, and the situation varies very much with geographical conditions. Due to the soaring population and the long history of cultivation in addition to the short-sighted approach in exploitation of natural resources, many natural hazards and disasters causing ecosystem degradation and destruction, soil erosion, desertification, pollution, species extinction and lowering of biodiversity, etc. have occurred. This situation is particularly severe since the last decades. Urgent need for ecological studies of China is aroused under such circumstances.

## General Review of the Development of Ecology in China

In the course of China's long history, wealth of knowledge and theories on the interactions between organisms and their environment have accumulated. Looking back at the development of ecology in China, we may roughly divide it into the following four phases, namely: primitive embryonic phase, fundamental ecological study phase, ecosystem study phase, and contemporary ecological development phase.

**Primitive Embryo Phase**
The traditional integrated philosophical thoughts to understand the world have long existed since time immemorial. Quite a number of ancient treaties and books reflect

the great achievements of the ancient Chinese as regards their production and life. For instance, they use the "San Cai" theory to explain the relationship between "the heaven", "the earth" and "human being". The theory of "Ying Yang" has been widely used to study the relationship between different components of the systems. Many excellent ideas in this aspect were recorded in some early outstanding works. For example, the four essays in Lü Shi Chun Qiu (Master Lü's Annals) including "Shang Nong" ("lay stress on agriculture"), "Ren Di" ("Capacity of soil"), "Bian Tu" ("Work and ground") and "Shen Shi" ("Fitness of the Season"), completed in the third century B.C., can be claimed as China's earliest agricultural treaties. The book Qi Min Yao Shu (Important Arts for People's Welfare) written by Jia Sixie of the sixth century, summarized the systematic knowledge of farming, forestry, animal husbandry, side-line production and fishery. Treaties and books on agriculture written over 2000 years have been handed down to the present day. According to incomplete figures, 376 essays and books on agriculture in those 2000 years were published, but many of them were scattered or lost. They form a large set of treatises with deep system and ecological thinking. Primitive but valuable thoughts are important parts of brilliant Huaxia civilization and left us many world important heritages. A few examples include rotation and intercropping systems in agriculture, Dujiangyan Water Conservancy Project, and karez and check dam engineering, dike-ponds, rice-fish integrated system, terrace planting, pastoral nomadism, etc. Ecological thinking and many practices were not only important in the ecological development of China, but also had important influence on the world. Although a wealth of knowledge related to ecology had been accumulated and scattered as a result of research carried out by individual scientists, ecology had been in its embryo phase until the establishment of People's Republic of China in 1949.

**Fundamental Ecological Research Phase**

Ecology as an independent branch of science obtained its rapid development after 1949. At the early stage of development, ecological research focused mainly on autoecology, population, and community ecology. From the very beginning of ecological research in China, Chinese scientists centered their research on practices to solve the crucial problems of the country. These include the exploitation and assessment of the suitability of habitat for cultivation of rubber plantation in tropical areas; the research on rational exploitation and regeneration of forests in northeastern and southwestern mountain regions; the combating with desertification in arid and semiarid land of northwestern China, and the construction of shelterbelt in the Northern China plain; and to establish nature reserves in various regions to conserve natural ecosystems, etc. Important contributions were made by the interdisciplinary scientific expeditions organized by the Chinese Academy of Sciences to carry out comprehensive survey to fill the gap of information including all disciplines of biological and ecological conditions. In this period, many studies were conducted at the community level. Based on the research, a great number of articles and monographs were published. The publication of "Vegetation of China" was a great event in the history of China's ecology. It sums up the achievements of vegetation research covering all the provinces of China and has made the classification of the vegetation systems and the Vegetation Map of China with 1:100 million scale.

During this period, the international exchange was limited to the Soviet Union and other socialist countries. It should be recognized that the joint expedition and training programs had played important roles in the initial stage of the development of ecology in China.

**Ecological Research Phase**

Since the 1960s, Chinese ecologists started their research on the ecosystem or biogeocenoses bases. With the help of experts from the former Soviet Union, China built its first biogeocenological experiment station in the tropical forest of Xishuangbanna, Yunnan Province in the early 1960s. With the suggestions of ecological scientists, China delineated 15 nature reserves. In some of these nature reserves, interdisciplinary observations and researches were conducted by experts of relevant Universities and research institutes. Since the late 1970s, experimental stations were established by the Academy of Sciences, Universities and some other organizations. Fragmental observations and systematical studies on the structure, function, and succession of these ecosystems were started. Although the development of such long-term research was carried out rather slowly and sometimes interrupted by some events, the ecosystem study did not stop and it was regenerated later in the 1980s by upgrading it into a series of nationwide networks to carry out observation and research on ecosystem. Special contribution was given to the construction of Chinese Ecosystem Research Network (CERN) under the Chinese Academy of Sciences. This program was started in 1988 and became prosperous in the 1990s. Now, CERN consists of 40 field research stations for various representative ecosystems involving agriculture, forest, grassland, lake, ocean, and city areas in different part of China. For effective coordination and management of the project, a general secretariat and several coordinating centers for water, soil, atmosphere, and biology and a synthesis center were established.

In 2006 the Ministry of Science and Technology of China began to establish the Chinese National Ecological Research Network (CNERN). Now a national network has been set up, which consists of 51 field observation stations for different types of ecosystems and one comprehensive research center. The project has carried out observations about ecosystem change covering the main region of our county and cross-department research and technology demonstration. Furthermore, the Chinese Forest Ecosystem Research Network (CFERN) has established 33 filed stations since March 2003. At the same time, works in the wetland and desert ecosystem station network also made some progress, relying on related state research projects. At present, the Chinese Wetland and Desert Ecosystem Research Network (CWERN) and CDERN have established five and four field stations respectively. The establishment of ecosystem network raised the study of ecology in China to a new level, which not only provides a long-term platform for the study of macro-scale ecological problems, but also promotes modernization of observation instruments and means. If during the period of IBP, China missed the opportunity of participation, then from the 1970s, during the implementation of the Man and the Biosphere Program (MAB) of UNESCO, China has already become a country with many active participants. The Cooperative Eco-Environmental Research Program (CERP) sponsored by UNESCO through the fund of BMFT of Germany was one of the biggest projects in MAB and played an important role in promoting the progress of ecology and training the personnels. Another important event was the establishment of Ecological Society of China in 1979.

**Ecology for Sustainable Development**

Since the second half of last century, especially during the first decade of twenty-first century, due to the drastic increasing population, escalating needs for consumption and short-sighted approach when exploiting natural resources, a long list of hazards and disasters became glaringly apparent. In exploring the solutions to the problems of sustainable development, contemporary ecological research experienced its revitalization and breakthrough, and a series of characteristic changes have occurred. These include:

- From natural ecosystems to nature-socio-economic-complex systems;
- Expansion of the scope of research both in space and in time;
- From short-term qualitative description to long-term quantitative and experimental research;
- From isolated plot study in limited areas to networking research covering many regions and ecological zones;
- Differentiation of disciplines and emerging new scientific branches in ecology;
- The modernization of research facilities and instruments.

In this period, remote sensing, geographic information system, global positioning system (3S system) are widely implemented for access to accurate information. Some ecological stations have been established as observation towers equipped with many automated continuous observation apparatus. Many manipulative experiments including free air $CO_2$ enrichment, Throughfall Displacement Experiment, Free air temperature enrichment, and ultraviolet-B variation are established to explore responses of ecosystem processes, structure and function to climate change. Isotope tracer method is used to study the rates of gross mineralization, nitrification, immobilization, and consumption of soil N. In addition, mini-rhizotron technique, dilution-plate method and denaturing gradient gel electrophoresis are used to measure root dynamics and species composition of microbial community in microbial ecology studies. Remote sensing satellites to detect the regional greenhouse gas concentrations, the lidar technique to measure atmospheric aerosol concentrations, as well as remote sensing inversion models are wide applied to research carbon cycle. These modern equipments lay a solid foundation for the continuous improvement of ecology.

In this phase, the hot spots of ecological research were concentrated on biodiversity conservation, global change, and sustainable development. These researches have been listed in the National Key Fundamental Research Program and were implemented by research institutes, universities, and related ministries/

administrations. With the advent of the twenty-first century, ecosystem rehabilitation, environmental protection, and afforestation had become an urgent and enormous strategic task and a fundamental plan for China in her seeking for survival and development for the nation and a series of key ecological construction programs had been implemented. These programs included the Natural Forestry Protection Program (NEPP), Conversion of Cropland to Forest Land, Sheltbelt Development Program in such regions as China's "Three-North Region" and the Yangtze River Basin areas; the Desertification Control Program in the vicinity of Beijing and Tianjin; The Wildlife Conservation and Nature Reserves Development Program; the Fast-growing and High-yield Timber Plantation Program. The above-mentioned programs have been incorporated into the national economic and social development plan and thus the ecology had increasingly become a bridge linking science and development. It is necessary to point out that the Chinese Society of Ecology, the National Natural Science Foundation of China, the National Committee of IGBP, and the National Committee of IHDP of China have played important roles in coordinating and promoting the ecological research in these fields.

## Brief Introductory Description and Analysis on Selected Areas of Ecological Research

In recent years, the ecological research developed rapidly and covered a wide range of topics. In this paper, we just select a few of them to give more detailed introduction. These include biodiversity conservation, ecological research on global change, restoration of degraded ecosystems, and desertification control and promotion of sustainable development from concept to action, etc.

### Biodiversity Conservation

China is a country with highly diversified fauna and flora. Since the funding of new China, much research work has been undertaken on domestication, cultivation, and breeding of wild plants and animals. The Chinese government attaches great importance to wildlife conservation and has organized a series of nationwide surveys focusing on conservation of wildlife and ecosystems. Much information and a great number of data have been accumulated. With the support of UNEP, and funded by GEF, a country-wide study on "China's Biodiversity" was carried out. It was a study of China's plants, animals, and microorganisms and their environment on the land and its marine waters. A comprehensive summary of accumulated information about biodiversity conservation related to agriculture, forestry, animal husbandry and fisheries, as well as the various activities involving many regions over China related to biodiversity conservation and evaluations of their economic, ecological, and social effects. More than 80 experts took part in the compilation and appraisal of the national report.

On the basis of long-term studies, a biodiversity-related database has been developed. In order to facilitate the study on flora and fauna, Chinese Academy of Sciences (CAS) has established a specimen collection system with 21 preservation units, containing 16.23 million specimen accessions. Based on the specimen data, CAS has established a comprehensive biodiversity information system (CBIS).

Since the beginning of the Eighth Five-year Plan, the Ministry of Science and Technology has taken the study of biodiversity conservation as a national key research program, and relevant departments have organized many researches on artificial of endangered animals and plants. During the period of the "Eighth Five-year Plan" (1991–1995), status quo and causes of damage to crucial ecosystems, including forest, grassland, freshwater, and coral reef ecosystems, have been carried out. A series of new methodologies of conservation biology such as population viability analysis, DNA sequence analysis, and others has been implemented in evaluating the threatened status of important species and its endanger mechanisms. The result of such research provided a scientific basis for biodiversity conservation especially for conservation of rare and endangered species as well as the life-supporting ecosystems. During the period of the "Ninth Five-year Plan" (1996–2000), the projects like "Study on Biodiversity Conservation in Key Areas of China" and "Biodiversity Changes, Sustainable Use and Regional Ecological Security in the Yangtze River Basin" have been included in the national key program of basic research.

Until 2008, China had founded 2538 natural reserves with a total area of 148.94 million ha, accounting for 15.13 % of the total national land territory of China. It has now formed a well-coursing national natural reserve network. The gross area of the protected areas reaches to 17 % of the total territory land of China which includes forest parks, cultural and natural heritages, scenery areas, wetland parks, geological parks. Besides, an integrated ex-situ conservation system has also been built in China, which includes zoos, botanical gardens, arboretums, wildlife-breeding bases, crop germplasm banks, and animal germ cell banks.

Based on the ecological research, China has preliminarily established a system of biodiversity conservation policies and regulations, including "Law of Wildlife Protection" (1988), "Regulation on Nature Reserves" (1994), "Regulation of Wild Plant Protection" (1997), "Regulations on Administration of Import and Export of Endangered Wild Animals and Plants" (2006) and so on. In order to implement "the Convention on Biological Diversity(CBD)", the State Council established the CBD Implementation Steering Committee in 1993, headed by the Ministry of Environmental Protection (MEP) with 24 ministerial members. It is followed by "the Joint-ministerial Conference for Species Resources Protection and Management" founded in 2003 with 17 ministries, coordinated by MEP. Meanwhile, an affiliated scientific body of "National Expert Commission for Biological Species Protection" was established. Besides, China National Biodiversity Protection Action Plan was issued by the Chinese government in 1994 and the updated National Biodiversity Strategy and Action Plan was officially issued in 2010.

**Global Change and Global Ecology**

Research on alleviation and adaptation to global change has been a key area of global change science since the end of the twentieth century. Chinese ecologists have also actively taken part in relevant researches, such as the responses and acclimation of ecosystem to global change, ecosystem C storage, processes involved in the C cycle and their mechanisms, the observation and simulation of ecosystem carbon, nitrogen and water fluxes, and the interaction between human activities and global change, etc.

Chinese scientists have suggested to set up the Northeast China Temperate-Grassland Transect in 1993, which was officially registered as the Fifth Transect of the Global Change and Terrestrial Ecosystems (GCTE) by IGBP in 1994. Later, an assumption of creating the transect from the polar tundra to rain forests along the east coasts of Eurasia Continent was promoted and its sector in China was named the North-South Transect of Eastern China (NSTEC). In 2002, the conception of China Grassland Transect (CGT) was proposed after the extensive researches on terrestrial ecosystem C cycle and its driving mechanism in China. It covers the main grassland types across the Northeast Plain, Inner Mongolia Plateau, Loess Plateau and hinterland of Qinghai-Tibet Plateau from Northeast to Southwest of China. Recently, an initiative has been made in cooperation with Euro-Asian Continental Grassland Transect (EACGT) with AsiaFlux. An integrated observation and research network platform for ecosystem and global change in Asia has been established, with the carrying out and achievement of the A3 foresight program (CarbonEastAsia) among ChinaFLUX, JapanFlux, and KoFlux.

Research on ecosystem C storage, process and mechanism of C cycle, and C budget in China started basing on the inventory of vegetarian productivity, biomass, soil census and experiments on organic fertility effect in the early 1960s. Feng Zongwei, for the first time, quantified the biomass yield of main forest ecosystem types in China and characterized its distribution pattern in 1999, by comprehensively summarizing and analyzing the research data accumulated since the 1960s. Hence, a large number of studies assessed the C budget and C storage of forests as well as grassland ecosystems in China by using the data of soil census and forest inventory data. Recently, Piao Shilong analyzed the current terrestrial carbon balance of China and the mechanisms of the involved processes during the 1980s and 1990s using three different methods: estimation of biomass and soil carbon inventories extrapolated by satellite greenness measurements, ecosystem models, and atmospheric inversions.

The Chinese Terrestrial Ecosystem Flux Observation and Research Network (ChinaFLUX) was established in 2001. A basic system of flux observations based on eddy covariance technique and gas chromatography has been developed. Continuous measurements on C and water vapor fluxes over typical terrestrial ecosystems have been made in the last 10 years. New efforts have also been made to make observations of atmospheric N deposition, biological N fixation, and the stable isotope nitrogen flux. ChinaFLUX has obtained long-term observation data of $CO_2$ fixation through photosynthesis, nitrogen fixation by bean root modules through biological processes, water evaporation through evapotranspiration, and energy fluxes in typical terrestrial ecosystems in China. The vegetation and soil inventory data around each flux site have also been collected. A series of datasets have been produced, including the spatiotemporal dynamic dataset and atlas of major climatic and meteorological factors, land resources, and LUCC since 1980 in China. Basing on the above datasets and modeling system, the spatiotemporal pattern of C sink or source of typical forest and grassland ecosystems in China were quantitatively evaluated, and the environmental and biological driving mechanisms of temperature, precipitation, radiation, and leaf area index on the C budget of different terrestrial ecosystems in the Asian Monsoon area were also investigated.

The research on C cycle simulation started later in China than that in European and American countries, but it progressed rapidly. In recent years, Chinese scientists have adopted and improved several terrestrial ecosystem C cycle models from abroad, such as CEVSA, GLO-PEM, BEPS, EALCO, etc. They also compared the performance of SIB2, BIOME-BGC, and BIOME3 models in simulating China's terrestrial C budget. Furthermore, Chinese scholars have also developed their own C cycle models for different ecosystem types; for example, the C budget model of forest ecosystems in China (FORCCHN), the biophysical process-based model for C budget in agro-ecosystem (Agro-C), the C budget model of grass ecosystems in China (DCTEM) based on IBIS model, and the atmosphere-vegetation interaction model version 2 (AVIM2). The rapid development and improvement of these models greatly promoted the simulation of C and water budget in China's terrestrial ecosystem.

Chinese scientists have carried out a series of experiments to test the ecosystem's responses to different warming, precipitation, $CO_2$ enrichment, and N deposition conditions. In 2007, National Natural Science Foundation of China (NSFC) launched a key research project focusing on the response and adaptation of China's typical terrestrial ecosystem to climate change and integrated the major observation sites in China. Important findings were obtained about the short-term response and long-term adaptation of ecosystem C and N processes, biological diversity, and ecosystem patterns to climate change. Currently, under experiments control were made focusing on the grasslands in Qinghai-Tibet Plateau and Inner Mongolia, where ecosystems are sensitive to climate change. Lately, simulation experiments on the effects of N deposition, precipitation change, and atmospheric $CO_2$ enrichment on plant growth have been carried out at the major forest ecosystems in eastern China.

**Combating Desertification and Erosion Control**

China is one of the countries most severely affect by desertification and soil erosion. According to preliminary estimation, 37.2 % of the country's territory—some 3.57 million km$^2$—is classified as drylands (including arid, semisarid and dry sub-humid arid areas). Of the drylands, 2.64 million km$^2$ falls under the category of desertified land in accordance with the definition of the UNCCD, and these desertified lands are distributed in 18 provinces and account for 27.5 % of the country's landarea. The desertified lands of China can be attributed to wind erosion, water erosion, salinization, and freezing-thawing processes. The life of over 400 million residents are affected by desertification, and the direct economic losses per year exceed 64 billion yuan.

China's desertification mitigation efforts began in the late 1950s. Through a number of high-profile programs, such as the Three-North's Shelterbelt Development Program was initiated in 1978, the National Program on Combating Desertification was initiated in 1990, the Sandification Control Program for Areas in the Vicinity of Beijing and Tianjin was launched in 2000, and the Conversion of Croplands to Forests and Grasslands Program initiated in 2000, the Government of China has poured on average 0.024 % of the country's annual gross domestic product (GDP) into desertification mitigation efforts and, as a result, some 20 % of the desertified lands have been brought under control.

Based on the result of scientific research, the scientists in field ecology put forward the following suggestions for further improvement the situation for policy making: (i) expanding the previous sectoral perspective to embrace a multi-stakeholder approach; (ii) setting priority zones within the restorable area, and establishing National Special Eco-Zones (e.g., forest farms, protected areas, and headwater areas); (iii) restructuring the state anti-desertification investment portfolio by changing the government direct investment in tree plantations to government acquisition of planted/greened areas; and (iv) introducing preferential policies against sandy-desertification, such as permitting land tenures for up to 70 years and compensating for ecological services.

China is one of the nations with most serious soil erosion in the world. According to the second national remote sensing soil erosion survey in the 1990s, the area of soil erosion and wind erosion reached 3.556 million $km^2$. About 50 million tons of soil would be lost every year.

China, as a country with an ancient civilization, has a long history of soil and water conservation. Having summarized the experiences in history, considering the current situation of the country, a soil erosion prevention system with Chinese characteristics was formulated, mainly based on small watershed comprehensive management. Small watershed management has been carried out in 27 provinces, autonomous regions and municipalities. More than 9800 small watershed areas are made control, with a total area of nearly 40 million $km^2$, of which 22 million $km^2$ are soil erosion area. Nearly 3000 small river basin control projects have been completed.

Although the worsening trend of soil erosion in China has been controlled, water and soil erosion areas are still facing high population, densities, and environmental pressures. Several countermeasures are being taken: (i) recognizing that "soil erosion is a serious ecological degradation" and soil and water conservation as a national policy; (ii) implementing integrated technology systems; (iii) strengthening scientific research on soil and water conservation; (iv) strengthening the legal construction of soil and water conservation; (v) enhancing soil and water conservation; and (vi) constructing soil and water conservation mechanisms, especially investment practices, compensation, incentive, and restriction mechanism.

**The Ecosystem Study on the Qinghai-Tibetan Plateau**

The Qinghai-Tibetan Plateau, also called "The Third Pole", is a unique geographic unit with an average altitude over 4000 m. The plateau is one of the biodiversity hotspots over the world; on the other hand, it is the most fragile region due to its extreme climate and habitats.

The ecological research dated back to the natural resources survey since the 1950s. The large-scale multi-disciplinary scientific expeditions in Chinese Himalayas and Hengduan Mountains, were organized and carried out by the Chinese Academy of Sciences from 1973 to 1979. A wealthy of knowledge obtained in understanding the ecological conditions and distribution of vegetation and the productivity of ecosystems. The results of such expeditions were summarized in the book Geological and Ecological Studies on the Qinghai-Xizang Plateau. In 1980, an International Symposium on the Qinghai-Tibetan Plateau was held in Beijing. It was a milestone of international cooperation in ecological research in the high mountain regions. The Commission of Integrated Survey of Natural Resources, Chinese Academy of Sciences, played an important role in ecological research on the plateau in collaboration with international organizations such as International Center for Integrated Mountain Development in Nepal and the Woodland Mountain Institute, West Virginia, USA and International Development Research Center in Canada. In the period of "the Eighth-Five Year Plan" (1991–1995), the national basic research program of "Formation, Evolution, Environmental Change and Ecosystem Research on the Qinghai, Tibetan Plateau" was initiated. Some long-term ecological research stations, for example Gonggashan Mountain Forest Station in west Sichuan, Haibei Alpine Grassland Ecosystem in Qinghai, and Lhasa Agricultural Ecosystem Research Station were founded as research bases for ecosystem study. As a consequence, a book Ecosystems of Qinghai-Xizang (Tibetan) Plateau and Approach for Their Sustainable Management was jointly published to summarize the results of the research in this period.

Through physiological studies it was discovered that the slow growth and small stature of alpine plants including crops is not associated with low rate of carbon uptake and high loss per unit of tissue. The carbon fixation rates were not lower than those at lower altitudes. On the contrary, the alpine plants have higher capacity for carbon assimilation in comparison with their lower populations or plant heights. The higher carbon investment into roots, but nothing significant difference in leaf mass ratio, guarantees carbon supply of alpine plants. The duration of active growing season appeared to be the overwhelming factor to constrain plant production. The net primary productions (NPP) of alpine plants in terms of production per growing day are not necessary lower than the low-altitude plants, even for the tropical or subtropical ones. However, NPP at the community level is lower than lower ecosystems due to lower leaf area index (LAI). Low temperature is the prominent factors limiting plant growth rather than photosynthesis, which is indicated by higher nonstructural carbohydrate concentration in alpine plant tissues in comparison with the growing at lower altitudes. Therefore, it is lower carbon use rather than poorer carbon source of alpine ecosystem boundaries such as treeline and grassland upper limit. The results of eddy covariance measurement of net ecosystem $CO_2$ exchange (NEE) indicate that alpine meadow and scrubs have weaker carbon sink or even source. Lower LAI is the key factor to cause lower carbon uptake. Rainfall patterns in the growing season and pulse rainfall in the beginning and at the end of the season affect the ecosystem respiration and thereby the carbon balance of ecosystem. Soil respiration is the main component of ecosystem respiration. Although temperature is the key factor controlling respiration, plant phenology modifies the temperature dependence of soil respiration in the season. Seasonal distribution of precipitation greatly affects the sink–source relation. Grazing and reclamation are the main disturbances that cause carbon loss due to enhancement of soil respiration. On the contrary, enclosure-pasturing system and recovery of degraded grassland might alleviate carbon emission and even accelerate carbon sequestration.

Global warming tended to be accelerated in the past decades in the high alpine area. Experimental warming caused rapid species extinction in alpine meadow, which was dampened by simulated grazing. Higher species losses occurred at the drier sites where N was less available. The indirect effect of climate warming on species richness was mediated by plant–plant interactions. Heat stress and warming-induced litter accumulation are potential causes. Grazing might reduce the risk of loss of species under the global warming scenarios.

Addition of nitrogen increases productivities of alpine N-limited ecosystems, but simulated nitrogen deposition often causes loss of species not only via increase of competition between plants species, but between plant and microorganisms. $CO_2$ enhancement ecosystem increases number of wheat tillers and stochastic photosynthetic rate of leaves. However, the wheat yields and biomass are not necessarily increased partly due to nitrogen supply shortage while the growing period.

**Promoting Sustainable Development from Concept to Action**

Unlike other biological communities, human society is a kind of artificial ecosystem dominated by human behavior, sustained by natural life support system, and vitalized by ecological process. It was named as Social-Economic-Natural Complex Ecosystem (SENCE) by Ma Shijun in 1984. Its natural subsystem consists of the five elements of Chinese tradition: metal (minerals), wood (living organism), water, fire (energy), and soil (nutrients and land). Its economic subsystem includes the essentials factors of production, consumption, reduction, transportation, and regulation. While its social subsystem includes technology, institution, and culture, the scientific study on sustainability in China is to coordinate the temporal, spatial, structural, and functional relationships among and within the three subsystems.

Grounded in ancient Chinese human ecological philosophy and SENCE approach, a campaign of Ecopolis development has been undergoing in some Chinese cities and towns since the 1990s. Ecopolis is a kind of administrative unit having productive and ecologically efficient industry, systematically responsible and socially harmonious culture, and physically beautiful and functionally vivid landscape. It is a kind of adaptive process toward sustainability through cultivating five facets of eco-sanitation, eco-security, eco-landscape, eco-industry, and eco-culture. The essential idea of ecopolis development is to plan, design, manage, and construct the ecosystem's function of production, living, and sustaining according to ecological cybernetics. It is a healthy process toward sustainable development within the carrying capacity of local ecosystem through changing the mode of production, consumption behavior, and decision instruments with three legs of Circular Economy, Harmonious Society, and Safe Ecology. Integration, demonstration, citizen's participation, and scientist's and technician's catalyzing are the keys in its development. The term "ecopolis" is used to imply an ecologically sound city region and its immediate periphery. The development of ecopolis needs five ways of motivation, i.e., administrative authorization, scientific supervision, industrial sponsorship, participation by the community, and motivation by media. There are four development stages: concept initiation and comprehensive planning, ecoscape planning and legislation, development through eco-engineering design, and ecosystem monitoring and management.

During the past two decades, 510 eco-demonstration zones have been appraised and named by the Ministry of Environmental Protection, 14 provinces/autonomous regions/directly governed city regions (Hainan, Jilin, Heilongjiang, Fujian, Zhejiang, Shandong, Anhui, Jiangsu, Inner Mongolia, Shanxi, Hebei, Guangxi, Sichuan, and Tianjin) were approved to carry on eco-province development. By February 2010, there were already 108 state comprehensive experimental districts toward sustainability approved by the Ministry of Science and Technology, including towns, counties, middle/small-size cities, and districts of large cities distributed in 93 % of China's provinces and autonomous regions.

Compared with foreign countries, China's ecopolis development is rather through top-down than bottom-up encouragement. The advantages of this way is that if the decision makers are smart enough, the ecopolis plan will be strongly implemented; otherwise, it will be just an oral promise or an utopian ideal. While main lessons and challenges are also gained such as institutional barrier, behavioral bottleneck, and technical malnutrition, the Sino-Singapore ecocity planning in Tianjin, and a quite few other cases of ambitious ecocity planning in China, show the public a dream of a sustainable city. To realize it, however, an adaptive process is needed to meet the local natural and human ecological condition needs, to reshape our production mode, consumption behavior, the goal of development, and the meaning of life, to reform the fragmented institution in legislation, organization, governance, decision making, planning and management, and to renovate the reductionisms based and chain-linked technology.

After entering the new century, eco-agricultural development entered into its new stage. There are some mile stones like the publication of Agro-ecological Farming Systems in China by Parthenon Publishing in 2001 and Eco-agriculture—Theories and Practices of China's Sustainable Agriculture by (Beijing) Chemistry Press. These two monographs, all compiled by Li Wenhua together with nearly 100 specialists in eco-agriculture, got excellent responses from both in China and abroad.

In 2005, The Food and Agriculture Organization (FAO) of the United Nations launched a program on the Conservation of Globally Important Agricultural

Heritage Systems (GIAHS). FAO defined GIAHS as "remarkable land use systems and landscapes which are rich in globally significant biological diversity evolving from the co-adaptation of a community with its environment and its needs and aspirations for sustainable development". The Traditional Rice-fish Culture in Qingtian County, Zhejiang Province, was selected by FAO as one of the first group of pilot sites of the GIAHS project. Up to now, some researches on biocultural diversity conservation, eco-tourism development, eco-compensation mechanism, organic agricultural development, and multi-participatory process on the pilot site were carried out and many papers published.

**Evaluation of Ecosystem Services and Eco-Compensation**

Since the middle 1990s, global assessment of ecosystem services and the launching of the Millennium Ecosystem Assessment (MA) on a regular basis at national and sub-national scales have attracted great attention and interest of Chinese scholars in joining the cooperation with international organizations and conducting the assessment work at home.

In China, since the late twentieth century, researchers have done some preliminary explorations on the evaluation theory, methods, and applications of ecosystem services. Especially after entering the twenty-first century, a lot of works on the evaluation of various ecosystems in different regions were done. Ecosystem services of different spatial scales such as national, watershed, provincial, municipal, and district have been studied, as well as different natural and artificial ecosystems such as forest, grassland, farmland, wetlands, desert, marine and reservoirs, etc. At the same time, much attention has been paid to some single functions and services like water and soil conservation, carbon sequestration and oxygen release, air purification and landscapes, too. These achievements have played important roles in decision-making process related to ecosystem management and environmental protection in China.

Recently, scientists in the field of forestry worked on the standardization of the indicators of ecosystem service assessment and the value of ecosystem services were calculated on the basis of the forest inventory data, which covers all the forests in China.

In 2008, a Joint US-China Center on Ecosystem Services (JUCCES) has been established. The establishment of the Center was based on a series of discussions and meetings between National Service Foundation of China (NSFC), Chinese Academy of Sciences (CAS), and National Science Foundation of USA (NSFA), aiming to establish a long-term mechanism for promoting cooperative research and exchange information in the field of eco-service and eco-compensation studies. The mission of JUCCES is to provide a platform for facilitating collaboration, communication, and coordination between scientists engaged in studies on natural resources, ecosystem services, eco-compensation, impacts of ecosystem changes on human well-being, and response options for developing a harmonious relationship between human being and nature.

Research and practice on eco-compensation mechanism are of great strategic significance in the new period of implementing full-scale scientific development approach and setting up of harmonious society in China, which is not only an effective way to deal with the severe environmental problems but an important measure to promote the environmental protection policy system, as well as to adjust the relationship of related stakeholders, to coordinate the regional development and to enhance the equity of the whole society. Since the 1970s, some theoretical and practical studies were carried out on eco-compensation in China, and now it is a hot issue attracting wide attention of all social circles. On the one hand, lots of local governments started to conduct pilot projects, actively exploring relevant experiences. On the other hand, the researches of the academic circles have entered a new era, transiting from researches on the theory and methodology of quantifying the value of eco-service function to the research on policy design of eco-mechanism, including the general framework, compensation principles, approaches and standards, and some pilot studies in key fields such as watershed eco-compensation, forestry eco-compensation, nature reserve eco-compensation, and mineral exploitation compensation.

## Challenges and Perspectives of Development of Ecology in China

### Challenges

Although much progress has been made in ecology, there are a number of weak-points in its development. These include:

- The insufficient basic research and lack of innovation in theoretical studies of ecology;
- Weakness in accumulation of knowledge to meet the immediate need to solve the ecological problems of the country;
- Weakness in participating international cooperation and active involvement in global issues;
- Ignorance of respect and development of indigenous knowledge in ecology;
- Separation in advanced concept with broad practices by people;
- Inadequate experience in integration of interdisciplinary research;
- Necessity for improvement of the mechanism system in scientific management.

### Strength

Having recognized the weakness in our ecological development, we are fully confident of the prosperity of its further development. This is due to the diversified natural conditions and rich biological resources; the Chinese have acquired the primary managerial experience of ecosystem, and accumulated a great amount of research data for ecology. Besides, the Chinese Ecosystem Research Network (CERN) was established, the research equipments have been greatly improved and international corporations have made great progress in ecology research in the recent years. More importantly, the enhancement of comprehensive national power, the increasing focus on the ecological constructions, and the public ecological awareness laid a solid foundation for the development of ecology in China.

**The Ways Forward**

The guiding thought of development of ecology has been formed from the mediums term plan as the following:

- Strengthening of the fundamental theoretical study focused on frontiers in ecology;
- Active participation in resolving the eco-environment problems in the social economic development and environment protection;
- Facilitating ecological study with modern scientific theories and methods; make discoveries and upgrade traditional and indigenous knowledge;
- Strengthening international cooperation and self-innovative creation in ecological study;
- Development of medium and long-term plan to coordinate and guide ecological research in China;
- Strengthening education and training of cultivate scientists in the field of ecology and popularize ecological knowledge to the public at various levels.

Based on the guiding principles listed above, considering the current situation of China, the following key problems should be taken as the priorities for further studies.

- Ecological assets accounting and ecological compensation;
- Repairing and reconstruction of degraded ecosystems;
- Comprehensive control and prediction of regional pollution;
- Ecological and environmental risk management;
- Research and development of circular economy;
- Compliance support for the control of global changes.

## Conclusion

In conclusion, I would like to say that the above-mentioned ecological research and achievements are just a small part of the various works done by Chinese scientists. I wish to express my sincere thanks to those who assisted me to prepare this paper. I would be satisfied if it can give you a general picture about what we have done and what we are going to do from my personal point of view. I do believe that with the unique and diverse biological and physical conditions in China and the rich experiences accumulated in thousands of years in ancient China civilization, with the joint efforts of our ecological research team at home in close cooperation with acknowledged scientists abroad, we are able to make responsible contribution to develop ecology with our own characters and for the welfare of human beings.

Beijing, China
March 2015

Wenhua Li

# Contents

# Contributors

**Shuqing An** School of Life Sciences, Nanjing University, Nanjing, China

**Yanying Bai** Institute of Geographic Sciences and Natural Resources Research, Chinese Academy of Sciences, Beijing, China

**Chao Bao** Institute of Geographic Sciences and Natural Resources Research, Chinese Academy of Sciences, Beijing, China

**Weikai Bao** Key Laboratory of Mountain Ecological Restoration and Bio-resource Utilization, Chengdu Institute of Biology, Chinese Academy of Sciences, Chengdu, Sichuan, China

**Lei Cai** Department of Nature and Ecology Conservation, Ministry of Environmental Protection, Beijing, China

**Jianhua Cao** Institute of Karst Geology, Chinese Academy of Geological Sciences, Guilin, China

**Shuyan Cao** Institute of Geographic Sciences and Natural Resources Research, Chinese Academy of Sciences, Beijing, China

**Guangting Chen** Research Institute, Chinese Academy of Sciences, Cold and Arid Regions Environmental and Engineering, Lanzhou, China

**Liding Chen** Research Center for Eco-Environmental Sciences, Chinese Academy of Sciences, Beijing, China

**Luzhen Chen** Key Laboratory of the Ministry of Education for Coastal and Wetland Ecosystems, College of Environment and Ecology, Xiamen, Fujian, China

**Panqin Chen** Bureau of Science and Technology for Resources and Environment, Chinese Academy of Sciences, Beijing, China

**Yaning Chen** State Key Laboratory of Desert and Oasis Ecology, Xinjiang Institute of Ecology and Geography, Chinese Academy of Sciences, Urumqi, China

**Bangbo Cheng** Institute of Geographic Sciences Natural Resources Research, Chinese Academy of Sciences, Beijing, China

**Peng Cui** Key Laboratory of Mountain Hazards and Earth Surface Process, Institute of Mountain Hazards and Environment, Chinese Academy of Sciences, Chengdu, China

**Limin Dai** State Key Laboratory of Forest and Soil Ecology, Institute of Applied Ecology, Chinese Academy of Sciences, Shenyang, China

**Yating Dai** Institute of Grassland Research, Chinese Academy of Agricultural Sciences, Hohhot, China

**Dewen Ding** The First Institute of Oceanography, State Oceanic Administration of China, Qingdao, China

**Hui Ding** Nanjing Institute of Environmental Sciences, Ministry of Environmental Protection, Nanjing, China

**Ming Dong** Institute of Botany, Chinese Academy of Sciences, Beijing, China

**Yunshe Dong** Institute of Geographic Sciences and Natural Resources Research, Chinese Academy of Sciences, Beijing, China

**Jie Fan** Resources Research, Institute of Geographical Sciences and Natural, Chinese Academy of Sciences, Beijing, China

**Chuanglin Fang** Institute of Geographic Sciences and Natural Resources Research, Chinese Academy of Sciences, Beijing, China

**Huajun Fang** Institute of Geographic Sciences and Natural Resources Research, Chinese Academy of Sciences, Beijing, China

**Jianguang Fang** Yellow Sea Fisheries Research Institute, Chinese Academy of Fisheries Sciences, Qingdao, China

**Jingyun Fang** Department of Ecology, College of Urban and Environmental Science, Peking University, Beijing, China

**Xiuqi Fang** Beijing Normal University, Beijing, China

**Bojie Fu** Research Center for Eco-Environmental Sciences, Chinese Academy of Sciences, Beijing, China

**Jixi Gao** Nanjing Institute of Environmental Sciences, Ministry of Environmental Protection of China, Nanjing, China

**Lijie Gao** Institute of Urban Environment, Chinese Academy of Sciences, Xiamen, China

**Shuqin Gao** Institute of Botany, Chinese Academy of Sciences, Beijing, China

**Yingzhi Gao** School of Life Sciences, Northeast Normal University, Changchun, China

**Feng Ge** Institute of Zoology, Chinese Academy of Sciences, Beijing, China

**Quansheng Ge** Institute of Geographic Sciences Natural Resources Research, Chinese Academy of Sciences, Beijing, China

**Lanping Guo** National Resource Center of Chinese Material Medical, China Academy of Chinese Medical Science, Beijing, China

**Xuebing Guo** Institute of Geographic Sciences and Natural Resources Research, Chinese Academy of Sciences, Beijing, China

**Zhaodi Guo** Department of Ecology, College of Urban and Environmental Science, Peking University, Beijing, China

**Xinmei Hao** Center for Agricultural Water Research in China, China Agricultural University, Beijing, China

**Daming He** Asian International Rivers Center, Yunnan University, Kunming, China

**Lu He** Institute of Geographic Sciences and Natural Resources Research, Chinese Academy of Sciences, Beijing, China

**Nianpeng He** Institute of Geographic Sciences and Natural Resources Research, Chinese Academy of Sciences, Beijing, China

**Qijin He** Chinese Academy of Meteorological Sciences, Beijing, China

**Yongtao He** Resources Research, Institute of Geographical Sciences and Natural, Chinese Academy of Sciences, Beijing, China

**Xiangyang Hou** Institute of Grassland Research, Chinese Academy of Agricultural Sciences, Hohhot, China

**Huifeng Hu** State Key Laboratory of Vegetation and Environmental Change, Institute of Botany, Chinese Academy of Sciences, Beijing, China

**Zhenqi Hu** Institute of Land Reclamation and Ecological Restoration, China University of Mining and Technology (Beijing), Beijing, China

**Fen Huang** Institute of Karst Geology, Chinese Academy of Geological Sciences, Guilin, China

**Jikun Huang** Center for Chinese Agricultural Policy, Chinese Academy of Sciences, Beijing, China

**Luqi Huang** National Resource Center of Chinese Material Medical, China Academy of Chinese Medical Science, Beijing, China

**Xianzhi Huang** State Key Laboratory of Silkworm Genome Biology, Southwest University, Chongqing, China

**Haiting Ji** School of Life Sciences, Nanjing University, Nanjing, China

**Xiangping Jia** Center for Chinese Agricultural Policy, Chinese Academy of Sciences, Beijing, China

**Jusheng Jiang** Hainan Academy of Agricultural Reclamation Sciences, Haikou, China

**Zhigang Jiang** Institute of Zoology, Chinese Academy of Sciences, Beijing, China

**Wenjun Jiao** Institute of Geographic Sciences and Natural Resources Research, Chinese Academy of Sciences, Beijing, China

**Shaozhong Kang** Center for Agricultural Water Research in China, China Agricultural University, Beijing, China

**Weijing Kong** Chinese Research Academy of Environmental Sciences, Beijing, China

**Fumin Lei** Key Laboratory of Zoological Systematics and Evolution, Institute of Zoology, Chinese Academy of Sciences, Beijing, China

**Chengyun Li** Yunnan Agricultural University, Kunming, China

**Jing Li** Institute of Land Reclamation and Ecological Restoration, China University of Mining and Technology (Beijing), Beijing, China

**Liang Li** Institute of Karst Geology, Chinese Academy of Geological Sciences, Guilin, China

**Long Li** College of Resources and Environmental Sciences, China Agricultural University, Beijing, China

**Nuyun Li** State Forestry Administration, Beijing, China

**Peicheng Li** Research Institute of Water and Development, Chang'an University, Xi'an, China

**Ping Li** Institute of Grassland Research, Chinese Academy of Agricultural Sciences, Hohhot, China

**Pingxing Li** Nanjing Institute of Geography and Limnology, Chinese Academy of Sciences, Nanjing, China

**Qiang Li** Institute of Karst Geology, Chinese Academy of Geological Sciences, Guilin, China

**Qilei Li** Research Institute of Water and Development, Chang'an University, Xi'an, China

**Shenggong Li** Institute of Geographic Sciences and Natural Resources Research, Chinese Academy of Sciences, Beijing, China

**Shidong Li** State Forestry Administration, Beijing, China

**Wenhua Li** Institute of Geographic Sciences and Natural Resources Research, Chinese Academy of Sciences, Beijing, China

**Xiliang Li** Institute of Grassland Research, Chinese Academy of Agricultural Sciences, Hohhot, China

**Xiubin Li** Institute of Geographic Sciences Natural Resources Research, Chinese Academy of Sciences, Beijing, China

**Yunhe Li** State Key Laboratory for Biology of Plant Diseases and Insect Pests, Institute of Plant Protection, Chinese Academy of Agricultural Sciences, Beijing, China

**Zhou Li** Rural Development Institute, Chinese Academy of Social Sciences, Beijing, China

**Biao Liang** Institute of Geographic Sciences and Natural Resources Research, Chinese Academy of Sciences, Beijing, China

**Guanghui Lin** Center for Earth System Science, Tsinghua University, Beijing, China

**Yongming Lin** College of Forestry, Fujian Agricultural and Forestry University, Fuzhou, China

**Zhihua Lin** Zhejiang Wanli University, Ningbo, China

**Guihuan Liu** Academy for Environmental Planning, Beijing, China

**Guofang Liu** Institute of Botany, Chinese Academy of Sciences, Beijing, China

**Haolong Liu** Institute of Geographic Sciences and Natural Resources Research, Chinese Academy of Sciences, Beijing, China

**Jiyuan Liu** Institute of Geographic Sciences and Natural Resources Research, Chinese Academy of Sciences, Beijing, China

**Junhui Liu** State Environmental Protection Key Laboratory of Regional Eco-Process and Function Assessment, Chinese Research Academy of Environmental Sciences (CRAES), Beijing, China

**Moucheng Liu** Institute of Geographic Sciences and Natural Resources Research, Chinese Academy of Sciences, Beijing, China

**Shirong Liu** Chinese Academy of Forestry, Beijing, China

**Shuzhen Liu** Institute of Mountain Hazards and Environment, Chinese Academy of Sciences, Chengdu, China

**Xiaohui Liu** Northeast Institute of Geography and Agroecology, Chinese Academy of Sciences, Changchun, China

**Xingtu Liu** Northeast Institute of Geography and Agroecology, Chinese Academy of Sciences, Changchun, China

**Xu Liu** Chinese Academy of Engineering, Beijing, China

**Xuelin Liu** Beijing Tourism Planning and Design Institute Davos Summit, Beijing, China

**Yan Liu** Nanjing Institute of Environmental Sciences, Ministry of Environmental Protection, Nanjing, China

**Zhijuan Liu** China Agricultural University, Beijing, China

**Baorong Lu** Ministry of Education Key Laboratory for Biodiversity and Ecological Engineering, Department of Ecology and Evolutionary Biology, Fudan University, Shanghai, China

**Chunxia Lu** Institute of Geographic Sciences and Natural Resources Research, Chinese Academy of Sciences, Beijing, China

**Fei Lu** State Key Laboratory of Urban and Regional Ecology, Research Center for Eco-Environmental Sciences, Chinese Academy of Sciences, Beijing, China

**Qi Lu** Institute of Desertification Studies, Chinese Academy of Forestry, Beijing, China

**Shiming Luo** Institute of Tropical and Subtropical Ecology, South China, Agricultural University, Guangzhou, China

**Shihai Lv** State Environmental Protection Key Laboratory of Regional Eco-Process and Function Assessment, Chinese Research Academy of Environmental Sciences (CRAES), Beijing, China

**Xianguo Lv** Northeast Institute of Geography and Agroecology, Chinese Academy of Sciences, Changchun, China

**Lin Ma** Centre for Resources, Environment and Food Security, China Agricultural University, Beijing, China

**Wenqi Ma** College of Resources and Environmental Sciences, Agricultural University of Hebei, Baoding, China

**Wei Meng** Chinese Research Academy of Environmental Sciences, Beijing, China

**Qingwen Min** Institute of Geographic Sciences and Natural Resources Research, Chinese Academy of Sciences, Beijing, China

**Songlin Mu** Beijing Agriculture Information Technology Research Center, Beijing, China

**Xiang Niu** Research Institute of Forest Ecology, Environmental and Protection, Chinese Academy of Forestry, Beijing, China

**Zhiyun Ouyang** Research Center for Eco-Environmental Sciences, Chinese Academy of Sciences, Beijing, China

**Tao Pan** Institute of Geographic Sciences and Natural Resources Research, Chinese Academy of Sciences, Beijing, China

**Shengji Pei** Kunming Institute of Botany, Chinese Academy of Sciences, Kunming, China

**Xia Pei** Institute of Geographic Sciences and Natural Resources Research, Chinese Academy of Sciences, Beijing, China

**Qin Peng** Institute of Geographic Sciences and Natural Resources Research, Chinese Academy of Sciences, Beijing, China

**Shaolin Peng** State Key Laboratory of Biocontrol, School of Life Sciences, Sun Yat-sen University, Guangzhou, China

**Yufa Peng** State Key Laboratory for Biology of Plant Diseases and Insect Pests, Institute of Plant Protection, Chinese Academy of Agricultural Sciences, Beijing, China

**Zongbo Peng** Hainan Academy of Agricultural Reclamation Sciences, Haikou, China

**Yuchun Qi** Institute of Geographic Sciences and Natural Resources Research, Chinese Academy of Sciences, Beijing, China

**Liu Qian** Wageningen University and Research Centre, Wageningen, The Netherlands

**Jian Qin** State Key Laboratory of Silkworm Genome Biology, Southwest University, Chongqing, China

**Pei Qin** Halophyte Research Lab, Nanjing University, Nanjing, China

**Huayong Que** Institute of Oceanology, Chinese Academy of Sciences, Qingdao, China

**Jizhou Ren** State Key Laboratory of Grassland Agro-ecosystems, College of Pastoral and Agricultural Science and Technology, Lanzhou University, Lanzhou, China

**Guofan Shao** Department of Forestry and Natural Resources, Purdue University, West Lafayette, IN, USA

**Haihua Shen** Department of Ecology, College of Urban and Environmental Science, Peking University, Beijing, China

**Yuancun Shen** Resources Research, Institute of Geographical Sciences and Natural, Chinese Academy of Sciences, Beijing, China

**Zhenxi Shen** Resources Research, Institute of Geographical Sciences and Natural, Chinese Academy of Sciences, Beijing, China

**Dongmei Shi** State Key Laboratory of Silkworm Genome Biology, Southwest University, Chongqing, China

**Honghua Shi** The First Institute of Oceanography, State Oceanic Administration of China, Qingdao, China

**Longyu Shi** Institute of Urban Environment, Chinese Academy of Sciences, Xiamen, China

**Peili Shi** Institute of Geographic Sciences and Natural Resources Research, Chinese Academy of Sciences, Beijing, China

**Changchun Song** Northeast Institute of Geography and Agroecology, Chinese Academy of Sciences, Changchun, China

**Changhong Su** Research Center for Eco-Environmental Sciences, Chinese Academy of Sciences, Beijing, China

**Xiaomin Sun** Institute of Geographic Sciences and Natural Resources Research, Chinese Academy of Sciences, Beijing, China

**Yucheng Sun** Institute of Zoology, Chinese Academy of Sciences, Beijing, China

**Lina Tang** Institute of Urban Environment, Chinese Academy of Sciences, Xiamen, China

**Qisheng Tang** Yellow Sea Fisheries Research Institute, Chinese Academy of Fishery Sciences, Qingdao, China

**Xiaoping Tang** Academy of Forest Inventory and Planning, State Forestry Administration, Beijing, China

**Ling Tong** Yellow Sea Fisheries Research Institute, Chinese Academy of Fishery Sciences, Qingdao, China; Center for Agricultural Water Research in China, China Agricultural University, Beijing, China

**FangHao Wan** Institute of Plant Protection, Chinese Academy of Agricultural Sciences, Beijing, China

**Bing Wang** Research Institute of Forest Ecology, Environmental and Protection, Chinese Academy of Forestry, Beijing, China

**Deli Wang** School of Life Sciences, Northeast Normal University, Changchun, China

**Gang Wang** School of Life Sciences, Lanzhou University, Lanzhou, China

**Jinfeng Wang** Research Institute of Water and Development, Chang'an University, Xi'an, China

**Jingsheng Wang** Resources Research, Institute of Geographical Sciences and Natural, Chinese Academy of Sciences, Beijing, China

**Ming Wang** Northeast Institute of Geography and Agroecology, Chinese Academy of Sciences, Changchun, China

**Peijun Wang** Institute of Land Reclamation and Ecological Restoration, China University of Mining and Technology (Beijing), Beijing, China

**Qingli Wang** Institute of Grassland Research, Chinese Academy of Agricultural Sciences, Hohhot, China

**R.L. Wang** Institute of Geographic Sciences and Natural Resources Research, Chinese Academy of Sciences, Beijing, China

**Rui Wang** Institute of Plant Protection, Chinese Academy of Agricultural Sciences, Beijing, China

**Sheng Wang** National Resource Center of Chinese Material Medical, China Academy of Chinese Medical Science, Beijing, China

**Tao Wang** Research Institute, Chinese Academy of Sciences, Cold and Arid Regions Environmental and Engineering, Lanzhou, China

**Wenqing Wang** Key Laboratory of the Ministry of Education for Coastal and Wetland Ecosystems, College of Environment and Ecology, Xiamen, Fujian, China

**Xiangrong Wang** Department of Environmental Science and Engineering, Fudan University, Shanghai, China

**Xiaodan Wang** Institute of Mountain Hazards and Environment, Chinese Academy of Sciences, Chengdu, China

**Xiaoke Wang** State Key Laboratory of Urban and Regional Ecology, Research Center for Eco-Environmental Sciences, Chinese Academy of Sciences, Beijing, China

**Jiangchun Wei** State Key Laboratory of Mycology, Institute of Microbiology, Chinese Academy of Sciences, Beijing, China

**Xuefa Wen** Institute of Geographic Sciences and Natural Resources Research, Chinese Academy of Sciences, Beijing, China

**Yihui Wen** Academy for Environmental Planning, Beijing, China

**Jianguo Wu** School of Life Sciences and Global Institute of Sustainability, Arizona State University, Tempe, AZ, USA

**Jianyong Wu** Nanjing Institute of Environmental Sciences, Ministry of Environmental Protection, Nanjing, China

**Jun Wu** Nanjing Institute of Environmental Sciences, Ministry of Environmental Protection, Nanjing, China

**Kongming Wu** State Key Laboratory for Biology of Plant Diseases and Insect Pests, Institute of Plant Protection, Chinese Academy of Agricultural Sciences, Beijing, China

**Shaohong Wu** Institute of Geographic Sciences and Natural Resources Research, Chinese Academy of Sciences, Beijing, China

**Wenliang Wu** College of Resources and Environmental Sciences, China Agricultural University, Beijing, China

**Cheng Xiang** Center for Chinese Agricultural Policy, Chinese Academy of Sciences, Beijing, China

**Zhonghuai Xiang** State Key Laboratory of Silkworm Genome Biology, Southwest University, Chongqing, China

**Yu Xiao** Institute of Geographic Sciences and Natural Resources Research, Chinese Academy of Sciences, Beijing, China

**Dong Xie** School of Life Sciences, Nanjing University, Nanjing, China

**Gaodi Xie** Institute of Geographic Sciences and Natural Resources Research, Chinese Academy of Sciences, Beijing, China

**Dingpeng Xiong** Institute of Geographic Sciences and Natural Resources Research, Chinese Academy of Sciences, Beijing, China

**Danghui Xu** School of Life Sciences, Lanzhou University, Lanzhou, China

**Haigen Xu** Nanjing Institute of Environmental Sciences, Ministry of Environmental Protection, Nanjing, China

**Ming Xu** Key Laboratory of Ecosystem Network Observation and Modeling, Institute of Geographic Sciences and Natural Resources, Chinese Academy of Sciences, Beijing, China

**Qiang Xu** Institute of Oceanology, Chinese Academy of Sciences, Qingdao, China

**Qinzeng Xu** Institute of Oceanology, Chinese Academy of Sciences, Qingdao, China

**Shixiao Xu** Northwest Institute of Plateau Biology, Chinese Academy of Sciences, Xining, China

**Zhiwei Xu** Institute of Geographic Sciences and Natural Resources Research, Chinese Academy of Sciences, Beijing, China

**Zhongqi Xu** Forestry College, Agricultural University of Hebei, Baoding, China

**Dayuan Xue** Nanjing Institute of Environmental Sciences, Ministry of Environmental Protection, Nanjing, China

**Xiwu Yan** Dalian Ocean University, Dalian, China

**Guangmei Yang** Shanghai Hongqiao New Energy Investment Corp, Shanghai, China

**Hongsheng Yang** Institute of Oceanology. Chinese Academy of Sciences, Qingdao, China

**Hui Yang** Institute of Karst Geology, Chinese Academy of Geological Sciences, Guilin, China

**Jing Yang** Yunnan Agricultural University, Kunming, China

**Li Yang** Urban Planning and Design Institute, Tsinghua University, Beijing, China

**Qingwen Yang** Institute of Crop Science, Chinese Academy of Agricultural Science, Beijing, China

**Shengchang Yang** Key Laboratory of the Ministry of Education for Coastal and Wetland Ecosystems, College of Environment and Ecology, Xiamen, Fujian, China

**Xiaoguang Yang** China Agricultural University, Beijing, China

**Zhaoping Yang** Nanjing Institute of Environmental Sciences, Ministry of Environmental Protection of China, Nanjing, China

**Zhifeng Yanng** School of Environment, Beijing Normal University, Beijing, China

**Buqing Yao** Northwest Institute of Plateau Biology, Chinese Academy of Sciences, Xining, China

**Zuofang Yao** Northeast Institute of Geography and Agroecology, Chinese Academy of Sciences, Changchun, China

**Xuehua Ye** Institute of Botany, Chinese Academy of Sciences, Beijing, China

**Guirui Yu** Institute of Geographic Sciences and Natural Resources Research, Chinese Academy of Sciences, Beijing, China

**Kongjian Yu** College of Architecture and Landscape Architecture, Peking University, Beijing, China

**Jianli Yuan** School of Life Sciences, Lanzhou University, Lanzhou, China

**TianXiang Yue** Institute of Geographic Sciences and Natural Resources Research, Chinese Academy of Sciences, Beijing, China

**Jinyan Zhan** Beijing Normal University, Beijing, China

**Chunlan Zhang** Key Laboratory of Zoological Systematics and Evolution, Institute of Zoology, Chinese Academy of Sciences, Beijing, China

**Fusuo Zhang** Centre for Resources, Environment and Food Security, China Agricultural University, Beijing, China

**Guofan Zhang** Institute of Oceanology, Chinese Academy of Sciences, Qingdao, China

**Huiyuan Zhang** Academy for Environmental Planning, Beijing, China

**Linbo Zhang** Chinese Research Academy of Environmental Sciences, Beijing, China

**Ming Zhang** Nanjing Institute of Environmental Sciences, Ministry of Environmental Protection, Nanjing, China

**Runzhi Zhang** Institute of Zoology, Chinese Academy of Sciences, Beijing, China

**Xianzhou Zhang** Resources Research, Institute of Geographical Sciences and Natural, Chinese Academy of Sciences, Beijing, China

**Yanmin Zhang** Academy for Environmental Planning, Beijing, China

**Yifeng Zhang** Institute of Geographic Sciences and Natural Resources Research, CAS, Beijing, China

**Yihui Zhang** Key Laboratory of the Ministry of Education for Coastal and Wetland Ecosystems, College of Environment and Ecology, Xiamen, Fujian, China

**Yong Zhang** Institute of Grassland Research, Chinese Academy of Agricultural Sciences, Hohhot, China

**Yuan Zhang** Chinese Research Academy of Environmental Sciences, Beijing, China

**Haijun Zhao** Foreign Environment Cooperation Center, Ministry of Environmental Protection, Beijing, China

**Hong Zhao** State Key Laboratory of Urban and Regional Ecology, Research Center for Eco-Environmental Sciences, Chinese Academy of Sciences, Beijing, China

**Jingzhu Zhao** Institute of Urban Environment, Chinese Academy of Sciences, Xiamen, China

**Liang Zhao** Northwest Institute of Plateau Biology, Chinese Academy of Sciences, Xining, China

**Xinquan Zhao** Northwest Institute of Plateau Biology, Chinese Academy of Sciences, Xining, China

**Lin Zhen** Institute of Geographic Sciences and Natural Resources Research, Chinese Academy of Sciences, Beijing, China

**Feimin Zheng** Research Institute of Water and Development, Chang'an University, Xi'an, China

**Jingyun Zheng** Institute of Geographic Sciences and Natural Resources Research, Chinese Academy of Sciences, Beijing, China

**Wei Zheng** The First Institute of Oceanography, State Oceanic Administration of China, Qingdao, China

**Yongyun Zheng** Marine Biology Institute of Shandong Province, Jinan, China

**Zhirong Zheng** State Environmental Protection Key Laboratory of Regional Eco-Process and Function Assessment, Chinese Research Academy of Environmental Sciences (CRAES), Beijing, China

**Linsheng Zhong** Institute of Geographic Sciences and Natural Resources Research, CAS, Beijing, China

**Xianghao Zhong** Institute of Mountain Hazards and Environment, Chinese Academy of Sciences, Chengdu, China

**Zhiwei Zhong** School of Life Sciences, Northeast Normal University, Changchun, China

**Guangsheng Zhou** Chinese Academy of Meteorological Sciences, Beijing, China

**Hangtao Zhou** Key Laboratory of the Ministry of Education for Coastal and Wetland Ecosystems, College of Oceanography and Earth Sciences, Xiamen, Fujian, China

**Hengjie Zhou** School of Life Sciences, Nanjing University, Nanjing, China

**Honghua Zhou** State Key Laboratory of Desert and Oasis Ecology, Xinjiang Institute of Ecology and Geography, Chinese Academy of Sciences, Urumqi, China

**Huakun Zhou** Northwest Institute of Plateau Biology, Chinese Academy of Sciences, Xining, China

**Ting Zhou** State Key Laboratory of Biocontrol, School of Life Sciences, Sun Yat-sen University, Guangzhou, China

**Youyong Zhu** Yunnan Agricultural University, Kunming, China

**Changxin Zou** Nanjing Institute of Environmental Sciences, Ministry of Environmental Protection of China, Nanjing, China

# Part I
# Biodiversity Conservation and Its Application

# Chapter 1
# Biodiversity Inventory and Researches

**Dayuan Xue, Jianyong Wu, Xu Liu, Baorong Lu and Shengji Pei**

**Abstract** Chinese authorities have organized many national or regional large-scale investigations, such as national key species resources, terrestrial wildlife resources, wetland resources, agricultural, and livestock genetic resources investigation. Currently, China has set up relevant database for biodiversity, published nearly 400 volumes of species cataloged annals, and established the monitoring system of forest resources, wetland resources monitoring centres, wildlife resources monitoring centres, and the Chinese ecosystem research network with dozens of ecological research stations. At the same time, China has gradually formed a set of perfect identification and evaluation technology on crop germplasm resources, formulated the unified germplasm resource description specification and data standard for more than 100 variety of crops, and established a set of scientific and standardized cryopreservation and monitoring technology system for genebanks, germplasm, and tube seedlings which is accord with the situation of China itself. Furthermore, the intimate connection between cultural diversity and biodiversity in China has been recognized and appreciated.

**Keywords** Biodiversity · Inventory · Crop germplasm resources · Monitor · Conservation · Genetic resources · Cultural diversity

D. Xue (✉) · J. Wu
Nanjing Institute of Environmental Sciences, Ministry of Environmental Protection, Nanjing 210042, China
e-mail: xuedayuan@hotmail.com

X. Liu
Chinese Academy of Engineering, Beijing 100088, China

B. Lu
Ministry of Education Key Laboratory for Biodiversity and Ecological Engineering, Department of Ecology and Evolutionary Biology, Fudan University, Shanghai 200433, China

S. Pei
Kunming Institute of Botany, Chinese Academy of Sciences, Kunming 650201, China



W. Li (ed.), *Contemporary Ecology Research in China*,
DOI 10.1007/978-3-662-48376-3_1

# 1 Progress on Biodiversity Inventory and Researches in China

## 1.1 Introduction

Biodiversity is defined as variability among living organisms from all sources, and it includes the diversities in three levels of gene, species, and ecosystem by the definition of the Convention on Biological Diversity (CBD) (Chandra and Idrisova 2011). Biodiversity plays an important role in sustaining human lives by providing different goods and services, and through its intrinsic, cultural, and socio-economic values. But many negative factors, such as habitat degradation and fragmentation, over-exploitation of biological resources, introduction of alien invasive species, and biodiversity-unfriendly agricultural practices, are threatening biodiversity's loss at an unprecedented level (Ayyad 2003; Ahrends et al. 2011).

China is one of the mega-biodiversity countries in the world. The main types of terrestrial ecosystems in China include forest, shrub, meadow, steppe, desert, and wetland; and the major marine ecosystems are the Yellow Sea, East China Sea, South China Sea, and Kuroshio Basin. China has known approximately 35,000 species of higher plants, ranking the third in the world. China also has 6445 vertebrate animal species, accounting for 13.7 % of the total in the world. The fungi species proven in China are about 10,000, making up 14 % of the total in the world. According to incomplete statistics, there are 1339 species of cultivated crops, 1930 species of wild relatives of crops and 567 varieties or breedings of domesticated animals in China. But at the same time, China is also one of the countries where biodiversity is under the most serious threat.

A basic element in the success of managing species for conservation concern is a full understanding of the country's biodiversity status. To halt the destruction of the diversity, it is essential to investigate, inventory, and assess the present biodiversity status and to forecast future change and trends. In 1950s and 60s, a large-scale biodiversity survey was carried out, laying down a solid foundation for composition of a series monographs of the vegetation, flora, and fauna of China.

## 1.2 Progress of the Study on Inventory of Biodiversity in China

Since 1990s, field surveys and inventory work have been conducting in some key areas to identify the important natural resources, as a result, categories and assessing criteria for Red List Species were developed, the endangered wild animal and plant species were evaluated and graded, and demonstrative evaluation for grading ecosystems was carried out. Great progress has also been made in ex situ and in situ conservation of biodiversity, and a series of relevant conservation planning have been developed. Although it is a little late for starting biodiversity conservation activities, great progress was made during the past 20 years in biodiversity researches, especially biodiversity's identification, and inventory.

### 1.2.1 Progress on Identification and Inventory of Biodiversity

A national-wide wildlife survey organized by the State Forestry Administration was carried out on the investigation to 189 nationally protected plant species and 252 wild animal species (State Forestry Administration 2009a, b). Another systematic survey for 191 wild species of crops has been undertaken in 2002–2009. Based on data collection, a monograph on national key protected wild plants was compiled (Wang et al. 2011). To achieve 2010 targets of CBD, in 2004, a project for biodiversity survey and inventory was launched by Ministry of Environmental Protection joined by other relevant ministries of forestry, agriculture, Chinese traditional medicines (TCMs), and so on. This survey was mainly focused on key species of plant, animal, and macro-fungi, mainly including rare, endemic, threatened and endangered, and economic species that are mostly involved in Red list and other lists for protection. Until 2010, "China National Biodiversity Conservation Strategy and Action Plan" was approved by the State Council. In this document, identification and inventory of biodiversity was listed as a priority strategy and action. Therefore, a new program for biodiversity identification and inventory was implemented since 2010. This program is targeted to inventory all species based on county level and to establish the biodiversity databases for counties in whole China. It means that this ambitious program is to identify and inventory biodiversity for more than 2000 counties in whole China during the next 10 years. To initiate the great job, a pilot project for biodiversity inventory in 26 selected counties in Southwest China (Guizhou. Guangxi and Yunnan) was implemented during 2010–2011. The project has succeeded in many aspects, for example, the databases established in 26 counties are not only limited in species, but also in ecosystems and genetic resources. It is expected that the experience achieved in this pilot project will play a demonstrative role for the next large scale of biodiversity investigation in the over 2000 counties in whole Chinese territory.

### 1.2.2 Inventory and Database of Biodiversity

Inventory and database are important parts of biodiversity investigation. China has completed all the 80 volumes of Flora and 125 volumes of Fauna of China, etc. Catalogue of Life (COL) China has an annual checklist. In September 2012, COL China Annual Checklist 2012 edition, compiled by Species 2000 China Node, was published by Science Press. The groups of species in 2012 Annual Checklist of

COL China and their numbers of accepted species names are: virus (348), bacteria (158), chromista (1540), fungi (140), protozoa (1291), plantae (35,487), and animalia (23,103). The numbers of total species, synonyms and aliases reached to 70,596, 92,635 and 32,120, respectively. Furthermore, a catalog for 60,000 species and 400,000 accessions of agricultural germplasm resources was completed, setting up a national database platform for biodiversity, including the databases of ecosystems, vegetations, species, genetic resources, case study reports, as well as a huge capacity for photos, video, and specimens.

In situ and ex situ conservation are common and effective approaches for biodiversity conservation. By the end of 2011, China had established 2640 nature reserves, accounting for 15 % of whole country's territory. However, ex situ conservation needs a strong management, and assessment for the management effectiveness of nature reserves is an effective way to promote management quality and achieve conservation targets. Chen et al. (2009) used geographic information system (GIS) to assess the conservation status of vegetation types, endangered plant and animal species, and biodiversity hotspots in China, based on the area, endangered species list, and geographic position of 2047 nature reserves. Yuan et al. (2009) investigated the protected conditions of state key protected wild plants within the national nature reserves as of 2008, based on the available data and by use of information collection and analysis. Quan et al. (2011) carried out a survey in 535 nature reserves in China to assess the management status quo in 2005 using a questionnaire. Then the state key protected wild plant in natural reserves and the wild animal species were displayed and assessed for their conservation capacities in the in situ and ex situ facilities in China.

### 1.2.3 Study of Categories and Criteria and Assessment of Endangerment Status of Biodiversity

It is an important task in biodiversity conservation to assess biodiversity status and determine protection priorities. IUCN Red list Criteria has been improved through periodical revision. In 2008, the International Union for Conservation of Nature (IUCN) established a working group at the Fourth World Conservation Congress to develop quantitative categories and criteria for assessing ecosystems' threat status. As an example, Chen and Ma (2012) illustrated the use of these criteria for assessing ecosystem threat status, and they evaluated the threat status of these four ecosystems by use of the existing literature data on the occupancy area for ecosystems in China's Liaohe Delta in 1988 and 2006. Under the funding support from Ministry of Environmental Protection of China, Jiang and Luo (2012) assessed the status of terrestrial vertebrates in China using IUCN method. The result showed that five species were listed in the category of extinct, 30 species were near extinct, 343 species were endangered, 459 species were threatened, 439 species were concerned, and 1032 species were least concerned. In addition, the development of the red list for higher plants was executed, joining by more than 100 specialists from different institutions and universities in China. Zhang et al. (2011) assessed the threatened status of the nationally protected wild plants in China, by use of IUCN Red List Categories and Criteria.

### 1.2.4 Policy Systematic on Biodiversity

China government has being focused on biodiversity conservation for a long time. Especially, since becoming the contracting party to CBD, China has taken a lot of actions to halt the loss of biodiversity. In 1994, China issued China Biodiversity Conservation Action Plan. In 2007, China launched the National Program for Conservation and Use of Biological Resources. In 2008, National Strategy for Plant Conservation was released as a response to the Global Strategy for Plant Conservation. In addition, the renewed China National Biodiversity Conservation Strategy and Action Plan was formulated during 2007–2010 and approved by the State Council in Sept 2010. Furthermore, the policy of access and benefit-sharing for genetic resources and associated traditional knowledge was presented in China National Intellectual Property Strategy Outlines, which was issued by the State Council in 2008.

## *1.3 Conclusion and Discussion*

Because of limited financial and expertise resources, biodiversity surveys and inventories in the past years were mainly concentrated in the important regions or aimed to the key species. These surveys and inventories are far from the target to know the biodiversity baselines in China. However, the baselines are significant for future monitoring and assessment for the country's biodiversity. Therefore, it is important to establish a national program for biodiversity inventory based on county unit level, which is helpful to set up a monitoring network based on over 2000 county nodes. Challenges for biodiversity conservation are not only limited in inventory and monitoring, but also in access and benefit-sharing for the utilization of genetic resources and associated traditional knowledge (Xue et al. 2012). However, a clear baseline and a sound monitoring network, generated by biodiversity survey and inventory, would be a solid foundation for implementation of access and benefit-sharing regime. Hence the authors suggest that the urgent work is to start the implementation of the national biodiversity inventory program identified in the China National Biodiversity Strategy and Action Plan (2011–2030).

# 2 Survey, Catalog, Evaluation, and Conservation of Crop Germplasm Resources

Crop germplasm resources are the basis for crop breeding and the development of modern seed industry. They are long-term, fundamental, and public work. The survey, catalog, evaluation, and conservation of crop germplasm resources have a far-reaching significance and realistic importance.

## *2.1 Survey and Collection of Crop Germplasm Resources*

Collection of germplasm resources is the basis of and first step to the conservation of crop germplasm resources. The main collecting methods include general investigation and specific investigation. When conducting the general investigation of crop germplasm resources, the related national research institutes draw up an investigation letter, formulate a collecting form, and send them to the relevant administrative departments at the provincial level. As required, the relevant local staff at district or county levels do the survey and collection for the germplasm resources in their areas. After putting them in order, the collected materials shall be sent to the agricultural department at the provincial level or the institutes that sent the letter and form. Finally, the collection of germplasm resources will be characterized, cataloged, multiplied, and preserved in the national genebank and/or field genebanks under the coordination of institute that leads the general investigation. In the 1950s, in order to avoid local germplasm resources getting lost in the process of promoting improved varieties, a general collection of germplasm recourses was conducted at the county level nationwide. In total, 200,000 accessions of 53 field crops, 17,000 accessions of 88 vegetable crops, and 12,000 accessions of introduced from foreign countries were collected. From 1979 to 1984, the general collection was re-conducted and 110,000 accessions of 60 crops were collected.

Besides the general investigation, China also conducted specific surveys and collecting missions of important crops and their wild relatives in key areas. Based on exhaustive literature and thorough understanding of geography, climate, social economy, cultural environment and crop species, diversity, and special resources of the areas, the work plan for targeted collections is developed, including the scope, time, personnel, equipment, management approach, and financial budget of the survey. The participating organizations and staff are determined. The collecting team will be composed of old, middle-aged, and young scientists with different expertise. The workshop and training will be conducted for enhancing the capacity for the surveys. The work plan shall be discussed between the investigation panel and the local government and related sectors. If the area to be surveyed covers a broad range and employs many staff, the collecting team could be divided into groups, each of which takes charge of part of the area. In the process, the staff needs to fill in the unified form for each accession of collected crop germplasm resources. If the tissues or organs of the crop can be preserved as germplasm resources, they should be collected on the site, numbered, and preserved temporarily. For wild species, the samples should be collected. But if it is not the right time to collect the species when the survey is conducted, the collecting team shall authorize the local team members to do it in its suitable time and send it to the preserving organization, making sure that the numbering of the samples are consistent.

From 1970s, China has already conducted specific survey of crop germplasm resources in Yunnan, Tibet, Shennongjia in Hubei Province, Three Gorges Area, Hainan, Daba Mountains, and the mountainous areas of southern Guizhou, western Guangxi, southern Jiangxi, and northern Guangdong. China has also surveyed and collected salt-tolerant crop germplasm resources in coastal areas of 11 provinces and drought-resistant germplasm resources in 7 provinces in the West part of China. By 2010, altogether 70,000 accessions of germplasm resources have been collected. Besides, wild soybean was surveyed and collected in 1020 counties nationwide, and 5000 accessions of soybean seeds and 4000 accessions of soybean plants were collected. Wild rice was investigated in Hainan, Guangdong, Guangxi, Yunnan, Fujian, Jiangxi, and Hunan provinces (Autonomous regions). It was discovered in 140 counties (cities) and 3800 seed samples and seedling stems were collected. During 1986–1990, 1000 accessions of 100 species in 10 genus of wild relatives of wheat were surveyed and collected in provinces, cities, and autonomous region in the Northwest and Southwest regions. Starting from 2002, the Ministry of Agriculture implemented a systematic investigation of wild plants related to agriculture as listed in "The List of Wild Plants under National Key Protection" (first and second parts), which checked up 30,000 distribution sites and collected 10,000 accessions of germplasm resources. Surveys have also been conducted on kiwi fruit nationwide, wild forage plants, mulberry plants in Sichuan, Hubei, Shanxi, and Guizhou provinces, cotton in Hainan, Guizhou and Guangxi provinces, and ramie in 14 provinces. The surveys made a thorough investigation of categories and distribution of the wild species and collected a number of precious germplasm resources.

## 2.2 *Evaluation and Catalog of Crop Germplasm Resources*

Over the past 30 years, China has formulated a very comprehensive evaluation techniques and methods and published "Descriptors and Data Standards for Crop Germplasm Resources" for 110 crops, which standardizes the experimental design, evaluation standards, data collection, and quality control of the evaluation of different species of crops, and therefore improves the credibility of the evaluation data.

### 2.2.1 Characterization and Evaluation

Characterization and evaluation of crop germplasm resources is mainly based on the principal theories and methods of agronomy, genetics, and statistics. It abides by the following three basic principles.

First, the basic agronomic traits shall be evaluated in places similar to the original ecological environment where the germplasm was collected. For example, it is suitable to evaluate indica rice in the South and japonica rice in the North. The process shall be carried out in similar ecological environment to its original places for different varieties of the same species, which are collected from different places. The scope of the evaluation of the basic agronomic traits differs for different crops. Taking winter wheat for instance, the evaluation shall include information of its sowing date, emergence date, the number of basic seedlings, returning green date, elongation date, heading date, flowering date, grain filling date, mature date, days of whole growth period, stem length, tillering number, number of ears per unit, grain number per spike, 1000-grain weight, yield per unit area, etc. The specific scope and standards for each crop shall follow the "Descriptors and Data Standards for Crop Germplasm Resources".

Second, biotic-stress and abiotic-stress resistance shall be evaluated in artificially controlled and natural environments. Characterization and evaluation of biotic-stress resistance (for example, disease and/or pest resistance) shall be conducted in controlled environment such as greenhouse and/or net houses. By means of artificial inoculation of disease and/or pest and in suitable temperature and humidity, the crops are characterized and evaluated for their biotic resistance, according to the standards and seriousness of being affected by the disease and/or pest. As to the specific techniques and methods, they shall be selected for different species of crops, diseases, or pests according to the relevant "Descriptors and Data Standards for Crop Germplasm Resources." Evaluation of abiotic-stress resistance (drought, salt, alkaline, cold, heat, low nitrogen, low phosphorus, barren, etc.) shall be conducted in natural environment with control, such as drought tents, salt/alkaline-resistant ponds, and plant growth chambers.

Third, nutrition and food processing traits shall be analyzed and evaluated in laboratories with equipment. The usual items include protein, starch, fat, amino acid content, and processing characteristics, for instance, sedimentation value and stabilization time of wheat flour, and the special-function nutrition factors, for example, isoflavone of soybean and $\beta$-glucan of oat. The specific evaluation techniques and methods shall accord with either national or ministerial standards for different species of crops.

China has completed evaluation of the basic agronomic traits (generally over 30 traits) of the crop germplasm resources as preserved in the national genebanks, field genebanks, and in vitro genebanks. For the evaluation of disease/pest resistance, salt/alkaline tolerance, and quality traits, the percentages of the germplasm resources evaluated so far are 50, 50, and 60 %, respectively. However, the evaluation of basic agronomic traits, disease/pest resistance, salt/alkaline tolerance, and quality traits of individual crops may be adjusted along with the change of breeding objectives. For instance, the evaluation of resistance to soybean phytophthora root rot and sclerotinia rot, and tolerance of low nitrogen/phosphorus of wheat are the emerging needs in breeding and agricultural production in recent years.

### 2.2.2 Unified Catalog

For the crop germplasm resources to be cataloged, they shall meet the following requirements: viability that meets the requirement to enter the long-term genebank/field genebanks; potential values in research and application; stable genetic characteristics; with observation and evaluation results and data on its agronomic traits for 2–3 years; with complete germplasm passport information.

The information needs to be cataloged includes the following items. (1) The purpose, background, process of the catalog, the categories, conditions, numbers of accessions cataloged, code and numbering method, and the institutes and staff produced the catalog. (2) The basic information of germplasm resources, i.e., the passport information, including scientific names of family, genus and species, national unique accessions number, germplasm name, the preservation institute name and number, country or region of origin, country or region of introduction, etc. (3) Botanical features and agronomic traits. (4) Biotic- and abiotic-resistance and quality traits. Key items shall be selected, identified, evaluated, and cataloged according to the needs of breeding and agricultural production. (5) Other important characteristics, such as the special traits of special germplasm resources. Each accession cataloged is designated a unique accession number. For individual crops, it shall accord with "National Accessions Number" standards in the "Descriptors and Data Standards for Crop Germplasm Resources."

By the end of 2010, there were 463,427 accessions of germplasm resources that are nationally uniquely numbered, in which 415,692 are in the seed banks, 44,724 in the field genebank, and 3011 in the in vitro genebanks. Catalogs for 370,846 accessions have been published. For the wild plant germplasm resources, experts organized by the Ministry of Agriculture have compiled "The Outline of Wild Agricultural Plants Under National Key Protection" covering the scientific names in Chinese and in Latin, geographic distribution, ecological environment, morphological traits, conservation value, and state of endangerment. It consists of 209 species, including subspecies and variants, in 72 families.

## *2.3 Conservation and Monitoring of Crop Germplasm Resources*

There are two main approaches to protect the crop germplasm resources, namely in situ and ex situ conservation. For the protection of the wild relatives of agricultural crops, the major approach is to establish the in situ conservation regions (or sites). The conservation site of wild rice, established in 1985 in Dongxing, Jiangxi Province, is the rudiment of the early stage of national protection of wild germplasm resources. By the end of 2010, 137 in situ conservation sites in 26 provinces, cities, and autonomous regions have been established for wild relatives of crops, including 26 species, namely, wild rice, wild soybean, wild relatives of wheat, wild lotus, coastal glehnia, wild buckwheat, cordyceps sinensis, wild apples, wild cherry-apples, wild sugarcane, wild citrus, Chinese ilex, wild kiwi fruit, Isoetes sinensis, wild tea, wild lychee, wild wolfberry, wild orchid, etc. The methods for ex situ conservation are seed banks, in vitro bank, cryopreservation, and field bank, among which the first one is of the most importance.

Low-temperature conservation of the germplasm in China can be traced back to the late 1970s. Over the past 30 years, it has witnessed significant achievement. First, a safe conservation system for crop germplasm resources, consisting of long-term genebanks, duplicate genebanks, and mid-term genebanks, has been set up. The system contains 1 national long-term genebank, 1 duplicate genebank, 10 national mid-term genebanks as well as about 30 provincial mid-term genebanks, which makes it possible for the safe storage and efficient utilization of national crop germplasm resources. Second, a technique system for conservation and monitoring has been established that is consistent with our national situations. The "Two Fifteen" seed drying system, where the temperature is controlled at 15 °C and the relative humidity is set at less than or equal to 15 %, has been developed. By the end of 2010, there were more than 360,000 accessions of germplasm resources conserved in the long-term bank, which belong to 735 species. The number of accessions in conservation ranks the second worldwide. In the national and provincial mid-term banks, there are around 600,000 accessions that can be distributed or are to be processed. The annual distribution of germplasm resources can reach 40,000 accessions, which provide a solid basis for the sustainable development of agriculture in China.

The viability of 14,000 accessions of 34 crops preserved in the national long-term banks for over 20 years is being monitored. Results show that germination rates of 92 % seeds monitored maintained above 85 %. However, 155 accessions, accounting for 1.1 % of the total, showed significant decline in germination rates, from above 80 % to below 70 %. China has started from 1978 conservation of germplasm resources of potatoes and sweet potatoes in vitro with tip meristem culture. In 1986, the national in vitro genebank of potato and sweet potato was established. By the end of 2010, 2204 accessions of potato and 1302 accessions of sweet potato were preserved. With SSR and morphological markers, the genetic stability of germplasm resources of potato and sweet potato was compared between preserved in the germplasm nursery gardens and in the in vitro genebanks. Results show that genetic stability is higher in the germplasm nursery gardens than in the in vitro genebanks. But germplasm materials in the nursery gardens are vulnerable to virus infection and thus are likely to degenerate. The variation of germplasm materials in the in vitro genebanks is temporary and can be restored by regenerating in the field.

Cryopreservation is usually used for preserving germplasm resources through its organs such as pollen, embryo, root tip, stem tip, sprout, branch, cell, and seed, etc., and considered as an ideal way of long-term preservation for vegetatively propagated germplasm resources. China has succeeded in cryopreservation of germplasm resources of many species in fruit trees, flowers, and vegetables, such as apples, pears, sweet cherries, plums, banana, etc. At present, large-scale cryopreservation for vegetatively propagated crop germplasm resources is being carried out in the country. Conservation of living plants in field genebanks is also an important way of germplasm preservation, with aims to conserve vegetatively propagated and perennial plants, including fruit trees like apple, pear, peach, and grapes; economic cops like tea, mulberry, and rubber; perennial herbaceous plants like wild rice, wild relatives of wheat and perennial forage grass; aquatic vegetables like lotus, cane shoot and yam; and tuberous root and stem like potato, sweet potato, and cassava. The construction of field genebanks in China has started in the 1970s. By December 2010, altogether 30 field genebanks have been constructed and 7 more are under construction. There are 47,805 accessions of germplasm resources belonging to 1098 species conserved in field genebanks in China. Monitoring of the germplasm resources in the nursery gardens show that natural disasters, including disease and pest, aging of the germplasm, and relocation are the main factors causing the loss of germplasm resources.

### *2.4 Outlook*

Survey, catalog, evaluation, and conservation of crop germplasm resources have been listed in the National Medium and Long-term Development Planning. In the coming 10–20 years, it is planned to add 100,000 accessions of crop germplasm resources into the strategic reserves, by introduction from foreign countries and domestic collection; to conduct deep phenotypic characterization for 30,000 elite accessions, and select a number of accessions that show eminent and stable elite characteristics for breeding; to expand 32 field genebanks and build a new national long-term genebank, enabling the capacity to reach 1.5 million accessions, which will meet the conservation needs for the coming 50 years; and to establish around 350 in situ conservation sites to improve the conservation of wild germplasm resources.

## 3 Conservation of Genetic Resources

### *3.1 Introduction*

The rapid increase in global population has posed a great challenge to the world food security. As indicated in the Declaration of the summary report from the World Summit on Food Security held at the Food and Agriculture Organization

(FAO) headquarters in Rome, Italy: "The number of people suffering from hunger and poverty now exceeds one billion, and that to feed a world population expected to surpass nine billion in 2050, agricultural output will have to increase by 70 %" (IISD 2009). This clearly suggests the severe situation of global food shortage and future food demand. The population increase is in parallel with the gradual decrease in arable land, shortage of agricultural sources such as water and minerals, loss of rural laborers, and global climate change. Such a situation arouses serious problems for the world sustainable development. Dramatically increase in the per unit crop production can relieve the pressure of world food security. The efficient utilization of genetic resources and new technologies in crop production can achieve this goal (Lu 2001).

Genetic resources have played an important role in the production and improvement of crop varieties. The transferring of useful traits such as those with enhanced yield and quality, and tolerance to biotic and abiotic stresses can improve greatly crop varieties. The well-known examples are the production of the "green revolution" rice, wheat, and maize varieties by introducing semidwarf (*SW-1*) gene to these crops and breeding of hybrid rice by introducing male sterility (*MS*) gene from wild rice to cultivated rice (Evenson and Gollin 2003). These examples demonstrate the importance of maintaining the long-term availability of genetic resources for the genetic improvement of crop varieties at present and in the future. However, due to great changes such as rapid economic development, urbanization, human population increase, changes in agricultural management and farming styles, and global climate change, the long-term availability of genetic resources is under great threats. Effective and strategic conservation of genetic resources is the guarantee for their stainable uses (Lu 2001).

## 3.2 The Concept and Category of Genetic Resources

According to the definition of "Convention on Biological Diversity (CBD)," biological resources include "genetic resources, organisms or parts thereof, populations, or any other biotic component of ecosystems with actual or potential use or value for humanity." Therefore, genetic resources represent a part of biological resources and can be defined as "genetic material of actual or potential value. Genetic material is any material of plant, animal, microbial or other origin containing functional units of heredity." Although genetic resources include a wide spectrum of living genetic materials such as medicinal plants, agricultural crops, and animal breeds, this article only emphasize those closely associated with agricultural genetic resources.

Agricultural genetic resources include: (1) domesticated crops; (2) wild species genetically closely related to crops; and (3) con-specific weeds of crops. Crop species are domesticated plants from wild species through long-term utilization and cultivation by human. Crop species have played extremely important roles in human civilization and the world food security. There are many cultivated types for each domesticated species that are usually divided into traditional crop varieties or landraces selected by farmers, modern or improved crop varieties bred by breeders, and genetic stocks from hybridization or genetic manipulation. All domesticated species are originated from their wild ancestors, and therefore any wild species that have a certain genetic relationship with the domesticated crops are referred to as crop wild relatives, including the direct ancestor of a crop.

Various types of genetic resources have different evolutionary relationships with crops, which reflect the accessibility to the genetic resources for utilization in crop breeding. This is due to compatibilities of a crop species that is involved in genetic improvement with other species used as genetic resources. Based on such compatibilities, Harlan and De Wet have categorized genetic resources into three gene pools: primary gene pool (GP-I); secondary gene pool (GP-II); and tertiary gene pool (GP-III) (Harlan and Wet 1971). GP-I includes genetic resources that belong to the same biological species as the target crop. Species including crop varieties, genetic stocks, and con-specific weeds in this gene pool can have sexual crosses and genetic recombination with the target crop species freely. GP-II comprises of genetic resources that have close relationships with the target crops. Species including wild species in the same genus of the crop species in this gene pool cannot easily hybridize with the target crop without manipulation. GP-II contains genetic resources that have distant relationships with the target crops. It is impossible to use species in this gene pool for breeding unless special technologies are applied (Fig. 1).

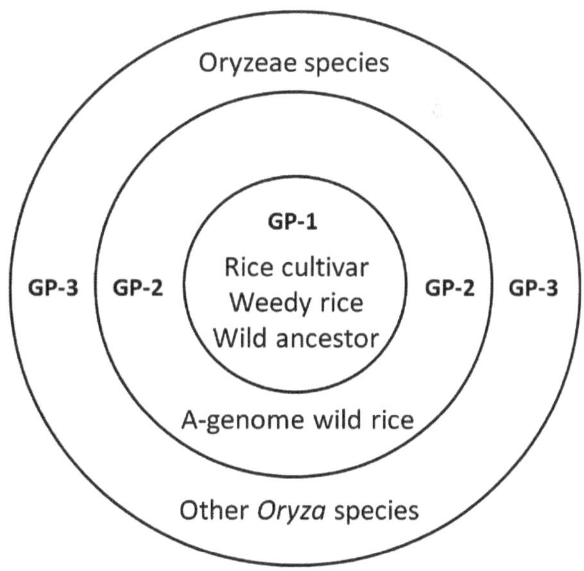


**Fig. 1** Categories of genetic resources into primary (GP-I), secondary (GP-II), and tertiary (GP-III) gene pools using cultivated rice as an example, based on the concept of Harlan and De Wet (1971)

## 3.3 Strategies for Genetic Resource Conservation

Given the importance of genetic resources, governments, nongovernment organizations, and international organizations have invested enormously for conserving genetic resources. Usually, there are three strategies for the effective conservation of genetic resources: ex situ conservation, in situ conservation, and on-farm conservation. Ex situ conservation, also referred to as off-site conservation, is the process of protecting a target species of plant or animal outside its natural habitats in a new location, for example, to store seeds of a target plant species in germplasm banks. In situ conservation, also referred to as on-site conservation, is the process of protecting a target plant or animal species in its natural habitats, for example, to protect plant or animal species in nature reserves and original conservation sites. On-farm conservation is a special conservation strategy for protecting landraces or traditional varieties of crop species. On-farm conservation of crop genetic resources is defined as the continued cultivation and management of a diverse set of crop populations by farmers in the agroecosystems where a crop has evolved (Lu et al. 2008; Zhu et al. 2003).

The three strategies provide complementary approaches to each other for the effective conservation of genetic resources because each of these strategies has its unique characteristics and limitations (Table 1). For example, ex situ conservation is static conservation in terms of evolution, and potential genetic variation from the changing environment will never occur in gene banks. In addition, genetic diversity may be compromised when collections are multiplied ex situ. In situ conservation can include great amount of genetic variation that always evolves under changing environment.

**Table 1** Different conservation strategies with their characteristics and limitations

| Strategy | Target species | Conservation site | Advantages and limitations |
|---|---|---|---|
| Ex situ conservation | Wild and cultivated species | In germplasm banks, gardens, or nurseries | Easy to access when genetic resources are need for utilization; evolutionary process of the conserved resources is stopped, no genetic variation can be generated during conservation |
| In situ conservation | Wild and cultivated species | All in natural habitats | Difficult to access when genetic resources are need for utilization; evolutionary process of the conserved resources is continuing; genetic variation can happen during conservation |
| On form conservation | Cultivated species (landraces and varieties) | All in farmer's fields | Easy to access when genetic resources are need for utilization; evolutionary process of the conserved resources is continuing; genetic variation occurs during conservation |

## 3.4 Challenge of Genetic Resource Conservation

Genetic resources are extremely important for human livelihood, proving fundamental materials such as food, clothes, and housing. However, the sustainable use and long-term availability of some genetic resources are under great threats due to many factors such as dramatic change of agricultural practices, expanding of farming land, extensive cultivation of only a few modern crop varieties in a huge area, and deterioration of agricultural and natural ecosystems (Bellon et al. 1998; Mack et al. 2000). The strategic conservation of genetic resources becomes increasingly important. Because of the above-mentioned factors, the conservation of genetic resources faces unprecedented challenges, which will significantly affect the efficient conservation of genetic resources. In addition, our understanding on genetic diversity of conserved genetic resources under different situation, relationships between conservation methodologies and effectiveness of conservation, for example, what are the sampling strategies for conserved resources of plant species, is still not perfectly known (Zhu et al. 2007). More scientific studies concerning the conservation methodology need to be carried out for ex situ as well as in situ conservation strategies.

The long-term availability of genetic resources of plant species is significantly influenced by the changing environment. For example, the habitat fragmentation and habitat losses will pose severe threat to the lone-term availability or even extinction of a plant species (Kiang et al. 1979). Factors such as extensive human disturbances to the habitats, expanding of farming land, changes in agriculture management styles, invasion of alien species to local habitats, and global climate change will exert significant threats to the existence of traditional crop varieties/landraces and wild relative species in many parts of the world. This situation is particularly true for the biodiversity-rich countries and regions located in the centers of origin for domesticated plant species. In addition, one important but neglected factor may significantly influence genetic diversity of traditional crop varieties and wild relative species. That is the potential impacts of continued introgression of crop genes, particularly transgenes from genetically engineered crops on conservation (Lu 2013).

## 3.5 Conclusion and Discussion

Genetic resources are essential for the production and improvement of crop species that are associated with the world food security. Given the increasing threats to the long-term availability and diversity of genetic resources, strategic and effective conservation of plant and animal species that harbor genetic diversity become very important. To achieve such objectives, we need to complete the follows tusks. (i) To strengthen scientific researches on conservation, such as sampling strategies, genetic diversity and structure of wild populations, and gene flow within and between populations, which will enable us to design strategies and measures for implementation and management of effective conservation, particularly for in situ conservation. (ii) To minimize genetic erosion during the conservation and management process, for example, contamination during seed multiplication in germplasm banks. (iii) To increase the utilization of genetic resources, which will promote the effective conservation of genetic resources, particularly for on-farm conservation of traditional crop varieties and in situ conservation of wild species. (iv) To strengthen scientific studies on the adaptive evolution of in situ conserved genetic resources, which can facilitate the effective conservation of genetic resources under changing environmental conditions. (v) To increase the public education on the biodiversity and conservation of genetic resources, which will mobilize resources from governments, nongovernment organizations, and the whole societies for the course of genetic resource conservation.

## 4 Cultural Diversity and Biological Diversity

### *4.1 Introduction*

With the conclusion of the CBD in 1992 by United Nations World Summit in Rio and the Convention on the Protection and Promotion of the Diversity of Cultural Expressions in 2005 by UNESCO in Paris, the conservation of biological and cultural diversity for environmental protection and sustainable development has now become a common concern of the international community. Since 1980s, ethno biologists, human ecologists, and conservationists around the world have been actively engaged in the research of indigenous community biodiversity conservation that lead to recognize of biodiversity and traditional culture is inseparable; genetic resources preservation and traditional knowledge are inseparable, and traditional cultural beliefs are not separated with ecosystem conservation. As McNeely (1993, 2003) pointed that cultural and biological diversity are intimately and inextricably linked. Case studies and reports from research have strongly supported the coevolutional relationships of biodiversity and cultural diversity which was first discussed in the Oxford University published book '*Our Common Future*' in 1987 and built up the theoretical basis for the study of BioCultural diversity conservation today (Pei 2006; Pei and Huai 2007).

China is a country of multicultural nationalities, 56 ethnic cultural groups have jointly together created the great Chinese civilization with which the root-base is agrocivilization covering agroculture; nomadic-patrolman culture; forest culture; medical culture; and other relevant cultures of utilization animals and plants throughout Chinese history. China has the richest northern temperate flora and fauna in the world. Known as one of the Mega-bio diversity countries of the world, for instance China has recorded 31,500 native species of vascular plants, around 8 % of the world's estimated total. China is the only country in the world that possess continue terrestrial ecosystems from tropical to cold rigid zones and from sea level to the highest peak of the world. The 5000 years noninterrupted Chinese civilization development has been developed from ancient time to present time is also very unique among nations of the world. Traditional knowledge on plants and animals in China is very rich that has been formed into the Chinese philosophy of 'Man and Nature are the One'; many plants and animals have been recorded in Chinese ancient literatures, for example, the 'Zhou Yi-Zhen-Yi' is one of the books over 2000 years. Different traditional forms and approaches of conservation ecosystems and biodiversity have been developed and maintained since ancient China which have made great contribution to the world civilization development and nature environment conservation (Pei 2006; Pei and Huai 2007).

## *4.2 The Relationship of Biodiversity and Cultural Diversity*

### 4.2.1 The Coevolutional Relationship of Biodiversity and Cultural Diversity

Biological resources are the basic nature resources for human life. Plants, animals, and associated ecosystem have been mixed together with human's material life and spiritual life. Today, human being is highly depended on and influenced on ecosystem even much stronger than any time in its history, at same time remarkable progress has been made in science and technology over the last half century, human have more powerful to change environment. However, culture and biodiversity are intimately and inextricably linked. The studies on coevolutionary relationship between biodiversity and cultural diversity are discussed in many ethnobotanical studies over the last three decades (Pei and Huai 2007; Pei 1987, 2000, 2001, 2002, 2006). The roots of appreciation of the value of biodiversity run very deep in many traditional world views connected to spiritual understanding and religions (McNeely 1993, 2003). The accelerated loss of biodiversity does not mean only the loss of gene, species, and ecosystem, but also destruction the unique structure relationships between all life forms and human cultures which further proves the importance to understand the coevolutional relationship of biodiversity and cultural diversity in biodiversity conservation.

### 4.2.2 The Interdependency of Biodiversity and Cultural Diversity

Without biodiversity, there is no cultural diversity to talk about. On the other hand, if there not have cultural diversity, biological diversity cannot be preserved effectively. Human society utilization and management of land and ecosystems are based on cultural values and arranged by the entitled social systems. The relationship of man and biodiversity is expressed through cultural expressions of human societies, which is the reflection of interdependency of plants, animals, and ecosystems and human being.

### 4.2.3 Culture Expresses Human Interactions and Biological Species

Human influence all aspects of biodiversity at different levels: species diversity, genetic diversity, ecosystem diversity, landscape diversity, etc.; man increases biodiversity throughout domestication and acclimatization biospecies, and decreases biodiversity through land use change, over-use biospecies, and introduce a certain economically high-value species (e.g., Rubber, coffee, cacao, tea, etc.) to replace native species resulting disappearing and extinction of hundreds of local species from local environment. On the other hand, man respect and protect some plants, animals, forest and ecosystems through cultural beliefs and religions systems, for instance Sacred Natural Sites (SNS) as cultural-landscape systems are well protected in southwest China through history, which greatly help biodiversity conservation in China (Pei 2002).

## *4.3 Biodiversity as the Material Basis for Building-up Human Cultures*

The relationship of biodiversity and traditional culture cannot be separated, the formulation, and development of human culture was influenced by biodiversity which is also the carrier of traditional culture. This chapter focuses on the analysis of human cultures that are directly related to biodiversity including linguistic culture, belief culture, landscape culture, agroculture, forest culture, medical culture, food culture, and folklore culture as discussed in below paragraphs:

### 4.3.1 Linguistic Culture and Biodiversity

The accumulated knowledge on biodiversity comes from leaning and exchange through language. In ancient Chinese literatures, there were massive records on names of plants, animals, and it distribution, usage, notes, and miscellanies. As early as the year 2000 BC, the Chinese Poem Collections ‘Shi Jing’ recorded more than 200 plants and more, including the famous aquatic weed *Potamogeton crispus* L. Linguistic culture contains very rich biological knowledge. There are 5000 and more ethnic groups in the world today; different culture have different names on plants, animals, and ecosystems; local names and nomenclature systems are even more details than modern scientific systems. For instance, the Hamunoo people in

Luzon Island of the Philippines have named 1600 plants in local flora but modern taxonomy could only identified 1200 species names (Cookling 1954).

### 4.3.2 Believe Culture and Biodiversity

Plants and animals play very important role in human belief systems; cultural believe regulates man's behavior and plays important role in biodiversity conservation, in particular religion belief; ancestor worships; and totem beliefs. Buddhism was introduced into China in Han dynasty (300 BC). Buddhist temples maintain many plants in the temple yard and protected significant nature forests in surroundings, case study shows that 58 plants are commonly cultivated in the temples in Xishuangbanna area, Yi people in Yunnan worship flowers of *Camellia reticulata* and *Rhododendron delavayi* both plants are well protected in forest area of Yunnan. SNS culture is one of the important cultural believes among ethnic minorities of China which includes the Dai's holly-hill forest; Tibetan's Sacred Mountains and Lakes; Miao's God-forests; Dong's village-protection forest, as well as the Han's 'Feng Shui' forests and 'Dragon-Pool' protection beliefs, which are not only important for biodiversity but also significant to protect ecological service functions of ecosystem (Pei and Huai 2007).

### 4.3.3 Landscape—Culture and Biodiversity

Many landscapes have been modified and influenced by human societies through human history. Since ancient time, landscape arrangement includes natural processes and man's activities that have shifted landscape in rural area and urban areas to meet their demanding for better life. Cultural landscapes contain rich biodiversity components. For instance, ficus trees are well protected by Dai people in Xishuangbanna, *Ficus religiosa*, *Ficus hookeriana*, *Ficus altissima* as well as *Ficus benjamina* are very attractive big trees in various cultural landscapes of south China and Tropical Asian countries. In Chinese Himalayas, the Chinese cypress tree (*Sabina chinensis*), Pine tree (*Pinus yunnanensis*), Oak tree (*Quercus pannosa*), Dog-wood tree (*Dendronbenthamis capitata*) among Naxi's landscape culture are important cultural plant species and well protected by local people in their surrounding landscapes (Pei 2006) (Fig. 2).

### 4.3.4 Agroculture and Biodiversity

Agro civilization was established on the success of domestication wild plants and animals by ancient farmers. China is one of the countries of the world agriculture origin in human history; paddy-rice cultivation can be traced back to 10,000 BC in China. China has extremely rich agrobiodiversity, from tropical agriculture, aquatic culture, and high plateau agriculture to Oasis agriculture, all can be found in China.

**Fig. 2** Traditional landscape in Xishuangbanna presents the Dai's landscape culture arrangement

Agro cultural diversity have benefited to agro biodiversity, contributing to maintain crop genetic at very high level. For instance, in the Hengduan Mountain's area of Northwest Yunnan where resides thirteen different ethnic groups (Tibetan, Naxi, Yi, Bai, Lisu, Lu, Jinpo, and others) landforms ranging from hot and dry valley to subalpine mountain lands distributed diverse agricultural patterns and maintains very high crop diversity, mountain crops such as buckwheat, belay, oat, potato, and maize are well maintained in farming systems of the region. In tropical region, shifting or swidden agriculture is a traditional agriculture practice that used to be a very common mode of production and way of life for mountain people in south China tropics. In 1980s, there was an argument on swidden agriculture among governments and societies in the region; however, better understanding the nature of swidden agriculture is the key. From scientific point of view, swidden agriculture is a traditional culture of millions people's way of life for their survivor; any change of the practice must be based on the swiddeners culture of their own. In fact, traditional swidden agriculture practice involves strategies for maintaining biodiversity and land-soil restorations in swidden farming lands, and crop diversity is highly maintained in swidden farming systems including hundreds of crop landraces of rice and wild edible plants (wild vegetable more than 20–30 species); even the Yak grass (*Imperata cymlindricd*) maintained in the fallow fields is used for construction houses and hunting sites. However, it is also revealed that swidden agriculture could be maintained only in areas where human population density is less than 15 people per km$^2$ (Ramboo 1983; Pei et al. 1997), but this old agroculture practice along with swidden agriculture are totally disappeared since 1990s in China due to rapid socio-economic development and cultural changes in the area.

### 4.3.5 Forest Culture and Biodiversity

Forest culture is the culture of traditional societies about forest management traditional knowledge and practices, forest traditional knowledge involves man's consciousness, belief, and values of forest; and management practice involves use of forest land, forest products harvesting, cultivation, management and protection methods, and technologies. In fact, forest culture is a complexity and precise knowledge body developed by people in forest areas. As forest culture is important for sustainable management of forest and at the time forest culture is rapid disappearing due to land use change and rapid urbanization development over the last two decades. In recent years, Chinese scholars (ethnobotanists, foresters, botanists, and ethnologists) have been working together to develop methods and approaches in forest culture studies and biocultural approaches for biodiversity conservation (Pei and Huai 2007). Therefore forest culture can be explained as a complexity of biocultural diversity with extensive application and significant role to play in biodiversity conservation. Traditional forest culture is not only consisting of belief, value, sacred forest, sacred land, and other spiritual cultures, but also involves forest utilization, management and protection rules, social organizations, community regulations, folk norms, technical standards, and other practical materialized standards. Globalization has come a time and modern industrial civilization was expanded into almost every corner of the world; traditional forest culture is being disappeared very rapidly, on the other hand, traditional forest management system is being transformed to adopt market economy model that threats biodiversity, for example, large tropical forest lands are converted into rubber plantations in Xishuangbanna over half century that is seen as a challenge to biodiversity conservation in the biodiversity hot-spot area, this was partially linked with forest land use from previous swidden agriculture into modern plantation economy and also transformation of mountain people livelihoods from forest people into plantation farmers (Pei and Huai 2007). We do not know what forest looks like in the future. Who will manage forest? How biodiversity is preserved? Can forest continue providing social and ecological service functions to us? So far there is no answer to these questions, but the facts are truly faced to us: nature forest area is continuing disappearing, forest structure and ecoservice functions are reduced rapidly, and people's forest value is being changed.

### 4.3.6 Medical Culture and Biodiversity

Use of plants and animals for medicine represents a long history of human interactions with the environment. In China, the first records herbal plants are the '*Shen-Nong Herbal Book*' in 3000 BC, it comprises 365 plants and animals. The updated inventory of TCM consists of 11,146 plants of which 10,654 are wild plants (Pei 2001). Today all knowledge about medicinal plants almost comes from traditional medical culture. Many modern medicine and new drugs are developed based on traditional medicine. Statistics show that there are 50,000 medicinal plants in the world, China ranks number one with 11,146 species; Europe holds about 1000 medicinal plants; North America about 1600 spp. Central and South America known for medicinal plants 5000 spp (Hamilton 2008). Therefore, as we can see even today plants are still the powerful weapons for human being to against disease. In fact all knowledge about medicinal plants and associated traditional medical culture are from traditional medicine of all nations in the world. Conservation of medicinal plants and all components of biodiversity have to be consulting with and working together with traditional medical culture in all nations.

### 4.3.7 Food Culture and Biodiversity

Food culture is directly linked to human health, food culture involves selection, harvesting, procession, preparation and storage of edible plants and animals, reflecting man's physical requirements (nutrient, ingredient, quantity, and quality), cultural demanding (color, fragrance, test, and appearance), and spiritual demanding (ritual, festival, ceremony, and worships). Chinese food culture is very diverse with long history; food sources include plants, animals, fungus, insects, and microorganism. Yunnan province is located in Southwest China that holds 50 % of plant and animal species of China, including edible plants 2000 spp.; edible fungus 150 spp.; edible insects over 100 spp. An unique food culture of flower-eating culture is remarkable among ethnic minority populations, number of edible flowers is over 150 plants such as flowers of *Gmelina arborea* in low-land tropics by the Dai people; and flowers of *Rhododendron decorum* in mountain forests by the Bai and Yi people; and flower of *Ottlea acuminate* from fresh waters in central Yunnan by all ethnic groups, the rich food culture of people in Yunnan is obviously supported by rich biodiversity in the mountainous province (Pei and Huai 2007).

### 4.3.8 Folklore Culture and Biodiversity

Plants and animals play an important role in folklore culture, since ancient time man have been established worships of nature and animal and plants totems. For instance, the Dragon totem, Tiger totem, Ox totem, Horse totem etc., are very old Chinese cultures among different ethnic groups. Many plants are worshiped by people in different areas, e.g., Calabash worship, worship of *Rhododendron delavayi*, *Camellia reticulata*, Artemisia, *Acorus calamus*, *Pyracantha fortuneana* etc., which are localized folklore cultures but with distinctive cultural symbolic meanings. Plants and animals are always used in folk music, arts, and writings; for example, plum, orchid, bamboo, and chrysanthemum are the 'Four Gentlemen'; Peony indicates indicating long life, firm, indomitable, etc. Biodiversity has been mixed up with people's daily life in every aspect including folklore culture, which can be explained not only that people love and enjoy biodiversity but also appreciate and preserve biodiversity.

## 4.4 *Trends and Progress on the Study of Biodiversity and Cultural Diversity*

### 4.4.1 Studies on Cultural Value and Ecosystem

Previous human ecology studies use cross-cutting culture methods to engage in a comparative study on the interactions of human culture and biophysical environments in Southeast Asia. In 1985 University of Michigan Ann Arbor published a book entitled 'Cultural Value and Human Ecology in Southeast Asia'(Hutterer et al. 1985), in which a number of case studies were presented using case study to analysis impact of cultural values on environment, one of the case study reports was 'Some Effects of Dai People's Cultural Beliefs and Practice upon Plant Environment of Xishuangbanna, Yunnan Province, SW China (Pei 1987)', this was first report on the interactions of culture and environment among ethnic minorities in China by a Chinese scholar.

In 1984–1996, Chinese ethnobotanists and ecologists in collaboration with American human ecologists and scientists from Southeast Asia countries working together in a regional network namely Southeast Asia University Agro-ecosystem Network (SUAN). The SUAN group was engaged in rural agroecosystem analysis and assessment throughout various rural ecosystem case studies in the region including Yunnan of China; the human ecology approach on rural ecosystem studies was based on a conceptual frame work of interactions between social system and forest-farming natural systems with five indicators of the hierarchical system: productivity, stability, sustainability, equilibrium, and resilience (Ramboo 1983) (Fig. 3).

During 1985–2009, Chinese scholars were starting to use ethnobotanical methodologies and approaches to study impact of human culture on biodiversity begun with ethnic minority culture and biodiversity in southeast China; hundreds of research papers and case study reports were published in this regard over the last

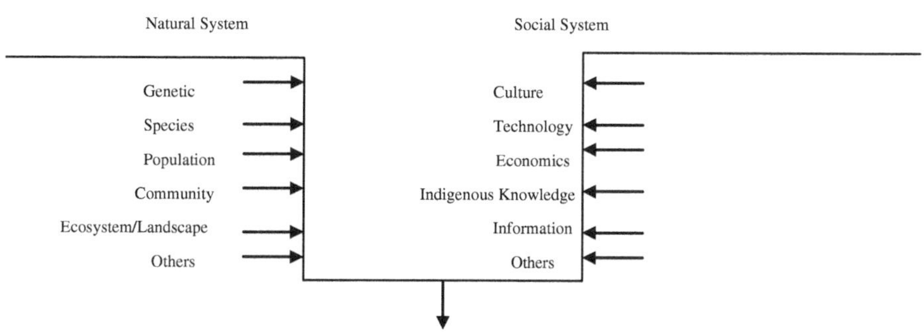


**Fig. 3** A theoretical modal for study social system and natural system (cited from: Pei and Sajise 1995)

three decades (Pei 2000, 2001, 2002, 2006, Pei and Huai 2007; Pei et al. 1997). Many of these studies were focused on the below aspects:

(i) Cultural value and biodiversity conservation
(ii) Traditional knowledge and biodiversity management
(iii) Adoptive technologies for biodiversity resource management
(iv) Natural and cultural heritage protection and biodiversity.

### 4.4.2 Culture-Disappearing Accelerates Biodiversity Lost

Over the last half century, globalization has brought the world into a new era, economic development model of China has been transformed from traditional agriculture model into industrial development model, traditional organic, biodiversity-based agriculture which is being transformed into intensive, mono-culture, petroleum–based, and gene transfer high-tech agriculture, which is seen as revolutionary change in Chinese history that has brought about unpredictable effects on biodiversity and cultural diversity. Anthropologists pointed that today in China the loss of 'Cultural Species' is much faster than the loss of biological species. At present, the loss of culture diversity in connection with biodiversity can be summarized into below six respects:

(i) Rapid loss of traditional knowledge
(ii) Change of cultural values
(iii) Modern development accelerates loss of traditional technologies
(iv) Dis-integrated the inheritance mechanism for traditional culture
(v) Dis-connection of biodiversity and cultural diversity
(vi) Misunderstanding traditional cultures.

## *4.5 Conclusion*

The modern human society has exerted massive impact upon all forms of life on earth, our social economy, development of science and technology, management of natural resources, cultural values, policy, and legislation all have impacts on nature and biodiversity. In fact, the conservation of biodiversity today is far more than a purely scientific issue. Globalization has accelerates the process of biodiversity loss. Many believe it is the economic drive to balance, actually the root-cause threat on biodiversity comes not only from the economy development, but also from the cultural values and the development model we use that decide the mode of economic development in a society. In fact, traditional knowledge has made great contribution to the development of science and technology in human history (ICSU 2002). Our study indicates that human culture is a critical factor for biodiversity conservation. Traditional cultures offer numerous examples of species and ecosystem presentations. By rescuing traditional cultures associated with cultural diversity and biodiversity coexist and coevolve with each other. Their interaction is strong and complicated, varying from one ethnic group to another ethnic group and from one region to another region. We believe that with 5000 years development of sophisticated Chinese culture and enormous rich traditional knowledge on biodiversity and environment, China will reach a new era of ecocivilization in which cultural diversity and biodiversity will be maintained in a harmony world to contribute to national social, economic, and cultural development.

## References

Ahrends, A., Rahbek, C., Bulling, M. T., Burgess, N. D., Platts, P. J., Lovett, J. C., et al. (2011). Conservation and the botanist effect. *Biological Conservation, 144*(1), 131–140.

Ayyad, M. A. (2003). Case studies in the conservation of biodiversity: Degradation and threats. *Journal of Arid Environments, 54*, 165–182.

Bellon, M. R., Brar, D. S., Lu, B. R., & Pham, J. L. (1998). Rice genetic resources. In N. G. Dwoling, S. M. Greenfield & K. S. Fischer (Eds.), *Sustainability of rice in the global food system, Chap. 16* (pp. 251–283). Davis, CA/Manila: Pacific Basin Study Center/IRRI.

Chandra, A., & Idrisova, A. (2011). Convention on biological diversity: A review of national challenges and opportunities for implementation. *Biodiversity and Conservation, 20*(14), 3295–3316.

Chen, G. K., & Ma, K. P. (2012). Criteria and methods for assessing the threat status of ecosystem. *Biodiversity Science, 20*(1), 66–75.

Chen, Y. H., Tang, Z. Y., & Fang, J. Y. (2009). Distribution of nature reserves and status of biodiversity protection in China. *Biodiversity Science, 17*(6), 664–674.

Cookling, H. C. (1954). *The relation of human culture to the plant world*. New Haven: Yale University.

Evenson, R. E., & Gollin, D. (2003). Assessing the impact of the green revolution, 1960–2000. *Science, 300*, 758–762.

Hamilton, A. C. (2008). *Medicinal plants in conservation and development: Case studies and lessons learnt*. Salisbury: Plantlife International.

Harlan, J. R., & De, Wet J. M. (1971). Toward a rational classification of cultivated plants. *Taxon, 20*, 509–517.

Huttterer, K., Ramboo, A. T., & Lovelace G. (1985). Cultural value and human ecology in Southeast Asia. In *Michigan Paper on Southeast Asia*. Ann Arbor: University of Michigan.

ICSU. (2002). ICSU series on science for sustainable development, no. 4, science. In *Traditional knowledge and sustainable development*. Paris: ICSU.

IISD (International Institute for Sustainable Development) and FAO (Food and Agriculture Organization). (2009). *A summary report from the world summit on food security*. World Report on Food Security Bulletin. http://www.iisd.ca/ymb/food/wsfs2009/html/ymbvol150num7e.html.

Jiang, Z. G., & Luo, Z. H. (2012). Assessing species endangerment status: Progress in research and an example from China. *Biodiversity Science, 20*(5), 612–622.

Kiang, Y. T., Antonvics, J., & Wu, L. (1979). The extinction of wild rice (Oryza perennis ormosa) in Taiwan. *Journal of Asian Ecology, 1*, 1–9.

Lu, B.-R. (2001). Conservation and sustainable use of biodiversity in wild relatives of crop species. In *Proceedings of 3rd IUPAC International Conference on Biodiversity, November 03–08, 2001*. Biodiversity: Biomoleculer aspects of biodiversity and innovative utilization (pp. 23–34).

Lu, B.-R. (2013). Introgression of transgenic crop alleles: Its evolutionary impacts on conserving genetic diversity of crop wild relatives. *Journal of Systematics and Evolution, 51*, 245–262.
Lu X.-X., Chen, S.-P., Liu, X., et al. (2008). *Technology regulations for crop germplasm resources preservation*. Beijing: Chinese Agriculture Press.
Mack, R. H., Simberloff, D., Lonsdale, W. M., Evans, H., Clout, M., & Bazzaz, F. A. (2000). Biotic invasions: Causes, epidemiology, global consequences, and control. *Ecological Applications, 10*, 689–710.
McNeely, J. A. (1993). Diverse nature, diverse culture, people and planet, *2*(3), 11–13.
McNeely, J. A. (2003). Biodiversity and cultural diversity, helex, in Darwin. In *International Symposium on Biodiversity and Human Health, Taichung, Taiwan*.
Pei, S. J. (1987). Some effects of the Dai People's cultural believe upon plant environment of Xishuangbanna, Yunnan Province, SW China. In K. Hutterer et al. (Ed.), *Michigan Paper on Southeast Asia*. Ann Arbor: University of Michigan.
Pei, S. J. (2000). Traditional culture and nature conservation. *Nationality Conservation, 8*, 34–36.
Pei, S. J. (2001). Ethnobotanical approaches of traditional medicine studies: Some experiences from Asia. *Pharmaceutical Biology, 39*(Supplement), 74–79.
Pei, S. J. (2002). Bio-Cultural diversity and development of West China. *Journal of Graduated School, Chinese Academy of Sciences, 2*, 1.
Pei, S. J. (2006). Sacred natural site and biodiversity conservation. *Science, 58*(6), 29–31.
Pei, S. J., & Huai, H. Y. (2007). *Ethnobotany*. Pudong: Shanghai Sciences and Technology Press.
Pei, S. J., & Sajise, P. (1995). *Regional study on biodiversity: Concepts, frameworks and methods*. Kunming: Yunnan University Press.
Pei, S. J., Xu, J. C., Chen, S. Y., & Long, C. L. (1997). *Collected regional papers on biodiversity in swidden agroecosystems in Xishuangbannan*. Kunming: Yunnan Education Publishing House.
Quan, J., Ouyang, Z. Y., Xu, W. H., & Miao, H. (2011). Assessment of the effectiveness of nature reserve management in China. *Biodiversity and Conservation, 20*, 779–792.
Ramboo, A. T. (1983). *Conceptual approaches to human ecology*. EAPI Research Report, No. 14, East-West Center, Honolulu, Hawaii, USA.
State Forestry Bureau. (2009a). *Investigation of key state protected wild plant resources*. Beijing: China forest press.
State Forestry Bureau. (2009b). *Investigation of key state protected wild animal resources*. Beijing: China forest press.
Wang, S. M., Li, L. H., Li, Y., Lu, X. X., Yang, Q. W., Cao, Y. S., et al. (2011). Status of plant genetic resources for food and agricultural in China (I). *Journal of Plant Genetic Resources, 12* (1), 1–12.
Xue, D. Y., Wu, J. Y., & Zhao, F. W. (2012). Actions, progress and prospects in implementation of the convention on biological diversity during the past 20 years in China. *Biodiversity Science, 20*(5), 623–632.
Yuan, H., Zhang, Y. B., Qin, H. N., Liu, Y., & Yu, M. (2009). The in situ conservation of state key protected wild plants in national nature reserves in China. *Biodiversity Science, 17*(3), 280–287.
Zhang, Y. B., Yuan, H., & Yu, M. (2011). Assessment of threaten status on the wild plants under state protection in China. *Biodiversity Science, 19*(1), 57–62.
Zhu, Y. Y., Wang, Y. Y., Chen, H. R., & Lu, B.-R. (2003). Conserving traditional rice through biodiversity management. *BioScience, 53*, 158–162.
Zhu, W. Y., Zhou, T. Y., Zhong, M., & Lu, B.-R. (2007). Sampling strategy for wild soybean (Glycine soja) populations based on their genetic diversity and fine-scale spatial genetic structure. *Frontiers of Biology in China, 2*, 397–402.

# Chapter 2
# Biodiversity Conservation and Its Research Process

**Zhigang Jiang, Fumin Lei, Chunlan Zhang and Moucheng Liu**

**Abstract** In this chapter, we mainly introduce the wildlife vertebrates and birds conservation and research. China has great habitat and biological diversity; however, many of wildlife are in peril due to habitat loss and over exploitation in the past few decades. Ever since, China had strengthened the protection of wild animals, especially the establishment of nature reserves. Here we take the protection of giant panda, Milu and Przewalski's wild horse as case studies. For birds, China has extremely rich bird resources, highly endemism, but faces high threaten for endangered species. In addition, the development of molecular biology, bioacoustics, a variety of analytical software, various analytical models, etc, have played an important role in promoting the scientific researches and conservation of birds in China.

**Keywords** Biodiversity · Vertebrates · Giant panda · Przewalski's wild horse · Birds · Endangered bird species · Wild bird diseases · Conservation

## 1 Wildlife Conservation and Research

### *1.1 Introduction*

China's fauna is divided by the Palaearctic and Indomalayan Realms with a boundary lies approximately along the Mt. Qinglin and Huihe River. The climate in

Z. Jiang
Institute of Zoology, Chinese Academy of Sciences, Beijing 100101, China

F. Lei · C. Zhang
Key Laboratory of Zoological Systematics and Evolution, Institute of Zoology, Chinese Academy of Sciences, Beijing 100101, China

M. Liu (✉)
Institute of Geographic Sciences and Natural Resources Research, Chinese Academy of Sciences, Beijing 100101, China
e-mail: liumc@igsnrr.ac.cn



W. Li (ed.), *Contemporary Ecology Research in China*,
DOI 10.1007/978-3-662-48376-3_2

China is diverse. Much of southern and southeastern China has a subtropical climate; which turns into more continental northwards and westwards. Monsoon rains affect the east coast, while the desert interior of the west has very low rainfall. The size of the country, its climate and topographical variety and its biogeographic position result in great habitat and biological diversity. Owing to the temperature difference, from the north to the south, there are Taiga, deciduous broadleaved forest, ever-green broad levered forest and rain forest in the Monsoon Zone. Wild animals in the zone are primarily forest inhabitants, like red and white giant flying squirrel *Petaurista alborufus*, sika deer *Cervus nippon*, and tiger *Panthera tigris*. Tibetan plateau characterized by the dry and cold alpine climate. Many wild animals live on the open plateau are endemic ungulates like Chiru *Pantholops hodgsonii*, Kiang *Equus kiang*, wild yak *Bos mutus*, white lipped deer *Przewalskium albirostris*, and Tibetan gazelle *Procapra picticaudata*, which are significantly different from the surrounding region. However, the differentiation of the fauna in the region is basically at the species or genus level, not the family level. Even though, Chen et al. (1996) claim the Tibetan region should be classified as a separate zoogeographic realm. Many researchers are exploring the ecological and evolutionary mechanism which determines the animal species richness pattern in China (Li et al. 2013; Luo et al. 2012).

## 1.2 Status of Terrestrial Vertebrates

As new species being discovered, new taxonomy is being adopted and new records are being added, and number of species in the country is increasing. Roughly, there are 2637 terrestrial vertebrate species in the country (Table 1). Endemic species varied from nearly 70 % in the amphibians to about 6 % in birds. During the last century, because of rapid economic growth, pressure on wildlife mounted in the country.

Generally, the species richness decreases from southeast to northwest in China. For an example, only 12.9 % of the 100 × 100 km grids contained more than 500 vertebrate species. These high vertebrate species richness (VSR) grids were mainly located in the southwestern areas, tropics, and sub-tropics of the country, which contained several hot spots, including the Hengduan Mountains, the Xishuangbanna region of Yunnan Province, the southeastern and southern coasts, Hainan Island, and Taiwan Island. The grids containing 200–500 species were mainly concentrated in

**Table 1** No. of species and endemics of mammals, birds, reptiles, and amphibians

| | No. of species | Endemics | Endemics (%) |
|---|---|---|---|
| Amphibians | 298 | 208 | 69.80 |
| Reptiles | 402 | 131 | 32.59 |
| Birds | 1330 | 82 | 6.17 |
| Mammals | 607 | 114 | 18.78 |
| Total | 2637 | 535 | 20.29 |

the vast eastern and northeastern plains of the country, which accounted for 49.2 % of the total of grid cells. The remaining grid cells (37.9 % of the total) had VSRs of <200 and they were mainly located in the northwestern areas and Qinghai–Tibetan Plateau (Luo et al. 2012).

Many of China's wildlife are in peril due to habitat loss and over exploitation. In 1989, The *Wild Animals Protection Law of PRC* was promulgated, which was a milestone for wildlife conservation in China. One hundred and one animal species are listed in the Category I of the National Key Protected Wild Animals. 84 species are listed in CITES Appendix I, including one amphibian, six reptiles, 34 birds and 43 mammals. One hundred and forty-eight species are listed in CITES Appendix II, including one amphibian, five reptiles, 99 birds and 43 mammals. Fifty-three species (10 amphibians, 15 reptiles, 4 birds and 24 mammals) are listed as CR in the IUCN Red List; 154 species (19 amphibians, 22 reptiles, 22 birds and 91 mammals) as En; 324 species (78 amphibians, 67 reptiles, 73 birds and 106 mammals) as VU (IUCN 2013).

Habitat management is one of the most important aspects of wildlife conservation, while establishing nature reserves is one of the most important measures to protecting habitat. Nature reserve in the country is following the MAB model, a nature reserve is divided into three function zones: core zone, buffering zone, and the experimental zone. Human activities are forbidden in the corn zone of reserves. Until the end of 2012, 2669 nature reserves of 1.50 million km$^2$ have been established, which accounted for 15 % of the territory of China.

## 1.3 Case Studies

### 1.3.1 Giant Panda

Giant panda *Ailuropoda melanoleuca* is a relic species which lives in fragmented habitats of Mt. Minshan, Mt. Chionglai, Mt. Major Xianglin, Mt. Minor Xianglin, Mt. Liangshan and Mt. Qinglin in central China. 1,569 pandas were estimated and lived in field at the end of twentieth century according to the Third Nationwide Giant Panda Survey. Long-term isolation caused genetic differentiation in panda population. It is reported that the giant panda of Mt. Qinglin is a subspecies of giant panda due to geographic isolation. Logging was a threat to the giant panda; five nature reserves were established in 1963 to protect the habitats of giant panda. Since 1970, the country has already finished three nationwide giant panda surveys; the fourth one is nearly finished. Since 1980, many researches on the field ecology of giant panda had been conducted (Schaller et al. 1985; Pan et al. 2001). Recently, development in molecular biology, like whole-genome sequencing of giant pandas provides insights into demographic history and local adaptation of the relic species (Li et al. 2010a; Zhao et al. 2013), means to assess implications for conservation of the drastic reduction of the smallest and most isolated giant panda population.

Nationwide survey, field and molecular ecological research identified key populations and habitats of giant panda and helped to understand behavior, ecology. and evolution of the giant panda; thus, set up scientific basis for field conservation. About 60 % of giant panda habitat, supporting over 70 % of the wild populations, is now protected in 64 nature reserves, most of them are national nature reserves. The ex situ population is demographically and genetically strong, and increased efforts are underway to develop an effective release program to reinforce wild populations. On the other hand, artificial breeding of giant panda has been carried out in the Wolong, Chengdu, Beijing and Fuzhou (Peng et al. 2001, 2009). Integrated global ex situ conservation strategy is supported by the Chinese Association of Zoological Gardens (CAZG) , the State Forestry Administration (SFA), and Conservation Breeding Specialists Group (CBSG) of IUCN. More than 300 giant panda have been bred in breeding centers and zoos in China. Global ex situ giant panda population is now 375 pandas, conservation breeding plan uses less genetically valuable females designated to produce offspring suitable for release training efforts. Recently a re-wild of giant panda project is well on the operation at Wolong Giant Panda Conservation Center.

### 1.3.2 Milu

After the last glacial period, Milu *Elaphurus davidianus* was restricted to swamp and wetland in the region south of 43°N and east of 110°E in China. Population of Milu declined because of human hunting and land reclamation as human population expanded in Holocene. Finally, Milu was extinct in the field (Cao 1992). The first conservation reintroduction of Milu into China included two groups of 20 (5♂: 15♀) and 18 (all♀) in 1985 and 1987, respectively. Beijing Milu Park (39°07′N, 116°03′E) was established. The second reintroduction of 39 Milu, selected from five UK zoos, was carried out in August of 1986. Dafeng Milu Natural Reserve (33°05′N, 120°49′E) was established to host the reintroduced Milu (Jiang et al. 2000). Further population growth in Beijing Milu Park was restricted by its limited size. Thus, more than 300 Milu were relocated to over 50 sites all over China. Ninety-one Milu were relocated to Shishou Milu Reserve, which was established in 1993 and 1995. A flooding of the Yangtze River in 1998 resulted in several cohorts of Milu leaving the initial release area and forming permanent herds in other parts of the province, as well as around Dongting Lake in Hunan province (Maddison et al. 2012). Researchers and graduate students conducted research projects including population monitoring on the introduced Milu in the country. Three international workshops on management and research on the reintroduced Milu were held at Beijing Milu Park in 2006 and Dafeng reserve in 2011 and 2012, respectively. Recently, a team is monitoring the field-released Milu in coast marsh of Dafeng with satellite collars. Many papers have been published in peer-reviewed journals (Zeng et al. 2013; Li et al. 2011a, b).

### 1.3.3 Przewalski's Wild Horse

Przewalski's wild horse *Equus przewalskii* is a fleet ship species in the Jungar Basin in northern Xinjiang where the first type specimen of the wild horse was collected in the 19th century. The wild horse was reintroduced to China in 1985. From 1985 to 2005, a total of 24 Przewalski's horses (14 males and 10 females) were transported to Jimsar Wild Horse Breeding Center in Xinjiang. The first foal was born at the breeding center in 1988. Since then, a total of 258 foals have been born, and the number of animals in the captive population continues to increase. On August 28, 2001, 27 wild horses were released into the Mt. Kalamaili Ungulate Nature Reserve. Studies on the acclimation, food habit, and community-based conservation were conducted (Chen et al. 2008). Because the first released site was close to a major highway, fatal vehicle collision caused several casualties in the released wild horses; the wild horses were then relocated to Qiaobaixili region of the reserve. After acclimating to the local habitat, the released wild horse families established territories and started to breed. Seventy-two foals were born, 7 families with 73 wild horses roamed in field in 2010. Number of re-wild wild horses reached 96 by the end of 2012 (Fig. 1).

### 1.3.4 Wildlife Trade

Commercial trade is identified as a threat to biodiversity. Unregulated international trade in wildlife not only devastates local ecosystems, but also poses threat to the

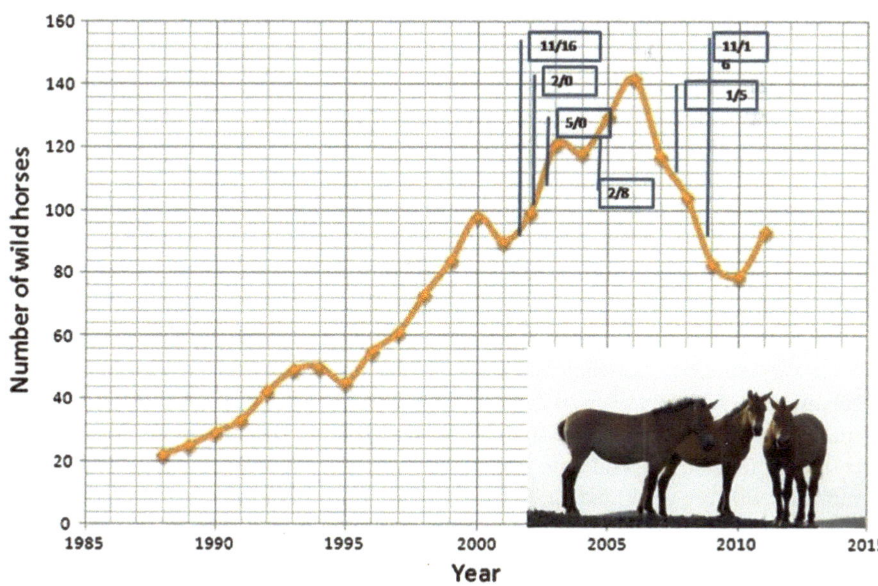


**Fig. 1** Population trends of the reintroduced wild horse in Xinjiang. Numbers in the flags indicate the numbers of wild horse (♂/♀) released into field

survival of individual species. Trade records show that since 1990, with respect to some species of snakes, China has changed from a net export country to a net import country. Importing snakes sharply increased in China before 2002 (Zhou and Jiang 2004). Since then, measures of suspending snake trade had been imposed by National Wildlife Management Authority in China. Jiang et al. (2013) found both import and export of all snakes in China recorded in the CITES Trade Database and the Wild Animal and Plant International Trade Database of China have sharply decreased since 2004. Li and Jiang (2014) also found that international trade of live birds in China peaked during the late 1990s; then decreased to the level before the surge of trade in a few years. The trade dynamics of wild birds may have been affected by governmental policy and the outbreak of avian influenza during the period.

## *1.4 Conclusion and Discussion*

Wildlife conservation is a branch of conservation science. It is to be noted that there are more conservation practices than theoretic research have been done in the field (Jiang et al. 2014, in press). The wildlife ecology should be the focus of the wildlife research in China; however, not until three decades ago, people had started to study live animals in the field. While the achievement in artificial propagation of endangered species and nature reserve construction in the country is encouraging, researches on the behavior, ecology, and management of the endangered wild animals in China have just been carried on. The reality is that many of China's wild animals either live in remote habitat, possess of cryptic nature or are of very low density; it is confronted with logistic problems to carry out field study, even though Chinese researchers are still working hard in field. Routine surveys monitor the population trends of wildlife. Progresses have been made in behavioral ecology. Radio and GPS collars and camera traps have been put in use in the field. GIS, in aid with GPS, is now widely in use for analyzing spatial data like GPS position data of animal movement and home range. Molecular ecology is developing rapidly in the country. With little samples, using PCR and computer software, people can infer the historical population trend, genetic landscape, gene flow, and population genetic structure of wild animals in laboratory, which becomes an indispensable tool in wildlife research. Nevertheless, as a country with mega animal diversity, field ecology study is not matched up with the number of species and degree of species in peril. Conservation of biodiversity, of which conservation of wildlife is a key part, is on the top agenda of the nation. There are gaps in scientific knowledge. Since many species are endemic which only live in the unique ecosystem in the country, the wildlife ecologists in the country need to desperately work more on the wild animals in the field in order to fill up the knowledge gaps. Younger field ecologists should be trained for qualifying the enormous work.

# 2 Biodiversity and Conservation of Birds in China

## 2.1 Introduction

As one of the "megadiversity" countries of the world, China has very rich bird resources; however, it has also been seriously threatened now. Increasing emergence of zoonotic diseases makes it even worse. China spans two ecozones, the Palearctic realm and the Oriental realm, which makes it an important region for biodiversity research and conservation. Benefited from new technologies in taxonomy, many new species and new records have been reported in China. The advance of phylogeography also increases our knowledge in understanding the formation of high endemism and species richness in China. This chapter therefore summarizes a brief review on the following five topics: biodiversity and distribution, endemism and conservation priority, phylogeography and molecular ecology, conservation and management for rare and endangered species, and avian diseases and eco-health.

## 2.2 Biodiversity and Distribution

New species or records of birds are discovered in China frequently because of the fast development of the research methodology and increasing numbers of ornithological researchers and bird watchers. During the past 50 years (1958–2008), six new species were reported by Yang and Lei (2009), e.g., *Bradypterus alishannensis*, *Phylloscopus emeiensis*, *Phylloscopus Hainanus*, *Seicercus omeiensis*, *Seicercus soror,* and *Certhia tianquanensis*. Currently, 1371 species are recognized by Zheng (2011), accounting for 13.12 % of the world total species.

China is characterized for holding both high species diversity and endemism of birds, thus has important role in conservation of world bird resources (Lei et al. 2003a, b). Some groups, such as cranes, pheasants, and babblers, have been attracting worldwide attention. Nine of 15 crane species of the world are distributed in China, while the black-necked crane is the only one in high altitude of Qinghai–Tibetan Plateau (QTP). There are 27 species of pheasants and 131 species of babblers in China, occupying more than half of the world species respectively. In China provinces, Yunnan has the highest species richness. Over 848 species account for over 62 % of the whole species recorded in China (Yang 2004), followed by Sichuan (683 species, Xu et al. 2008). Geographically, Hengduan Mountain areas harbor the highest richness, which has also been suggested as original center for many taxa. Nearly half of pheasant species and babblers are distributed here with high subspecies and population differentiation. Qinling Mountains lies in central China. It is an important boundary between the Palearctic Realm and the Oriental Realm.

China also has varied habitats, especially wetlands, and thus it has rich waterfowls. On the East Asian-Australian Flyway and Central Asian-Indian Flyway, there are the most important wintering and breeding grounds as well as stopovers for wild bird migration, for example *Grus leucogeranus*, *Grus monacha,* and *Grus grus* wintering in Poyang lake; *Anser indicus* and *Tadorna ferruginea* breeding in Qinghai lake. Climate warming was recently reported influencing the geographical distribution range of birds. One hundred and twenty species in China were reported to extend their distribution range northward, e.g., *Egret tagarzetta*, *Ardeola bacchus*, *Pycnonotus sinensis* (Du et al. 2009). Wang et al. (2010) analyzed the change of species richness patterns from 1976 to 2005 based on breeding birds distribution database and found that the richness increased in all zoogeographic subregions. Southern Yunnan Hilly subregion, Qiangtang Plateau subregion, and East Meadow subregion had the most obviously increasing but Hainan and Taiwan subregions are relatively more stable.

## 2.3 China Avian Endemism and Biodiversity Conservation Priority

Endemism is the most interesting question in biogeography and biodiversity conservation (Crisp et al. 2001). China is one of the most important countries in global biodiversity and biogeography. The distribution of endemic species has been considered very important for China avifaunal regionalization (Cheng et al. 1997). It has also been suggested in setting priorities for biodiversity conservation (Lei et al. 2003a, b).

The peak of richness distribution of endemic species was found in three areas, including Hengduanshan Mountains, mountain areas of western Qinling, north Sichuan province and South Gansu province, as well as Taiwan Island (Lei et al. 2003a). For endemic genera, the northern and eastern Hengduanshan Mountains, and the Qinling, Dabashan and Minshan Mountain regions have been found with the highest richness. Obviously, both endemic species and endemic genera have high richness in east edge of QTP, which was considered because of the uplift of QTP. Of these areas, Taiwan Island has the highest narrow distributed species than mainland, which implies that island isolation has great contribution to differentiation for Chinese avifauna. By comparing subregional distribution of overall endemic species, narrow distributed range species (EOSR), monotypic species and subspecific diversification, Lei et al. (2007) concluded that this pattern might reflect the avifaunal evolutionary and ecological isolation results from the highly diversified habitats and geographical environments as well as the historical effects from the primitive avifauna, inferring this "ecological island effect" hypothesis for explaining what driving this pattern. Huang et al. (2010a, b) using Parsimony analysis of endemicity recognized four Areas of Endemism (AOE): Qinghai-Zangnan Subregion, the Southwest Mountainous Subregion, the Hainan

Subregion, and the Taiwan Subregion). All these four AOEs located at the mountainous habitats, which implied the hypothesis that "mountainous environment may act as "historical and ecological barriers" preventing population gene flow, promoting speciation and maintaining a high endemism in explaining China avian endemism."

Based on endemicity of birds, biodiversity conservation hot spots and priority have been proposed. BirdLife International has ever suggested the Endemic Bird Area (EBA) in explaining avian endemism (Stattersfield et al. 1998). Lei and Lu (2006) also argued that it is wise, reasonable and practical to determine the priority for biodiversity conservation for China, a developing country. Lei et al. (2003a, b, 2007) proposed the distribution center of endemic species as the "biodiversity hotspots" in reference to set the priority of biodiversity conservation. By considering the distribution patterns of both endemic species and endemic genera, the southeastern peripheral areas of the QTP (e.g., Qinling-south Gansu mountainous region, Henduanshan Mountain areas), being considered as refugia in Pleistocene because of the high endemism and genetic diversity, should have the highest priority for conservation (Lei et al. 2003a, b; Lei and Lu 2006).

The study on distribution patterns by using GIS in China avian endemism was referenced by other taxa researches, and these algorithms for site prioritization have also been cited to identify indicative sets of potential conservation areas (Solymos and Feher 2005). The studies from the distribution pattern of endemic species and genera have inferred the "historical and ecological barrier" hypothesis, the "ecological island effect" hypothesis, and "evolutionary powerhouse" in particularly explaining the hot spot in Hengduan Mountain areas. These macro-ecology based findings have provided scientific questions and testable hypotheses for explaining the underlying mechanisms of formation of China avifauna and biogeographical distribution patterns, meanwhile the regional endemism properties have also provided new scientific questions for interpreting the global scenario of all animal endemicity.

## 2.4 Phylogeography and Molecular Ecology

Phylogeography is in understanding the principles and processes of forming the geographic distributions of genealogical lineages (Avise et al. 1987). Phylogeographical studies on birds obtained significant improvement by understanding the genetic consequences of the geological events on birds' population structure. Particularly, in North America, the recent population expansion was revealed in many passerine species, indicating that the current distribution patterns may be the consequences of the post-glacial population expansion during the late Pleistocene (Spellman et al. 2007). In Europe, many species experienced population expansions after the Last Glacial Maximum (LGM), which was believed profoundly affecting population dynamics (Hewitt 2000). The glacial cycles in Asia differ from those in Europe, even though the two continents spread similar latitudinal belts (Hewitt 2000). Post-glacial population expansion was less diagnosed in Asian birds, and the populations were rather stable through the late Pleistocene, inducing unique phylogeographic patterns in Asian birds. The phylogeographical researches about birds in China disclosed multiple phylogeographical patterns and glacial refuges and indicated that population divergences were much affected by Pleistocene climate changes earlier than LGM (Yang et al. 2009).

Phylogeographic patterns and its driving factors of birds in China have been well studied during the last 10 years. Phylogeographical studies on birds in Qinghai–Tibetan Plateau (QTP) revealed population dynamics and geographical distribution shifts response to Pleistocene glacial oscillations. The uplift of QTP greatly impacted the phylogeographic structure of the Plateau species, while most of the QTP species underwent population expansion after glacial movements. The eastern margin of the QTP area was detected as refugia for many plateau species during the Pleistocene glaciations. e.g., Onychostruthus taczanowskii, Pyrgilauda ruficollis, and Pseudopodoces humilis experienced rapid population expansion (0.07–0.19 Ma) from the eastern "refugia" to the platform of the plateau after the retreat of the extensive glaciers (Qu et al. 2005; Yang et al. 2006). But most species have not shown any deep phylogeographical structures, with the "no divergence" pattern, while Pseudopodoces humilis has distinct phylogeographical structure with a "north-south divergence" pattern. A "platform and edge" phylogeographical divergence was also detected by other plateau species, e.g., Carduelis flavirostris. (Qu et al. 2010). The QTP platform populations were derived from a single refuge at the eastern edge of the plateau. No bottleneck effect or population expansion was found at the lower altitude edge populations (Qu et al. 2010). The results implicates that plateau birds experienced population expansions from edge to the platform of the plateau around 0.17–0.50 mya, after the glacial extension in QTP. Elliot's laughing thrush is endemic to the Hengduan Mountain. It was isolated in different areas during the interglacial periods but connected again when they expanded to suitable habitats at low elevation during glacial periods; these repeated population isolation and extension are occurring in the spatial "sky island" pattern (Qu et al. 2011). The pre-LGM population expansions are rather earlier than post-LGM expansion scenarios common in European and North American birds. The studies shed lights on evolutionary history and formation of the avian fauna of the QTP. The refuges revealed by the phylogeographical studies are coincident with endemic hot spots detected by biogeographical analyzes, implying fundamental processes and mechanisms of avian fauna dynamics response to climate changes.

Phylogeographical structures and population divergence of south China species is different from the QTP species in many cases. Bambusicola thoracica, Alcippe morrisonia, Stachyridopsis ruficeps, and Parus monticolus distributed in Southern China are all detected with multiple phylogeographical breaks, but different species have different lineage structures. This population divergence was considered to be related to the uplift of the QTP, topographic complexity, and mountain system. Populations of most species have experienced the expansion events much earlier than the Last Glacial Maximum (LGM). The consistent patterns of this "pre-LGM" population expansion imply less impact of climate changes on birds in South China (Dai et al. 2011; Song et al. 2009; Huang et al. 2010a, b).

Remarkable achievements have been done for avian phylogeography in China within the last 10 years. The preliminary findings shed lights on the mechanism and processes of avian speciation and diversification of birds in China. The results from phylogeographcial studies also help us, from the evolutionary viewpoint, to explore the history and underlying mechanism in forming the diversity pattern as well as endemic pattern of birds in China, which enhance our knowledge in understanding the formation and evolution of the global biodiversity and endemicity.

## 2.5 Conservation and Management of Rare and Endangered Species

China has very rich rare and endangered bird species. From IUCN report in 2011, 124 species are threatened, 7 of them are critical endangered, e.g., *Eurynorhynchus pygmeus, Garrulax courtoisi, Pseudibis davisoni, Grus leucogeranus, Fregata andrewsi, Sarcogyps calvus,* and *Sterna bernsteini*, while 17 species are endangered, e.g., *Gyps bengalensis, Nipponia nippon, Arborophila rufipectus, Aythya baeri, Ciconia boyciana, Grus japonensis, Mergus squamatus, Platalea minor*, etc. (http://www.iucnredlist.org). In regard to conservation, China has contributed great efforts and obtained important achievements for these species. Here we have only taken fewer examples, e.g., Cabot's Tragopan (*Tragopan caboti*), Crested Ibis (*Nipponia nippon*), Black-faced Spoonbill (*Platalea minor*), etc.

Cabot's Tragopan is an endangered and endemic pheasant in China and listed as national key protected animal species (class I) by law. Since 1980, Chinese researchers have been working on the study and conservation and have now successfully solved the breeding, feeding, and conservation problems. A series of key techniques referring to the artificial sperm collection and insemination has been grasped. Based on these basic research results, Chinese scientists have now successfully established the artificial breeding population for over 100 individuals (Zhang 2005). The restoration of the habitat is the key to protect these endangered species. Since 1990, lots of habitat restoration programs have been conducted in Wuyanling Nature Reserve. The results of basic researches have been used for managing conservation strategy for the field population. The research team lead by professor Zheng Guangmei have obtained many national prizes in honor of their great contributions for the endangered species research and conservation. Crested Ibis was historically widely distributed in East Asia in China, Russia, Japan, and Korea Peninsula (Bird Life International 2001). Since 1950, the population has decreased dramatically and locally extinct in Russian, Japan and Korea Peninsula due to the shortage of foods, illegal hunting, loss of breeding trees and wetland habitats. This species was once declared to be extinct (Yu et al. 2006). In May 1981, researchers from Institute of Zoology, Chinese Academy of Sciences found a small population of only 7 birds in Yang County, Shaanxi Province. After 30 years of research and conservation efforts, the population has now increased to 1100. 600 of them are wild population distributed in 11 counties of Shaanxi province. The rest artificial populations are distributed mainly in Yangxian, Zhouzhi, and Ningshan counties of Shaanxi, in Beijing Zoo, Henan and Zhejiang provinces (Ding 2004). Because of their great achievements in research, rescuing, restoration, and management for the species, the Chinese team has received the national second prize. IUCN specialists suggested this as a successful example to the world for conservation of endangered species. The black-faced spoonbills had ever been very common in Asia in 1950s, but, the population decreased to only 288 in 1988. However, no one knows where the breeding population is. Researchers first discovered the small breeding colony in Xingren Islet, Changhai County, Liaoning Province in June 15th, 1999 (Yin et al. 1999). More breeding birds were reported and found after that, and the wild breeding area has been protected as a nature reserve.

Chinese colleagues have also actually achieved great success in protecting these rare and endangered birds but are still far from the final objectives. Here, we cannot list all these achievements from different researchers and for different species. However, conservation of birds in China also faces severe challenge as the global climate change and increasing disturbance from human activities, and especially when compared with the developed countries. The conservation program is a long-term task and thus needs a long-term effort.

## 2.6 Wild Bird Diseases and Eco-health

Birds are important hosts and vectors for many zoonotic diseases. Some of the diseases are fatal, e.g., High Pathogenic Avian Influenza (HPAI). As the most focused infectious zoonose, HPAI has tremendously threatened the economic development and eco-health of our society. Wild birds are considered natural reservoir of all known avian influenza (AI) virus subtypes, and Anatidae species are identified as the major vectors. However, whether wild birds are the hosts or vectors of H5N1 virus has been widely debated (Altizer et al. 2011; Normile 2005). H5N1 virus have ever been isolated from the Peregrine Falcon (*Falco peregrinus*), Grey Heron (*Ardea cinerea*), and other wild birds, however, these are sporadic cases, isolated viruses from migratory birds in Qinghai Lake was the first case report from wild bird population over the world (Liu et al. 2005). This finding supposed the relationship between migratory birds and global circulation of H5N1 virus. The latter on AI surveillance in wild birds in China isolated 17 strains in corresponding to five clades in the genomic phylogenetic tree, which suggested that high genetic diversity existed among H5N1 viruses (Kou et al. 2009). After the extensive surveillance programs, HPAI H5N1 viruses were isolated from diversified wild bird species, which strongly support the association between viral transmission and bird movements (Kou et al. 2005, 2009). The satellite-tracking studies on bird migration further indicated the preservation possibility of H5N1virus in the Qinghai lake areas, and spread through wild bird migration (Cui et al. 2011a, b; Li et al. 2010b, c). Moreover, resident birds could also be a potential vector to carry and spread H5N1 virus (Kou et al. 2005). However, some new genotypic H5N1 strains emerged in Qinghai Lake after 2009, which was suspected to be related with bird migration and the viruses reassortment from migratory birds and poultries (Li et al. 2011a, b; Hu et al. 2011). The breakouts of H5N1 have threatened eco-health and human society. The global epidemic of H5N1 virus has killed lots of wild bird species and caused thousands of migratory birds death, including very endangered species, e.g., Black-necked crane (*Grus nigricollis*) (Chen et al. 2006). We have also estimated the risk of H5N1 virus introduction to Qinghai Lake by using a quality–quantity integrative method and indicated that ducks and geese were most likely the vectors to introduce HPAI H5N1 into the lake through migration. The studies highlighted the importance of eco-health and AI surveillance activities around the lake, which could be guided to AI monitoring program in the whole Central Asian Flyway (Cui et al. 2011b). The finding of H5N1 outbreak in wild bird population in Qinghaihai Lake was published in Science in 2005. The paper was elected to get the "Thomson Reuters Research Fronts Awards" by Thomson Reuters in 2008, to recognize the outstanding contributions to the frontier scientific field worldwide. The significance of above researches on AI has also lies in confirming the role of wild birds in dissemination of AIV, so as to guide national AI prevention and control management, and global response to AI outbreaks; to supply crucial scientific recommendations on the establishments of the national monitoring system of AI and other emerging infectious zoonotic diseases.

## 2.7 Conclusion and Discussion

China has extremely rich bird resources, high endemism, but faces high threaten by endangered species. The uplift of the QTP, the Himalayan orogeny, geographical isolation of islands, and Pleistocene glaciations have greatly impacted on the endemism, distribution, and phylogeographical divergence, in facilitating the forming process of regional endemism and speciation. One hundred and tenty-four species are highly threatened, of them 7 are critical endangered; in addition, currently varied zoonotic avian diseases have also threatened the biodiversity and ecological security.

The technological developments have brought innovative opportunities for scientific researches. The development of molecular biology, bioacoustics, a variety of analytical software, various analytical models, and so on, all these have played an important role in promoting the scientific researches and conservations of birds in China and even the whole world. On morphology basis, to us multiple gene markers including mitochondrial and nuclear genes, sonograph analysis have played great role in bird taxonomy and species recognition. In particular, entering the next generation of genomics, a large number of sequences into mechanization at large scale, will dramatically reduce the financial and human resources invested, which make it possible to base on the whole genome, multiple gene combination in studying evolution, genetic diversity, phylogeny, and phylogeography researches. Despite birds are well studied among all animal taxa, there are still lack of basic knowledge in making conservation and management measures for endangered species. So, researches of behavior, ecology, geographical distribution, and biodiversity conservation are still a long way to go; therefore, basic biology in filling necessary data gap is still important. As global climate changes, illegal wildlife trade, the burst of human movements, will promote and rise the emerging of potential zoonotic diseases, so important pathogen surveillance and studies on co-evolution between pathogen and host of birds are still hot topics for human society.

## References


Altizer, S., Bartel, R., & Han, B. A. (2011). Animal migration and infectious disease risk. *Science, 331*, 296.

Avise, J. C., Arnold, J., Ball, R. M., et al. (1987). Intraspecific phylogeography: The mitochondrial DNA bridge between population genetics and systematics. *Annual Review of Ecology and Systematics, 18*, 489–522.

Bird Life International. (2001). *Threatened birds of Asia: The bird life international red data book.* Cambridge, UK: BirdLife International.

Cao, K. (1992). Selection of suitable area for re-introduction of wild Père David's deer in China. In N. Ohtaishi & H. L. Sheng (Eds). *Deer of China, biology and management*. B. V. Amsterdam, The Netherlands: Elserver Science.

Chen, Y., Chen, Y., & Liu, H. (1996). Studies on the position of Qinghai-Xizang Plateau region in zoogeographical division and its eastern demarcation line. *Acta Hydrobiologica Sinica, 20*, 97–103.

Chen, H. L., Li, Y. B., Li, Z. J., et al. (2006). Properties and dissemination of H5N1 viruses isolation during an influenza outbreak in migratory waterfowls in western China. *Journal of Virology, 80*(12), 5976–5983.

Chen, J., Weng, C., Chao, J., Hu, D., & Taya, K. (2008). Reproduction and development of the released Przewalski's horses (Equus przewalskii) in Xinjiang, China. *Journal of Equine Science, 19*, 1–7.

Cheng, T. H., Zheng, G. M., & Zhang, F. Y. (1997). *Fauna sinica: Aves* (Vol. 1, pp. 1–199). Beijing: Science Press.

Crisp, M. D., Laffan, S., Linder, H. P., et al. (2001). Endemism in the Australian flora. *Journal of Biogeography, 28*, 183–198.

Cui, P., Hou, Y. S., Tang, M. J., et al. (2011a). Movements of the bar-headed geese Anser indicus in relation to H5N1 spread during breeding and post breeding periods in Qinghai Lake, China. *Journal of Ornithology, 152*, 83–92.

Cui, P., Hou, Y. S., Xing, Z., et al. (2011b). Bird migration and risk for H5N1 transmission into Qinghai Lake, China. *Vector-Borne and Zoonotic Diseases, 11*(5), 567–576.

Dai, C. Y., Zhao, N., Wang, W. J., et al. (2011). Application of phylogeographic and ecological niche models to an East Asian subspecies group of black-throated tit (Aegithalos concinnus talifuensis–concinnus): Novel patterns, complex dynamics, multiple refugia and conservation insights. *PLoS ONE, 6*(12), e29329.

Ding, C. Q. (2004). *Research on the crested ibis* (pp. 1–388). Shanghai: Education Press of Shanghai Science and Technology.

Du, Y., Zhou, F., Shu, X. L., et al. (2009). The impact of global warming on China avifauna. *Acta Zootaxonomica Sinica, 34*(3), 664–674.

Hewitt, G. M. (2000). The genetic legacy of the quaternary ice ages. *Nature, 405*, 907–913.

Hu, X. D., Liu, D., Wang, M. Y., et al. (2011). Clade 2.3.2 avian influenza virus (H5N1), Qinghai Lake region, China, 2009–2010. *Emerging Infectious Diseases, 17*, 560–562.

Huang, Z., Liu, N., & Liang, W. (2010a). Phylogeography of Chinese bamboo partridge, Bambusicola thoracica thoracica (Aves: Galliformes) in south China: Inference from mitochondrial DNA control-region sequences. *Molecular Phylogenetics and Evolution, 56*, 273–280.

Huang, X. L., Qiao, G. X., & Lei, F. M. (2010b). Use of parsimony analysis to identify areas of endemism of Chinese birds: implications for conservation and biogeography. *International Journal of Molecular Sciences, 11*(5), 2097–2108.

IUCN. (2013). *IUCN red list of threatened species*. Version 2013.2. www.iucnredlist.org. Downloaded on February 07, 2014.

Jiang, Z., Feng, Z., Yu, C., Zhang, L., Xia, J., Ding, Y., & Lindsay, N. (2000). Reintroduction and recovery of Père David's deer in China. *Wildlife Society Bulletin, 28*(3), 681–687.

Jiang, Z., et al. (2014). *Principles in conservation biology*. Beijing: Science Press (In Press).

Jiang, Z., Zhou, Z., Meng, X., Meng, Z., Li, L., Ping, X., et al. (2013). Domestic and CITES regulations controlling the international snake trade in China. *Oryx, 47*(4), 532–534.

Kou, Z., Lei, F. M., Yu, J., et al. (2005). New genotype avian influenza H5N1 viruses were isolated from tree sparrows in China. *Journal of Virology, 79*(24), 15460–15466.

Kou, Z., Li, Y. D., Yin, Z. H., et al. (2009). The survey of H5N1 flu virus in wild birds in 14 provinces of China from 2004 to 2007. *PLoS ONE, 4*(9), e6926.

Lei, F. M., & Lu, T. C. (2006). *China endemic birds* (pp. 1–640). Beijing: Science Press.

Lei, F. M., Qu, Y. H., Lu, J. L., et al. (2003a). Conservation on diversity and distribution patterns of endemic birds in China. *Biodiversity and Conservation, 12*(2), 239–254.

Lei, F. M., Qu, Y. H., Tang, Q. Q., et al. (2003b). Priorities for the conservation of avian biodiversity in China based on the distribution patterns of endemic bird genera. *Biodiversity and Conservation, 12*(12), 2487–2501.

Lei, F. M., Wei, G. A., Zhao, H. F., et al. (2007). China subregional avian endemism and biodiversity conservation. *Biodiversity and Conservation, 16*, 1119–1130.

Li, L., & Jiang, Z. (2014). International trade of CITES listed bird species in China. *PLoS ONE* (In press).

Li, R., et al. (2010a). The complete genome sequence of the giant panda. *Nature, 463*, 311–317.

Li, Y. D., Li, P., Lei, F. M., et al. (2010b). Persistent circulation of highly pathogenic influenza H5N1 virus in Lake Qinghai area of China. *Avian Diseases, 54*(2), 821–829.

Li, Y. B., Shi, J. Z., Zhong, G. X., et al. (2010c). Continued evolution of H5N1 influenza viruses in wild birds, domestic poultry, and humans in China from 2004 to 2009. *Journal of Virology, 84*, 8389–8397.

Li, Y., Liu, L., Zhang, Y., et al. (2011a). New avian influenza virus (H5N1) in wild birds, Qinghai, China. *Emergency Infectious Diseases, 17*, 265–267.

Li, C., Yang, X., Ding, Y., Zhang, L., Fang, H., Tang, S., & Jiang, Z. (2011b). Do Père David's deer lose memories of their ancestral predators? *PLoS ONE, 6*(8), e23623.

Li, L., Wang, Z., Zerbe, S., Abdusalih, N., Tang, Z., et al. (2013). Species richness patterns and water-energy dynamics in the drylands of Northwest China. *PLoS ONE, 8*(6), e66450.

Liu, J. H., Xiao, H. X., Lei, F., et al. (2005). Highly pathogenic H5N1 influenza virus infection in migratory birds. *Science, 309*, 1206.

Luo, Z., Tang, S., Li, C., Fang, H., Hu, H., Yang, J., et al. (2012). Environmental effects on vertebrate species richness: Testing the energy, environmental stability and habitat heterogeneity hypotheses. *PLoS ONE, 7*(4), e35514.

Maddison, N. R., Jiang, Z., & Boyd, M. (2012). The Père David's Deer: Clinging on by a cloven hoof. *WAZA Magazine, 13*, 29–32.

Normile, D. (2005). Are wild birds to blame? *Science, 310*, 426–428.

Pan, W., et al. (2001). *A chance for lasting survival*. Beijing: Beijing University Press.

Peng, J., Jiang, Z., & Hu, J. (2001). Status and conservation of giant panda. *Folia Zoologica, 50*, 81–88.

Peng, J., Jiang, Z., Qin, G., Huang, Q., Li, Y., Fang, H., et al. (2009). Mate choice in giant panda (Ailuropoda melanoleuca). *Belgian Journal Zoology, 139*, 87–92.

Qu, Y. H., Ericson, P. G., Lei, F. M., et al. (2005). Postglacial colonization of the Tibet plateau inferred from the matrilineal genetic structure of the endemic red-necked snow finch, Pyrgilauda ruficollis. *Molecular Ecology, 14*, 1767–1781.

Qu, Y. H., Lei, F. M., Zhang, R. Y., et al. (2010). Comparative phylogeography of five avian species: Implications for Pleistocene evolutionary history in the Qinghai-Tibetan plateau. *Molecular Ecology, 19*, 338–351.

Qu, Y. H., Luo, X., Zhang, R. Y., et al. (2011). Lineage diversification and historical demography of a montane bird Garrulax elliotii—implications for the Pleistocene evolutionary history of the eastern Himalayas. *BMC Evolutionary Biology, 11*, 174.

Schaller, G. B., Hu, J., Pan, W., & Zhu, J. (1985). *The giant panda of wolong*. Chicago: University of Chicago Press.

Solymos, P., & Feher, Z. (2005). Conservation prioritization based on distribution of land snails in Hungary. *Conservation Biology, 4*, 1084–1094.

Song, G., Qu, Y. H., Yin, Z. H., et al. (2009). Phylogeography of the Alcippe morrisonia (Aves: Timaliidae): Long population history beyond late Pleistocene glaciations. *BMC Evolutionary Biology, 9*, 143.

Spellman, G. M., Riddle, B., & Klicka, J. (2007). Phylogeography of the mountain chickadee (Poecile gambeli): Diversification, introgression, and expansion in response to Quaternary climate change. *Molecular Ecology, 16*, 1055–1068.

Stattersfield, A. J., Crowby, M. J., Long, H. J., et al. (1998). *Endemic bird areas of the world: Priorities for biodiversity conservation: conservation series 7*. Cambridge, UK: Birdlife International.

Wang, K. F., Zhang, J. R., & Lei, F. M. (2010). Geographical distribution pattern and its temporal and spatial variation of birds breeding in zoogeographical subregion of China. *Acta Zootaxonomica Sinica, 35*(1), 145–157.

Xu, Y., Ran, J. H., & Yue, B. S. (2008). The updated number of speciesof birds in Sichuan. *Sichuan Journal of Zoology, 27*(3), 429–431.

Yang, L. (2004). *The avifauna of Yunnan China* (pp. 1–1056). Yunnan: Yunnan Science and Technology Press.

Yang, L., & Lei, F. M. (2009). Morphological taxonomy, fauna and zoogeography of birds in China. *Zootaxonomica Sinica, 34*(2), 316–328.

Yang, S. J., Lei, F. M, & Dong, H. L. (2009). Phylogeography of regional fauna on the Tibetan Plateau. Progress in Natural Science, pp. 789–799.

Yang, S. J., Yin, Z. H., Ma, X. M., et al. (2006). Phylogeography of ground tit (Pseudopodoces humilis) based on mtDNA: Evidence of past fragmentation on the Tibetan Plateau. *Molecular Phylogenetics and Evolution, 41*, 257–265.

Yin, Z. H., Lei, F. M., Ding, W. N., et al. (1999). The new discovery of breeding site of the black-faced spoonbill (Platalea minor) in China. *Chinese Journal of Zoology, 33*(6), 30–31.

Yu, X. P., Liu, N. F., Xi, Y. M., et al. (2006). Reproductive success of the crested ibis nipponia nippon. *Bird Conservation International, 16*, 325–343.

Zeng, Y., Li, C., Zhang, L., Zhong, Z., & Jiang, Z. (2013). No correlation between neonatal fitness and heterozygosity in a reintroduced population of Père David's deer. *Current Zoology, 59*(2), 249–256.

Zhang, Y. Y. (2005). A review on the research of the Cabot's tragopan. *Chinese Journal of Zoology, 40*(1), 104–107.

Zhao, S., Zheng, P., Dong, S., et al. (2013). Whole-genome sequencing of giant pandas provides insights into demographic history and local adaptation. *Nature Genetics, 45*, 67–71.

Zheng, G. M. (Ed.). (2011). *A checklist on the classification and distribution of the birds of China (second edition)* (pp. 1–444). Beijing: Science Press.

Zhou, Z., & Jiang, Z. (2004). International trade status and crisis of snake species in China. *Conservation Biology, 18*, 1384–1396.

# Chapter 3
# Biodiversity Evaluation and Monitoring

**Haigen Xu, Xiaoping Tang, Jiyuan Liu, Hui Ding, Jun Wu, Ming Zhang, Qingwen Yang, Lei Cai, Haijun Zhao, Yan Liu, Rui Wang and FangHao Wan**

**Abstract** Biodiversity evaluation is the basic work and an important means to objectively know the status and trend changes in biodiversity and scientifically conducts the protection of biodiversity. Here we take the 2010 global biodiversity target as an example which develop national indicators to successfully evaluate China's progress toward the 2010 target. But the lack of a national biodiversity monitoring system still hinders timely and accurate assessment of biodiversity. In addition, invasive alien species (IAS) has worsened it in recent years. Because of the severe situation of biological invasions in China, the theoretical and applied research on IAS has gained great importance since 1990s. Main research of Chinese scientists focused on eco-impact mechanisms and management basis of important

H. Xu (✉) · H. Ding · J. Wu · M. Zhang · Y. Liu
Nanjing Institute of Environmental Sciences, Ministry of Environmental Protection, Nanjing 210042, China
e-mail: xhg@nies.org

X. Tang
Academy of Forest Inventory and Planning, State Forestry Administration, Beijing 100714, China

J. Liu
Institute of Geographic Sciences and Natural Resources Research, Chinese Academy of Sciences, Beijing 100101, China

Q. Yang
Institute of Crop Science, Chinese Academy of Agricultural Science, Beijing 100081, China

L. Cai
Department of Nature and Ecology Conservation, Ministry of Environmental Protection, Beijing 100035, China

H. Zhao
Foreign Environment Cooperation Center, Ministry of Environmental Protection, Beijing 100035, China

R. Wang · F. Wan
Institute of Plant Protection, Chinese Academy of Agricultural Sciences, Beijing 100193, China

 
W. Li (ed.), *Contemporary Ecology Research in China*,
DOI 10.1007/978-3-662-48376-3_3

invasive species in China. And research models on biological invasions and a discipline of invasion biology are gradually formed with Chinese characteristics.

**Keywords** Biodiversity · Indicators · Evaluation · Threats · Monitoring · Invasive alien species (IAS) · Ecological impact · Management

# 1 Indicators and Assessment of 2010 Global Biodiversity Target

## *1.1 Introduction*

Over the past hundred years, humans have caused species extinction rates to increase 1000 times as much as the background rates that were typical over Earth's history (Millennium Ecosystem Assessment 2005; Pimm et al. 1995). Parties to the Convention on Biological Diversity (CBD) adopted in April 2002, the 2010 biodiversity target "to achieve by 2010 a significant reduction of the current rate of biodiversity loss." We developed national indicators to evaluate China's progress toward the 2010 target.

## *1.2 Indicators*

Indicators should reflect loss of biodiversity, cover main elements of biodiversity, be sensitive to change, and be acceptable by decision-makers and public. We developed indicators in three aspects: state, threats (including driving force and pressure), and response (Table 1).

## *1.3 Results*

### 1.3.1 Status and Trends of the Components of Biological Diversity

In terms of change in land coverage, the area of cultivated lands, inland waters, and residential quarters increased, but the area of forests, grasslands, and undeveloped lands decreased from the late 1980s to 2000; the area of cultivated lands, grasslands, and undeveloped lands decreased while the area of forests, inland waters and residential quarters increased between 2000 and 2005.

**Table 1** National indicators for assessment of the 2010 biodiversity target

| Indicators |
| --- |
| *Status and trends of the components of biological diversity* |
| (1) Change in land cover |
| (2) Net primary productivity (NPP) |
| (3) Total growing forest stock and annual net increase of growing forest stock |
| (4) Area of desert land |
| (5) Marine trophic index |
| (6) Water quality in marine ecosystems |
| (7) Water quality in freshwater ecosystems |
| (8) Change in status of threatened species |
| (9) Genetic diversity of domesticated animals, cultivated plants, fish species of major socioeconomic importance |
| *Threats to biodiversity* |
| (10) Discharge of major pollutants |
| (11) Difference between nitrogen input and output |
| (12) Density of railroad and expressway |
| (13) Trends in invasive alien species |
| (14) Impact of climate change on biodiversity |
| *Response* |
| (15) Number and coverage rate of nature reserves |
| (16) Status of access and benefit-sharing of genetic resources and traditional knowledge |
| (17) Financial resources for biodiversity conservation |

Net primary productivity (NPP) plays a significant role in global carbon balance. In the last two decades, the NPP of China showed an increasing trend (Gao and Liu 2008).

Forest area, forest coverage rate, and forest growing stock are indicators for the ranges and functions of forest ecosystems. Over the last two decades, forest resources have been increasing. China has become a country with the fastest growth in forest resources in the world (Xu et al. 2009).

Marine Trophic Index (MTI), or mean trophic level, is the mean position in a food web that an organism occupies, and it indicates the integrity of a marine ecosystem (Pauly et al. 1998; Pauly and Watson 2005). From the early 1980s to the mid-1990s, overfishing led to a significant decline in the MTI. However, statistics showed a steady increase in the MTI from 1997 to 2006 (Fig. 1). This may be attributed to the implementation of summer fishing ban on all marine waters of China.

Water quality of inland waters has been improving in mainland of China since 2001. According to the Third National Monitoring of Land Desertification (State Forestry Administration 2005), the area of desertified land decreased by 6416 $km^2$ from 1999 to 2004 in mainland of China, a drift from an average annual expansion of 3436 $km^2$ to an average annual reduction of 1283 $km^2$. Loss of grasslands

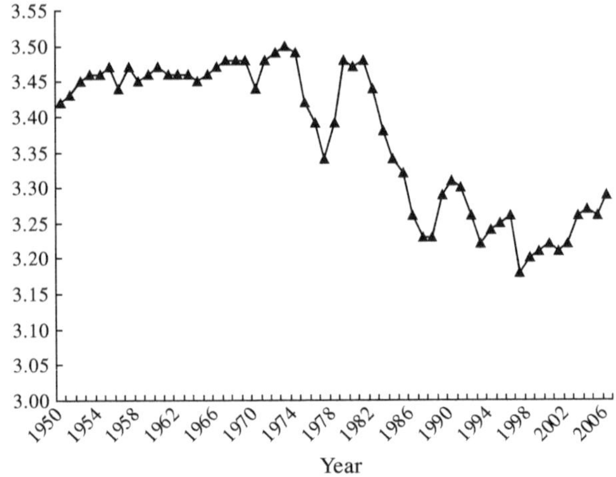


**Fig. 1** Marine trophic index of all marine waters of China

continues, and 33 % of natural grasslands are overloaded (Ministry of Agriculture 2007), although grassland conservation programs have been initiated in recent years.

Red List Index (RLI) illustrates the relative rate at which a particular set of species change in overall threat status (Butchart et al. 2004, 2005; Lamoreux et al. 2003). RLI value of mammals and freshwater fish decreased from 1998 to 2004, which showed a continuing deterioration in the threat status of mammal and fish species. RLI value of bird species showed a decreased deterioration in the threat status based on equal-steps approach (Butchart et al. 2004); if the weight of critically endangered species is higher, the RLI value of bird species would show a continuing deterioration in its threat status. China's wetland conservation has contributed to better conservation of bird species.

### 1.3.2 Threats to Biodiversity

The total discharge of chemical oxygen demand, and sulfur dioxide in waste gas, two national-targeted pollutants, first decreased in 2007, down 3.14 and 4.66 % from the previous year, respectively. There were decreases in the discharge of toxic and harmful pollutants (mercury, cadmium, hexavalent chromium, plumbum, arsenic) in wastewater, the emission intensity of chemical oxygen demand of key industries, the emission of soot dust and industrial dust in waste gasses, and discharge of solid wastes. However, China still faces severe environmental pressure. The total emissions of pollutants are still high and especially wastewater discharge is increasing at a 3.90 % average annual rate.

Agrochemicals have been the key driver for the remarkable increase in food production. The annual application of fertilizers has increased by 5.48 % on average since 1980, and that of pesticide has also increased by 4.50 % on average since 1991, but as much as 60 % of the nitrogen fertilizer and 70 % of the pesticide

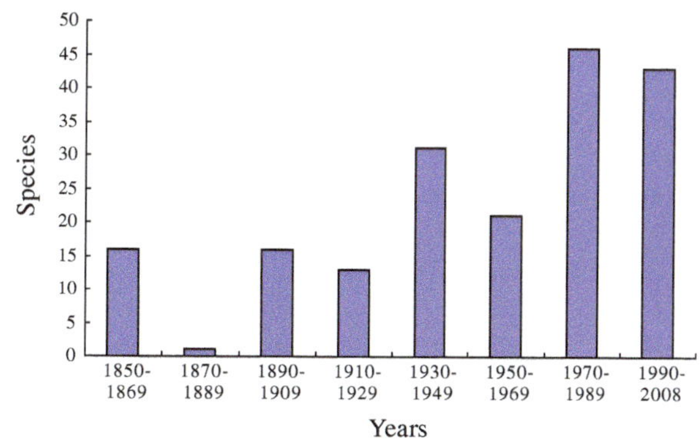


**Fig. 2** Number of invasive alien species newly discovered at 20 years' interval

applied may be lost to the environment (China EPA 2005), which leads to soil pollution, lake eutrophication, ground water pollution, etc.

Invasive alien species (IAS) expedite the losses of biodiversity (Xu et al. 2006a, b). The annual economic losses caused by IAS to China accounts for 1.36 % of its GDP (Xu et al. 2006a). The number of IAS newly discovered in 20 years' interval was between 1 and 16 before the 1930s. The figure rose to 21 and 32 from 1930s to 1960s and exceeded 40 after the 1970s, showing a tremendous growing trend (Fig. 2).

### 1.3.3 Response

As of the end of 2007, the number of nature reserves was 2531, 32 times of the number in 1978; its area was 151.88 million hectares accounting for 15.2 % of the national territory (China EPA 2008) and 120 time of the area in 1978. The Chinese government has initiated environmental pollution control programs and forest conservation programs. Investment in pollution control increased at an average annual rate of 40.35 % and that of forest conservation increased at 40.83 %.

## *1.4 Discussion*

It is crucial that progress toward the 2010 biodiversity target and beyond can be monitored. The lack of a national biodiversity monitoring system still hinders timely and accurate assessment of biodiversity more. Owing to limited monitoring capacity and data availability, the indicators "trends in extent of selected biomes, ecosystems and habitats" and "trends in abundance and distribution of selected species," were not included in China's national indicator framework. The establishment of this system calls for properly planned partnerships, development of sampling regimes, design of data collection programs, and statistical analyses.

# 2 Invasive Alien Species in China

## 2.1 *Introduction*

China is one of the world's hotspots of biodiversity. However, about 5000 higher plants are at the edge of extinction or nearly extinct, which account for 12–20 % of all the higher plants in China (Lin 2008). This phenomenon has been worsened by IAS in recent years. There were about 520 IAS invaded into China last century (Fig. 3), in which 50 species were listed among "100 of the world's worst IAS" published by the World Conservation Union (IUCN). IAS invaded almost all the ecosystems in China (Wan et al. 2009). The spatial distribution of abundances of IAS in China indicates a significant variation among provinces and territories. These invaders severely threaten China's biodiversity and protection of genetic resources (Wan et al. 2002; Xu et al. 2006). With economic globalization and rapid development of international trades, challenges posed by biological invasions will be very serious (Lin et al. 2007; Ding et al. 2008).

Because of the severe situation of biological invasions in China, the theoretical and applied research on IAS has gained great importance since 1990s (Wan et al. 2009). Main research progress by Chinese scientists was summarized in this section, with focus on eco-impact mechanisms and management basis of important invasive species in China.

## 2.2 *Eco-impact Mechanisms and Management Basis of IAS*

### 2.2.1 Ecological Impacts of Important IAS

IAS has caused huge economic losses and ecological disasters in various ecosystems of China (Xu et al. 2006). After an alien species successfully established its

**Fig. 3** Numbers of various categories of invasive alien species in China

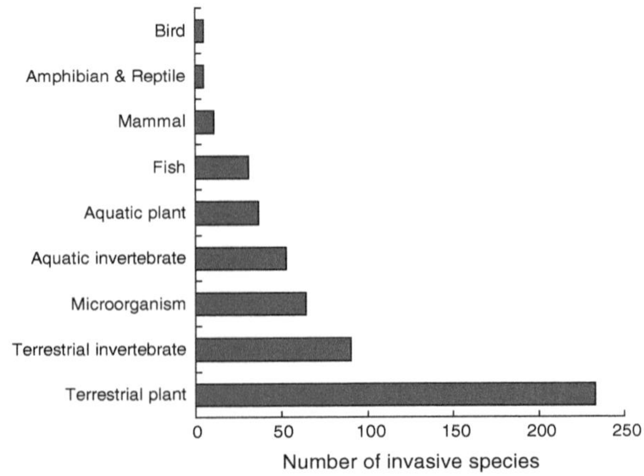

population in one area, it will squeeze the local species out through competition and occupation of native species' ecological niche, resulting in decrease of species diversity and changes in population structure, thus affecting ecosystems' material cycles, energy flows and the whole service function in the end (Wan et al. 2002, 2009). Above all, the recession of ecosystems service function is persistent and irreversible. Some important IAS (e.g., smooth cordgrass, alligator weed, and Crofton weed) usually create mono-dominant community and threaten native species, resulting in vanishing and extinction of native species, especially leaving the precious and rare species at endangered or extinct risk status (Wan et al. 2005) (Figs. 4 and 5). For example, extensive spread of smooth cordgrass has not only destroyed habitat of neritic organisms, but also competed with native plants for growing space, which finally led to disappearance of mangrove trees and impossible restoration of habitat conditions in such areas that had been invaded (Chen et al. 2004).

**Fig. 4** Single dominance community of Crofton weed in southwest of China

**Fig. 5** Alligator weed in canal

### 2.2.2 Invasion and Expansion Mechanisms of Major IAS

Biological invasions could be viewed as an orderly ecological process with phases of introduction, establishment, dispersion, and outbreak (Wan et al. 2009, 2011a). According to the sequential process with different core scientific issues during different stages, the research on biological invasions in China covered biological intrinsic features like introduction and population establishment, survival and adaption, development and evolution and interspecies interaction, as well as extrinsic characteristics like response and resistance of ecosystem and prevention and management technology.

From the above studies, population establishment and expansion mechanisms, genetic and ecological adaption mechanism of important alien species, co-adaption between IAS and its host, competition and replacement between IAS and native species, synergetic effects between IAS and other biological factors were revealed. Some new viewpoints, hypotheses and theories were brought forward, such as the intrinsic propagation potential under competition or adverse conditions (Cheng et al. 2008; Wan et al. 2010a), asymmetric mating interactions driving widespread invasion and displacement in a whitefly (Liu et al. 2007), interaction between plant virus and virus vectors (Luan et al. 2013), semiochemical regulation of invasion behavior (Lu 2008; Zhao et al. 2013), synergetic effects between IAS and native species (Lu et al. 2010), hypothesis of the evolution of nitrogen allocation (Feng et al. 2009), feedback of soil biota and self-reinforced invasion mechanisms of invasive plant (Niu et al. 2007), and multiple mechanisms underlie the spread of invasive plant (Wang et al. 2011). These achievements contribute immensely for the construction and development of invasion biology in China and provide scientific basis for the IAS management strategies and techniques.

### 2.2.3 Management Strategies of Invasive Alien Species

According to the invasion process of the alien species, we have established related prevention and control technology (Wan et al. 2009). We first established risk assessment and early warning techniques for potential and local-level invasive species (Wan et al. 2010b). The potential geographical distribution areas and high risk area for 99 IAS (including TCK, pine wood nematode, Crofton weed, etc.) were analyzed (Wan et al. 2010b). Based on the potential distribution area and spread pathway of newly arrived and established local-level invasive species, blocking zones were established on the front edge and potential spread pathway. Spread interdiction and construction of blocking zone have been carried out in China on 12 IAS, such as red imported fire ant, Colorado potato beetle, grape phylloxera, pine wood nematode, and crofton weed, etc. (Wan et al. 2009, 2010b). Second, we developed rapid detection and monitoring technology of 35 important IAS, based on rapid molecular detection techniques of invasive diseases and small insects, chemical and physical monitoring techniques of invasive insects (Wan et al. 2011c). All these techniques, methods, and standards provide powerful technological support to deal with emergency situations caused by IAS. Third, sustainable management systems of important IAS based on classical biological control and ecological restoration were developed. Sustainable management technologies of common ragweed, alligator weed, coconut leaf beetle, and tobacco whitefly have been constructed and demonstrated in Hunan, Fujian, Hainan, and Zhejiang, respectively (Wan et al. 2008).

## 2.3 Conclusion and Prospect

The research on biological invasions in China has achieved great progresses under the support and R&D investment from government departments. Gradually, research models on biological invasions and a discipline of invasion biology are formed with Chinese characteristics (Wan et al. 2011b). The research on biological invasions is a persistent task in the long term and requires multidisciplinary approach to study the subject. On the basis of the discipline framework of invasion biology and development of current research results, we should first conduct innovative cutting-edge research, upgrade our research level at different scales (Wan et al. 2009, 2011b). At the global dimension, the future research should focus on the effects of global changes (including changes of climate, environment and land use) on biological invasions, especially the influences of ecological factors on the colonization, dispersal and damage of IAS, thus revealing the correlations between different geographic environments and invasion success. At specific ecosystem level, the research should focus on the relations between biological invasions and changes of agricultural and forest ecosystems services and functions to reveal influence processes and mechanisms of biological invasions on native species losses, erosion of species' genetic resources, changes and function loss of community structure and food chain, and the decline of ecosystems services and functions, etc. At the inter-species relationship dimension, the research should focus on IAS' population control imbalance and the adjustment mechanism, analyze the ecological factors of natural control imbalance of invasive species, investigate the ecological processes of population construction and collapse of invasive species, etc. Second, we should be more innovative and develop more effective prevention and control technologies, establish four-technology systems on prevention and early warning, detection and monitoring, eradication and containment, and control and management, especially to develop innovative quantitative risk analysis methods, construct the technological system for quick detection of invasive species; investigate the epidemic monitoring and emergency eradicating technology for important invasive species. Third, we should develop management model in accordance with China's actual conditions, upgrade obligation mechanism in accordance with international trading rules and international conventions, as well as lay a foundation for empowering China's international negotiations on relevant issues.

# References


Butchart, S. H. M., Stattersfield, A. J., Baillie, J. E. M., et al. (2005). Using red list indices to measure progress towards the 2010 target and beyond. *Philosophical Transactions of the Royal Society of London B, 360*, 255–268.

Butchart, S. H. M., Stattersfield, A. J., Bennun, L. A., et al. (2004). Measuring global trends in the status of biodiversity: Red list indices for birds. *PLoS Biology, 2*, e383.

Chen, Z. Y., Li, B., & Chen, J. K. (2004). Ecological consequences and management of Spartina spp. invasions in coastal ecosystems. *Biodiversity, 12*(2), 280–289.

Cheng, X. Y., Cheng, F. X., Xu, R. M., & Xie, B. Y. (2008). Genetic variation in the invasive process of *Bursaphelenchus xylophilus* (Aphelenchida: Aphelenchoididae) and its possible spread routes in China. *Heredity, 100*(4), 356–365.

China EPA. (2008). China Environment Status Communique. Available at http://www.sepa.gov.cn

Ding, J. Q., Mack, R. N., Lu, P., Ren, M. X., & Huang, H. W. (2008). China's booming economy is sparking and accelerating biological invasions. *BioScience, 58*, 317–324.

Feng, Y. L., Lei, Y. B., Wang, R. F., Callaway, R. M., Valiente-Banuet, A., Inderjit, Li Y. P., & Zheng, Y. L. (2009). Evolutionary tradeoffs for nitrogen allocation to photosynthesis versus cell walls in an invasive plant. *Proceedings of the National Academy of Sciences of the United States of America, 106*, 1853–1856.

Gao, Z. Q., & Liu, J. Y. (2008). Comparative study on net primary productivity of China. *Chinese Science Bulletin, 53*(3), 317–326.

Lamoreux, J., Akcakaya, H. R., Bennun, L., et al. (2003). Value of the IUCN red list. *Trends in Ecology and Evolution, 18*, 214–215.

Lin, L. (2008). Discussion on the deficiencies of legal protection of wild plant resources in China and the countermeasures. *Journal of Northwest A&F University, 28*(1), 109–113.

Lin, W., Zhou, G. F., Cheng, X. Y., & Xu, R. M. (2007). Fast economic development accelerates biological invasions in China. *Plos One, 2*(11), e1208.

Liu, S. S., de Barro, P. J., Xu, J., Luan, J. B., Zang, L. S., Ruan, R. M., & Wan, F. H. (2007). Asymmetric mating interactions drive widespread invasion and displacement in a whitefly. *Science, 318*(5857), 1769–1772.

Lu, M. (2008). Invasiveness of Dendroctonus valens LeConte (Coleoptera: Curculionidae: Scolytinae) based on the inter specific facilitation between this invasive bark beetle and its related species. *Ph.D. Dissertation*. Beijing: Chinese Academy of Sciences.

Lu, M., Wingfield, M. J., Gillette, N. E., Mori, S. R., & Sun, J. H. (2010). Complex interactions among host pines and fungi vectored by an invasive bark beetle. *New Phytologist, 187*, 859–866.

Luan, J. B., Yao, D. M., Zhang, T., Walling, L. L., Yang, M., Wang, Y. J., et al. (2013). Suppression of terpenoid synthesis in plants by a virus promotes its mutualism with vectors. *Ecology Letters, 16*, 390–398.

China EPA. (2005). Investigation and assessment of the status of the environment in China: Synthesis. Beijing: Chinese Environmental Science Press.

Millennium Ecosystem Assessment. (2005). Ecosystems and human well-being: Biodiversity synthesis. Washington, DC: World Resources Institute.

Ministry of Agriculture. (2007). Report on national monitoring of grasslands in China. Available at http://www.agri.gov.cn

Niu, H. B., Liu, W. X., Wan, F. H., & Liu, B. (2007). An invasive aster (*Ageratina adenophora*) invades and dominates forest under stories in China: Altered soil microbial communities facilitate the invader and inhibit natives. *Plant and Soil, 294*, 73–85.

Pauly, D., Christensen, V., Dalsgaard, J., Froese, R., & Jr Torres, F. C. (1998). Fishing down marine food webs. *Science, 279*, 860–863.

Pauly, D., & Watson, R. (2005). Background and interpretation of the 'Marine Trophic Index' as a measure of biodiversity. *Philosophy Transactions Royal Society B, 360*, 415–423.

Pimm, S. L., Russell, G. J., Gittleman, J. L., & Brooks, T. M. (1995). The future of biodiversity. *Science, 209*, 347–350.

State Forestry Administration. (2005). Report on the status of land desertification in China. Available at http://www.forestry.gov.cn
Wan, F. H., Feng, J., & Xu, J. (2011a). *Biological invasion: Detection, surveillance and monitoring*. Beijing: Science Press.
Wan, F. H., Xie, B. Y., & Yang, G. Q. (2011b). *Invasions biology*. Beijing: Science Press.
Wan, F. H., Yan, Y., Wang, R., & Yang, G. Q. (2011c). Invasion biology: Development and perspective in China. *Journal of Biosafety, 20*(1), 1–19.
Wan, F. H., Guo, J. Y., & Wang, D. H. (2002). Alien invasive species in China: Their damages and management strategies. *Chinese Biodiversity, 10*, 119–125.
Wan, F. H., Guo, J. Y., & Zhang, F. (2009). *Research on biological invasions in China*. Beijing: Science Press.
Wan, F. H., Li, B. P., & Guo, J. Y. (2008). *Biological invasion: Biological control theory and practice*. Beijing: Science Press.
Wan, F. H., Liu, W. X., Guo, J. Y., Qiang, S., et al. (2010a). Invasive mechanism and control strategy of *Ageratina adenophora* (Sprengel). *Science China-Life Sciences, 53*, 1291–1298.
Wan, F. H., Peng, D. L., & Wang, R. (2010b). *Biological invasion: Risk analysis and early prevention*. Beijing: Science Press.
Wan, F. H., Zheng, X. B., & Guo, J. Y. (2005). *Biology and management of alien invasive species in agriculture and forestry*. Beijing: Science Press.
Wang, R., Wang, J. F., Qiu, Z. J., Wan, F. H., Meng, B., & Wang, Y. Z. (2011). Multiple mechanisms underlie rapid expansion of an invasive alien plant. *New Phytologist, 191*, 828–839.
Xu, H. G., Ding, H., Li, M. Y., Qiang, S., Guo, J. Y., Han, Z. M., et al. (2006a). The distribution and economic losses of alien species invasion to China. *Biological Invasions, 8*, 1495–1500.
Xu, H. G., Ding, H., Li, M. Y., Qiang, S., et al. (2006b). The distribution and economic losses of alien species invasion to China. *Biological Invasions, 8*, 1495–1500.
Xu, H. G., Qiang, S., Han, Z. M., et al. (2006c). The status and causes of alien species invasion in China. *Biodiversity and Conservation, 15*(9), 2893–2904.
Xu, H. G., Tang, X. P., Liu, J. Y., et al. (2009). China's progress toward the significant reduction of the rate of biodiversity loss. *BioScience, 59*(10), 843–852.
Zhao, L. L., Lu, M., Niu, H. T., Fang, G. F., Zhang, S., Sun, J. H. (2013). A native fungal symbiont facilitates the prevalence and development of an invasive pathogen-native vector symbiosis. *Ecology, 94*(12), 2817–2826. http://www.esajournals.org/doi/abs/10.1890/12-2229.1

# Chapter 4
# Technology and Perspective of Sustainable Biodiversity Utilization

**Luqi Huang, Lanping Guo, Sheng Wang, Qisheng Tang, Ling Tong, Long Li, Jing Yang, Chengyun Li, Youyong Zhu, Runzhi Zhang, Wenhua Li, Qingwen Min and Lu He**

**Abstract** The combination of traditional knowledge and modern biological technology promotes the biodiversity conservation and sustainable utilization in China, which embodies in biological pharmaceutical, healthcare, crop breeding and cultivation, control of disease and insect pest, forest management and administration, ecological protection, restoration and construction, and so on. Here we list the following aspects. First, the conservation and utilization of medicinal bioresources have been gradually formed a multilevel, multifaceted, and multidisciplinary characteristic and achieved remarkable results. Second, China is one of the early countries that conduct the ocean ecosystem dynamics research, including from the structure and process of ocean ecosystem to the ecosystem service and production function, which laid the groundwork for the sustainable development of the ocean ecosystem. Third, intercropping, as one of the multiple cropping systems, is the essence of Chinese traditional agriculture. We have conducted plenty of intercropping practices about increasing production and efficient use of resources,

---

L. Huang · L. Guo · S. Wang
National Resource Center of Chinese Material Medical,
China Academy of Chinese Medical Science, Beijing 100700, China

Q. Tang · L. Tong
Yellow Sea Fisheries Research Institute, Chinese Academy of Fishery Sciences,
Qingdao 266071, China

L. Li
College of Resources and Environmental Sciences, China Agricultural University,
Beijing 100193, China
e-mail: lilong@cau.edu.cn

J. Yang · C. Li · Y. Zhu
Yunnan Agricultural University, Kunming 650224, China

R. Zhang
Institute of Zoology, Chinese Academy of Sciences, Beijing 100101, China

W. Li (✉) · Q. Min · L. He
Institute of Geographic Sciences and Natural Resources Research,
Chinese Academy of Sciences, Beijing 100101, China
e-mail: liwh@igsnrr.ac.cn



W. Li (ed.), *Contemporary Ecology Research in China*,
DOI 10.1007/978-3-662-48376-3_4

including legumes and nonlegumes intercropping, single cotyledon and dicotyledonous intercropping, gramineae and nongramineous intercropping, and so on. In addition, intercropping can increase the species diversity of farmland ecosystem, improve utilization of system resources for biology, sufficiently protect and use natural enemy, and inhibit pest outbreaks. Finally, agricultural heritage, as a traditionally sustainable agriculture practices, also plays an important role on biodiversity protection and utilization. Currently, this research on agricultural heritage mainly concentrated on the theoretical consideration of agricultural heritage, agrobiodiversity characteristics of agricultural heritage, multivalues of agricultural heritage dynamic conservation, substitute industry, and development law and policy on conservation of agricultural heritage.

**Keywords** Medicinal bioresources · Conservation · Utilization · Ocean ecosystem · GLOBEC · Intercropping · Root interaction · Nutrient acquisition · Disease management · Ecological management pest insects · Agricultural heritage · GIAHS

# 1 Conservation and Utilization of Medicinal Bioresources

## *1.1 Introduction*

Medicinal bioresources refer to biological resources which have medical or health effects for human diseases, or have therapeutic effects on domesticated animals, as well as those have insecticidal, fungicidal, herbicidal, and other effects in general (Li 2007). Medicinal bioresources are the basis for the sustainable development of Chinese medicine and the material basis of human health. Along with social progress and development, the awareness of self-care continues to improving. People pay more and more attention to seek medical treatment from natural plant and animal which biological resources of medicinal increasing pressure.

As the core country of origin, distribution, and production of medicinal plants and animals in the world, China have 11,146 kinds of medicinal plants and 2215 kinds of medicinal animals, in approximately 80 % of wild medicinal plants and animals. Although enrich in resources of medicinal plants, China faced an increasing pressure from conservation and sustainable utilization of medicinal plant resources with the expansion of demand due to population growth and industrial development of traditional Chinese medicine. The disorder and unrestraint utilization of traditional Chinese medicine resources induced a great threat for the survival of many wild medicinal plants, so far, there are 168 kinds of medicinal plants have listed in the *Rare and Endangered Plants List of China*. Some high economic value medicinal plant and animal resources (especially Daodiherbs) have been seriously undermined. *China Plant Red Data Book* released 398 kinds of endangered plants in 1992, including 168 kinds of medicinal plants.

Since the fifties of last century, a large number of researchers dedicated to conservation and sustainable development of medicinal bioresources research, and the research and development medicinal bioresources has been gradually formed a multilevel, multifaceted, and multidisciplinary characteristic and achieved remarkable results.

## *1.2 Advances*

Faced on the drawbacks of irrational development and utilization of medicinal bioresources, a large number of scholars dedicated to conservation and utilization of medicinal bioresources and have initially established a system of medicinal bioresources research, conservation, and sustainable utilization, including the following aspects.

### 1.2.1 Investigation and Collation of Medicinal Bioresources

In this regard, China carried out the fourth National Investigation of CMMR which will provide a strong foundation for the protection and utilization of medicinal bio resources. This will help our nation ascertaining the real situation of Chinese material medica resources fully, grasping impact of environmental changes on Chinese material medica resources, analyzing reasons of changes in Chinese material medica resources, strengthening protection, research, utilization of protected wild endangered species, and establishing Chinese material medica resources database. It is of great significance for the protection of genetic resources and associated intellectual property rights.

### 1.2.2 Protection of Endangered Medicinal Bioresources

Many scholars studied the protection strategies of endangered medicinal bioresources and endangered mechanisms. There are many reasons for decreased wild and endangered Chinese material medica resources, but it mainly due to the human impacts and ecological environment changes. Protected methods of endangered medicinal bioresources include in situ conservation and ex situ conservation, germplasm repository of Chinese material medica resources, gene banks, and wild medicinal materials tending. We have progressively realized multifaceted protection of medicinal species biodiversity, ecosystem diversity, and genetic diversity.

### 1.2.3 Mechanism Researches About Environmental Impact on Quality Formation of Medicinal Bioresources

Since the growth and quality formation of medicinal bioresources has a very close relationship with the surrounding natural environment, thus resulting in a conception of "Geoherbs," also known as "Daodi-herb" (Guo et al. 2002; Lu et al. 2006). Research about environmental impact on quality formation of medicinal bioresources is specific in Ecology of Chinese Material Medica Resources. With the development of "Ecology of Chinese Material Medica Resources" and "Medicinal Plant Physiological Ecology," researches about environmental impact on quality formation of medicinal bioresources have been made significant progress.

### 1.2.4 Regionalization of CMM for Cultivation

Regionalization of CMM for Cultivation is the basis for introduction and cultivation of Chinese herbal medicine. For full and rational use of medicinal bioresources and high-yield production of Chinese material medica, it is of very significance. In recent years, technologies and methods of regionalization of CMM for cultivation are constantly updated. Based on the traditional method, some modern technology such as GIS and remote sensing analysis combined with regionalization of CMM method will become the main method in this field (Guo et al. 2005). Using GIS technology to achieve storage, management, analysis, and display of daodiherbs spatial data can overcome the current prevalence problem of sample representation in Ecology of Chinese Material Medica Resources researches. And it will provide scientific basis for the introduction and standardized cultivation of Chinese herbal medicines (Guo et al. 2007).

### 1.2.5 Ecological Farming of Medicinal Plants

At present, about 40 % of Chinese herbal medicines on the market come from the cultivation plants and breeding animals. In more than 200 kinds of commonly used Daodiherbs, 25 % of them completely rely on cultivation, and 60 % is cultivated and wild resources coexist. In 1998, China proposed GAP in standardized medicinal planting and officially launched countries in June 2002. With the promotion of Chinese herbal medicine GAP standardized planting, our government have established large-scale modernization of Chinese material medica industry bases in Yunnan, Sichuan, Ningxia, Jilin, Zhejiang Province and supported and funded standardized cultivation of dozens medicinal species. At present, China has built more than 50 kinds of commonly used Chinese herbal medicines GAP planting bases; some of the base have been great success, such as Notoginseng base in Yunnan Wenshan, Salvia base in Shangluo, wolfberry base in Zhongning, and so on.

### 1.2.6 Bioengineering Researches of Medicinal Bioresources

Bioengineering researches of medicinal bioresources include: rapid propagation technology to achieve rapid production of quality seeds and seedlings of medicinal plants; cultivation of medicinal plants cells, tissues, or organs in bioreactor to get the active ingredients of medicinal plants directly and quickly (Gao 2008). And others like transgenic organ culture, Genetic Engineering breeding, biotransformation, and biosynthesis of the active ingredient associated function genes colon. Among them, study about biosynthesis of the medicinal bioactive components related genes has become a research hotspot this year (Zeng 2011), and biotransformation technology for developing new medicine, reducing toxicity of the active ingredient of Chinese medicine, improving the activity of the active ingredient, and transforming the void ingredients into active ingredient and other aspects of a good application.

## *1.3 Discussion and Conclusion*

In order to achieve the goal of sustainable development of medicinal bioresources, with the system of sustainable development and utilization of medical bioresources established gradually, researches will show the following trends:

(i) The evaluation and monitoring of endangered medicinal bioresources will be standardized and normalized gradually (Li 2007);
(ii) Fixed-point cultivation based on daodiherbs and directed breeding targeted on the active ingredient maturing;
(iii) Soil microbial ecology becomes a hotspot in environment basic researches of quality formation of medicinal bioresources (Huang et al. 2011);
(iv) Research areas of medicinal bioresources will expand in both micro and macro directions supported by modern molecular biology and 3S technology.

# 2 Global Ocean Ecosystem Dynamics Research in China

## *2.1 Introduction*

The ocean is a key component of the earth system and playing a major role in regulating the earth's climate and biogeochemical cycling of key elements. Ocean ecosystem dynamics is one of the most active front fields of ocean science research related with the global change and marine sustainable studies. Global Ocean Ecosystem Dynamics (GLOBEC) and Integrated Marine Biogeochemistry and Ecosystem Research (IMBER) are two projects of International Geosphere-Biosphere

Programme (IGBP) core element structure related with ocean research. Both projects operate together and continuously to develop an integrated understanding of the linkages, interactions, and food-banks between physical forcing, biogeochemical cycles, and food webs of ocean. Chinese marine scientists involved GLOBEC and IMBER studies focus mainly on the scientific issues of the shelf ecosystem dynamics in the characteristics of shallow sea in North Pacific.

The GLOBEC program is an important part of the global change and marine sustainable studies established by the Scientific Committee on Oceanic Research (SCOR) and IOC in late 1991 and incorporated into the IGBP Core Element structure as first ocean project in 1995. The GLOBEC Science Plan published in 1997 set out the GLOBEC goal as *To advance our understanding of the structure and functioning of the global ocean ecosystem, its major subsystems, and its response to physical forcing so that a capability can be developed to forecast the responses of the marine ecosystem to global change* (Harris et al. 1997). GLOBEC research was organized around four foci: retrospective analyses in the context of large-scale climatic changes, process studies, predictive modeling capacity, and feedbacks from changes in marine ecosystem structure (Aksnes et al. 1999). The research was initially developed within the regional projects and a series of national projects.

## 2.2 *Development of Global Ocean Ecosystem Dynamics with Scientific Questions in China*

A working group of the national strategic research on marine ecosystem dynamics development was founded under the support of National Nature Science Foundation of China in 1994 (Tang et al. 1995). One result from the group is to identify four main themes of China GLOBEC study: (1) influencing mechanism of key physical process on biological production, (2) nature of bio-elements recycles and settlement input, (3) primary production process and manipulation of zooplankton, and (4) trophic dynamics of food web and alternation principle of dominate resources. Multidisciplinary and synthesis studies are encouraged to provide breakthrough in understanding ecosystem dynamics and recruitment mechanism of living resources in continental shelf areas.

The Chinese GLOBEC studies focus mainly on the scientific issues of the shelf ecosystem dynamics in the shallow sea and shelf of Chinese waters which are an area affected heavily by global change and human activities. The experimental fields should be characteristics in typically physical, chemical, and biological environment and sensitive to global change so the Bahia Sea, Yellow Sea, and the East China Sea are determined as the field study of China GLOBEC. The China national strategic research of GLOBEC brings forward a goal to study the functioning of ecosystem dynamics in coastal ocean. The process studies are most important part directly implemented in the research. China GLOBEC studies developed six scientific questions related closely with the ecosystems in continental shelf. The questions are (i) energy flow and shift of key species, (ii) recruitment of zooplankton population, (iii) recycle and renewal of bio-elements, (iv) ecological effect of key physical process, (v) coupling of pelagic and benthic systems, and (vi) microbial food loop contribution in ecosystem. All the questions emphasize the interaction and coupling of physical and biological progress in the continental shelf (Tang et al. 2000).

## 2.3 General Review of the Development of GLOBEC Studies in China

China is a country to develop GLOBEC and IMBER research early in the world. Chinese scientist acted as member of GLOBEC Scientific Steering Committee in 1991 and involved in the framing the GLOBEC Science Plan and Implementation Plan. In 2004, IMBER Scientific Steering Committee came into existence. One Chinese scientist become the member of the first committee group and joined to prepare the IMBER Science Plan and Implementation Strategy. China GLOBEC/IMBER has been promoting through the program "Ecosystem Dynamics and Sustainable Utilization of Living Marine Resources in China Coastal Seas." Three phases of the program, including China GLOBEC-I Project, China GLBOEC-II Project, and China GLBOEC-III/IMBER-I Project, are being implemented from 1997 to 2010. The Chinese GLOBEC and IMBER studies are regarded as a regional contribution to the international research to providing a case study of coastal ecosystem and its living resources dynamics.

### 2.3.1 China GLOBEC-I Project

China GLOBEC-I Project entitled "Bohai Sea Ecosystem Dynamics and Sustainable Utilization of Living Resources" (BoSEC 1997–2000) has four major themes: early life history of the Bohai prawn and critical processes in its habitat, zooplankton population dynamics, and its role in the Bohai Sea productivity, trophodynamics of the food web, and the mechanism of the dominant species shift in Bohai Sea ecosystem and Bohai Sea ecosystem dynamics modeling (Su et al. 2002). The achievements of project study are mainly on:

(i) Environmental processes of the habitat of *Penaeus chinensis* and its biomass change of the early life, including stock dynamics of *P. chinensis*, relevant physical and biogeochemical processes in its habitat, long-term variations of atmospheric parameters, and hydrographic properties and their influence on the marine ecosystem.
(ii) Population dynamics of zooplankton and its controlling effects in the marine ecosystems, including phytoplankton composition, primary productivity and new productivity, bacteria production, community structure and population

dynamics of zooplankton, feeding pressure on phytoplankton, ecological conversion efficiency and secondary production, and benthos and benthic productivity.

(iii) Trophodynamics of food web and species change, including feeding relationship and food web structure, trophodynamics in higher trophic level, community structure and biological productivity, and influence of human activities on living resources.

(iv) Ecosystem models of the Bohai Sea, including 3-D primary productivity models and a box model of pelagic–benthic in ecosystem dynamics study.

### 2.3.2 China GLOBEC-II Project

China GLOBEC-II Project, entitled "Ecosystem Dynamics and Sustainable Utilization of Living Resources in the East China Sea and the Yellow Sea (EYSEC 1999–2004)," is funded through the National Key Basic Research and Development Program of China (973 Program). The scientific objectives of the project are to determine:

(i) Impacts of key physical processes on biological production,
(ii) Cycling and regenerations of biogenic element,
(iii) Basic production processes and role of zooplankton in the ecosystem,
(iv) Food web trophodynamics and shifts of dominant species.

The goals of EYSEC project are to understand the function, production, and the critical ecosystem-relevant physical mechanisms of the coastal sea, as part of the knowledge basis for the formulation of strategy to achieve sustainable use of the marine resources. It regards the interaction and coupling of physical and biological progress happening in the shelf as its main efforts. The key target species of the research in the East China Sea and Yellow Sea include zooplankton species (*Calanus sinacus*), small pelagic fish (anchovy, *Engraulis japonicus*), and large commercial species (Spanish mackerel, *Scomberomorus niphonius*). These key species form a main linkage for the study of all the identified major scientific questions of the project. All studies also carry out in three stratums of key species, important species and bio-community. The dominant physical forcing mechanisms, biogenic elements, and its transfer mechanism, and ecological characteristics are examined to find potential linkages between the mechanisms and the change of the living resources. The major themes of this study in the East China Sea and the Yellow Sea as one integral region are: trophic dynamics of key species; recruitment of zooplankton; recycle and regeneration of biogenic elements; critical physical processes in high-productivity areas; coupling of pelagic and benthic systems; and microbial contribution to secondary production. One of the main results commonly in the coastal ocean of China is the academic frame of China coastal ocean ecosystem dynamics research based on the six scientific questions.

### 2.3.3 China GLBOEC-III/IMBER-I Project

China GLBOEC-III/IMBER-I Project entitled as Key Processes and Sustainable Mechanisms of Ecosystem Food Production in the Coastal Ocean of China from 2006 to 2010 funded by 973 Program. The project carries out integrated studies among multidisciplinary subjects by focusing on the coupling mechanism of the marine biogeochemical cycles of biogenic elements and the end-to-end food web in the China seas to comprehend the supporting, regulating, and producing functions of food production and to understand the sustainable mechanisms in the coastal ocean ecosystems of China seas from the perspectives of both anthropogenic impacts and natural changes.

The major scientific questions to be dealt with are: the biogeochemical processes of food production, the physical mechanisms of biogenic element cycle and supplement, coupling mechanism of primary production with major biogeochemical processes, and food web trophodynamics of major biological functional groups. The research activities mainly aim at some unique subecosystems in the Yellow Sea and the East China Sea with studies on ecological capacity. The following four foci will be deployed: the supporting role of main biogeochemical processes in food production, key physical processes of biogenic element cycle and supplement, primary production coupling with main biogeochemical processes, and food production processes of biological function groups together with their sustainable models.

One main scientific achievement is on the spring phytoplankton bloom in the cold mass area of the Yellow Sea. Phytoplankton bloom ecosystem process and its various trophic levels and trophodynamic interaction in the food web are studied (Sun and Song 2009). The spring bloom happening and developing are founded composing by a series of subprocesses. The different type blooms are actually diverse phases of bloom. Phytoplankton biomass increased during spring bloom is following a parabola shape so that the highest biomass occurs in the metaphase. The biomass of subprocesses and spatial–temporal change lie on environment and phytoplankton species of bloom. The highest biomass of subprocesses is happen from south forward north with patch distribution in the region of Yellow Sea Cold Water Mass. Wind is the key control factor of bloom coming into being, and wind and water circumfluence play an important effect to carry through and develop the bloom. Nutriment is the most control factor but dust storms swept over the Yellow Sea from Northern China also effect the bloom in the continental shelf of the in the Central Yellow Sea.

## *2.4 Perspectives of the Development of GLOBEC*

After the completion of the GLOBEC research, SCOR and IGBP set up a working Group of Transition Task Team (TTT) in 2008 on how the second phase of the IMBER program proceeding to accommodate new developments in marine ecosystem research that needs addressing. Key aspects of IMBER research, as only one IGBP program conducted on ocean research, will be the seamless integration of biogeochemical and ecosystem research in a truly trans-disciplinary approach and the incorporation of social science research to enable the investigation of options for mitigating or adapting to the impacts of global change. This integration is also important because feedbacks are critical. Marine biogeochemical and ecosystem responses to global change are complex and diverse, and can only be evaluated through integrated interdisciplinary studies that allow observation and analysis of the target process in the context of the system and its feedbacks. Such studies will include targeted field-based process studies, in situ mesocosm studies and laboratory experiments, and comprehensive observation and modeling of biological, chemical and physical processes.

Through discussion of GLOBEC and IMBER activities, the TTT has identified some emerging scientific issues that are recommended to be addressed in IMBER-II (John et al. 2010). These issues are $CO_2$ enrichment and ocean acidification, new metabolic and biogeochemical pathways, role of viruses, coupled biogeochemical-ecosystem model projections. The term presents research approaches as following:

(i) Innovative approaches.
(ii) Innovative technologies.
(iii) Process studies.
(iv) Sustained observations.
(v) Palaeo-oceanography.
(vi) Molecular genetics and functional groups.
(vii) Integration of human dimensions in ecosystem models.
(viii) Comparative approach among ecosystems.
(ix) Synthesis and modeling.

## 3 Overyielding and Efficient Resource Utilization in Intercropping via Interspecific Below-Ground Interactions

### 3.1 Introduction

Intercropping, as one of the multiple cropping systems, has been practiced in China for thousands years as well as all over the world for many years (Li et al. 2013). Intercropping is a productive and sustainable system due to its effective resource utilization (water, light, and nutrients), and especially symbiotic nitrogen fixation into the cropping system when legumes involved in an intercropping system (Zhang and Li 2003). Most studies on intercropping have focused on crop combination and above-ground resource use; little attention has been paid to below ground interactions and their effects on yield advantage and resource-use efficiency.

Since the 1990s, we have conducted a series of field and greenhouse experiments to investigate the role of interspecific root interactions and rhizosphere processes in yield advantage in intercropping systems. The objectives of these studies were to examine the complementarity N utilization between intercropped nonlegume/legume species, interspecific facilitation of phosphorus uptake by intercropped P-efficient/P-inefficient species, microelement acquisition by intercropped species, and the root distribution relative to interspecific interactions in various intercropping.

## 3.2 Interspecific Root Interactions Play an Important Role in the Yield Advantage of Intercropping

Yield advantages of various intercropping within China were examined by literature study and our field experiments. We found that there was more than one of land equivalent ratio (LER) for the most intercropping practiced by farmers in China. The crop combinations included cereals/cereals (i.e., wheat/maize, barley/maize, spring maize/summer maize, etc.), cereals/legumes (i.e., maize/soybean, maize/peanut, maize/pea, maize/faba bean, wheat/faba bean, wheat/soybean, etc.), and some cereals/vegetables intercropping (Li et al. 1999, 2001a, 2013; Li 2013).

Interspecific below-ground interactions play an important role in overyielding of intercropping. A significant positive yield effect on maize was found when the root systems intermingled freely (no root barrier) or partly (400 mesh nylon mesh barrier) compared with no interspecific root interaction (plastic sheet root barrier) in a microplot experiments (Li et al. 1999; Zhang and Li 2003; Li et al. 2007). When the roots of two species intermingled, LER values based on total yields and grain yields were 1.21 and 1.34, respectively, but when the roots of the two species were separated completely, the intercropping advantage was greatly diminished and the LER values were reduced to 1.06 and 1.12 (Li et al. 1999). In wheat/maize and wheat/soybean intercropping, we also found that almost half of yield increases in wheat were derived from below-ground interspecific interactions between intercropped species by similar root barrier experiments under field conditions (Zhang et al. 2001).

## 3.3 Complementary Utilization of Nitrogen Between Intercropped Nonlegume/Legume Species

Nitrogen transfer from faba bean to wheat was estimated by the indirect $^{15}N$ isotope dilution technique and by direct plant labeling via petiole injection of a $^{15}N$ solution in intercropping of faba bean and wheat. With the indirect method, N transferred from faba bean to the associated wheat was 2 mg with a mesh barrier and 6 mg without barrier, which accounted for 1.2 and 2.4 % of total N of wheat plant, respectively. Using direct labeling method, N transferred from faba bean to companion wheat was 7 mg, equal to 15 % of total N in wheat (Xiao et al. 2004).

On the other hand, intercropping increased the percentage of N derived from air (%Ndfa) of the wheat/faba bean intracropping, but not that of the maize/faba bean system when no N-fertilizer was applied (Fan et al. 2006). When receiving 120 kg N/ha, however, intercropping did not significantly increase %Ndfa either in the wheat/faba bean system or in the maize/faba bean system in comparison with faba bean in monoculture. The amount of shoot N derived from air (Ndfa), however, increased significantly when intercropped with maize, irrespective of N-fertilizer application. NDFA decreased when intercropped with wheat, albeit not significantly at 120 kg N/ha (Fan et al. 2006).

Intercropping alleviated the inhibitory effect of N fertilization on nodulation and $N_2$ fixation of legumes. Symbiotic nitrogen fixation is usually inhibited by N fertilization in intensive farming systems (Salvagiotti et al. 2008). We tested the effect in two years of field experiments, where different N-fertilizer rates (0, 75, 150, 225, and 300 kg N $ha^{-1}$) were applied to faba bean (*Vicia faba* L.)/maize (*Zea mays* L.) intercropping and corresponding sole cropping systems in the north-western part of China (Li et al. 2009d). The nodule biomass and nitrogen derived from the atmosphere (Ndfa) in intercropped faba bean were increased by 7–58 % and 8–33 % at the start of flowering, 8–72 % and 54–61 % at peak flowering, 4–73 % and 18–50 % at grain filling, and 7–62 % and 7–72 % at maturity, respectively, compared with sole faba bean (Li et al. 2009d).

## 3.4 Interspecific Facilitation of Phosphorus Uptake by Intercropped Species

Interspecific facilitation of P uptake by intercropped species was examined under inorganic P and organic P supplies. Phosphorus in soil includes the inorganic P and the organic P which is mostly not available to general plants. However, some plant species had greater capability to acquire insoluble inorganic P (i.e., white lupin and faba bean) and organic P (i.e., chickpea). Mechanisms behind interspecific facilitation on P uptake by intercropped crops included (i) rhizosphere acidification by P-efficient species resulted in a pH decrease in the rhizosphere, which increased the availability of insoluble inorganic P in soil, such as $Ca_{10}$–P, $FePO_4$, and $AlPO_4$, and benefited associated P-inefficient species that has less ability of mobilizing insoluble soil P; (ii) carboxylates from root exudation of one species chelated Ca, Fe, and Al, consequently mobilizing insoluble soil P, which will benefit the species and other species grown together with it; (iii) greater phosphatase activity in the rhizosphere decomposed soil organic P into an inorganic form, which can be used by both species, such as wheat/chickpea and maize/chickpea (Li et al. 2003a, b, c, 2004, 2007; Zhou et al. 2009)

Maize/faba bean intercropping with rhizobia inoculation enhances productivity and recovery of fertilizer P in a reclaimed desert soil. The apparent P recovery of faba bean/maize intercropping system was much greater ($P < 0.001$) than that of sole cropping systems (weighted means) and was highest at the intermediate P application rate on average. Moderate fertilizer P application enhanced productivity and noclulation of the intercropped faba bean in a reclaimed desert soil, and P deficiency was ameliorated to some extent (Mei et al. 2012).

### 3.5 Maize Improves Iron Nutrition of Intercropped Peanut

Iron deficiency chlorosis frequently occurs in peanut (*Arachis hypogaea* L.) grown on calcareous soils, especially in north China. Interestingly, iron deficiency chlorosis of the crop is more severe in sole cropping systems than in the intercropping system of peanut/maize on these soils. The studies showed that the improvement in the Fe nutrition of peanut intercropped with maize was mainly caused by rhizosphere interactions between peanut and maize (Zuo et al. 2000).

### 3.6 Yield Advantage, Nutrient Acquisition and Competition-Recovery Production Principle

Except for interspecific facilitation, there is another mechanism behind overyielding of intercropping and competition—recovery production principle (Li et al. 2001a, b). In wheat/maize or wheat/soybean intercropping, wheat is grown with maize or soybean for 80 days and afterward maize grows alone for around two months after wheat harvest. During the co-growth period, lasting for about 80 days from maize or soybean emergence to wheat harvesting, yield and nutrient acquisition by intercropped wheat increased significantly, while those by maize or soybean intercropped with wheat decreased significantly (Li et al. 2001a, b). Both aggressivities and nutrient competitive ratio of wheat relative to either maize or soybean revealed the greater competitive ability of wheat than either maize or soybean (Li et al. 2001a).

The biomass and nutrient accumulation in either intercropped soybean or intercropped maize were significantly smaller than in sole soybean or maize before wheat harvest but thereafter increased sharply. The rates of dry matter accumulation in the intercropped maize (10.0–20.1 g/m$^2$ per day) were significantly lower than those in the sole maize (17.1–34.8 g/m$^2$ per day) during the early stage from 7 May to 3 August, while mostly intercropped with wheat. After 3 August, however, the rates of intercropped maize, increasing to 58.9–69.9 g/m$^2$ per day, were significantly greater than in sole maize (22.7–51.8 g/m$^2$ per day), and nutrient acquisition showed the same trends as growth (Li et al. 2001b). It was concluded that then was indeed recovery of growth after wheat harvesting in wheat/maize and wheat/soybean intercropping (Li et al. 2001b).

These findings explained well-yield advantage of cereals/cereals intercropping by temporal niche complementarity.

### 3.7 Root Distribution and Interactions Between Intercropped Species

In wheat/maize intercropping, interspecific interactions lead to an increase for wheat, but a decrease for maize, in terms of yield and nutrient acquisition during the co-growth stage (Li et al. 2001a). In maize/faba bean intercropping, increases in plant growth and nutrient acquisition of both faba bean and maize have been observed, and the interspecific below-ground interactions contributed more to yield advantage in this case than in that of wheat/maize intercropping (Li et al. 1999, 2003a, b, c). We investigated the root distribution of the contrast intercropping to define the relationship between the interspecific interactions and the root distribution (Li et al. 2006, 2011a, b; Xia et al. 2013).

Our studies showed that the roots of intercropped wheat spread under maize plants and had much greater root length density (RLD) at all soil depths than sole wheat. The roots of maize intercropped with wheat were limited laterally but had a greater RLD than sole-cropped maize. The RLD of maize intercropped with faba bean at different soil depths was influenced by intercropping to a smaller extent compared to maize intercropped with wheat (Li et al. 2006).

Faba bean had a relatively shallow root distribution, and the roots of intercropped maize spread underneath them. The results support the hypotheses that the overyielding of species showing benefit in the asymmetric interspecific facilitation results from greater lateral deployment of roots and increased RLD, and that compatibility of the spatial root distribution of intercropped species contributes to symmetric interspecific facilitation in the faba bean/maize intercropping (Li et al. 2006).

Intercropping with wheat leads to greater root weight density and larger below-ground space of intercropped maize at late growth stages (Li et al. 2011a, b). In addition, the roots of intercropped maize have a longer life-span than the roots of sole maize at later growth stage (Li et al. 2011a, b). The result may further understand the recovery growth of later mature species at later growth stage (Li et al. 2001b).

## 4 Intercropping and Disease Management

### 4.1 Introduction

The abundant biodiversity in agricultural ecosystem can be achieved through increasing crop species, including cover crops, intercropping, changing cultivating spatial patterns, applying microbial manure and pesticides, using organic fertilizer to improve soil microorganisms, building the marginal zone and buffer zone in agricultural ecosystem, etc. It is widely approved that intercropping not only effectively manage the disease occurrence and prevalence, but also increase ecosystem diversity, enhance stability, increase crop yield ,and improve the quality of the product. To grow crop varieties with rich genetic backgrounds can effectively defense against potential dangerous of pathogen and insects, decrease the risk of alien invasive species as well. Innovative cultivation technologies make the function of the genetic resources in farmland ecosystem can be more effectively exploited. Lots of practices suggest that strategy of introducing single-resistance gene into elite line can be overcomed by pathogens and pests rapidly, and pyramiding a number of resistance genes into one line is technologically difficult. Exploration and utilization of combination of different genes are alternative method to exert excellent germplasm resources. Intercropping of crops is the essence of Chinese and Asian traditional agriculture, but principles underlying was undiscovered recently. Patterns of intercropping include combination of differences of plant species, growth period, nutrient characteristics existed in different crops to effectively utilize nutrient and heat resources, importantly, the interaction between different species or genotypes of same species was revealed also.

## *4.2 Effects and Mechanisms on Rice Genetic Diversity for Blast Disease*

### 4.2.1 Rice Genetic Diversity for Preventing Prevalence of the Disease

Rice blast disease, caused by *Magnaporthe oryzae* (Hebert) Barr, is one of the most serious fungal diseases worldwide (Ou et al. 1985), and the disease seriously limits rice yield annually. Chemical control and utilization of resistant varieties are main measure to manage the blast disease, but pesticide is of higher cost and easily pollutes the farmland ecosystems, in addition, rice varieties carrying single disease-resistant gene are easily loss the resistance within a few years because of complexity and diversity of physiological races of pathogens. Many production practices have proved that varieties with single-resistant gene are prone to disease in the short term when they are deployed in large-scale areas, namely the blast resistance will quickly loss once these varieties continuously cultivate in large-scale areas for three or five years (Ou 1985). In order to extend cultivation life of resistant varieties to achieve purpose of sustainable management of blast disease, plant pathologists are making lots of exploration and research in the aspects of utilization of rice genetic diversity to manage the disease. Zhu et al. (2000) uses hybrid rice varieties of Shanyou 63 (or Shanyou 22) to intercrop with two good quality glutinous varieties such as Huangkenuo and Zinuo in Yunnan in 1998 and 1999; the results show that rice yield increase by 89 % and blast disease severity decrease by 94 % when intercropping two good quality glutinous varieties, comparing with monoculture of two glutinous varieties.

According to crop genetic diversity research and practical application, mechanism on crop genetic diversity for managing disease can be summarized as follows.

One is the dilution of source amount of compatible races of pathogen. Two is the barrier effect of resistant plants. Three is induction resistance, e.g., nonpathogenic strain or weak pathogenic strain preinoculate plants, the plants can achieve the function that defense against strong pathogenic strain, so that reduce occurrence of leaf blast and panicle blast. Four is microecological effects such as modern variety with high-yield and short-stalk intercropped with glutinous varieties with good quality and tall stalk in the same field, which decrease relative humidity on parts of panicles and shorten dew duration on panicle neck of glutinous varieties, then reduced suitable condition of blast disease occurrence.

### 4.2.2 The Effect of Rice Genetic Diversity on Population Genetic Structure of *M. oryzae*

Because of complexity and diversity of *M. oryzae* races formed in the long period of evolution, the pathogenicity also appear variability, and thus cultivars are often eliminated due to occurrence and prevalence of rice blast. One of the measures to solve this problem is to continuously breed the new resistant varieties, but these varieties will be quickly eliminated when they continuously monoculture in large-scale areas for several years, which lead to a vicious spiral such as breeding speed do not keep pace with resistant loss of varieties. In view of the above reason, it necessary to mine-resistant resources, avoid-resistant varieties carrying single same resistant gene to deploy in large-scale areas for a few years and alternate cultivate different cultivars with different genetic backgrounds, which prevent formation of single variety carrying single-resistance gene and dominant races, stabilize composition of physiological races of pathogen and lessen pesticide application. Through these measures, the ultimate objective is to reduce damage of ecological environment.

Zhu et al. (2003) analyze population genetic structure of blast strains isolated from samples collected from different rice varieties intercropping fields in Shipping County between 1999 and 2000 using Pot2-rep-PCR and pathogenicity test, the result confirm that rice genetic diversity favor to stabilizing selection, keeping genetic diversity of pathogen population, limiting the development of dominant races, and leading significantly to decreasing of disease severity in the field.

## *4.3 Multi-effects and Mechanisms of Maize Intercropping with Soybean*

Intercropping of different crops makes full use of agricultural production resources and reaches the goal for the intensive utilization of lands, soil nutrients, natural resource including moisture, light, heat etc. Intercropping can increase solar energy utilization efficiency, improve the soil quality, reduce diseases and pests, and increase yield significantly. Intercropping maize with soybean is a typical cultivation pattern of long-stalked crops and dwarf-stalk crops. Maize variety of 'Yunrui 88' and soybean variety of 'Nandou 12' was intercropped as different rows (monoculture of maize, intercropping 2 rows of maize with 2 rows of soybean, an intercropping 2 rows of maize with 4 rows of soybean) in 2009 and 2010. The maize and soybean yield, stomata characteristics, photosynthetic characteristics, and the change of field microclimate were systematically studied in Yunnan, China. The results show briefly as follow.

### 4.3.1 Change of Light Intensity and Humidity

Light intensity and humidity in different parts of plants in different periods of crop growth and development in fields were measured, and the result showed that the light intensity and humidity were higher in intercropping field than that in monoculture field. In small bell stage, large-bell stage and anthesis, the humidity in middle part of maize plants was higher than that in intercropping field.

### 4.3.2 Stomata Characteristics of Leaf Lower Epidermis

Stomata characteristics of leaf lower epidermis showed that the length of stomata decreased in the upper part of plants, but increased in the middle and lower parts in maize-intercropping field, comparing with in maize-monoculture field. As for the width and density of stomata, in maize-intercropping field, maize leaves in the upper and middle parts had smaller stomata but bigger ones than in lower parts, more density of stomata in the upper part but less density in the middle and lower parts in intercropping field, comparing with those in monoculture field. There were significant differences in stomatal characteristics parameters between intercropping and monoculture.

### 4.3.3 Photosynthetic Characteristics

Photosynthetic characteristics showed that the photosynthetic rate of the upper part of plant (above 250 cm from ground), the middle part of plant (above 120 cm from ground), the lower part of plant (above 30 cm from ground) in intercropping fields increased comparing with that in monoculture, and the increase of the photosynthetic rate had a positive relations with soybean sow number. There were great differences in photosynthetic rate between maize intercropping and monoculture. According to the two-year data, there were random changes in transpiration rate and stomatal conductance in different parts of maize plants in intercropping fields of maize and soybean.

Comparing with height of maize plant in monocropping field, there was a tendency such as plant height shortened, leaf number, and leaf area increased in intercropping field, and the tendency enhanced along with increase of soybean rows number. The data of maize yields showed that seed setting rate significantly increased, which leading to increase of maize yield.

### 4.4 Summary and Future Prospects

Construction of crop genetic diversity for effectively managing diseases requires conservation of more bioresources and keeps higher genetic heterogeneity within crop varieties. Zhu et al. (2000) found that varieties with different genotypes were mixture cropped in the same field, which resulted in decreased incidence of blast disease due to increased genetic diversity, comparing with monoculture of one variety. Based on recent research and application, the mechanisms on crop genetic diversity for effectively managing disease can summarized as follows: (1) dilution: the dilution of infection source amount of compatible race of pathogen; (2) isolation: the barrier effect of the resistant varieties can be the barrier of pathogen, increasing distance of compatible hosts; (3) induced resistance for host plants, and (4) variation of microecological condition: humility, dewing time, and temperature can be changed to not favorable for pathogen infection and disease development. In addition to above-mentioned factors, traditional varieties with good quality are higher price than that of hybrid ones, which make more favor market competition of the products and income increase. Actually, farmers have more choice of crop combination both for marketing needs and cultivation practices, and more effective for pests management and yield increasing (Li et al. 2009a, b, c) than in paddy field. The mechanism is also more complex the paddy field system, which need investigating in detail.

Due to different crops or varieties intercropping in the same plot of field, species diversity and genetic diversity are increased, and disease, insect, and weed can be effectively controlled without chemical pesticides, which decrease the risks and pressures of agropractice to ecological environments. These technologies can play the important roles in sustainable development of agriculture.

## 5 Ecological Management of Agricultural Pest Insects

### 5.1 Introduction

Agriculture is the foundation of China's national economy with agricultural development increasing production to meet demand for agricultural products. However, ensuring food safety standards and maintaining ecosystem function require constant vigilance. China's environmental, social and economic development faces a resource bottleneck. Ecological agriculture needs to incorporate more sustainable rural development to ensure high-quality industrial practices, multifunctional systems and the integration of modern technology (Li et al. 2011a, b).

Agricultural diversification measures including leisure farm landscapes, intercropping and no-tillage practices could reduce the occurrence of pests. Wheat and cotton aphid [*Aphis gossypii* Glover; *Sitobion miscanchi* (Takahashi), *Schizaphis graminum* (Rondani), and *Rhopalosiphum padi* (L.)] are the main agricultural pests in northern China, requiring the intensive application of chemical pesticides to control them, causing increasingly serious impacts on the environment and food chain. Farmland incorporating ecological systems has inherent pest control factors including the pests' natural enemies, allowing crop plants to produce higher yields. Understanding the importance of biological regularity (including the pest) using various methods will achieve long-term sustainable agricultural development.

## *5.2 Ecological Management Methods to Control Aphids*

### 5.2.1 Cotton Aphid

The cotton aphid, *A. gossypii*, is the first major pest of cotton in Xinjiang, China. The cost of controlling the cotton aphid contributes to more than 20 % of the annual crop loss and is considered a serious threat to cotton production in the Xinjiang Autonomous Region, the largest cotton production area in China (Zhang and Zhang 1998a). In Xinjiang, the cotton aphid emerges and begins damaging cotton plants in mid-June. Therefore, the optimal stage to control this pest is mid-to late June. The natural enemies of cotton aphid in order of their effectiveness are: ladybird beetles, lacewings, hoverflies, Syrphus bugs, and spiders. To encourage natural enemies into this system requires plants that provide food and shelter to the predators. Alfalfa has been used in conservation biological control, benefiting natural enemies, and can be used as a food source because it develops earlier than cotton, allowing sufficient natural enemy populations to establish to control aphids in cotton (Zhang et al. 2004). The planting of a marginal alfalfa zone near cotton fields, followed by cutting the alfalfa while the aphids are entering the cotton, resulted in the natural enemies moving into the cotton crop and controlling aphid populations (Fig. 1).

The main pest species may differ from one plant to another, but they usually share common natural enemies. Alfalfa and cotton have different insect pests, but these pests have similar natural enemies, including ladybird beetles, lacewings, and hoverflies (Fig. 2). During the first ten-day period in June, while cotton aphids started invading the cotton field, the total amount of insects on alfalfa showed a higher density compared with the cotton crop. Insect abundances on the alfalfa crop were approximately 6.94 times higher compared with the cotton crop. The cotton aphid was the dominant species in the cotton fields accounting for 67.77 % of the total insect count. During the second ten-day period in June, insect numbers in cotton fields increased rapidly. The total number of insects in the cotton crop was

**Fig. 1** The arrangement of plants for cotton aphid management in Xinjiang, China

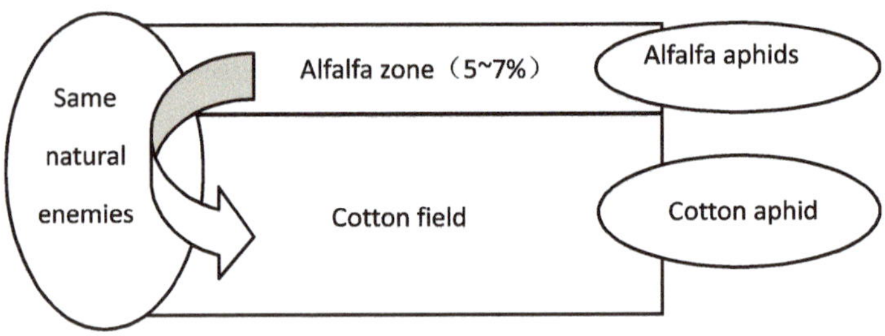


**Fig. 2** The principle of cotton aphid ecological management

approximately 1.42 times higher compared with the alfalfa crop. Concurrently, the number of major natural enemies on alfalfa was 13.65 times higher than in cotton. The natural enemy densities on alfalfa were 27.89, 11.69, and 3.16 times higher (ladybird beetles, lacewings, and hoverflies, respectively) than those in the cotton field (Zhang et al. 2000a).

This pest control technology has the advantage of simple operation, relatively long lasting, and is environmentally friendly (Zhang et al. 2000b). Other advantages are: (i) alfalfa is a high-quality forage species, suitable for the adjustment of agricultural structure and the development of animal husbandry in traditional rural areas; (ii) farmland protection forest provides shading, adversely affecting crop growth, whereas planting alfalfa can significantly improve the use of limited croplands; (iii) alfalfa can improve soil fertility.

### 5.2.2 Wheat Aphids

Wheat aphids are a significant pest on wheat crops, potentially spreading viruses and leading to decreased wheat yields and quality. Aphid species on wheat include *S. miscanchi* (Takahashi), *S. graminum* (Rondani), and *R. padi* (L.). The natural enemies of wheat aphids can be divided into two types, either predatory or parasitoids (Huang et al. 2008). The main predators are: *Harmonia axyridis* Pallas, *Propylaea japonica* (Thunberg), *Coccinella septempunctata* L., *Episyrphus balteata* De Geer, *Lasiopticus pyrastri* (L.), *Syrphus corollae* F., *Chrysoperla sinica* (Tjeder), *Chrysopa formosa* Brauer,and *Eringonidium graminicolum* (Sundevall), with the main parasitoids being *Apridius avenae* Haliday and *Apridius gifuensis* Ashmaed.

Huang et al. (2008) investigated aphid populations on wheat in Yucheng, Shandong Province, showed that populations peaked in mid-May with parasitoids and ladybirds peaking 5–10 and 15–20 days later (respectively). Aphid density was negatively correlated with their natural enemy density, including a significant negative correlation with ladybird density. At the wheat heading stage in late April and early May, aphid populations exponentially increased because of the low density of natural enemies'. From late April to early May, predator density on *Rumex* sp. and alfalfa was 2–3 times higher compared with wheat. This suggests that these plant species have potential applications in regulating aphid populations.

In the wheat/cotton farmland ecosystem, the main problems for insect pest management were: (i) prior to wheat harvest, a large concentration/population of parasitoids and predators is required in the area. However, following the wheat harvest the natural enemies of aphids will move to find new hosts and habitat; (ii) after the wheat harvest, a natural enemy population could be introduced into grazing type plants such as rumex or alfalfa (with a large number of aphids) (Fig. 3). When the cotton aphid begins to damage the crop, their natural enemies could be introduced to the crop as biological controls. This concept of ecological management using mutual plant (grass) species to mediate the natural enemy population and control pest insects is critical in improving environmental impacts.

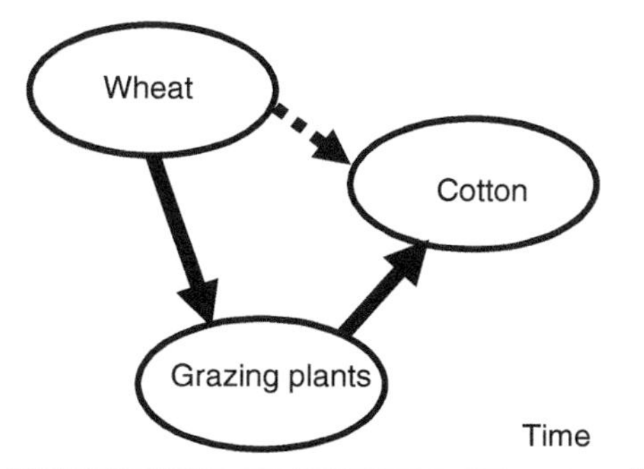


**Fig. 3** Transfer of natural enemies from wheat to cotton via grazing plants

**Fig. 4** The use of *Rumex* sp. (**a**) and alfalfa (**b**) for rearing natural enemies to control wheat aphids

When rumex and alfalfa species are at early growth stages compared with wheat, aphids on early stage species could be used as food to attract ladybirds from wheat, especially during the wheat harvest (Fig. 4). The most effective period to control aphids with natural enemies is during the aphids' population growth stage. At this time, it is important to assist predator transfer from the mutual plant species to target the protective crops. The planting of forage grasses along the crop boundary or intercropping the grasses is relatively simple, ensuring the natural enemies will enter the crops while the pests are emerging. Conversely, weed management along the field boundary is critical to increase plant diversity within the crop field. This can provide suitable habitat for the natural enemies of agricultural pests using ecological management techniques.

## 5.3 *Discussion*

At the annual meeting of China's Ecology Society, Li (2011) suggested that the field of ecology in China developed relatively late compared with western countries. This development can be divided into five stages according to the research direction during different periods (Li 2011). Agricultural pest control has been developed to prevent/reduce biological invasion, emphasizing the protection of biodiversity and the use of green control technologies. The ecological management of agricultural pests includes the mutual plant protection concept (MPP). This concept emphasizing the relationship between pests and their natural enemies regulated by mutual plants will become increasingly important (Zhang and Zhang 1998b). Frank (2010) developed a similar concept known as the 'banker plant' (Frank 2010). The historical experiences of Europe prove that ecological farming system optimization is crucial to the control of agricultural pests (Vasileiadis et al. 2011). Therefore, MPP is an important developmental direction in agriculture and forestry pest management and a necessary pathway for sustainable agricultural and forestry development.

## 6 Agricultural Heritage Research in China: Progresses and Perspectives

Agriculture has flourished in China from time immemorial. With a long history of thousands years of agricultural development, Chinese farmers have been developing for many agricultural practices adaptive to different natural conditions; these traditional practices or models are not only the synthetic application of traditional Chinese philosophy but also the foundation of modern ecological agriculture, which have much positive influence on the sustainable agriculture movement throughout the world (Li 2001). However, because of the population growth and economic driving, more and more farmers gave up the traditional way and adopted modern agricultural techniques. The intensive production practices in modern agriculture resulted in significant negative externalities (Pimentel et al. 1992).

This paper reviewed the progresses of agricultural heritage research in China, which can be divided into two aspects, documentary-based traditional agricultural heritage research and practice-oriented research of dynamic conservation and adaptive management. The former one including China's history of traditional agricultural heritage and archeological have laid a solid foundation for starting the latter one and further study. Dynamic conservation and adaptive management of agricultural heritage was promoted by the Globally Important Agricultural Heritage Systems (GIAHS) project initiated by FAO in 2002. Until now, this research mainly concentrated on the theoretical consideration of agricultural heritage, agrobiodiversity characteristics of agricultural heritage, multivalues of agricultural heritage dynamic conservation, substitute industry and development law and policy on conservation of agricultural heritage. At last, the authors indicated that some greater efforts should be made in the future to advance the study of agricultural heritage: further enrichment for the content of investigation, more creation in the methods of research, deeper research in conservation and utilization of agricultural heritage.

In fact, researches on China's agricultural heritage have been started since the end of 19th century, focusing on agricultural archeology, agricultural history, traditional Chinese philosophy, and agricultural folklore, etc. (Wang et al. 2010). Prior to the intervention of GIAHS project in 2005, Chinese researchers had already done plenty of work on traditional agricultural heritage with emphasis on agricultural history, agricultural archeology, ecological ideas of traditional agriculture, and agriculture custom. These studies tried to discover and identify the agricultural heritages through history records. The results have laid the foundation for any further study.

Compared to the systematic collection and documentation of agricultural history materials and agro-archeological achievements, the research on other aspects of agricultural heritage had been neglected before. Since 2005, based on previous researches, many Chinese scientists have studied on China's agricultural heritage from different perspectives, using modern methods. These would be very helpful for the agricultural heritage conservation and management. These studies can be divided into five aspects.

The first aspect is theoretical developments. FAO defines GIAHS as "Remarkable land use systems and landscapes which are rich in globally significant biological diversity evolving from the co-adaptation of a community with its environment and its needs and aspirations for sustainable development." However, controversies exist regarding the concept of agrocultural heritage, mainly focusing on its interpretation from English and its exact connotation (Min 2007; Han et al. 2007; Zhang et al. 2008; Wang et al. 2010).

The second one is agrobiodiversity evaluation. One of the salient features of agricultural heritage is their high degree of agrobiodiversity. The early studies on agrobiodiversity of agricultural heritage focused on the performance information gathering and compilation. Later researches on agrobiodiversity of agricultural heritage employed more quantitative research methods (Zhang et al. 2010b) and focused on the mechanisms of ecological effect, such as diseases, pests, and weeds control in species coexistence systems (Wang et al. 2007; Zhang et al. 2010a).

The third one is multivalues analysis. Due to specific natural conditions and human activities, agricultural heritage sites always have a fragile ecological environment, rich cultures, undeveloped economy as well as multifunctionality in agriculture. The Multi-values of agricultural heritage have been gradually recognized and many efforts were put on it, from qualitative research (Sun et al. 2008; Li et al. 2009a, b, c; Gao et al. 2010) to quantitative research (Qin et al. 2010; Zhang et al. 2009a, b, 2010; Liu et al. 2010).

The fourth one is dynamic conservation approaches. The agricultural heritage and its dynamic conservation and adaptive management have attracted increasing attention in recent years and have been becoming an emerging field of inquiry. It has become a consensus that agriculture heritage conservation should be integrated with the local socioeconomic development. This requires the development of alternative income sources for local farmers so that dynamic conservation of agricultural heritage will be possible (Liu et al. 2008; Zhang and Tang 2008; Cui 2008). Suggestions range from developing ecotourism (Min et al. 2007; Yan et al. 2008; Chang et al. 2008), developing organic agriculture (Liang et al. 2010; He et al. 2009), production of high-quality agricultural products, development and industrialization of characteristic agriculture, to promoting ecological agriculture and related industries.

The last one is legislation and policy suggestions. The current legal framework for the protection of agricultural heritage is sporadic. The weaknesses and problems concerning the existing legislations and policies lay in the fact that the national interest, especially the interest of traditional communities are not taken consideration during the implementation of the international conventions (Li 2007; Xue et al. 2009). There is no specific law on the maintenance, protection, and utilization of the agricultural heritage. And, the law of protection of agricultural culture is absent. In short, legislations and policies at both international and national levels should be better coordinated and integrated for effective protection of agricultural heritage (Wu et al. 2010).

Although the promotion of GIAHS has received great enthusiasm and support from related governmental departments and organizations, agricultural heritage is still a new concept. Greater effort should be made in the future to advance the study of agricultural heritage. Future research agenda should shift from static literature studies to dynamic explorations and focus on agricultural heritage conservation and utilization, which has greater concern about people's livelihood. The ultimate goal is to achieve the objectives for both protection of traditional agricultural systems and the sustainable development of the local population.

## References

Aksnes, D., & The members of the GLOBEC Scientific Steering Committee. (1999). *Global ocean ecosystem dynamic (GLOBEC) implementation plan*. IGBP Report No. 47. Stockholm, Sweden: IGBP Secretariat, Royal Swedish Academy of Sciences.

Chang, X., Wu, D. T., & Qiao, N. (2008). Ecotourism development of agricultural heritage. *Journal of Beijing Forestry University (Social Sciences), 7*(4), 33–38. (in Chinese).

Cui, F. (2008). On the conservation-oriented tourism development of agricultural heritage. *Journal of Nanjing Agricultural University (Social Sciences Edition), 8*(4), 103–109. (in Chinese).

Fan, F. L., Zhang, F. S., Song, Y. N., Sun, J. H., Bao, X. G., Guo, T. W., & Li, L. (2006). Nitrogen fixation of faba bean (*Vicia faba* L.) interacting with a non-legume in two contrasting intercropping systems. *Plant and Soil, 283*(1–2), 275–286.

Frank, S. D. (2010). Biological control of arthropod pests using banker plant systems: Past progress and future directions. *Biological Control, 52*, 8–16.

Gao, W. y., & Xiao, P. g. (2008). Application of bioengineering technology to protection of endangered medicinal plant resources. *Chinese Traditional and Herbal Drugs, 39*(07), 961–964.

Gao, Z., & Chen, J. (2010). Analysis of the effect of rice-fish system on controlling agricultural non-point source pollution. *Anhui Agricultural Science Bulletin, 16*(9), 162–164. (in Chinese).

Guo, L. p., Huang, L. q., Lv, D. m., Shao, A. j., & Wang, J. (2007). Construction and application of special analysis database of geoherbs based on 3S technology. *China Journal of Chinese Materia Medica, 17*, 1821–1824.

Guo, L. p., Huang, L. q., Yan, H., et al. (2005). Habitat characteristics for the growth of *Atractylodes lancea* based on GIS. *China Journal of Chinese Materia Medica, 30*(8), 565–569.

Guo, L. p., Huang, L. q., & Yan, Y. n. (2002). The influences of inorganic elements in soil on the geolism of *Atractylodes lancea*. *China Journal of Chinese Materia Medica, 27*(4), 5–10.

Han, Y. P., & Liu, J. P. (2007). Some concepts related to agricultural heritage. *Ancient and Modern Agriculture, 3*, 111–115. (in Chinese).

Harris, R., & The members of the GLOBEC Scientific Steering Committee. (1997). *Global ocean ecosystem dynamics (GLOBEC) science plan*. IGBP Report No. 40. Stockholm, Sweden: IGBP Secretariat, Royal Swedish Academy of Sciences.

He, L., Min, Q. W., Zhang, D., et al. (2009). Study on agricultural development mode in traditional agricultural areas: A case study of Congjiang County of Guizhou Province. *Resources Science, 31*(6), 956–961. (in Chinese).

Huang, D., Zhang, R., Dong, Z., Gao, F., Pan, G., & Ouyang, Z. (2008). Multi-population dynamics of aphids and natural enemies on winter wheat and forages and their relationships. *Journal of Environmental Entomology, 30*(4), 325–330.

Huang, J. h., Tan, J. f., Jie, H. k., et al. (2011). Effects of inoculating arbuscular mycorrhizal fungi on *Artemisia annua* growth and its officinal components. *Chinese Journal of Applied Ecology, 22*(06), 1443–1449.

Field, J., & The members of the GLOBEC-IMBER Transition Task Team. (2010). *Supplement to IMBER the science plan and implementation strategy*. IGBP Report No. 52A. Stockholm, Sweden: IGBP Secretariat, Royal Swedish Academy of Sciences.

Li, C. Y., He, X. H., Zhu, S. S., et al. (2009a). Crop diversity for yield increase. *PLoS ONE, 4*, e8049. doi:10.1371/journal.pone.0008049.

Li, G. (2007). Discussion on legal protection for legacy of agricultural culture. *Journal of Beijing University of Agriculture, 22*(4), 46–49. (in Chinese).

Li, L. (2013). *The ecological principles and applications of biological N2 fixation in legumes-based intercropping systems*. Beijing: China Agricultural University Press. (in Chinese).

Li, L., Li, S. M., Sun, J. H., Zhou, L. L., Bao, X. G., Zhang, H. G., & Zhang, F. S. (2007). Diversity enhances agricultural productivity via rhizosphere phosphorus facilitation on phosphorus-deficient soils. *Proceedings of National Academy of Sciences USA, 104*(27), 11192–11196.

Li, L., Sun, J. H., & Zhang, F. S. (2011a). Intercropping with wheat leads to greater root weight density and larger below-ground space of irrigated maize at late growth stages. *Soil Science and Plant Nutrition, 57*(1), 61–67.

Li, L., Sun, J. H., Zhang, F. S., Guo, T. W., Bao, X. G., Smith, F. A., & Smith, S. E. (2006). Root distribution and interactions between intercropped species. *Oecologia, 147*(2), 280–290.

Li, L., Sun, J. H., Zhang, F. S., Li, X. L., Rengel, Z., & Yang, S. C. (2001a). Wheat/maize or wheat/soybean strip intercropping. II. Recovery or compensation of maize and soybean after wheat harvesting. *Field Crops Research, 71*(3), 173–181.

Li, L., Sun, J. H., Zhang, F. S., Li, X. L., Yang, S. C., & Rengel, Z. (2001b). Wheat/maize or wheat/soybean strip intercropping. I. Yield advantage and interspecific interactions on nutrients. *Field Crops Research, 71*(2), 123–137.

Li, L., Tang, C., Rengel, Z., & Zhang, F. S. (2003a). Chickpea facilitates phosphorus uptake by intercropped wheat from an organic phosphorus source. *Plant and Soil, 248*(1–2), 297–303.

Li, L., Yang, S. C., Li, X. L., Zhang, F. S., & Christie, P. (1999). Interspecific complementary and competitive interactions between intercropped maize and faba bean. *Plant and Soil, 212*(2), 105–114.

Li, L., Zhang, F. S., Li, X. L., Christie, P., Sun, J. H., Yang, S. C., & Tang, C. (2003b). Interspecific facilitation of nutrient uptakes by intercropped maize and faba bean. *Nutrient Cycling in Agroecosystems, 65*(1), 61–71.

Li, L., Zhang, L. Z., & Zhang, F. S. (2013). Crop mixtures and the mechanisms of overyielding. In S. A. Levin (Ed.), *Encyclopedia of biodiversity* (2nd ed., Vol. 2, pp. 382–395). Waltham, MA: Academic Press.

Li, S. M., Li, L., Zhang, F. S., & Tang, C. X. (2004). Acid phosphatase role in chickpea/maize intercropping. *Annals of Botany, 94*(1), 297–303.

Li, W. (2011). *China's ecological researches have the turning point*. http://www.cast.org.cn. December 1, 2011.

Li, W., Liu, M., & Min, Q. (2011b). China's ecological agriculture: Progress and perspectives. *Journal of Resources and Ecology, 2*(1), 1–7.

Li, W. H. (Ed.). (2001). *Agro-ecological farming systems in China* (pp. 1–433). Paris: UNESCO and USA & UK: Parthenon Publishing Group.

Li, W. H., Liu, M. C., & Zhang, D. (2009b). Tradeoff analysis on comprehensive valuation of traditional agriculture and rice monocropping in Zhejiang. *Resources Science, 31*(6), 899–904. (in Chinese).

Li, W. X., Li, L., Sun, J. H., Zhang, F. S., & Christie, P. (2003c). Effects of nitrogen and phosphorus fertilizers and intercropping on uptake of nitrogen and phosphorus by wheat, maize and faba bean. *Journal of Plant Nutrition, 26*(3), 629–642.

Li, Y. Y., Sun, J. H., Li, C. J., Li, L., Cheng, X., & Zhang, F. S. (2009c). Effects of interspecific interactions and nitrogen fertilization rates on the agoronomic and nodulation characteristics of intercropped faba bean. *Scientia Agricultura Sinica, 42*(10), 3467–3474.

Li, Y. Y., Yu, C. B., Cheng, X., Li, C. J., Sun, J. H., Zhang, F. S., et al. (2009d). Intercropping alleviates the inhibitory effect of N fertilization on nodulation and symbiotic N-2 fixation of faba bean. *Plant and Soil, 323*(1–2), 295–308.

Liang, Z. Y., & Chen, E. H. (2010). Conservation of traditional agricultural techniques and development of eco-agriculture: A case study of the Wanjiang region. *Resources, Science, 32*(6), 1026–1034. (in Chinese).

Liu, M. C., Zhang, D., & Li, W. H. (2010). Evaluation of comprehensive benefit of rice-fish agriculture and rice monocropping—A case study of Qingtian County, Zhejiang Province. *Chinese Journal of Eco-Agriculture, 18*(1), 164–169. (in Chinese).

Liu, P. F., Gao, Q. J., & Xu, W. S. (2008). An analysis on the relationship between agricultural heritage conservation and socio-economic development. *Ancient and Modern Agriculture, 4*, 89–98. (in Chinese).

Lu, S. p., Sui, X. x., Sun, Q., & Sun, B. q. (2006). Biological functions of secondary metabolism of medicinal plants and influences of ecological environment. *Natural Product Research and Development, 18*(06), 1027–1032.

Mei, P. P., Gui, L. G., Wang, P., Huang, J. C., Long, H. Y., Christie, P., & Li, L. (2012). Maize/faba bean intercropping with rhizobia inoculation enhances productivity and recovery of fertilizer P in a reclaimed desert soil. *Field Crops Research, 130*, 19–27.

Min, Q. W., Sun, Y. H., Cheng, S. K., et al. (2007). Primary study on the features and development of GIAHS's tourism resources. *Economic Geography, 27*(5), 856–859. (in Chinese).

Min, Q. W. (2007). Explanation of GIAHS and other related questions. *Ancient and Modern Agriculture, 3*, 116–120. (in Chinese).

Ou, S. H. (1985). *Rice disease* (2nd ed.). Kew, UK: Common-wealth Mycological Institute.

Pimentel, D., Acquay, H., Biltonen, M., et al. (1992). Environmental and economic costs of pesticide use. *BioScience, 42*, 750–760.

Qin, Z., Zhang, J. E., Luo, S. M., et al. (2010). Estimation of ecological services value for the rice-duck farming system. *Resources Science, 32*(5), 864–872. (in Chinese).

Salvagiotti, F., Cassman, K. G., Specht, J. E., Walters, D. T., Weiss, A., & Dobermann, A. (2008). Nitrogen uptake, fixation and response to fertilizer n in soybeans: A review. *Field Crops Research, 108*(1), 1–13.

Su, J., Tang, Q., et al. (2002). *Study on ocean ecosystem dynamics in coastal ocean II-processes of the Bohai Sea ecosystem dynamics*. Beijing: Science Press. (in Chinese).

Sun, J., & Song, S. (2009). Phytoplankton growth and micro-zooplankton herbivory during the spring phytoplankton bloom period in the East China Sea. *Acta Ecilogica Sinica, 29*(12), 6429–6438. (in Chinese).

Sun, Y. H., Min, Q. W., & Cheng, S. K. (2008). Value of the GIAHS-China traditional rice-fish system. *Chinese Journal of Eco-Agriculture, 16*(4), 991–994. (in Chinese).

Tang, Q., et al. (1995). *The strategy study of Chinese ocean ecosystem dynamics development*. Beijing: Earth/Life Science Departments of National Natural Science Foundation of China. (in Chinese).

Tang, Q., Su, J., et al. (2000). *Study on ocean ecosystem dynamics in coastal ocean I-key scientific questions and development strategy*. Beijing: Science Press. (in Chinese).

Vasileiadis, V. P., Sattin, M., Otto, S., Veres, A., Pálinkás, Z., Pons, X., et al. (2011). Crop protection in European maize-based cropping systems: Current practices and recommendations for innovative integrated pest management. *Agricultural Systems, 104*, 533–540.

Wang, J., & Su, X. M. (2007). The conservation and sustainable development of ecological museum and agricultural heritage. *Chinese Museum, 1*, 91–96. (in Chinese).

Wang, S. M., & Lu, Y. (2010). China's agricultural heritage research: Progress and change. *Agricultural History of China, 1*, 3–11. (in Chinese).

Wu, L., & Jiao H. T. (2010). Study on globally important agricultural heritage system, from law perspective. In J. M. Yi (Eds.), *China yearbook of science & technology law (2008)* (pp. 183–219). (in Chinese).

Xia, H. Y., Zhao, J. H., Sun, J. H., Bao, X. G., Christie, P., Zhang, F. S., & Li, L. (2013). Dynamics of root length and distribution and shoot biomass of maize as affected by intercropping with different companion crops and phosphorus application rates. *Field Crops Research, 150*, 52–62.

Xiao, Y. B., Li, L., & Zhang, F. S. (2004). Effect of root contact on interspecific competition and N transfer between wheat and fababean using direct and indirect N15 technique. *Plant and Soil, 262*(1–2), 45–54.

Xue, D. Y., & Guo, L. (2009). On protection and benefit-sharing of genetic resources and associated traditional knowledge in the ethnic areas of China. *Resources Science, 31*(1), 10–20. (in Chinese).

Yan, J., Wu, D. T., & Xiao, M. (2008). Ecotourism: Balance between protection and development in agricultural heritage site—taking "traditional rice-fish agriculture" of Qingtian County as an example. *Rural Economy, 2*, 74–77. (in Chinese).

Zeng, Q. P., & Bao, F. (2011). Research achievements in the synthetic biology and metabolic engineering of artemisinin. *Chinese Science Bulletin, 27*, 2289–2297.

Zhang, D., Liu, M. C., Min, Q. W., et al. (2009a). Comparison of ecosystem services of rice-fish agriculture in Qingtian and Congjiang. *China Population Resources and Environment, 19*(6), 30–36. (in Chinese).

Zhang, D., Min, Q. W., Cheng, S. K., et al. (2009b). Ecosystem services evaluation of traditional agricultural regions: A case study of Congjiang county. *Guizhou Province. Resources Science, 31*(1), 31–37. (in Chinese).

Zhang, D., Min, Q. W., Cheng, S. K., et al. (2010a). Effects of different rice farming systems on paddy field weed community. *Chinese Journal of Applied Ecology, 21*(6), 1603–1608. (in Chinese).

Zhang, D., Min, Q. W., He, L., et al. (2010b). The value of ecosystem services in conventional and organic rice paddies: A case study in Wannian, Jiangxi, China. *Chinese Journal of Population Resources and Environment, 8*(2), 47–54.

Zhang, F. S., Li, L., & Sun, J. H. (2001). Contribution of above- and below-ground interactions to intercropping. In W. J. Horst, et al. (Eds.), *Plant nutrition—food security and sustainability of agro-ecosystems* (pp. 979–980). Dordrecht: Kluwer Academic Publishers.

Zhang, F. S., & Li, L. (2003). Using competitive and facilitative interactions in intercropping systems enhances crop productivity and nutrient-use efficiency. *Plant and Soil, 248*(1–2), 305–312.

Zhang, R., Liang, H., Tian, C., & Zhang, G. (2000a). Biological mechanism of controlling cotton aphid (Homoptera: aphididae) by the marginal alfalfa zone surrounding cotton field. *Chinese Science Bulletin, 45*(4), 355–357.

Zhang, R., Ren, L., Wang, C., Lin, R., & Tian, C. (2004). Cotton aphid predators on alfalfa and their impacts on cotton aphid abundance. *Applied Entomology and Zoology, 39*(2), 235–241.

Zhang, R., & Zhang, G. (1998a). The importance of mutual plants in biological control of pests. *Chinese Journal of Biological Control, 14*(4), 176–179.

Zhang, R., & Zhang, G. (1998b). The issues and research topic on cotton planting in Xinjiang, China. *Bulletin of Chinese Academy of Sciences, 12*(1), 50–58.

Zhang, R., Tian, C., & Wang, D. (2000b). The development of rural grass industry and realize the cotton pest ecological control. *Bulletin of Chinese Academy of Sciences, 5*, 346–348.

Zhang, W. Y., & Tang, S. (2008). On agricultural heritage and value judgment. *Journal of Anhui Agricultural, Science, 36*(25), 11041–11042. (in Chinese).

Zhou, L. L., Cao, J., Zhang, F. S., & Li, L. (2009). Rhizosphere acidification of faba bean, soybean and maize. *Science of the Total Environment, 407*(14), 4356–4362.

Zhu, Y. Y., Chen, H. R., Fan, J. H., et al. (2000). Genetic diversity and disease control in rice. *Nature, 406*, 718–722.

Zhu, Y. Y., Leung, H., Chen, H., et al. (2003). Using genetic diversity to achieve sustainable rice disease management. *Plant Disease, 87*, 1155–1169.

Zuo, Y. M., Zhang, F. S., Li, X. L., & Cao, Y. P. (2000). Studies on the improvement in iron nutrition of peanut by intercropping with maize on a calcareous soil. *Plant and Soil, 220*(1–2), 13–25.

# Chapter 5
# Case Study of Species and Population Conservation

**Guanghui Lin, Luzhen Chen, Yihui Zhang, Shengchang Yang, Wenqing Wang, Hangtao Zhou, Jian Qin, Dongmei Shi, Xianzhi Huang, Zhonghuai Xiang, Jusheng Jiang, Zongbo Peng and Zhiyun Ouyang**

**Abstract** In this chapter, we listed three important species and population as typical cases of species and population conservation. Mangrove, as a special ecosystem in the edge of the land and sea, has been conducted plenty of in-depth researches for a long time. Here, we tease and describe the historical perspectives and current status of mangrove ecological research. In addition, the protection consciousness of the ecological environment and biodiversity has gradually improved during the development process of biological industry. Sericulture, as a traditional industry, played an important role in the economic, culture, and social development in China at one time. In recent years, researches have been focused on the multiple ecological functions of mulberry, such as windbreak and sand-fixation, soil and water conservation, etc. Rubber is a typical tropical cash crop, and natural rubber is an important industrial raw materials and strategic resources. The sustainable management on rubber ecosystem has been reinforced at the same time of natural rubber industry development.

---

G. Lin
Center for Earth System Science, Tsinghua University, Beijing 100084, China

L. Chen · Y. Zhang · S. Yang · W. Wang
Key Laboratory of the Ministry of Education for Coastal and Wetland Ecosystems, College of Environment and Ecology, Xiamen 361005, Fujian, China

H. Zhou
Key Laboratory of the Ministry of Education for Coastal and Wetland Ecosystems, College of Oceanography and Earth Sciences, Xiamen 361005, Fujian, China

J. Qin · D. Shi · X. Huang · Z. Xiang
State Key Laboratory of Silkworm Genome Biology, Southwest University, Chongqing 400715, China

J. Jiang · Z. Peng
Hainan Academy of Agricultural Reclamation Sciences, Haikou 570206, China

Z. Ouyang (✉)
Research Center for Eco-Environmental Sciences, Chinese Academy of Sciences, Beijing 100085, China
e-mail: zyouyang@rcees.ac.cn

 
W. Li (ed.), *Contemporary Ecology Research in China*,
DOI 10.1007/978-3-662-48376-3_5

**Keywords** Mangrove · Historical perspectives · Recent developments · Mulberry · Ecosystem function · Rubber · Ecosystem management

# 1 Historical Perspectives and Current Status of Mangrove Ecological Research in China

## *1.1 Introduction*

Mangrove species in China covered >50,000 ha in 1950s (Lin 1999), but reduced to about 23,000 ha now (Chen et al. 2009). Since 1995, the majority of natural mangroves have been protected as part of the national wide mangrove nature reserves. Chinese scientists have conducted a great deal of research on mangroves since 1950s (Lin 1999). Chen et al. (2009) provided one in-depth reviews on Chinese mangrove current status and research. Since then, increased government investments have greatly improved the research on mangroves in China. In this paper, we provided some historical perspectives and reviewed the rapid developments in the mangrove conservation, restoration, and researches in China after 1990s.

## *1.2 Historical Perspectives*

Mangrove research in China can be divided into the following five stages:

(i) Early awareness and taxonomy study period (before 1955): There was only 38 reports, most descriptions and taxonomy of mangrove species, on mangroves in China between 1600 and 1975 (Rollet 1981), including «Fujian Flora and Plant Communities», «Chinese Mangrove Forests», etc. Under the supervision of Prof. Jing He, Peng Lin started his mangrove community field surveys in 1953.

(ii) Community ecology study period (1955–1965): Mangrove research in China shifted from taxonomy studies to more on plant ecology, and yielded some milestone publications such as «Mangrove Forests» in Taiwan, «Mangrove Ecology» in Fujian and «Mangrove Communities on Leizhou Peninsula» in Guangdong. Meanwhile, *Kandelia candel* was successfully introduced in 1952 to Ruian, Zhejiang from Guangdong and Fujian.

(iii) Mangrove resource and conservation study period (1976–1985): As vegetation science progressed in China, especially publication of «Vegetation of Guangdong» in 1976 and «Vegetation of China» in 1980, mangrove values and related studies received more and more attentions during this period. Prof. Peng Lin and his research group at Xiamen University made significant progresses in mangrove ecological studies, and published some significant books such as «Mangrove Communities and Distributions along Chinese Southeastern Coastlines» (1981) and «Mangrove Forests» (1984). Yunzhang Gao in Botany Institute of South China and Hongda Zhang at Zhongshan University conducted some excellent work on mangroves in Hong Kong, Macao and Guangdong. In Taiwan, a few papers on mangrove ecology were published in 1982, including famous authors Ming-yi Chen, Chang-hung Chou, etc.

(iv) Mangrove ecosystem ecology research period (1985–2000): This was the peak period for mangrove ecosystem ecology research, with strong focus on the structure and functions of mangrove ecosystems in China, including energy flow and nutrient cycle processes, biodiversity of benthic animals, algae, phytoplankton, microbes, and birds, and trophic relations among primary producers, consumers, and decomposers. Mangrove physiological ecology and molecular ecology were also received great attentions (e.g., Ge and Sun 1999). Influential research groups included the Mangrove Research group of Prof. Peng Lin at Xiamen University (members: Changyi Lu, Guanghui Lin, Ronghua Chen, Wenjiao Zheng, Yiming Lin, Wenqing Wang, etc.), the research group of Prof. Hongda Zhang at Zhongshan University (members: Guizhu Chen, Shenyu Liao, etc.), the research team led by Yushan Wong and Nora FY Tam at City University of Hong Kong, and the research team led by Hanqing Fan at Center for Mangrove Research of Guangxi (Chen et al. 2009). In 1992, Research Group of Mangrove Ecologists was established within Ecology Society of China, and the first national symposium on mangrove ecosystems in China were organized in Beihai, Guangxi in 1993, followed by the second symposium in Guangzhou in 1996 and third symposium in Xiamen Fujian in 1996, respectively. These meetings greatly promoted the academic exchanges among Chinese mangrove ecologists and stimulated Chinese mangrove research.

(v) Mangrove-global change interaction study period (2001–present): During recent years, Chinese mangrove researchers paid great deal to possible interactions between mangrove ecosystems and global change. In addition to field or experimental studies on methane emission and mangrove responses to sea level rise (e.g., Lu et al. 1999; Ye et al. 2003, 2004), several long-term field research stations were established, respectively, in Dongzhaigang National Mangrove Nature Reserve by Bao Liao's group at Institute of Tropical Forestry, Chinese Academy of Forestry and in Zhangjiangkou National Mangrove Nature Reserve and Zhanjiang National Mangrove Nature Reserve by Guanghui Lin's group at Xiamen University (now moved to Tsinghua University). Another threat of global change on Chinese mangrove wetlands came from the fast spread of *Spartina alterniflora* and introduction of exotic *Sonneratia apetala*, which received intensive studies during recent years (e.g., Chen et al. 2008, 2012, 2013; Zhang et al. 2012). However, few studies are available on the effects of elevated $CO_2$, nitrogen deposition, and land-use change on mangroves in China (Chen et al. 2009).

## 1.3 Recent Developments in Mangrove Research

Since 1990s, significant amount of books and scientific papers have been published, indicating the rapid development in this research field (Chen et al. 2009). In total, there were 1473 papers published in domestic and international journals from 1990 to 2007, according to the records in the Weipu Database of China and the Web of Science Database (Table 1). Although the number of annual published papers on mangroves by Chinese scientists increased exponentially from 1990 to 2007, the proportion of the peer review journal papers written by Chinese scientists to the world mangrove publications after 2000 was 10.2 %, which was lower than that in 1990s (Table 1).

Between 1990 and 2007, the mangrove research in China focused on a dozen areas (Fig. 1), which included remote sense and modeling, aquaculture, global ecology, geography and hydrography, energy flow, morphology and anatomy, molecular ecology, pharmaceutics and active material exploitation, silviculture, community and population ecology, biodiversity, pollution ecology, ecophysiology, conservation and management. Among them, five research areas increased most rapidly, including molecular ecology, pollution ecology, biodiversity, conservation and management, silviculture and pharmaceutics, and active material exploitation (Chen et al. 2009).

Extensive researches on mangrove ecosystem structure and function revealed extremely high biomass and primary production for the mangrove forests in China (see Chinese literature cited by Chen et al. 2009). Based on these results, a "Three-High" or "3-H" theory on mangrove communities, i.e., high productivity, high return ratio, and high decomposition ratio, was later proposed (Lin 1999). A great deal of field and greenhouse studies pointed to great challenges in selecting

**Table 1** Number of papers on China's mangroves published from 1990 to 2007 (Chen et al. 2009)*

| Year | No. of publications in Weipu database of domestic journals in Mainland China | No. of publications in SCI journals of web of science database | Total No. of world mangrove publications SCI journals of web of science database | Proportion of China's SCI publications to world mangrove publications (%) |
|---|---|---|---|---|
| 1990–99 | 307 | 104 | 679 | 15.3 |
| 2000–7 | 780 | 282 | 2752 | 10.2 |
| Total (1990–2007) | 1087 | 386 | 3431 | 11.3 |

*Officially published papers of Mainland China, Hong Kong, Taiwan and Macao mangroves were included. While the abstracts submitted to conferences and book chapters were not included

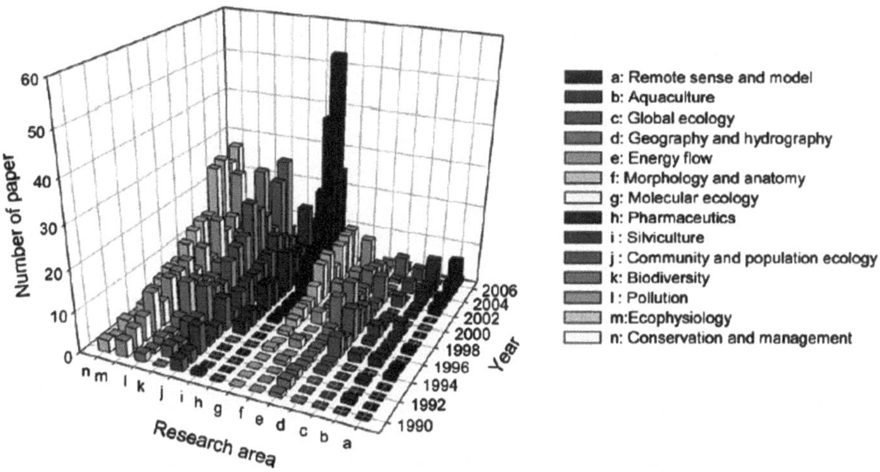


**Fig. 1** Distribution of published papers in 14 major research areas from 1990 to 2007 (Chen et al. 2009)

suitable tidal flats for the mangrove afforestation efforts in China (Fan and Li 1997; Mo and Fan 2001). Physiological studies showed that *Bruguiera gymnorrhiza* had lower tolerance to soil flooding than *K. candel* (Ye et al. 2003), while the optimal tidal inundation period for *Kandelia obovata* growth and photosynthesis was 2–4 h per tidal cycle (Chen et al. 2004, 2005; Wang et al. 2011a, b). Studies on mangrove management and new techniques in silviculture developed rapidly after 2000 (Fig. 1). The exotic *Sonneratia apetala* was shown to have good tolerance of high tide and chilling conditions (Liao et al. 2003), which may explain why it was considered as one of the best mangrove afforestation species and used almost in all mangrove afforestation projects in China.

Researches on the potentials of mangrove wetlands for wastewater treatments and pollutant degradation have been also greatly promoted in China since 1990s (Tam and Wong 1997; Wong et al. 1997), and it was estimated that more than 70 % of dissolved organic carbon, ammonia, and total Kjeldahl nitrogen; 50 %of inorganic nitrogen and many polycyclic aromatic hydrocarbons could be removed or degraded by a constructed mangrove wetlands (Zhong et al. 2007; Yang et al. 2008). Furthermore, medicinal applications of Chinese mangrove plants were known for a long period of time (Lin and Fu 1995), which stimulated great interests in the studies on the sources, compound structures, and bioactivities of natural products from mangrove materials after 2000 (Li et al. 2007; Tao et al. 2007; Wu et al. 2008). However, the direct utilization of mangrove materials for medicine production will likely reduce mangrove resources and should be avoided. A better way for this application is to formulate new medicines through chemical synthesis base on the compound configurations of related compounds found in certain mangrove materials (Chen et al. 2009).

Great progresses have been made in the field of molecular ecology since 2000, especially in the areas on the geographical distances and species relationships of Chinese mangroves. The genomic basis of the adaptive evolution and speciation in mangrove was established (Zhong et al. 2002; Shi et al. 2002, 2005), and several molecular biomarker, including chloroplast DNA, mitochondrial DNA and inter-simple sequence repeat, were used to identify the gene flows between South China Sea and nearby regions (Tan et al. 2005). Molecular markers were also used as taxonomic evidences for several mangrove species classifications (Sheue et al. 2003). These studies illustrated the values of using modern technologies in resolving long-standing ecological or evolutionary issues in mangroves (Chen et al. 2009).

## 1.4 Conclusion and Discussion

Because of the apparent success in mangrove conservation and reforestation during last two decades, the loss of mangrove habitats was halted or even reversed, with about two third of mangrove forests being protected in more than 30 natural reserves of national, provincial and county-level as well as over 2000 ha mangrove forests being restored in southern China (Chen et al. 2009). However, there are still many threats to Chinese mangroves, including urban and aquaculture wastewater discharge, oil pollution, biological invasion, insect outbreak and the influence of water transportation remain serious threats to mangroves in China. For example, human destruction of mangrove mud flats or forest canopy may facilitate fast invasion of *S. alterniflora* into mangrove forests (Zhang et al. 2012).

A large number of case studies on Chinese mangroves have significantly increased our understanding of the structure and functions of the mangrove ecosystems as well as the values of mangrove wetlands over the past two decades (Chen et al. 2009). However, there are still many areas needed to be strengthened in the future, including (i) the role of keystone species in determining Chinese mangrove ecosystems and the relationships between species diversity and ecosystem functions; (ii) the consequence of large scale sea wall construction on the fate of mangrove forests in China under rising sea levels in coming decades or century; (iii) the invasive mechanisms and the efficient measures for controlling biological invasions caused by exotic species such as *S. alterniflora*; and (iv) universal standard system for evaluating such efforts and achievements and enforcement of such standard system by governmental agencies (such as State Administration of Forestry), research institutions and local communities, etc.

In conclusion, the population boom and rapid economic developments have greatly reduced mangrove areas in China since 1980s, leaving only about 23,000 ha mangroves in Mainland China (Chen et al. 2009). Chinese government has launched a series of programs to protect mangroves since 1980s and has established mangrove ecosystems as high-priority areas for improving environmental and living resource management. During last three decades, a total of 34 natural mangrove conservation areas have been established, which accounts for 80 % of the total existing mangroves areas in China. Mangrove restoration areas in Mainland China accounted for <7 % of the total mangroves areas in 2002. A great deal of research papers on Chinese mangroves has been published in international journals. However, more systematic protection strategies and active restoration measurements are still urgently needed in order to preserve these valuable resources in China.

## 2 Ecological Issues of Mulberry and Sustainable Development

Mulberry trees have long been planted solely for silkworm rearing. Its ecological roles were neglected. In recent years, as people paid more and more attention to the increasingly deteriorated ecological environment, the roles of mulberry trees in prevention and control of desertification, control of stony desertification, water and soil conservation, saline land management, and returning farmland to forestry have received renewed cognition. Meanwhile, multiple values of mulberry trees as source of other products other than as silkworm food have also been gradually explored, leading to an innovation and transformation on management of mulberry industry. Mulberry ecological industry, its core technologies, and its ecological governance and industrialization mode in a typical fragile ecological zones, have become the focuses for academia and industry.

Mulberry trees are good carbon sink plants. It is primarily estimated that 1 mu mulberry trees could absorb about 4162 kg of $CO_2$ (equivalent to 135 kg of carbon) and release 3064 kg of oxygen each year. In addition, mulberry leaves have high endurance and certain absorption and purification function to air pollutants, such as chlorine, hydrogen fluoride, and sulfur dioxide. Under 6 h of fumigation at $0.79 \times 10^{-6}$ concentration of sulfur dioxide, 1 kg dry mulberry leaves could absorb 5.7726 g sulfur dioxide and one cubic meter mulberry forest could absorb 20 mL sulfur dioxide gas each day. Mulberry has strong resistance to sulfur dioxide pollution, being a first-class resistant tree species against sulfur dioxide. Mulberry also has high resistance to chlorine dominant air pollution. Under such air pollution, mulberry leaves remained undamaged or area of damage was below 20 % of total leaf area, and trees grew and developed normally, being a first-class resistant tree species which can be used as resistant plants. No matter what is considered, mulberry is an ideal tree species for city landscape due to its excellent features in tree form, leaf color, growth vigor, tenacity, and resistance. As a deciduous arbor species, mulberry is resistant against drought and flood. It can be planted along riversides, at field edges, on slopes, at garden corners, along roadsides, and in public parks and other recreation places. It is an excellent tree species for city afforestation at "four sides".

Mulberry has very strong root system. Its roots form a greatly tangled and densely network in the soil. Mulberry plantation is highly capable of suppressing sandstorm and conserving water and soil. According to experiments at Natuo High-efficiency Agri-ecological Garden, Fuling District of Chongqing City, anti-erosion of soil under mulberry tree hedgerow was considerably improved. The aggregation degree and aggregation status of top soil was increased by 25.2 and 50.6 %, respectively than that under traditional planting pattern, while dispersion rate was lowered by 3.7 %, water stability index was 1.9 times to that of traditional down slope planting pattern, and the destruction rate of water stable soil aggregates within the first 10 min was only 57.2 % to that of the traditional planting pattern. Compared to traditional planting pattern, runoff volume, and runoff coefficient were reduced by 10.34–20.00 %, erosion was lowered by 55.23–67.84 %, and sand content of runoff was lowered by 48.60–59.80 %. Under heavy rain, mulberry tree hedgerow had significant effect in reducing total runoff and enrichment ratio of nutrients as well.

Mulberry root system occupies a room bigger than aerial part of the plant. In loess plateau of Shaanxi, China, length of the root system of a 1-year-old mulberry sapling could reach 1000 m. That of a 10-year-old mulberry tree could reach 10,000 m. The longest taproot was 8 m deep in the soil. The longest lateral root was over 9 m. The distribution area of underground root system could be 4–5 times to the projected area of tree canopy. Compared to un-afforested land, water conservation was 20 $m^3$ higher and mud and sand runoff was 3 t less in 1 mu mulberry field per year. While no other vegetation was present, the agricultural land with mulberry plantation was capable of stabilizing 3067.8 $kg/km^2$ sand a year. The erosion ditches were 57 % less, 59.2 % shorter, 61 % narrower, and 64 % shallower than those in agricultural land without mulberry plantation. After establishment of ecological mulberry plantation for water and soil conservation on agricultural land with 40° steep slope, reduction of rainfall runoff could reach 70 % and reduction of soil erosion could reach 79.7 %.

Mulberry trees have high adaptability to soil. Under soil pH value of 4.5–8.5 and slat content of 0.2 %, they could still grow normally (Hu and Zhou 2010). Mulberry trees could grow in barren soil with poor nutrients. In arid or semiarid desert area with annual rainfall less than 300–600 mm, they could still grow well under natural condition. Even in desert area of Xinjiang with annual rainfall below 150 mm, they could also grow and develop normally. Under abundant supply of soil moisture, the transpiration coefficient of mulberry trees was 350–450. Under drought stress, net photosynthetic rate, transpiration rate and water utilization efficiency of mulberry leaves would decline. The transpiration coefficient of some drought resistant mulberry varieties in the north of China was as low as 274, being lower than that of *Populus diversifolia* (300), *Elaeagnus angustifolia* (383), seabuckthorn (483), and poplar (513). Wilting coefficient of northern drought resistant mulberry varieties was 9, being lower than wild apricot (13), white elm (13), poplar (15) and other tree species, demonstrating that mulberry could adapt to dry climate. At present, mulberry has been used as an ecological tree species for water and soil conservation of loess plateau in Shaanxi, desert control in Xinjiang, control of stony desertification in Chongqing, sand control in Beijing, control of stony desertification in Guangxi, and construction of ecological environment in many other provinces.

Mulberry trees can resist chilling and freezing of −30 °C and endure high temperature of 40 °C. Dormant mulberry trees have the highest resistance against chilling. Mulberry trees also have certain resistance against chilling at growth stage. Investigations showed that adult mulberry trees could survive from an inundation of 20 days during their growth period. This is very rare among other xerophytic plants. Mulberry trees have very strong endurance to waterlogging in dormant stage. Mulberry trees in hydro-fluctuation belt of the Three Gorges Reservoir Area which had undergone inundation of 150 days in over 10 m deep water could still germinate and grow, being a tree species with best growth after emergence of the hydro-fluctuation belt.

As an ecological tree species, mulberry trees not only bring good ecological benefit but also yield high economic income. This is a prominent characteristic of mulberry ecological industry. Mulberry leaf, fruit, stem, and bark can be utilized and easily integrated with other advantageous industries. All these utilizations are beneficial and sustainable, having caused wide attentions of researchers at home and abroad.

Mulberry leaf is a kind of forage resource with complete nutritional composition. Mulberry tea has the effects of calming liver, improving eyesight, and evacuating wind-heat, being a healthcare drink suitable to all ages. Mulberry fruit is a kind of good taste, tonic fruit. Mulberry tree has very high medicinal value. It has multiple medicinal parts. Its root, root bark, leaf, twig, and fruit can all be used as medicine. In recent years, many clinic application studies have discovered that the chemical components of mulberry tree have various pharmacological functions, such as reducing blood sugar level, reducing blood lipid level, lowering blood pressure, antibacteria, antivirus, antitumor, delaying aging, anti-filariasis, spasmolysis, and antiulcer, having important clinic application value.

Mulberry trees can also be utilized in food and chemical industries. For examples, food, food additives, and cosmetics made from mulberry leaf and fruit have been available in the market. A special edible carbon made from carbonization of mulberry root and stem has been widely used as food additive. Recombined wood boards and bars made from mulberry branch can be used to substitute wood logs for making floor boards, door jambs, and wooden models of buildings, furniture, and parts of wooden structures. Mulberry branches can substitute miscellaneous woods or cotton seed hulls for producing various edible fungi such as mushroom, agaric, *Ganoderma lucidum*, *Flammulina velutipes*, and *Pleurotus ostreatus*. Meanwhile, it is an excellent raw material for paper making and active carbon production. Mulberry branch has quite high unit combustion value, being an excellent biomass fuel. Mulberry twigs can be used to make skin care agent. Mulberry bark fiber has excellent characteristics of natural fiber, being good raw material for making high rank paper.

Mulberry is a fast-growing woody plant characterized by deep roots, flourish leaves, high resistance to pollution, high resistance to wind and sand, high resistance to drought, high endurance to salinity, strong adaptability, and easy cultivation. The broad ecological adaptability of mulberry to light, temperature, water, soil, and other natural conditions objectively enables it to have multiple ecological protective functions in water and soil conservation, wind resistance and sand consolidation, water source preservation, and air refreshment. As a traditional economic woody plant, mulberry also has great values for comprehensive development and utilization. It is an excellent tree species with both ecological and economic benefits for water and soil conservation and ecological environment construction in China.

To sum up, mulberry tree has a strong adaptability, strong ecological function, high economic efficiency, can grow in different ecological fragile areas. At the meanwhile, the economic use of the diversity for mulberry ecological industry, make it a good coupling in various kinds of ecological industrial system, to achieve a sustainable development.

## 3 Ecosystem Management in the Natural Rubber Industry

### *3.1 Introduction*

Ecosystem management originated from management and utilization of the traditional forest resources (Christensen et al. 1996). Forest ecosystem management has formed a robust theoretical framework in China (Liao and Zhao 1999; Yang and Jiang 2003; Zhou et al. 2007a, b).This management method has been widely recognized because of growing awareness of environmental degradation and resource depletion. However, ecosystem management is not only applicable to the management of natural resources, but has also developed into an integrated resource management system targeting the structure, function, and processes of sustainability. This approach ensures that ecological services and biological resources have not been irreversibly consumed due to human activities, and achieves long-term sustainability (Zhao et al. 2004; Yang et al. 2004).

The natural rubber ecosystem has been degenerated due to a long-term focus on economic benefits, and a new method for the management of natural rubber is needed. The natural rubber industry ecosystem is a typical compound ecosystem, and relationships between ecosystem elements are complex (Jiang and Wang 2004). So its management can not merely focus on the economic benefits but should take into full account the coordination of ecological, economic, and social benefits for sustainability, health, and industry development. The ecosystem management provides a new approach for natural rubber industry ecosystem management. This method can guide production and management of the natural rubber industry, and this will benefit the healthy and sustainable development of the natural rubber industry.

Hainan is a tropical island and the second largest island in China, the environment is good for growing rubber trees. After decades of effort, Hainan has become the largest production and supply base of natural rubber in China and rubber forest is one of the largest artificial ecosystems on the island (Jiang and Wang 2004). So Hainan is selected as an example to present a basic model of natural rubber industry ecosystem management in this paper.

## 3.2 Ecosystem Management in the Natural Rubber Industry

Natural rubber industry ecosystem management mainly reflected in the ecological management engineering of rubber plantation, industrial ecosystem regulation, ecosystem evaluation, and sustainable management, Eco-management information (Fig. 2).

### 3.2.1 Ecological Engineering of Rubber Plantation

Environmental Selection Engineering

Agro-climate analysis and regionalization are most important in selection and evaluation of rubber planting environments. Two main principles should be followed: (i) more than 30 years of meteorological data should be considered, since the rubber tree is a long-term crop; (ii) meteorological indices of selected planting sites should be as similar as possible to the place of origin. Temperature and moisture related to chilling injury are the main indicators for climatic regionalization of the rubber tree when selecting appropriate sites.

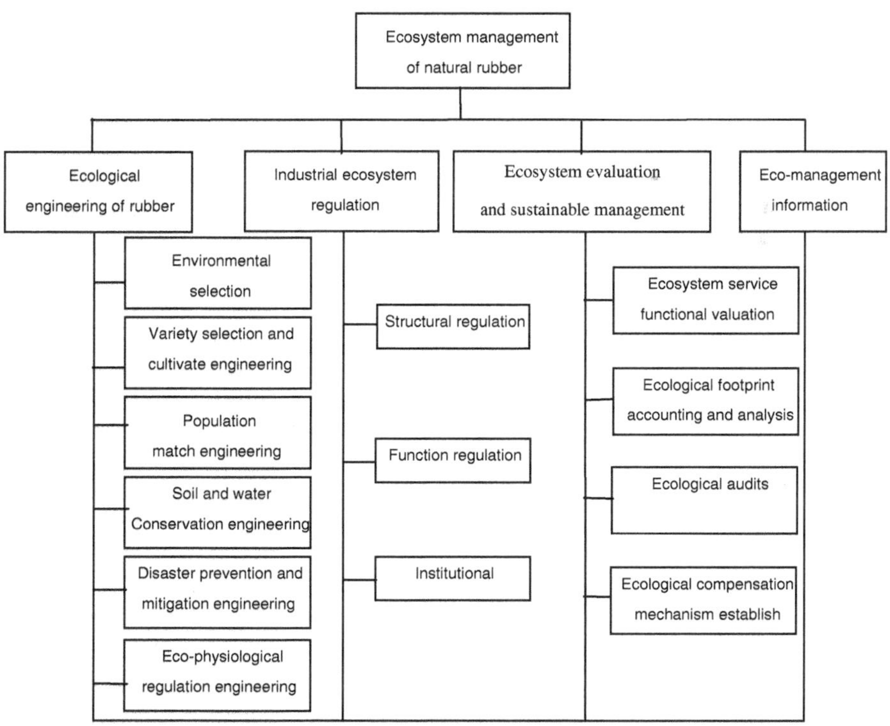


**Fig. 2** Ecosystem management of natural rubber industry

### Variety Selection and Cultivate Engineering

Selection and cultivating of desirable rubber clones is one of the main ways for China's rubber plantation industry to succeed in a relatively short period of time. The specific measures include: local field trial and local extension of the clones; cultivation of multi-clones, planting clones in the line of local conditions; rubber planting recommendation; and implementation of appropriate agricultural technical standards and measures.

### Population Match Engineering

Many years of scientific research and production practice proves that planting other crops between the rows in the rubber tree was able to conservation soil and water, increase soil nutrient content, maintain and improve biological diversity of rubber plantation, etc., and ultimately achieve ecological, economic, and social benefits. Common cover crops include cereal, oil crops, potatoes, economic crops, and green manure crops. The common cropping system includes a one-year, two-year, or three-year rotation system.

### Soil and Water Conservation Engineering

Annual rainfall is high and concentrated in Hainan's rubber plantations, and hard to be absorbed completely by the soil in a short time, causing soil erosion. We must focus on water and soil conservation engineering when the rubber tree planted. The main technical measures for soil and water conservation engineering in rubber plantation include: contoured planting on slopes, building benches and terraces on gentle slopes.

### Disaster Prevention and Mitigation Engineering

Many years of rubber planting have shown that shelter-forest can reduce natural disasters and protect rubber plantations, so it is essential to construct wind belt network in rubber plantations areas. The protective effect of the shelter-forest belt depends on the structural configuration and the direction, width and distance between the shelter-forest.

### Eco-physiological Regulation Engineering

Eco-physiological regulation engineering of rubber plantation divided into biological regulation and chemical regulation, biological regulation mainly include Intercropping, chemical control including nutritional diagnostic fertilization, stimulation tapping. Intercropping in rubber plantation has numerous advantages, such as conserving soil and water, increasing soil nutrient content, maintaining and improving biodiversity, promoting growth and increasing yield of the rubber tree. Fertilization management significantly affects growth of the rubber trees, so the rubber tree needs supplementary fertilizer through nutrient diagnosis. Since stimulated tapping has been practiced, it is very important to good tend and enhance fertilizer for the rubber trees.

### 3.2.2 Natural Rubber Industry Ecosystem Regulation

Structural Regulation

Structural regulation of the natural rubber industry ecosystem includes: regulate spatial structures of the rubber planting acceleration replanting of the old rubber plantation to adjust time structure of rubber plantation; adjust variety structure of the rubber planting; make full use of existing factories, and adjust processing layout of the natural rubber.

Institutional Regulation

Institutional regulation of the nature rubber industry ecosystem includes: establish scientific and rational structural framework of industrial business entities on the basis of reforming existing systems; emphasis on building the institutional network, establish and improve market circulation, finance and social service institutions, and develop the market of futures transaction and e-commerce transaction; build the prototype of targeted institutional system model, adjust targets and measures in the next phase, improve the institutional system of the natural rubber industry.

Functional Regulation

After new tapping system was adopted, physiological processes and material flow processes changed significantly, and material circulation was significantly faster, accelerating nutrient depletion of the rubber plantation ecosystem. This forces us to seek a new functional regulation strategy, and to improve the fertilizer formula and fertilizing mode. In addition, biological regulation should also adopted, and strengthen research in ecological adaptability and service function of the inter-crops, and other areas of the biological regulation should also to research.

### 3.2.3 Ecosystem Evaluation and Sustainable Management

Ecosystem Service Functional Valuation

The artificial forest ecosystem replaced of the natural forest ecosystem conforms to the law of value transfer in ecosystem service functions. Ecosystem service functional value in economic and social systems will be amplified several times or even scores of times through input of human labor, material and energy. The total value of ecosystem service function in rubber plantations in Hainan State Farms are lower than those of montane rainforest, the lower part of which can be regarded as natural input and bring about powerful economic and social service function value.

Ecological Footprint Accounting and Analysis

Accounting and analysis of ecological footprint to evaluate regional sustainable development has become central in ecological economics. The results of accounting and analysis of the natural rubber industry ecosystem shows that the Hainan State Farms had a per capita ecological footprint of 0.44038 ha and per capita ecological capacity of 9.11196 ha, available per capita ecological footprint of 8.01852 ha, per capita ecological capacity of 7.57814 ha, and had a per capita ecological footprint surplus of 5.71852 ha. This indicates that the natural rubber industry ecosystem in Hainan state farms is healthy and sustainable in economic development terms.

Ecological Audits

Ecological audits reflect relationships between the economic growth of natural rubber industry with the resource environment and social progress. Research shows that natural rubber plantations have strong ecological service functions in soil conservation, water saving, fixing of carbon dioxide, and the release of oxygen. The natural rubber industry in Hainan State Farms mainly relies on energy input and affected by the environment at lower levels, and the ratio of the energy input to currency is lower than in other areas of the country. This is to say that the natural rubber industry has relatively good economic benefits (Yu 2007).

Ecological Compensation Mechanism Establish

Ecological compensation mechanism is environmental economic policies that integrated use of administrative and market instruments to adjusting ecological protection and construction interests. The ecological compensation mechanism has begun to explore in Hainan and Yunnan rubber planting area. The fees of the ecological compensation through government to engage in the production of rubber processing enterprises imposed was used to transformation low-yielding rubber plantation, and to compensate for the environment losses in rubber production process, regulate coordination of the natural rubber industry development and ecological environment.

# 4 Conclusion and Discussion

Ecosystem management research has mainly focused on the theoretical study in ecological economics and policy. Little practical study into the specific technical and management measures has been done. According to the actual situation of natural rubber industry ecosystem in Hainan, we initially explored a set of natural rubber industry ecosystem management approaches and methods experienced many years of effort. However, ecosystem management of natural rubber plantations is a huge systematic project involving ecological, economic and social factors, long-term and large-scale substitution of natural vegetation with rubber plantations and change of global climate will give some impact on the structure, function and ecological process of the regional ecosystem, and the regional ecological environment will also evolved. Therefore, we need to carry out long-term monitoring and basic experiments of the ecological environment and explore the effective management of natural rubber industry ecosystem to promote the coordination and long-term sustainable development of the natural rubber industry.

# References

Chen, L., Tam, N. F. Y., Huang, J., Zeng, X., Meng, X., Zhong, C., et al. (2008). Comparison of ecophysiological characteristics between introduced and indigenous mangrove species in China. *Estuarine, Coastal and Shelf Science, 79*, 644–652.

Chen, L. Z., Tam, N. F. Y., Wang, W. Q., Zhang, Y. H., & Lin, G. H. (2013). Significant niche overlap between native and exotic Sonneratia mangrove species along a continuum of varying inundation periods. *Estuarine, Coastal and Shelf Sciences, 117*, 22–28.

Chen, L. Z., Wang, W. Q., & Lin, P. (2004). Influence of waterlogging time on the growth of *Kandelia candel* seedlings. *Acta Oceanology Sinica, 23*, 149–158.

Chen, L. Z., Wang, W. Q., & Lin, P. (2005). Photosynthetic and physiological responses of *Kandelia candel* L. Durce seedlings to duration of tidal immersion in artificial seawater. *Environmental and Experimental Botany, 54*, 256–266.

Chen, L. Z., Wang, W. Q., Zhang, Y. H., & Lin, G. H. (2009). Recent progresses in mangrove conservation, restoration and research in China. *Journal of Plant Ecology-UK, 2*, 45–54.

Chen, L. Z., Zeng, X. Q., Tam, N. F. Y., et al. (2012). Comparing carbon sequestration and stand structure of monoculture and mixed mangrove plantations of *Sonneratia caseolaris* and *S. apetala* in Southern China. *Forest Ecology and Management, 284*, 222–229.

Christensen, N. L., Bartuska, A. M., & Brown, J. H. (1996). The report of the ecological society of America committee on the scientific basis for ecosystem management. *Ecological Applications, 6*(3), 665–691.

Fan, H. Q., & Li, G. Z. (1997). Effect of sea dike on the quantity, community characteristic and restoration of mangroves forest along Guangxi coast. *Chin Journal of Appllied Ecology, 8*, 240–244. (in Chinese with English abstract).

Ge, X., & Sun, M. (1999). Reproductive biology and genetic diversity of a cryptoviviparous mangrove *Aegiceras corniculatum* (Myrsinaceae) using allozyme and intersimple sequence repeat (ISSR) analysis. *Molecular Ecology, 8*, 2061–2069.

Hu, J., & Zhou, J. X. (2010). *Development and utilization of mulberry eco-industry in sand field* (pp. 10–11). Beijing: Chinese Forestry Press (in Chinese).

Jiang, J. S., & Wang, R. S. (2004). *Ecology of natural rubber industry in Hainan* (pp. 36–44). Beijing: China Science and Technology Press.

Li, M. Y., Wu, J., Zhang, S., et al. (2007). Xylocarpins A and B, two new mexicanolides from the seeds of a Chinese mangrove *Xylocarpus granatum*: NMR investigation in mixture. *Magnetic Research Chemistry, 45*, 705–709.

Liao, B. W., Li M, M., Zheng, S. F., et al. (2003). Study on intraspecific and interspecific competition in exotic species *Sonneratia apetala*. *Fororest Research, 16*, 418–422. (in Chinese with English abstract).

Liao, L. P., & Zhao, S. D. (1999). Theoretical and practical approaches to ecosystem management of Chinese fir plantations. *Resources Science, 21*(4), 1–6.

Lin, P. (1999). *Mangrove ecosystem in China*. Beijing: Science Press.

Lin, P., & Fu, Q. (1995). *Environmental ecology and economic utilization of mangroves in China*. Beijing: Higher Education Press. (in Chinese).

Lu, C. Y., Wong, Y. S., Tam, N. F. Y., et al. (1999). Methane flux and production from sediments of a mangrove wetland on Hainan Island, China. *Mangroves and Salt Marshes, 3*, 41–49.

Mo, Z. C., & Fan, H. Q. (2001). Comparison of mangrove forestation methods. *Guangxi for Science, 30*, 73–581. (in Chinese with English abstract).

Rollet, B. (1981). Bibliography on mangrove research 1600-1975. Unesco.

Sheue, C. R., Liu, H. Y., & Yong, J. W. H. (2003). *Kandelia obovata* (Rhizophoraceae), a new mangrove species from Eastern Asia. *Taxon, 52*, 287–294.

Shi, S., Huang, Y., Zeng, K., Tan, F., He, H., Huang, J., et al. (2005). Molecular phylogenetic analysis of mangroves: Independent evolutionary origins of vivipary and salt secretion. *Molecular Phylogenetics and Evolution, 34*, 159–166.

Shi, S., Zhong, Y., Huang, Y., Du, Y., Qiu, X., & Chang, H. (2002). Phylogenetic relationships of the Rhizophoraceae in China based on sequences of the chloroplast gene matK and the internal transcribed spacer regions of nuclear ribosomal DNA and combined data set. *Biochemical Systematics and Ecology, 30*, 309–319.

Tam, N. F.·Y., & Wong, Y. S. (1997). Accumulation and distribution of heavy metals in a simulated mangrove system treated with sewage. *Hydrobiologia, 352*, 67–75.

Tan, F., Huang, Y., Ge, X., Su, G., Ni, X., & Shi, S. (2005). Population genetic structure and conservation implications of *Ceriops decandra* in Malay Peninsula and North Australia. *Aquatic Botany, 81*, 175–188.

Tao, S. H., Wu, J., & Qi, S. H. (2007). Scyphiphorins A and B, two new iridoid glycosides from the stem bark of a Chinese mangrove *Scyphiphora hydrophyllacea*. *Helvetica Chimica Acta, 90*, 1718–1722.

Wang, Z. H., Wu, Y. Y., & Zhang, Y. Z. (2011a). Analysis on organoleptic quality and nutrient active ingredients of mulberry-leaf teas made by different processing techniques. *Science of Sericulture, 37*(2), 272–277. (in Chinese).

Wang, W. Q., Yan, Z. Z., You, Y. S., Zhang, Y. H., Chen, L. Z., & Lin, G. H. (2011b). Mangroves: Obligate or facultative halophytes? A review. *Trees-Structure and Function, 6*, 953–963.

Wong, Y. S., Tam, N. F. Y., & Lan, C. Y. (1997). Mangrove wetlands as wastewater treatment facility: A field trial. *Hydrobiologia, 352*, 49–59.

Wu, J., Xiao, Q., Xu, J., Li, M. Y., Pan, J. Y., & Yang, M. (2008). Natural products from true mangrove flora: Source, chemistry and bio activities. *Natural Product Reports, 25*, 955–981.

Yang, R. J., Fu, B. J., Liu, G. H., et al. (2004). Principles and methods of sustainable management of ecosystem. *Chinese Journal of Ecology, 23*(3), 103–108.

Yang, X. M., & Jiang, Z. L. (2003). Forest ecosystem management and its relationship with traditional forest management. *Journal of Nanjing Forestry University, 27*(4), 91–94.

Yang, Q., Tam, N., Wong, Y., Luan, T., Su, W., Lan, C., et al. (2008). Potential use of mangroves as constructed wetland for municipal sewage treatment in Futian, Shenzhen, China. *Marine Pollution Bulletin, 57*, 735–743.

Ye, Y., Tam, N. F. Y., Wong, Y., & Lu, C. (2003). Growth and physiological responses of two mangrove species (*Bruguiera gymnorrhiza* and *Kandelia candel*) to waterlogging. *Environmental and Experimental Botany, 49*, 209–221.

Ye, Y., Tam, N. F. Y., Wong, Y., & Lu, C. (2004). Does sea level rise influence propagule establishment, early growth and physiology of *Kandelia candel* and *Bruguiera gymnorrhiza*? *Journal of Experimental Marine Biology and Ecology, 306*, 197–215.

Yu, W. (2007). Ecological-audit of Hainan state farm bureau's rubber industry. *Master's thesis*, pp. 15–46.

Zhang, Y., Huang, G., Wang, W., Chen, L., & Lin, G. (2012). Interactions between mangroves and exotic Spartina in an anthropogenically-disturbed estuary in southern China. *Ecology, 93*, 588–597.

Zhong, Y., Zhao, Q., Shi, S., Huang, Y., & Hasegawa, M. (2002). Detecting evolutionary rate heterogeneity among mangroves and their close terrestrial relatives. *Ecology Letters, 5*, 427–432.

Zhou, Y. M., Yu, X. B., & Yu, G. R. (2007a). Ecosystem approach for natural resources and ecosystem management: Concept, principle and application. *Advances in Earth Science, 2*, 171–178.

Zhou, R., Zeng, K., Wu, W., et al. (2007b). Contrasting ancestral and extant polymorphisms in non-model organisms: I Mangroves. *Molecules Biology Evolution, 24*, 2746–2754.

# Part II
# Ecosystem Management

# Chapter 6
# Classification and Research Methods of Ecosystem

**Jixi Gao, Shihai Lv, Zhirong Zheng, Junhui Liu, Changxin Zou, Zhaoping Yang, Liding Chen, Bojie Fu, Changhong Su and Wenhua Li**

**Abstract** The vast territory, diverse climate, rich landscape types, the numerous lakes, and vast sea in the east and south, relatively superior natural, historical and geographical conditions from tertiary and quaternary, make China one of the richest countries in ecosystem in the world. In this chapter, our topic include the main types of ecological system, crisscross band of ecological system, ecological security, ecological function area division, landscape pattern, ecological process, microbial ecology, and so on. Because rapid population growth and large-scale rapid urbanization process bring huge pressure on resources and environment, combined with the impact from the global climate warming and unreasonable land-use activities, ecological security has become a key focus areas of national security and sustainable development strategy. At present China still has not formed a unified evaluation index system and standard of evaluation methods to evaluate ecological security, and the exact meaning of ecological security is a lack of unified understanding. We still use static evaluation in most work at present, short in researching dynamic change, and the evolution trend of ecological security. Ecological security properties, assessment scales of time and space of theoretical

---

J. Gao · C. Zou · Z. Yang
Nanjing Institute of Environmental Sciences, Ministry of Environmental Protection of China, Nanjing 210042, China

S. Lv · Z. Zheng · J. Liu
State Environmental Protection Key Laboratory of Regional Eco-Process and Function Assessment, Chinese Research Academy of Environmental Sciences (CRAES), Beijing 100012, China

L. Chen · B. Fu · C. Su
Research Center for Eco-Environmental Sciences, Chinese Academy of Sciences, Beijing 100085, China

W. Li (✉)
Institute of Geographic Sciences and Natural Resources Research, Chinese Academy of Sciences, Beijing 100101, China
e-mail: liwh@igsnrr.ac.cn



W. Li (ed.), *Contemporary Ecology Research in China*,
DOI 10.1007/978-3-662-48376-3_6

basis, evaluation index system, and evaluation method are an important part of the study of ecology in China at present. To develop regional ecological function zoning plan, clarifying the main ecological problems of different ecological function areas has very important practical significance. Ecological regionalization has become the current macro and ecosystem ecology research hot spot. Using the theory of modern ecology, with full consideration regional ecological processes, ecosystem service functions, and ecological sensitivity to human activity intensity relationship on the basis of comprehensive functional zone is one of the ecology theories in practice. Ecological function regionalization is based on partition method and consolidation method. The relationship between landscape patterns and ecological processes and scale features is complex as landscape itself, and the simple concept of causality is not enough for quantitative research. The main research methods include landscape pattern index analysis, spatial autocorrelation analysis, and landscape model. The research contents involved in the evolution of the landscape pattern and drive process.

**Keywords** Classification of ecotones · Forest–steppe ecotone · Farming–pastoral ecotone · Desert–oasis ecotone · Geographical distribution · Forest–steppe · Desert–oasis ecotone · Landscape pattern analysis

# 1 Typical Ecotones in China

## *1.1 Introduction*

Ecotone refers to a transition area between two biomes or two different ecosystems (Odum 1971). Since the ecotone was first proposed by Clements in 1905, many have contributed to ecotone research and theory. A focus on ecotones has also played an important role in protecting ecological environments and enhancing industrial and agricultural development. Since the 1970s, especially, the study of the ecological ecotone has increasingly gained notoriety as an important facet of ecological research.

The ecotone possesses many unique natural attributes such as the distinctiveness of edge effects (Clements 1905; Hardt 1989), noncontinuity of vegetation distribution, heterogeneity in landscape structure (Walker 1979, 1985), and the fragility of the ecological environment. These attributes guide the study of ecological ecotones and play an important and irreplaceable role in the exploration of natural ecological laws and protection of the ecological environment. For these reasons, the ecotone has increasingly received attention from scientists and governments (Di Casstri and Hansen 1992; Kevin and Thomas 2006; Temuulen 2005; Wang et al. 2000).

China has a vast territory with a complex and changing geology, climate, and vegetation, and is consequently home to a variety of ecological ecotones. This variety is witnessed in breeds, scales, and sizes. This vast array has not only allowed for a solid foundation of scientific research but has also enhanced China's economic development, social development, and cultural diversity conservation. The rational selection of a typical ecological ecotone, the in-depth study of its spatial distribution laws, characteristics of natural environment, as well as the fragility of the ecological environment will lead to a comprehensive understanding of the evolutionary patterns of ecotones. This study will play a significant role in protecting regional ecological environments and optimizing economic structures.

## *1.2 Classification and Basic Characteristics of Ecotones*

### 1.2.1 Classification of Ecotones

According to the geographical distribution and ecological features of China's large-scale ecotone and the national land type classification standard, this thesis uses system ecology, agricultural climatology, and landscape ecology theory to establish "The trinity index system" including land-use types, natural climate characteristics, and agricultural economic development. Based on this index system, we define the spatial distribution of typical ecotones in China (Table 1).

In accordance with the professional division standards, the indicators for national ecological ecotone, and the basic characteristics of the fragile ecological environment, we adopted techniques such as remote satellite sensing, satellite interpretation, geographic information system, and graphics overlay to classify ecotones in China into five different classes, eighteen sub-ecotones, and a number of units. We classify localized ecotones if provided with regional vegetation types, characteristics of plant communities, and other detailed data. The classification only applies to the ecological transition zone in the ecosystem scale.

### 1.2.2 Basic Characteristics

Forest–Steppe Ecotone in Northern China

The distribution of the forest–steppe ecotone is concentrated in the semiarid and sub-humid temperate zone of northern China, which refers to the area among the outer Daxinganling Forest, south of the Yanshan Mountain, west of the Bashang Plateau, and part of the Loess Plateau. This area can be roughly divided into the northern temperate forest–steppe ecotone and the northern warm-temperate forest–steppe ecotone.

**Table 1** The spatial definition indices of typical ecotones in China

| Ecotone types | Land-use patterns | Natural climate characteristics | Agricultural economy development |
|---|---|---|---|
| Forest–steppe ecotone in Northern China | Forest and grassland mosaic distribution, and woodland area ratio is less than 30 % | Annual precipitation 350–500 mm, aridity 0.6–1.0 | Animal husbandry accounted for more than 60 % of agricultural economy |
| Farming–pastoral ecotone in Northern China | Dominated by grass, and farmland area less than 30 % | Annual precipitation 300–450 mm, aridity 1.0–2.0 | Farming herd coexists, cultivate industry accounted for more than 60 % of agricultural economy |
| Desert–oasis ecotone in Northwest China | Dominated by desert, oasis distribution was mosaic and the area is less than 30 % | Annual precipitation <150 mm, aridity >4.0 | Farming herd coexists, crop occupies 60 % above |
| The Chuan-dian farming–pastoral ecotone in Southwest China | Altitude 2500–4500 m in southwest mountain area, vertical distribution of forest and grassland, ravine area of farmland area less than 30 % | Annual precipitation >1000 mm, above 10 °C accumulated temperature 500–4000 | Agriculture forestry animal husbandry to coexist, animal husbandry, animal husbandry accounted for more than 30 % of agricultural economy |
| Eastern marine-terrestrial interlaced zone | Eastern marine-terrestrial interlaced zone, there is significant tidal wetland | Annual precipitation >800 mm, across temperate, subtropical and tropical | Fishing accounted for more than 60 % of agricultural economy |

### Farming–Pastoral Ecotone in Northern China

We redefined the exact location of the farming–pastoral ecotone in northern China from the two perspectives of land-use and climatic elements. The results show that climate changes are very obvious in the farming–pastoral ecotone in northern China, which covers a total area of 621,000 km$^2$, and spans over 10° in latitude and 20° in longitude. Located in 154 counties (autonomous counties and county-level city), it is shared by nine provinces or autonomous regions (banners, cities). The boundaries of land use have been extended frequently, growing to include an additional 23,000 km$^2$ between 1986 and 2000 (Fig. 1).

### Desert–Oasis Ecotone in Northwest China

The distribution region of desert–oasis ecotone in northwest China mainly includes the outer Hetao Plain, the west of the Helan Mountain, Hexi Corridor, the north and south of Tianshan Mountains, Wushaoling, Qilian Mountains, Altun Mountains,

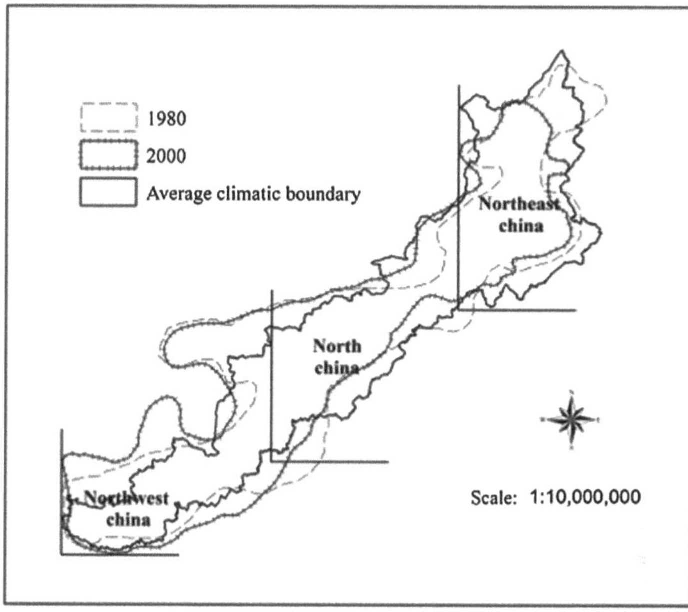


**Fig. 1** The boundary changes of the land use in the farming–pastoral ecotone in northern China

and the large marginal areas in the north of the Kunlun Mountains. Such areas are located in Inner Mongolia, Gansu, Xinjiang Uygur Autonomous Region, Qinghai, and Ningxia Autonomous Region. The desert–oasis ecotone in China has the following characteristics: circular or horseshoe shape, dominated by desert vegetation, simple structure, extremely fragile environment, etc. Its area is about 1490,000 km$^2$, accounted for about 13.5 % of total area.

### The Chuan-Dian Farming–Pastoral Ecotone in Southwest China

We have defined the Chuan-dian farming–pastoral ecotone in southwest China following three aspects of agro-climate, agricultural economy, and land use. The results show that the elevation of the ecotone ranges from 2500–2800 to 4200–4500 m, the accumulated temperature isocline above 10 °C ranges from 500 to 4000 °C, and the average temperature of the hottest period is above 10 °C, while the average temperature of the coldest period is above -10 °C. Based on county-level administrative boundaries, the ecotone is located in a total of 40 county-level administrative units including Ganzi County, Aba County, and Liangshan County in Sichuan, Diqing County, Nujiang County, and Lijiang County, and covers an area of about 280,000 km$^2$.

Eastern Marine-Terrestrial Interlaced Zone

The large-scale marine-terrestrial interlaced zone is the junction between terrestrial ecosystems and marine ecosystem, located in eastern China. It is a sensitive zone and transition zone. The total length of the Chinese eastern coastline, comprising the main body of the marine-terrestrial interlaced zone, is 32,000 km, of which 18,000 km is mainland coastline and 14,000 km is island coastline. The total area of the marine-terrestrial interlaced zone is approximately 100,000 $km^2$.

### *1.3 Conclusion and Discussion*

Based on existing problems such as excessive deforestation, a northward moving tree line, grassland degradation, and biodiversity loss in the northern forest–steppe ecotone, effective actions are needed. This action may include strengthening regional ecological construction and building a reasonable vegetation protection system and environmentally friendly industrial system.

In the northern farming–pastoral ecotone, attention should be paid to prevent land boundaries from spreading to grasslands. This spread is due to over-grazing, over-cultivation, and land desertification, and should be avoided wherever possible. In addition, more effort should be directed toward the construction of ecological projects and the establishment of farming–pastoral industry structure, with specific focus on the maintenance of ecological balance. In so doing, hazards from sandstorms may be effectively curbed and a state ecological protective screen built.

As for ecosystem deterioration caused by climate acidification, and over-exploitation of soil and water resources in the northwestern desert–oasis ecotone, the strategy of "determining production by water" should be implemented. An industrial structure and development mode based on the assessment of water resources should be established in order to recover natural vegetation. This action will help to maintain ecological security in the oasis.

In the Chuan-dian farming–pastoral zone in southwest China, ecology is rather fragile and people live on little economic resources. Since soil erosion has been caused mostly by human activities, including farming, afforestation, and grazing, programs should be established to ensure that human activity is suited to local conditions. Advantage-oriented ecological industries should be encouraged and a mutually beneficial relationship between environment protection and economic development is required.

## 2 Eco-environmental Security Assessment

Ecological and environmental problems have become more and more serious and the international society has attached high importance to eco-environmental security. To ensure regional eco-environmental security and to support sustainable development, it is urgent need to build ecological security assessment system, early warning system, and environmental management system. Eco-environmental security is also called environmental security. General concept of eco-environmental security means a complex ecosystems including resource security, environmental security, biological security, and ecosystem security. Limited concept of eco-environmental security means natural (seminatural) ecosystem health and integrity.

The concept of the eco-environmental security is rich in content. The evaluation of eco-environmental security has a variety of scales which vary from individual, population, community, ecosystem, landscape, and catchment to the whole earth. From the perspective of nature protection, the objective of eco-environmental security evaluation varies from single ecosystem, complex ecosystem, region, watershed ecosystem, and eco-region (biogeographic region) to the ecosphere. From the perspective of social system and development, the objective of eco-environmental security evaluation varies from village, town, county, city, country, and multi-countries, to the whole world, which restricted by administrative boundaries.

The evaluation of eco-environmental security is confide in a spatial scale or concentrated on natural–social-economic complex system. Based on the theories of ecosystem hazard, ecosystem health, landscape ecology, and sustainable development, the purpose of eco-environmental security evaluation is to realize sustainable development, identify the eco-environmental security status, and predict and control the risk with the trend. The selection of evaluation indexes system complies with the principles of science, integrity, layer-divisibility, easy-to-operation, and dynamic. General steps of eco-environmental evaluation are identifying the target (region), building evaluation indexes system, establishing the standards of each index, selecting the evaluation model, generating the result, analyzing, and discussing the process and conclusion.

Theoretically, the eco-environmental security is evaluated by various models such as mathematic model, ecological model, landscape model, and digital terrain model. The common used models are ecological carrying capacity-based model, pressure-state-response assessment model, eco-environment management model, and risk prevention/control model. The followings are summaries of these models:

1. Ecological carrying capacity refers to the ecological system of self-maintenance and self-adjustment. It also means the capability of resource storage and environmental containment that supporting the economic growth. The concept of ecological carrying capacity contains three components, namely, resource carrying capacity, environment carrying capacity, and ecological elasticity. The assessment of ecological carrying capacity includes three aspects. First, the environmental pressure should not surpass the ecological edacity. Second, the demand of economic development should not surpass the provision of resources. Third, the discharge of waste should not surpass the decomposition of pollutants.

2. In the model of pressure-state-response assessment, pressure means the unsustainable human activities including the consumption discharge of wastage or any other related factors in economic system, state reflects the situation of each economic and environmental sub-systems, and response means the effective measures or investment taken for sustainable development and pollutant control.
3. The eco-environment management concentrates on the ecosystem structure, the process of succession, and the function of ecological service. To realize sustainable development, the targets of eco-environment management are to protect, to restore, and to rebuild the ecosystem. Measures of protection, restoration, and reconstruction should be made in order to keep the equilibrium of various types of ecosystems.
4. Generally, ecological risk assessment is a special evaluation in the scopes of biological engineering, ecological invasion, and natural/human disaster. In a regional scale, ecological risk assessment contains the describing and assessing the consequence of adverse effects such as environment pollution and natural/human disaster. Generally, the possibility of natural/human disasters and environmental pollution and the extent of the adverse influences on ecosystem (or its components) are the objectives of the ecological risk assessment.

The evaluation of eco-environment security is complicated. Up to now, systematic evaluation methods and theories have not established. Same indexes could not be applied to another region and determining the threshold of ecosystem security meets difficulties. Moreover, the eco-environmental security observation system and management platform have not established, which retard the disaster prediction and pollution control. In future, the construction of observation system and management platform needs to be strengthened.

# 3 Landscape Ecological Studies in China

## *3.1 Introduction*

Landscape ecology grows rapidly since it was introduced into China in the early 1980s (Fu 1983; Xiao 1991). The scientists engaging in landscape ecological studies have increased from tens to thousands in past three decades. Until today, seven national conferences on landscape ecology have been organized since 1989 when the first national conference was held in Shenyang of northeastern China (Xiao 1991). In addition, two international conferences on landscape ecological studies circum-pacific regions and the 8th World Congress of IALE (2011) were organized. After 30 years growth, a theoretical framework on landscape ecological study is gradually developed in China by digesting the academic achievements on landscape ecological studies of both North-America and Europe, and considering the specialty of landscape in China.

## *3.2 Recent Progresses on Landscape Ecological Studies in China*

Although many achievements on landscape ecological studies were made in China, most of them are the consequences of following the leading groups in the world. The main academic merits made in past decade in China include the following fields: landscape pattern change and its multiple scale effects (Li et al. 2006, 2007), landscape pattern analysis and optimization of landscape eco-network (Cheng et al. 2009; Guo et al. 2010), allocation of ecosystem service land and urban eco-security pattern design (Yu et al. 2009a, b), forest landscape modeling and ecosystem management for insect pest and fire control (Chang et al. 2008; Liu et al. 2009), source-sink landscape pattern analysis and soil erosion risk assessment (Chen et al. 2008a, 2009), landscape pattern index and soil erosion risk identification at multiple scales (Fu et al. 2006, 2013), etc.

### 3.2.1 Landscape Pattern Change and Its Multiple Scale Effects in the Loess Plateau

Landscape change and its driving force is the hot topic in landscape ecological studies, and many works are conducted in China (Chen et al. 2008b; Fu et al. 2008). An interested work was completed by Zhang et al. (2012) who investigate land-use/land-cover change in Yan'an region of northern loess plateau using survey data on cropland, orchard, woodland, pasture land, and residential land in 1980, 1986, 1996, 2000, and 2001. In this study, the index describing the dynamic degree of land-use change was employed to explore the characteristics of land-use/land-cover change. In detail, the characteristics of land-use/land-cover change in Yan'an region at different periods of 1980–1986, 1986–1996, 1996–2000, and 2000–2001 was examined, as well as the features of driving force of landscape pattern evolution in spatial scale. It was found that the index of dynamic degree of land-use change is becoming smaller as the spatial scale increase. Furthermore, distinct temporal scale effects are presented on land-use/land-cover change besides the spatial scale effects. Generally, the land-use types with an annual growth period such as cropland and pasture grassland have the rule that the index of dynamic degree is increased with temporal scale becoming smaller. However, the index of dynamic degree of the other land-use types, for example, woodland, orchard, and residential land, is varied, and the peak index of dynamic degree appears at the temporal scale of 5 years. It becomes smaller with the temporal scale increases or decreases.

Usually, landscape pattern is the consequence of the driving forces composed of natural and socio-economic factors. It was found that the natural factors may function predominantly on landscape pattern in long term or large scale, and the response is slow and uncertain. However, the socio-economic factors are the main driving force of landscape change and may affect landscape pattern by determining the trend of landscape change.

In the loess plateau, land-use/land-cover change has distinct spatial scale effects, and the driving force in the large scale is normally the combined impacts of all driving forces in the small scale. In addition to the effects of spatial scale, clear effects of temporal scales are also displayed on the driving force of landscape pattern change. In short period, the effects of natural factors on landscape pattern are slight; however, it may contribute more in the long term.

### 3.2.2 Landscape Pattern Analysis and Optimization of Landscape Eco-Network

To explore methodology and means for landscape pattern optimization in large scale by considering the effects of landscape pattern on ecological processes and ecosystem services is highly addressed in China and some achievements were realized.

*Landscape concentricity and eco-network optimization.* To build a high-connectivity, landscape eco-network is a useful method in biodiversity conservation, ecosystem management, and environmental hazard control (Fu et al. 2008). Landscape planning and management thus received many attentions from landscape ecologists and planners. However, how to seek out the key points in a heterogeneous landscape is critical for landscape eco-network planning and management. For such purpose, an index called landscape concentricity was proposed by Teng et al. (2010) based on graph theory and network analysis principles. The significance of landscape elements in the heterogeneous landscape and ecological processes are emphasized in the model. The index thus can be applied in landscape eco-network planning and management for biodiversity conservation and landscape sustainability.

*Urban expansion model and optimization in spatial scale.* The spatial models of urban expansion and eco-network optimization are important topics in landscape ecological studies. It aims to establish a sustainable urban landscape planning based on spatial model of landscape eco-network. In such purpose, Wang and Liu (2009) classified the urban landscape eco-network into four types after investigating the features of landscape pattern and ecosystem service land in the urban area of China, i.e., Wetland-scenery-woodland landscape pattern, Woodland-road-dwelling landscape pattern, Plain-city-farm-shrub-river landscape pattern, and Island-city-green-land-road landscape pattern. And the features, functions, and ecological significance of these landscape patterns were further elaborated. The result can be used as guidance to urban planning and urban ecosystem management.

*Landscape planning and nature reserve network design.* The relationship between landscape pattern and animal movement is one of important research fields in landscape ecology. Many studies indicate that landscape pattern may produce crucial impacts on animal movement, as well as biodiversity conservation. Thus, to study the effects of landscape pattern on the movement and conservation of endangered species is highlighted. However, it sometimes may bring adverse effects on biodiversity conservation due to inappropriate distribution of nature reserves, particularly in spatial allocation from large scale. Thus to establish a fit landscape network from large scale is required. Many Chinese landscape scientists have conducted researches addressing such issues. For example, Xu et al. (2010) proposed an optimal network of nature reserves in the purpose of protecting Giant Panda in Qinling mountainous areas based on habitat suitability evaluation and accessibility analysis. Based on the network planning, two new nature reserves and three habitat corridors are to be built for a safe pattern for Giant Panda living in the concerned regions. Zhang et al. (2011) delineated out the most crucial areas to be protected for biodiversity conservation in Hainan Island by habitat suitability assessment as for 140 endangered animal species.

### 3.2.3 Ecosystem Service Land and Landscape Security Design

The issues such as extremely crowed living, traffic congestion, water shortage and pollution, ecosystem degradation, etc., due to rapid urban expansion, are becoming social hot topics, especially in the developing countries. How to solve the practical problems appeared during urbanization by integrating landscape pattern analysis has been paid much attention in China (Li et al. 2011b; Yu et al. 2009a, b). One of them is about ecosystem service land and landscape eco-security design.

*Ecosystem service land and urban security pattern*. How to determine the minimum quota of land to be used for ecosystem services in the metropolitan area is of high significance, particularly in the case of reducing land resource and increasing demand on land resource due to economic development. Yu et al. (2009a, b) proposed a fundamental and safe landscape pattern for sustainable development in Beijing using landscape eco-security pattern theory and GIS techniques. In this study, a comprehensive and safe landscape pattern of Beijing at minimum level is given by considering the effects of urban landscape pattern on hydrological process, environmental hazards, biodiversity, and ecosystem services such as cultural landscape protection and recreation for local people. As well, the environmental effects of urban expansion were further investigated.

*Urban expansion based on the distribution of ecosystem service land*. The level of landscape eco-security pattern of Nanchong of Sichuan in southwestern China was estimated by Li et al. (2011a) after taking landform, flooding, soil erosion, vegetation pattern, geological hazards, and biodiversity conservation using RS/GIS techniques. Ecological corridors and critical landscape points were determined using minimum distance resistance model in the purpose of ensuring urban eco-security by reasonable urban expansion both in area and spatial configuration. An index system for landscape eco-security evaluation was set up by Song and Cao (2010) by integrating landscape pattern, function, vitality, ecological sensitivity, and landscape stress. And landscape eco-security level of Beijing in 1988 and 2004 was compared.

### 3.2.4 Forest Landscape Modeling for Pest/Fire Control

How to control fire/pest burst in forest area is a worldly concerned issue. As for this topic, many researches were carried out in northeastern China, particularly in the Daxinganling mountainous areas.

*Dynamic modeling for forest landscape management and fire disturbance control.* An important work conducted by Chang et al. (2008) is that a comparison on spatial pattern of fire frequency, burned area, intensity, and patch of forest fires under two scenarios in Huzhong region of Daxinganling region was explored using LANDIS model. Another research finished by Liu et al. (2009) is modeling the effects of 10 scenarios on treatment of combustible under forest, including five human involved treatments and five combined treatments of human and natural disturbance. The long-term effects of such treatments were compared using LANDIS model in term of burned area and fire intensity.

*Dynamic modeling of forest landscape and pest burst control.* As the most two important disturbances in forest evolution, fire and insect pest, have been paid much attentions in forest ecosystem management, however, how to control them in heterogeneous forest landscape is still not clear. Chen et al. (2011) simulate the interaction of fire and insect pest in forest evolution in future 300 years using LANDIS model. It was found that fire frequency of forest may be reduced in the early and middle stage of forest evolution due to combustible removal with the occurrence of frequent insect pest in the forest landscape. It helps to keep a stable and healthy forest landscape at low-level pest occurrence.

### 3.2.5 Location-Weighted Landscape Index and Soil Erosion Risk Assessment

How to integrate landscape pattern analysis with ecological process is a key and tough work for landscape ecologists, and has been paid much attention in China (Chen et al. 2008; Wu et al. 2012). The construction of the landscape index Location-Weighted Landscape-contract Index (IWLI) may give a new concept on this field in landscape ecology.

*Location-weighted landscape contract index (IWLI).* Understanding the relationship between landscape patterns and ecological processes is a central yet challenging research theme in landscape ecology. Over the past decades, many landscape metrics were proposed but few of them directly incorporated ecological processes. The landscape index, i.e., location-weighted landscape-contract index, is developed by Chen et al. (2008a) to link landscape pattern analysis with the ecological processes such as soil erosion or nutrient loss. In this index, relative distance, relative elevation, and slope of landscape units located were employed to indicate the importance of landscape to the targeted ecological process at the outlet in watershed scale. It can be used to characterize the contribution of landscape pattern to a targeted ecological process (e.g., nutrient losses) with respect to a specific monitoring point in a watershed. In 1999, it was improved by Chen et al. (2009) as

$$\mathrm{LWLI} = \sum_{i=1}^{m} A_{\mathrm{source}\,i} * W_i * \mathrm{AP}_i \Bigg/ \left[ \sum_{i=1}^{m} A_{\mathrm{source}\,i} * W_i * \mathrm{AP}_i + \sum_{j=1}^{n} A_{\mathrm{sink}\,j} * W_j * \mathrm{AP}_j \right]$$

where $A_{\mathrm{source}\ i}$ is the area of the $i$th source landscape type versus the distance, the relative elevation, or the slope gradient; and $A_{\mathrm{sink}\ j}$ represents the area of the $j$th sink landscape type versus the distance, the relative elevation, or the slope gradient. $W_i$ and $W_j$ are the weights of the $i$th source and the $j$th sink landscape type, respectively; $m$ is the number of source landscape types, while n is the number of sink landscape types; $\mathrm{AP}_i$ and $\mathrm{AP}_j$ are the area percentages of the $i$th source landscape type and the $j$th sink landscape type in a watershed, respectively. In the equation, when function of source and sink landscape is in balance, the value of LWLI would be 0.5. This landscape pattern in a watershed would produce little nutrient losses or soil erosion.

The highlighted significance of IWLI foundation is the integration of spatial pattern of landscape types ($A_{\mathrm{source}}$) using Lorenz Curve, the quantitative attributes of landscape types ($\mathrm{AP}_i$) in terms of landscape percentage, and the importance of landscape types at a watershed scale by weight assigning. It can be employed to compare the effects of landscape pattern on ecological processes.

*Source-Sink landscape pattern analysis and soil erosion risk assessment.* After corrected by Xu (2009) with considering the contribution of rainfall, landforms, and soil erosivity to soil erosion, LWLI can be used to estimate the risk of soil erosion or nutrient loss at watershed scale by calculating IWLI value. It can also be used to determine which landscape pattern is better in a watershed by comparing LWLI index at different periods, or be used to determine which watershed is the potential area on soil erosion by comparing LWLI index among different watersheds in same period. In general, the bigger the LWLI index in a watershed, the higher the risk of soil erosion occurs. Apart from evaluating the potential risk of nutrient losses or soil erosion, LWLI can be used to characterize the effects of landscape pattern on ecological processes, such as meta-population and wildlife conservation and urban heat island effect.

### 3.2.6 Landscape Pattern Index for Soil Erosion Assessment at Multiple Scales

To quantify the relationship between landscape pattern and soil erosion is a highlighted issue in landscape ecological studies (Fu et al. 2009, 2013). However, how to define the effect of landscape pattern on soil erosion in different scales is a hard work. In China, a framework on evaluating soil erosion risk at multiple scales was proposed by Fu et al. (2006) at considering the effects of land use, terrain, soil, and rainfall on soil erosion and using scale–pattern–process principles in landscape ecology. The scales involved include slope transact, small watershed, and regional scale. This model can be used not only in exploring the relationship between landscape pattern and soil erosion, but also in identifying the potential risk of soil erosion in different landscape patterns. It is a useful tool to land-use planners and decision-makers on sustainable land-use planning and ecosystem restoration.

### 3.2.7 Rural Landscape Analysis and Environmental Conservation

Landscape analysis and environmental conservation in the rural areas of China is an important field for landscape ecologists, and many researches have been conducted. Agricultural landscape pattern in Beijing at both patch and landscape level was investigated by Zhao and Zhang (2008) using Fragstats3.3 based on land-use map of 1993 and 2004. Based on the results, four functional zones in Beijing, i.e., urban agro-environmental regulation area, suburb agro-environmental protection area, plain agro-production aggrandized area, and hilly agro-ecosystem service maintaining area, and suggestions on landscape pattern optimization in each area were given. Moreover, a comprehensive agro-landscape quality assessment index was set up by Pan et al. (2009) based on field survey and expert knowledge by considering the aesthetic values of agro-landscape, wildness, openness, diversity, pollution risk, regularity, etc. The quality and spatial difference of agro-landscape in Beijing was addressed based on land-use data and remote sensing images.

### 3.2.8 Landscape Health Assessment

As a new trend of ecosystem health study, landscape health focuses on the health value in landscape scale. An important indicator of landscape health is to see whether the landscape pattern is sustainable. A case study carried out by Li et al. (2010) is about the spatial difference of landscape health in Xixi Wetland Park in Hangzhou of Jiangsu Province using an index developed based on socio-economic factors and ecosystem protection. Another case study conducted by Shuo et al. (2011) is about the evaluation of the response of ecosystem services to landscape change in Liaohe River delta using remote sensing images of 1990–2010.

### 3.2.9 Oasis Landscape and Sustainable Development

Oasis landscape pattern analysis and function dynamic are also concerned in China. The hydrological response to the landscape pattern change in the Oasis area in northwestern China has been paid a lot attention. The important works mainly cover the following areas: (1) landscape change in oasis area and its driving force (Li et al. 2007). Generally, both natural and human affecting factors that driving landscape change were addressed using remote sensing images and GIS techniques. Recently, the effects of snow smelting due to global warming and water shortage due to human activity increase in the upper river on landscape health are becoming a hot topic; (2) Oasis landscape stability assessment and regional eco-security. Oasis landscape stability is closely related to eco-security of oasis (Li et al. 2007). Many researches have been conducted on landscape stability in oasis region using the methodology with weight assigning by experts to the proposed index system; (3) Landscape change and hydro-ecological balance in oasis area. Oasis landscape in the lower river and the water source in the upper river compose a complex system in the arid regions. However, the change in landscape pattern will result in hydrological reaction in the drainage area and further water shortage, and thus the hydro-ecological cycle and balance in regional scale are to be explored.

#### 3.2.10 Landscape Fragmentation and Biodiversity Conservation

In landscape fragmentation and biodiversity conservation, a new research field concerned by landscape ecologists is the hereditary effects of landscape fragmentation due to natural or human disturbance. The spatial variation of gene diversity was affected by landscape pattern, as well as the gene flow between meta-population using landscape pattern index (Shen and Ji 2010). As for the plant meta-population due to deforest, Jian et al. (2008) found that the pattern of hereditary structure is determined by landscape fragmentation and gene flow among different populations. Therefore, to investigate the response of gene flow to landscape fragmentation requires to study further the reproductive strategy (Wang et al. 2009).

### *3.3 Conclusions and Remarks*

By integrating the advantages of both North-American landscape ecological studies and European landscape ecological studies, a Chinese framework on landscape ecological studies is gradually developed. Some original works both in theories and methodologies have been done in China after the concept of landscape ecology introduced into China since 1980s. They are mainly focused on the following fields: multiple scale effects of landscape change, landscape eco-network design for wildlife conservation, and spatial arrangement of ecosystem service land for urban eco-security. Further, seeking new landscape index is also addressed, such as LWLI index (location-weighted landscape contrast index) by integrating landscape pattern and ecological processes, traverse-slope landscape pattern index, or longitudinal-slope landscape pattern index (You and Li 2005), and the multiple-scale soil erosion index based on the correlation between landscape pattern and soil erosion(Fu et al. 2006). All the above-mentioned indexes are new and valuable metrics to be employed in describing the relationship between landscape pattern and ecological processes (Chen et al. 2008b). Although many achievements are made on landscape ecological studies in China, there is still much work to be done. The highly geographical difference both in cultural and human-nature coupling relationship resulted from long history of human activities and complicated environmental conditions in China call for a methodology to seek a solution to the increasing environmental issues (Jiao et al. 2012; Qin et al. 2010). Additionally, the intensity, diversity, complexity, and typicality of the interaction between natural ecosystem and social system in China are quite different from that in the other countries. All above-said provides a chance for Chinese landscape ecologists for acquiring much more achievements, and also raise a challenge for them to resolve all the problems faced to realize sustainable and healthy landscape in China.

# 4 Coupling Analysis of Landscape Pattern and Ecological Processes

## *4.1 Introduction*

Landscape ecology has been applied widely for its theories and approaches. The link between landscape patterns and ecological processes forms the foundation of landscape ecology, understanding which is key to further promote the study of landscape ecology (Fu et al. 2001; Wu 2007; Wu and Hobbs 2002). Exploration of the coupling relation between landscape pattern and ecological processes is conducive to understandings of ecosystem function mechanisms and prudent land-use policies.

## *4.2 Quantification of Landscape Pattern*

Landscape pattern metrics, spatial statistical analysis, and dynamic models constitute the three types of primary methods for quantifying landscape patterns. Landscape pattern metrics are used most widely in analyzing landscape structural composition and spatial configurations. Unfortunately, most existing landscape metrics fail to capture the fundamental patterns in landscapes important to ecosystem processes, due to lack of or confusion in ecological meanings (Jones et al. 2012), which makes it difficult to explain the underlying mechanism of the landscape pattern change. Real-world landscapes are always typified by spatial autocorrelations which reflected the spatial gradient variations. Studies on these gradient variations lay the foundations for regional ecological process researches and the future directions of exploration on spatial evolutions of landscape patterns, albeit subjected to various natural factors. Dynamic models of landscape pattern consist primarily of spatial Markov (Aaviksoo 1995), cellular automata (Wu 2002), and agent-based models (Bithell and Brasington 2009). Markov model calculates the land-use transfer probability matrix table based on historic land-use images, and simulates the future landscape patterns. Cellular automata simulate the complex spatial structure through simple transferring or domain rules. Agent-based models are robotic in depicting the influences of complex human decision on landscape pattern. Integrating the agent-based models into cellular automata is a key future direction for landscape study.

## 4.3 Scale Dependence of Landscape Evolution

Along with the spatial scale increase, the magnitude of land-use change will decrease, and the complexity of land-use transfer will increase. (i) Slope is the smallest spatial scale, which is characterized by land use, hydrology, topology, soil, etc. In addition, physical and anthropogenic processes are directly interacted at slope scale. The landscape evolution at the slope scale in the Loess Plateau during the Grain for Green program period is characterized by slope farmland conversion to nonfarm land use based on the criteria of slope degree. (ii) Watershed/sub-watershed is a basic unit key to understand land use and hydrological response. At watershed/sub-watershed scale, in the Loess Plateau, patches become more regular and aggregated. Landscape becomes more fragmented with higher landscape diversity and complexity. (iii) County is the basic unit for regional sustainable development. As a relatively independent unit, county is a scale at which natural resources and social statistical data are easy to be obtained. At county level, the Loess Plateau underwent frequent land-use transfer characterized by decreased farmland, and expanded forest, grassland, and residential land (Zhang et al. 2004; Wang et al. 2006). (iv) Region is a unit characterized by correlations between population, resources, and environment. Take Yan'an district as an example, from 1980 to 2001, land-use types of farmland, water area, and waste land decreased, whereas vegetable land, forest, grassland, and industrial and mining land increased.

Generally, vegetation changed more abruptly at shorter temporal scale. The studies on the Loess Plateau show that, since 1930, Zhifanggou watershed has undergone four stages, i.e., sudden destroy, slow recovery, stable, and quick recovery (Zhang et al.'s 2004); studies show that along with the shortening of the temporal scale, land-use dynamic index (LUDI) of farmland showed a monotonic decreasing tendency, LUDI of vegetable yard and industrial/mining land showed a "V"-shaped trajectory, LUDI of forestry showed a invert "U"-shaped trajectory, and LUDI of grassland showed a first-fluctuate-then-surge tendency.

## 4.4 Driving Mechanisms of Landscape Pattern Change

The driving system of landscape pattern consisted dominant and nondominant driving factors. At larger scale, natural factors of topology and climate and anthropogenic factors of population, culture, and regional social and economy take the leading function. At medium/small scale, vegetation, soil, and technological renovation exert the pivotal effects. Generally, studies on larger scale are conducive to exploration of the overall tendency of the relations between landscape variation and the multiple factors, but it is not advantageous for grasping the landscape pattern and driving factors at smaller scale. At 1970–1990, the major driving factors of land use in Zhifanggou sub-watershed include population policy, land-use policy, reforming of economic system, economic development, agricultural technology advancement, comprehensive harnessing program, and factors of slope and soil. Whereas, at county (Anshai) scale and region (Yan'an District) scale, the driving factors are more comprehensive, anthropogenic factors of population policy, economic policy and technological renovation, and physical factor of topology became the dominant factors. Some factors insignificant at shorter temporal scale, e.g., temperature and precipitation, possibly become influential at longer temporal scale (Zhang et al. 2004).

## 4.5 *Coupling of Landscape Pattern and Ecological Processes*

### 4.5.1 The Theoretical Framework of Coupling Study

At finer scales, in situ observation and experimentation are used widely in studying landscape pattern and ecological processes. High controllability and accuracy makes in situ observation good verification for researches of landscape pattern and ecological processes at larger scale. Land unit provides 'bricks' for coupling studies at various scales. Hierarchical patch theory and scale transition strategy provide the theoretical bases for establishing spatial explicit model linking landscape pattern and ecological processes. We provided a framework for coupling landscape pattern and ecological processes (Fig. 2).

### 4.5.2 Models for Landscape Pattern and Ecological Processes

According to the directions of the interactions between landscape patterns and ecological processes, the models coupling landscape pattern and ecological processes are generally divided into models that analyze the effect of landscape patterns on ecological processes (Hattermann et al. 2006), models that analyze the influence of ecological processes on landscape patterns (Jeltsch et al. 1999), and models coupling landscape patterns and ecological processes.

### 4.5.3 Influences of Landscape Pattern on Various Ecological Processes

Landscape pattern is a key factor affecting hydrological cycling and water resources. Land-use configuration pattern exerts profound influences on the feature

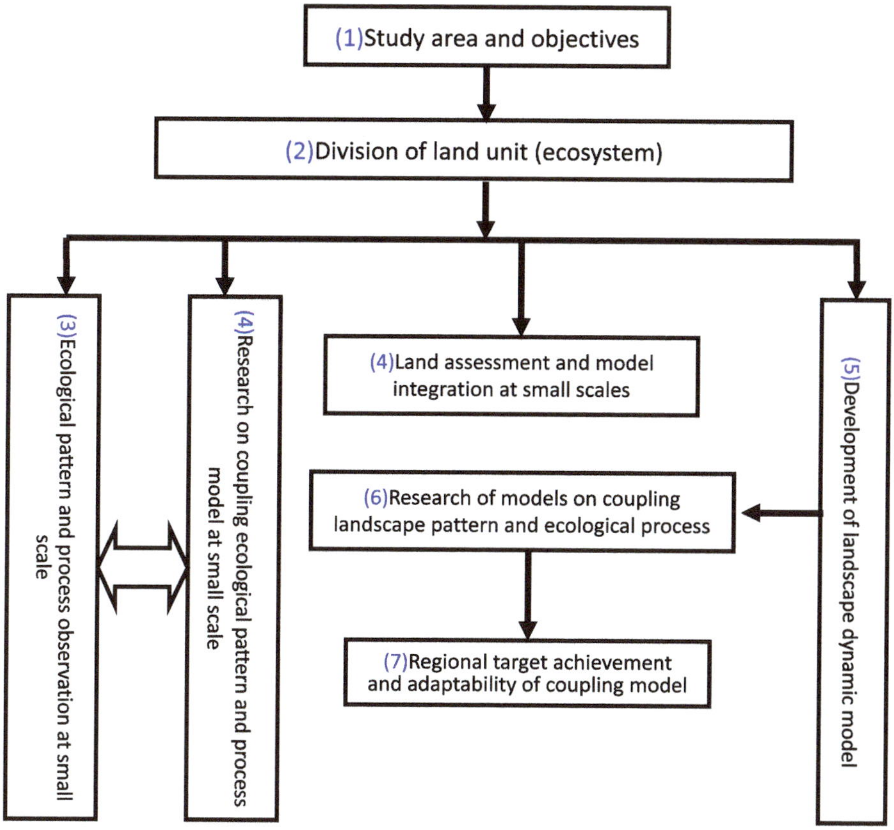


**Fig. 2** Framework on coupling landscape pattern and ecological process (Lü et al. 2012)

of the substrate and the physical/chemical characteristics of soils. At sub-watershed scale, configuration of slope farmland–grassland–forest has higher nutrition retention ability than other patterns (Fu et al. 1999). Biodiversity and environment have formed relatively stable co-adaptive relations during the long-period evolution. Generally, land-use pattern close to natural state is conducive to biodiversity conservation. Landscape pattern directly influences the soil carbon cycling. Farmland ecosystem generally is regarded as an atmospheric carbon sink. Conservation farming practice is effective in improving soil structure and soil organic carbon. Conversion from farmland to forest/grass or fallowing farmland can improve soil organic carbon. Logging or weeding accelerates carbon released from plant residues and reduces the carbon content in the soil.

## 4.6 Conclusion and Discussions

Given the awareness and understanding of landscape ecology, the mere description of landscape patterns can no longer meet the needs of academic pursuits. Developing landscape pattern indices that reflect ecological processes has much theoretical and practical significance for future study.

Due to the complexity and abstract nature of ecological processes, most current studies are confined to small/medium scales. It is necessary to test the cross-scale relationships between landscape patterns and ecological processes, and to reveal the scaling characteristics of ecological processes.

Long-term ecological research (LTER) is significant in exploring dynamic and periodic ecological processes. One of the main purposes for landscape model is to establish quantifiable and repeatable methods studying landscape pattern and ecological processes. Parameters and functions need to be verified by in situ observation and controllable experiment which necessitate the LTER.

So far, research on the influences of landscape pattern on ecological processes is relatively mature. However, research on the effects of ecological processes on landscape patterns has not aroused sufficient attention. Ecological processes are subtle and long lasting, which call for long-term and unremitting study.

## References

Aaviksoo, K. (1995). Simulating vegetation dynamics and land use in a mire landscape using a Markov model. *Landscape and Urban Planning, 31*(1/3), 129–142.

Bithell, M., & Brasington, J. (2009). Coupling agent-based models of subsistence farming with individual based forest models and dynamic models of water distribution. *Environmental Modelling and Software, 24*(2), 173–190.

Chen, L., Fu, B., & Zhao, W. (2008a). Source-sink landscape theory and its ecological significance. *Frontiers Biological of China, 3*(2), 131–136.

Chen, H., Hu, Y., Chang, Y., Bu, R., Li, Y., & Liu, M. (2011). Interaction between pest and fire disturbances in Huzhong area of Great Xingan Mountains: Long-term simulation. *Chinese Journal of Applied Ecology, 22*(3), 585–592.

Chen, L., Liu, Y., Lu, Y., Feng, X., & Fu, B. (2008b). Landscape pattern analysis in landscape ecology: Current, challenges and future. *Acta Ecologica Sinica, 28*(11), 5521–5531.

Chen, L., Tian, H., Fu, B., & Zhao, X. (2009). Development of a new index for integrating landscape patterns with ecological processes at a watershed scale. *Chinese Geographical Sciences, 19*(1), 37–45.

Cheng, H., Zheng, H., Wang, Z., & Jian, X. (2009). The type and pattern characteristics of green land and its relationship with surface land temperature. *Acta Scientiarum Naturalium Universitatis Pekinensis, 45*(3), 495–501.

Clements, F. E. (1905). *Research methods in ecology*. Nabraska: University Publishing Company.

Di Casstri, F., Hansen, A. J. (1992). The environment and development crises as determinants of landscape dynamics. In Hansen, A. J., Castris, F. D. (Eds.), *Landscape boundaries* (pp. 3–18). New York: Springer.

Fu, B. (1983). Landscape ecology-a new study field of geography. *Chinese Journal of Ecology*, 60.

Fu, B. J., Chen, L. D., & Ma, K. M. (1999). The effect of land use change on the regional environment in the Yangjuangou Catchment in the Loess Plateau of China. *ACTA Geographica Sinica, 54*(3), 241–246. (in Chinese).

Fu, B. J., Chen, L. D., Ma, K. M., & Wang, Y. L. (2001). *The principles and application on landscape ecology* (pp. 1–14). Beijing: Science Press. (in Chinese).

Fu, B., Lv, Y., Chen, L., Su, C., Yao, X., & Liu, Y. (2008). The latest progress of landscape ecology in the world. *Acta Ecologica Sinica, 28*(2), 798–807.

Fu, B., Wang, Y., Lv, Y., He, C., Chen, L., & Song, C. (2009). The effects of land use combination on soil erosion-a case study in Loess Plateau of China. *Progress in Physical Geography, 33*(6), 793–804.

Fu, B., Wang, S., Su, C., & Forsius, M. (2013). Linking ecosystem processes and ecosystem services. *Current Opinion in Environmental Sustainability, 5*, 4–10.

Fu, B., Zhao, W., Chen, L., Lu, Y., & Wang, D. (2006). A multi-scale soil loss evaluation index. *Chinese Science Bulletin, 51*(4), 448–456.

Guo, H., Huang, Y., Ye, G., Li, C., & Chen, L. (2010). A study on the evaluation and optimization of urban ecological function network in Xiamen. *Journal of Natural Resources, 25*(1), 71–79.

Hardt, R. A. (1989). Boundary from effects on woody colonization of reclaimed surface mine. *Ecology, 70*(5), 1252–1260.

Hattermann, F. F., Krysanova, V., Habeck, A., & Bronstert, A. (2006). Integrating wetlands and riparian zones in river basin modelling. *Ecological Modelling, 199*(4), 379–392.

Jeltsch, F., Moloney, K., & Milton, S. J. (1999). Detecting process from snapshot pattern: Lessons from tree spacing in the southern Kalahari. *Oikos, 85*(3), 451–466.

Jian, Y., Shi, L., Li, D., Zhang, C., Shi, M., Wang, R., & Chen, X. (2008). Effects of deforestation on the genetic structure of Castanopsis sclerophylla. *Acta Ecologica Sinica, 28*(12), 6228–6234.

Jiao, Y., Li, X., Liang, L., Takeuchi, K., Okuro, T., Zhang, D., & Sun, L. (2012). Indigenous ecological knowledge and natural resource management in the cultural landscape of China's Hani Terraces. *Ecological Research, 27*, 247–263.

Jones, K. B., Zurlini, G., Kienast, F., Petrosillo, I., Edwards, T., Wade, T. G., Li, B. L., Zaccarelli, N. (2012). Informing landscape planning and design for sustaining ecosystem services from existing spatial patterns and knowledge. *Landscape Ecology, 28*(6), 1175–1192 (in press).

Kevin, L., & Thomas, V. (2006). Climatic variability and episodic Pinus ponderosa establishment along the forest-grassland ecotones of Colorado. *Forest Ecology and Management, 228*(1), 98–107.

Li, Y., Liu, H., Cao, X., & Zheng, N. (2010). Spatial differences of landscape health in urban wetland parks: A case study of Xixi Wetland Park in Hangzhou. *Acta Geogrphica Sinica, 65* (11), 1429–1437.

Li, Y., Liu, H., Zheng, N., & Cao, X. (2011a). A functional classification method for examining landscape pattern of urban wetland park: A case study on Xixi Wetland Park, China. *Acta Ecologica Sinica, 31*(4), 1021–1028.

Li, S., Tiemao, S., Fu, S., Zhou, L., Liu, M., & Wang, W. (2011b). Landscape ecological security pattern during urban expansion of Nanchong City. *Chinese Journal of Applied Ecology, 22*(3), 734–740.

Li, X., Xiao, D., He, X., & Chen, W. (2007). Factors associated with farmland area changes in arid regions: A case study of the Shiyang River basin, northwestern China. *Frontiers in Ecology and the Environment, 5*(3), 139–144.

Li, X., Xiao, D., He, X., Chen, W., Jiao, Y., Song, D., & Hu, Z. (2006). Integrated assessment of oases maturity in Inland River Basins of China. *Acta Geogrphica Sinica, 61*(8), 855–864.

Liu, Z., Yu, C., Hu, H., Hu, Y., & Wnag, W. (2009). Effects of different forest fuel treatments on potential forest fire regimes in Great Xingan Mountains: a simulation study. *Chinese Journal of Ecology, 28*(8), 1462–1469.

Lü, Y., Fu, B., Feng, X., Zeng, Y., Liu, Y., Chang, R., et al. (2012). A policy-driven large scale ecological restoration: Quantifying ecosystem services changes in the loess plateau of China. *PLoS ONE, 7*(2), e31782. 10.1371/journal.pone.0031782.

Odum, E. P. (1971). *Fundamentals of ecology* (2nd ed.). Pennsylvania: Saunders W B Company.

Pan, Y., Xiao, H., & Yu, Z. (2009). Spatial evaluation on ecological and aesthetic quality of agricultural landscape. *Chinese Journal of Applied Ecology, 20*(10), 2455–2460.

Qin, Z., Zhang, J., Luo, S., Xu, H., & Zhang, J. (2010). Estimation of ecological services value for the rice-duck farming system. *Resources Science, 32*(5), 864–872.

Shen, Z., & Ji, C. (2010). Landscape genetics: Principles and its applications for the genetic effects of habitat fragmentation. *Acta Ecologica Sinica, 30*(18), 5066–5076.

Shuo, A., Yu, Y., & Han, F. (2011). Response of ecosystem service value to wetland landscape pattern change in Panjin region of Liaohe delta. *Ecological Economy, 6*, 147–151.

Song, Y., & Cao, M. (2010). Landscape ecological security assessment of Beijing city based on RS and GIS. *Chinese Journal of Applied Ecology, 21*(11), 2889–2895.

Temuulen, T. S. (2005). *20th Century forest ~ grassland shift and effects of livestock herbivory*. Montana: Montana State University Dissertation.

Teng, M., Zhou, Z., Wang, P., Xu, Y., & Wu, C. (2010). Landscape centrality and its applications in ecological network planning and management. *Chinese Journal of Applied Ecology, 21*(4), 863–872.

Walter, H. (1979). *Vegetation of the earth and ecological systems of the geo-biosphere* (2nd ed.). New York: Springer.

Walter, H. (1985). *Community and ecosystems* (2nd ed.). New York: Macmillan Publishing Co., Inc.

Wang, R., Ai, B., Gao, B., Yu, S., Li, Y., & Chen, X. (2009). Spatial genetic structure and restricted gene flow in a functionally dioecious fig, *Ficus pumila* L. var. pumila (Moraceae). *Plant Ecology, 51*, 307–315.

Wang, Q., Feng, Z., & Luo, J. (2000). Study on biodiversity of forest-grassland ecotones in the north of Hei Bei and the east of Inner Mongolia. *Plant Ecology, 24*(2), 141–146.

Wang, Y., & Liu, Y. (2009). Researches on spatial model application of urban landscape eco-network planning. *Resource and Environment in the Yangtze Basin, 18*(9), 819–824.

Wang, S., Zhang, Z., Sun, G., Zhang, M., & Yu, X. (2006). Effects of land use change on hydrological dynamics at watershed scale in the Loess Plateau—A case study in the Luergou watershed, Gansu Province. *Journal of Beijing Forestry University, 28*(1), 48–54. (in Chinese).

Wu, F. L. (2002). Calibration of stochastic cellular automata: The application to rural urban land conversions. *International Journal of Geographical Information Science, 16*(8), 795–818.

Wu, J. (2007). *Landscape ecology: Pattern, process, scale and hierarchy* (2ed ed., pp. 1–23). Beijing: Higher Education Press. (in Chinese).

Wu, J., & Hobbs, R. (2002). Key issues and research priorities in landscape ecology: An idiosyncratic synthesis. *Landscape Ecology, 17*(4), 355–365.

Wu, P. F., Zhou, D. M., & Gong, H. L. (2012). A new landscape expansion index: Definition and quantification. *Acta Ecologica Sinica, 32*(13), 4270–4277.

Xiao, D. (1991). *Application, methodology and theory of landscape ecology*. Beijing: China Forestry Press.

Xu, S. (2009). Landscape pattern analysis and soil erosion evaluation based on the source-sink process. A Ph. D. thesis of the Graduate School of Chinese Academy of Science, Beijing.

Xu, W., Luo, C., Ouyang, Z., & Lu, Z. (2010). Designing regional nature reserves group: the case study of Qinling Mountain Range, China. *Acta Ecologica Sinica, 30*(6), 1648–1654.

You, Z., & Li, Z. (2005). The effect of landscape pattern on soil erosion in Loess Plateau catchment. *Journal of the Graduate School of the Chinese Academy of Sciences, 22*(4), 447–453.

Yu, C., He, H., Hu, Y., Bu, R., & Li, X. (2008). Historic and current fire regimes in the great Xing'an mountains, northeastern China: Implications for long-term forest management. *Forest Ecology and Management, 254*, 445–453.

Yu, K., Qiao, Q., Li, D., Yuang, H., & Wang, S. (2009a). Ecological land use in three towns of eastern Beijing: A case study based on landscape security pattern analysis. *Chinese Journal of Applied Ecology, 20*(8), 1932–1939.

Yu, K., Wang, S., Li, D., & Li, C. (2009b). The function of ecological security patterns as an urban growth fra mework in Beijing. *Acta Ecologica Sinica, 29*(3), 1189–1204.

Zhang, Q. J. (2004). Study of the characteristics of land use change and it driving forces at multiple scales in the hilly area of the Loess Plateau. Doctoral dissertation of Research Center for Eco-Environmental Sciences, Chinese Academy of Sciences (in Chinese).

Zhang, Q. (2012). The scale effects of land use change (chapter 4). In L. Chen & C. Lu (Eds.), *Study on the scale effects of land use change and soil and water loss in the loess hilly areas.* Beijing: Science Press.

Zhang, L., Ouyang, Z., Xiao, Y., Xu, W., Zheng, H., & Jiang, B. (2011). Priority areas for biodiversity conservation in Hainan Island: Evaluation and systematic conservation planning. *Chinese Journal of Applied Ecology, 22*(8), 2105–2112.

Zhao, H., & Zhang, F. (2008). Agricultural landscape pattern change and functional zonation in Beijing. *Transaction of the Chinese Society of Agricultural Engineering, 24*(Supply 1), 78–84.

# Chapter 7
# Ecosystem Service Evaluation

**Gaodi Xie, Shuyan Cao, Yu Xiao, Xia Pei, Yanying Bai, Wenhua Li, Bing Wang, Xiang Niu, Xiaohui Liu, Zhongqi Xu, Qingwen Min, Chunxia Lu, Honghua Shi, Wei Zheng, Dewen Ding, Jiyuan Liu, Jinyan Zhan, Lin Zhen, Li Yang, Xuelin Liu and Moucheng Liu**

**Abstract** With the development of the ecology research, the in-depth understanding of the characteristics of different types of ecological systems, ecosystem observation techniques progress, all kinds of ecosystem observation data to accumulate, an value analysis and evaluation of the ecosystem make it possible for understanding the ecological system evolution and discussing the sustainable development of ecosystem. All parts of the ecosystem including humans and their environment, ecosystem provide all kinds of ecological services to human through the process of its function. Ecosystem services (ESs) is considered as no value by the existing economic model and theory for a long time, only part of the ecological

G. Xie · S. Cao · Y. Xiao · X. Pei · Y. Bai · W. Li · Q. Min · C. Lu · J. Liu · L. Zhen · M. Liu (✉)
Institute of Geographic Sciences and Natural Resources Research, Chinese Academy of Sciences, Beijing 100101, China
e-mail: liumc@igsnrr.ac.cn

B. Wang · X. Niu
Research Institute of Forest Ecology, Environmental and Protection, Chinese Academy of Forestry, Beijing 100091, China

X. Liu
Northeast Institute of Geography and Agroecology, Chinese Academy of Sciences, Changchun 130102, China

Z. Xu
Forestry College, Agricultural University of Hebei, Baoding 071000, China

H. Shi · W. Zheng · D. Ding
The First Institute of Oceanography, State Oceanic Administration of China, Qingdao 266061, China
e-mail: shihonghua@fio.org.cn

J. Zhan
Beijing Normal University, Beijing 100875, China

L. Yang
Urban Planning and Design Institute, Tsinghua University, Beijing 100084, China

X. Liu
Beijing Tourism Planning and Design Institute Davos Summit, Beijing 100101, China



W. Li (ed.), *Contemporary Ecology Research in China*,
DOI 10.1007/978-3-662-48376-3_7

products have the market price, in order to obtain tangible ESs, human damage or even destroy some invisible ecological services at the same time, led to a decline in ESs. Since the mid-1990s, the Chinese ecologists widely study the ESs and its value assessment research, contenting forest, grassland, wetland, farmland, and marine ecosystems. Through the general laws of ecological service value method and the time spatial heterogeneity of ESs space model method, we assess the ESs value in whole and reflect its spatial and temporal heterogeneity. Value assessment method of ESs is developing constantly, widely used in the main are valued based on the price per unit area, based on the function of laws of value and dynamic laws of value based on the single function time three types. Three methods are widely adopted, they are based on the price per unit area, based on the function of laws of value, dynamic laws of value based on the single function time. In a new paradigm to investigate the relationship between ecosystem and economic system, so as to promote an efficient economic decisions for resource allocation in the economic system and ecological system of integrated system framework. To cultivate and develop effective market mechanism for ecological service has become one of the hot spot of ecological service research and ecosystem management. Researching on ESs consumption and building ecological system efficient continuous consumption mode, we can hold the direction of the rational utilization of ecological system, and it is of great practical significance to safeguard the ecological safety. On the basis of understanding the principle of ecosystem consumption and consumption measurement model, analyzing consumption process, analyzing consumption utility function (UF), and analyzing the consumption process, we can realize ESs, and this provide a scientific basis to optimize the structure of consumption of ESs, and to build efficient continuous consumption patterns (CPs).

**Keywords** Valuation methods · Spatial heterogeneity · Ecological compensation · Forest ecosystem service · Monetary value of FES · Wetland ecosystem service · Grassland ecosystem service · Agroecosystem service · Marine ecosystem service

# 1 Theory and Method

## *1.1 Introduction*

Economists and ecological economists now recognize that nature is providing a wide range of services that were previously ignored, and have proposed the term ecosystem services (ESs) to encompass these services. The widely accepted definition for ES is the environment and its effectiveness that human beings must rely on, maintained by ecosystems or formed by ecological processes (Daily 1997), or the benefits people obtain from ecosystems (Costanza et al. 1997; WGMEA 2003a, b, c). Obviously, the economies of every country are all entirely based on the goods and services provided by ecosystems, and human life itself depends on the ability of ecosystems to continuously provide multiple benefits.

Scientists have increasingly recognized that both economic goods and services and ecological goods and services are equally important to human welfare. In order to achieve environmental sustainability in policy and management practice, it is very important to determine ESs and evaluate these. The main purpose of this section is to summarize valuation methods for ES and the progress and contributions of Chinese ecologists in this field.

## *1.2 Valuation Methods of Ecosystem Services and Progress*

### 1.2.1 Valuation Methods of Ecosystem Services

The value basis of ESs includes: (i) the value of ES which is a utility value; (ii) consumer surplus and producer surplus; and (iii) the willingness to pay or willingness to accept compensation.

ESs have multiple values. The value of ESs includes two parts: the use value and nonuse value. The use value includes direct use value and indirect use value; the nonuse value includes the heritage value and existence value. Except for the values mentioned above, there is an option value, which can be classified as either a use or nonuse value.

Valuation methods for ESs are challenging and there remains no internationally recognized or standardized valuation methodology. Existing valuation methods include Avoided Cost (AC), Replacement Cost (RC), Factor Income (FI), Travel Cost (TC), Hedonic Pricing (HP), Contingent Valuation (CV), Group Valuation (GV), and Marginal Product Estimation (MP).

### 1.2.2 Dynamic Change and Spatial Heterogeneity of Ecosystem Service Value

Valuation methods for ESs can assess the eco-services value of a certain region for a given period. Therefore, as research in this area deepens, temporal and spatial changes in ES values are receiving increased attention. Dynamics of ES value can be expressed by ES flow. The spatial heterogeneity of ES can be expressed by a space model.

### 1.2.3 Example Valuation of Ecosystem Services in China

Since the mid-1990s, Chinese ecologists have extensively studied and assessed ESs and accomplished outstanding achievements. For example, Li et al. (2002, 2008) organized experts to compile and publish a series of monographs including the

*Study of Ecosystem Services, Theory, Methods and Applications of Ecosystem Services Valuation*, and *Ecological Compensation Mechanism and Its Policy in China*. These books systematically summarize the fundamental theories and methods underpinning ES valuation and establish an ecological compensation system and policy for China. Ouyang at al. (1999a, b) evaluated ESs in China quantitatively and others have systematically explored ESs for different ecosystems, including forest (Zhao et al. 2003), grassland (Xie et al. 2001), river and farmland (Xiao et al. 2004). As to valuation methods for ESs, and Xie et al. (2003, 2008) posited an equivalent factor method for ESs. The development of these methods has played an important role in ES valuation in China. Unlike in other countries, ecological service valuation research has been directly used to promote the establishment of ecological compensation mechanism in China. In fact, some research results have been adopted directly by national authorities, for example, the *Evaluation Standards for the Ecosystem Services of Forest Ecosystems* has become the forestry industry standard in China and was officially issued by the State Forestry Administration (SFA) of China in 2008. ES valuation methods remain a sphere of development, and widely used methods can be divided into three types: (i) valuation methods based on the price of unit area; (ii) valuation methods based on the function value; and (iii) valuation methods based on temporal changes in a single function.

### 1.2.4 The Mechanism of ES Value

Most ESs are public goods; however, since there is no market or market development is far from perfect, they are always forgotten in the resource allocation of the social-economic system. In essence, the coordination between economy and ecology is a large challenge faced by many countries in the world. Although ecological degradation has an indivisible relationship with the market economy, market mechanisms are not the cause of ecological degradation but only intensify market failure for ESs. The cause of ecological problems lies in economic activity, rather than the market mechanism itself. Compared with nonmarket mechanisms, market mechanisms are more effective for resource allocation and remain a fundamental way to ease ecological degradation and protect and restore ecosystems. ESs provide a new paradigm to evaluate the relationship between the ecological and economic systems, and promote effective allocation of resources within a comprehensive framework composed of economic and ecological systems according to social and economic policies.

Trade form of ES: whether the supply and demand of ESs in the market are real or not, as long as they exist and the quantity or value of ES transfer occurs in time or space, it can be regarded as ES trade. There are three main forms of trade for ESs: (i) economic trade; (ii) the spatial transfer of ESs; and (iii) the transfer of ESs between different generations.

ES payment mechanisms have the potential to promote the formation of ecological service protection funds by creating new demands for ecological products and services. The ES payment system is developing rapidly around the world and accumulating a lot of experiences in promoting the marketization and value realization of ESs. However, these are mainly case studies only and exploring the 'market' for ESs remains at an early stage and far from a 'real' ES market. Market mechanisms will fully play a role in resource allocation only under an 'ideal' market. Therefore, the fair, impartial, sustainable ES trade requires effective intervention from governments and robust international cooperation.

### *1.3 Conclusion and Discussion*

ESs have emerged as a kind of theory and thought area within ecological economics. The theory and method of ecological service valuation is still being developed. The supply, consumption, and value of ESs will be gradually merged into the theory of ecological service valuation, and ES payment mechanisms will be established in a gradual manner.

ESs provided a new paradigm through which to evaluate the relationship between ecological and economic systems, promote the cultivation and establishment of ES markets in many countries, and accumulate experience in coordinating the relationship between economic development and ecological protection. However, the market for ESs development remains local and at a pilot stage, and effective government intervention is necessary for continued development. Legal and economic systems must be brought into the concept and framework of ESs. As a society, specialized departments and policy are needed to ensure the reasonable management and utilization of ESs.

## 2 Assessment of Forest Ecosystem Services in China

### *2.1 Introduction*

Coupled human and natural systems (CHANS) are systems in which human and natural components interact (Liu 2007). According to Mooney and Ehrlich (1997), the idea that humans depend on natural systems dates back as far as Plato, but the first modern publication that addresses this issue is *Man and Nature* by George Perkins Marsh in 1864. Today, interactions between human and natural systems have emerged as concerns because human activities are globally connected. At the same time human societies and globally interconnected economies rely on ecosystems services and support (Millennium Ecosystem Assessment 2005). ESs are the conditions and processes through which natural ecosystems, and the species that make them up, sustain, and fulfill human life (Daily 1997), including provision services, regulation services, cultural services, and support services. It is now clear that patterns of production, consumption, and well-being develop not only from economic and social relations within and between regions but also the capacity of other regions' ecosystems to sustain them (Arrow et al. 1995).

Valuation of ecosystems has continued (de Groot et al. 2002), but research and attention has expanded greatly since the estimation of the value of ESs and natural capital (Costanza et al. 1997). The development of science-based policy has been increasingly recognized as an method for protecting and managing the environment in the context of global change (Sun and Chen 2006; Daily and Matson 2008; Fisher et al. 2008; Mäler et al. 2008; Carpenter et al. 2009). The recent Millennium Ecosystem Assessment (MEA) provided a new general conceptual framework for estimating the value of ESs at the regional, national, and global scales (MEA 2005). In April 2011, a UK National Ecosystem Assessment (UK NEA) report was published and it was the first and relatively complete assessment of ESs at a national scale (UK NEA 2011). The UK NEA included four recognized services provided by all the main ecosystems: supporting, regulating, provisioning, and cultural services.

Here we focus on valuing China's forest ecosystems services (FES). Forests cannot only provide timber, but also critically represent important habitats and ESs (Miller and Tangley 1991; Mendelshon and Balick 1995; Pearce 1998, 1999). The ESs provided by forest ecosystems are diverse and difficult to quantify accurately at a national level. In the last two decades, the estimation of the value of FES at the national or regional scale has been the focus of ES research. For example, economic techniques for estimating the total economic value (TEV) of forests in Mexico was proposed (Adger et al. 1995), however, only a proportion of this value can feasibly be 'captured' within Mexico: much of the benefits of Mexico's forests fall outside its borders and is therefore not considered by forest users or national policy-makers. These benefits include maintaining water quality, reducing storm water runoff and erosion, improving air quality, regulating climate and carbon sequestration, providing habitat for wildlife, maintaining biodiversity, and providing a destination for recreation and tourism in addition to providing timber and non-timber resources, which were estimated in America (Krieger 2001). The economic value of Mediterranean forests, brought together forest valuations at the national level from 18 countries, is based on extensive local data and research findings in the context of institutions and new policy approaches for improving management at national, regional, and local levels (Merlo and Croitoru 2005). A study on the TEV of Amazonian deforestation during the period of 1978–1993 also suggested the value of FES from different points of view (Torras 2000), and scientists combined the green income accounting and TEV approaches and applied the new framework to Brazil in order to assess the foregone economic benefits resulting from Amazonian deforestation. Canada also assesses the real value of its boreal ecosystems (Anielski and Wilson 2005). Assessment work was carried out in 1972, 1991, and 2000 in Japan (Wang 2005), and more recently in the UK (UK NEA 2011). All these studies indicate the high value of FES, which has important implications in the development of policy to protect and manage forests via ecological compensation.

In the last 30 years, forest resources in China have rapidly increased along with its economy. In 2008, the total area of forests was 195 million ha with a growing stock of more than 13 billion cubic meters (State Forestry Administration 2009a). Forested areas covered 20.36 % of the land base of China in 2008, a value that has tripled from 8 % 60 years ago. The Chinese government has announced a plan to expand forested areas by 40 million ha between 2005 and 2020 with carbon sequestration as the main policy objective. China is an important country in the world in terms of the importance of forests. Without a doubt, the benefits of afforestation and reforestation make an important contribution to environment improvement and economic development. The value of China's FES in forest ecology and forestry economics is an essential issue to consider. Based on the latest national forest resources survey (the seventh) and socioeconomic data, this section aims to show the monetary value of China's FES at a national scale in 2008, and to discuss the characteristics and implications of these assessments in an international context.

Research is committed to supporting policy action toward a sustainable use of forest resources nationwide, and the forest economic evaluation challenge has gradually reached the national policy agenda. The methodology and approaches for assessing FES and applying results are currently being developed in China, but uncertainty factors when assessing FES remain. Therefore, sharing research methods and results among scientists, forest managers, policy-makers, and the public in different countries is important. China has an exceptionally diverse climate, geography and hydrology and forest vegetation, and social and economic conditions differ between regions. While Chinese scientists and policy-makers desire to learn from the work of other countries, China's diverse conditions favor the development of a methodology that is applicable to other countries with different climatic, geographical, and hydrological conditions. China should play an important role in the development of environmental and forest management.

## 2.2 Data and Methods

### 2.2.1 Data Sources

Field measurement data consisting of ecological properties (e.g., net primary productivity (NPP), water, and soil conservation) and characteristic parameters were obtained based on 50 long-term research stations (consisting of 286 supplement stations) in the China Forest Ecosystem Research Network (CFERN). This network covers almost all forest ecosystems of all dominant trees in the country. The forest inventory dataset was also used. Although SFA data are available from 1973 to 2008 via seven national forest resources inventories (NFI), (1973–1976) (the 1st NFI), 1977–1981 (the 2nd NFI), 1984–1988 (the 3rd NFI), 1989–1993 (the 4th NFI), 1994–1998 (the 5th NFI), 1999–2003 (the 6th NFI), and 2004–2008 (the 7th NFI), only the period 2004–2008 (the 7th NFI) is complete and reports forest area, timber volume, and some ecological parameters. Data recorded included the forest group (planted and natural forests), dominant tree species and age-class and so on. We used data from the 7th NFI. Social-economical public data released by authorities was also used in this assessment.

### 2.2.2 A Framework for the Ecological Valuation of Forest Ecosystem Services in China

We evaluated the valuation of forest ESs based on the framework in Fig. 1. The framework comprises four steps: (i) selection of indicators for the FES assessment; (ii) identification of units for assessing FES; (iii) calculation; and (iv) synthesis of the results.

#### Selecting Items for the FES Assessment

Forests are amongst the most biologically rich terrestrial systems in the world and provide us with a wide variety of ESs. Six items have been selected to assess the FES, including water conservation, soil conservation, carbon sequestration and oxygen release, nutrient accumulation, atmosphere environmental purification, and biodiversity conservation. Data from China are available for all the six measures. Water conservation service contains water quantity regulation and water quality purification, reflecting the role forests play in mitigating natural disasters such as droughts and floods, as well as clean drinking water. Water and soil erosion is widespread and affects all natural and human-managed ecosystems. It often causes soil deterioration (Marques et al. 2008), decline in land productivity (Pimentel and Kounang 1998), and is caused by a lack of vegetation protection (Canton et al. 2001; Ludwig et al. 2005). Soil conservation is an indicator to demonstrate the interaction between human and natural systems. The reasons for considering carbon sequestration and oxygen release are that forest ecosystems are important carbon sinks and have a close relationship with climate change. Forests account for around 50 % of total aboveground terrestrial organic carbon, as well deforestation and forest degradation are estimated to cause 20 % of annual greenhouse gas emissions (SCBD 2008). So this indicator can better understand the effects of human activities on natural systems and the responses of natural systems to human activities. Key publications such as the MEA (2005) and Red List of Threatened Species (IUCN 2004) indicate that a large and increasing number of forest ecosystems, populations, and species are threatened globally due to the loss and degradation of forest habitat, and thus this indicator is needed to assess the value of forest biodiversity conservation. With industrialization and increasing pollution, human living environmental issues are more and more remarkable. Forest functions by absorbing air pollutants and biochemistry cycles, and thus the indicators nutrient accumulation and atmosphere environment purification are therefore required.

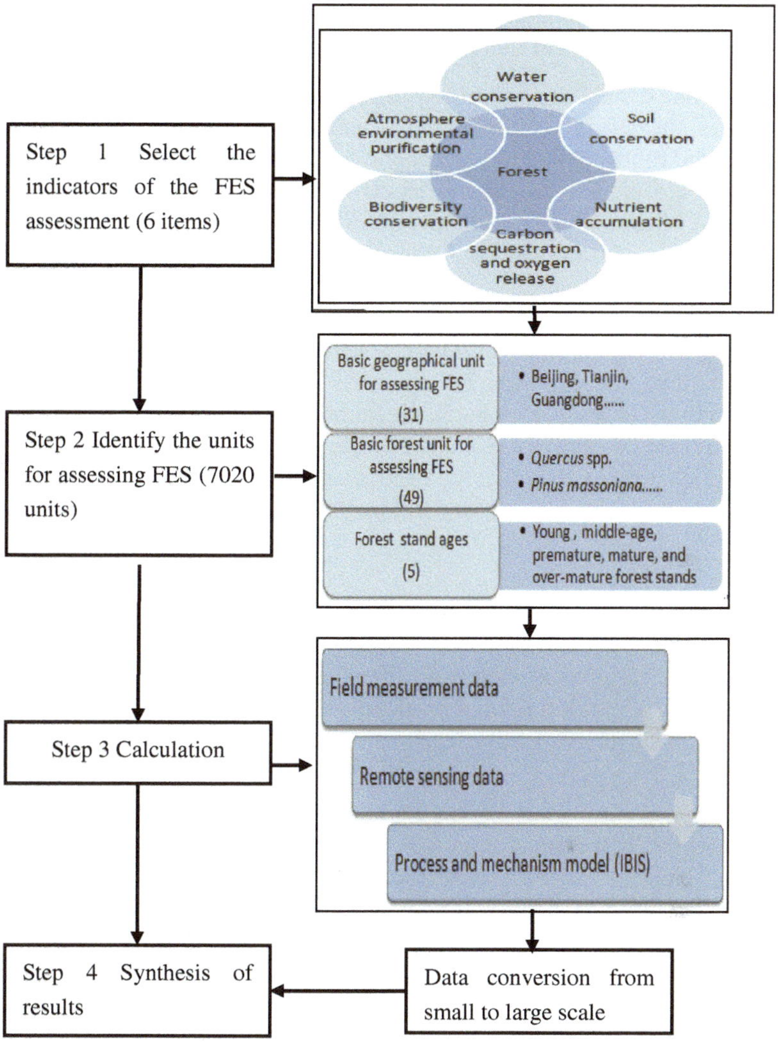


**Fig. 1** Framework for evaluating FES in China

## Identifying the Units for Assessing FES

*Basic geographical unit* China is a large country of varied topography, climate, forest vegetation, forest management systems, and social-economic development levels. In order to minimize differences in natural conditions and maximize the implication of the assessment results in policy-making, 31 administrative regions (provinces and municipalities directly under the Central Government) across the

Chinese mainland were used as assessment units. The value of FES in Hong Kong, Macau, and Taiwan was not included here.

*Basic forest unit* Forest vegetation was divided into 46 kinds of forests based on dominant tree species. In addition, a large area of economic forest (fruit trees, cork plantations, and rubber plantations), bamboo forest, and shrubbery woodland exists. These forests are difficult to classify based on dominant trees and are treated as three kinds of forests. In total, 49 kinds of forests were used in this assessment. In view of the effect of stand age on FES, each kind of forest was divided into five age-classes: young, middle-age, premature, mature, and overmature stands. Similarly, each kind of forest was divided into five age-group stands in a province based on the NFI in 2008 (SFA 2009a). When all age-class stands in all kinds of forests were found in a province, 245 age-class stands were involved in accounting FES at the provincial level. In general, the number of age-class stands was less than 245 in a province because some kinds of forests or some age-groups in a kind of forest were not present. Ultimately, 7020 assessment units with homogenization were used as accounting units at the national level.

### Calculation Process

Based on field measurement data from forest ecosystem research stations, remote sensing data, and using process and mechanism models such as IBIS, we were able to convert data conversion across scales.

### Synthesis of FES Results

An age-group stand was used as the basic forest unit for accounting the value of each FES. For each age-class stand in a province, the annual quantity of each FES provided by the stand was estimated based on local research station observations or published data. The unit price of each FES was determined based on relevant Chinese yearbooks or published data. For example, the price of nitrogen conserved in the stand was determined by referring to the current market price of nitrogen fertilizer (Wang and Yang 2008). For an age-class stand, the annual monetary value of each FES was obtained by multiplying the annual amount provided by the stand with the price per unit amount. For each FES, the total values for quantity and monetary could be obtained for a kind of forest, forest type, and province by summing the quantity and monetary value of each FES in different age-class stands within a province.

## 2.3 Conclusion and Discussion

### 2.3.1 Monetary Value of FES at a National Level

In 2008, the total value of the six FES (water conservation, soil conservation, carbon sequestration and oxygen release, nutrient accumulation, atmosphere environmental purification, and biodiversity conservation) was estimated to be 10.01 trillion CNY (1.48 trillion USDs) per year. The contribution of each ES to the total FES from highest to lowest was water conservation (40.51 %), biodiversity conservation (24.01 %), carbon sequestration and oxygen release (15.57 %), soil conservation (9.92 %), atmosphere environmental purification (7.92 %), and nutrient accumulation (2.07 %) (Fig. 2).

### 2.3.2 Monetary Value of FES at the Provincial Level

Sichuan provided the largest FES value (1059 billion CNY) and Shanghai was the least (2.31 billion CNY) (Fig. 3). The values for water conservation, biodiversity conservation, and carbon sequestration and oxygen release at the provincial level are the three largest fractions, and the value of nutrient accumulation was the smallest. The value of FES showed the different traits in spatial distribution. In general, southwestern provinces (e.g., Sichuan 10.57 %, Yunnan 10.24 %, and Guangxi 7.73 %) and northeastern provinces (e.g., Heilongjiang 8.57 % and Inner

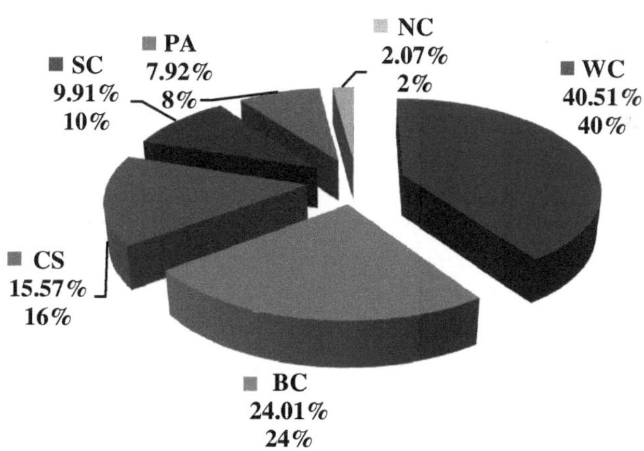


**Fig. 2** Proportions of water conservation (*WC*), biodiversity conservation (*BC*), carbon sequestration (*CS*), soil conservation (*SC*), purification of the atmosphere (*PA*), and nutrient conservation (*NC*) of total value of FES (10.01 trillion CNY, 1.48 trillion USD) in China

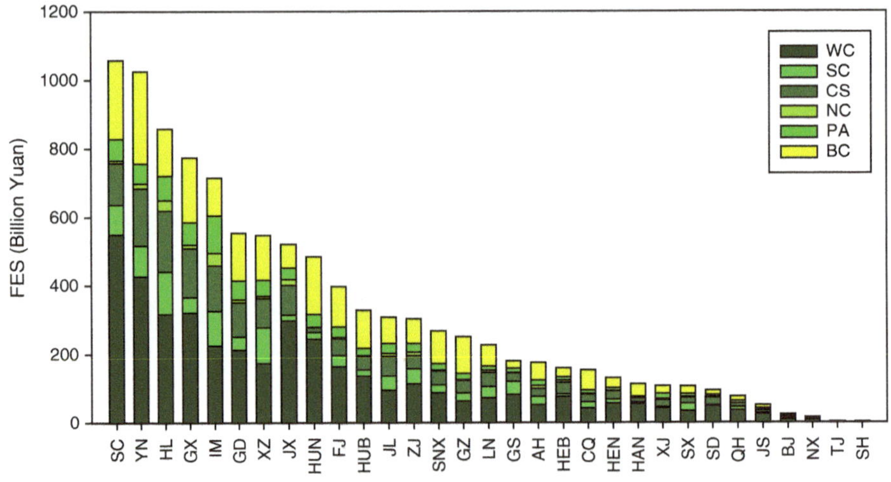


**Fig. 3** Monetary value of water conservation (*WC*), biodiversity conservation (*BC*), carbon sequestration (*CS*), soil conservation (*SC*), purification of the atmosphere (*PA*), and nutrient conservation (*NC*) (billion CNY) provided by forests in 31 provinces. *AH* Anhui; *BJ* Beijing; *CQ* Chongqing; *FJ* Fujian; *GS* Gansu; *GD* Guangdong; *GX* Guangxi; *GZ* Guizhou; *HAN* Hainan; *HEB* Hebei; *HL* Heilongjiang; *HEN* Henan; *HUB* Hubei; *HUN* Hunan; *IM* Inner Mongolia; *JS* Jiangsu; *JX* Jiangxi; *JL* Jilin; *LN* Liaoning; *NX* Ningxia; *QH* Qinghai; *SD* Shandong; *SH* Shanghai; *SNX* Shannxi; *SX* Shanxi; *SC* Sichuan; *TJ* Tianjin; *XJ* Xinjiang; *XZ* Xizang (Tibet); *YN* Yunnan; *ZJ* Zhejiang

Mongolia 7.15 %) had higher contributions to the total national FES; Central and Eastern provinces (e.g., Jiangsu 0.51 % and Shanghai 0.02 %) had lower contributions (Fig. 4).

### 2.3.3 Per Unit Area FES Value at the Provincial Level

Variation in provinces exists in terms of mean FES per unit forest area ($\times 10^3$ CNY/ha). For instance, the highest FES per unit forest area was in Hainan ($64 \times 10^3$ CNY/ha), and the lowest one in Xinjiang ($16 \times 10^3$ CNY/ha). The mean FES per unit forest area for the whole country was $46 \times 10^3$ CNY/ha.

### 2.3.4 FES of Different Forest Types

Forty-nine kinds of forests were classified into four groups according to their contribution to the total FES: >10 % group (shrubbery and broadleaf forest), 5–10 % group (*Quercus* spp. forests, economic forests, conifer and broadleaf mixed forests, and *Pinus massoniana* forests), 1–5 % group (13 kinds of forests), and <1 % group (30 kinds of forests). The first two groups displayed a substantial contribution

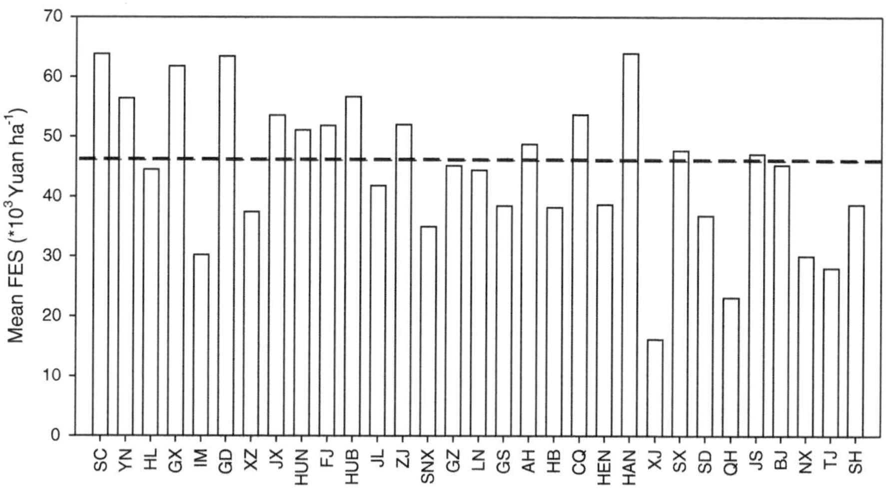


**Fig. 4** The mean FES per unit area ($\times 10^3$ CNY/ha) for each province (*column*) and entire country (*dash line*). For province acronyms see the legend to Fig. 3

to total FES (57.79 %). As a single genus, *Quercus* spp. forests had the largest contribution to total FES (11 %), and as a species, *P. massoniana* had the largest contribution (5.00 %)

### 2.3.5 Characteristics of China's FES Value

*High spatial heterogeneity of FES value and decoupling from GDP* China has an unbalanced development for the economy and environment protection. Like economic production (e.g., GDP as an indicator), FES varied greatly among provinces (Fig. 4). The contributions of the FES value of provinces were decoupled from the contribution of GDP (in 2009) at the national level (Fig. 4). Large FES contributions arose from Sichuan, Yunnan, Guangxi, Heilongjiang, and Inner Mongolia, whereas large GDP contributions arose from Guangdong, Jiangsu, Shandong, Zhejiang, and Henan. The smallest FES contributors were from Shanghai, Tianjin, Ningxia, Beijing, and Jiangsu. In general, the largest FES contributors were undeveloped provinces and remote provinces are located in the northeast and southwest China. Large GDP contributors were the Eastern and Central provinces and metropolitan regions. As an extreme example, Shanghai is the most developed region in terms of economy, society, and technology, but its FES contribution is negligible at a national scale. Tibet, in southwest China, is an opposite example (Fig. 4). Given the pattern of decoupled FES and GDP across the country, special importance for economic compensation from higher GDP provinces to higher FES provinces in China is apparent.

*Rapid increase in future FES* Changes in ESs in the UK were investigated, showing that a decline or increase in ESs was dependent on the ecosystems of the previous years (UK NEA 2011). In contrast, China's FES value will rapidly increase in the coming decades as a large proportion of forests are young plantations (33.8 %) (SFA 2009a). These young plantations have great potential to grow in size and ecological function, and will likely result in a net increase in FES values. The Chinese government has announced a plan to expand the forested areas by 40 million ha between 2005 and 2020 because of the huge potential for sequestering atmospheric $CO_2$ (Pan et al. 2011). Therefore, forests in China are expected to increase in stock (stock density and total area) and ecological functions in all key aspects of FES, including carbon sequestration, nutrient accumulation, and water and soil conservation in the future.

*Parts of plantations and economic forest in providing FES* China has the largest area of plantation in the world, accounting for 38 % of its total forest area; more plantations are being planned over the next 10 years. Plantations play essential roles in providing FES and materials. This is a distinct feature compared to Canada, Russia, and the USA (FAO 2010). In particular, the plantation area of China is about 62 million ha (38 % of national forest areas), of which half is economic forest (32 million ha) (SFA 2009b). For economic forest, the annual direct economic income was 3.56 trillion CNY in 2008 (SFA 2009b) and provided 1.4 trillion CNY FES value, about 40 % of direct economic income.

*Differences in FES demands by region* Different regions and provinces have distinct demands for FES because of differences in geography, climate, and social development. For example, in the eastern plains of China, where the economy and society are developed, the function of forest ecosystems in the purification of air pollution is more favorable, while in the northwestern mountainous region, water conservation is vital. Similarly, such differences exist among countries. Relative to the larger need for the roles of water and soil conservation in China and Japan, the value of protecting biodiversity, landscape, culture, and tourism in the UK (UK NEA 2011) and Canada (Anielski and Wilson 2005) may be more important.

### 2.3.6 Policy Development for Ecological Compensation at a National Level

Relative to countries with balanced economic and environment development (e.g., Canada, USA, and UK), developing ecological compensation policy is especially important in China because of its distinct decoupled GDP and FES. In fact, much discussion in China has focused on how to determine economical compensation for underdeveloped regions with high FES from low FES but economically developed regions (Fei et al. 2004; Sun and Chen 2006). For establishing such a framework, quantifying the monetary value of ESs is important. In this sense, the assessed value of FES could be used as a basis for establishing ecological compensation at a national level. For example, Sichuan, Yunnan, Heilongjiang, Guangxi, Inner Mongolia, and Tibet are the top five provinces (from 7.15 to 10.17 %) (Fig. 4) in terms of FES contribution to the national total, but are all located in undeveloped regions. In general, eastern provinces have higher GDP contributions, whereas northeastern and western provinces have higher FES contributions (Fig. 4).

Further, 58 % of forests fall under collective ownership in China, with the state owning remaining forests. Reforms to change the collective ownership forest system to private ownership in China are ongoing. This is considered a way to efficiently manage existing forests and increase new plantations in non-forest mountains (Zhao et al. 2010) and these private forests can be traded in a forest ownership trading center. This ongoing area of reform is considered a significant change in China's forestry system and the FES value in a stand should be taken into account in future trading.

Although current forest coverage across China is 20.36 % and it has 286.6 million ha of forested land (SFA 2009b), establishing forests in mountainous areas where commercial forests for timber production are not suitable is in high demand. Farmers are more interested in planting trees in those areas if their investments deliver through the consideration of FES values on the forest trading market, in addition to the direct benefits of forests.

### 2.3.7 Improving Assessment Methods

Barriers to the complete assessment of FES across an entire country remain (e.g., methodology) despite that a theoretical framework has been proposed to account the value of all ESs (e.g., supporting, provisioning, regulating, and cultural services) (Costanza et al. 1997). UK NEA (2011) and Merlo and Croitoru (2005) exercised a relatively more complete FES assessment at the national scale in UK and in some Mediterranean countries. However, the current assessment in China only includes parts of the FES. Specifically, the aesthetic values, cultural heritage, and educational values of ecosystems are not considered because of a lack of methods for assessing the FES values of ecosystems, other than water conservation, soil conservation, carbon sequestration and oxygen release, nutrients accumulation, atmosphere environment purification, and biodiversity conservation.

International timber trading may transfer some of the FES observed in a country to losses of forests in other countries (Mayer et al. 2005). China is not self-sufficient in wood products and needs to import round wood and pulp from other countries. Thus, the effects of international trade should be considered when calculating the net FES value when accounting FES in a country.

An important feature of China's FES assessment is its close combination with data from national forest resources surveys conducted nationally at 5-year intervals. This means it is possible to produce a regular assessment of FES at a provincial and national level into the future.

# 3 Assessment of Wetland Ecosystem Services in China

## *3.1 Introduction*

It is well known that wetlands are an important component of the terrestrial landscape, performing significant ESs such as climate regulation, flood storage, water supply, and biodiversity conservation. Increasing human land use has put wetlands at risk. The Organization of Economic Cooperation and Development (OECD) estimates that the world may have lost 50 % of its wetlands since 1900, and land conversion for agriculture is the principal cause (OECD/IUCN 1996). Once thought to be wastelands, wetlands have been extensively drained for economic development. Direct land conversion for agricultural drainage, forestry, and urban construction has caused widespread degradation and destruction of wetlands.

In the mid-1990s, an article titled *The Value of the World's Ecosystem Services and Natural Capital* represented the beginning of TEV estimation of our planet. The services of ecological systems and natural capital stocks that produce them are critical to the functioning of the earth's life support system (Costanza et al. 1997).

Loss or degradation of wetland habitats can result in a loss of biodiversity, reduction in water supply and water storage, and increased soil erosion. Additionally, wetland conversion for industrial and agricultural purposes has directly or indirectly contributed to an increase in atmospheric concentrations of major greenhouse gases. How to recognize or understand these functions generally remains a challenge and it is very important to introduce value estimation. Restoration actions that enhance both biodiversity and ESs are necessary worldwide. In order to achieve the goal of 'no net loss' of wetland function, we should establish wetlands compensation accounts to balance any loss.

## *3.2 Wetland Ecosystem Services and Value Estimation*

### 3.2.1 Wetland Ecosystem Services

Ecosystem functions refer variously to habitat, biological or system properties or processes of ecosystems. Ecosystem goods (such as food) and services (such as waste assimilation) represent the benefits human populations derive, directly or indirectly, from ecosystem functions. For simplicity, we will refer to ecosystem goods and services together as ESs. A large number of wetland services have been identified such as carbon sequestration and greenhouse gases emissions, flooding regulation, biodiversity, nutrient cycling, water supply, waste treatment, food production, and sediment retention.

Wetlands regulate biogeochemical cycling, play an important role in the global carbon budget and exchange greenhouse gases such as carbon dioxide ($CO_2$) and methane ($CH_4$) with the atmosphere. According to the Chinese Second Soil Survey, wetland soil organic carbon (SOC) density ranges from 14.1 to 60.0 kgC/m$^2$ (Pan 1999). Wetland SOC density ranged from 13.9 to 47.3 kgC/m$^2$ in the Sanjiang Plain, northeastern China (Ma et al. 1996). Recent global concerns over increased atmospheric $CO_2$ have furthered interest in SOC changes and the carbon sequestration capacity of various ecosystems, especially wetlands. $CH_4$ is a very potent greenhouse gas, and $CH_4$ emissions from natural wetlands account for 20 % of global emissions (Liu 2004).

Flooding regulation, storm protection, or drought recovery of wetland habitat responding to environmental variability are mainly controlled by wetland vegetation structure. For example, the porosity of vegetation roots was up to 71–93 % and maintains flooding. The mean depth of marshlands is 30 cm, so natural marshlands can store $17.15 \times 10^8$ m$^3$ of water. Taking into account the Sanjiang Plain, Liu et al. (2007) indicated that maximum soil water storage was $46.97 \times 10^8$ m$^3$.

Wetland habitats are keystones of biodiversity reservation. Patchy shapes, area and corridor length influence species migration. Endangered waterfowl and other species that rely on wetlands have become threatened or extinct in areas where wetland habitat has been destroyed (Liu 2005).

### 3.2.2 The Coupling of Wetland Structure, Process, and Service

Land use changes impact GHG emissions because of the growing human population. Although the greenhouse effect is a global issue and under global influence, one should consider the effect of these gases on our local and regional climates. Wetland ESs such gas regulation have become one of the key issues in environmental and ecological scopes (Liu et al. 2013). What influence do changes in ecosystem structure and process have on services? This is a key question when trying to improve wetland services.

Flood pulse supports floodplain biological productivity functions, but under disturbance by people, hydrological regime changes may impact the flood pulse. After wetlands are drained, they are converted from $CO_2$ sinks to $CO_2$ sources, leading to carbon cycles. In the long run, there is a potential to influence global warming (Lu 2004).

### 3.2.3 Wetland Ecosystem Service Estimation and Dynamic Change

The value coefficients of different wetlands ESs are listed in Table 1. There is much research on the types of natural wetland services and their values, but little research on the ESs of constructed wetlands.

Compared to the value coefficients of different services, we must consider regional heterogeneity when estimating ESs. Services are affected by different geography, ecology, and climates, so the values of these services also differ.

**Table 1** The value coefficient of different wetland ecosystem services (CNY/(ha year)

| | Gas regulation | Climate regulation | Water storage | Soil formation and conservation | Waste treatment | Biodiversity maintenance | Food production | Raw materials | Recreation and culture | Sum |
|---|---|---|---|---|---|---|---|---|---|---|
| Nation | 1573 | – | – | 2 | – | – | – | – | – | 1575 |
| Poyang Lake | 598 | – | 301 | 2008 | – | – | – | – | – | 2906 |
| Yangtze River Estuary | – | – | 716 | – | 1585 | 1328 | – | – | 2235 | 5863 |
| Wetland Reserve | – | – | – | – | – | 5063 | – | – | – | 5063 |
| Rice paddy | – | – | – | 16 | – | – | – | – | – | 16 |
| Global | 1104 | 37,674 | 31,665 | – | 34,171 | 2523 | 2125 | 880 | 12,077 | 122,218 |

*Note* China's wetland area is $6594 \times 10^7$ ha; this table is from Liu et al. (2008)

Worldwide, wetlands have had the fastest rate of loss amongst all ecosystem types. Many studies have investigated marshland loss and landscape changes in some portions of the Sanjiang Plain or over the whole region. Land use changes have spatial and temporal patterns, and correspondingly, ESs exhibit dynamic change. The value of wetland ESs may decline when wetlands shrink.

### *3.3 Conclusion and Discussion*

Ecological economic methods could be an effective way to value wetland ESs. These methods clarify the conceptual difference in the functions and services of ecosystems and reveal the relationship among structure, process, and service. This approach also identifies the different services of different wetland types and means the indirect value of ESs are easily understood. In the future, dynamic changes in service estimation should stimulate wetland conservation and restoration.

Considering environmental and human factors, we should give priority to dominant services across different wetland types. Ecological compensation should be implemented for loss of wetland and other ecosystems. Monitoring station network data are critical in wetland service estimation, such as for the service of biodiversity.

## 4 Assessment of Grassland Ecosystem Services in China

### *4.1 Introduction*

ESs are the foundation that sustain and fulfill human life and development (Daily 1997; Ouyang et al. 1999a, b). Grasslands are the largest and most important terrestrial ecosystem in China and play a major role in maintaining ecological safety, forage production, soil erosion prevention, biodiversity conservation, and carbon sequestration (Xie et al. 2003). Costanza et al. (1997) valued grassland ESs at a global scale and showed that the valuation of global grassland is $9.06 \times 10^{10}$ USD, accounting for 7.3 % of total terrestrial ESs. Zhao and Ouyang (2004) and Min et al. (2004) studied the ESs of Chinese grasslands and the Inner Mongolia steppes, respectively, and found that grassland ESs are important for maintaining ecological conditions and the development of society (Zheng et al. 2009). However, several grassland ecosystems and their services have been degraded across China because of the rapid development of local economies, increased population, and animal husbandry (Xu et al. 2005; Yu et al. 2005). It is important to understand how human actions impact the ESs of grasslands so that these precious resources can be managed in a sustainable way. Here, I report on a study of the impacts of human disturbance on ESs in the Inner Mongolian steppes region, including grazing, non-grazing, and reclamation.

## 4.2 Site Description

The study area is located in experimental pasture at the Meteorological Bureau of Xilinhaote City, China (116°04′N–117°05′N, 43°26′E–44°08′E). The region has a temperate semiarid continental climate, with a windy and dry spring, a warm and rainy summer, a very short autumn, and a long and cold winter. Precipitation is 300 mm and the annual average air temperature is −0.1 °C. The site is located in the central region of Inner Mongolia typical steppe, composed of the following species: *Stipa krylovi*, *Leymus chinensis*, *Cleistogenes squanosa*, *Anemarrhena asphodeloides*, *Allium anisopodium*, *Allium ramosum*, *Salsola collina,* and *Carex duriuscula*. The soil is characterized as chestnut soil. There are experimental plots for non-gazing of different times (2, 7 and 17 years), and grazing plots are located outside non-grazing plots.

## 4.3 Conclusion and Discussion

### 4.3.1 Impact of Disturbance on Biomass

Cultivation and grazing decrease biomass in the ecosystem and non-grazing increases it. The net productivity of the ecosystem is highest for farmland and smallest for grassland under grazing; grassland under non-grazing for 7 and 2 years is higher than that for 17 years. The valuation of production of goods is biggest for the farmland ecosystem from grassland. Non-grazing can raise the valuation of production of goods, and the longer the banned-grazing period, the larger the valuation.

### 4.3.2 The Impact of Disturbance on Carbon Storage

Cultivation leads to a sharp decrease in soil carbon storage; non-grazing distinctly leads to increases in the carbon storage of soil, litter, and aboveground plants and roots. Topsoil (0–20 cm) carbon storage is affected strongly by human disturbance rather than deep soil. Topsoil (0–20 cm) organ carbon densities are ranked in the descending order as follows: grassland non-grazing for 17 years (4.47 kg/m$^2$) > grassland banned-grazing for 7 years (4.23 kg/m$^2$) > grassland banned-grazing for 2 years (4.01 kg/m$^2$) > grazing grassland (3.70 kg/m$^2$) > farmland (2.27 kg/m$^2$). Total carbon storage is ranked in the descending order as follows: grassland non-grazing for 17 years

(9.44 kg/m$^2$) > grassland non-grazing for 7 years (9.09 kg/m$^2$) > grassland non-grazing for 2 years (8.82 kg/m$^2$) > grazing grassland (8.78 kg/m$^2$) > farmland (4.72 kg/m$^2$).

### 4.3.3 The Impact of Disturbance on Anti-wind Erosion

Wind erosion rates (WER) of grassland soil increase as wind speed rises, but are limited by the amount of erodible soil. In general, the WER of farmland soil is unlimited because of its deep plough layer. The WER and wind erosion amount (WEA) in banned-grazing grasslands are lower than that under grazing, and the longer the banned-grazing period the less the WEA and WER. In cultivated land the WEA and the WER are more than for natural grasslands and the differences between them increase with wind speed. Non-grazing reduces the loss in valuation due to wind erosion, and cultivation and grazing can increase this. Value loss rates ranked in the descending order are: farmland (initial stages of cultivation) > farmland > grassland under grazing > grassland non-grazing for 2 years > grassland non-grazing for 17 years.

### 4.3.4 The Impact of Disturbance on Soil Nutrient Content

Cultivation and grazing lead to a decrease in soil nutrient content, especially topsoil, and cultivation affects it more significantly than grazing. Non-grazing increases soil nutrient content, which increases with the time of banned-grazing. Using valuation by biology nutrient pools, the value of the nutrient circle maintaining service is 1302.54 CNY/(ha year) for farmland, 571.87 CNY/(ha year) for grassland non-grazing for 2 years, 556.09 CNY/(ha year) for grassland non-grazing for 7 years, 551.65 CNY/(ha year) for grassland non-grazing for 17 years, and 441.03 CNY/(ha year) for grassland under grazing. Subtracting the cost for farmland, the value of farmland is only 2284.37 CNY/(ha year). Using valuation by soil nutrient pools, the value of the nutrient circle maintaining service is 3911.52 CNY/(ha year) for grassland non-grazing for 17 years, 3813.43 CNY/(ha year) for grassland non-grazing for 7 years, 3665.09 CNY/(ha year) for grazing grassland, 3648.77 CNY/(ha year) for grassland non-grazing for 2 years, and 2893.43 CNY/(ha year) for farmland.

### 4.3.5 The Impact of Disturbance on Biodiversity Nutrient Content

Non-grazing does not affect the composition of dominant species markedly, but does affect the dominance of species. Non-grazing enhances the biodiversity of grasslands at an early stage, but as non-grazing time extends, biodiversity and the richness of grasslands decline. Water conditions affect biodiversity distinctly and appropriate non-grazing times and better water conditions favor the restoration of grassland biodiversity. The decline in biodiversity of grasslands in this region can be ascribed to over grazing and drought.

# 5 Assessment of Agroecosystem Services in China

## 5.1 Introduction

Agroecosystems have become integrated crop production systems with significant human disturbance. Compared to natural ecosystems, the crop production function of agroecosystems has intensified while impairing other ESs such as gas and water regulation, soil conservation, and biodiversity maintenance (Fowler and Mooney 1990; Wood et al. 2000; Tilman et al. 2002). In recent years, agro-ESs have become more and more scarce because of demands from rapid global development (Tilman et al. 2001; Foley et al. 2005). This has resulted in increasing concern about the values and services of agroecosystems across the world. The biodiversity of agroecosystems was the physical basis for ESs (Wall et al. 2010). There have been many studies on agro-ESs, including production supply, carbon sequestration, soil conservation, nutrient cycling and water regulation, and comprehensive assessments of agro-ESs (Xie et al. 2005; Swinton et al. 2007; Sandhu et al. 2008; Porter et al. 2009; Yoshikawa et al. 2010). The passive effects (e.g., nonpoint pollution, $CH_4/N_2O$ emissions and heavy metal pollution) of agricultural production on human society and the environment have also been explored (Liang et al. 2007; Köel-Knabner et al. 2010; Xiao et al. 2010). Farmers and government organizations have realized the importance of the tradeoff between the advantages and disadvantages of agricultural production. The various effects of agricultural production systems (e.g., integrated, conventional and organic agroecosystems, and combined food/forest systems) on the services of different agroecosystems have been compared (Yang et al. 2007; Xiao et al. 2011). Only multifunctional agricultural production systems maximize welfare supported by agro-ESs. To manifest the effects of ESs of agroecosystems to human well-being, three main studies have been conducted and will be discussed in this section.

## 5.2 Assessments of Ecosystem Services by Agroecosystems in China

### 5.2.1 Ecosystem Services by Wheat-Maize Croplands on the North China Plain

Field investigations were conducted at Luancheng Agro-Ecosystem Experimental Station of the Chinese Academy of Sciences in Hebei from 2006–2007 (Xiao et al. 2011). Field data were then used to evaluate the ESs of wheat-maize croplands in the North China Plain. ESs analyzed included primary products, gas regulation, soil organic matter (SOM) accumulation, water regulation, and nitrogen transformation. The results showed that primary products from croplands accounted for 5.04–5.71 t/(ha year) of wheat grain, 6.69–8.24 t/(ha year) of maize grain, 8.58–9.72 t/(ha year)

**Table 2** Assessment of ecosystem services by wheat-maze agroecosystems CNY/(ha year)

| Services types | 2006 | | 2007 | |
|---|---|---|---|---|
| | N1 | N2 | N1 | N2 |
| Primary production | 19,679 | 21,210 | 19,267 | 19,233 |
| Gas regulation | 37,346 | 41,384 | 36,256 | 36,141 |
| SOM accumulation | 2033 | 967 | 2033 | 967 |
| Water regulation | −215 | −230 | −173 | −188 |
| N transformation | 2977 | −962 | 2900 | −1424 |
| Integrated values | 61,897 | 62,484 | 60,356 | 54,809 |

*Note* In the integrated values of ecosystem services of the wheat-maize agro-ecosystem, the value for $N_2O$ emissions was just counted once, though it was calculated both in gas regulation and nitrogen transformation. $N_2O$ emission values were −78 CNY/(ha year) (N1), −114 CNY/(ha year) in 2006 and −73 CNY/(ha year) and −79 CNY/(ha year) (N2) in 2007

of wheat straw, and 6.97–8.58 t/(ha year) of maize straw. As for cropland gas regulation: $O_2$ and $N_2O$ emissions were 24.99–28.64 t/(ha year) and 0.72–1.13 kg/(ha year), whereas $CO_2$ and $CH_4$ assimilations were 34.23–39.22 t/(ha year) and 3.39–5.70 kg/(ha year), respectively. While cropland SOM accumulation was 1.13–2.39 t/(ha year), that of water consumption was 2890–3830 $m^3$/(ha year). Soil nitrogen content dropped considerably at a rate of −107.73 to 5.33 kgN/(ha year) after one crop rotation. The TEV of cropland ESs was estimated at $5.48 \times 10^4$ to $6.25 \times 10^4$ CNY/(ha year), three times the value of food production. Based on these results, the effects of nitrogen fertilizer on the welfare of cropland ESs appear complicated. Nitrogen application led to economic loss due to increasing nitrogen transformation, simultaneously, increased economic value of primary production, gas regulation, $CO_2$ fixation, and $O_2$ release. Most ES studies have focused on the positive effects of ecosystems on human welfare, but a balanced and reasonable approach is to analyze the positive and negative effects of cropland ESs on human welfare (Table 2).

### 5.2.2 Ecosystem Services by Rice Paddy Ecosystems in Suburban Shanghai

The rice paddy is one of the most important farmland systems in China, responsible for more than one-third of total food production. Rice paddy ecosystems cannot only provide food, but also support many ESs such as gas regulation, water regulation, flood controlling, SOM accumulation, nutrient transformation, and environment purification. In 2002, we executed a field experiment in a rice paddy ecosystem at Wusi Farm, Shanghai to examine the ESs of rice paddies under different levels of nitrogen application (Xiao and Xie 2009). We investigated primary production, gas regulation, nutrient transformation, SOM accumulation, water regulation, and environment purification and their values. The results of primary production indicated that the rice seeds and stalks and their values increased with N application. Gas regulation showed that rice paddies absorbed $CO_2$ from the atmosphere; $O_2$ emissions and $N_2O$ emissions by rice paddies increased with N addition, but $CH_4$ emissions decreased with N addition. The economic value of gas regulation by rice paddies with 375 kgN/ha was the highest. Nitrogen transformation indicated that among different sources of nitrogen inputs, nitrogen application provided the most nitrogen input and the harvest and ammonia volatilization were the most important nitrogen outputs. According to this study, nitrogen transformation resulted in economic loss and these increased with nitrogen addition. SOM showed that rice planting increased the content of SOM, and the quantities of SOM accumulation and their values increased with N application. Rice cultivating consumed water resources, and the ridge of the rice paddies controlled flooding as a reservoir with a height of 5 cm. Comprehensive evaluation indicated that ESs by rice paddies in Wusi Farm provided many benefits to society in the range of 3.83–4.85 × $10^4$ CNY/ha. The ESs by rice paddies with 375 kgN/ha of N application yielded the highest economic value. However, the economic values of the ESs by these rice paddies did not have significant advantage over those without N application. The economic loss of environmental damage by N application comprised the benefits supported by rice paddies. Therefore, to maximize the benefits of ESs and the sustainability of rice production, the government should control the release of nitrogen fertilizer and improve nitrogen application technologies to improve the efficiency of nitrogen application (Table 3).

**Table 3** Assessment of ecosystem services from a rice paddy agro-ecosystem at Wusi Farm in 2002 CNY/ha

| Services types | N0 | N225 | N375 | N525 |
|---|---|---|---|---|
| Primary production | 6469 | 7694 | 11,769 | 13,181 |
| Gas regulation | 36,858 | 34,586 | 37,810 | 30,808 |
| N transformation | 628 | −5264 | −9956 | −14,992 |
| SOM accumulation | 3160 | 4073 | 8731 | 8936 |
| Water regulation | 1891 | 1891 | 1891 | 1891 |
| Environment purification | 941 | 941 | 941 | 941 |
| Integrated values | 48,322 | 41,250 | 48,518 | 38,290 |

*Note* In the integrated values of ecosystem services of the rice paddy agroecosystem, the value for $N_2O$ emissions was counted once, though it was calculated both for gas regulation and nitrogen transformation

### 5.2.3 Evaluation of Ecosystem Services Provided by 10 Typical Rice Paddies in China

Based on a reference review, this study investigated ESs supported by 10 typical rice paddies in six rice planting regions of China (Xiao et al. 2011). The services were primary production, gas regulation, nitrogen transformation, SOM accumulation, and water regulation and flood control. The results indicated that grain production of the 10 rice paddies was between 4.71 and 12.18 t/(ha year); straw production was 4.65–9.79 t/(ha year); gas regulation was calculated to emit $O_2$ ranging from 8.27 to 19.69 t/(ha year) and to assimilate greenhouse gases ranging from −2.13 to 19.24 t/(ha year) (in $CO_2$ equivalent); nitrogen transformation was estimated as nitrogen input from 209.70 to 513.93 kgN/(ha year) and nitrogen output of 112.87–332.69 kgN/(ha year); SOM accumulation was between 0.69 and 4.88 tC/(ha year); water regulation was estimated to consume water resources of 19,875 $m^3$/(ha year) and to support water resources of 6430 $m^3$/(ha year); and flood control of several of the rice paddies was calculated to be 1500 $m^3$/(ha year) (Table 4). The integrated economic value of ESs of these rice paddies was estimated at 8605–21,405 USD per ha per year, of which 74–89 % of the value can be ascribed to ESs outside primary production. The integrated economic value of the ESs of the 10 rice paddies was higher when nitrogen fertilizer was applied in the range of 275–297 kgN/(ha year). Until now, the economic value of the rice paddy ecosystem has been underestimated as only the economic value of grain and straw production were previously calculated. As more and more forestland and grassland

**Table 4** Integrated economic values per unit area of ecosystem services for 10 rice paddies in China

| Sites | Primary production | Gas regulation | Nitrogen transformation | SOM accumulation | Water regulation | Integrated economic values |
|---|---|---|---|---|---|---|
| Guangzhou | 3013 | 11,557 | 287 | 2538 | 444 | 16,525 |
| Changshu | 2665 | 15,095 | 317 | 852 | 387 | 18,351 |
| Chengdu | 2219 | 11,654 | −71 | 2063 | 385 | 15,407 |
| Taoyuan | 3564 | 14,735 | 264 | 3487 | 367 | 21,405 |
| Yingtan | 2556 | 4610 | 58 | 2509 | 359 | 9677 |
| Bijie | 2166 | 5486 | 497 | 1145 | 414 | 8605 |
| Fengqiu | 1484 | 7532 | −1498 | 494 | 387 | 8657 |
| Shenyang | 2622 | 14,353 | 381 | 739 | 32 | 17,002 |
| Hailun | 1810 | 13,687 | 632 | 979 | −22 | 16,338 |
| Lingwu | 2283 | 16,719 | −906 | 1213 | −204 | 18,878 |

*Note* All values are USD/(ha year). As values of grain and straw harvest in nitrogen transformation and primary production were recalculated, only the economic value of primary production was calculated in the integrated value. As the economic value of $N_2O$ emission was also recalculated in gas regulation and nitrogen transformation, only the economic value of $N_2O$ emission in gas regulation was calculated in the integrated value is lost to urban and industrial uses, cropland, especially rice paddies, will become more ecologically important to society. The economic value of ESs supplied by rice paddies, outside primary production, are therefore worthy of increased research attention.

### *5.3 Conclusion and Discussion*

In this section, we reviewed research on the ESs of agroecosystems and estimated the economic value of ESs of (1) wheat-maize croplands in Luancheng County, Hebei, (2) rice paddy fields in suburban Shanghai, and (3) 10 typical rice paddy fields across China. The results indicate that primary production was the core service delivered by these agroecosystems. Gas regulation, water conservation, nutrient maintenance and transformation, and environment purification also contribute to human well-being.

Agroecosystems support food and other critical ESs to people, but impair human well-being by the addition of chemical fertilizers, pesticides, and herbicides. This tradeoff between food production and the impassive effects remain a major challenge for the implementation of sustainable agriculture. Compared with natural ecosystems, intensive agro-systems supply much more food and relatively fewer ESs, such as gas regulation, water regulation, and nutrient transformation. The impassive effects of intensive agriculture on social-economic systems counteract the human well-being supplied by food and fiber production in the long run. Agroecosystems with a natural buffer belt and those that compromise primary production and other ESs are better choices for achieving primary products, other ESs and sustainable agriculture. The future direction of sustainable agriculture will be to manage agroecosystems by optimizing human well-being supported by both primary production and other ESs.

## 6 Assessment of Marine Ecosystem Services in China

### *6.1 Introduction*

ESs refer to the natural environmental conditions and their benefit formed by ecosystems through ecological processes, on which humankind relies on for survival (Daily 1997). Since the early 1990s, studies on ESs have rapidly progressed and become one of the dominant research topics and frontlines (Daily 1997; Costanza et al. 1997). The promulgation of the MEA by the United Nations in 2005 has drawn broad attention and concern from the governments of many countries and a wide variety of sectors of society. To date, almost every type of ecosystem on earth has been assessed and we have gradually recognized the great value of natural ecosystems to our long-term survival and development.

Marine ESs refer to the products and services provided for the survival and development of human society by specific marine ecosystems within a given period of time and through certain ecological processes. Compared to our knowledge of land systems, our recognition and understanding of seas and oceans are relatively poor. In particular, the complexity and uniqueness of marine ecosystems and the applicability of assessment methodologies means assessments of marine ESs and their values are extremely difficult.

### 6.1.1 Progress in Basic Research on Ecological Theories

The formation and realization of ESs requires the support of complex ecological processes. Changes in the size of an ecosystem are influenced by changes in biodiversity. Many ecologists have conducted in-depth research on relationships between ESs and biodiversity (Naeem and Li 1997; Loreau et al. 2001; Hemminga and Duarte 2000). A number of researchers have also focused their efforts on the formation mechanisms of ESs, including sources of ESs and methods for realizing their value (Zhang et al. 2006; Wu 2006) and the influence of human activity on ESs (Holmlund and Hammer 1999).

Naeem and Li (1997) found that biodiversity can strengthen the stability of ecosystems and that biodiversity can prevent population loss and functional reduction: the higher the species number, the more stable the ecological system. Hemminga and Duarte (2000) stated that the correlations between biodiversity and ecosystem functions/services are very complex. To study these correlations is very important to elucidate the impacts of biodiversity on ecological system function, and is also helpful when explaining reasons for protecting biodiversity rather than protecting individual species alone. Using seaweed as a case study, Hemminga and Duarte (2000) analyzed the supporting function of biodiversity for the function and services of various ecosystems. Holmlund and Hammer (1999) defined fish ESs as the maintenance of functions and elasticity of ecosystems and called these ecological services derivatives of human demands. Holmlund and Hammer (1999) stated that overexploitation of fishes globally not only reduces the total allowable catch and the capability of population regeneration but also threatens ecological services provided by fish populations, reduced biodiversity, ecological functions and our welfare.

Zhang et al. (2006) analyzed ESs supported by major ecological processes, including photosynthesis, respiration, biological pump functions, decomposition, mineralization, calcification, bioturbation processes, nitrogen cycle processes, biotransport, and bioabsorption. Referring to the classification methods of ESs developed by Costanza et al. (1997), Wu (2006) divided ecological service functions of a mangrove ecosystem in Guanxi, China into three types (resource function, environmental function, and humanity function) and analyzed in the formation processes for each service function.

### 6.1.2 Progress in Basic Research on Economic Theories

The evaluation of the value of ESs requires the support of economic theory. Costanza et al. (1997) analyzed the characteristics of ESs and stated that the total value of resources was the sum of producer surplus and consumer surplus. Due to the requirement of curve estimation for ESs and the difficulty in calculating consumer surplus, a contingent valuation method (CVM) based on public surveys on the willingness to spend on environmental quality improvement or willingness to accept the tolerance of environmental loss, plays an important role in evaluation of the nonmarket values of ESs (Loomis et al. 2000).

Stephen et al. (2002) expounded the economic and ecological meaning of values and corresponding methods for evaluation techniques. Values based on utility commodities and services reflect people's willingness to acquire them or the willingness of accepting compensation for giving them up, whereas values based on trade reflect the values of commodity and trade services. In the evaluation of market values, this value reflects the marginal values of the commodities or services. However, in the evaluation of the nonmarket values, indirect estimation methods are required. This method is new and a supplement method for the classical value theory or a replacement of energy value theory, and is also one of the methods for value evaluation on the natural assets recommended by ecologists and economists. Time can be also used as a means for value evaluation (Farber et al. 2002). The ecological footprint (EF) method based on material flow within ecosystems can be used to carry out calculations and assessments by converting a variety of human consumption and activity into land area (Rees 1992). Thus, evaluation methods for monetization are not the only standard method for the value evaluation of ESs. Due to convenience, this is the most common method used to evaluate the value of ESs.

Chinese researchers have conducted some research on economic theories for the assessment of marine ESs. Based on labor theory of value and utility theory of value, Wang et al. (2005) investigated the value theory of marine ecological resources. Zheng et al. (2006) analyzed the role of ESs in meeting human requirements from the angle of welfare economics. They stated that both the ESs and human welfare influenced and interacted mutually. Through definition and content analysis of ecological assets, Zheng et al. (2007) analyzed asset attributes from the aspects of revenue, scarcity, and rights.

### 6.1.3 Progress in Research on Value Assessment Methods

Types of Marine Ecosystem Service Values

Analysis and classification of the components of marine ESs are the basis for carrying out value evaluation of ESs. Pearce et al. (1989), McNeely et al. (1990), Turner et al. (2001) have conducted substantial research on the components of ESs and methods for assessment. Pearce et al. (1989) proposed a theory of the TEV of natural resources that includes utilization values, existence values, and selection values. The utilization values can be divided into direct utilization values and indirect utilization values. The selection values include the individual's future utilization values, other peoples' future utilization values and future generations' utilization values. McNeely et al. (1990) stated that biological resources can be divided into direct values and indirect values. The former can be further divided into consuming utilization values and producing utilization values, whereas the later can be further divided into nonconsuming utilization values, selection values, and existence values. When Turner et al. (2001) discussed the utilities of wetlands and their management, he divided the TEV of wetlands into utilization values and non-utilization values. The utilization values included direct utilization values, indirect utilization values, and selection values. The non-utilization values included existence values and heritage values. Pearce et al., McNeely et al., and Turner laid the foundation for theoretical research on the classifications of natural assets and the values of ESs. It is generally accepted that the TEVs of ESs can be divided into utilization values and the non-utilization values. The TEVs of marine ESs can be correspondingly divided into utilization values and non-utilization values also. Currently available assessment technologies can easily distinguish direct utilization values from nondirect utilization values; however, due to overlaps among selection values, heritage values and existence values, it is difficult to clearly distinguish these from one another.

Economic Assessment Methods for the Value of Ecosystem Services

The value assessment methods for marine ESs are mainly divided into three types: the conventional market assessment methods, replacement market assessment methods, and proposed market assessment methods. For each marine ES, several methods can be used to conduct assessments. Selection of the assessment method is determined by the characteristics, application scope of the assessment methods, and obtainability of data.

### 6.1.4 Applied Research

Research on the assessment of marine ESs has mainly been static assessments and few studies have addressed dynamic assessments. More research has been conducted on the specific types of ecological systems than on ecosystems at a regional scale. More applied research has been conducted on direct assessments than applied research on conclusive assessments.

Taking studies conducted by Chinese researchers as examples, in recent years Chinese scientists have begun to pay attention to applied research on the management of functional assessments of offshore-coastal ESs. For example, Peng and Hong (2006) established a series of ecological-economic models and conducted an assessment and applied research on the value evaluations of ESs for the Xiamen.

Shi et al. (2008, 2009a, b) examined bay and island ESs and methods of assessment for Sangguo bay (Shi et al. 2008) and the Miaodao Islands (Shi et al. 2009a, b). They constructed assessment models for detecting the sensitivity and uncertainty of ES values (Shi et al. 2009a, b). Zheng et al. (2009a, b) constructed cost–profit analytic models for marine aquaculture patterns based on the value theory of marine ESs, and provided a feasible framework for incorporating ESs and their value assessment into marine management.

## 6.2 *Characteristics of Marine Ecosystem Services and Their Values*

### 6.2.1 The Alien Land Realization of Marine Ecosystem Services

Compared to land ESs, the alien land realization of marine ESs is especially prominent. Due to the connectivity and dynamics of marine ecosystems, the values of marine ESs are not usually realized locally. For example, climate regulation and oxygen generation provided by marine ESs are usually realized at a global scale. Furthermore, organisms within marine ecosystems can swim and migrate at a larger scale and these characteristics result in marine ESs follow an obvious alien land realization (Zhang et al. 2006).

Alien land realization makes the value assessment of marine ESs difficult. Due to alien land realization of marine ESs, it is difficult to evaluate precisely what proportion is contributed by alien land during the specific assessment process. It is also difficult to determine the range of temporal and spatial scales for certain services.

### 6.2.2 Temporal and Spatial Scales of Marine Ecosystem Services and Regional Dependence

The formation of marine ESs is dependent on ecosystem structure and processes at certain spatial and temporal scales. Only within certain spatial and temporal scales can the ESs play a leading role and cause significant impacts (Zhang et al. 2007). Different types of ESs confer different utilities for humans in different temporal and spatial scales. For example, the material supply service of an ecosystem is usually more closely related to the interest of local residents, whereas the regulatory services and supporting services of the ecosystem are usually related to the interests of people at regional, national, and global scales. Cultural services are closely related to interest-related aspects at the local-global scale. How large the temporal scale the formation of marine ESs depends on is a difficult point that has attracted research attention. For example, ESs such as marine tourist/entertainment and climate regulation can be realized at regional and global scales, but the size of the selected scope of the scale of the assessment is an important factor that influences the size of the assessment results.

Marine ESs are displayed as utilities that marine ecosystems provide for people. However, the value realization of some marine ESs relies on local social and economic conditions. For example, different coastal leisure/entertainment values (an ES) at the same period of time differ. The leisure/entertainment values in the regions where population density is high and the economy is well developed are usually higher than for regions where the population density is low and the economy is less developed. Even for the same coast, due to changes in the levels of economic and social development over different periods of time, leisure/entertainment values also change. Thus, marine ESs have obvious characteristics of regional dependence.

## 6.3 *Problems with Value Assessments of Marine Ecosystem Services*

### 6.3.1 Deficiencies in Systematics, Theory, and Methods

Scientific systems for the theory and methods of value assessment of marine ESs have not been established. Currently, there is a deficiency in systemic studies on the formation mechanisms of marine ESs and routes for realizing service values. A majority of applied research in this area comprises simulated applications of theories and methods for land ESs. However, due to the systematics and dynamics of marine ecosystems they differ greatly from land ecosystems and the application of these methods influences the accuracy of the value assessment of marine ESs. Taking the value assessment of the nonmarket ESs as an example, due to differences in social and economic development status, public psychological characteristics, recognition level, understanding level and exploration and utilization intensity, hypothetical market assessment technologies (such as conditional value methods) widely applied to assessments on land ecosystems are less commonly applied to marine ecosystems due to large errors. This influences the value assessment of nonmarket ESs.

### 6.3.2 Further Improvements

Due to a lack of background multidisciplinary knowledge, our recognition of marine ecosystems is not adequately deepened. Additionally, it is difficult to acquire data and we usually place emphasis on assessment of the utilization values (the direct utilization values and the indirect utilization values) and not the assessment of non-utilization values (selection values, heritage values, and existence values). Thus, our value assessment of marine ESs is neither complete nor perfect.

Due to different regions, differences in possession of research data and differences in the understanding of the contents of marine ESs and the application of different assessment methods, the comparability of assessment results across studies is greatly reduced.

## *6.4 Conclusions and Discussion*

### 6.4.1 Basic Research on Mechanisms of Marine Ecosystem Services and Theories for Value Realization

The reason why current research on marine ESs is not adequately deepened is mainly due to the complexity of marine ecological processes and the relatively low level of marine exploration and utilization. We are still not clear what the sources of many marine ESs are and this leads to incorrect evaluations or omission of parts of ESs. Due to the characteristics of alien land realization and regional dependence, when we conduct value evaluation we must have a clear understanding of the routes and processes of realization of their values so that we can accurately grasp evaluation criteria. Thus, strengthening basic research on the formation mechanisms of marine ESs and theories for realization of their values is the only method for deepening and maturing our understanding of marine ESs.

### 6.4.2 The Uncertainty of Value Assessments

The accuracy of value assessments of marine ESs is consistently a tough problem in this field of research. In reviewing the entire assessment process, the main errors are outlined below.

#### Limitation of Assessment Methods

We do not have mature methods for the assessment of nonmarket values. Methods such as substitute cost, conditional value, shadow project, and TC are usually applied to evaluate nonmarket values. These methods have their own applicability scope and prerequisites. For example, when the substitute cost method is applied to assess the value of marine ESs and when certain ecosystems have two or more substitute products or services, determining which product or service is the most suitable becomes critical for whether the evaluation is accurate or not. When the conditional value method is applied to study a person's willingness to pay, the interviewed person's level of the recognition and understanding of ESs usually influence their willingness to pay. We always place our hope on reducing errors through larger samples; however, these errors cannot be eliminated entirely this way.

The Weak Reliability of the Selected Assessment Criteria

Spatial heterogeneity exists between the supply of marine ESs and value realization. During the assessment process, due to the limitation of available data, criteria used for assessment of ecosystems in other regions are frequently cited. Attention should be paid to conversion at certain temporal and spatial scales to avoid errors.

Differences in Researcher Recognition of Marine Ecosystem Services

Differences in researcher recognition of marine ESs are an important factor causing assessment errors. Different researchers have conducted classifications and defined assessment methods based on their own understanding of the ecosystem and no uniform criteria for value assessment exist.

We must quickly establish the accuracy of value assessments for marine ESs, perform comprehensive assessments on the distribution characteristics of a variety of key factors and conduct research into uncertainty, so that we can provide a relatively complete description for regional marine ESs and their value.

### 6.4.3 Applied Research on Management Based on the Values of Marine Ecosystem Services

Research on marine ESs has mainly been static assessments at different spatial scales, which do not truly reflect the responses and feedback of marine ecosystem to man-made interference. An assessment of ESs forms a bridge between our recognition of nature and ecosystems and management decisions. Assessments are also the basis for ecological protection, ecological restoration, and ecosystem management (Ouyang 2007). The goal of marine management based on ecosystems is to make the supply capability of the ESs optimal through standardization and optimization of regional human activity. Thus, to conduct research on the responses of, and feedback from, marine ESs to human interference is needed to transition research into a decision-making process.

Due to the limitation of our recognition of marine ecosystems and assessment methods, our current attempt to conduct a complete and reliable value evaluation on ESs is impossible to realize. The purpose of assessment on marine ESs is mainly to uniformly incorporate nature and ecosystems into our economic system; the ultimate goal is to apply the values of ESs. Often, we study only a few ESs that have already been influenced by human activities and this could be a focal point for our research on marine ESs into the future.

# 7 Assessment of Regional Ecosystem Services in China

At the beginning of the twenty-first century, the impact of ecological degradation on human welfare and economic development increased at a global scale. The MEA was an important international cooperation plan initiated by the United Nations at a critical turn of the century aimed at depicting the health condition of the earth's ecosystem through international cooperation and meeting the scientific requirements of decision-makers. The MEA was announced by the Secretary General of the United Nations, Kofi Annan, in June 2001 and the assessment report was officially released on March 30, 2005.

The MEA is an integrated assessment on ecosystem and human well-being at multiple spatial scales, such as the community, basin, nation, region, and globe. Due to high variability of ecosystems across space and time, the MEA deployed a series of Sub-Global Assessments at all scales to ensure the evaluation results were helpful for the effective management of ecosystems.

The Chinese Ministry of Science utilized the MEA scientific conceptual framework and ecosystem assessment methods during the successful implementation of the Western China Development Strategy and consequently launched the Western China Ecosystem Assessment (MAWEC) as part of a Sub-Global Assessment of the MEA. In April 2001, the MAWEC project was officially recognized by the MEA as one of five first draft Sub-Global Assessment projects. In March 2005, the MAWEC and MEA reports were simultaneously released in Beijing.

The integrated ecosystem assessment of western China was based on the framework of the MEA and proved to be a comprehensive assessment of the status quo, evolution law and future scenarios of macrostructures and service functions of ecosystem in western China. This platform was also a driving mechanism of ecosystem change and the relationship between ecosystems and human well-being.

The MAWEC developed an integrated ecosystem process simulation and earth information science system using multitier ecological zoning and information systems to develop models for analyzing ESs. Data collected from nine areas representative of 'typical' ecologies and socioeconomic conditions were used for in-depth assessment and modeling at the local level. Trends and scenarios for water and food supply services of various ecosystems, carbon storage, and biodiversity were analyzed by combining developed models with geographical information systems (GIS). Additionally, a method of surface modeling population spatial distribution (SMPD) was developed to analyze the relationship between ESs and human well-being. Integration of multi-scale information and discrepancies with GIS data were handled through the development of high precision surface modeling (HPSM).

Through in-depth research in these nine typical areas, MAWEC revealed key conflicts between ESs and human well-being in different ecological zones and systems, refined some effective human–ecosystem relationship optimization models, and offered good examples to guarantee the sustainability of western China ecosystems.

The results of the integrated ecosystem assessment of western China show that due to the driving forces of climate change and human activities, in the past 20 years all kinds of ecosystems in western China have experienced different degrees of degradation. For example, the area of permanent snow and ice continued to decrease and desert area increased, but ecosystem diversity increased (important carbon sequestration in China). As for the carrying capacity of ecosystems, the surrounding metropolitan areas (parts of Gansu, Shaanxi, Ningxia, and Guizhou) showed signs of overload, but the western area on the whole exhibited marginal potential carrying capacity.

The results of the future scenario analysis showed that in the next 50–100 years, due to increasing biological temperature in western China, rainfall will increase. In addition to human-driven systematic ecological restoration and reconstruction, ecosystem diversity and forest cover will increase, productivity for all kinds of ecosystems will increase and carbon sequestration will strengthened. However, rising temperatures will lead to a constant shrinking of the permanent snow zone on the Qinghai Tibet Plateau and northwest China and the area of desert will expand at a low rate.

The food supply function of ecosystems in western China will rise on the whole. Food area in the northwest increases greatly. The food supply potential of eastern parts of Inner Mongolia, Guangxi, Yunnan, and the Qinghai Tibet Plateau will fall slightly or experience less obvious change. Though there are increases in other regions, the increase was significantly lower than that of northwest China. According to analysis of population growth, the carrying capacity of ecosystems in western China guarantees population development over the next century.

In the next 50–100 years, if China engages in effective ecosystem protection and restoration according to western China development plans, ecosystems in western regions will show a benign development trend. If the intensity of human activities goes beyond the regulation range of ecosystems, western ecosystems will maintain their current trend of degradation and this may even accelerate.

In response to global environmental change and intensified human activity affecting ecosystems in western China, the following policy stages are necessary:

(1) According to ecological function zoning, conduct regional targeted ecological construction, guide the behavior of people in particular ecological zones and ecological systems, and avoid blind development.
(2) Adopt building a water-saving society as a basic strategy. Adapt measures to local conditions to develop the utilization and protection of water resources. For example, develop and utilize water resources of international rivers properly; develop soil reservoirs, forest reservoirs, and reservoirs; change slopes to terraces or flat farmland; conserve water through farmland; and conduct afforestation and water conservation.
(3) Implement desertification control projects and protect vegetation. Protect existing vegetation and strictly prohibit random deforestation, disordered mining,

farming, and grazing. Define ecological imbalanced areas which are difficult to restore or are sharply deteriorating as 'Depopulated zones' or 'No animal areas' in order to facilitate ecological balance and recovery in these regions.

(4) Ecological environmental construction is a long-term and complicated systematic project. Given the scale of households over a vast area, simple government investment and fiscal subsidy policy is insufficient for change: there is an urgent need to establish a complete policy system that ensures ecological environmental construction in western China. In particular, all economic activity in the desertification area requires judicial management and strict monitoring.

(5) Taking water and soil resources carrying capacity and ecological environment capacity as the premise and the structure-function-balance-efficiency principle of the ecosystem, optimize system structure and enhance system function; combine ecological construction with economic development; vigorously promote industrialization of agriculture; and cultivate ecological economic systems to break the vicious spiral of vulnerability-poverty and fragility.

(6) Implement ecological compensation mechanisms across China. Long-term investment of people, money and materials are needed in the middle and upper reaches of rivers to implement ecological management and protect vegetation and water and enhance ecological environmental security downstream; sacrifice economic interests to some degree in these areas. The nation should take responsibility, duty and authority into consideration. Establish comprehensive economic compensation mechanisms in ecosystem function benefit areas and ecosystem function protection and provision zones, and realize common prosperity in eastern and western China.

# 8 Consumption of Ecosystem Services

## *8.1 Introduction*

According to the MEA, approximately 60 % of ESs are degraded or used unsustainably and human activity is a major driver of these changes (MEA 2005). In recent decades, research has focused on ES studies and especially the concepts and theories underpinning this concept (Daily et al. 2000; Ouyang et al. 1999a, b) and economic assessments of global, regional, and typical ES (Costanza et al. 1997; Turner et al. 2003; Ouyang and Wang 2000; Zhao et al. 2000; Xie et al. 2001). The MEA has addressed the interrelationship between ES changes and human well-being and this has become a way forward for ES studies (MEA 2003).

Given recent population growth, rapid economic growth and associated changes in lifestyle, the human utilization and consumption of goods and services provided by natural ecosystems are changing rapidly. Here, we aim to summarize research progress on human consumption of ES and highlight major challenges into the future.

## 8.2 Conceptualization and Measurement of Ecosystem Service Consumption

Consumption is a process whereby humans use goods and services to meet basic needs and demands for survival (Shen, 2008). Ecosystem service consumption (ESC) is defined as the human utilization and occupation of natural ecosystems' goods and services in order to meet demands for production and living (Zhen et al. 2010). Therefore, ESC includes individual consumption and production consumption; the latter refers to the utilization of other ES in order to produce certain kinds of ES, for instance, food provision service depends on soil formulation, soil and water conservation, and other services. Individual consumption refers to our daily consumption, utilization, and occupation of ES. Consumption behavior significantly influences human consumption features, which can be divided into the six aspects of who, when, what, where, why, and how (i.e., 5W1H). Due to differences in consumption behavior, different CPs can be identified. CP reflects main consumption features over a certain time period, and includes components of consumption, level of consumption, structure of consumption, approach of consumption and trends in consumption, expressed as follows (Zhen et al. 2008):

$$E_c = E_{dc}\left(f(X_a, X_{hz}, X_{gen}, X_{edu}, X_{bev}, X_{cmr}, X_{inc}, \ldots, X_n) + E_{idc}\right)$$

where $E_c$ = total ES consumption; $E_{dc}$ = direct ES consumption, of which, $X_a$: accessibility to ES, $X_{hz}$: household size; $X_{gen}$: sex; $X_{edu}$: educational attainment; $X_{bev}$: consumption behavior; $X_{cmr}$: type of consumers; $X_{inc}$: income of consumers; and $E_{idc}$ = indirect consumption, which includes consumption of regulation and supporting services like water and soil conservation and soil formation for the production of goods and services.

Both physical and monetary assessments can be used to evaluate ESC. (1) Physical measurements directly assess the actual quantity consumed and are commonly used for measuring food, fuel wood, water, raw materials, medicine, and other goods consumed. Some ES are indirectly consumed for producing directly consumed goods and services and therefore the total consumption of ES should include both of direct and indirect consumption. For example, for food consumption, we need to calculate both the food quantity consumed and other ES such as the water and soil nutrients necessary for food production. (2) Monetary assessment is normally used for marketable goods where market prices are available and values can be calculated. For ES that do not yet have market value, alternative methods can be used such as the CVM to determine willingness to pay and willingness to accept by stakeholders through intensive questionnaires measuring landscape amenity. Those are only few of the commonly used methods for ESC, however, specific methods for ESC quantification would need to be identified and applied during actual case studies.

## 8.3 Case Studies of the Consumption of Ecosystem Services

### 8.3.1 Consumption Pattern and Accessibility

CPs are affected by the availability and accessibility of ES. Based on household questionnaires in the Mongolian Plateau in 2006, we found that herders relied heavily on ES produced locally such as meat, milk, and products (Zhen et al. 2010) (Table 5). To reduce risks, herders normally combine different consumption items and diversify their consumption. Mongolian herders tend to accept diversified meat consumption like mutton, beef, and horse meat, while Inner Mongolian herders used different fuels such as fuel wood, crop straw, and animal dung. Accessibility to ES is affected by distance, time, and the costs involved in obtaining the ES. We selected fuel wood consumption to investigate the influence of accessibility to fuel wood resources on consumption in the upper stream of the Jinghe watershed on the Loess Plateau. Through surveys, we found that fuel wood collection intensity decreased from 2000 to 2005 as some households had given up fuel wood collection due to poor accessibility.

In recent years, an increasing number of Chinese studies have focused on relationships between ES production and consumption. The EF has been widely applied to the calculation of land resources required for producing consumption items (Xu et al. 2002; Jiao et al. 2010). Dong et al. (2006) estimated the balance between ES

**Table 5** Consumption patterns of ecosystem services on the Mongolian Plateau

| Consumption pattern | | Proportion of household surveyed (%) | |
|---|---|---|---|
| | | Mongolia ($N$ = 150) | Inner Mongolia ($N$ = 102) |
| Meat | Mutton | 4.7 | 8.3 |
| | Beef | 0.7 | 13.9 |
| | Mutton + beef | 15.4 | 75.0 |
| | Mutton + horse meat | 13.4 | 0 |
| | Beef + horse meat | 0.7 | 0 |
| | Mutton + beef + horse meat | 64.4 | 0 |
| Dairy | Milk | 33.6 | 83.3 |
| | Butter | 2.0 | 0 |
| | Milk + butter | 58.4 | 0 |
| Fuel | Fuel wood | 18.8 | 0 |
| | Dried grass | 0.7 | 0 |
| | Animal dung | 38.9 | 27.8 |
| | Fuel wood + straw | 0 | 5.6 |
| | Fuel wood + animal dung | 19.5 | 44.4 |
| | Fuel wood + coal | 2.0 | 0 |
| | Dried grass + animal dung | 0 | 11.1 |
| | Straw + animal dung | 2.7 | 8.3 |

*Source* Zhen et al. (2010)

supply and consumption in Taicang city, Jiangsu using the methods of alternative cost, shadow price, and protection cost and found that total production and consumption remained balanced from 1996 to 2001, but after 2001 consumption increased quickly leading to a rapid decline in net value between production and consumption. Consequently, total consumption exceeded total production.

### 8.3.2 Utility Function Derived from ES Consumption

The ultimate purpose of consumption is to achieve maximum utility function (UF). In ES consumption studies, UF refers to the consumption behavior, preferences, and choices of consumers (Lui 2009). The Cobb–Douglas function is adopted to analyze UF of imperfect substitution consumption types (Varian and Repcheck 2006) and the main variables used for the analysis include quantity of the items consumed and preference (represented by cost). This method identifies the appropriate and alternative combinations of different consumption items if certain UF is desired (Fig. 5). Based on household survey data in the Jinghe watershed, we found that when the consumption of grains, vegetables and fruits, and meat, egg, and milk were 200, 200, and 26.6 kg, respectively, the UF was 123. However, when grain consumption remained unchanged, vegetable and fruit consumption declined by 100 kg and an additional 102.4 kg of meat, egg, and milk was needed in order to maintain the same UF, meaning the total cost will be increased by 546 CNY.

Expenditure on food is important for UF. An ideal UF can be reached through readjustment of the quantity of each of the food items consumed (Table 6) and this is useful when identifying a rational consumption structure with high UF but with lower cost and resource use.

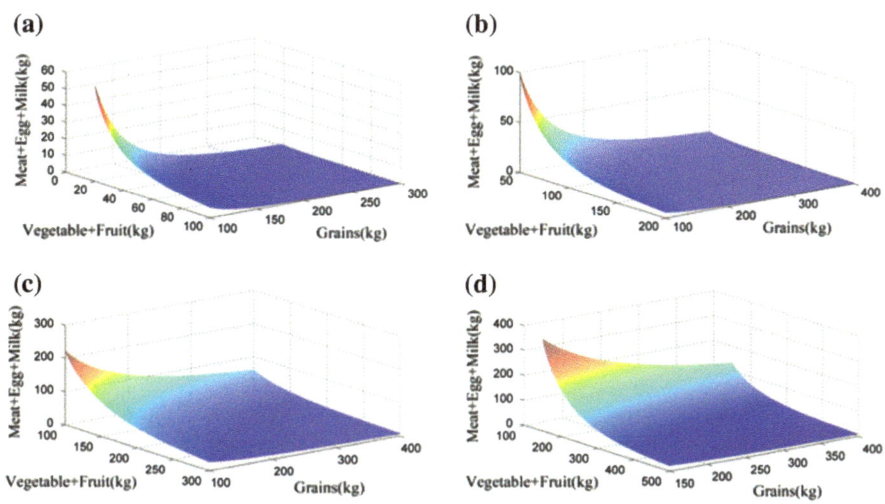


**Fig. 5** The utility function

**Table 6** Utility function and associated consumption patterns under different expenditure levels

| Expenditure (CNY/year) | ≤500 | | 500–1000 | | 1000–1500 | | >1500 | |
|---|---|---|---|---|---|---|---|---|
| UF/person | 60 | | 78 | | 125 | | 179 | |
| Consumption/person year | AC[a] | MUFC[b] | AC | MUFC | AC | MUFC | AC | MUFC |
| Grains (kg) | 162 | 151 | 200 | 184 | 221 | 221 | 275 | 246 |
| Vegetable + fruit (kg) | 30 | 27 | 95 | 76 | 194 | 177 | 309 | 283 |
| Meat, egg and milk (kg) | 8.4 | 6.4 | 17.9 | 17.6 | 37.2 | 32.7 | 46.9 | 43.2 |
| Actual expenditure (CNY/year) | 393 | 353 | 726 | 639 | 1216 | 1127 | 1727 | 1573 |

*Source* Household survey in the Jinghe watershed 2008
[a]Actual consumption
[b]Consumption under MUF

## 8.4 Conclusion and Discussion

Human consumption of ES significantly impacts natural ecosystems, and it is important to examine interactions between ESs and socioeconomic systems (Li 2008). Here, we conclude that humans directly (e.g., provisioning services) and indirectly (e.g., regulation and cultural services) consume ES and that both physical and monetary methods are being used for these assessments. CPs are affected by the availability and accessibility of ES. The UF can be achieved through different combinations of ES types and quantities consumed.

Many challenges exist in this field of study. For example, there would be cross-calculations on indirect consumption when considering food consumption for items such as water and soil conservation, soil formation, primary production, and nutrition circulation. It is necessary to explore the relationship between production and consumption regions when consumed ES are transported from other regions and in-depth case studies are required here. The UF needs to be further investigated to establish functional relationships between consumption and its utility in order to maximize resource use while meeting our basic consumption needs.

## References


Adger, N., Brown, K., Cervigni, R., & Moran, D. (1995). Total economic value of forests in Mexico. *Ambio, 24*(5), 286–296.

Anielski, M., & Wilson, S. (2005). *Counting Canada's natural capital: Assessing the real value of Canada's boreal ecosystems*. Ottawa, Canada: Canadian Boreal Initiative and Pembina Institute.

Arrow, K., Bolin, B., Costanza, R., et al. (1995). Economic growth, carrying capacity and the environment. *Science, 268*, 520–521.

Canton, Y., Domingo, F., Sole-Benet, A., & Puigdefábregas, J. (2001). Hydrological and erosion response of a badlands system in semiarid SE Spain. *Journal of Hydrology, 252*, 65–84.

Carpenter, S. R., Mooney, H. A., Agardc, J., Capistranod, D., DeFriese, R., Díazf, S., et al. (2009). Science for managing ecosystem services: Beyond the millennium ecosystem assessment. *Proceedings of the National Academy of Sciences, 106*, 1305–1312.

Costanza, R., d'Arge, R., de Groot, R., Farber, S., Grasso, M., Hannons, B., et al. (1997). The value of the world's ecosystem services and nature. *Nature, 387*, 253–260.
Merlo M., & Croitoru, L. (2005). *Valuing mediterranean forests: Towards total economic value*. Wallingford: CABI.
Daily, G. C. (1997). *Nature's services: Societal dependence on natural systems*. Washington, D. C.: Island Press.
Daily, G. C., & Matson, P. A. (2008). Ecosystem services: From theory to implementation. *Proceedings of the National Academy of Sciences, 1*(05), 9455–9456.
Daily, G. C., Soderqvist, T., Aniyar, S., Arrow, K., Dasgupta, P., Ehrlich, P., et al. (2000). The value of nature and the nature of value. *Science, 289*(5478), 395–396.
De Groot, R. S., Wilson, M. A., & Boumans, R. M. J. (2002). A typology for the classification, description, and valuation of ecosystem functions, goods, and services. *Ecological Economics, 41*, 393–408.
Dong, J., Bao, C., & Shu, T. (2006). Analysis on the balance between supply and consumption of ecosystem services. *Acta Ecologica Sinica, 26*(6), 2001–2010.
FAO (Food and Agriculture Organization). (2010). *Global forest resources assessment*. Rome: FAO.
Farber, S. C., Costanza, R., & Wilson, M. A. (2002). Economic and ecological concepts for valuing ecosystem services. *Ecological Economics, 41*(3), 375–392.
Fei, S., Peng, Z., & Zhou, J. (2004). A discussion on the compensation of forest ecological benefit. *Scientia Silvae Sinicae, 40*, 171–179.
Fisher, B., Turner, K. M., Zylstra, M., Brouwer, R., de Groot, R., Farber, S., et al. (2008). Ecosystem services and economic theory: Integration for policy-relevant research. *Ecological Applications, 18*, 2050–2067.
Foley, J. A., DeFries, R., Asner, G. P., Barford, C., Bonan, G., Carpenter, S. R., et al. (2005). Global consequences of land use. *Science, 309*, 570–574.
Fowler, C., & Mooney, P. (1990). *Shattering: Food, politics and the loss of genetic diversity*. Tucson, USA: University of Arizona Press.
Hemminga, M.A., & Duarte, C.M. (2000) *Seagrass ecology*. New York, USA: Cambridge University Press.
Holmlund, C., & Hammer, M. (1999). Ecosystem services generated by fish populations. *Ecological Economics, 29*, 253–268.
International Union for the Conservation of Nature (IUCN). (2004). *2004 IUCN red list of threatened species: A global species assessment*. Gland, Switzerland and Cambridge, UK: IUCN.
Jiao, W., Min, Q., Cheng, S., Zhen, L., & Liu, X. (2010). Measurement of ecosystem services consumption: a case study of the traditional agricultural area in Congjiang County of Guizhou Province. *Acta Ecological Sinica, 30*(11), 2846–2855.
Köel-Knabner, I., Amelung, W., Cao, Z., Fiedlerd, S., Frenzele, P., Jahnf, R., et al. (2010). Biogeochemistry of paddy soils. *Geoderma, 157*, 1–14.
Krieger, D. J. (2001). *The economic value of forest ecosystem services: A review*. Washington, D. C., USA: Wilderness Society.
Li, W. (2008). *The assessment theory, methods and application of ecosystem service value*. Beijing: Renmin University of China Press.
Li, W., Ouyang, Z., & Zhao, J. (2002). *Study of ecosystem services*. Beijing: Meteorological Press.
Liang, X., Chen, Y., Li, H., Tian, G., Ni, W., He, M., & Zhang, Z. (2007). Modeling transport and fate of nitrogen from urea applied to a near-trench paddy field. *Environmental Pollution, 150*, 313–320.
Liu, Z. (2004). Carbon stock and GHG emission of wetland ecosystem. *Scientia Geographica Sinica, 24*(5), 634–639.
Liu, H. (2005). *Wetlands landscape changes and the environmental effects*. Beijing: Science Press.
Liu, X. (2007). Water storage and flood regulation functions of marsh wetland in the Sanjiang Plain. *Wetland Science, 5*(1), 64–68.

Liu, J., Dietz, T., Carpenter, S. R., Folke, C., Alberti, M., Redman, C. L., et al. (2007). Coupled human and natural systems. *AMBIO: A Journal of the Human Environment, 36*(8), 639–649.
Liu, X., Dong, G., Zhang, Y., Lu, X., & Jiang, M. (2013). Contribution to the global warming mitigation of marshlands conversion to croplands in the Sanjiang Plain, Northeast China. *Clean-Soil Air Water, 41*(4), 319–324.
Liu, X., Lu, X., Jiang, M., Shang, L., & Wang, X. (2008). Research on the valuation of wetland ecosystem services. *Acta Ecologica Sinca, 28*(11), 5625–5631.
Loomis, J., Kent, P., Strange, L., Fausch, K., & Covich, A. (2000). Measuring the total economic value of restoring ecosystem services in an impaired river basin: results from a contingent valuation survey. *Ecological Economics, 33*(1), 103–117.
Loreau, M., Naeem, S., Inchausti, P., Bengtsson, J., Grime, J. P., Hector, A., et al. (2001). Biodiversity and ecosystem functioning: Current knowledge and future challenges. *Science, 294*, 804–808.
Lu, X. (2004). *Wetlands ecosystem conservation and management*. Beijing: Chemical Industry Press.
Ludwig, J. A., Wilcox, B. P., Breshears, D. D., Tongway, D. J., & Imeson, A. (2005). Vegetation patches and runoff-erosion as interacting ecohydrological processes in semiarid landscapes. *Ecology, 86*(2), 288–297.
Lui, X. (2009). *Consumption of ecosystem goods and services: With a case study in Jinghe Watershed of northwestern China*. Beijing: Graduate University of Chinese Academy of Sciences.
Ma, X., Lu, X., & Yang, Q. (1996). The carbon cycle of Sanjiang Plain. *Scientia Geographica Sinica, 16*(4), 323–330.
MälerK, G., Aniyar, S., & Jansson, Å. (2008). Accounting for ecosystem services as a way to understand the requirements for sustainable development. *Proceedings of the National Academy of Sciences, 105*, 9501–9506.
Marques, M. J., Bienes, R., Perez-Rodriguez, R., & Jimenez, L. (2008). Soil degradation in central Spain due to sheet water erosion by low-intensity rainfall events. *Earth Surface Processes and Landforms, 33*(3), 414–423.
Mayer, A. L., Kauppi, P. E., Angelstam, P. K., Zhang, Y., & Tikka, P. M. (2005). Importing timber, exporting ecological impact. *Science, 308*, 359.
McNeely, J.A., Miller, K.R., Reid, W.V., Mittermeier, R.A., & Werner, T.B. (1990) *Conserving the World's Biological Diversity*. Washington, D.C.: IUCN, WRI, WWF and World Bank
Mendelsohn, R., & Balick, M. (1995). The value of undiscovered pharmaceuticals in tropical forests. *Economic Botany, 49*(2), 223–228.
Millennium Ecosystem Assessment (MEA). (2003). *Ecosystems and human well-being: A framework for assessment*. Washington, D.C.: Island Press.
Millennium Ecosystem Assessment (MEA). (2005). *Ecosystems and human well-being: Synthesis*. Washington, D.C.: Island Press.
Miller, K., & Tangley, L. (1991). *Trees of life: Saving tropical forests and their biological wealth*. Boston: Beacon Press.
Min, Q., Xie, G., Hu, D., Shen, L., & Yan, M. (2004). Service valuation of grassland ecosystem in Qinghai province. *Resource Science, 26*(3), 56–60.
Mooney, H. A., & Ehrlich, P. R. (1997). Ecosystem services: A fragmentary history. In G. C. Daily (Ed.), *Nature's services: Societal dependence on natural ecosystems*. Washington, D. C.: Island Press.
Naeem, S., & Li, S. (1997). Biodiversity enhances ecosystem reliability. *Nature, 390*, 507–509.
OECD/IUCN. (1996). *Guidelines for aid agencies for improved conservation and sustainable use of tropical and sub-tropical Wetlands*. Paris: OECD.
Ouyang, Z. (2007). *China's ecological construction and sustainable development (in general program for China's sustainable development)* (Vol. 11). Beijing: Sciences Press.
Ouyang, Z., & Wang, R. (2000). Ecosystem services and their economic valuation. *World Sci-tech Research and Development, 22*(5), 45–50.

Ouyang, Z., Wang, X., & Miao, H. (1999a). A primary study on Chinese terrestrial ecosystem services and their ecological-economic values. *Acta Ecologica Sinica, 19*(5), 607–613.
Ouyang, Z., Wang, R., & Zhao, J. (1999b). Ecosystem services and their economic valuation. *Chinese Journal of Applied Ecology, 10*(5), 635–640.
Pan, (1999). Study on carbon reservoir in soils of China. *Bulletin of Science and Technology, 15* (5), 330–332.
Pan, Y., Birdsey, R. A., Fang, J., Houghton, R., Kauppi, P. E., Kurz, W. A., et al. (2011). A large and persistent carbon sink in the world's forests. *Science, 333*, 988–993.
Pearce, D., Markandya, A., & Barbier, E. (1989). *Blueprint for a Green Economy*. London: Earthscan.
Pearce, D. W. (1998). Can non-market values save the tropical forests? In B. Goldsmith (Ed.), *Tropical rain forest: A wider perspective*. London: Chapman and Hall.
Pearce, D. W. (1999). Can non-market values save the world's forests? In S. Roper & A. Park (Eds.), *The living forest: The non-market benefits of forestry*. London: The Stationery Office.
Peng, B., & Hong, H. (2006). *Evaluation on coastal ecosystem services-theory and applied research*. Beijing: China Ocean Press.
Pimentel, D., & Kounang, N. (1998). Ecology of soil erosion in ecosystems. *Ecosystems, 1*, 416–426.
Porter, J., Costanza, R., Sandhu, H., Sigsgaard, L., & Wratten, S. (2009). The value of producing food, energy, and ecosystem services within an agro-ecosystem. *Ambio, 38*, 186–193.
Rees, W. E. (1992). Ecological footprint and appropriated carrying capacity: What urban economic leaves out. *Environment and Urbanization, 4*(2), 121–130.
Sandhu, H. S., Wratten, S. D., Cullen, R., & Case, B. (2008). The future of farming: The value of ecosystem services in conventional and organic arable land: An experimental approach. *Ecological Economics, 64*, 835–848.
SCBD. (2008). *Subsidiary body on scientific, technical and technological advice (13/3): In-depth review of the expanded programme of work for forest biological diversity*. Accessed at http:// www.cbd.int/doc/meetings/sbstta/sbstta-13/official/ sbstta-13-03-en.pdf
Shen, M. (2008). *Eco-economics*. Beijing: China Environmental Science Press.
Shi, H., Zheng, W., Ding, D., & Lv, J. (2009a). Valuation and ecosystem services of typical island. *Marine Environmental Sciences, 28*(6), 743–748.
Shi, H., Zheng, W., Ding, D., Lv, J., & Zhang, X. (2008). Valuation of typical marine ecosystem services—A case study in Sanggou Bay. *Marine Environmental Sciences, 27*(2), 100–104.
Shi, H., Zheng, W., Wang, Z., & Ding, D. (2009b). Sensitivity and uncertainty analysis of regional marine ecosystem services value. *Journal of Ocean University of China (Oceanic and Coastal Sea Research), 8*(2), 150–154.
State Forestry Administration. (2009a). The seventh national forest inventory (2004–2008). Beijing: China Forestry Press.
State Forestry Administration. (2009b). *China forestry statistical year book 2008*. Beijing: China Forestry Press.
Stephen, C., Lester, S., Black, W., Fyfe, M., & Raverty, S. (2002) Multispecies outbreak of cryptococcosis on southern Vancouver Island, British Columbia. *Canadian Veterinary Journal, 43*, 792–794.
Sun, C., & Chen, L. (2006). *A study of policies and legislation affecting payments for watershed services in China*. London, UK: Research Center of Ecological and Environmental Economics Beijing, and International Institute for Environment and Development.
Swinton, S. M., Lupi, F., Robertson, G. P., & Hamilton, S. K. (2007). Ecosystem services and agriculture: Cultivating agricultural ecosystems for diverse benefits. *Ecological Economics, 64*, 245–252.
Tilman, D., Cassman, K. G., Matson, P. A., Naylor, R., & Polasky, S. (2002). Agricultural sustainability and intensive production practices. *Nature, 418*, 671–677.
Tilman, D., Fargione, J., Wolff, B., D'Antonio, C., Dobson, A., Howarth, R., et al. (2001). Forecasting agriculturally driven global environmental change. *Science, 292*, 281–284.

Torras, M. (2000). The total economic value of Amazonian deforestation, 1978–1993. *Ecological Economics, 33*, 283–297.
Turner, R. K., Bateman, I. J., & Adger, W. N. (2001). *Economics of coastal and water resources: Valuing environmental functions*. Dordrecht: Kluwer.
Turner, R. K., Paavola, J., Cooper, P., Farber, S., Jessamy, V., & Georgiou, S. (2003). Valuing nature: Lessons learned and future research directions. *Ecological Economics, 46*(3), 493–510.
Uk, N. E. A. (2011). *National ecosystem assessment: Synthesis of the key findings*. Cambridge: UNEP-WCMC.
Varian, H. R., & Repcheck, J. (2006). *Intermediate microeconomics: A modern approach* (6th ed.). Shanghai: Shanghai People's Publishing House.
Wall, D. H., Bardgett, R. D., & Kelly, E. F. (2010). Biodiversity in the dark. *Nature Geoscience, 3*, 297–298.
Wang, D. (2005). Japanese compensation systems for forest ecological benefits and recent implementation. *World Forestry Research, 18*, 65–70.
Wang, M., Liu, X., Li, H., & Duan, Z. (2005). Discussion on the price theory of marine ecological resources. *Marine Sciences, 29*(1), 43–47.
Wang, B., & Yang, F. (2008). *Specification for assessment of forest ecosystem services in China*. Beijing: Chinese Standard Publisher.
WGMEA (Working Group of the Millennium Ecosystem Assessment). (2003a). *Ecosystems and human well-being: Biodiversity synthesis*. Washington, Covelo, London: Island Press.
WGMEA (Working Group of the Millennium Ecosystem Assessment). (2003b). *Ecosystems and human well-being: Wetlands and water*. Washington, Covelo, London: Island Press.
WGMEA (Working Group of the Millennium Ecosystem Assessment). (2003c). *Ecosystems and human well-being: Desertification synthesis*. Washington, Covelo, London: Island Press.
Wood, S., Sebastian, K., & Scherr, S. J. (2000). *Pilot analysis of global ecosystems: Agroecosystems*. Washington, D.C.: International Food Policy Research Institute and World Resources Institute.
Wu, S. (2006). *The functional assessment on mangrove ecosystem services in Guanxi province and their value evaluation*. Guilin, Guangxi: Guanxi Normal University.
Xiao, Y., An, K., Xie, G., & Lu, C. (2011). Evaluation of ecosystem services provided by 10 typical rice paddies in China. *Journal of Resources and Ecology, 2*, 328–337.
Xiao, Y., & Xie, G. (2009). Comprehensive valuation of the ecosystem services of rice paddies in Shanghai. *Resources Sciences, 31*, 38–47.
Xiao, Y., Xie, G., Lu, C., Ding, X., & Lv, Y. (2004). The gas regulation function of rice paddy ecosystems and its value. *Journal of Natural Resources, 19*(5), 617–623.
Xiao, T., Yang, Q., Ran, W., Xu, G., & Shen, Q. (2010). Effect of Inoculation with arbuscular mycorrhizal fungus on nitrogen and phosphorus utilization in upland rice-mung bean intercropping system. *Agricultural Sciences in China, 9*, 528–535.
Xie, G., Lu, C., & Leng, Y. (2003). Ecological assets valuation of the Tibetan Plateau. *Journal of Natural Resources, 18*(2), 189–196.
Xie, G., Xiao, Y., & Zhen, L. (2005). Study on ecosystem services value of food production in China. *Chinese Journal of Eco-agriculture, 3*, 10–13.
Xie, G., Zhang, Y., Lu, C., Zheng, D., & Cheng, S. (2001). Study on valuation of rangeland ecosystem services of China. *Journal of Natural Resources, 16*(1), 47–53.
Xie, G., Zhen, L., Lu, C., Xiao, Y., & Chen, C. (2008). Expert based valuation method of ecosystem services in China. *Journal of Natural Resource, 23*(5), 911–919.
Xu, Z., Li, W., Min, Q., & Xu, Q. (2005). Research on change in value of ecosystem services in Xilin river basin. *Journal of Natural Resources, 20*(1), 99–104.
Xu, Z., Zhang, Z., Cheng, G., Su, Z., Lu, A., Lin, Q., & Zhang, H. (2002). Measuring the total economic value of restoring Ejina Banner's ecosystem services. *Acta Geographica Sinica, 57* (1), 107–116.
Yang, Z., Zheng, D., & Feng, S. (2007). Value evaluation of negative external effect in Beijing city farmland production. *China Environmental Science, 27*, 29–33.

Yoshikawa, N., Nagao, N., & Misawa, S. (2010). Evaluation of the flood mitigation effect of a paddy field dam project. *Agricultural Water Management, 97*, 259–270.
Yu, G., Lu, C., & Xie, G. (2005). Progress in ecosystem services of grassland. *Resource Science, 27*(6), 172–179.
Zhang, H., Ouyang, Z., & Zheng, H. (2007). Spatial scale characteristics of ecosystem services. *Chinese Journal of Ecology, 26*(9), 1432–1437.
Zhang, Z., Shi, H., Jiang, Z., & Ding, D. (2006). Sources and implementation of marine ecosystem services. *Chinese Journal of Ecology, 25*(12), 1574–1579.
Zhao, T., & Ouyang, Z. (2004). Evaluation of forest ecosystem services and their value in China. *Journal of Natural Resources, 19*(4), 480–492.
Zhao, T., Ouyang, Z., Jia, L., & Zheng, H. (2003). Ecosystem services and their valuation of China grassland. *Acta Ecology Sinica, 24*(6), 1101–1110.
Zhao, Y., Wen, Q., & Ai, J. (2010). Ecosystem service value of forests in Yunnan Province. *Forest Research, 23*, 184–190.
Zhao, J., Xiao, H., & Wu, G. (2000). Comparison analysis on physical and value assessment methods for ecosystems services. *Chinese Journal of Applied Ecology, 11*(2), 290–292.
Zhen, L., Lui, X., & Wei, Y. (2008). Consumption of ecosystem services: Models, measurement and management framework. *Resources Science, 30*(1), 100–106.
Zhen, L., Ochirbat, B., Lv, Y., Wei, Y., Liu, X., Chen, J., et al. (2010). Comparing patterns of ecosystem service consumption and perceptions of range management between ethnic herders in Inner Mongolia and Mongolia. *Environmental Research Letters, 5*(1), 1–11.
Zheng, W., Shi, H., Chen, S., & Ding, D. (2006). Perception to ecosystem service function on the view point of welfare economics. *Ecological Economy, 5*, 78–81.
Zheng, W., Shi, H., Chen, S., Zhang, Z., Wang, Z., & Ding, D. (2007). Perception of the characters of marine ecological asset and its value. *Marine Environmental Sciences, 26*(4), 393–396.
Zheng, W., Shi, H., Chen, S., & Zhu, M. (2009a). Benefit and cost analysis of mariculture based on ecosystem services. *Ecological Economics, 6*(8), 626–1632.
Zheng, S., Wang, K., Zhao, M., Han, G., & Feng, Y. (2009b). Primary evaluation of the indirect value on rangeland ecosystem services in Northern agro-pastoral ecotone: A case study in Taipusi banner and buan league. *Pratacultural Science, 26*(9), 18–23.

# Chapter 8
# Ecosystem Management

**Limin Dai, Guofan Shao, Jizhou Ren, Tao Wang, Guangting Chen, Xianguo Lv, Qinzeng Xu, Hongsheng Yang, Qiang Xu, Shiming Luo, Lin Ma, Liu Qian, Wenqi Ma, Fusuo Zhang, Jikun Huang, Xiangping Jia and Cheng Xiang**

**Abstract** Ecosystem management is to make adaptive management strategy based on full understanding the process of composition, structure and function of ecological system, and to restore or maintain ecosystem integrity and sustainability. Forest ecosystem management attempts to maintain forest ecosystem complex process, path and the interdependent relationship, and maintaining their function good, for sustainable forest management and forest ecological system, and establishing and developing comprehensive theoretical system of forest ecosystem and sustainable forestry management system and decision support system, method and technology system. Grassland ecosystem is the prairie land biological coupling with

---

L. Dai
State Key Laboratory of Forest and Soil Ecology, Institute of Applied Ecology, Chinese Academy of Sciences, Shenyang 110164, China
e-mail: lmdai@iae.ac.cn

G. Shao
Department of Forestry and Natural Resources, Purdue University, West Lafayette, IN 47907, USA
e-mail: shao@purdue.edu

J. Ren
State Key Laboratory of Grassland Agro-ecosystems, College of Pastoral and Agricultural Science and Technology, Lanzhou University, Lanzhou 730020, China
e-mail: renjz@vip.sina.com

T. Wang · G. Chen
Research Institute, Chinese Academy of Sciences, Cold and Arid Regions Environmental and Engineering, Lanzhou 730000, China

X. Lv (✉)
Chinese Academy of Sciences, Northeast Institute of Geography and Agroecology, Changchun 130102, China
e-mail: luxg@neigae.ac.cn

Q. Xu · H. Yang · Q. Xu
Chinese Academy of Sciences, Institute of Oceanology, Qingdao 266071, China

S. Luo
Institute of Tropical and Subtropical Ecology, South China, Agricultural University, Guangzhou 510642, China

 
W. Li (ed.), *Contemporary Ecology Research in China*,
DOI 10.1007/978-3-662-48376-3_8

its survival in nature. Absorbing new achievements of science and technology, to build and to make the new grazing system unit perfect, and make it as the indispensable important component of modern agriculture. Desert ecosystem management needs to solve problems of harmonious interaction and the sustainable development of river basin, as well as the ecological water requirement of river basin security problems. Desert ecosystem management must adopt comprehensive ecological management measures in the process of management, give attention to utilization and protection to optimize economic, social, and environmental benefits. Wetland surface interaction process and the system response, the succession of wetland ecosystem and its ecological effect, and the further research of wetland ecosystem function evaluation becomes scientific foundation to explain wetland process route and wetland ecosystem management. Lake ecosystem is affected by human activities. Controlling food chain control of biological on the basis of the classic and nonclassical biological control theory is one of the important measures to control eutrophication in lake. Aquatic plant restoration is also very important in the process of eutrophication in lake. We developed the marine ecosystem management system in China, including ocean space planning and Marine functional zoning planning, Marine nature reserve construction, fishing intensity control and the closed fishing rule, ecological restoration and resource conservation, maintenance management information system construction, and emphasize more on the ocean in terms of ecosystem management in new pattern. Driven by the interests of the market due to the economic benefit and production effect, ecological environmental benefits are neglected in agricultural ecological system. Using the ecology principle "prevention and control of integrated system" is the fundamental way for agricultural pest control.

**Keywords** Grassland ecosystem management • Forest ecosystem management • Desert ecosystem management • Wetland ecosystem management • BKF restoration measures

---

L. Ma · F. Zhang
Centre for Resources, Environment and Food Security, China Agricultural University, Beijing 100193, China
e-mail: Zhangfs@cau.edu.cn

L. Qian
Wageningen University and Research Centre, P.O. Box 47, 6700 AA Wageningen, The Netherlands

W. Ma
College of Resources and Environmental Sciences, Agricultural University of Hebei, Baoding 071001, China

J. Huang · X. Jia · C. Xiang
Chinese Academy of Sciences, Center for Chinese Agricultural Policy, Beijing 100101, China

# 1 Forest Ecosystem Management in Northeast China

## *1.1 Introduction*

About one-third of China's natural forests are found in its northeastern region, including the Daxing'anling, Xiaoxing'anling, and Changbai Mountains (Tang et al. 2009). Northeast (NE) China is not only an important national base for wood products, but an ecologically significant region as well.

Broadleaved-Korean pine mixed forest (BKF) is the native climax forest vegetation in the eastern part of NE China and accounts for a large proportion of the temperate forest zone both in the Xiaoxing'anling and Changbai Mountains. BKF is primarily comprised of *Pinus koraiensis* and broadleaved tree species such as *Tilia amurensis*, *Betula costata*, *Fraxinus mandshurica*, *Juglans mandshurica*, *Quecus mongolica*, *Acer* spp., and others (Dai et al. 2003). This forest type, thus, constitutes not only a huge gene pool and a rich reservoir of biodiversity, but also has great economic and societal values. However, due to long-term human disturbance in the region, including extensive logging in the latter half of the last century, the area of primary BKF forests has decreased substantially, accompanied by severe disruption of stand structure and serious degradation of overall forest quality and function (Pei 2000; Hao et al. 2000). In effect, the history of forest logging and management in BKF mirrors the developmental history of forestry in NE China. Addressing the challenges in forest management of BKF in NE China is thus essential for the sustainability of these forests in the future.

## *1.2 1949–1998: Human Disturbance*

During the latter half of the past century, especially in the early years of the People's Republic of China, the main goal of forestry in NE China was to produce timber in order to meet the exigent demands of economic construction and social development. As a result, excessive harvesting of timber was encouraged (Zhou 2006a, b; Zhang et al. 2000).

In 1950–1956, diameter-limit harvesting of BKF forests was the rule. Although, this promoted forest regeneration to some extent, because most of the young trees were broadleaved species of low quality, this guideline led to large areas of substandard young broadleaved mixed forest in NE China (Liu 1963, 1973).

During 1957–1965, clear-cutting with artificial regeneration was the predominant harvesting regime (Wang et al. 1997). Cutting in narrow, broad, and unequal strips were all adopted (Liu et al. 1965). As a result, large areas of primary BKF disappeared and were replaced with equally large expanses of Korean pine plantations (Dai et al. 2004).

The most drastic phase of this early exploitation of the country's timber resources occurred from 1966 to 1976 during the 10-year period of the "Cultural Revolution," when neither clear-cutting with artificial regeneration nor selective harvesting with other silvicultural treatments were employed by most of the forest bureaus in NE China. These were replaced by heavy selective cutting methods with intensities of 70–90 % and no artificial regeneration (Dai et al. 2003; Deng 2008). Thus, forest landscapes were fragmented (Shao and Zhao 1998), and much of the remaining primary BKF were replaced by large areas of low-quality secondary forests (Shao et al. 2001).

From 1978 to 1998, although some protective measures and forestry laws were adopted, they were not sufficient to sustain the nation's forest resources (Zhao and Shao 2002). Excessive logging of natural forests in NE China was not effectively prevented, and small-area clear-cutting was a common practice until 1996 (Shao and Zhao 1998). In addition, serious problems of illegal timber cutting and resultant excessive deforestation were also rampant (Xu et al. 2004).

## *1.3 Conservation Efforts*

During the last half century of extensive logging of BKF in NE China, some efforts at resource conservation were also attempted. Thus, for example, several nature reserves in which harvesting of BKF was prohibited were established, beginning in the 1960s with Fenglin and Liangshui reserves in Xiaoxing'anling and the Chagbai Mountain Nature Reserve in 1960. Although not seeing fruition as implemented polices, attention was given in the scientific community to exploring ways of formulating and adjusting forest laws and regulations in seeking a balance between forest utilization and conservation, forest growth and harvesting, and between developing forest plantations and harvesting natural forests (Wang et al. 2004).

Forestry scientists did begin to consider the relationship between BKF harvesting and its restoration as early as end of the 1950s. First, the biological characteristics of Korean pine came under debate with the publication of three papers by a Japanese ecologist in the 1950s (Guo et al. 1995). With respect to natural regeneration, researchers who viewed Korean pine as a shade-intolerant species insisted that clear-cutting was the best approach to managing BKF forests, while others who considered Korean pine to be a shade-tolerant species insisted that it should be selective harvesting (Liu et al. 1965). At the same time, forestry scientists who claimed that Korean pine is shade-intolerant investigated areas of cutover lands and found that the natural regeneration of Korean pine was unsatisfactory (Wang 1957; Liu 1957).

Thus the situation regarding Korean pine natural regeneration and forest restoration after harvesting became a dilemma for researchers. As debates continued, new perspectives eventually emerged. From the perspective of botanical dynamics, Liu concluded that the tolerance of Korean pine varies with age, in that it requires shade as a seeding followed by sunlight thereafter. It was thus proposed that selective harvesting combined with subsequent silvicultural treatments was the appropriate management approach for BKF (Liu 1963; Wang et al. 1997). This helped clarify the perplexing problems regarding the biological characteristics of Korean pine as well as the dynamics of Korean pine regeneration. As a result, prescriptions for appropriate harvesting methods, including a core of postharvest silvicultural treatments, gradually became more crystallized.

In the 1980s, based on years of research, the management theory of "Plant conifers and reserve young broad-leaved trees" for BKF was put forward (Zhou 1982). This method offered a much more effective approach to mixed forest management in the region by integrating conifer planting with a protective strategy regarding the general problems of plantation management. In effect, it signaled that forest management of BKF in NE China was maturing.

During this time forest researchers in China also continued making efforts to explore ways of formulating and adjusting forest laws and regulations (Wang et al. 2004). In 1981, the "Resolution on Issues Concerning Forest Protection and Development" was issued; and in 1984, the Forest Law of China was passed. In the same year, forest management strategies based on a five-category classification system of forest resources was also put into effect (Dai et al. 2009, 2011). In 1987, "cutting quotas," which regulated the annual timber, output, were introduced (Liu 2001; Dai et al. 2013). All of the above effects, while only marginally successful in slowing the enormous pressure on BKF as a source of timber for economic development, helped lay the foundation for more permanent changes in the future.

Many feel that the corner was turned in 1998, when China broadened the focus of forest management in the country and launched the Natural Forest Conservation Program (NFCP) aimed at protecting and restoring natural forests (Zhang et al. 2000; Dai et al. 2009). At the same time, China's forest law was amended to reclassify the five forest categories into two—Commodity Forest (CoF) and Ecological Welfare Forest (EWF) (Cai et al. 2003). These changes have provided unique opportunities to pursue ecosystem management of forests in NE China (Zhao and Shao 2002). To aid in this process, state financial funding for forestry was further enlarged, and the forest harvesting quota in NE China was further decreased. Unlawful timber cutting and excessive logging were also effectively contained. In addition, a digital decision support system was used for managing and protecting primary forests in NE China (Shao et al. 2003, 2005). These changes and innovations have provided formal recognition and official status to the many other functions and services of forests in addition to timber production. Moreover, with respective to Korean pine, this species may no longer be harvested for timber in NE China, but only utilized as a seed-tree species (Yu et al. 2011).

All of the above measures had immediate and significant effects on forest resource protection and restoration in NE China. Since their adoption, forest area of BKF has slowly increased (Wang et al. 2004).

## 1.4 Current Problems and Challenges in BKF Sustainability

The practice of monoculture forestry for several decades in NE China not only contributed to many economic difficulties in forested regions, but also to a series of ecological disasters for BKF (Shao et al. 2003).

(i) The area of primary BKF decreased substantially and was replaced with low-quality secondary forests. Even within the existing primary BKF, tree species composition was modified with only Korean pine being reserved and associated tree species such as Manchurian ash (*F. mandshurica*), Amur cock tree (*Phellodendron amurense*), and Manchurian walnut (*Juglans maudsurica*) declining or disappearing.
(ii) The sharp decline in the quantity and quality of primary BKF has resulted in the loss and fragmentation of natural habitats, along with the decline and disappearance of ecological functions and services.
(iii) Forest age structure of BKF was suboptimal while it was dominated by old-growth forest, with a much smaller proportional representation of middle-aged and young age classes, which contributed to the poor regeneration of BKF in NE China.

On an overall basis, ecosystem management remains a primary philosophical concept underlying the management of China's forests, especially for dealing with large spatial scales and long time frames. It also represents the foundation for addressing the array of challenges faced by forest managers and forestry scientists in meeting the country's demands for sustainable multiple-use management of the nation's forests.

## 1.5 BKF Restoration Measures

In view of the problems for BKF outlined earlier and with the aid of lessons learned during the country's tumultuous history of forestry development, it is worthwhile to consider some restoration measures for BKF in NE China.

First, the protection of existing primary old-growth and mature BKF both in the center and buffer zones of nature reserves is paramount for the restoration of biodiversity and ecological functions. This includes the prohibition of timber harvesting and any other human activities that would damage the integrity of core ecosystems, as well as reasonable protection of buffer areas to ensure the caliber of ecological attributes and processes.

Second, the quality of the large areas of natural secondary forest should be improved through effective silvicultural practices and management regimes in a multiple-use framework that encompasses both restoration of regenerative capacities and utilization of timber resources.

Third, the development of plantations for large diameter tree should be accelerated to replace components of existing plantations with low quality and short rotations so that an effective mix of both short and longer rotations can better address societal demands for wood production.

## 2 Grassland Ecosystem Management

Grassland ecosystems are an important component of terrestrial ecosystems. According to FAO's statistics, 69 % of agricultural land is permanent grazing land, where Oceania, sub-Saharan Africa, South America, and East Asia were 89, 83, 82, and 80 %. Grazing systems provide mankind with more than half of meat, 1/3 of the milk, fur, and other animal products.

Grazing is the human agricultural behavior on the aim of grassland management as well as utilization through livestock grazing. Grazing has two aims. One aim is to keep pasture healthy via managing grassland. Another aim is to produce animal products, which can achieve sustainable economic benefits.

Grazing contains dual combination with interface: grassland—livestock—habitat which constitute factor groups, it is the initial combination of grazing behavior; based on this foundation, spatial and temporal combination of grassland—livestock, that is to say, secondary portfolio of grazing behavior is implemented, and further contribute grassland and livestock cooperative development.

Grazing go through different historical stages, a dramatic change had taken place especially at the late nineteenth to early twentieth century, which was a critical period, and finally in 1930 completed a grazing modernization. Modern grazing is still prevalent in North America, Australia, New Zealand, Western Europe, and Northern Europe.

Grazing is a "double-edged sword," it is our powerful tool for grassland management as well as to achieve livestock products with appropriate use. But if misused, too light or overgrazing, for example, it will damage the grass as well as lead to low productivity. The key is to make good use of grazing. We do not treat grazing as a tool for grassland management, instead, grazing has been demonized, and grassland ecological degradation, pastoral poverty, and cultural backwardness ascribe to the grazing. Many districts forbid grazing in order to highlight grassland management reform efforts. For now grazing forbidden is prevalent all over the country, even called for more than 90 % of the grassland should be completely abolished grazing in two or three years, this should arouse us great concerns.

Grazing is a system constituted of three elements which are human settlements, grassland, and livestock in grassland ecosystems. The grassland ecosystem maintains healthy development status only under the correct use of grazing and coexist the three elements in harmony.

The grassland carrying capacity shows dimensionless of carrying capacity under grazing pressure, it contains three elements: yields of grassland which is denoted by $a$, grazing duration which is denoted by $b$, and the number of grazing animals which is denoted by $c$. So for the measurements of carrying capacity, there are three formulations: area, time (duration), and livestock.

Carrying capacity area unit load method: $a = b \times c$, expressed as a paddock, for given animals, how much time grazing in the area.

Carrying capacity time (duration) unit load method: $b = a \times c$, expressed as a paddock, allowing livestock grazing how much time.

Carrying capacity animal husbandry unit load method: $c = a \times b$, expressed as a paddock, a certain time to allow how many animals grazing.

There can be mutual conversion among the three units, in order to facilitate the calculation, often based on one kind of livestock by habit, European and American adopt "cow units," while China, New Zealand, and Australia adopt "sheep units." Because they differ in livestock species, body sizes, production levels and management, productivity denoted by carrying capacity which expressed the grassland conversion efficiency is also different. The real carrying capacity is grassland and livestock husbandry system coupling reach close to a perfect desired level, in other words, it depends on the overcoming process and overcoming degree of time contrary, species contrary and space contrary in interface system coupling between grassland and herbivore.

The most simple and accurate metering system is animal production unit (APU). One APU is equal to a moderate fertility grazing cattle weight gain 1 kg, which is equivalent to 111 MJ digestible energy (DE), or 94.2 MJ metabolizable energy (ME), or 58.2 MJ net consumption growth energy (NE) as well. APU of variety grazing livestock is just compared to the energy consumption of fertility grazing cattle weight gain 1 kg, in this way, the APU coefficient of livestock can obtain. Any livestock production in grassland anywhere can convert to APU by the coefficient (Table 1). APU directly describes the final grassland productivity, in addition, it reflects comprehensive scientific and technological level of grassland management indirectly, it is the most accurate way to measure grassland load as well. APU measures the true grass loads simultaneously avoiding deviation caused by livestock species, numbers, management, and other factors since it is according to the principles of energetic grassland production.

Grazing went through a long process of historical development. After a primitive nomadic stage, which last from the late Pleistocene to early Holocene (8000 aB.P–6000 aB.P.), after the global flood subsiding. Nomad is the survival forms of early humankind, for they moved from place to place in search of water and grass. In East Asia, as represented by the Hongshan culture, creating a golden era when there is harmony between man and nature. Herbivore has the instinct of looking for grass and water, human trail the animals, it is so-called "grazing." The herds lead to migration of human, its essence is human tracking food source, just as wolves following behind the deer to hunt for food at any time. Grass—herds—human food chain, which forming a ring of ecosystems. When people put this grazing system from passive to active and fixed, to be manipulated on the completion of the first human bionics, it was a big step forward of human civilization.

With the formation of human habitats, there were organized grazing order according to the natural watershed, the key point is grazing system unit of habitat–

**Table 1** The conversion of animal product unit and different animal products

| Livestock production | Animal production unit |
|---|---|
| Weight gain 1 kg | 1.0 |
| Carcass of 50 kg sheep | 22.5 (Slaughter rate 45 %) |
| Carcass of 280 kg cattle | 140.0 (Slaughter rate 50 %) |
| Edible offal | 1.0 |
| 1 kg standard milk (Fat content 4 %) | 0.1 |
| 1 kg various clean fur | 13.0 |
| One three-year-old serving horse | 500.0 |
| One three-year-old serving cattle | 400.0 |
| One four-year-old serving Camel | 750.0 |
| One four-year-old serving donkey | 200.0 |
| Serving horse work for a year | 200.0 |
| Serving cattle work for a year | 160.0 |
| Serving camel work for a year | 300.0 |
| Serving donkey work for a year | 80.0 |
| One sheep skin(fur sheep breed) | 13.0 |
| One fur coat | 15.0 |
| One cowhide | 20.0 (or count as 7 % live weight) |
| One horse skin | 15.0 (or count as 5 % live weight) |
| One sheep skin | 4.5 (or count as 9 % live weight) |
| One weed out mutton sheep(live weight 50 kg) | 34.5 (or count as 69 % live weight) |
| One weed out mutton cattle (live weight 280 kg) | 196.0 (or count as 70 % live weight) |

grassland–livestock. Its basic elements were seasonal pastures within one year or between years, in other words, it contains grazing pastures of different seasons, which can constantly provide or offer rotary grazing fields, this ensure relative stability of the grazing system unit. In the long history of the social development, with combination of human habitats, social may change significantly, but we should keep stability of grazing system unit, and its managing model of habitat–grassland–livestock has no essential alteration. *Grazing system unit*—the symbiosis of habitat–grassland–livestock is an enduring legacy of primitive society, it finally, inevitably will find and create its own grazing system unit no matter how social order turbulent and change, otherwise, it will lead to the collapse of nomadic society. Grassland animal husbandry management system gradually transmute on the premise of stability of the grazing system unit and complete modernization of livestock grazing finally, its key aspect is transformation of grazing management. From original nomadic stage to livestock modern grazing phase, which can be called "pre-modernization grazing" period. This process may be long or short, early or late due to social conditions divergence, but from original nomadic stage to livestock modern grazing phase which is a necessary historical process. In western countries, with the industrializing, industrial civilization penetrates to grassland management, which built a more scientific and meticulous management system and completed the transformation of grazing management as well. The grazing development history of North America has typical sense to grazing history, this article takes Americans for example. America is in state of primitive nomadic under American Indian before colonialists invaded the American continent. After American civil war, with arrival of the fast economic development, "westward movement" completely broke the grazing system of native Indians. The government established Indian reservation, and then forced aboriginal moved to protected areas, and allocated most of grassland to new immigrants. It provided great favorable conditions for the development of grazing livestock of western United States. A lot of social capital put in grassland animal husbandry of western United States, new cattle nurture companies emerged massively and got climax of foreign investments, which constituted "Livestock Kingdom."

America's "westward movement" lasts three centuries from 1607 to 1890, when black storm swept prairie. The United States gradually formulated relevant laws and regulations in the context of this disorder situation. The modern transformation of grassland management had been basically completed at 1934 when promulgated the "Taylor grazing act." Its iconic features can be summarized as follows: (i) ranch boundaries clear, grassland property rights was protected by law; (ii) with development of the industrial production of fence, grazing grassland gradually improved which focus on rotational; (iii) establishing good drinking water system, it was to support rotational grazing; (iv) after the gradual development of ultra-biological production level, market disciplined; (v) the modern grassland management scientific system initially formed, the textbook "grassland management" published, and published professional monograph; (vi) making full use of system coupling, combined with the different ecological and economic zones, in which rotational grazing as the core, at last established grassland—beef cattle systems, grassland—cow system, grassland—sheep systems. In summary, the human's grazing techniques went through three stages of development in transition which are original grazing, extensive grazing and intensive grazing, lasting three centuries. Our country is totally unaware of transformation of grazing, treating original grazing as the only type and behind things, in this case, grazing has been repeatedly abandoned over the years. This article criticizes several misunderstanding on grazing cognitive: "grazing is a backward, primitive mode of production"; "grazing should be replaced by soiling"; "grazing damages grassland"; " to replace all natural grassland with cultivated grassland"; " the way of saving grassland degradation is to break up into small pieces, subcontracting home"; "grazing is harmful to afforestation, it should be forbidden."

# 3 Desert Ecosystem Management

## 3.1 Introduction

Desert is a kind of geographic landscape developed in arid climate conditions. Desert ecosystems mainly distribute in subtropic and temperate extreme arid and rainless regions. Zonal desert derived planetary wind system distribute around Tropic of Cancer or Tropic of Capricorn. In Asia–Europe and Africa continents, an obvious desert belt formed in the North Hemisphere, from Sahara Desert at the western coast of North Africa, to Arab Peninsula, to temperate deserts in the Central Asia and Sino-Mongolia eastward. Desert in China mainly distributes in the northwest arid region to the west of Helan Mountains, and its east border is basically consistent with the total annual precipitation isocline of 200 mm. This is closely related to the impact of geographical location in the hinterland of continent, downward current of air caused by landforms, especially Tibetan lift blocking warm wet air from Indian Ocean.

Desert ecosystem is very fragile. In case that desert ecosystem was disturbed or destroyed by natural disasters or human activities, its rehabilitation is very difficult and slow, and even often impossible. Water is the limit factor of desert ecosystem. Those places with water supply will develop oases, otherwise will become deserts.

Oasis is a kind of unique geographical landscape, and is a sub-ecosystem with super productivity supported by water in the desert regions (Shen et al. 2001). We must be very careful and discreet to develop and utilize desert. The oasis size in arid region must be restricted to the carrying capacity of water resources. To explore desert ecosystem, especially management and control of oasis ecosystem, we should make full use of the function of oasis ecosystem, natural environment and resources. During seeking development of human society and economy, eco-environmental construction and protection, and sustainable development of oases, the harmonious relation between mankind and natural environment, improvement of human livelihood should be emphasized.

## 3.2 Basics of Desert Ecosystem Management

Extreme continental climate is the basic characteristic in the desert region. Annual precipitation is less than 200 mm in most places of temperate desert in China and its variation is very huge. Potent evaporation is several dozens or even one hundred times of the precipitation, which caused extremely lack of water in air and soil. Temperature enormously changes and daily temperature difference is particularly obvious. Strong reflection of desert surface often destroy radiation balance and thermal balance, and result in horizontal and vertical air current and induce wind-blown sand or dust storms.

### 3.2.1 Structure and Function Features of Desert Ecosystem

In desert community, there are poor plant types, simple plant structure and lower vegetation cover, even complete bare land. Nutrient materials in soil also are very little. Due to barren food supply, animal types and quantities are very less. However, there are longer sunshine and abundant quantity of heat in growing season. In some places with better conditions, such as oases, where can become local high-productivity areas by irrigation and fertilization.

According to differences in physical features and cause of formation, desert ecosystem can be divided into the following five sub-ecosystems (Table 2).

### 3.2.2 Several Problems in Desert Ecosystem Management

With the development of society and economy, a kind of Social-economic-natural compound ecosystem had been built surrounding those areas in arid region where generate water resources and produce runoff and been exploited and used, in which inland river basin was taken as a unit (Qin et al. 1998). This system is a semi-artificial ecosystem. In which, human activities were taken as the dominant, and oasis ecosystem was taken as the support, and water resource was taken as the key, and society system was taken as the frame.

The obvious contradict of water resources shortage in the northwest arid region of China, mainly show in: (1) shrinkage of surface water, serious decline of underground water level, increasing mineralization degree. The important reason is lack of effective and reasonable allocation and management of water resources; (2) lack of effective protective measures for natural vegetation resulted in the whole ecosystem facing collapse; (3) soil salinization extending in the farmlands of oases.

## *3.3 Desert Ecosystem Management*

### 3.3.1 Oasis Ecosystem Management

Oasis is the most harmonious ecosystem unit in arid region, with the better match among climate resource, water resource, land resource, biological resource and eco-environment. Modern oasis is supported by irrigation. Thus, at certain degree, oasis ecosystem management actually is water resources management, including: (i) to integrate and plan water resources taking river basin as a unit; (ii) to develop saving-water irrigation agriculture and to enhance utilization ratio of water resources; (iii) to determine farmland area according to water resources quantity and to improve water transferring ditch net; (iv) to implement market-oriented irrigation management mechanism; (v) to strengthen union management and associated utilizing of surface water and ground water, and to prevent water pollution; (vi) to

**Table 2** Classification of desert ecosystem in China

| Sub-ecosystem | Subsystem | Type | Structure | Utilizing status |
|---|---|---|---|---|
| Oasis ecosystem | Artificial oasis | Irrigation farmland system | | Agriculture tillage |
| | | Shelter forest system for farmland | | Prevent wind activities |
| | Natural oasis | Woodland–grassland–wetland system | Riparian sparse-tree shrub meadow | Protection forest or season pasture |
| | | | *Tamarix chinensis* shrub meadow around oasis | Season pasture |
| | | | *Haioxyion ammodendron* and weeds in lake basin | |
| | | | Lakes or marshes with *Phragmites australis* | Fisher or weaving |
| | | Fixed, semi-fixed dune fields | Shrubs or shorter arbors | Fuel-wood forest or grazing |
| Gobi ecosystem | Desert grassland | Sandy desert grassland | *Haloxylon Persicum Bunge ex Boiss. Et Buhse* and weeds | Grazing |
| | | | *Calligonum mongolicunl* and weeds | |
| | | Gobi desert grassland | Shorter semi-arbor or shrub | |
| | | | Semi-shrub or shorter semi-shrub | |
| | | | Succulent and salt shorter semi-shrub and herb | |
| | Bare gobi | Deflation gobi | | Not utilizing |
| | | Accumulation gobi | | |
| Sandy desert ecosystem | Sandy desert | Fixed or semi-fixed dune field | | |
| | | Shifting dune field | | |
| Salty desert ecosystem | Salty desert | | | |
| Arid middle or lower hill ecosystem | | Lower Loess hills | *Artimisia*-type desert grassland | Grazing |
| | | Middle or lower stony hills | Salt semi-shrub grassland | |

reasonably develop and utilize land resources according to ecological rules; (vii) to protect natural vegetation, and to actively build artificial vegetation.

### 3.3.2 Oasis-Desert Transition Belt Ecosystem Protection

In natural conditions, there is a transition belt between oasis and desert. An oasis-desert transition belt is the place that oasification and desertification begin and frequently confront, where supports stability of inner oasis and agriculture sustainable development, and where is also residents making cutting and grazing, and where is a very significant part of oasis safe and stability.

Basic rule and approaches to restore oasis-desert transition belt ecosystem include: (i) correctly recognizing ecological status of oasis-desert transition belt; (ii) prohibiting to blindly expand artificial oasis under lack of enough irrigation; (iii) making full use of high scientific and technical means to solve residents' fuel and to enhance their livelihood standards for avoiding woodcutting, gathering and excessive human activities; (iv) increasing vegetation cover of oasis-desert transition belt and to extend its scopes, through air seeding or manual sowing.

### 3.3.3 Desert Ecosystem Protection and Reconstruction

Desert ecosystem is the most fragile one in land ecosystem, developing under the conditions of rigorous climate and extreme imbalance of water and heat supply. The fragility of desert ecosystem determines the place of desert ecosystem distribute as the frequent area of natural disasters, and is also one of the most easily prone to degradation (Liang et al. 2003).

Generally speaking, to restore degraded ecosystem should follow the three basic principles: (i) natural restoring rule; (ii) scientific technology, feasible economy, and acceptant measures rule; (iii) aesthetic rule (Peng and Lu 2003).

As far as desert ecosystem restoration is concerned, it is different from other ecosystems to follow principles (Jiao 2003). They can be summarized: (i) coordination between ecology and economy rule; (ii) dominant preventing and combination with control; (iii) mutual complementarities among many measures rule; (iv) coordination between long-term and short-term profit rule; (v) suitable measures for local conditions rule; (vi) combination between scientific–technological support and administration rule; (vii) combination of biological, ecological, and engineering technique rule (Zhao et al. 2009).

Protection methods of desert vegetation (including microbiotic crust) include: (i) protecting sand for grass propagation and returning farmland to woodland or grassland; (ii) controlling livestock quantities and conducting reasonable rotation grazing. In addition, during comprehensively using desert industry, clear energy, and developing travel industry, it should be insisted to protect environment.

## 3.4 Outlook

Aiming at desert ecosystem in northwest arid region of China, it needs to adopt comprehensive ecological management for optimizing economy, society and environment profits, through improving available ecological service level from all types of support, regulation, supply, and culture provided by natural ecosystem, and preventing continuous deterioration of ecosystem.

# 4 Wetland Ecosystem Management

## 4.1 Introduction

Wetlands ecosystems are around all over the world and are an ecological landscape that has the most biodiversity in nature, as well as being one of the most important environments necessary for humans to prosper (Mitsch and Gosselink 2000). Wetlands are located in the interface of the atmosphere, terrestrial, and aquatic ecosystems, they play an important role in water, nutrient, organic, sediment, and pollutant transportation. The wetland is also called the cradle of life, the kidney of the earth, species genes reserve, and so on. Since 1971, since the Ramsar Convention was established, there has been an increase in awareness by international communities of the importance and urgency of fostering wetland protection, ecological restoration, and promoting reasonable and continuous wetland exploitation (Liu et al. 1999). From initially emphasizing the wetlands' function as waterbird habitats, on the focus on wetlands has expanded to various aspects of protecting and reasonably utilizing wetland ecosystems, wetland ecosystem management being a hot topic today.

Today, crucial hydrological, biological, chemical, and physical processes maintaining ecological stability have been drastically changed on the global scale, and rates of change are increasing in general, which results in ecological degradation, sustainability weakness, biological diversity loss, and ecological function reduction (Lu 2001).

Wetland ecological management styles emphasize driving and stressing factors of ecological structure and function, as well as focusing on the interactions between management activities and systemic structure and function, this includes wetland ecosystem processes and natural and human factors which cause ecological process change (Yuan and Lu 2004). Administrating human activities are the main parts of wetland ecological management because it is more practical than adjusting natural factors. The manager must learn the objective laws of wetland ecosystems, to clear wetland processes and their mechanisms, making certain the purpose, problems, methods, and routes of management activities to achieve the optimum configuration of systemic structure and function. To realize these goals, natural laws of wetlands must be followed (Zhao and Gao 2007).

## 4.2 Wetland Amphibious Mutual Process and Ecological Response

Wetlands are located in the transitional zones between terrestrial (e.g., forest, grassland, and so on) and aquatic ecosystems (e.g., marine), and contain multiple ecological functions due to the interaction of terrestrial and aquatic ecosystems. Wetlands possess a unique structure, process and function, and are sensitive to environmental change.

Wetland processes are the occurrence and development of dynamic characteristics of events and phenomenon in wetlands. Wetlands are interracially affected by multiple earth spheres and are among the most complex ecosystem on the Earth's surface. Wetlands often contain many physical, chemical and biological processes, and wetland functions are the outside presentation of various processes.

## 4.3 Wetland Ecological Succession Processes and Ecological Effects

Natural disturbances play important roles in maintaining wetland structures and functions. Disturbances to wetlands due to human activities include both positive and the negative effects, affecting wetland succession processes in various aspects. Wetland ecosystem succession characteristics include changes of species compositions, diversity indices, and life-forms of plants at different succession stages (Yao et al. 2009). Wetland succession process changes would bring great ecological fluctuations. First, wetland environment changes result in wetland fragmentation and degradation, wetland types change and biodiversity vary. On the other hand, wetland ecosystem succession would also cause plant biomass and soil biogeochemical nutrient levels variation (He and Zhao 2001; Zhang et al. 2013), and lead to wetland function changes at last in the aspect of materials circulation, biodiversity protection, water purification, climate adjustment, and so on.

## 4.4 Wetland Ecological Function Assessment

Wetland function assessment methods include two types, which are energy analytic approach and the price evaluation approach. Due to the variety of wetland ecological service functions, its evaluation is also complex. Different wetlands require different methods according to functions supplied by wetlands (Xu et al. 2006).

By the middle of the twentieth century, although some wetlands were protected for hunting, fishing and water birds, wetland management policies both in China and abroad mainly focus on agricultural and other uses, which limited the understanding of wetland value. Due to unreasonable utilization, key biological, chemical, and physical processes maintaining wetland ecosystems were as disturbed, which resulted in wetland degradation, biodiversity loss, and service functions reduction.

Until the middle of the twentieth century, some managers of fishing and entertainment industries realized the value of wetlands as wildlife habitats, and they promote the maintenance of hydrological conditions of wetlands to protect fish and waterbird communities. Other functions of wetlands were understood in the late twentieth century, such as flood regulation and water quality promotion, and the management of wetlands saw an increase in quality and purpose.

The wetland ecosystem is hierarchical and with different characteristics and functions at different gradations (Chen and Lu 2003). Wetland management often starts from wetland structure to reveal their ecosystem processes and functions based on the understanding of material and energy circulation within the wetland. Based on the results of the monitoring, changes of wetland functions, information integration, adjustment and prediction were performed so as to provide scientific policies to the government.

### *4.5 Wetland Management Practice in China*

Since 1992, when China officially joined in the Ramsar Convention, The Chinese government has been strengthening the intensity of wetland protection. In December 2000, China's wetland protection plan was established by the State Forestry Bureau and other 17. In August 2008, the State Forestry Bureau and the Nation Development and Reform Commission (NDRC) of China finished the national wetland protection engineering plan (2002–2030), having it approved by the State Council in September. Wetland protection and management in China is in the rapid development period. There are three main models of wetland protection in China, which are natural reserve, wetland parks, and wetland protection cells. Up to now, a lot of work has been done for wetland restoration and reestablishment.

## 5 Marine Ecosystem Management

### *5.1 Introduction*

The ocean can provide many resources, ecosystem services and functions, such as seafood, energy, oil and natural gas. Marine ecosystems are very complex due to multiple scales caused by flow of water and species movements, so marine ecosystem are vulnerable to rapid social and environmental changes. It is necessary for government and related stakeholders to apply more efficient solutions to manage marine ecosystem.

## 5.2 Concept, Features, and Principle of Marine Ecosystem Management

Marine ecosystem has many characteristics that are unique and are different with terrestrial ecosystem. The first factor is its openness and large scale; the second one is the difficulties to collect ecosystem information; and the final one is the complex impacts from human activities.

The marine ecosystem management concept is based on comprehensive coordination among different ecosystem services in specific space and time range, and following the inner structure and rule of the ecosystem, the administrations create scientific and reasonable adjusting policy by motivating different stakeholders to reach a healthy marine ecosystem.

The management of marine ecosystem also has three factors. First, there are several variable (spatial and time) management targets. Second, there are many management agents with different administrative implementations. Finally, the decision-making process and the evaluation are complex.

The marine ecosystem management should follow some principles: sustainable development, maintenance of ecosystem health, prevention, and early warning and public participation.

## 5.3 Methods for Marine Ecosystem Management

### 5.3.1 Marine Spatial Planning and Marine Functional Zoning

Marine spatial planning (MSP) is a process about analyzing and allocating the spatial and temporal distribution of human activities in the marine areas to achieve ecological, economic, and social objectives.

Marine functional zoning (MFZ) is a management tool about utilizing sea space areas in China based on its geographical and ecological features, natural resources, current usage, and socioeconomic development needs. MFZ has been characterized as a practice of MSP in China.

### 5.3.2 Marine Protected Areas Establishment

The marine protected areas (MPAs) in China started to be built from the 1960s. A total of 201 MPAs which scattered in 11 coastal provinces have been declared in China and total area amount is up to 3,300,000 $hm^2$. The MPAs contain the main habitats of mangrove, coral, coastal wetland, island, estuary, bay, and other areas with fragile but important marine ecosystem.

### 5.3.3 Fishing Restriction and Fish-Ban System Reform

Fish-ban system has been implemented since 1995. The fish-ban period and the category of restricted fishing instruments have been enlarged. Chinese administration has controlled the license of marine fishing and encouraged the fishermen to work in other industry of Mariculture such as aquatic product processing, recreational fishery, and other relative industries.

### 5.3.4 Ecological Restoration and Marine Living Resources Maintenance

Marine ranching is the best way to restore the impaired ecosystem and depleted marine living resources. However, such method in China is premature and it mainly focuses on the implementation of artificial reef as the primary step to carry out the marine ranching method. Fisheries resources proliferation has been implemented in China. Many species such as crab *Portunus trituberculatus*, jellyfish *Rhopilema esculentum*, fish *Paralichthys olivaceus*, *Pagrosomus major,* and other demersal fishes have been proliferated and released to the suitable ocean habitat. Ministry of agriculture detailed the specification and standard about proliferation.

### 5.3.5 Establishment of Information System for Marine Ecosystem Management

Establishment of information system is crucial to the marine management. The European and American countries have established complete and effective management system but few effective marine manage systems exist. The National Marine Data and Information System (NMIS) was built in 1997. Marine Scientific Data Sharing Project System was activated in 2003.

### 5.3.6 Comprehensive Method for Marine Ecosystem-Based Management

In fact, marine ecosystems are presently problematic despite of many attempts to manage individual threats in the absence of a system-wide approach. Therefore, ecosystem-based management (EBM) is a promising approach for managing the sea areas proved by government, scientists, and others.

There are four steps to implement the EBM, the first is to determine the regulatory boundaries of EBM, and the second is to fix the critical management objectives, optimize management methods, and create the public feedback mechanism.

### 5.4 Prospects on Marine Ecosystem Management

In the future, the marine ecosystem management would develop in the following aspects: encouraging management consciousness about the whole ecosystem; straightening out the model and system of marine ecosystem management; improving the system about marine monitoring and evaluation; implementing comprehensive management system.

## 6 Agroecosystem Management

### 6.1 Introduction

There are three interacting levels of agroecosystem regulation and control. The first level regulation and control is provided by natural processes. The second level is direct regulation and control exerted by human activities. The third level is indirect regulation and control provided by social and economical background which deeply affects the human behavior (Fig. 1).

### 6.2 Direct Intervention by Strengthening Natural Processes

The first-level regulation and control of agroecosystem is the natural processes. Direct intervention exerted by people can make good use of the natural processes and hence achieve ecologically sound results with much less negative impacts on resources and environment. The direct intervention includes (i) regional and landscape level land use planning, (ii) restoration of ecosystem recycling structure and function, and (iii) the use of biodiversity to increase natural resource efficiency and to deal with biology and physical stresses (Fig. 2).

The regional land use planning relates to the development strategy in the region. It should carefully balance the relation of agricultural land use and nonagricultural land use, the relation among residential, production, and ecological needs. The optimum land use arrangement within agriculture production along different positions of a watershed according to physical and social economic situation should be carefully considered. Natural vegetation within cropland including buffering zone along the drainage channel, wind belt system, and cover on marginal space of cropland are important landscape structures.

Agriculture and forestland can play an important role in the restoration of ecosystem recycling structure and function. Six scales of recycling structure base on agriculture land can be identified and should be strengthened (Fig. 3). They are

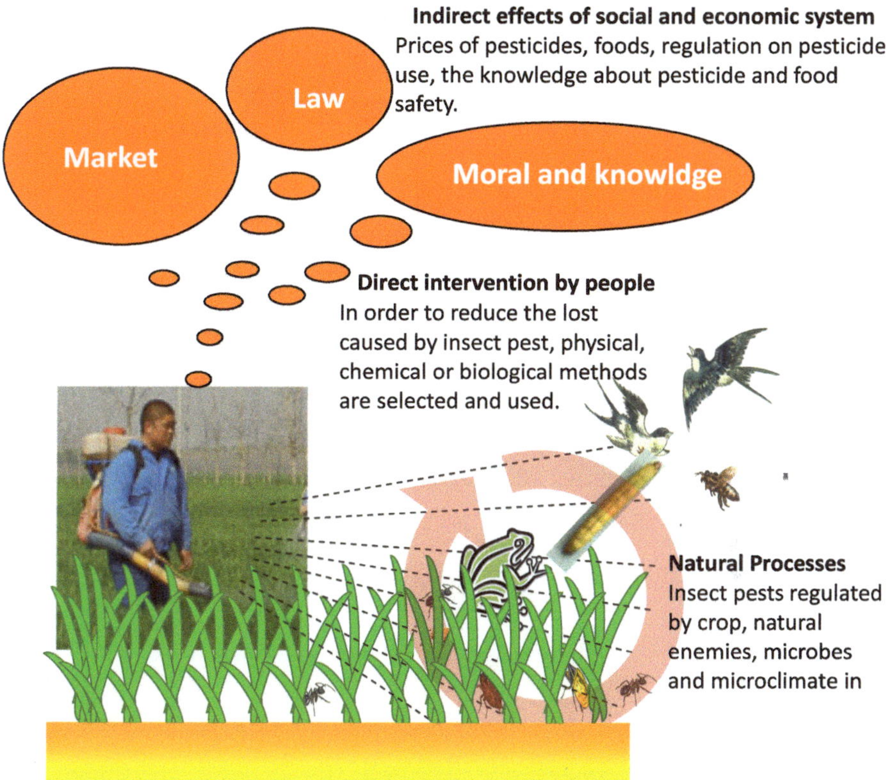


**Fig. 1** The three levels of agroecosystem control and regulation demonstrated by pest control (Luo 2010a, b)

(i) recycle of crop residue within field level, (ii) recycle of animal waste between animal production and crop production, (iii) recycle of village waste, (iv) recycle of processing waste from agro-industry, (v) recycle of organic waste from cities and towns, and (vi) recycle of $CO_2$ in a global scale.

The use of biodiversity in agriculture includes community, population, and genetic diversity organization. Ground cover in orchard or woodland, and intercropping or relay cropping system in the field are examples of biodiversity atcommunity level. Niche differentiation among community species is an important rule in the design of community structure. Good community diversity not only is able to make good use of natural resources, but also can resist and withstand both physical and biological stresses. Research by Zhu (2007) showed that the intercropping of maize with potato, corn with sugarcane, corn with yam could gain higher yield and less pest outbreak. Li et al. (2007) showed that the intercropping of

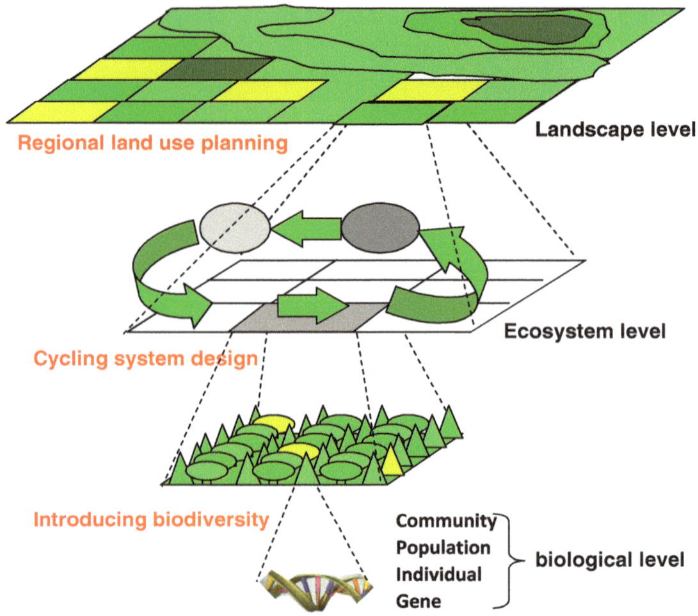


**Fig. 2** The three scales of direct intervention by strengthening natural processes

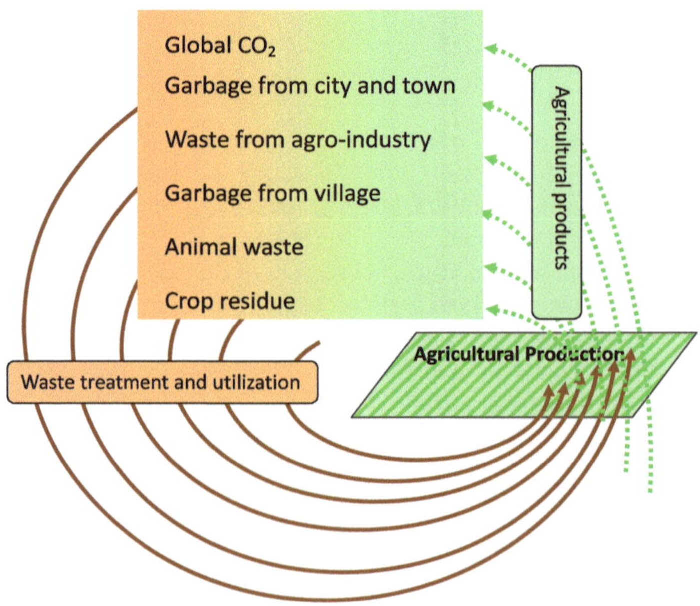


**Fig. 3** The restoration of recycling processes base on agriculture in different scales

corn and broad bean could increase the nitrogen fixation of bean and increase the availability of soil P, Fe, Zn, etc. In population diversity organization, research by Quan et al. (2008) showed that rice duck mixed system could reduce field pests, increase lodging resistance and rice quality. Xie et al. (2009) showed the effect of rice fish system could reduce weed, rice plant hopper, rice sheath blight, and rice blast disease. Releasing natural enemies of pests is another important method to increase population diversity in the field. Genetic diversity use including the use of high-yielding gene, resistant gene, and high-quality gene is very important. Zhu Youyong's research (2000) showed that intercropping of traditional rice variety which was sensitive to rice blast disease with hybrid rice which was resistant to rice blast disease in the field with ratio of 1–4 lines could control blast disease very well without chemical use.

### *6.3 Social Economic Regulation and Control for Agroecosystem*

In order to encourage the implementation of ecologically sound practices mentioned above, social and economic measures should be worked out to regulate the behavior of farmers and farming enterprises. Motivation of people mainly drives by their judgment on whether the action is righteous or beneficial. Education through school and public media will deeply influence the view of people about what behavior is right or wrong. The technical standard and assessment standard for ecologically sound practices will guide the right direction of framer and farming enterprises. Law can let the ecologically wrong actions become illegal and hence can guide the general direction of social behavior. In order to internalize the external economic effect of ecological sound practices, on the one hand ecological compensation measures such as high tax for the use of resources, high penalty for discharge of pollutant, can be taken by government. On another hand, the method to clearly define the ownership of natural resources and encourage the growth of market for the exchange of natural resources can be adopted. The progress of the management system for agroecosystem in China is significant in the past 30 years. New regulation including the protection of agro-resources, rural planning and agricultural regionalization, agro-environment protection, quality requirement for agricultural input products, quality requirement for agricultural output products, and ecological agricultural have been made. However, there are still a lot of serious challenges faced by Chinese agriculture. An important consultation project "Strategy research on the optimum resource allocation, integrated environment management and regional agriculture development" was launched by the Chinese Academic of Engineering and deal with this challenge in a comprehensive way (Shi 2008; Dai and Hu 2008).

# 7 Nutrient Management in the Food Chain

## 7.1 Introduction

Chinese food security and ecological sustainability remain among the greatest challenges for the twenty-first century and both are significantly affected by the way we manage nutrients for agricultural systems, particularly nitrogen (N) and phosphorus (P). Critical to the growth of plants and animals and ultimately to human nutrition and health, N and P are also known to be major contributing factors for ecological deterioration that can negatively impact atmosphere, ground, and surface waters (Galloway et al. 2008). Nutrient management at farm and field levels has been defined as achieving agronomic and environmental objectives through an iterative series of six consecutive steps: analysis, decision-making, planning, execution, monitoring, and evaluation (Oenema and Pietrzak 2002). Such a concept has not been defined and described at regional levels yet. Managing nutrients in the food chain at regional level provide a safe and secure food supply while protecting the ecological system.

Here we summarize developments of nutrient management in the food chain in China in our group during the last 10 years. We begin with a brief introduction of food chain model. We then summarize the main findings of nutrient management in the food chain. We close by the prospects of the future research.

## 7.2 Development of Food Chain Model

Nutrient flows in Food chains, Environment and Resources use (NUFER) is a deterministic model with large databases that calculate the flows, use efficiencies, and emissions of N and P in the food chain of 31 regions and China on an annual basis. It uses a mass balance approach with detailed accounts of the partitioning of N and P inputs and outputs. It makes a distinction between "new" N and P (from bio-fixation and imported fertilizers, natural grassland and fish), and "recycled" N and P (from recycled material such as manure, crop residues, wastes, etc.).

NUFER consists of an input module with activity data and transformation and partitioning coefficients, a calculation module with equations, an optimization module and an output module. NUFER allows assessment of the N and P flows in the pyramid in two directions, viz., from the food production side and from the consumption side (Ma et al. 2010).

The progresses and compositions of NUFER model are shown in Fig. 4 and the detailed information are presented in the references (Liu et al. 2007, 2008a, b; Ma et al. 2008, 2009, 2010, 2011; Wang et al. 2007a, b, 2010, 2011; Wei et al. 2008).

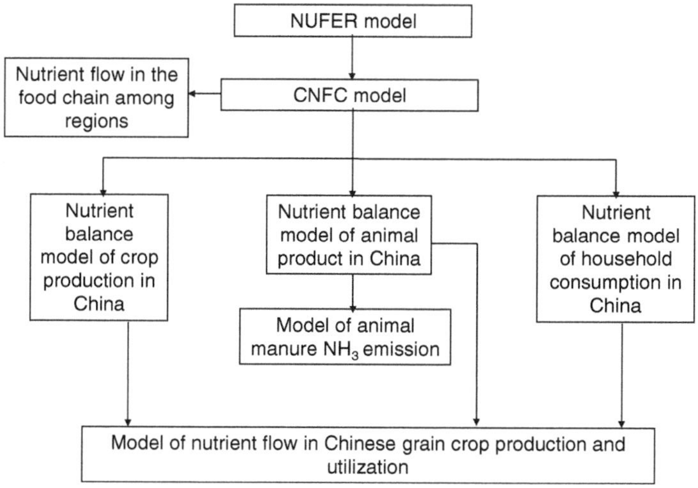


**Fig. 4** Models of nutrient flows in the food chain in China

## *7.3 Main Findings*

The main achievements of this thesis are:

- The major developments in nutrient management in China during the last 50 years were reviewed, and the future challenges for the next decades, considering nutrient management in the whole food chain of "crop production–animal production–food processing–food consumption" were addressed (Ma et al. 2013b).
- NUFER model was developed, as the first model that allows quantitative assessment of the N and P cost of food chain and losses to air and waters at regional and national level in China (Ma et al. 2010).
- Between 1980 and 2005, the mean N cost of 1 kg N in food in China increased dramatically from 6 to 11 kg $kg^{-1}$. Mean P cost increased even more, from 5 to 13 kg $kg^{-1}$ (Ma et al. 2012).
- The N (P) use efficiency in the food chain (NUEf and PUEf) decreased from ~20 % to less than 10 % in most of the regions between 1980 and 2005, and the regional variations were large (Ma et al. 2012).
- Quantitative analysis of driving forces indicated that a larger impact of changing nutrient management practice than population growth on elevated nutrient flows in China's food chain (Hou et al. 2013).
- Losses of $NH_3$ and $N_2O$ to air and N and P to groundwater and surface waters increased greatly in China between, especially in the Beijing and Tianjin metropolitans, Pearl River Delta, and Yangzi River Delta (Ma et al. 2012).
- In 2030 (the business as usual scenario), N and P fertilizer consumption will both increase by 25 %, and N and P losses will increase by 44 and 73 %,

respectively, relative to 2005, driven by population growth and diet changes in China (Ma et al. 2013a).

- Implementation of a package of integrated nutrient management measures, including balanced fertilization, precision feeding, and improved manure management, may roughly nullify the increase in the efficiency of N and P throughout the whole food chain in China at national level (Ma et al. 2013a).

### 7.4 Concluding Remark and Prospects

A food chain approach for managing nutrient flows at regional and national level is a new and evolving concept. The NUFER model is the first model that allows a quantitative assessment of the N and P cost of food, and N and P losses to air and waters at the food chain at national and regional level in China. Our researches are the first integrated assessment of changes of NUE and PUE in the food chain in China, and of N losses via $NH_3$, and $N_2O$ emissions to air and N and P leaching to groundwater and surface waters during 1980–2005 and future scenario in 2030.

For China, further research needs to close the knowledge gaps among researchers, farmers, and policy-makers. Food production technologies improvement should focus on, not only technologies themselves, but sound extension for farmers practices. Further research is needed to develop the food chain approach not only for nutrient management, but for water management, land management, and energy management and integration of these aspects. Upscaling of insight in nutrient management in the food chain to global scale is needed to understand the effects of globalization on nutrient cycling, soil fertility and NUE and PUE in food production—consumption chains (Sutton et al. 2013).

## 8 Fertilizer and Greenhouse Gas Management in Agroecosystem

### 8.1 Introduction

While chemical fertilizer is important for China's crop production, there have been increasing concern on its intensive use and environmental consequences. Chemical fertilizer application per hectare has grown rapidly after 1960s in China. The previous studies have showed that rising application of chemical fertilizer has significantly lowered the efficiency of fertilizer use (Wang 2007; Zhang et al. 2008; Zhu 1998) and also resulted in serious environmental stresses by increasing greenhouse gas (GHG) emissions and polluting ground and surface water through nitrogen leaching (Izaurralde et al. 2000). It is estimated that the manufacture and use of N fertilizer contributed to approximately 30 % of agricultural GHG emissions and about 5 % of China's total GHG emission in 2007 (SAIN 2010).

Recent studies have also shown that excessively high uses of chemical fertilizer have mainly occurred in N fertilizer. Several studies observed that N fertilizer used in wheat and maize production was often overused by farmers in North China Plain (Chen 2003; Cui 2005). Large number of studies also observed massive N fertilizer use in the rice production and approximately 30 % of N fertilizer can be reduced without lowering (and even increasing) rice yield in major rice production regions (Hu et al. 2007; Peng et al. 2006; Want et al. 2007). Some scientists believe that there is a lack of technology to improve the efficient use of N fertilizer for farmers, therefore substantial efforts have been made to identify efficient N fertilizer use methods in the field (Chen et al. 2006; Cui et al. 2008a, b; Zhao et al. 2006).

However, despite great efforts made by scientists to improve the efficiency of fertilizer use, N fertilizer use continues to rise while its efficiency remains low, which has brought economists' attention to the rationality of overusing N fertilizer. Huang et al. (2008) argued that farmers' lack of knowledge and information on crop responses to N fertilizer are the primary reason for its overuse. Chinese farmers have been used their experience from the previous Green Revolution in 1960s–1970s that significantly increased agricultural productivity by adopting high-yield varieties and using highly responsive chemical fertilizer.

If the above arguments are valid for N fertilizer use in rice production, several questions are raised. Can N fertilizer use also be reduced significantly in other major crops such as maize in China? Will reducing N fertilizer in crop production lead to declines in crop yields? How to deliver appropriate knowledge and information on the efficiency of fertilizer use to millions of small farmers in China? These questions are crucial, not only for the fertilizer industry, given the size of China's fertilizer market, but also for China's public agricultural extension system, which has a mandate to deliver technologies but has faced great challenges in offering appropriate technology and knowledge to millions of farmers (Hu et al. 2007; Huang et al. 2009).

The overall goals of this study are to provide empirical evidence for the above-raised questions by investigating the impacts of delivering information and knowledge on the efficiency of nitrogen fertilizer use in maize production through training. The following section will introduce the research design, the data collection, and investigate the impacts of N fertilizer application training on farmers' overall use of N fertilizer in the study sites. Then policy recommendations about reducing nonpoint pollution through crop production and China's public agricultural extension system are drawn in the last.

## 8.2 Method and Results

### 8.2.1 Research Design and Data Collection

The experimental research was conducted in Shouguang (SG) County of Shandong Province in the North China Plain (NCP) in 2009. In the county, 3 townships and 5 villages from each township, were randomly selected. In each township, we divided the 5 villages into 2 groups: 3 treated villages and 2 non-treated villages. In total, there were 9 treated villages and 6 non-treated villages. Villages were randomly selected to ensure these two groups of villages were comparable before the N fertilizer use training course was conducted in the treated villages.

For each of the treated villages, a training course on nitrogen fertilizer use in maize production was offered to farmers by the trained extension staff in April 2009, prior to maize being planted. The extension staffs were selected from local townships and were trained by China Agricultural University. Extension staffs came to each of the treated villages and randomly selected farmers in group for training.

Information provided during the training was based on the local soil situation and the results of N fertilizer management experiments in maize production conducted by soil scientists from China Agricultural University. Farmers were advised to use N fertilizer in the following ways: (1) controlling the total amount of N fertilizer use between 150 and 180 kg/ha; and (2) applying N fertilizer during maize growing season twice—once before the 10-leaf stage and once after the 10-leaf stage.

In November 2009 after the maize was harvested, we randomly selected farmers from both treated and non-treated villages for a face-to-face questionnaire-based household survey. In total, 226 farmers (66 trained and 160 non-trained) were selected from treated villages. In each of the non-treated villages, we randomly selected 20 farmers for a face-to-face questionnaire-based household survey, just as we did for those farmers from treated villages. At the end, we surveyed a total of 116 farmers from non-treated villages.

### 8.2.2 Fertilizer Use and N Fertilizer Application

Table 3 reports average fertilizer use by trained and non-trained farmers in 2009, which provides several interesting observations.

First, it seems that farmers were not interested in applying fertilizers twice during one maize growing reason, one of two key findings provided in the training course. Farmers might be used to their conventional fertilizer application practices, and as such it was difficult to change their fertilizer use behaviors in the short run. Some farmers might also be concerned about increased labor input by changing their one-time fertilizer application to a two-time application.

Second and most importantly, while the application time does not change with the introduction of training, trained farmers used much less N fertilizer than

**Table 3** Average fertilizer use by trained and non-trained farmers in maize production and the yield of maize in 2009

| | Trained farmers | Non-trained farmers | |
|---|---|---|---|
| | | Treated villages | Non-treated villages |
| Fertilizer use | | | |
| Total number of application | 1.23 | 1.36 | 1.32 |
| Number of N application | 1.20 | 1.36* | 1.32 |
| Amount used (kg/ha)[a] | | | |
| N | 179 | 221** | 217* |
| P | 101 | 106 | 94 |
| K | 56 | 50 | 58 |
| Average yield (kg/ha) | 8512 | 7991** | 7890** |

*Source* CCAP's survey

* and ** Denotes statistical significance of the mean different from the trained farmers in treated villages at 1 %. There is no case with statistical significance at 5 %

[a]The figures indicate pure content of N, P, and K

non-trained farmers (row 3, Table 3). Non-trained farmers applied an average of 221 kg/ha of inorganic N fertilizer in treated villages and 217 kg/ha in non-treated villages, while the trained farmers applied an average of 179 kg/ha of N fertilizer which was in the amount recommended by scientists (150–180 kg/ha).

Third, training in the improved N management practice for maize production had no impacts on farmers' potassium and phosphorus application.

Lastly, notwithstanding that trained farmers reduced overall N fertilizer use in maize production, the yield was not affected. As shown in Table 1, the average yield of maize for trained farmers in treated villages did not decrease when comparing the figures for farmers in non-treated villages.

### 8.2.3 Multivariate Analysis on the Impacts of Knowledge Training on Farmers' N Fertilizer Use

Model Specification

Based on the survey data, we created a cross-section dataset consisting of 342 farmers from 15 villages studied. To estimate the impacts of training on maize farmers' N fertilizer use, the following empirical model is specified:

$$\mathrm{N}_i = a + b \cdot \mathrm{TF}_i + c \cdot \mathrm{NTF}_i + \varphi \cdot X + \varepsilon_i \tag{1}$$

$$\mathrm{Freq}_i = a + b \cdot \mathrm{TF}_i + c \cdot \mathrm{NTF}_i + \varphi \cdot X + \varepsilon_i \tag{2}$$

where $\mathrm{N}_i$ is the $i$th household's N fertilizer use per hectares, and $\mathrm{Freq}_i$ is the number of N fertilizer applications. The key independent variable of interest is TF, trained farmers in treated village; it is a binary variable that equals 1 if a household attended the N fertilizer application training in the treated village, otherwise it equals 0. NTF indicates non-trained farmers in treated village; it is designed to catch the likely spillover effect within a treated village. The bases for comparison are those households from non-treated villages.

As a set of control variables, $X$ includes a household's demographics (for example, land area, age of household head, education of household head, female headed household, share of off-farm labor before the maize season, and consumption asset per capita in 2009), and regional characteristics (for example, access to the nearest fertilizer shop and county dummy variable). We specify an ordinary least squares estimator (henceforth, OLS) and OLS with logarithmic transformation to estimate Eq. (1), and use Poisson model to estimate Eq. (2). While the linear model presents marginal effects directly, the logarithmic functional specification directly provides coefficients with percentage effect interpretations. When the dependent variable is discrete and satisfy Poisson distribution, the method of Poisson model estimation is regarded more efficient than the OLS.

#### Estimation Results

The modeling performs well and the results are presented in Table 4. The estimated coefficients of variables of interest and for control have intuitive signs. The statistical significance for all estimated coefficients is also robust and consistent in both OLS and logarithmic transformation specifications. Results are summarized as follows:

First, the training led to a significant reduction—23 % (34.82 kg/ha)—of chemical N fertilizer use by trained farmers in treated villages. The reduction rate related to knowledge training is 23 % (column 2, Table 4); the training was thus effective in reducing maize farmers' N use.

Second, there are no spillover effects in the treated village. As shown in Table 4 (row 2), the coefficients of non-trained farmers in treated villages in Eqs. (1) and (2) are both insignificant, implying that, when holding all else constant, there is no significant difference between farmers in non-treated villages and farmers who did not receive the direct training in the treated villages.

Third, the rising household land area or farm size could result in a significant reduction on N fertilizer use per hectare (row 3, Table 4). Double farm size could reduce N fertilizer use by 15 % (column 2). As expected, the more aged farmers tended to apply more N fertilizer.

## 8.3 Conclusion and Discussion

Chemical fertilizer plays an important role in increasing food production in China. However, farmers in China use much more fertilizer per hectare than do farmers in

**Table 4** Estimated results of farmers' N fertilizer use in maize production in 2009

| | Total N fertilizer use [a] (kg/ha) | | Number of N fertilizer application (3) |
|---|---|---|---|
| | N (1) | Ln (N) (2) | |
| Trained farmers in treated villages (Yes = 1; No = 0) | −34.82** (2.12)[b] | −0.23*** (3.03) | −0.14 (0.77) |
| Non-trained farmers in treated villages (Yes = 1; No = 0) | 13.21 (0.95) | −0.00 (0.02) | 0.02 (0.13) |
| Household land area (ha) | −31.24*** (2.63) | −0.15*** (4.05) | −0.02 (0.16) |
| Age of household head (years) | 1.37** (2.23) | 0.00 (1.40) | 0.00 (0.22) |
| Education of household head (years) | −0.84 (0.39) | −0.01 (0.94) | 0.02 (0.66) |
| Female headed household (Yes = 1; No = 0) | 18.53 (0.82) | 0.02 (0.22) | −0.02 (0.09) |
| Share of off-farm labor before maize season (%) | −0.07 (0.38) | 0.01 (0.70) | −0.00 (0.04) |
| Consumption asset per capita in 2009 (1000 yuan) | 0.09 (0.31) | 0.05 (1.57) | −0.00 (0.10) |
| Distance to nearest fertilizer shop (km) | −8.93 (1.61) | −0.00 (0.02) | 0.02 (0.31) |
| Intercept | 170.62*** (3.81) | 4.82*** (22.12) | |
| Total samples | 342 | 342 | 342 |
| $R^2$ | 0.08 | 0.10 | |

*Source* CCAP's survey
[a]The figures indicate pure N content
[b]The figures in the parentheses are absolute t ratios of estimates
*, **, and *** Represent the statistical significance at the 10, 5 and 1 %, respectively

many other countries. The overuse of nitrogen fertilizer has resulted in serious environmental stress and increased greenhouse gas emissions. This study seeks to find appropriate measures to reduce excess fertilizer use through a training program. The results show that delivering information and knowledge on the efficiency of nitrogen fertilizer through the public agricultural extension system can significantly lower inorganic N fertilizer use by 22 % in maize production in the NCP; knowledge training indeed matters. Farm size is negatively associated with per hectare N fertilizer use.

Training for farmers on nitrogen management has great potential on reducing agricultural GHG emissions. In China 30 % of agricultural GHG emissions come from N fertilizer production and utilization, the total emissions generated by N fertilizer have exceeded 500 million ton every year (SAIN 2010). Moreover, the environmental damage resulted from the overuse of nitrogen fertilizer has also been significant. SAIN (2010) showed that through appropriate use of N fertilizer, China's GHG emissions could reduce by 320 million ton without lowering the grain yield. However, such estimation has not fully considered the real world or farmers' behaviors. The results of this study suggest that, under the current farming system, farmers' N fertilizer use could reduce by about 20 %, which would result in the reduction of 100 million ton GHG emissions. Although this figure is much less than the optimal emission-cutting scheme designed by the agricultural scientists, N fertilizer management in crop production can significantly contribute to a reduction of China's GHG emissions in the coming years.

The findings of this study also have other important policy implications. First, policies on land rental markets or land consolidation programs that aim to expand farm size can also help Chinese farmers to reduce N fertilizer use in crop production. Second, despite significant reductions of N fertilizer use by farmers through training program, training more than 200 million small farmers is not without cost. How to deliver appropriate information and knowledge on the efficiency of N fertilizer to millions of farmers is an issue that requires further study because the current agricultural extension system also faces great challenges related to providing technology services to farmers (Huang et al. 2009). Finally, the recommended technologies to farmers should meet farmer's demand for labor-saving technologies. With rising wages and off-farm employment opportunities, and given the predominance of small-scale farms in China, advising farmers to use higher frequency but less-intensive fertilizer technologies seems not to appeal by farmers. New technologies (e.g., slow release fertilizer and nitrification inhibitors) that are less labor-demanding may fit with farmers' habits and strategies of optimizing household welfare.

## References


Cai, T., Kong, F., & Jiang, D. (2003). Present situation, problem and countermeasure of forest classified management in China. *Journal of Northeast Forestry University, 31*, 42–44. (in Chinese).

Chen, Y. Y., & Lu, X. G. (2003). The wetland function and research tendency of wetland science. *Wetland Science, 1*(1), 7–11.

Chen, X. (2003). *Optimization of the N fertilizer management of a winter wheat/summer maize rotation system in the Northern China Plain*. Unpublished doctoral dissertation, Universitat Hohenheim.

Chen, X., Zhang, F., Römheld, V., Horlacher, D., Schulz, R., Böning-Zilkens, M., et al. (2006). Synchronizing N supply from soil and fertilizer and N demand of winter wheat by an improved Nmin method. *Nutrient Cycling in Agroecosystems, 74*(2), 91–98.

Cui, Z. (2005). *Optimization of the nitrogen fertilizer management for a winter wheat-summer maize rotation system in the North China Plain from field to regional scale*. Unpublished doctoral dissertation, China Agricultural University, Beijing.

Cui, Z., Chen, X., Miao, Y., Zhang, F., Sun, Q., Schroder, J., et al. (2008a). On-farm evaluation of the improved soil Nmin-based nitrogen management for summer maize in North China Plain. *Agronomy Journal, 100*(3), 517–525.

Cui, Z., Zhang, F., Miao, Y., Sun, Q., Li, F., Chen, X., et al. (2008b). Soil nitrate-N levels required for high yield maize production in the North China Plain. *Nutrient Cycling in Agroecosystems, 82*(2), 187–196.

Dai, J., & Hu, Y. (Eds.). (2008). *Agriculture structure adjustment and regional planning*. Beijing: China Agriculture Press.

Dai, L., Gu, H., Shao, G., & Wang, Q. (2004). *The Broadleaved-Korean pine mixed forest on Changbai Mountain of China*. Shenyang: Liaoning Science and Technology Publishing House. (in Chinese).

Dai, L., Shao, G., & Xiao, B. (2003). Ecological classification for mountain forest sustainability in northeast China. *Forestry Chronicle, 79*(2), 233–236.

Dai, L., Wang, Y., Su, D., Zhou, L., Yu, D., Lewis, B., & Qi, L. (2011). Major forest types and the evolution of sustainable forestry in China. *Environmental Management, 48*, 1066–1078.

Dai, L., Zhao, F., Shao, G., Zhou, L., & Tang, L. (2009). China's classification-based forest management: procedures, problems, and prospects. *Environmental Management,*. doi:10.1007/s00267-008-9229-9.

Dai, L., Zhao, W., Shao, G., Lewis, B., Yu, D., Zhou, L., & Zhou, W. (2013). The progress and challenges in sustainable forestry development in China. *International Journal of Sustainable Development and World Ecology,*. doi:10.1080/13504509.2013.775193.

Deng, H. (2008). *Sustainable forest management in China*. Beijing: Science Press. (in Chinese).

Galloway, J. N., Townsend, A. R., Erisman, J. W., Bekunda, M., Cai, Z., Freney, J. R., et al. (2008). Transformation of the nitrogen cycle: recent trends, questions, and potential solutions. *Science, 320*, 889–892.

Guo, H., Ge, J., & Li, J. (1995). Outline of ecological literatures on Korean pine forest in China. *Journal of Northeast Forestry University, 23*(3), 57–62. (in Chinese).

Hao, Z., Dai, L., Wang, Q., Huang, N., & Dong, S. (2000). Relationship of natural forest conversation and biodiversity protection. In P. C. Zhang (Ed.), *Discussion on natural forest conservation programme* (pp. 294–301). Beijing: Chinese Forestry Press. (in Chinese).

He, C. Q., & Zhao, K. Y. (2001). The accumulation, allocation and biological cycle of the nutrient elements in *Carex lasiocarpa* weltand. *Acta Ecologica Sinica, 21*(12), 2074–2080.

Hou, Y., Ma, L., Gao, Z. L., Wang, F. H., Sims, J. T., Ma, W. Q., & Zhang, F. S. (2013). The driving forces for nitrogen and phosphorus flows in the food chain of China, 1980 to 2010. *Journal of Environmental Quality, 42*, 962–971.

Hu, R., Cao, J., Huang, J., Peng, S., Huang, J., Zhong, X., et al. (2007). Farmer participatory testing of standard and modified site-specific nitrogen management for irrigated rice in China. *Agricultural Systems, 94*(2), 331–340.

Huang, J., Hu, R., & Zhi, H. (2009). The 30 years' development and reform for the grass- roots agricultural technology extension system: Policy assessments and recommendations. *Journal of Agrotechnical Economics, 1*, 4–10. (in Chinese).

Huang, J. K., Hu, R. F., Cao, J. M., & Rozelle, S. (2008). Training programs and in-the-field guidance to reduce China's overuse of fertilizer without hurting profitability. *Journal of Soil and Water Conservation, 63*(5), 165A–167A.

Izaurralde, R. C., McGill, W. B., Rosenberg, N. J., & Schlesinger, W. H. (2000). Carbon cost of applying nitrogen fertilizer. *Science, 288*, 811–812.

Jiao, J. (2003). Keys and considers on ecological restore. *Soil and Water Conservation in China, 2*, 1–2.

Li, L., Li, S., Sun, J., Zhou, L., Bao, X., Zhang, H., & Zhang, F. (2007). Diversity enhances agricultural productivity via rhizosphere phosphorus facilitation on phosphorus-deficient soils. *PNAS, 104*(27), 11192–11196.

Liang, Z., Zuo, C., & Jiao, J. (2003). The role of ecology restore on soil and water consternation in Loess Plateau. *Journal of Northwest Forestry University, 18*(1), 20–24.

Liu, D. (2001). Tenure and management of non-state forests in China since 1950: A historical review. *Environmental History, 6*(2), 239–263.

Liu, H. Y., Zhao, Z. C., & Lv, X. G. (1999). A study on wetland resources and protection in China. *Resources Science, 21*(6), 34–37.

Liu, L. L., Wang, R., Duan, Y., Wang, X., Wang, S., & Wang, Y. (2008a). Status and effect of enhancement release of marine fisheries resource in Shandong. *Transactions of Oceanology and Limnology, 2008*(2), 91–98.

Liu, S. (1957). Problems of forest regeneration on Mount Da Xinganling and Xiao Xinganling. *Forestry Science, 3*, 263–280. (in Chinese).

Liu, S. (1963). Korean pine forest must choose selecting cut and natural regeneration. In Z. Wang (Ed.), (1985) *Liu shen'e collect works* (pp. 291–294). Beijing: Science Press. (in Chinese).

Liu, S. (1973). Comprehending the forest cutting and regeneration regulation from the situation of cutting and regeneration of Korean pine in Northeast China. In Z. Wang (Ed.), (1985) *Liu shen'e collect works* (pp. 314–326). Beijing: Science Press. (in Chinese).

Liu, S., Zhu, J., & Cao, X. (1965). Advice on change large area successive clear cutting in Korean pine forest. In Z. Wang (Ed.), (1985) *Liu shen'e collect works* (pp. 295–299). Beijing: Science Press. (in Chinese).

Liu, D., Ma, L., Wang, F. H., Ma, W. Q., & Zhang, F. S. (2007). Rate of nitrogen excretion and ammonia emission from pig production: A case study of Hebei province. *Journal of Agricultural University of Hebei, 30*, 5–10. (in Chinese).

Liu, D., Wang, F. H., & Ma, L. (2008b). Evaluation of $NH_3$ emission from pig manure in China I $NH_3$ emission factor. *Transactions of the Chinese Society of Agricultural Engineering, 24*, 57–61. (in Chinese).

Lu, X. G. (2001). Wetland protection and water resources security in China. *China Water Resrouces, 11*, 26–27.

Luo, S. (2010a). A system approach toward ecoagriculture in China. In *The Proceedings of the 9th European IFSA Symposium* (pp. 2108–2115). Vienna, Austria: IFSA.

Luo, S. (2010b). On the technical package for eco-agriculture. *Chinese Journal of Eco-Agriculture, 18*(3), 453–457.

Ma, L., Ma, W. Q., Velthof, G. L., Wang, F. H., Qin, W., Zhang, F. S., & Oenema, O. (2010). Modeling nutrient flows in the food chain of China. *Journal of Environmental Quality, 39*, 1279–1289.

Ma, L., Velthof, G. L., Wang, F. H., Qin, W., Zhang, W. F., Liu, Z., et al. (2012). Nitrogen and phosphorus use efficiencies and losses in the food chain in China at regional scales in 1980 and 2005. *Science of the Total Environment, 434*, 51–61.

Ma, L., Wang, F., Zhang, W., Ma, W., Velthof, G., Qin, W., et al. (2013a). Environmental assessment of management options for nutrient flows in the food chain in China. *Environmental Science and Technology, 47*, 7260–7268.

Ma, L., Wei, J., Wang, F. H., Gao, L. W., Zhao, L., Ma, W. Q., & Zhang, F. S. (2009). Studies on nitrogen flow in food chain among regions based on MFA and model-A case of Huang-Huai-Hai plain. *Acta Ecologica Sinica, 29*, 475–483. (in Chinese).

Ma, L., Zhang, W. F., Ma, W. Q., Velthof, G. L., Oenema, O., & Zhang, F. S. (2013b). An analysis of developments and challenges in nutrient management in China. *Journal of Environmental Quality, 42*, 951–961.

Ma, W. Q., Li, J. H., Ma, L., Wang, F. H., Sisak, I., Cushman, G., & Zhang, F. S. (2008). Nitrogen flow and use efficiency in production and utilization of wheat, rice, and maize in China. *Agricultural Systems, 99*, 53–63.

Ma, W. Q., Ma, L., Li, J. H., Wang, F. H., Sisak, I., & Zhang, F. S. (2011). Phosphorus flows and use efficiencies in production and consumption of wheat, rice, and maize in China. *Chemosphere, 84*, 814–821.

Mitsch, W. J., & Gosselink, J. G. (2000). *Wetlands* (3rd ed.). New York: Wiley.

Oenema, O., & Pietrzak, S. (2002). Nutrient management in food production: Achieving agronomic and environmental targets. *Ambio, 31*, 159–168.

Pei, T. (2000). Natural forest conversation and flood control and treatment in Northeast China. In P. C. Zhang (Ed.), *Discussion on natural forest conservation programme* (pp. 275–293). Beijing: Chinese Forestry Press. (in Chinese).

Peng, S.-L., & Lu, H.-F. (2003). Some key points of restoration ecology. *Acta Ecologica Sinica, 23*(7), 249–257.

Peng, S., Buresh, R. J., Huang, J., Yang, J., Zou, Y., Zhong, X., et al. (2006). Strategies for overcoming low agronomic nitrogen use efficiency in irrigated rice systems in China. *Field Crops Research, 96*(1), 37–47.

Qin, P., An, S., & Yan, J. (1998). *Eco-engineering*. Nanjing: Nanjing University Press.

Quan, G.-M., Zhang, J., Yang, J., Chen, R., & Xu, R.-B. (2008). Impacts of integrated rice-duck farming system on rice quality. *Acta Ecologica Sinica, 28*(7), 3475–3483.

SAIN. (2010). *UK-China sustainable agriculture innovation network policy brief.* From http://www.sainonline.org/English.html.

Shao, G., Dai, L., Li, Y., Liu, Y., & Bai, G. (2003). FORESTAR: A decision-support system for multi-objective forest management in Northeast China. *Journal of Forestry Research, 14*(2), 141–145.

Shao, G., Wang, H., Dai, L., Wu, G., Li, Y., Lang, R., & Song, B. (2005). Integrating stand and landscape decisions for multi-purposes of forest harvesting. *Forest Ecology and Management, 207*, 233–243.

Shao, G., & Zhao, G. (1998). Protecting versus harvesting of old-growth forests on the Changbai Mountain (China and North Korea): a remote sensing application. *Natural Areas Journal, 18*, 334–341.

Shao, G., Zhao, P., Bai, G., Zhao, G., & Wang, Z. (2001). Ecological classification system for China's natural forests: protection and management. *Acta Ecologia Sinice, 21*, 1564–1568. (in Chinese).

Shen, Y., Wang, J., Wu, G., & Han, D. (2001). *Oases in China*. Kaifeng: Henan University Press.

Shi, Y. (2008). *Reasonable agro-resources allocation and integrative productivity increase in agriculture*. Beijing: China Agriculture Press.

Sutton, M. A., Bleeker, A., Howard, C. M., Bekunda, M., Grizzetti, B., de Vries, W. et al. (2013). *Our nutrient world: the challenge to produce more food and energy with less pollution*. Global Overview of Nutrient Management.

Tang, L., Shao, G., & Dai, L. (2009). Roles of digital technology in China's sustainable forestry development. *International Journal of Sustainable Development and World Ecology, 16*(2), 94–101.

Wang, F., Sims, J. T., Ma, L., Ma, W., Dou, Z., & Zhang, F. (2011). The phosphorus footprint of China's food chain: Implications for food security, natural resource management, and environmental quality. *Journal of Environmental Quality, 40*, 1081–1089.

Wang, F. H., Dou, Z. X., Ma, L., Ma, W. Q., Sims, J. T., & Zhang, F. S. (2010). Nitrogen mass flow in China's animal production system and environmental implications. *Journal of Environmental Quality, 39*, 1537–1544.

Wang, J. (2007). *Analysis and evaluation of yield increase of fertilization and nutrient utilization efficiency for major cereal crops in China*. Unpublished doctoral dissertation, China Agricultural University, Beijing.

Wang, G., Zhang, Q. C., Witt, C., & Buresh, R. J. (2007a). Opportunities for yield increases and environmental benefits through site-specific nutrient management in rice systems of Zhejiang Province. *China Agricultural Systems, 94*(3), 801–806.

Wang, J. Q., Ma, W. Q., Jiang, R. F., & Zhang, F. S. (2007b). Development and application of nitrogen balance model of agro-ecosystem in China. *Transactions of the Chinese Society of Agricultural Engineering, 23*, 210–215. (in Chinese).

Wang, S., van Kooten, G., & Bill, W. (2004). Mosaic of reform: forest policy in post-1978 China. *Forest Policy and Economics, 6*, 71–83.

Wang, Z. (1957). On the forest regeneration and cutting system of Korean pine forest of Xiao Xinganling Mountain. *Forestry Science, 3*, 281–288. (in Chinese).

Wang, Z., Fu, P., & Deng, Y. (1997). Founder and pioneer of China's botany and ecology—in memory of 100th anniversary of Liu Shen's birth. *Chinese Journal of Ecology, 16*(5), 77–80. (in Chinese).

Wei, J., Ma, L., & Lu, G. (2008). The influence of urbanization on nitrogen flow and recycling utilization in food consumption system of China. *Acta Ecologica Sinica, 28*, 1016–1025. (in Chinese).

Xie, J., Liu, L., Chen, X., Chen, J., Yang, X., & Tang, J. (2009). Control of diseases, pests and weeds in traditional rice-fish ecosystem in Zhejiang. *China Bulletin of Science and Technology, 25*(6), 801–805.

Xu, J., Tao, R., & Gregory, S. (2004). An empirical analysis of China's state-owned forests. *Forest Policy and economics, 6*, 379–390.

Xu, S. G., Guo, H. J., & Tian, K. (2006). Progress of wetland functional evaluation. *Environmental and Sustainable Development, 5*, 12–14.

Yao, C., Wan, S. W., Sun, D. L., & Qin, P. (2009). Ecological mechanisms of vegetation succession of coastal wetland in Yancheng Nature Reserve. *Acta Ecologica Sinica, 29*(5), 2203–2210.

Yu, D., Zhou, L., Zhou, W., Ding, H., Wang, Q., Wang, Y., et al. (2011). Forest Management in Northeast China: History, Problems, and Challenges. *Environmental Management,*. doi:10.1007/s00267-011-9633-4.

Yuan, J., & Lu, X. G. (2004). Progress in Research o Functional Assessment of Wetlands. *Wetland Science, 2*(2), 153–160.

Zhang, P., Shao, G., Zhao, G., Le Master, D., Parker, G., Dunning, J., & Li, Q. (2000). China's forest policy for the 21st century. *Science, 288*, 2135–2136.

Zhang, Z. S., Song, X. L., Lu, X. G., & Xue, Z. S. (2013). Ecological stoichiometry of carbon, nitrogen, and phosphorus in estuarine wetland soils: influences of vegetation coverage, plant communities, geomorphology, and seawalls. *Journal of Soils and Sediments, 13*, 1043–1051.

Zhang, F., et al. (2008). *China's fertilizer industry and scientific fertilization strategy research report*. Beijing: China Agricultural University Press.

Zhao, G., & Shao, G. (2002). Logging restrictions in China: a turning point for forest sustainability. *Journal of Forestry, 100*, 34–37.

Zhao, H., Zhao, X., Zhang, T., & Li, Y. (2009). *Restoration ecology*. Beijing: Science Press.

Zhao, Q. G., & Gao, J. F. (2007). Ecosystem services of wetlands and their delineation in China. *Chinese Journal of Eco-Agriculture, 15*(1), 1–4.

Zhao, R. F., Chen, X. P., Zhang, F. S., Zhang, H., Schroder, J., & Römheld, V. (2006). Fertilization and nitrogen balance in a wheat-maize rotation system in North China. *Agronomy Journal, 98*(4), 938–945.

Zhou, G. (2006a). *The rise and fall of the Cattle Kingdom*. Beijing: People's Publishing House.

Zhou, S. (2006b). *Forestry in China: historical transitions and industry developments*. Singapore: Thomson.

Zhou, X. (1982). Restoration way of Korean pine forest-planting conifer and protecting broadleaved tree. *Journal of Northeast Forestry University, 10*(8), 18–28. (in Chinese).

Zhu, Y. (2007). *Genetic diversity for crops diseases' sustainable management*. Beijing: Sciences Press.

Zhu, Y., Chen, H., Fan, J., Wang, Y., Li, Y., Chen, J., Fan, J., Yang, S., Hu, L., Leung, H., Mew, T. W., Teng, P. S., Wang, Z., Mundt, C. C. (2000). Genetic diversity and disease control in rice. *Nature, 406*(6797), 718–722.

Zhu, Z. (1998). Fertilizer, agriculture and environment. *Exploration of, Nature, 17*(4), 25–28.

# Chapter 9
# Water Management

**Wei Meng, Yuan Zhang, Weijing Kong, Qingwen Min, Wenjun Jiao, Yaning Chen, Honghua Zhou, Shaozhong Kang, Ling Tong, Xinmei Hao and Shirong Liu**

**Abstract** Fresh water area in global surface is less than 1 % in the area of global land. Human are connected to freshwater ecosystems closely, but also affects the freshwater ecosystems. Water ecological degradation of river basin becomes the main problems of the current global water ecosystem. River basin water ecological management and quality management of water environment are the basis of water environment management. Ecological water requirement is brought up under the background of ecosystem degradation caused by extrusion ecological water from water resources development and utilization of activity. Configuration of ecological water requirement study will be integrated with the mechanism research and model research in future, forming a complete organic research system. The response between ecological process and hydrological processes is an important part of ecological hydrology study in arid areas. We carry out dual research of the management of water resources in arid areas from the macro-scale and micro-scale level. Saline-alkali land management and development has always been the hot spot of the domestic and foreign research theory and method innovation, and it is the important content of saline-alkali land management development. Minjiang river basin is an important large-scale, complex ecological transition zone, and the complexity of the natural environment, the borderline of economic development

W. Meng · Y. Zhang · W. Kong
Chinese Research Academy of Environmental Sciences, Beijing 100012, China

Q. Min · W. Jiao
Institute of Geographic Sciences and Natural Resources Research, Chinese Academy of Sciences, Beijing 100101, China

Y. Chen · H. Zhou
State Key Laboratory of Desert and Oasis Ecology, Xinjiang Institute of Ecology and Geography, Chinese Academy of Sciences, Urumqi 830011, China

S. Kang · L. Tong · X. Hao
Center for Agricultural Water Research in China, China Agricultural University, Beijing 100193, China

S. Liu (✉)
Chinese Academy of Forestry, Beijing 100091, China
e-mail: liusr@forestry.ac.cn

 
W. Li (ed.), *Contemporary Ecology Research in China*,
DOI 10.1007/978-3-662-48376-3_9

and social cultural transition are typical in china. Hydrologic cycle of forest ecosystem dynamics model and coupling of multiple scale water circulation regulation theory has not yet been established, deforestation, and the effects of climate change on river basin ecological hydrology is yet to be evaluated.

**Keywords** Eco-function zones · Freshwater eco-function management zone · Eco-hydrological processes · Ecological safety and water demand · Water-heat balance · Limiting crop water consumption

# 1 Water Eco-function Zones

## *1.1 Introduction*

Although freshwater habitats occupy less than 1 % of the Earth's surface, they host nearly 10 % of all animal species found globally; hence, freshwaters are considered biodiversity hotspots (Strayer and Dudgeon 2010). Freshwater ecosystems also provide extensive resources and ecosystem services for humans. Human activities, however, have resulted in point and non-point source pollution, overfishing, land reclamation, invasive species, flow regulation, and excessive water withdrawals that negatively affect the physical, chemical, and biological integrity of freshwater ecosystems (Strayer and Dudgeon 2010). As a result, freshwater species are increasingly threatened and endangered. Humans are also widely affected by ecological degradation, with nearly 80 % of the global population now faced with issues of water security (Vörösmarty et al. 2010). The current state of freshwater resources demonstrates the need for comprehensive management of aquatic ecosystems.

Within the current mode of economic development, environmental management of freshwaters in China has undergone two stages: control of pollutant concentrations, and control of the total amount of pollutant discharge. Since these pollutant control measures were not linked to overall water quality targets, protection of the aquatic ecosystem has not been ensured (Meng et al. 2011). Taking into account the challenges for environmental management in China, Meng (2007) developed a new strategy for pollution control done by categories, zones, and stages, with water quality targets based on different levels.

In 2006, the National Long-Term Science and Technology Development Plan and Outline included a provision for delineation of freshwater ecological function management zones. In 2008, the National Major Science and Technology Program for Water Pollution Control established several projects to delineate the freshwater eco-function management zones in the main river and lake basins of China. These programs will ensure the transition of freshwater management from water quality to ecosystem management.

## *1.2 Freshwater Eco-function Management Zones*

### 1.2.1 Current Methods Used for Eco-Function Delineation

Eco-function zones are widely used to guide ecological management throughout the world, including in the United States, European Union, Australia, and New Zealand (Maxwell et al. 1995; European Union Commission 2000; Higgens et al. 2005). Eco-function zones, or eco-regions, are typically arranged as hierarchically nested systems, based on the extent of influence each factor exerts on the system (Sorrano et al. 2010). Different classification systems exist for different regions, ecosystems, and contexts. The system developed by The Nature Conservancy has four levels, from broadest to narrowest, of aquatic zoogeography, ecological drainage, ecological system, and macrohabitat units (Higgens et al. 2005). Aquatic ecological units for North America include 11 levels extending from the very large continental scale, to the coarse basin scale, down to the channel unit scale (Maxwell et al. 1995). The number and type of levels is determined by the scale and context of management objectives.

### 1.2.2 Freshwater Environment Management Zones in China

Since the 1980s, several ecological regionalization schemes have been conducted in China, including eco-function zoning and eco-geographical zoning. These zonings, however, were based on terrestrial ecosystems, and were therefore not appropriate for use in management of freshwater ecosystems. Delineation of freshwater eco-regions in China started in the 1980s, with the first fish-based zoning scheme completed in 1981. Zoning schemes were also completed to control the discharge of pollutants into freshwaters. While such zonings work for management of pollutant discharge based on human use of water resources, they are not sufficient for ensuring ecological condition. Eco-function zoning that incorporates measures of ecological integrity is now needed for effective and comprehensive management of freshwater resources in China.

### 1.2.3 Definition of Freshwater Eco-function Management Zone

Freshwater eco-function zones are delineated based on hierarchical elements that govern different levels of the system, resulting in the partitioning of a watershed into spatial units based on environmental factors, ecosystem features, and ecological functions. The zones represent the distribution of freshwater ecosystems and habitats, which include key and endangered species and their habitats. The zones also include aspects of ecosystem function, with targets for ecological security according to management guidelines. The overall aim of the freshwater eco-function zones is to protect the integrity of the freshwater ecosystem and provide the basis for ecological management, conservation, and rehabilitation of freshwater resources (Meng et al. 2013).

### 1.2.4 Framework of Management Zone Systems

The current system of freshwater eco-function zones in China has four hierarchical levels. Each level represents different ecosystem features and meets different management demands. Levels I and II reflect the general ecosystem features of the watershed, while Levels III and IV classify river types and ecological function. Levels I and II were delineated based on overarching climatic and terrestrial factors, which set the physical and chemical background of the rivers. Level III was classified based on integrated aspects of the river system. Level IV was based on ecological function in smaller units of river segments. Each level is designed to meet different management demands, such as river health assessment in Level II, and standards for water quality in Level IV.

### 1.2.5 Management Use of Freshwater Zones

The freshwater eco-function zone provides the basic unit for ecosystem management. Meng et al. (2011) discussed a framework for the use of eco-function zones to determine water quality targets. Eco-function zones are mainly used in ecosystem management to: (i) assess freshwater ecosystem health, (ii) determine sites for monitoring, and (iii) set water quality standards and benchmarks (Meng et al. 2008a, b). These three uses for eco-function zones have been extensively studied and widely applied in China (Fig. 1).

## *1.3 Discussion and Conclusions*

Levels I and II of the freshwater eco-function zoning were completed in 2010 for watersheds of the Songhua, Liao, Hei, Hai, Huai, Dong, and Gan Rivers, as well as the Dian, Er, and Chao Lakes. Reference sites have been selected based on the zones used in river health assessment. Level III and IV classification will be completed in 2015 for the 10 main river and lake watersheds. These eco-function zones will play an important role in management of water quality targets for freshwater ecosystems.

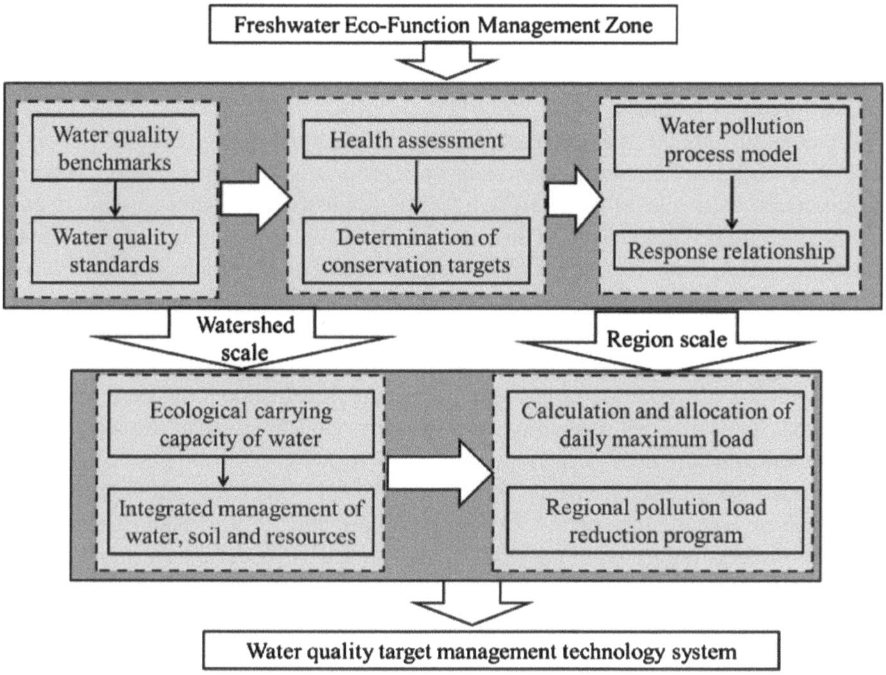


**Fig. 1** Framework for determining freshwater eco-function management zones for use in setting water quality targets (Meng et al. 2011)

## 2 Water Quality Target Management

China has been faced with water problems in many aspects for quite a long time, such as water pollution, water shortage, ecological degradation, and flood. Water pollution has in effect aggravated to the other three water problems to a certain degree. Work related to water pollution prevention and control has been extensively carried out across major river and/or lake basins of the country and has achieved some success for these years. However, water pollution will remain for quite a long period as the key factor constraining economic and social sustainable development of China.

Reasons are manifold. A very important aspect lies in the mismatches between application details and setting targets in the "total quantity control of pollutant emission" which is the major system implemented for water pollution prevention and control in China. For instance, pollution control techniques mismatched the conservation of aquatic ecosystems; emission control standards mismatched the improvement of water quality; the district-based water environment function zoning mismatched the watershed-based water pollution regulation.

The traditional ideas have been unable to meet China's current requirements in water environment management which is now in urgent need of reform and innovation. The whole country needs, as soon as possible, to change the target from pollution control to ecological protection, to transform the mode from administrative management to basin management, and to improve the technique from target control to capacity control

Water environment management is a complex system engineering. Even in developed countries, water environment management is in a process of continuous improvement as the level of awareness continues to rise and the demand is constantly changing. The TMDL (Total Maximum Daily Loads) Plan of the United States, the EU Water Framework Directive, and the Total Amount Control Plan of Japan, are all very representative. After several years of development and improvement, the TMDL program has gradually formed a complete system of strategies, technologies, and methodologies for the total amount control of pollutant emission. It has played a very important role in the improvement of water quality of the United States and has represented the development direction of the world water environment management.

A framework of water quality target management has been put forward after learning from experiences of the TMDL technique and fully considering practical problems such as policy continuity, measure operability, and data availability. This framework regards the health of aquatic ecosystems as the objective of water environment management, takes small watershed units that are called control units as the foundation of water environment management, and considers the water environment capacity as the key of water environment management.

When it comes to practical application, a complete system of the water quality target management consists of six major parts. They are, respectively, defining the protective function; determining the water quality target; calculating the water environment capacity (that is, the total maximum loads the ecosystem can assimilate); estimating existing pollution sources; allocating maximum loads among sources; and formulating pollution reduction plans.

The water quality target management has been attemptedly applied to the Taihu Lake Basin, which is facing serious water pollution and ecological degradation. The upstream area of Taihu Lake Basin was selected as the case study area, which includes Changzhou and Huzhou, two prefecture-level cities, and Yixing, one county-level city, with the total area of about 12,242 km$^2$. The study area was divided into 31 control units, and for each unit the total maximum monthly loads (TMML) were calculated, the existing pollution sources and their monthly loads were estimated. Finally, the TMML were allocated among the pollution sources in each control unit and pollution reduction plans for each control unit were formulated.

The application in the Taihu Lake Basin indicates that, although it is difficult to consider the daily loads for the current water environment management in China, the amount of pollutants and the water environment capacity can be fully taken into account on a month-by-month basis. That is why the water quality target management in the Taihu Lake Basin can also be named as the TMML. It is demonstrated that the water quality target management helps to realize the reform and innovation of the water environment management in the Taihu Lake Basin and it also provides theoretical support and methodological guidance for similar basins.

## 3 Eco-hydrological Processes in Arid Region

### 3.1 Introduction

Arid area in northwestern China, account for 1/4 of land area of China, is a very fragile region due to the dry climate and little precipitation, and is very sensitive for the global climate change. Changes in water cycle process are the direct driving force for ecological environment evolution of arid area. So, the interaction between ecological and hydrological processes is an important part of the eco-hydrological issues in the arid area. As the main producers of the arid inland river basin, desert riparian forest vegetation is the most important part of arid ecosystems. Water, which directly affects the desert riparian forest vegetation's growth and development, and controls plant communities' composition and succession, is the key ecological factor to control the composition, development, and stability of the ecosystem in arid inland basin. Therefore, hydrological processes is the main line of the eco-hydrological researches, and physiological and ecological responses and adaptation of desert riparian forest vegetation is the main content of the eco-hydrological researches (Chen et al. 2010).

In recent decades, ecological processes and hydrological mechanism in arid inland basin always has been the priorities and hot issues of ecology and hydrology researches. In China, researches conducted comprehensive, multi-angle and in-depth eco-hydrological researches, and they have made many important results about the eco-hydrological processes of desert riparian forest vegetation in the arid inland basin. These results provide the theoretical and technological supports for restoring the damaged ecosystem of arid inland river basin in northwestern China. This section provides an overview of the researches of eco-hydrological processes and mechanism in arid inland river basin, northwestern China.

### 3.2 Eco-hydrological Processes and Mechanism in Arid Inland River Basin

#### 3.2.1 Physiological Response of Desert Riparian Forest Vegetation in Arid Inland River Basin

Under the drought stress, desert riparian forest vegetation could reduce the harm caused by the drought though their physiological activities. Physiological responses of the desert riparian forest vegetation to different groundwater depths were sensitive (Chen et al. 2003, 2010). With the increasing of groundwater depth, drought stress became serious, and the leaves of *Populus euphratica*, Tamarix ramosissima, and reeds could actively accumulate soluble sugar, proline, abscisic acid, MDA, and SOD; in the other hand, their POD reduced with the increasing groundwater depth (Chen et al. 2003, 2004).

Likewise, Photosynthesis, chlorophyll fluorescence parameters, and water potential of *P. euphratica* changed with the increasing groundwater depth (Chen et al. 2003, 2004, 2006, 2011; Zhou et al. 2010; Fu et al. 2006; Zhuet al. 2010). According to the analysis between water potential of plant leaves in the arid area and groundwater depths, −6.5 and −7.12 MPa, respectively, were critical water potential to determine whether *P. euphratica* and Tamarix encounter drought stress (Fu et al. 2006, 2007).

### 3.2.2 Water Use Strategy of Desert Riparian Forest Vegetation in Arid Inland River Basin

In the lower reaches of Heihe River, the water resources of trees and shrubs in the desert riparian forest mainly came from groundwater, but the water resources of herbs mainly was from the surface water in soil. For example, Isotope analysis showed that groundwater accounted for 71.5–97 % water resources used by *P. euphratica*, and deep soil moisture (200–300 cm layer) accounted for 1.5–18.6 % water resources by *P. euphratica*, but surface soil water did not significantly affect the water use of *P. euphratica* (Hao et al. 2013). Similarly, 90 % water used by Tamarix came from groundwater. But 97 % water used by *Sophora alopecuroides* L. came from the soil water in 0–80 cm layers, and water used by *sonchus oleraceus* L. and Herba Taraxaci mainly came from the soil moisture in 0–20 cm and precipitation (Zhao et al. 2008).

Further analysis indicated that diurnal variation of sap flow of *P. euphratica* showed a ‘s’ shape, that is, in the morning (08:00–10:00), its transpiration was weak, and its water consumption and velocity of sap flow were small; during 10:00–20:00, the velocity of sap flow and the water consumption increased; after 20:00, the velocity of sap flow and the water consumption decreased. The seasonal changes of transpiration of *P. euphratica* also showed a regular change, that is, its transpiration mainly occurred in June–September, which accounted for 87.5 % total transpiration in a year, and the transpiration peak presented in July (Ma et al. 2010).

### 3.2.3 Ecological Water Depth for the Desert Riparian Forest in Arid Inland River Basin

For the *P. euphratica*, its physiological characteristics did not showed a significant change within groundwater depth 4 m, but soluble sugar and proline content in leaves significantly increased when the groundwater depth increased, especially proline in leaves rapidly increased from 11.28 μg·g-1.DW to 16.28 μg·g-1.DW when groundwater depth was more than 9 m, which suggested *P. euphratica* encountered the serious water stress. So, the rational and critical groundwater for *P. euphratica* were 4 and 9 m, respectively. Similarly, the analysis of changes in physiological characteristics with the change groundwater depths showed the rational and critical groundwater for Tamarix were 6 and 9 m, respectively; the rational groundwater depth for reed was 3.5 m (Chen et al. 2007).

For the plant community of desert riparian forest, species diversity was highest when the groundwater depth was 2–4 m, followed by it in the groundwater depth was 4–6 m, then it was in the groundwater depth was 0–2 m. Species diversity sharply reduced and Hill diversity index tended to be straight when the groundwater depth was more than 6 m. Combined with the changes in plant communities niche and species diversity with the changes in groundwater depth, rational groundwater for restoring plant community in desert riparian forest in the lower Tarim River should be 2–6 m (Hao et al. 2008). Likewise, according to the analysis the relationship between groundwater depth and species diversity, including richness, Shannon-Wiener, Simpson and Pielou, the rational groundwater depth to keep species diversity in plant communities at oasis-desert ecotone in the lower reaches of Tarim River should be less than 4 m, and the critical groundwater depth for vegetation survive in the lower reaches of Tarim River should be not more than 9 m (Zhou et al. 2008a).

### 3.2.4 Ecological Safety and Water Demand in the Arid Inland River Basin

Minimum ecological flow for river course in the Lower reaches of the Tarim River was showed in the Table 1, and ecological water demand for the river course in the Lower reaches of Tarim River was $0.79 \times 10^8$ m$^3$ without considering evaporation and seepage (Ye et al. 2008). Ecological water demand for the ecological system safety in the main stream of Tarim River was $31.74 \times 10^8$ m$^3$, in which ecological water demand in the upper, middle and lower reaches were $9.95 \times 10^8$ m$^3$, $18.47 \times 10^8$ m$^3$ and $3.32 \times 10^8$ m$^3$, respectively (Chen et al. 2008). Moreover, the

**Table 1** The minimum river flow calculation in the lower reaches of Tarim River

| Items | Qiala-Yingsu | Yingsu-Aalagan | Alagan-Yiganbujima |
|---|---|---|---|
| Length of river/km | 210 | 128 | 96 |
| Water consumption per unit river length/($10^8$m$^3$ km$^{-1}$) | 0.0111 | 0.0061 | 0.0041 |
| Multi-average river flow/ (m$^3$ s$^{-1}$) | 29.31 | 26.82 | 25.63 |
| Minimum ecological flow/ (m$^3$ s$^{-1}$) | 2.85 | 3.76 | 1.76 |
| Percentage of minimum flow and multi-average flow/% | 9.1 | 14.0 | 6.9 |

critical groundwater depth for the phreatic evaporation was 5 m in the lower Tarim River (Ye et al. 2007). So, according to the distribution of plants and phreatic evaporation, annually minimum ecological water demand for the desert riparian forest vegetation was $3.2 \times 10^8$ m$^3$, and main period needed ecological water was from April to September, which accounted for 81 % of annual water ecological demand of plants, especially for the water demand in May to July, which accounted for 47 % of annual water demand (Ye et al. 2007).

The annually average ecological water demand in the middle reaches of Heihe River was $(11.16 \pm 2.67) \times 10^8$ m$^3$, in which it was $(9.13 \pm 2.29) \times 10^8$ m$^3$ for the oasis ecosystem; the annually average ecological water demand in the lower reaches of Heihe River was $(16.16 \pm 4.04) \times 10^8$ m$^3$, in which it was $(11.06 \pm 2.77) \times 10^8$ m$^3$ for the present status of oasis ecosystem (Wanget al. 2005).

### 3.3 Conclusion and Discussion

Vegetation obviously affected ecosystem structure and function of arid inland river basin, while water significantly influences physiological characteristics of individual plants, which further influences the composition and structure of plant communities. Therefore, ecosystem safety researches focused on ecosystem and plants stability under water process. Although there were many researches on plant and ecosystem in arid inland area, changes in composition, structure, distribution and succession of plant communities as well as their ecological water demand with different groundwater depths was main researches direction in future. Moreover, adaptation and survival strategies of desert riparian forest vegetation to drought stress still need further study.

## 4 High-Efficient Water Use Mechanisms in Arid Region

### 4.1 Introduction

Water shortage and deteriorated ecological environment have become the two biggest obstacles that hinder social-economic sustainable development in inland river basins of northwest China. How to rationally appropriate and effectively protect the limited water resources, as well as to adopt high-efficient water use schemes, therefore, has become the most urgent task and is expected to have great impacts in promoting high-efficient use of the limited water resources in the region.

At present, related research work has been carried out primarily in Hei River Basin, Shiyang River Basin, Shule River Basin of Gansu Province, and Tarim River Basin of Xinjiang Uygur Autonomous Region. Shiyang River Basin is the most populated region with the greatest extent of development and utilization of water resources, the fiercest competition for water resources by different sectors, the most severe eco-environmental problems, and the great hindrance to economic and social development by water shortage in inland river basins of Hexi region, Gansu Province. This paper focuses on research work conducted in Shiyang River basin in recent years, specifically, regional water resource transformation theories and water resource system modeling, estimation methods, and spatiotemporal variations of different crops' water consumption, water-heat balance, water transfer processes and modeling for typical farmlands in northwest arid region, ecosystem-based regional water resource rational allocation theories and optimization and regulation schemes, water-saving, and quality-regulating irrigation technologies based on limiting crop water consumption.

## 4.2 *Regional Water Transformation Theories and Water Resource System Modeling*

Water resource and hydrological processes are the most important factors determining the regional economic structure, land use, and eco-environment of inland river basins. Multiple transformations between surface and subsurface waters within the region enable water reuse multiple times, which alters water transformation process and utilization structure, and meantime affects water-dependent ecosystems. To tackle the key research questions associated with regional water transformation and system modeling, studies on regional water transformation, parameters calculation associated with soil properties, soil water flow modeling, effects of changing environment on runoff and simulations, groundwater modeling have been carried out (Lai and Ren 2007; Zhou et al. 2008a; Huo et al. 2011; Sun et al. 2009; Wang et al. 2008; Ma et al. 2008). Zhou et al. (2008a) developed a two-dimensional dynamic root water uptake model for grapevine, and a dynamic soil water flow model under alternate partial root-zone drip irrigation for the vineyard was built based on the root water uptake model; Huo et al. (2011) proposed the ANN-FEFLOW model, which extended the applications of regional groundwater level models to dynamic boundary conditions; Wang et al. (2008) modified the evapotranspiration estimation module in a regional water transformation model (SWAT model) through relating seasonal potential evapotranspiration (PET) to regional digital elevation model (DEM), and established a SWAT-based distributed hydrological model for Zamu River Basin; Ma et al. (2008) proposed an evaluation method which can quantify the effects of climate change and human activity on regional runoff, and the results showed that the contribution percentage from human activity on runoff at regional mountain outlet was from 12.1 to 35.5 %.

## 4.3 Estimation Methods and Spatiotemporal Variations of Different Crops' Water Consumption

At present, there are no uniform standard or module on estimation and measurement of crops' water consumption. To explore appropriate estimation and measurement methods of crops' water consumption grown in different types of farmlands in northwest arid region, various methods including water balance, Bowen ratio-energy Balance, eddy covariance, sap flow + micro-lysimeter, crop coefficient, and theoretical modeling have been evaluated based on major crops of Shiyang River Basin, such as corn, cotton, potato, grape, apple-pear, water melon, hot pepper, and tomato in green house. The applicability and application conditions of different measurement methods have been elucidated through systematic experimental studies and comprehensive comparisons of those methods (Du et al. 2006; Li 2009; Zhang et al. 2007; Hou et al. 2010; Liu et al. 2011; Wang et al. 2011; Ding et al. 2010). Ding et al. (2010) monitored the temporal changes of ET during the whole corn growing season using the large-scale lysimeter and eddy covariance system, and their results showed that half-hour average ET values measured by eddy covariance system ($ET_{EC}$) during daytime were 21.8 % lower than those measured by lysimeter ($ET_L$). Average $ET_{EC}$ would be very similar to average $ET_L$ over different growth stages after adjustment using forced Bowen ratio-energy balance and filtering-interpolation methods. Zhang et al. (2009) developed a water consumption estimation model for uneven wetting and sparse canopies. Zhang et al. (2011), Tong et al. (2007), and Zhang et al. (2010) conducted studies on spatial up-scaling of crop water consumption estimation. Using the scale-transfer function which is set up from elevation (H), aspect (A), and latitude (V) which derived from DEM, the spatial distribution model of ETc and net irrigation water requirement (I) of different crops in the basin were obtained through GIS software, based on the DEM and land use map (Tong et al. 2007).

## 4.4 Water-Heat Balance, Water Transfer Processes and Modeling for Typical Farmlands in Northwest Arid Region

Zhang et al. (2007) studied daily, seasonal, and yearly variations and changing patterns of water and heat fluxes in a vineyard, and found that daily variations of components of water and heat fluxes, and Bowen ratios displayed a typical single-peak pattern during the entire growing season. Guo (2010) monitored water and carbon fluxes of the vineyard and reported that daily change of $CO_2$ flux showed a typical two-peak curve with two peaks taking place, respectively, at about 10 and 14 o'clock, and the lowest flux around noon. Li (2009) investigated the changing patterns of different components of corn energy, and quantify the percentage of each component over the whole energy. Their studies showed that daily average net radiation tended to increase at the earlier stage, then decrease over time during the whole growing season, soil heat flux fluctuated greatly at the earlier growth stage, and then became relatively stable during the middle and late growth stages, and latent heat changed as a convex-shape parabolic curve during the growing season.

## 4.5 Ecosystem-Based Regional Water Resource Rational Allocation Theories and Optimizing and Regulating Scheme

Water consumption changing patterns and calculation model for wild wind prevention and sand fixation species, such as *Haloxylon ammodendron*, *Caragana korshinskii*, and *Hedysarum scoparium*, and artificially planting species such as *Tamarix elongata ledeb*, *Caragana korshinskii kom*, and *Populus alba* L. var. *Pyramidalis* were achieved based on experimental studies on water consumption of typical vegetation at Shiyanghe Experimental Station for Water-saving in Agriculture and Ecology of China Agricultural University (CAU) in Wuwei, Gansu (Xia et al. 2008; Xu 2010). Their results showed that *Haloxylon ammodendron* consumed the least amount of water among wild species, with May—October total sap flow of 515.58 and 445.65L in 2004 and 2005, respectively. Among the three artificial species, *Tamarix elongata ledeb* was the one saving most water, the next saving-water species was *Caragana korshinskii kom* while *Populus alba* L. var. Pyramidalis consumed the largest amount of water with May—October total sap flow of 532.3–910.7 L, 587.3–922.6 L, and 1371.5–2375.2 L for 2005–2008 respectively. Xu (2010) developed an evapotranspiration estimation model for *Populus alba* L. var. *Pyramidalis,* which extended the single-layer canopy resistance model to dual-source model and improved canopy resistance and soil resistance calculation methods in the dual-source model.

Methods for better estimating ecological water demand at watershed scale in northwest arid region, and for dynamic evaluation of ecological value for inland river basins based on coefficient of social development stages and degree of resource shortage have been proposed. A water resource rational allocation model accounting for ecological water demand for Shiyang River Basin was developed, and allocation scheme for up, middle, and low reaches of Shiyang River Basin, allocation scheme for ecological and economic water demands, and optimized cropping structure for the region were achieved. The optimized irrigated farmland would be within 213800–240300 ha, and the optimized crop acreage ratio for grain—economic—forage crops would be around 54:31:15 in Shiyang River Basin. Zeng et al. (2010) built a multi-objective irrigation decision model with uncertain parameters and obtained optimized crop structure under different water-saving irrigation methods in Shiyang River Basin.

## 4.6 Water-Saving and Quality-Regulating Irrigation Technologies Based on Limiting Crop Water Consumption

Kang et al. (2009) first proposed the concept of water-saving and quality-regulating high-efficient irrigation, and put forward theories and implementation method of water-saving and quality-regulating high-efficient irrigation based on water-quality interrelationships for typical economic crops. The effects of water control at different growth stages and at different parts of roots on crop growth were investigated for cotton, apple, grapevine, tomato, and green pepper in greenhouse. The sensitivity of different quality components to water stress at different growth stages were studied, the relationships between quality and water control were quantified, and the threshold values of different quality indices to soil water regulating were determined (Du et al. 2006, 2008; Wang et al. 2011; Kang et al. 2009). Du et al. (2008) discovered that alternate partial root-zone irrigation could increase Vitamin C content, reduce tartaric acid content, and substantially increase total soluble solid content. They also proposed water-saving, quality-regulating, and high-efficient integrative technological irrigation systems for wine grape based on water-comprehensive quality-yield-economic benefit interrelationships under furrow and drip irrigations. The systems could save 40–60 $m^3$ irrigation water during a growing season. Wang et al. (2011) found that water shortage during blossom and fruit-development stages could increase yield of favorable fruits and improving the fruit uniformity for greenhouse tomatoes. Regulated Irrigations with 1/3 potential ET during blossom and fruit-development stages, and with 1/3 or 2/3 potential ET at ripe and harvest stages could enhance substantially fruit total soluble solid content, soluble sugar, organic acids, Vitamin C and lycopene, improve fruit hardness, and reduce fruit water moisture content.

## 4.7 Prospects

Coordinating multiprocesses and comprehensively studying mass exchange mechanisms between interfaces will help to understand water-heat-carbon circulation mechanisms in soil-plant-atmosphere-continuum. Regional spatiotemporal crop water use optimization technologies under water shortage will be a new research topic. Decision systems on optimizing water-saving and quality-regulating irrigation module based on crop water demand information and water-quality-yield-revenue model need to be explored. Water-saving and quality-regulating irrigation module for major cereal and economic crops and associated operation guidelines are needed.

## References

Chen, Y. N., Chen, Y. P., Li, W. H., et al. (2003). Response of the accumulation of proline in the bodies of Populus euphratica to the change of ground water level at the lower reaches of Tarim River. *Chinese Science Bulletin, 48*(18), 1995–1999.

Chen, M., Chen, Y. N., Li, W. H., et al. (2007). Drought resistance mechanism of three plant species in the middle reaches of Tarim River China. *Acta Botanica Boreali-Occidentalia Sinica, 27*(4), 747–754.

Chen, Y. P., Chen, Y. N., & Xu, C. C. (2011). Effects of groundwater depth on the gas exchange and chlorophyll fluorescence of Populus euphratica in the lower reaches of Tarim River. *Acta Ecologica Sinica, 31*(2), 344–353.

Chen, Y. N., Chen, Y. P., Xu, C. H., et al. (2010). Effects of ecological water conveyance on groundwater dynamics and riparian vegetation in the lower reaches of Tarim Rive, Chinar. *Hydrological Processes, 24*(2), 170–177.

Chen, Y. N., Hao, X. M., Li, W. H., et al. (2008). An analysis of the ecological security and ecological water requirements in the inland river of Arid Region. *Advanced in Earth Science, 23*(7), 732–738.

Chen, Y. N., Li, W. H., Chen, Y. P., et al. (2004). Physiological response of natural plants to the change of groundwater level in the lower reaches of Tarim River, Xinjiang. *Progress in Natural Science, 14*(11), 975–983.

Chen, Y. N., Wang, Q., & Li, W. H. (2006). Rational groundwater table indicated by the eco-physiological parameters of the vegetation: A case study of ecological restoration in the lower reaches of the Tarim River. *Chinese Science Bulletin, 51*(Supp. I), 8–15.

Ding, R., Kang, S., Li, F., Zhang, Y., Tong, L., & Sun, Q. (2010). Evaluating eddy covariance method by large-scale weighing lysimeter in a maize field of northwest China. *Agricultural Water Management, 98*, 87–95.

Du, T., Kang, S., Zhang, J., Li, F., & Hu, X. (2006). Yield and physiological response of cotton to partial root zone furrow irrigations in the oasis region of northwest China. *Agricultural Water Management, 84*, 41–52.

Du, T., Kang, S., Zhang, J., Li, F., & Yan, B. (2008). Water use efficiency and fruit quality of table grape under alternate partial root-zone drip irrigation. *Agricultural Water Management, 95*(6), 659–668.

European Union Commission. (2000). *The EU water framework directive integrated river basin management for Europe. 2008-12-14.* http://ec.europa.eu/environment/water/water-framework/index_en.html

Fu, A. H., Chen, Y. N., & Li, W. H. (2006). Analysis on water potential of Populus euphratica oliv and its meaning in the lower reaches of Tarim River Xinjiang. *Chinese Science Bulletin, 51* (Supp. 1), 221–228.

Fu, A. H., Chen, Y. N., & Li, W. H. (2007). Stems water potential of *Tamarix ramsissima* Lbd. and influencing factors in the lower reaches of Tarim River, Xinjiang. *Arid Land Geography, 30*(1), 108–114.

Guo, W. (2010). Research on variation of vineyard's carbon, water fluxes and modeling the exchange of carbon flux in arid northwestern china. *Thesis for Master's Degree Northwest A&F University.*

Hao, X. M., Chen, Y. N., Guo, B., et al. (2013). Hydraulic redistribution of soil water in *Populus euphratica* Oliv. in a central Asian desert riparian forest. *Ecohydrology, 6*(6), 974–983.

Hao, X. M., Li, W. H., & Chen, Y. N. (2008). Water table and the desert riparian forest community in the lower reaches of Tarim River China. *Journal of Plant Ecology, 32*(4), 838–847.

Higgins, J. V., Bryer, M. T., Khoury, M. L., & Fitzhugh, T. W. (2005). A freshwater classification approach for biodiversity conservation planning. *Conservation Biology, 19*(2), 432–445.

Hou, X., Wang, F., Han, J., Kang, S., & Feng, S. (2010). Duration of plastic mulch for potato growth under drip irrigation in an arid region of Northwest China. *Agricultural and Forest Meteorology, 150*, 115–121.

Huo, Z., Feng, S., Kang, S., Mao, X., & Wang, F. (2011). Numerically modelling groundwater in an arid area with ANN-generated dynamic boundary conditions. *Hydrological Processess, 25* (5), 705–713.

Kang, S., Su, X., Du, T., Feng, S., Tong, L., Shen, Q., et al. (2009). *Water resource transformation and efficient water-saving pattern in arid northwest China—A case study in Shiyang River Basin*. Beijing: China Water Resources and Hydro Power Press.

Lai, J., & Ren, L. (2007). Assessing the size dependency of measured hydraulic conductivity using double-ring infiltrometers and numerical simulation. *Soil Science Society of America Journal, 71*(6), 1667–1675.

Li, S. (2009). *Variation and simulation of water, heat and carbon fluxes above typical fields in oasis region of northwest China*. Dissertation for Doctor Degree China Agricultural University.

Liu, C., Kang, S., Li, F., Li, S., Du, T., & Tong, L. (2011). Seasonal variation of maximum daily stem shrinkage in apple tree and its relationship with environmental factor in arid region of northwest China. *Scientia Horticulturae, 130*, 118–125.

Ma, J. X., Chen, Y. N., Li, W. H., et al. (2010). Response of sap flow in *Populus euphratica* to changes in groundwater depth in the middle and lower reaches of the Tarim River of northwestern China. *Chinese Journal of Plant Ecology, 34*(8), 915–923.

Ma, Z., Kang, S., Zhang, L., Tong, L., & Su, X. (2008). Analysis of impacts of climate variability and human activity on streamflow for a river basin in arid region of northwest China. *Journal of Hydrology, 352*, 239–249.

Maxwell, J. R., Edwards, C. J., Jensen, M. E., Paustian, S. J., Parrott, H., Hill, D. M. (1995). *A hierarchical framework of aquatic ecological units in North America (Nearctic Zone)*. U.S. Department of Agriculture, Forest Service, North Central Forest Experiment Station, General Technical Report NC-176.

Meng, W. (2007). The strategy of comprehensive pollution prevention and cure of water environment in Chinese watershed. *China Environmental Science, 27*(5), 712–716. (in Chinese with English abstract).

Meng, W., Liu, Z. T., Zhang, N., & Hu, L. L. (2008a). The study on technique of basin water quality target management II: Water environmental criteria, standard and total amount control. *Research of Environmental Sciences, 21*(1), 1–8. (in Chinese with English abstract).

Meng, W., Qin, Y. W., Zheng, B. H., Zhang, L., & Zhang, N. (2008b). The study on technique of basin water quality target management III: Basin monitor technique for water environment. *Research of Environmental Sciences, 21*(1), 9–16. (in Chinese with English abstract).

Meng, W., Zhang, Y., Zhang, N., Cai, M. T., & Huang, Y. (2013). Concept, characteristics, and implementary policy of freshwater ecosystem function management regionalization. *Research of Environmental Sciences, 26*(5), 465–471. (in Chinese with English abstract).

Meng, W., Zhang, Y., Zhang, N., Tian, Z. Q., Liu, Z. T., & Lei, K. (2011). Study on aquatic ecological function regionalization and water quality target management in the river basin. *Acta Scientiae Circumstantiae, 31*(7), 1345–1351. (in Chinese with English abstract).

Soranno, P. A., Spence Cheruvelil, K., Webster, K. E., Bremigan, M. T., Wagner, T., & Stow, C. A. (2010). Using landscape limnology to classify freshwater ecosystems for multi-ecosystem management and conservation. *BioScience, 60*(6), 440–454.

Strayer, D. L., & Dudgeon, D. (2010). Freshwater biodiversity conservation: Recent progress and future challenges. *Journal of the North American Benthological Society, 29*(1), 344–358.

Sun, Y., Kang, S., Li, F., & Zhang, L. (2009). Comparison of interpolation methods for depth to groundwater and its temporal and spatial variations in the Minqin oasis of northwest China. *Environmental Modelling & Software, 24*, 1163–1170.

Tong, L., Kang, S., & Zhang, L. (2007). Temporal and spatial variations of evapotranspiration for spring wheat in the Shiyang river basin in northwest China. *Agricultural Water Management, 87*, 241–250.

Vörösmarty, C. J., McIntyre, P. B., Gessner, M. O., Dudgeon, D., Prusevich, A., Green, P., et al. (2010). Global threats to human water security and river biodiversity. *Nature, 467*, 555–561.

Wang, F., Kang, S., Du, T., Li, F., & Qiu, R. (2011). Determination of comprehensive quality index for tomato and its response to different irrigation treatments. *Agricultural Water Management, 98*(8), 1228–1238.

Wang, S., Kang, S., Zhang, L., & Li, F. (2008). Modelling hydrological response to different land-use and climate change scenarios in the Zamu River basin of northwest China. *Hydrological Processes, 22*, 2502–2510.

Wang, G. X., Zhang, Y., Liu, G. M., et al. (2005). Water demand evaluation of riverside ecosystem in arid inland river basin: the case of Heihe River Basin. *Acta Ecologica Sinica, 25* (10), 2467–2476.

Xia, G., Kang, S., Li, F., Zhang, J., & Zhou, Q. (2008). Diurnal and seasonal variations of sap flow of Caragana korshinskii in the arid desert region of north-west China. *Hydrological Processes, 22*, 1197–1205.

Xu, X. (2010). *Models in evapotranspiration and root water uptake of irrigated plants for wind sheltering and sand fixation in arid region*. Dissertation for Doctor Degree China Agricultural University.

Ye, Z. X., Chen, Y. N., & Li, W. H. (2007). Ecological water demand of vegetation based on eco-hydrological processes in the Lower Reaches of Tarim River. *Acta Geographica Sinica, 62* (5), 451–461.

Ye, Z. X., Chen, Y. N., Li, W. H., et al. (2008). Minimum ecological flow of river course in the lower reaches of Tarim River. *Progress in Natural Science, 18*(5), 531–537.

Zeng, X., Kang, S., Li, F., Lu, Z., & Guo, P. (2010). Fuzzy multi-objective linear programming applying to crop area planning. *Agricultural Water Management, 98*, 134–142.

Zhang, Y., Kang, S., Ward, E. J., Ding, R., Zhang, X., & Zheng, R. (2011). Evapotranspiration components determined by sap flow and microlysimetry techniques of a vineyard in northwest China: Dynamics and influential factors. *Agricultural Water Management, 98*(8), 1207–1214.

Zhang, B., Kang, S., Zhang, L., Du, T., Li, S., & Yang, X. (2007). Estimation of seasonal crop water consumption in a vineyard using Bowen ratio-energy balance method. *Hydrological Processes, 21*(26), 3635–3641.

Zhang, X., Kang, S., Zhang, L., & Liu, J. (2010). Spatial variation of climatology monthly crop reference evapotranspiration and sensitivity coefficients in Shiyang river basin of northwest China. *Agricultural Water Management, 97*, 1506–1516.

Zhang, B., Kang, S., Zhang, L., Tong, L., Du, T., Li, F., & Zhang, J. (2009). An evapotranspiration model for sparsely vegetated canopies under partial root-zone irrigation. *Agricultural and Forest Meteorology, 149*, 2007–2011.

Zhao, L. J., Xiao, H. L., Chen, G. D., et al. (2008). A preliminary study of water sources of riparian plants in the lower reaches of the Heihe Basin. *Acta Geoscientica Sinica, 29*(6), 709–718.

Zhou, H. H., Chen, Y. N., & Li, W. H. (2008a). Response of species diversity to water resources in Oasis-desert Ecotone. *Progress in Natural Science, 18*(7), 789–794.

Zhou, H. H., Chen, Y. N., Li, W. H., et al. (2010). Photosynthesis of *Populus euphratica* in relation to groundwater depths and high temperature in arid environment, northwest China. *Photosynthetica, 48*(2), 257–268.

Zhou, Q., Kang, S., Li, F., & Zhang, L. (2008b). Comparison of dynamic and static APRI-models to Simulate soil water dynamics in a vineyard over the growing season under alternate partial root-zone drip irrigation. *Agricultural Water Management, 95*(7), 767–775.

Zhu, C. G., Li, W. H., Ma, J. X., et al. (2010). Effects of groundwater level on chlorophyll fluorescencecharacteristics of Tamarix hispida inlower reaches of Tarim River. *Chinese Journal of Applied Ecology, 21*(7), 1689–1696.

# Part III
# Restoration of Degraded Ecosystem

# Chapter 10
# Restoration of Degraded Ecosystem

**Shaolin Peng, Ting Zhou, Deli Wang, Yingzhi Gao, Zhiwei Zhong, Dong Xie, Hengjie Zhou, Haiting Ji, Shuqing An, Ming Dong, Xuehua Ye, Guofang Liu and Shuqin Gao**

**Abstract** Ecosystem is the most complete basic structure and functional unit for the research of ecology and other branches, so ecosystem restoration is the basis of different levels of ecological restoration research, and the research and practice of ecological restoration should take ecosystem as the basic object. The theme for the International Conference of Restoration Ecology in 2003 was "understand and restore ecosystem", which indicated the basic significance of ecosystem restoration. There are many types of ecosystems, including forest, grassland, desert, ocean, lake and river. They are not only different in appearance, but also composed with separately unique biotic components. For different ecosystem types, the theory and methods of recovery and reconstruction are different. In the first three sections of this chapter, we discussed the ecological restoration for degraded forest ecosystem, grassland ecosystem, and wetland ecosystem. The fourth section described protection and restoration of sandland ecosystem. In the restoration research of each ecosystem type, the chapter emphasized the degradation actuality of different ecosystems, its reasons, corresponding restoration ways, and benefits and evaluation. There have been great achievements have been made in the ecosystem restoration research field in China, and some have reached the advanced international level. The authors of all sections are experts, who have spent years researching the ecosystem restoration, thus whether their discourses were based on themselves research, or their articles integrated peer research nationwide, they summarized the front research which reflected the overall level of this field in current in China.

---

S. Peng (✉) · T. Zhou
State Key Laboratory of Biocontrol, School of Life Sciences, Sun Yat-sen University, Guangzhou 510275, China
e-mail: lsspsl@mail.sysu.edu.cn

D. Wang · Y. Gao · Z. Zhong
School of Life Sciences, Northeast Normal University, Changchun 130024, China

D. Xie · H. Zhou · H. Ji · S. An
School of Life Sciences, Nanjing University, Nanjing 210093, China

M. Dong · X. Ye · G. Liu · S. Gao
Institute of Botany, Chinese Academy of Sciences, Beijing 100093, China

 
W. Li (ed.), *Contemporary Ecology Research in China*,
DOI 10.1007/978-3-662-48376-3_10

**Keywords** Ecological restoration · Degraded ecosystem · Forest ecosystem · Grassland ecosystem · Wetland ecosystem · Sandland ecosystem

# 1 Ecological Restoration of Degraded Forest

## *1.1 Introduction*

Forest ecosystem contributes to the maintenance of biodiversity and regional eco-balance, thus the restoration of degraded forest ecosystem is crucial for regional sustainable development. On the other hand, we are all in the benefits of ecological restoration against the background of resource exhausted and increasing global change(Alcoze et al. 2000). Ecological restoration influences the environment through regulating the global carbon dynamic and biogeochemical cycling (Stone 2009). In view of this, ecological restoration becomes a global issue. Bowers, the chairman of the International Society for Ecological Restoration, proposed that ecological restoration in the changing world is to restore the world's future (Peng and Hou 2007). Prof. Dixon, the chairman of the 19th International Conference of Restoration Ecology, also pointed out that restoration ecology might be the only choice in a changing world (Peng and Zhou 2009).

Many processes in restoration ecology have made in China, such as in the eroded soils in tropical region (Li et al. 1996), desertification management (Li et al. 2005), mangrove restoration (Peng et al. 2008), control of water loss and soil erosion in loess plateau (He and Lang 2009). In this study, we reviewed the theory and practice of the restoration of degraded forest ecosystem based on our long-term researches.

## *1.2 Theoretical and Technological Background of Ecological Restoration of Degraded Forest*

### 1.2.1 The Basic Theory of Forest Ecosystem Restoration-Successional Theory

Full-scale understanding of the dynamic principles increases the efficiency of forest vegetation restoration and reconstruction. All successful artificial forest ecosystems are the simulation of natural ecosystems and based on the principles of dynamics and ecology, which conform to the successional law. Therefore, when restoring and reconstructing degraded forest ecosystem we rely on restoration reference, which is the representative community and species combination in a successional chronology.

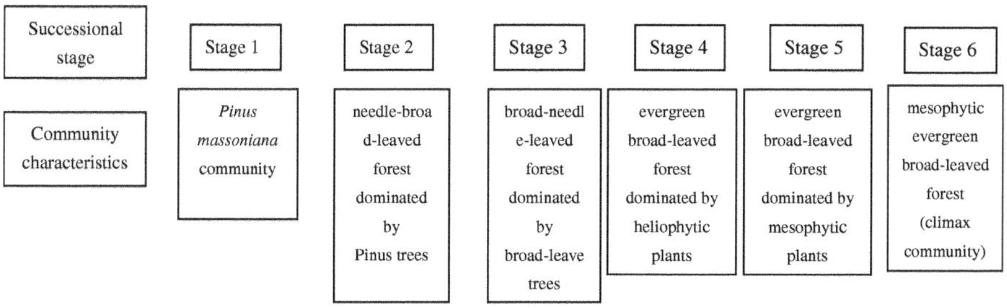


**Fig. 1** Successional process of subtropical forest

Subtropical forest succession region generally follow the process in Fig. 1 and Table 1 with the exclusion of anthropogenic interference and this restoration reference was proved as an effective practical guidance on afforestation and forest vegetation transformation (Peng et al. 2010).

### 1.2.2 Technological System of Forest Ecological Restoration

The technological system mainly includes three aspects: the recovery technology involving water, soil, air and other environmental factors; the recovery technology involving species, populations and community; the overall planning and assembly technology involving ecosystems and landscapes. And these technical methodologies of restoration and reconstruction of degraded ecosystems continue to develop and improve.

## *1.3 Restoration and Reconstruction of Forest Vegetation of Degraded Ecosystem*

### 1.3.1 Degraded Ecosystems

Spontaneous restoration of forest vegetation in extremely degraded land under natural conditions is infeasible and need the manual startup process.

Taking southern extremely degraded bare land as example, we should first take engineering measures and biological measures to control the ecological factors causing extreme degradation. Second, we should choose fast-growing, drought resistance, barren pioneer species to reconstruct pioneer community. Third, we should carry out stand improvement and accelerate the recovery process of succession according to local restoration reference. Finally, utilization of ecosystems can be considered to improve the economic benefits of restoration of degraded ecosystems.

**Table 1** Time division of subtropical forest communities at different successional stages

| Forest age/year | 0 | 0–25 | 25–50 | 50–75 | 75–150 | 150–∞ |
|---|---|---|---|---|---|---|
| Community characteristics | Pioneer forest community (e.g. *Pinus massoniana*) | Needle-broad-leaved forest dominated by Pinus trees | Broad-needle-leaved forest dominated by broad-leave trees (e.g. *Castanopsis fagaceae*, *Schima superba*) | Evergreen broad-leaved forest dominated by heliophytic plants (e.g. *Castanopsis fagaceae*, *Schima superba*) | Evergreen broad-leaved forest dominated by mesophytic plants (e.g. *Cryptocarya chinensis*, *Cryptocarya concinna*) | Mesophytic evergreen broad-leaved forest |

In practice, ecological restoration can be obtained very significant ecological, economic and social benefits (Palmer and Filoso 2009). It is proved by numerous researches conducted in Chinese Academy of Sciences Heshan positioning station.

### 1.3.2 Restoration of Secondary Forest Ecosystem

The habitats of secondary forest ecosystems are generally well, so the restoration of the ecosystem is primarily in line with restoration reference, and artificially promote development along the line of succession. The main stages of secondary forest ecological restoration include closing forest, stand improvement, increasing the light transmission and post-secondary forest conservation.

## *1.4 Eco-benefit of Restoration of Degraded Forest*

### 1.4.1 Biological Benefit of Restoration of Degraded Forest

The biological benefit is first reflected in the woodland biomass accumulation. Because there are sufficient good light, heat, water and other conditions in tropical and subtropical area, vegetation recovery and development could be very fast with appropriate choice of species. Heshan degradation slope has been in re-vegetation for seven years and the existing capacity has reached 100–150 t $hm^{-2}$, which accumulation process is very fast. During that recovery process, solar energy utilization of vegetation grows, and it promotes the accumulation of forest biomass and primary productivity, thereby increasing the functional intensity of forest vegetation (Peng et al. 2003).

Second, the biological benefit is reflected in the increasing biodiversity, especially the plant diversity. Ecosystem biodiversity is built on the basis of plant diversity. Different plant species have varied biological and ecological characteristics and plant diversity will lead to community complex, which means that more vertical stratification, higher levels of plague pattern and more complicated root system. These diverse microhabitats would be easier for reside more living organisms, such as insects, birds, microorganisms and animals.

### 1.4.2 Environmental Benefit of Restoration of Degraded Forest

In addition to the improvement of biological benefit, the eco-benefit of restoration of degraded forest is largely reflected on the marked improvement of environmental benefits, mainly in the aspects of controlling of soil erosion, soil quality improvement in woodland and forest microclimate optimization and so on. Researches by Chinese Academy of Science Xiaoliang Tropical Plantation Research Station suggested that Bare land erosion is the most serious of 52.3 t $hm^{-2}$ $a^{-1}$; followed by eucalyptus, 10.79 t $hm^{-2}$ $a^{-1}$; Mixed lowest, 0.18 t $hm^{-2}$ $a^{-1}$. Soil and water conservation capacity of artificial broad-leaved forest in Xiaoliang Station close to natural mixed forest (Peng et al. 2003). Soil organic matter content increased from 1.34 to 2.68 % in 5–25 years old artificial forest and the average annual growth was 0.067 %. Nitrogen increased from 0.076 to 0.0135 % and the annual average growth was 0.003 %. In spite of the fast recovery, the simulation results predicted that it would take 148 years for the soil organic matter to achieve zonal natural forest vegetation level (Yu et al. 1996).

During forest vegetation restoration, forest development exerts high ecological effects on forest soil, water and woodland microclimate. Vegetation restoration not only affects the woodland itself, but also affects the surrounding environment, and the regional and even global ecological balance. Since 1976, the new forest in Guangdong Province, annually fixed more than half of industrial C emissions (Peng et al. 2009). We should note that ecological restoration of vegetation not only release the global increasing of $CO_2$, but also important to the amelioration of other global change issues, such as temperature rise, UVB increases and reduction of biodiversity.

# 2 Grassland Degradation and Ecological Restoration

## *2.1 Introduction*

As one of the most important terrestrial ecosystem types on earth, grassland plays a key role in preserving water and soil, reducing pollution and purifying the air, preventing desertification and maintaining ecological balance, the ecological functions, and resource value of grasslands is unique and cannot be replaced by other ecosystems. There are about 400 million $hm^2$ of natural grasslands in China, which account for 41 % of the total land area of the country. It is the biggest terrestrial ecosystem in China, however, a majority of grasslands are now considered degraded due to the human utilization such as long-term overgrazing, and the lack of adaptive management and capital investment. Grassland degradation has a "bottleneck effect" on the sustainability of livestock husbandry and pastoral social economy in China. Therefore, it is a critical issue to carry out the long-term planning and comprehensive management or utilization strategies for those grasslands, especially for degraded grasslands.

Grassland degradation usually refers to the degradation of functions for grassland ecosystems, reflecting the decreases in forage yield, vegetation productivity and animal performance, and the deterioration of soil conditions due to overgrazing, high frequent cutting or reclaiming. Furthermore, grassland degradation includes two basic aspects: one is the degradation of vegetation, which mainly refers to the changes in various vegetation characteristics in grasslands, such as decreases in vegetation coverage, productivity and plant diversity; the second is the degradation of soil, which refers to the changes in soil physical and chemical properties that are on longer suitable for plant growth (Wang et al. 1996; Gao et al. 2004). In generally, natural grassland ecosystems exhibit many kinds of degradation as desertification, and salination or alkalization. In Songnen grasslands of northeast China, a large lands of salt and alkali were found to accumulate in the topsoil owing to the special terrain and geological process and overgrazing which can make the grasslands become salinization and alkalization. Again, as a result of overgrazing and continued drought, Maowusu sandy grasslands in Inner Mongolia of China is now experiencing more and more serious desertification. Given different drivers and changing processes for grassland degradation, it needs to have various techniques with simple manipulation and low cost for combating the grassland degradation.

## 2.2 *Causes of Grassland Degradation*

Grassland degradation is due mainly to the integrated effects of natural causes and human activities. Some scientists argued that grassland degradation should be attributed largely to natural factors, and they believed that the changes in grassland ecosystems are highly associated with the changes of geographical features, climatic conditions, soil properties and natural disaster events in a region. The climatic factor, especially the changes of annual precipitation, could largely determine the developing direction of grasslands. By contrast, other scientists believed that human activities such as excess reclamation, overgrazing and unreasonable use of water resources are the main reasons of grassland degradation.

### 2.2.1 Natural Factors

It is no doubt that the ecosystems including grasslands are strongly affected by a wide variety of natural factors on the earth. In fact, the formation and dynamics of ecological systems are driven by various environmental factors such as physical conditions-light radiation, temperature, precipitation, soil properties, fire, and other geological activities. Empirical work clearly show that the dramatic changes in these environmental factors may directly or indirectly cause the changes or degradation in ecosystems. For a specific ecosystem, biological factors such as exotic invasion, outbreaks of insects or rodents can also result in the significant changes or degradation of grasslands.

Climate change is one of the main causes of grassland degradation. In recent years, global change mainly refers to three aspects: first, global average temperature rises as atmospheric $CO_2$ concentration increases; second, the climate in most part of the world, especially in grassland regions, will tend to be drier; third, the variation in climatic patterns such as precipitation and temperature. For grassland ecosystems, vegetation production may be limited by drought and the uneven temporal and spatial distribution of precipitation. Besides, the decreased

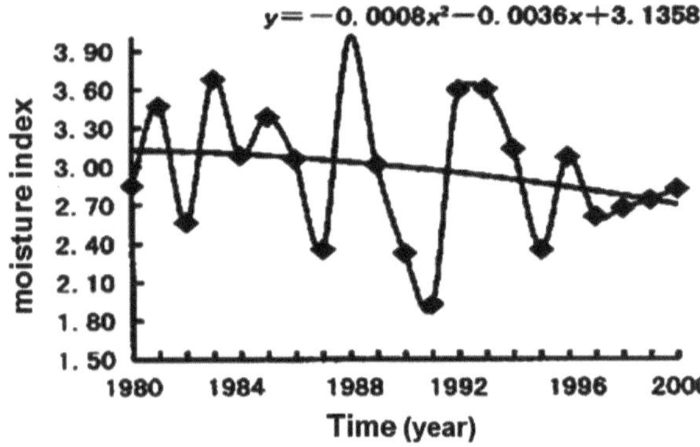


**Fig. 2** The variation dynamic of grassland precipitation in Jin Qiang district, Gansu province in twenty years (Bai and Pu 2003)

precipitation accompanying with the increased temperature, leads to higher soil evaporation. Therefore, such changes often limit plant growth or production, and thereby cause the degradation in grasslands. Lü and Lü (2002) reported a significant increase in temperature and evaporation but a significant decrease in precipitation in pasturing areas of northeast Tibet Plateau since the 1980s. Climatic warming-drying trend has led to a shortage of water and the degradation of grassland in this region. Again, Bai and Pu (2003) fund that grassland average annual temperature in Jinqianghe area of Gansu Province has increased during the past twenty years. Average annual precipitation in this region varied largely, average annual precipitation from 1991 to 2000 was 4.7 % lower than that from 1980 to 1990 (Fig. 2).

Pest insects and rodents are also the main causes of grassland degradation. Nowadays, more than 28 million $hm^2$ grassland areas in China are suffering from various degree of rodent pest, which account for 10 % of the available grasslands in the country (Zhong and Fan 2002). In the Qinghai-Tibet alpine meadow ecosystem, rodent problem is particularly serious. Besides, by consuming a large amount of above-ground biomass, pest insects can also lead to the degradation of grasslands. There are two main groups of pest insects, namely grasshoppers and caterpillars. Under suitable climatic conditions, it is particularly easy to provoke the outbreaks of pest insects. For example, as the temperature in grassland regions is relatively high during June and July each year, abundant summer precipitation may cause the plague of locusts. In 1998, it was reported that the grassland of approximately 2 million $hm^2$ in Qinghai Province were suffered from insect pest (Ma et al. 2002), and it is a common phenomenon in the Inner Mongolian grasslands.

### 2.2.2 Human Activities Disturbance

As the same as the other parts in the world, the human activities such as overgrazing, frequently mowing, reclamation are often considered as the main causes of the degradation in grasslands. Overgrazing is the primary factor influencing the sustainable development of grassland ecosystems. Livestock grazing is the main way of human grassland use, and grassland resources were directly transformed into animal products through livestock grazing. However, it should be noted that both of the effectiveness of grassland use and the ecosystem health of grasslands depend on the rationality of grazing regime, and the balance between grassland production and animal foraging. With increasing grazing intensity, grasslands will fail to maintain high and sustainable vegetation production, and thereby restrain animal production. This may lead to serious consequences, as degradation of vegetation may develop into soil degradation in grasslands (Wang et al. 1996, 2005). Under the overgrazing, through the effect of trampling (also called hoof effect), livestock grazing can exert strong impacts on grasslands. Previous studies showed that livestock trampling can cause forage losses in grassland, for example, as grazing time increases, the falling rate of *Leymus chinensis* has risen from 4.2 to 11.0 % (Table 2) (Teng and Wang 2002). Meanwhile, intensive grazing may change soil physical structure and chemical properties by trampling (Teng et al. 2006).

The intensity and frequency of mowing can directly influence grassland vegetation dynamics, including plant species composition and plant productivity (Wang and Yang 2003). In the Songnen grasslands, previous studies showed that mowing significantly changed the composition of plant community (Fig. 3). As the intensity of mowing increases, *L. chinensis* + forbs communities will turns into *L. chinensis* + *Carex duriuscula* communities, finally leads to a decline in the number of plant species. Therefore, just like overgrazing, intensive mowing will also exert a strong negative impact on grasslands, and cause the degradation of grassland vegetation.

**Table 2** The foraging loss and falling rate of plants under different grazing rates (g DM $m^{-2}$, %)

| Item | Grazing time 0.5 h | | Grazing time 1.0 h | | Grazing time 1.5 h | |
|---|---|---|---|---|---|---|
| | Loss | Falling rate | Loss | Falling rate | Loss | Falling rate |
| *Leymus chinensis* | 3.5 | 4.2 | 5.0 | 9.4 | 1.6 | 11.0 |
| *Phragmites communis* | 0.6 | 13.4 | 0.6 | 8.5 | 0.2 | 2.6 |
| *Kalimeris integrifolia* | 0.8 | 3.3 | 0.3 | 10.0 | 1.0 | 6.6 |
| *Apocynum venetum* | 0.1 | 0.0 | 0.0 | 14.8 | 0.0 | 10.7 |
| *Puccinellia tenuiflora* | 1.1 | | 1.3 | | 0.4 | |
| Total and average | 6.1 | 5.2 | 7.2 | 8.5 | 3.2 | 6.8 |

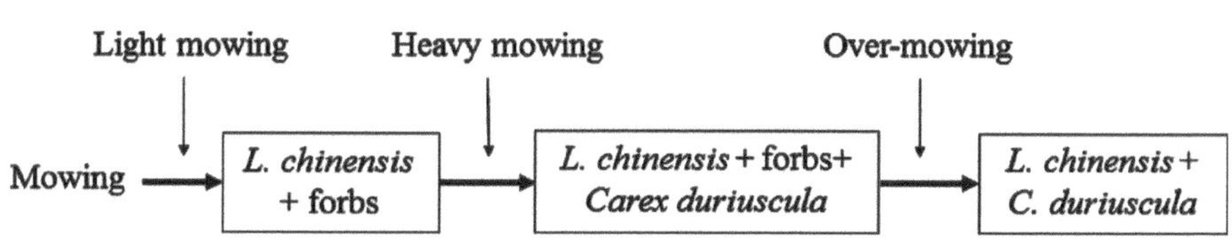


**Fig. 3** The vegetation degradation of Songnen grassland under different mowing intensity

Reclamation is still a driven factor leading to grassland degradation in northeaster China. Crop cultivation in the reclaimed grasslands can damage or decrease native grassland vegetation productivity and soil fertility, and thus reduce the stocking capacity, which naturally cause the degradation in grasslands. After the founding of the country, the United States had conducted a large scale of reclamation on Great Plains of North America, and a huge black storm broke out in Great Plains of North America in 1934, and resulted in huge damage. Two-thirds of the land areas of the United States were suffered from the storm, and 15–85 % of vegetation in these regions was destroyed, 45 million $hm^2$ reclaimed farmland was disappeared, and 60,000 farmers and herdsmen were bankrupt.

Except for the impact factors mentioned above, low investment is also another cause of grassland degradation. According to the statistics of relevant departments, during 1949–1986, the total investment in grasslands was 8.73 billion RMB in China, which accounted for 3.4 % of the total investment in agriculture and 1.6 % of the gross output in animal husbandry, the average investment per hectare per year in grasslands was only 0.58 yuan (Sun et al. 1997).

## *2.3 Techniques for the Restoration of Degraded Grassland Ecosystems*

Techniques for the restoration of degraded grasslands mainly include physical and chemical technology, engineering technology, and biological technology. In practice, in order to restore the degraded grasslands more effectively, these techniques can be used separately or conjunctively.

### 2.3.1 Physical and Chemical Technology

'Sand bedding' is an effective physical technique for the restoration of degraded saline-alkaline grasslands. The basic principle of this technique is to incorporate sands into grassland soils. This technique could effectively change the physical and chemical properties of these saline-alkali soils, and finally lead to a decrease in soil salinity and alkalinity in the degraded grasslands. In Tongyu country of Jilin Province, 'sand bedding' significantly improved soil physical structure and properties in a saline-alkali grassland after 4 years of large scale experiment (Table 3; Jiang 1978). 'Sand bedding' technique mainly changes the physical structure of soils, and thus improves the aeration and absorption of soils. However, it should be noted that there are some limitations of this technique. First, sand sources should be available for applying this technique into practice, but there are not sand sources in many grassland regions; second, the cost of this technique is a little bite high, it is estimated that the cost of this technique can be as high as 3000 yuan per $hm^2$; third,

**Table 3** The variation of saline-alkali soil properties before and after sand bedding

| Treatment | Item | | | | | |
|---|---|---|---|---|---|---|
| | Sampling depth (cm) | Physical clay (<0.01 mm) | Physical sand (>0.01 mm) | Soil texture | Salt content (%) | Note |
| Before sand bedding | 0–20 | 52.07 | 47.93 | Light clay | 0.157 | |
| After sand bedding | 0–20 | 23.23 | 76.77 | Light loam | 0.062 | Four years continuous sand bedding |

Jiang (1978)

when the sand is taken from one place to grasslands, vegetation within source region could be damaged.

Chemical technology is often considered as an experimental or practical way to improve degraded grasslands. Indeed, this technique can effectively improve the conditions of degraded soils, and then make the grassland vegetation easier to recover. Chemical agents are the key of chemical technology. Usually, those chemicals that used to improve degraded soils can be called soil conditioner. According to the sources of raw materials, soil conditioner can be divided into four types, namely natural conditioner, synthetic conditioner, natural-synthetic copolymer conditioner and biological conditioner.

Xiao et al. (2009) conducted a field experiment testing the improvement of sodic soils with desulphurization gypsum in a Latin square design in Ningxia Xidatan and studied the application amounts of desulphurization gypsum in cropping rice (Oryza sativa). Results showed that the application of desulphurization gypsum reduced soil alkaline, total alkalinity and pH, and increased seedling rate and yield of rice. However, different application amounts of desulphurization gypsum led to different reductions of soil alkaline, total alkalinity and pH. According to the simulation curve of the application amount of desulphurization gypsum with the reduction of soil alkalization, total alkalinity and pH, the reduction of alkaline soil, total alkalinity and pH reached the maximum when the application amount of desulphurization gypsum was 2.8–3.1 kg $m^{-2}$. Meanwhile, according to the simulation curve of the application amount of desulphurization gypsum with rice seedling rate and yield, rice seedling rate reached the maximum of 84.7 % when the application amount of desulphurization gypsum was 2.86 kg $m^{-2}$; when the application amount of desulphurization gypsum was 2.79 kg $m^{-2}$, the highest rice yield was up to 0.75 kg $m^{-2}$. Therefore, the optimal application amount of desulphurization gypsum was 2.8–3.1 kg $m^{-2}$ to improve sodic soils for planting rice.

### 2.3.2 Engineering Technology

According to the formation of saline soils, an engineering technology for grassland restoration was developed in China. Establishment of irrigation and drainage facilities is the main engineering technique for the restoration of degraded saline-alkali grasslands. Effective drainage systems can decrease the salinity and alkalinity of grassland soils, and make the degraded grassland move back to normal state. Generally, soils contain a variety of alkali and salt ions in the upper layers. Some of these ions are the essential nutrients for plant growth. However, they are also poisoned for plants, especially when their concentrations are high. Li and Zheng (1997) showed that the as groundwater rose, soluble salinity will transfer from deep soils to the surface soils, and resulted in salinization or alkalinization in soils.

The establishment of irrigation and drainage facilities can effectively inhibit the processes of salinization: First, drainages facilities can improve the drainage speed, decrease the level of groundwater, and inhibit the transfer of soluble salinity from deep soils to the surface soils; Second, drainages facilities enhance the leaching effects, which can effectively decrease the concentrations of alkali and salt in soil surface. To restore a desertification grassland, one of the first needs is to fix the mobile sand dunes, and then to restore the grassland vegetation. By establishing a various kinds of sand barriers, engineering technique plays a key role in changing soil property, reducing the wind speed and preventing wind erosion in desertification grasslands (Feng et al. 1994). Sand barriers can be made of concretes, soils, woods and grasses. Grasses are the ideal materials to produce sand barriers in degraded grasslands, because they are cheap and the supplies are abundant. More importantly, grasses within the sand barriers could decompose quickly and then enhance the fertility of grassland soils (Fig. 4).

**Fig. 4** A framework of sand barriers in the Shabotou sand dunes

### 2.3.3 Biological Technology

Except for the technologies mentioned above, we can also use bio-ecological technique to restore the degraded grasslands. According to the principles (or theories) of ecology and the resilience stability of grassland vegetation, bio-ecological techniques such as grass planting, plant transplanting (Zhang and Wang 2009), establishment of the litter layer (Guo et al. 1998), fertilization and irrigation can significantly improve the conditions of degraded grasslands (Wang and Yang 2003).

(i) **Grassland fencing** Grassland fencing and planting grasses are the traditional biological techniques to restore degraded grasslands. Generally, grassland fencing is considered to be an effective bio-ecological technique in grassland management and restoration. After long-term grazing or mowing, vegetation and soil fertility in grasslands will decrease significantly. Moreover, under intensive utilization, both natural and artificial grasslands will degrade within a relatively short period of time. Before grasslands were completely destroyed, if we immediately stop using the grasslands and further carry out grassland fencing, grassland vegetation and soils will recover quickly. Chinese ecologists have previously conducted grassland fencing experiments in multiple grasslands, and have achieved very good results. In Erdos of Inner Mongolia, grassland fencing significantly changed plant species composition: the proportion of fine quality forage such as legume and grasses increased significantly, whereas the proportion of poisonous plants declined. In Songnen grassland of northeast China, 5 years grassland fencing caused a strong expansion of the dominant species *L. chinensis*, which significantly increased grassland vegetation cover by approximately 50 % (Table 4; Li and Zheng 1997). In the Qinghai-Tibet alpine meadow ecosystem, the above-ground biomass of fine quality forage has risen from 174.0 to 325.0 g $m^{-2}$, and the below-ground biomass was also significantly increased after 3 years of grassland fencing (Li 1994).

(ii) **Seeding** Planting grasses is the major technique in grassland restoration, and this technique has been widely accepted and applied in China. Using the theory of plant succession, ecologists from Northeast Normal University has previously conducted many experiments to explore how to restore the degraded saline-alkali grasslands, and they have proposed an empirical mode for the

**Table 4** The development of *L. chinensis* community (area) fenced after five years

| Time | Communities No. ($m^2$) | | | | | | | | | | Total |
|---|---|---|---|---|---|---|---|---|---|---|---|
| | 1 | 2 | 3 | 4 | 5 | 6 | 7 | 8 | 9 | 10 | |
| Before fencing | 0.50 | 0.4 | 4.2 | 2.1 | 1.2 | 1.3 | 51 | 0 | 0.8 | 12.4 | 73.9 |
| Fened for 5 years | 1.4 | 1.9 | 8.1 | 3.6 | 1.5 | 2.4 | 72 | 0.4 | 1.7 | 16.7 | 109.7 |
| Extended area | 0.9 | 1.5 | 3.9 | 1.5 | 0.3 | 1.1 | 21 | 0.4 | 0.9 | 4.3 | 35.8 |

Li and Zheng (1997)

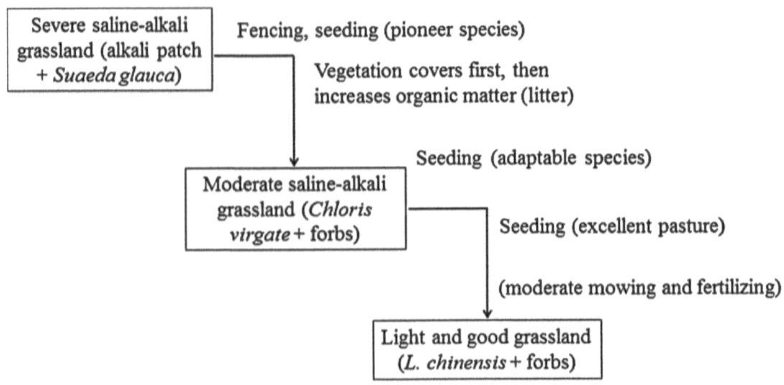


**Fig. 5** Seeding patterns of saline-alkali grassland restoration

restoration of these grasslands (Fig. 5). To restore the salinized grasslands, we should carry out the restoration step by step and closely according to the plant succession theory. There are large areas of salinized soils in Songnen grasslands, as those soils are unsuitable for plant growth, directly plant grasses on these soils is obviously not a viable alternative. Thus, we should firstly increase the accumulation of organic matter in grassland soils before planting grass. Generally, the establishment of litter layer may be a good way to improve the organic matter accumulation in soils. Guo et al. (1998) reported that after the establishment of litter layer, ground surface temperature has decreased by 7 °C, and soil water content has increased by 27 %. The establishment of litter layer can also decrease soil bulk density, soil pH and soil salt content, whereas increase the content of soil organic matter and available nitrogen (Table 5). The amount of litter could significantly influence the recover progress of grassland vegetation and soils. For example, once litter mass is below 500 g $m^{-2}$, the effects of litter layer on the restoration of salinized grasslands are not obvious. When litter mass is up to 1.5–2.0 kg $m^{-2}$, the density, cover, height, production and survival of plants will increase significantly. Compared with other types of lands, it is more difficult to establish artificial grasslands in salinized lands. Therefore, it is important to choose the native plant species for grassland restoration in this region. Zhang and Wang (2009) reported that the transplantation of *L. chinensis* to salinized grasslands could strongly improve the conditions in these regions. The transplantation of *L. chinensis* may strongly affect soil pH and electric conductivity in soils; likewise, as the physiological tolerance to salinity of *L. chinensis*, this plant species will expanse quickly in salinized lands, and thus can effectively restore this type of degraded grassland.

**Table 5** The effects of litter layer on physical and chemical properties of salinization soil

| Community type | Plot | Surface temperature (°C) | Soil water content (%) | Soil bulk density (g cm$^{-3}$) | SOM (%) | Available nitrogen (mg g$^{-1}$) | pH | Salt content (mmol kg$^{-1}$) |
|---|---|---|---|---|---|---|---|---|
| Saline-alkai | CK | 35 | 18 | 1.73 | 0.75 | 0.13 | 10.2 | 0.838 |
| patch | Treatment | 28 | 23 | 1.68 | 1.05 | 0.22 | 9.4 | 0.765 |
| Suaeda | CK | 34 | 19 | 1.72 | 0.76 | 0.21 | 10.0 | 0.807 |
| community | Treatment | 28 | 24 | 1.65 | 1.10 | 0.31 | 9.0 | 0.668 |
| Chloris | CK | 32 | 20 | 1.68 | 1.32 | 0.23 | 9.5 | 0.664 |
| community | Treatment | 27 | 24 | 1.52 | 1.63 | 0.28 | 8.6 | 0.581 |
| Puccinellia | CK | 33 | 18 | 1.61 | 1.34 | 0.25 | 9.6 | 0.754 |
| community | Treatment | 28 | 22 | 1.48 | 1.67 | 0.27 | 8.7 | 0.612 |

Guo et al. (1998)

### 2.4 Outlook

Grassland degradation is a serious problem in China. At the dawn of the 21st century, grassland ecologists in our country have put great efforts into the restoration of degraded grasslands, and they have made a lot of progress and achievements in this area. However, as the ecological and technological support systems of grasslands are still relatively incomplete in our country, it is clear that we need more studies on the restoration and reconstruction of degraded grasslands.

## 3 Ecological Restoration of Degraded Wetlands in China

### 3.1 Introduction

The wetland is one of the most important ecosystems on the earth, and an important place for the world's biological and genetic material (gene pool) storage. However, the rapid growth of human population and economy, together with the long-term over-exploitation, has made the global wetland ecosystems overwhelmed and on the verge of collapse. More than 50 % of the global wetlands have been lost today, and the situation of the Chinese wetlands' damage is also not optimistic, as more than 60 % of the wetlands have been damaged in varying degrees (Liao et al. 2006). Therefore, improving and restoring wetlands' various functions, such as slowing the runoff, flood control and drought prevention, water purification and restoring and rebuilding wetlands in the appropriate geographical places are the pressing issues human face today (An 2003).

Today, the restoration of the degraded wetlands is one of the most important issues of ecological research and has developed rapidly with the development of wetland restoration theories. Researches on the degraded wetland restoration were started early abroad: The United States invested huge funding for wetland restoration which is organized and implemented by the Clean Lake Program (CLP) since 1970; the U.S. National Research Council (NRC) commissioned the Committee on Restoration of Aquatic Ecosystems (CRAE) to carry out water ecosystem restoration and the overall evaluation in 1989. Also, some European countries have worked on water environmental governance and the restoration since 1970s. Since 1980, Japan has actively promoted and restored the natural state of the riparian environment construction for a beautiful natural environment on the basis of ensuring the function of the flood control and water utilization (Zhang et al. 2003).

Chinese researchers set out on the wetland restoration in 1980s, carrying out water environmental restoration research in different regions such as Chaohu Lake and Taihu Lake, and successful experience has been gained (Hu et al. 2008; Yang and Liu 2010; Zhai et al. 2010). Based on the fundamental concepts of the wetland restoration, this article expounded its principles and process, illustrated three basic modes in China, and summarized the main goal and basic strategy. The assessment and future trends of the wetland restoration were also discussed.

## 3.2 *The Basic Principles and Processes of the Degraded Wetland Ecological Restoration*

### 3.2.1 Basic Principles of Ecological Restoration of Degraded Wetlands

(i) Holism principle. Broadly speaking, the wetland ecosystem is regarded as the ecotone between typical terrestrial ecosystems and aquatic ecosystems. Wetland ecosystems integrate hygrophyte, helophytes, hydrophyte, mesocole, micro-organisms, some biological factors and other abiological factors closely related to the sun and soil. It releases and absorbs the nutrients and other chemicals through the hydrological pathways (such as rainfall, surface runoff, groundwater flow and tidal exchange). Therefore, the various components of the wetland ecosystem are a unified and indivisible organic whole.

(ii) Harmony and balance principle. The ecosystem is full of relative stability and coordinated internal structure and function inside certain period in the long-term process of evolution and development. Therefore, wetland ecosystem restoration should focus on restoring ecosystems function rather than the simple recovery and reconstruction of the structure, especially for the disappeared wetlands. Meanwhile, the restoration of the wetland ecosystem should balance the natural, economic and social factors.

(iii) Self-resiliency principle. Wetland ecosystem restoration should fully consider the wetland hydrology, biogeochemistry, ecosystem dynamics and the species adaptation and so on, restoring and rebuilding a self-organizing, self-maintaining, and self-design function for reducing unnecessary human participation. However, since the self-regulating mechanism of the wetland ecosystem is limited, over-reliance on the self-regulating mechanisms of the ecosystem is bound to have an adverse impact.

(iv) Circulation principles. Ecological restoration adjusts all aspects of the circulation, coordinating the amount of substance of the input, output and transformation of these aspects. Ecological restoration aims at achieving the ecological, economic, social benefits in the removal of endogenous and exogenous pollutants.

### 3.2.2 The Guidelines of the Degraded Wetland Restoration

The overall goal of wetland restoration is to restore degraded wetland ecosystem structure and function through the appropriate biological, ecological and engineering, and ultimately reach a state of the wetland ecosystem's self-sustaining

(Sun et al. 2007). But the focus and requirements of the restoration will be varied as the degraded wetland ecosystems are different.

(i) A comprehensive analysis about the original wetland ecological system should be carried out before planning the restoration.
(ii) Determining the restriction factor of the ecological system is very important to the restoration of wetland ecosystems.
(iii) As for different regions and types of the degraded wetlands, the corresponding strategies and the index system are different.
(iv) Appropriate monitoring plan should be made for monitoring the wetland ecosystem regularly and promptly.

## *3.3 Modes of Degraded Wetland Restoration*

### 3.3.1 Function-Oriented Model

(i) Restoration of habitat function. Various measures are taken to protect and improve the diversity and stability of damaged habitats and gradually restore the function, such as building the wild feeding point, covertly or biological wall, nest boxes or nest, ecological corridor, berm and gentle slope (Zhang and Wang 2001; Zhang et al. 2009).
(ii) Restoration of biogeochemical function. An important task of ecological restoration is to artificially change the dominant factor or terminate the process of ecosystem degradation, adjusting and optimizing the inner system as well as the flow and its temporal and spatial order with the outside substances to restore the ecological system to a certain or even higher level. However, concrete measures should be taken carefully to avoid secondary damage or pollution by over-simplified physical and chemistry methods.
(iii) Restoration of water resources protection. Socio-economic developments and lax environmental oversight have increased water demand and pollution (An et al. 2007). Restoring the wetland through restoring water function for the sustainable development and solving the contradictions between the development and utilization of water resources are indispensable.
(iv) Restoration of social service function. The landscape design aims at overall harmony in the wetland restoration, considering various factors such as the harmony between the form and the internal structure, and the environmental and social service functions. Based on the original body of water, plants, topography and other constituent landscape factors, the design avoids the destruction of the ecological integrity and the imbalance of the environment.

### 3.3.2 Policy-Oriented Model

(i) Restoration of hydrology and water resources. The primary goal of the restoration is to maximally avoid evaporation and the loss of water resources. So far, establishing the drainage system and the dam (raising the water level) are the main measures for the replenishment and water diversion of the wetlands.
(ii) Restoration of habitat diversity. Habitats' diversity is fundamental to biological diversity. In given geographical area, the diversity of the habitat and its constituent elements greatly influence the biological diversity. The ecological restoration such as flood storage, drought prevention and water conservation lays foundation to biodiversity.
(iii) Restoration of biodiversity. Biodiversity restoration must first adjust the types and structure of aquatic plants. For example, the aquatic plants provide a good living environment for other organisms and improve the structure and function of aquatic ecosystems in the degraded wetland restoration.

## *3.4 Type-Oriented Modes*

(i) Restoration of swamps. The hydrological regime of the swamp ecosystem is mainly determined by the replenishment (e.g., precipitation) and the amount of water loss (e.g., evaporation, surface and underground runoff). Therefore, wetland restoration first restores vegetation and habitat, reduces the surface water loss (Labadz et al. 2002).
(ii) Restoration of rivers. The river restoration aims at restoring the healthy river system, but it is a multi-objective, multi-level and multi-constrained issue. The river ecosystem restoration should be set out from restoring the function (Li and Ju 2005; Ni and Liu 2006).
(iii) Restoration of lakes. An effective way of the lake governance is restoring aquatic plants to control the eutrophication and water purification. The aquatic plants restoration needs certain preconditions. For instance, the eutrophic lake ecological restoration needs moderate nitrogen and phosphorus concentration, wind and waves and the water depth, as well as the mainly carnivorous fish structure and no organic-rich sediments (Qin 2007, 2009).
(iv) Restoration of coastal wetlands. Although the overall reasons of the degraded coastal ecosystems have known, the relationship among various components and their interaction mechanism still needs furthered. Coastal ecosystem restoration and pilot demonstration study are still stuck in the small, local

region or centralized to a single biome or vegetation type, lacking comprehensive regional research and demonstration. The deficient study on the functional indicators of sound coastal ecosystem blinds the restoration.

(v) Restoration of fishery ponds. Conversion of fisheries into lakes and wetland and developing the ecological fishery is critical to the ecological restoration.

### 3.5 Evaluation of the Ecological Restoration of Degraded Wetlands

The restoration program should be assessed after its completion for the construction and environmental management, with the perspective of the ecological and environmental protection and sustainable development. The influence of the restoration construction could be predicted based on the full investigation of ecological environment of regional and the surrounding area during and after the restoration. Integrating the engineering and regional ecological environmental characteristics, reasonable and feasible ecological protection and mitigation measures can be taken to minimize the adverse effects of the restoration for ensuring the effective protection and sustainable development of the wetland ecosystems and biodiversity. The assessment of degraded wetland restoration effect can be considered from the following ecological, social and economic aspects.

### 3.6 Conclusion and Discussion

The continuous deterioration of the wetland seriously constrains the rapid socio-economic development, so how to restore the degraded wetland ecology and maintain the sound structure and function of the wetland is a major issue for human. Thus some aspects of the future study on the degraded wetland ecological restoration should be emphasized as follows:

(i) Tolerance of the aquatic plant to pollutants. At present, most studies domestic and abroad focus on the role of absorption and degradation of the major aquatic plants to the pollutants. However, studies of the variability of the tolerance of the different aquatic plants in different concentrations and types of pollutants are less involved and need to be further studied. Additionally, for the specific circumstances of the restoration in different regions, setting different planting density and species composition is also very necessary.

(ii) The integration of various wetland ecosystem restoration modes. It is difficult to achieve the desired results by a single restoration mode for the water body pollution is complicated. The complex ecosystem together with a variety of ecological recovery modes can make up for its shortcomings and make the system run more stable.

(iii) Complete the technical specifications and evaluation system and establish a typical demonstration project of the degraded wetland ecological restoration.

(iv) Enhance the management of degraded wetland post-restoration. The influence of the restoration could be predicted based on the full investigation of regional ecological environment pre-restoration and post-restoration. Also, combining with the engineering and regional ecological environmental characteristics, reasonable and feasible ecological protection and mitigation measures can be taken to minimize the adverse effects of the restoration for ensuring the effective protection and sustainable development of the wetland ecosystems and biodiversity.

## 4 Protection and Restoration of Sandland Ecosystems in China

### *4.1 Introduction*

Since 1960s, the term "sandland" or "sandy land" has been using to describe inland dune ecosystems in North China, because they are in semiarid region, being distinguished from "desert" in arid region. The four most representative and largest sandland ecosystems in China are: (1) Mu Us Sandland (107° 20′E–111° 30′E; 37° 30′N–39° 20′N), located in south-central of Ordos Plateau in Inner Mongolia and northern Shaanxi province, with total area 32,100 $km^2$ in the end of 1950s and 38,940 $km^2$ in the middle of 1990s; (2) Otindag Sandland (112° 22′E–117° 57′E; 41° 56′N–44° 24′N), located in southern of Xilingol League and northwest Chifeng city in Inner Mongolia, with total area 21,400 $km^2$ in the end of 1950s and 29,220 $km^2$ in the middle of 1990s; (3) Horqin Sandland (118° 35′E–123° 30′E; 42° 41′N–45° 15′N), located in Xiliaohe river in western Chinese Northeast Plain, with total area 42,300 $km^2$ in the end of 1950s and 50,440 $km^2$ in the middle of 1990s; and (4) Hulunbeir Sandland (117° 10′E–121° 12′E; 47° 20′N–49° 59′N), located in Hulunbeier High Plain in northeastern Inner Mongolia, with total area 7200 $km^2$ in the end of 1950s and 6410 $km^2$ in the middle of 1990s (Zhu et al. 1980; Zhong 1998). The four major sandlands cover in north-south direction about 10° of latitude and in east-west direction about 16° of longitude. All of them exist in the semiarid temperate steppe region of northern China, where were historically covered by highly productive grasslands (Chinese Academy of Sciences 1985); currently, however, they are covered by aeolian mobile dunes, semi-fixed dunes and fixed dunes.

The early study on the sandlands focused on status, monitoring and control of desertification. Since 1990s, more researches had been done on the sandlands, distinguished from the deserts. And a number of ecological research stations were established in the four major sandlands for their scientifical importance and their importance for national economic development, including one in Ordos Sandland (Ordos Sandy Grassland Ecological Research Station, which was established in 1990), three in Horqin Sandland (Naiman Desertification Research Station, established in 1985; Daqinggou Ecosystems Research Station, established in 1988; and Wulanaodu Desertification Experimental Station, established in 1975), three in Otindag Sandland (Xilingol Grassland Ecosystem Research Station, established in 1979; Duolun Restoration Ecology Research and Demonstration Station, established in 2000; and Otindag Sandland Ecological Research Station, established in 2002), and one in Hulunbeier Sandland (Hulunber Grassland Ecosystem Research Station, established in 1997). After years of observation and study, these ecological stations make great achievements not only in basic researches, but also in application of the science and technology.

## 4.2 Protection and Restoration of Sandland Ecosystems

### 4.2.1 Ecosystem Characteristic in Sandlands

Sandy matrix is the major ecological feature of the sandland ecosystems. On the one hand, the sand coverage is the cause of drastic temperature changes, poor water and fertilizer retention ability of soil, and the instability of soil matrix. Sandy matrix has a major role in ecological processes of sandland ecosystems. It even may cause serious land desertification in the case of runaway operation. On the other hand, the sand coverage can prevent and reduce evaporation in the arid climate conditions, and it is conducive to store water in the deep layer. The "underground reservoir" formatted by sand coverage prevents the occurrence of land salinization, while results in richer biological and ecosystem diversity of sandland ecosystems, as comparing to grassland ecosystems.

Highly environment heterogeneity is another feature of sandland ecosystem. The terrain in sandland is characterized by alternative distribution of sandy dunes and lowland (dune interphase). According to Mu Us sandland landscape ecological processes and factors, there are 4 categories and 10 kinds of landscape ecological types.

#### Composition

Producers in sandland ecosystem are mainly xerophytes, which are often sparse and short. Animal species, as the consumer in sandland ecosystems, are few due to the lack of food sources. The decomposers in sandland ecosystem mainly are soil microbes, like bacteria, fungi and actinomycetes.

Plant Diversity

The four major sandlands are all located in ecotone between agriculture and animal husbandry, with rich landscape and plant diversity. For example, there are total 1083 vascular plant species in Otindag Sandland, belong to 85 families and 392 genus; Mu Us Sanland, named as "the shrub kingdom" in northern China for its most abundant shrub species, has 92 shrub species, including 22 semi-shrubs species, belonging to 25 families and 50 genus.

Ecosystem Stability and Function

Just like the forests and grassland ecosystems, sandland ecosystems play a big role in the carbon cycle. Studies have shown that the average biomass and productivity in Otindag Sandland, are 90 and 59 % higher than in typical steppe respectively (Li 2011). But sandland ecosystems suffered from greater threat by global changes and human interferences, for its relative simple structure and poor ecosystem stability.

Ecosystem Services

Sandlands' ecosystem services are mainly reflected in the soil formation and protection, water conservation, climate mediation, biodiversity conservation and other indirect ecosystem services. The value of these services accounts for 88.9 % of the total value, while the production of food and raw materials are only 7.3 %, and recreational and cultural value are 3.8 % (Zhang et al. 2007).

### 4.2.2 Ecosystem Succession in Sandlands

Cycle of Vegetation Succession

Due to instability of sandy matrix, specificity of regional climate (drought, strong wind) and poor adaptable ability of shallow root species, the climatic climax, Stipa communities, cannot exist stably for long-term in sandlands. Without shrubs with strong soil fixation ability like Artemisia, sandy soil activate again and irreversible desertification happen, resulting climatic climax succeed to the dune stage, then a cycle of vegetation succession is formed (Guo 2000).

Plant Functional Types (PFT) in Different Succession Stage

In sandlands, it is often that annual pioneer species settled first in the early stage of succession, and then perennial grasses became the dominant species in the later succession stage. Perennial $C_3$ clonal shrubs (PFT of perennial-$C_3$-clonal-shrub) showed a more stable performance with higher important values and percentage of species number in all succession stages, and play a key role in the early succession stages. Along with the dune fixation from mobile, semi-fixed to mixed dune, PFT-diversity and ecosystem primary productivity increased (Qiao et al. 2012).

Driving Force of Succession

Difference of adaptability of the dominant species to the sandy soil matrix in different successional stages may be an important driving force in sandlands, and global change, such as increased precipitation and nitrogen deposition, can modify the succession process (Qiao 2009).

### 4.2.3 Degradation of Sandland Ecosystems

Many factors may have contributed to the sandland ecosystems' degradation, including climate warming and drying, decline of underground water level caused by coal mining, irrational exploitation of vegetation such as overgrazing and firewood collection. Overgrazing is considered as the direct cause of sandland ecosystems' degradation. Desertification in sandland ecosystems can be manifested in the biodiversity, soil and ecosystem productivity.

Biodiversity

The process of ecosystem degradation is a process of decay of species diversity, degradation causes most native species disappear (Jiang et al. 2008), species diversity and evenness decreased (Zhang and Zhao 2010). In the four major sandland, species diversity and functional diversity in mobile dunes were significantly lower than which in the fixed dunes.

Soil

With the soil degradation process, soil becomes coarse-grained while nutritious tiny particles in soil are eroded by wind, resulting in loss of soil organic carbon, nitrogen, phosphorus and microbial activity. Vegetation restoration can produce significant sequestration and protection on soil nutrient and water, promoting the healthy development of soil-vegetation system.

Ecosystem Productivity

Poor soil nutrient of degraded sandland ecosystems resulting in the change of vegetation composition and structure, then in low ecosystem productivity. Study showed that aboveground biomass in mobile and semi-mobile dunes in Horqin Sandland were 48.2 and 60.4 g $m^2$, respectively much lower than those in the semi-fixed and fixed dunes, as 152.8 and 167.6 g $m^2$, respectively.

### 4.2.4 Restoration of Sandland Ecosystems

Principles

The protection and restoration of sandland ecosystems is a systems engineering, needs to follow principles as follows: (i) moderate development; (ii) landscape heterogeneity and multiple scales; (iii) integration of ecological, economic and social benefits (Ye and Liang 2004).

Paradigms

Based on the long-term observation, experiments and demonstration, aiming at the four major sandlands, a lot of eco-productive paradigms were constructed to protect and restore the degraded sandland ecosystems (Zhang 1994; Ye and Liang 2004; Jin et al. 2006; Ci et al. 2007; Yan and Zhang 2008), including the follows: (i) "3-circles" eco-productive paradigm in Mu Us Sandland; (ii) Restoration paradigm based on "10 % principle" in Otindag Sandland; (iii) "family ranch" complex ecosystem restoration paradigm in Horqin Sandland; and (iv) comprehensive management of agriculture, animal husbandry and forestry paradigm in Hulunbeir Sandland.

### 4.2.5 Resources Available in Sandland Ecosystems

Biotic Resources

There are rich plant resources in sandlands, according to different uses they can be divided into the following categories: (i) oil plants, such as *Xanthoceras sorbifolia*; (ii) fiber plants, as *Apocynum venetum*; (iii) medicinal plants, as *Radix Glycyrrhizin*; (iv) edible plants, including potherb and grass crops; (v) sand-fixing plant, as *Artemisia ordosica*; (vi) ornamental plants as *Sabina vulgaris*; (vii) bio-energy plants, as *Salix cheilophila*; and (viii) rare and protected plants, as *Picea mongolia* in Otindag Sandland.

Abiotic Resources

In sandland ecosystems, with sparsely populated and conventional energy shortage, it is imperative and feasible to utilize new energy sources.

Wind energy: as an important natural resource, with no pollution, the total reserves of wind power was estimated 1010 GW in Inner Mongolia, among which 101 GW were available for development and utilization (Zang and Feng 1997).

Solar energy: there are abundant solar energy resources in sandlands, especially in the summer season. It is estimated that the potential of solar power generation exceeds 3000 GW in Inner Mongolia.

Tourism resources: the unique natural landscape in sandland ecosystems is a good tourist resource, such as grassland, sand ridges and honeycomb dunes, dotted with numerous small lake and sand springs.

## *4.3 Conclusion and Discussion*

For the particularity of sandland ecosystems, people pay more and more attention on it, and more and more studies about sandland ecosystems are carried out. Most of earlier studies (i) focused on combating desertification; (ii) faced to climate change; and (iii) payed attention to combining theory with practice.

Further researches about sandland ecosystems are needed to focus on: (i) eco-hydrological processes; (ii) biological functional traits; (iii) adaptation to global climate change; (iv) ecosystem services and biological resources, and (v) sandland conservation model based on process and mechanism.

# References

Alcoze, T. A., Covington, W. W., & Fule, P. Z. (2000). Restoration ecology. *Science, 287*, 2159.

An, S. Q. (2003). *Wetland eco-engineering-optimization model of utilization and protection of Wetland resources*. Beijing: Chemical Industry Press.

An, S. Q., Li, H. B., Guan, B. H., et al. (2007). China's natural wetlands: Past problems, current status, and future challenges. *AMBIO A Journal of the Human Environment, 36*(4), 317–324.

Bai, X., & Pu, X. (2003). The analyses of climate change of alpine meadow grassland in Jinqianghe, Gansu province. *Prataculturał Science, 20*(7), 51–53.

Chinese Academy of Sciences. (1985). *Vegetation of Inner Mongolia*. Beijing, China: Scientific Press.

Ci, L. J., Yang, X. H., & Zhang, X. S. (2007). The mechanism and function of "3-Circles"—An eco-productive paradigm for desertification combating in China. *Acta Ecologica Sinica, 27*, 1450–1460.

Feng, L., Lu, J., & Di, Y. (1994). Review on the prevention of sand damages to railway line in desert areas of China. *Journal of Desert Research, 14*(3), 47–53.

Gao, Y., Han, X., & Wang, S. (2004). The effects of grazing on grassland soils. *Acta Ecologica Sinica, 24*, 790–797.

Guo, K. (2000). Cyclic succession of *Artemisia ordosica* Krasch community in the Mu Us sandy grassland. *Acta Phytoecologica Sinica, 24*, 243–247.
Guo, J., Shun, G., & Jiang, S. (1998). Ecological effect of litter layer on vegetation recovery of saline-alkali grassland. *Journal of Northeast Normal University, 3*, 107–110.
He, Y. T., & Lang, H. L. (2009). Significance and strategies of vegetation construction on the Loess Plateau. *Research of Soil and Water Conservation, 16*(4), 30–38.
Hu, W. P., Zhai, S. J., Zhu, Z., et al. (2008). Impacts of the Yangtze River water transfer on the restoration of Lake Taihu. *Ecological Engineering, 34*(1), 30–39.
Jiang, Y. (1978). *Saline-alkali soil improvement*. Changchun: Jilin People's Publishing House.
Jiang, D. M., Cao, C. Y., Toshio, O., A, L. M. S., & Li, M. (2008). Characteristics of the vegetation and soil changes in the process of desertification in Horqin sandy land. *Journal of Arid Land Resources and Environment, 22*, 156–161.
Jin, M., Jia, Z. Q., & Lu, Q. (2006). Integrated desert control pattern in Hunshandaka sand land and its benefit evaluation—A case study of duolun county. *Forest Research, 19*, 321–325.
Labadz, J. C., Butcher, D. P., & Sinnott, D. (2002). Wetlands and still waters. In M. R. Perrow & A. J. Davy (Eds.), *Handbook of ecological restoration—Principles of restoration* (Vol. 1, pp. 106–132). Cambridge: Cambridge University Press.
Li, X. (1994). Phytomass changes of alpine meadow in closed condition. *Chinese Qinghai Journal of Animal and Veterinary Sciences, 24*(4), 9–11.
Li, H. Y., & Ju, M. T. (2005). *Principles and practice of ecological restoration*. Beijing: Chemical Industry Press.
Li, G., Li, Y. G., Liu, M. Z., & Jiang, G. M. (2011). Vegetation biomass and net primary production of sparse forest grassland in Hunshandake sandland. *Science & Technology Review, 29*, 30–37.
Li, Z. A., Wang, H., & Yu, Z. Y. (1996). Effect of man-made forests on eroded soils in tropical region in Guangdong. *Tropical Geography, 16*(4), 310–318.
Li, S. Z., Xiao, H. L., Luo, F., Song, Y. X., Liu, L. C., & Li, S. L. (2005). Regulation effect of microbiotic crusts on soil hydrological process in Shapotou vegetated sanddunes. *Journal of Desert Research, 25*(2), 228–233.
Li, J., & Zheng, H. (1997). *The management and biological and ecological mechanisms of saline-alkali grassland in Songnen plain*. Beijing: Science Press.
Liao, W. G., Du, Q., Tan, H. W., et al. (2006). Application Status and trends of water ecological restoration techniques. *China Water Resources, 17*, A61–A63.
Lü, X., & Lü, X. (2002). Climate tendency analysis of warming and drying in grassland of northeast Qingzang plateau of China. *Grassland of China, 24*(4), 8–13.
Ma, M., Hou, X., & Zhang, H. (2002). Pest prevalence and its control on the Qinghai pasture. *Qinghai Prataculture, 11*(3), 24–27.
Ni, J. R., & Liu, Y. Y. (2006). Ecological rehabilitation of damaged river system. *Journal of Hydraulic Engineering, 37*(9), 1029–1037.
Palmer, M. A., & Filoso, S. (2009). Restoration of ecosystem services for environmental markets. *Science, 325*, 575–576.
Peng, S. L., & Hou, Y. P. (2007). Ecology-based restoration in a changing world. *Acta Ecologica Sinica, 27*(8), 3536–3537 (Review on the 18th SER conference).
Peng, S. L., Hou, Y. P., & Chen, B. M. (2010). Establishment of Markov successional model and its application for forest restoration reference in southern China. *Ecological Modelling, 221*, 1317–1324.
Peng, S. L., Yang, L. C., & Lu, H. F. (2003). Environmental effects of vegetation restoration on degraded ecosystem in low subtropical China. *Journal of Environmental Sciences, 15*(4), 514–519.
Peng, S. L., & Zhou, T. (2009). Making change in a changing world. *Acta Ecologica Sinica, 29*(9), 5161–5162 (Review on the 19th SER conference).
Peng, Y. S., Zhou, Y. W., & Chen, G. Z. (2008). The restoration of mangrove wetland: A review. *Acta Ecologica Sinica, 28*(2), 786–797.

Qiao, J. J. (2009). *Analysis of PFT-spectra and dominant species' adaptation at successional stages of steppe sandlands*. PhD degree paper, Institute of Botany, Chinese Academy of Sciences, Beijing.

Qiao, J., Zhao, W., Xie, X., Liu, G., Ye, X., Chu, Y., et al. (2012). Variation in plant diversity and dominance across dune fixation stages in the Chinese steppe zone. *Journal of Plant Ecology, 5* (3), 313–319.

Qin, B. Q. (2007). Principles and approach for lake ecological restoration. *Acta Ecologica Sinica, 27*(11), 4848–4858.

Qin, B. Q. (2009). Progress and prospect on the eco-environmental research of Lake Taihu. *Journal of Lake Science, 21*(4), 445–455.

Stone, R. (2009). Nursing China's ailing forests back to health. *Science, 325*, 556–558.

Sun, X. (1997). The production status and potential of livestock husbandry of grassland in China. *Inner Mongolia Prataculture, 4*, 30–35.

Sun, Y., Guo, J. B., Dang, P. X., et al. (2007). Research on everglade wetland remediation. *Journal of Inner Mongolia Forestry Science & Technology, 33*(3), 33–38.

Teng, X., & Wang, D. (2002). The positive effects of grazing sheep on grasslands of the Songnen Pains. *Journal of Northeast Normal University, 34*, 34–35.

Teng, X., Wang, D., & Zhang, B. (2006). The grazing sheep characters in rainy day and rainless day on *Leymus chinensis* grassland. *Acta Ecologica Sinica, 26*(3), 762–767.

Wang, D., Han, G., & Bai, Y. (2005). Interaction between foraging behaviour of herbivores and grassland resources in the eastern Eurasian steppes. In D. A. McGilloway (Ed.), *Grassland: A globe resource* (pp. 97–110). The Netherlands: Wageningen Academic Publishers.

Wang, D., Lu, X., & Luo, W. (1996). Analysis to effects of different grazing density on characteristics of rangeland vegetation. *Acta Prataculturae Sinica, 5*(2), 28–33.

Wang, D., & Yang, L. (2003). *Grassland ecology and its management and utilization*. Beijing: Chemical Industrial Press.

Xiao, G., Luo, C., Bai, H., et al. (2009). Research on the amount of desulfurized gypsum from the coal-burning power plant applied to improve the alkalized soil for paddy rice. *Ecology and Environmental Sciences, 18*(6), 2376–2380.

Yan, D. R., & Zhang, B. Z. (2008). Research on Hulunbeier sandy land. *Journal of Inner Mongolia Forestry Science & Technology, 34*, 34–39.

Yang, S. Q., & Liu, P. W. (2010). Strategy of water pollution prevention in Taihu Lake and its effects analysis. *Journal of Great Lakes Research, 36*(1), 150–158.

Ye, X. H., & Liang, S. C. (2004). A review on the study of optimized eco-productive paradigm in the farming pastoral zone of north China. *Acta Ecologica Sinica, 24*, 2878–2886.

Yu, Z. Y., & Peng, S. L. (1996). *Ecological restoration of degraded ecosystem in tropical and subtropical*. Guangzhou: Guangdong science and Technology Press.

Zang, R., & Feng, S. Z. (1997). Evaluation of wind energy resources and development of wind power in Inner Mongolia. *International Power, 4*, 17–20.

Zhai, S. J., Hu, W. P., & Zhu, Z. C. (2010). Ecological impacts of water transfers on Lake Taihu from the Yangtze River, China. *Ecological Engineering, 36*(4), 406–420.

Zhang, X. S. (1994). Principles and optimal models for development of Maowusu sandy grassland. *Acta Phytoecologica Sinica*, 1–16.

Zhang, M. X., Liu, G. Q., & Tang, X. P. (2009). Techniques and methods for wetland restoration. *Wetland Science & Management, 5*(3), 12–15.

Zhang, Y. Z., & Wang, X. (2001). A review of ecological restoration studies on natural wetland. *Acta Ecologica Sinica, 21*(2), 309–314.

Zhang, B., & Wang, D. (2009). A technique of rapidly restoring alkali-steppe with directly cultivating *Leymus chinensis* on the Songnen Plains. *Journal of Northeast Normal University, 41*(3), 97–100.

Zhang, W. H., Zhang, X. H., Xiao, B. D., et al. (2003). Progress on restoration techniques for inland aquatic environment. *Shanghai Environmental Sciences, 22*(11), 811–816.

Zhang, H., Zhang, A. P., & Yang, J. (2007). Effects of land use change on ecosystem service values of Horqin sandy land. *China Population, Resources and Environment, 17*, 60–65.
Zhong, D. C. (1998). *The dynamic evolution of China sand sea*. Lanzhou: Gansu Culture Press.
Zhong, W., & Fan, N. (2002). The cause of rodent pests occurring on the grassland in china and the strategies of ecological management. *Bulletin of Biology, 37*(7), 1–4.
Zhu, Z. D., Wu, Z., & Liu, S. (1980). *The desert of China*. Beijing: Science Press.

# Chapter 11
# Ecological Restoration in the Typical Areas

**Yuancun Shen, Xianzhou Zhang, Jingsheng Wang, Peili Shi, Yongtao He, Zhenxi Shen, Xinquan Zhao, Huakun Zhou, Shixiao Xu, Liang Zhao, Buqing Yao, Ting Zhou, Shaolin Peng, Jianguo Wu, Jianhua Cao, Fen Huang, Hui Yang, Liang Li, Qiang Li, Weikai Bao, Zhenqi Hu, Peijun Wang, Jing Li, Pei Qin, Jie Fan and Pingxing Li**

**Abstract** Human activities are strongly regional and targeted, and usually tightly bound to the regional natural resources, the needs of social economic development, and ecological fragility, which cause typical regional ecosystem degradation problems. For serious degradation areas, it is an effective way to carry out regional

Y. Shen · X. Zhang · J. Wang · P. Shi · Y. He · Z. Shen · J. Fan
Resources Research, Chinese Academy of Sciences, Institute of Geographical Sciences and Natural, Beijing 100101, China

X. Zhao · H. Zhou · S. Xu · L. Zhao · B. Yao
Chinese Academy of Sciences, Northwest Institute of Plateau Biology, Xining 810001, China

T. Zhou · S. Peng
State Key Laboratory of Biocontrol, School of Life Sciences, Sun Yat-sen University, Guangzhou 510006, China

J. Wu
School of Life Sciences and Global Institute of Sustainability, Arizona State University, Tempe, AZ 85287-4501, USA

J. Cao · F. Huang · H. Yang · L. Li · Q. Li
Chinese Academy of Geological Sciences, Institute of Karst Geology, Guilin 541004, China

W. Bao (✉)
Key Laboratory of Mountain Ecological Restoration and Bio-resource Utilization, Chengdu Institute of Biology, Chinese Academy of Sciences, Chengdu 610041, Sichuan, China
e-mail: baowk@cib.ac.cn

Z. Hu · P. Wang · J. Li
Institute of Land Reclamation and Ecological Restoration, China University of Mining and Technology (Beijing), Beijing 100083, China

P. Qin
Halophyte Research Lab, Nanjing University, Nanjing 210093, China

P. Li
Chinese Academy of Sciences, Nanjing Institute of Geography and Limnology, Nanjing 210008, China

 
W. Li (ed.), *Contemporary Ecology Research in China*,
DOI 10.1007/978-3-662-48376-3_11

ecological restoration and construction for restoration ecology study, which is the key to promote sustainable development and ecological security, and also the urgent requirement for the current reality. According to the demand of regional sustainable development, this chapter focused on ecological deterioration, and restoration and construction in typical areas. The chapter specially made thematic discussions on new progression in degradation problems of developing and utilizing the resource, ecological restoration practices, theory, techniques, mode, and management. This chapter was trying to provide experience and lessons, macro guidance and decision-making reference for ecological restoration in typical areas. From the regional cases of this chapter, we can see that regional ecological restoration and construction is intricate. Finally, it is necessary to point out that there are many regional ecological restoration cases, here just introduce several cases in western fragile region. The ecological restoration cases revealed in this chapter are just the first attempt to conclude the past experiences in China, which are initial identification for correlated theories and methods.

**Keywords** Ecological construction · Loess plateau · Tibet plateau · Grassland restoration · Vegetation restoration · Karst ecosystem · Hengduan Mountains · Arid valley · Mine land · Coastal ecosystem

## 1 Ecological Construction on Loess Plateau

The loess plateau is wide, which is surrounded by the Taihang Mountain range in the east, the Riyue mountain—Helan mountain in the west, the Qinling range in the south, and Yinshan range in the north. It covers an area of some $64.2 \times 10^4$ km$^2$, about 83 % of Yellow River basin. The loess plateau is the key area of ecological construction in China, due to its serious water loss and soil erosion.

### *1.1 Physical and Ecological Character of Loess Plateau*

The terrain is slightly sloping, high in the northwest and low in the southeast with elevation of 2000–500 m above sea level. There are sand loess, loess, and argillic loess zones from northwest to southeast (Liu et al. 1985), which form sandstorm in the north and loess plateau in the south. The climate is warm, with 2500–4500 °C yearly accumulated temperature above 10 °C, frost-free period of 150–250 days. The south belongs to warm temperate zone, while the north belongs to temperate zone. The rainfall is 700–200 mm. From southeast to northwest, the climate, vegetation, and ecosystem have zonal differentiations: semi humid—semiarid—

arid, forest—forest and grassland—grassland—desert grassland, broad-leaved deciduous forest in warm temperate climate—forest and grassland in semi humid and semiarid warm temperate climate—grassland in semiarid warm temperate climate—desert grassland in arid temperate climate (Huang 1983; Synthetical expedition team on loess plateau and Chinese academy of sciences 1990a, b, c; Yang 1990; Shen and Hong 2003; Shen and Yang 2006; The ministry of water resources of the people's republic of China et al. 2010).

## 1.2 The Status, Problem, and Harm of Ecology Environment

There is fragile ecology environment and frequent ecological disasters in loess plateau. In detail, the following are the main contents.

### 1.2.1 The Status and Harm of Soil Erosion

The quantity of soil erosion has been $1.6 \times 10^9$ t/a on loess plateau in recent 100 years, among which the quantity caused by nature has been $9.76 \times 10^8$ t/a (Ke et al. 1983). The soil erosion has made tremendous harm, especially it made the terrain fragmentized. In half a century, arable land has been eroded more than $2.67 \times 10^6$ $hm^2$, evenly over 6.7 $hm^2$ per year (The ministry of water resources of the people's republic of China et al. 2010). The total quantities of lost nitrogen, phosphorus, and kalium were more than $3.52 \times 10^{10}$ kg in each year (Zhang 1992). The quantity of silt piled up on the downriver bed was about $4 \times 10^8$ t. The riverbed in lower Yellow river became higher and higher by 10 cm per year, which has been higher than the ground, named "overground river". There has been frequent floods, which led to tremendous loss to the people (Li 2008). The silt made by soil erosion can accumulate in the reservoir, which cut down the capacity of reservoir and weaken its adjustment for the floods. Sandy area in the north of loess plateau is the main source of the sandstorm in our country. The sandstorm can harm Beijing, Tianjin, and the whole north of China.

### 1.2.2 Arid Climate and Severe Floods

The rainfall is lower than the transpiration in loess plateau (Zhang 1992). The scarcity of water becomes more and more serious from southeast to northwest. So there is severer arid in northwest than in southeast (Li 2008). In history, the arid made less production in agriculture and results in famine. The drought with over 10,000 died happened 16 times from B.C. 180–1949, which made total $1.7988 \times 10^7$ persons die (Chen and Gao 1984). Averagely, severe drought happened once per 2 or 3 years (Wang 1988). There was always heavy rain in loess plateau, which made flood and secondary disasters, such as collapsed reservoir, submersed field, road, and village.

#### 1.2.3 Damage and Harm of Vegetation

The native vegetation has been seriously destroyed. Now the natural vegetation coverage is only 41.42 % of total land area. The forest has been largely turned into bushwood, meanwhile grassland degradation and desertification have become serious. So disasters become common, such as floods, Debris flow, sandstorm, silt accumulate in the riverbed.

#### 1.2.4 Degeneration and Harm of Slope Farmland Ecosystem

The agricultural development in loess plateau began in the 8000 years ago (Synthetical expedition team on loess plateau and Chinese academy of sciences 1991). During nearly 50 years, the population has increased from 36.395 million in 1949, 88.52 million in 1999, to more than 100 million in 2010 (Chen 1990; Zhao 1991). Due to serious shortage of food and fuel woods, excessive deforestation and reclamation on the steep slope became more and more, which resulted in the sloping farmland ecosystem seriously deteriorated (Institute of soil and water conservation et al. 1990; Synthetical expedition team on loess plateau and Chinese academy of sciences 1990a, b, 1991).

### *1.3 Strategy of Ecological Construction and Setting up the Integrated Management System*

#### 1.3.1 Strategy on Ecological Construction

Strategic Concept

Based on the law of nature, we should set the target for eliminating human erosion and establish long-term, continuous, and steady governance.

Strategic Goal

(i) Ecological strategic goal

The quantity of soil erosion should be decreased to 975 million t/a only caused by nature after implementing ecological construction.

(ii) Goal of ecological industry

The goal of ecological industry should be moderate on loess plateau. The elementary goal is to realize self-supplying food production, for-sale fruit and drug production, sheltered forest and stockbreeding of farming and grazing.

Strategic Measure

(i) A series of dams should be constructed to protect the cropland and realize food self-supplying. High-production croplands should distribute on the gently slope less than 7° and alluvium with dams in the valley. To construct high-production croplands with dams is the premise and key to ensure people survive and realize ecological construction. Now the quantities of this kind of land resource is 3.36 μ/person, which shows there is steady basic for realizing food self-supplying and good ecological environment (Synthetical expedition team on loess plateau and Chinese academy of sciences 1992).

(ii) System of special industry should be set up to protect the ecology and soil on the slope. It can improve local economic steady development and make the people rich. This area is the most suitable place for developing the warm temperate fruits and drugs. So it is suitable to develop ecologic and economic industry on the slope, for example, crop-wood, crop-fruit tree, crop-drug, crop-vegetable, crop-fence or crop-grass can get ecologic and economic benefit. The system has had the case in the representative area on loess plateau.

(iii) The special resource should be integrated and developed to improve economic development in the sand area. The sand area is mainly distributed on the north of loess plateau. On the premise of ecological protection, special biology resources should be planted and efficiently developed with new technology. Developing the economic in the sand area is believed to be the soul of sand production. Now the special biology resources have been planted and developed in Yanchi county, which is succeeded in both ecology and economic.

(iv) Planting the vegetation according to the quantity of water and natural restoration capability. The vegetation can be naturally restored with close cultivation and management. The ecology and environment will go on well to change the strategy of "suitable land, suitable tree" to "suitable water, suitable tree".

(v) Small drainage area should be the fathering unit in order to get the whole development of regional ecology, social, and economic. Small drainage area is the basic unit in geography. There are many typical and successful cases, where all kinds of fathering methodologies and designs on developing industry are based on the small drainage area to improve the ecology—production—economic.

### 1.3.2 Building the System of Synthetic Management to Improve the Ecology and Environment on Loess Plateau

To make a synthetic science plan on ecology construction is the premise of realizing the good ecology and environment. Setting up an organization doing synthetic management and making decisions can ensure realization of the goal—effective ecology and environment construction. To widen the financing channel and built omnidirectional financing system can ensure long ecology and environment construction. It is necessary to construct monitoring system of the law and criterion.

## 2 Causes of Grassland Degradation and Grassland Restoration Technology in the Northern Tibet Plateau

The Chang Tang Plateau is main body of Tibetan Plateau, with three grassland types, i.e., alpine meadow, alpine steppe, and alpine desert grassland. The Chang Tang Plateau is sensitive to global change and human activities due to its high altitude, cold and drying climate condition, and fragile ecosystems. Many facts indicate that the grassland degradation trend of the northern Tibetan Plateau generally turns for the better since 1980s, although local region shows more severe degradation.

Climate change is one of the main factors causing grassland degradation. Annual air temperature on the northern Tibetan Plateau has increased about 1.4 °C during the past 50 years and the warming accelerates since 1980s. The warming magnitudes in fall (1.6 °C) and winter (1.8 °C) are significantly larger than the annual average. Annual precipitation has increased about 55 mm, approximately 50 mm of which occurs since 1980s. Overall, the warming and wetting trend increases ecosystem productivity of grassland on the northern Tibetan Plateau, whereas the positive effect of climate change is mainly concentrated on the eastern region. By contrast, the decrease trend of net primary production results from the warming and drying trend in the western region during the recent years (Ali region is one of the typical area). The long-term overgrazing intensifies severe degradation on the central and eastern of Chang Tang Plateau. In addition, the insect and mouse damages caused by climate change should not be ignored.

In recent years, several grassland protection measures and techniques have been continuously conducted on the Chang Tang Plateau. The enclosure mode, one main method adopted by the grazing ban project of the Ministry of Agriculture in China, increases coverage by 5 % and aboveground biomass by 42–65 % for the moderately and severely degraded grassland. The enclosure and fertilizer mode is better than the enclosure mode. N and P addition (5 $g/m^2$) increases average coverage by about 20 % and grass yield by 2.2–3.7 times in *Stipa purpurea* grassland. Mixture of N and P addition significantly enhances plant biomass, with 24.1 % increase of total biomass, 164.8 % increase of aboveground biomass, and 4.7 % decline of belowground biomass, and the fertilizing amount is positively correlated with plant biomass in alpine meadows. The enclosure, fertilizer, and resowing mode is fit for overgrazing-induced severe degradation grassland and increases aboveground biomass by 163.2–348.6 %, being 2–4.5 times equivalent to control. However, this mode requires higher input cost and technique measures. Therefore, it has certain difficulty to expand this mode.

The Tibetan Plateau is the most significant area in regional differentiation in China and even the world. The Tibetan Plateau has three ecological function area: (1) the northern Tibetan alpine grassland and desert region with an altitude of above 4000 m, (2) the Brahmaputra River and its two tributaries with an altitude of 3000–4000 m, and (3) the southeast Tibetan forest region with an altitude of below 3000 m. Considering the ecological security barrier construction, grassland degradation and recovery on the northern Tibetan alpine grassland region and the high land productivity on the Brahmaputra River and its two tributaries, the Tibet has requirement conditions of transporting artificial forage grass in the wastelands and middle-and-low yielding fields on the agricultural region to the northern Tibetan pasturing region. This can not only adjust industrial structure on the agricultural region, but also alleviate feed-animal imbalance status and prevent grassland degradation. Consequently, this eventually reaches the win–win situation of ecological environment protection and increment of farmers and herdsmen income.

# 3 The Restoration of Degraded Alpine Grassland Ecosystem and the Sustainable Development Patterns of Pastoral Livestock Production

## 3.1 Introduction

The alpine grassland is mainly distributed on Qinghai–Tibetan Plateau (QTP) in China and Central Asia, the Andes mountains in South America, and the Alps in Europe, but the QTP is the most typical representative of alpine grassland. The most concentrated distribution of alpine meadow is in the region of the QTP in China's western area. With the different hydrothermal conditions within the plateau, the alpine grassland ecosystem comprises alpine shrub, alpine meadow, alpine steppe, alpine desert, and alpine wetland, and cushion vegetation and talus vegetation distributing in high-altitude locales. In accordance with the "National Long-term Scientific and Technological Development (2006–2020)", the alpine grassland belongs to the typical ecological fragile area. In recent years, due to the impact of human disturbance and climate warming, alpine grassland degradation is serious, with a sharp deterioration of the ecological environment. After several years of large-scale ecological restoration engineering, grassland degradation trend has been curbed, but the overall deterioration situation is not reversed, especially at the alpine grassland in the headwater region of three rivers, degradation situation is becoming increasingly grim. According to the survey, 90 % of the alpine grassland in the headwater region of three rivers appeared in different degrees of degradation. At present, area of moderate degraded grassland is of 187 million mu (1 $hm^2 = 15$ μ), accounting for 58 % of the useful grassland area in the region. Compared with that of recent 50 years, yield per unit area decreased by 30–50 %, high quality forage proportion decreased by 20–30 %, poisonous weeds increased by 70–80 %, grassland vegetation coverage reduced by 15–25 %, dominant grass height decreased by 30–50 %, and grass height decreased by more than 20 % (He et al. 2008). Only in the case of the source of the Yellow River, grassland degradation rate from 1980 to 1990 was more than double than the average in 1970s . The headwater region of three rivers "black soil beach" area has reached to 42.44 million mu, accounting for 15 % of the total area of the grasslands. The desertification area has reached to 44 million mu, and still expand in the speed of 78,000 mu annual. The average rate of desertification increased from 3.9 % in the 1970s–1980s to 20 % in 1980s–1990s. The fragmentation of the original ecological landscape, vegetation succession showed the reverse succession patterns from alpine meadow by degraded alpine meadow to the desert. Because of the conflicts of grass and livestock, the number of livestock has been maintained over the grassland carrying capacity, grassland degradation, the number of livestock will continue to decline, entered the vicious circle of "the overgrazing of grassland degradation intensified conflicts between livestock and herbage—the deterioration of the ecological environment—seriously affected the healthy development of the life and pastoral livestock economy of the headwater region of three rivers".

The restoration to the degraded alpine grassland is a complex system engineering, and the effective measures of comprehensive improvement of degraded alpine grassland are to reduce the natural grassland grazing pressure, strengthen the grazing management, make ecological environment not suitable for harmful biology thriving, ecological control, and the construct degraded grassland ecology, and protect and restore of organically structure and ecological function of alpine grassland ecosystems. Degraded alpine grassland restoration and governance are a long process, "prevention and control" combined with the specific ways and measures, has been widely reported (Ma et al. 2007; Zhao 2011). In general, adopting and carried out steadily what kind of control strategy and technical measures, are rest with the strength of grazing system, causes of the local climate, soil and grassland degradation, and other factors. Through the studies of different degraded grassland, ecological restoration technology integration and demonstration of the alpine grassland, and highly targeted remediation technology of the severe degradation grassland: "black soil beach" degraded alpine grassland restoration process, there will provide a significance great demonstration and radiation effect to prevent the "black soil beach" degraded alpine grassland further degradation and desertification.

In recent years, the ecological restoration of damaged environment has attracted more and more attention and got rapid development, ecological degradation of the environment has become one of the hotspots in the field of ecology, and restoration of degraded ecological system has become a hot issue of restoration ecology. America is one of the world's first country of ecological restoration research and practice. At the beginning of the twentieth century, the Europe and the United States have practiced under drought stress recovery in agricultural ecosystem (Peng and Ren 2003). A lot of works have been done by ecological restoration in degraded artificial grassland at home and abroad. Earliest research and deep restoration were carried in large area of ecological mining that industrial revolution left in British and European heath. Very fruitful research results have been obtained in cold temperate coniferous forest stub land vegetation in the Nordic countries. In Australia, African continent, and Mediterranean Europe, the focus of the study is related to artificial restoration of degraded arid land. In foreign countries, the methods of fertilization and irrigation are relatively common in the cultivation and management of artificial grassland. The additions of trace elements were performed according to the conditions of soil fertility difference. Guillaume (1986) and Chambers (1997) restored and studied the degradation alpine ecosystem, respecting to recovery of the vegetation succession characteristics and corresponding environmental stress factors, in Western America using the methods of sowing, plant grass transplanting, and turf transplantation, and received better results.

Because of the far transportation distance between the alpine grassland area and county, high cost management, no relevant coordination mechanism, lack of innovation mechanism in aspects of characteristic resources mining, marketing, production organization, and horizontal cooperation, local resources cannot be fully utilized. For example, animal husbandry was a characteristic resource in the headwater region of three rivers, but only as a cheap raw material due to the need of the backward development of animal husbandry and industrialization. After hundreds of years of efforts, Europe and the United States and other developed countries make full use of advanced scientific technology and management experience, the layout of the area, factory farming, automation, specialization, scale, intensive production, standardized management, realize the modernization of animal husbandry to obtain high profits. Therefore, in guidance of Scientific Outlook on Development, by speeding up the transformation of animal husbandry production from traditional to the modern mode of production, promoting the healthy development of animal husbandry and animal products, and increasing the added value of alpine grassland distribution, is the route that must be taken to develop the animal husbandry of the alpine grassland area.

Practice has proved that popularization and application of sci-technology on alpine grassland distribution area of animal husbandry economic growth rate is the driving force for the healthy development of animal husbandry and stable place, such as cattle, sheep-fattening techniques, so that each sheep meat production increased about 5 kg, cattle carcass weight increase about 40 kg. Application of warm-shed feeding technology in pastoral industry has marked effect on improving survival rate of lamb and numerous living rate, reducing death rate, and preventing the fat loss in winter and spring, so that benefits are very obvious with propagation rate increasing 13 % in average, adult livestock loss dead dropping 2.2 % (Zhao 2011). Ga and Qing (2004) think that the way of development of animal husbandry in southern Qinghai area and actively exploring the development path of industrialization should accord to following methods: first, to vigorously support from the capital, policies for development of animal products processing enterprises, through the "ten households", "base the household" business model, develop the order animal husbandry industry; second, to support a number of large farm households of cow, cattle and sheep fattening and trafficking, and animal husbandry; third, to cultivate pastoral intermediary organization. It is necessary to use the scientific methods to improve the overall production efficiency. High yield animal husbandry and increase of herdsmen income would be achieved by shortening the breeding cycle through the methods of "warm sukkah, improving variety of livestock, and ammoniated grass", to realize the transformation of the way of breeding, increase rate of households, the contribution rate of science and technology, and accelerate the development of animal husbandry science and technology.

Northwest Institute of Plateau Biology, CAS, and Qinghai Academy of Animal Husbandry and Veterinary (QAAHV) have explored the experience, and broke an ecological Qinghai characteristic animal husbandry industry road in the headwater region of three rivers in recent years (Ma et al. 2007; Zhao 2011), made a complete Qinghai alpine grassland ecological animal husbandry technology system and promotion modes, such as grassland ecological animal husbandry model suitable for poor natural conditions of the natural grassland, with premise of ecology protection, and development mode of the ecological animal husbandry of livestock and modern organic ecological mode suitable for agricultural and pastoral areas, returning farmland to grassland, and conditional planting artificial grassland area, with the goal of resource recycling, have achieved good ecological, economic, and social benefits.

## *3.2 Body Part*

### 3.2.1 Alpine Grassland Degradation Classification and Restoration

Characteristics of Grassland Degradation and Its Index System

*Degraded Grassland Assessment and Grades*

The comprehensive evaluation was carried out according to some scholars to formulate the evaluation index system of grassland, the vegetation fractional coverage of grassland, forage yield, the biomass proportion and height of edible forage, and compared to the native vegetation on alpine grassland quality. According to this standard, the appropriate classification standard of the alpine meadow in the headwater region of three rivers was developed, and alpine meadow in the headwater region of three rivers can be divided into five grades: the original vegetation, lightly degraded grassland, moderate degraded grassland, severe degraded grassland, and extremely degraded grassland (black soil beach) (Table 1). Such classification criteria can also be used in Tibet, Gannan, northwest Sichuan, northwest Yunnan, and similar areas in alpine meadow (Ma et al. 2007).

### *"Black Soil Beach" Degraded Grassland Division Standards, Type, and Size*

In the degraded grassland classification, according to the terrain conditions and engineering management, the extreme degradation grassland "black soil beach" is divided into three types: namely, the beach slope between 0 and 7°; the gentle slope ≥ between 7 and 25°; the steep slope ≥ 25°. The main index and then through field investigation and questionnaire survey and application of principal component analysis to optimize classification, each type is divided into three grades such as mild, moderate, and severe one (Ma and Li 1999).

Take the headwater region of three rivers as an example, there are a total area of 73.6 million mu of different grades of "black soil beach" degraded grassland, including mild "black soil beach" degraded grassland area of 29.0 million mu, moderate "black soil beach" degraded grassland area of 25.4 million mu, and severe "black soil beach" degraded grassland area of 19.2 million mu. In addition, the different types of "black soil beach" degraded grassland (<7° slope) cover an area of

**Table 1** The evaluation grade standard of grassland degradation in the headwater region of three rivers

| Degradation level | Coverage (%) | Aboveground biomass proportion (%) | Proportion of palatable herbage (%) | Height of palatable herbage (cm) | Grassland quality |
|---|---|---|---|---|---|
| Nondegraded grassland | 80–95 | 100 | >70 | 25 | Standard |
| Lightly degraded grassland | 70–85 | 50–75 | 50–70 | −(3–5) | Fall 1 level |
| Moderately degraded grassland | 50–70 | 30–50 | 30–50 | −(5–10) | Fall 1 level |
| Heavily degraded grassland | 30–50 | 15–30 | 15–30 | −(10–15) | Fall 1–2 level |
| Extremely degraded grassland | <30 | <15 | 0 | – | Very bad |

46.8 million mu, the gentle "black beach" degraded grassland (7 ≤ slope < 25°) covers an area of 23.0 million mu, and steep slopes to the "black beach" degraded grassland (≤25° slope) cover an area of 3.8 million acres (Zhao 2011).

## The Analysis Way to Restore Degraded Alpine Meadow

### *Analysis of Restoration of Degraded Grassland*

Management regulation of degraded alpine meadow ecosystem can be divided into two levels: the implementation of degraded grassland degradation reasons and solutions. The previous level includes many factors that lead to the degraded alpine meadow, which is the level of governance measures after recovery. Alpine meadow degradation levels include the following six factors: F1—the long-term overgrazing, F2—the harm of mouse insect, weed, F3—the human unreasonable interference (shovel turf, herb digging, gold road, dredging, quarrying, deforestation, etc.), F4—herd structure unreasonable, F5—warm and dry climate, F6—soil erosion, freeze–thaw action. Restoring measures level consists of the following eight factors: the E1—control to reduce grazing intensity, E2—fencing and grazing rotation, E3 —the establishment of artificial grassland, E4—rat and insect pest and weed control, E5—stop unreasonable human interference, E6—fertilization, sowing, harrowing, E7—optimizing livestock structure, E8—excellent forage breeding.

Grassland management objective is to achieve the best economic conditions and protecting grassland vegetation, maintain the ecological balance of the total target, and achieve sustainable development. Suppose the general goal of C management problems, it can be formulated in an alpine meadow ecosystem for the hierarchy below (Fig. 1).

Although the alpine grassland distribution areas of low population density, grassland area is large, but because the grassland contracting sharply increasing livestock and grassland, seasonal distribution is not balanced and artificial pasture irrational use, resulting in most of the winter spring pasture overgrazing serious and degradation of vegetation. In addition, according to investigation of Wang and Cheng (2001) from the Cold and Arid Regions Environmental and Engineering

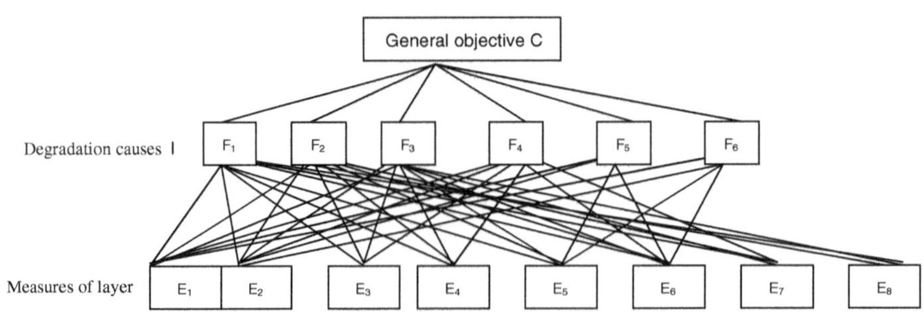


**Fig. 1** Hierarchical structure of alpine meadow grazing ecosystem

Research Institute, the Chinese Academy of Sciences, in the headwater region of three rivers, grassland rodent and pest hazard area is up to 6–21 % of the total area of grassland, degradation area ratio can reach to 13–58.25 % because of pest damage, the proportion varies across different regions. Under normal circumstances of grazing intensity, the pests and rats and grazing intensity in alpine meadow in the headwater region of three rivers is directly related to the more serious pests and rats destruction.

In general, the reasons and interaction of alpine region to grassland degradation ecological environment deterioration can be expressed in Fig. 2.

It can be clearly seen from Fig. 2 that climate change and human activities are the two major factors leading to the deterioration of the ecological environment of the study area. Rodent and pest were incidental to bring grassland overloaded, produced on these two factors playing a very important role during grassland restoration.

In the local natural conditions, fencing and grazing rotation (E2) governance best benefit (Fig. 3), followed by the establishment of artificial grassland (E3), control the reasonable grazing intensity (E1), rat pest and weed (E4), reducing the irrational human interference (E5), harrowing, sowing, fertilization of improved grassland (E6). These good governances were efficient. Because the optimizing livestock structure (E7) and excellent forage breeding (E8) in the treatment of degradation have indirect effects of grassland, their benefits and contributions are relatively low. For mild degraded grassland, based on the protection, by reducing grazing pressure, control of grazing intensity and fenced grazing, and other measures, can prevent the further degradation, and succeed to the direction of native vegetation. For moderate degraded grassland, we should take the harrowing, sowing, fertilizing, fencing, and other measures, at the same time, for rodent control. These measures will effectively curb the continued degradation of grassland, and improve soil fertility, good ecological and economic benefits. For the severe degradation and extremely degraded

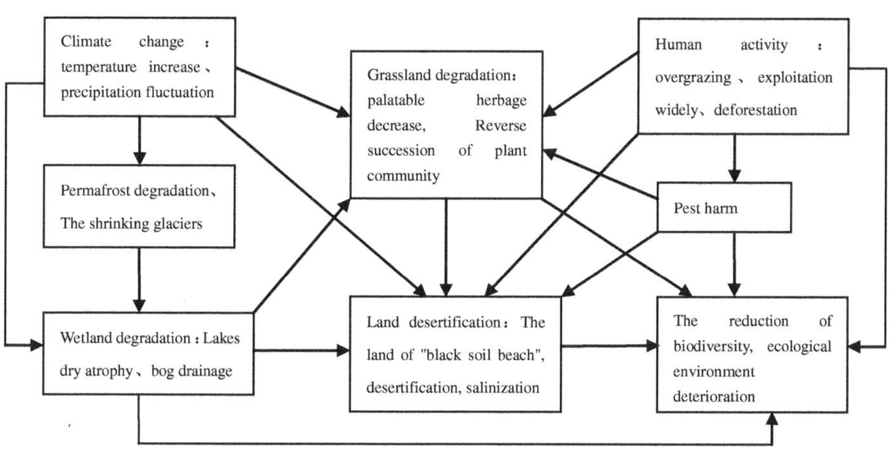


**Fig. 2** The causes and its interactions of the ecological environment deterioration in alpine region

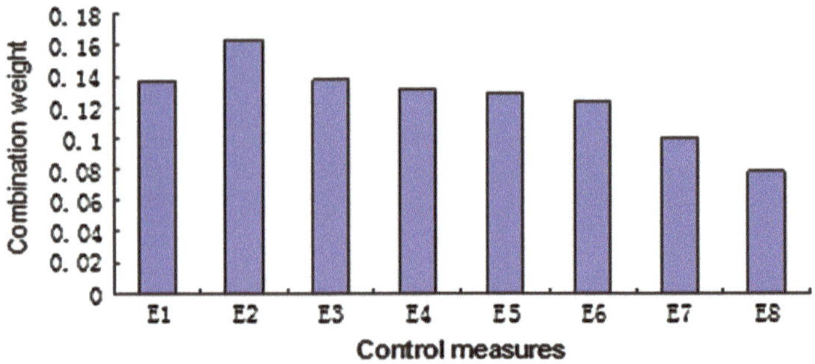


**Fig. 3** The distribution of combination weights under different controlling measures. $E_1$ grazing intensity decrease, $E_2$ fence, $E_3$ artificial grassland construction, $E_4$ pest control, $E_5$ stop unreasonable human interference, $E_6$ fertilization, sowing, harrowing, $E_7$ optimizing livestock structure, $E_8$ excellent forage breeding

grassland, natural recovery is more difficult to the restoration and reconstruction. Comprehensive measures should be adopted to establish artificial or semi-artificial grassland to rapid restoration of degraded grassland vegetation, so as to become the ecosystem reconstruction and achieve a new ecological balance.

### *Different Types of Vegetation After the Destruction of the Resilience and Recovery Rate*

To take the Maqin alpine meadow (I), Maduo alpine grassland (II), and Tongde steppe (III) in the headwater of three rivers in different years borrow natural recovery sequence, as the comparative study of dynamic object, the restoration of natural ability (SL) in Maqin alpine meadow (I), Maduo alpine steppe (II), and Tongde with temperate grassland (III) were 0.41, 1.33, and 1.50, respectively. The vegetation that completely destroyed after about 50, 40, and 30 years of natural recovery can achieve near the top of plant community types, their natural recovery (SL) and recovery rate are III > II > I.

Damage recovery capability of grassland community was quantitatively evaluated with $S_L = H/H_0\ (1-\Delta F)\ I$ (Wang 2010), in which: SL for the comprehensive index of damaged grassland ecosystem vegetation restoration ability was dimensionless, the greater values means the greater recovery capacity. I as the top species invasive index, refers to function of dominant species number and the frequency of occurrence in the natural conditions: $I = 1 + \sum_{i=1}^{s} P_i$, where $P_i$ and $s$ denote the dominant frequency of occurrence (or dominant species relative abundance of $I$ relative importance value, $P_i = N_i/N$. $N_i$ $I$ absolute importance value, $N$ plot all kinds of important value of the absolute) and species number; $H$ and $H_0$ represent the Shannon–Wiener indexes of sampling site and natural grassland, respectively; $\Delta F$ is the difference of cover degree between the sampling site and natural grassland vegetation.

### The Overall Design of Grassland Management Technology of Degraded Alpine Region

With the foundation of the ecology and system science, and restoration ecology and sustainable development theory, using multidisciplinary cross-combining theory with practice, combined supporting technology project is implementing ecological engineering, science and technology training and the rule of law education methods, based on the existing technology and research to develop the practical technology of ecological construction and environmental protection, to achieve ecological development between economic and social and environment; comprehensive controlling measures of ecological control are widely used. The overall framework of technical line is shown in Fig. 4.

The main innovations of this design are: (i) together with the technology demonstration engineering and application promotion, driven technology development and integration by the construction of the demonstration zone, achieving the feasibility and practicability of inspection technology demonstration project. So, this not only shortens the research cycle, but also can speed up the popularization and application of comprehensive ecological treating technologies. (ii) through the

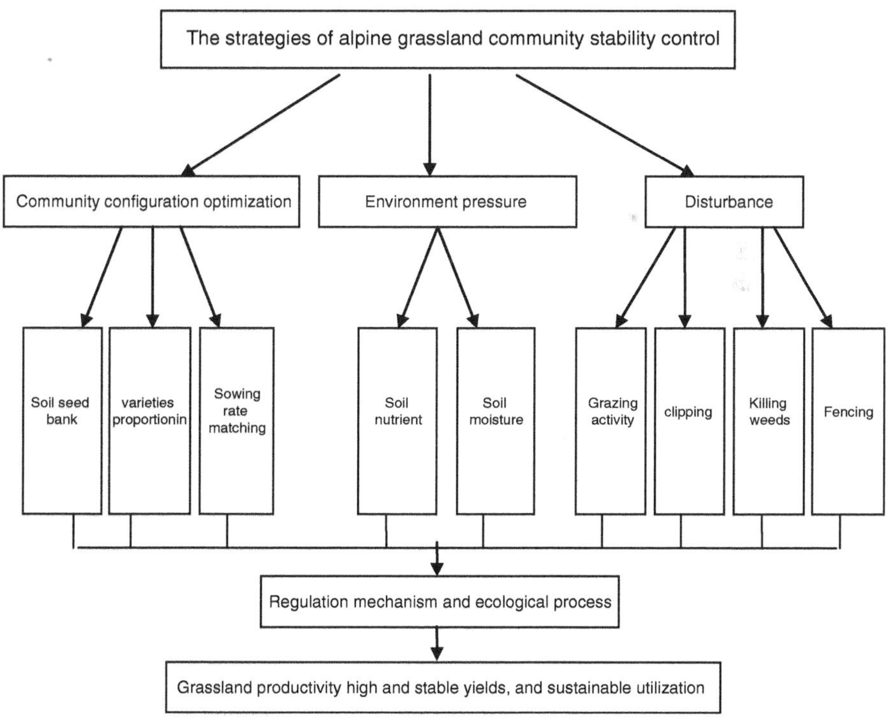


**Fig. 4** Integrated management and sustainable utilization of grassland degradation in alpine region

combination of scientific demonstration and the existing ecological treatment project phase, and solved the key technology urgently that needs to solve the problems in the ecological engineering, the research results can be directly applied to the practice of the ecological environment governance. (iii) the degraded grassland management and sustainable development of regional economy are combined, the overall optimization can not only realize the alpine grassland ecosystem structure and function, but also can achieve both in animal husbandry efficiency and the income of farmers and herdsmen target. (iv) the degraded grassland vegetation restoration combined with rodents ecological control can accelerate the restoration of vegetation, also can inhibit the mice growth and invasion. (v) combining ecological animal husbandry management mode will be harnessing degenerated grassland and intensive, which can keep the rectification results, achieve the coordinated development of grassland animal husbandry and ecological environment of the target.

## The Degraded Grassland Management Mode and Technology Integration

### *Mode of Governance and the Main Technical Characteristics*

Measures were carried according to the stage of succession and ecological degradation of natural grassland in alpine area, integrated by techniques of fencing, harrowing, sowing, fertilization, weed and pest control (Table 2), to restore rapidly degraded grassland vegetation and improve the primary productivity, curb the development and spread of degraded grassland.

(i) Application range and main technical means of mild degraded grassland management mode

**Table 2** The degraded grassland management technologies and modes

| The degree of degradation succession | Technical measures | Technical standard |
|---|---|---|
| Lightly degraded grassland | Fencing; rodent control; fencing + fertilize | The vegetation restoration technology rules of the lightly and moderately degraded alpine meadow |
| Moderately degraded grassland | Fencing; fencing + sowing herbage; killing weeds + fertilize | The vegetation restoration technology rules of the lightly and moderately degraded alpine meadow |
| Heavily degraded grassland | Fencing; harrowing + sowing herbage; artificial, semi-artificial grassland | The rules of classification and integrated control technology of black soil type degraded meadow |
| Extremely degraded grassland | Reconstruction of artificial community | Artificial vegetation construction and utilization of grassland management specification of black soil beach |

Primary grassland is generally applied to mild degradation of grassland plant communities, the species composition is higher, and richness is more than 30 kinds of species. Vegetation coverage is more than 70 %, excellent forage proportion is larger, and poisonous weeds accounted for less than 30 % in the composition of plant community biomass. Grassland landscape neat, turf layer remained intact; plant community has self-repair ability when grazing pressure is reduced, and recovers soon under enclosure condition.

(ii) Application range and main technical means of moderately degraded grassland management mode
The item is for the moderate degraded grassland restoration so that the moderate degradation grassland in Qinghai distributed at an altitude from 3500 to 4500 m, with an average annual temperature of 0 °C. The method is suitable for the thin soil layer or gentle hillside, where the total vegetation coverage is 50–70 %, the proportion of forage grass is 30–50 %.

(iii) Application range and main technical means of severe degradation grassland management mode
According to the degree and the local climate, topography and land use of degraded grassland, severe degradation grassland management mode is always used, such as enclosure and reseeding + fertilization or artificial, semi-artificial grassland planting pattern, etc. The treating methods were applied to the degraded grassland of altitude from 3500 to 4500 m, with an average annual temperature of 0 °C, flat terrain, slope less than 25°. The former is applicable to the native vegetation coverage of 30–50 %, forage yield is 15–30 % with soil of thin flat or gentle hillside. The technical measures of sowing time, sowing quantity and seed selection, and sowing method are same as the moderate degraded grassland management mode.
The construction of artificial, semi-artificial grassland model suitable for native vegetation is that the plant cover is less than 30 %, plant species composition of 60–80 % is the poisonous weeds, the hydrothermal conditions are better, the economic value and ecological service function of the grassland are poor, and the natural recovering ability is poor, and must use artificial intervention measures in order to achieve the purpose of restoration.

(iv) Application range and main technical means of the extreme degradation grassland management mode

Excellent forage rapid decreased with grassland degradation succession process. In the extreme degradation as the "black soil type degraded grassland", the proportion of forage grass is almost zero, the grassland grazing value almost decline. The "black soil type degraded grassland" is a severely damaged ecological system, which has lost the capacity of natural recovery, or in the short term cannot naturally recover, must through the ways of artificial rebuilding of damaged ecosystems restore the original ecological and production function. To reconstruct the artificial community, the forage species suitable for alpine meadow growing, the top grass such as *Elymus nutans*, *Festuca Sinensis* Keng, *Elymus sibiricus* Linn, *E. sibiricus* Linn var. Tongdeensis et al. were chosen, and the low grass such as *Poa crymophila* Keng, *Poa Pratensis* L,

*P. Pratensis* var. anceps Gaud cv. Qinghai, *Puccinellia tenuiflora, Festuca krylo-viana, Festuca kirilowii, Poa poophagorum* Bor at. al. were always chosen. Agronomic measures and its process include the deratization, plowing, harrowing, soil preparation, fertilizing, sowing (or row), covering by soil, and repression.

### 3.2.2 Process and Mechanism of the Degraded Alpine Meadow

Different Distribution Characteristics of Plant Communities in Degraded Stages and Carbon, Nitrogen, and Reserves

*Characteristics of Plant Communities in Alpine Meadow with Different Degradation Stages*

With degradation of alpine grassland, vegetation coverage, grassland quality index and good aboveground biomass, and the similarity index between grassland decreased, while the index of plant community diversity and evenness index in the moderate degraded steppe reach the highest stage. With the degradation degree increase, the plant community diversity and evenness index show a single peak curve. The total aboveground biomass in the lightly degraded stage is the highest, the lowest in the extreme degradation stage, with the degradation, weed biomass increased significantly, and sedges and grasses biomass decreased significantly (Table 3; Zhou et al. 2012). The 0–40 cm total root biomass, underground biomass of sedges and grasses with aggravation of degradation decline, changes of under-ground biomass of forbs is increased gradually, and the extreme degradation decreased. With the degradation degree aggravate, root volume and distribution in each layer are less, the underground root system has the characteristics of the shallow layer trend. Among all kinds of ground, underground biomass was positively correlated with aboveground biomass significantly.

*Different Succession Stages of Vegetation Carbon and Nitrogen Distribution*

The total carbon, total nitrogen concentration, and C:N ratio on the main functional groups of lightly degraded alpine Kobresia meadow were significantly higher than those in severe degradation grassland (Table 4). In the mild degraded grassland, the group's order of carbon and total nitrogen concentration were forbs > grass > sedges, C/N ratio was grass > sedge > forbs. In the severe degraded grassland, the group's order of carbon concentration was forbs > sedge > grasses, total nitrogen concentration was forbs > sedge > grasses; C /N ratio was grass > forbs > grass. The carbon concentration between grasses and sedges was different significantly to the lightly degraded grassland and moderate degraded grassland; the carbon concentration among forbs, sedges, and grasses was different significantly ($P < 0.05$) at the severe degradation grassland.

The total carbon and total nitrogen concentrations of plant root at severe degraded grassland are higher than those at the mild degraded grassland (Table 5).

**Table 3** Properties and qualities of plant community at different degraded Alpine steppe ($n = 6$)

| Degraded level | Nondegraded alpine steppe | Lightly degraded steppe | Moderately degraded steppe | Heavily degraded steppe |
|---|---|---|---|---|
| Dominant species of plant community | *Stipa purpurea* | *Stipa purpurea*, *Leontopodium nanum*, Carex sp. | *Lanceoleaf thermopis*, *Stipa purpurea*, *Leontopodium nanum*, Carex sp., *Potentilla bifurca* | *Ajania tenuifolia*, *Saussurea arenaria* Maxim, *Roegneria thoroldiana* (Oliv.) Keng |
| Aboveground biomass (0.25 g/m$^2$) | 15.67 ± 4.11 | 14.91 ± 4.84 | 25.83 ± 9.35 | 4.84 ± 0.98 |
| Species number/m$^2$ | 6.17 ± 1.94 | 8.50 ± 1.87 | 11.50 ± 1.97 | 8.67 ± 2.16 |
| Diversity index | 0.67 ± 0.31 | 1.13 ± 0.29 | 1.59 ± 0.22 | 1.30 ± 0.51 |
| Evenness index | 0.37 ± 0.12 | 0.52 ± 0.09 | 0.65 ± 0.08 | 0.60 ± 0.21 |
| Index of grassland quality | 2.80 ± 0.06 | 2.29 ± 0.43 | 0.70 ± 0.40 | 0.43 ± 0.87 |
| Biomass percent of *Stipa purpurea* (%) | 79.64 ± 12.15 | 63.45 ± 18.46 | 18.62 ± 8.53 | 7.66 ± 7.37 |
| Belowground biomass at 0–20 cm soil layer (g/0.25 m$^2$) | 198.26 ± 13.04 | 230.51 ± 30.60 | 218.97 ± 59.67 | 102.78 ± 35.80 |

The ratio of C/N in mild degraded grassland was significantly higher than that in severe degradation grassland. With the increase of soil depth, the total carbon, total nitrogen concentration of plant root decreased. Analysis of variance showed mild

**Table 4** The C, N concentration of main function in different successional stages of alpine Kobresia pygmae meadow (standard error)

| | Lightly degraded meadow | | | Heavily degraded meadow | | |
|---|---|---|---|---|---|---|
| | Total carbon (%) | Total nitrogen (%) | C/N | Total carbon (%) | Total nitrogen (%) | C/N |
| Grass | 42.072 (0.615)a | 1.335 (0.070)A | 31.515 | 37.354 (1.752)b | 1.310 (0.112)A | 28.515 |
| Forbs | 42.544 (0.338)a | 1.416 (0.079)A | 30.045 | 40.488 (1.278)a | 1.384 (0.063)A | 29.254 |
| Sedges | 40.772 (1.875)a | 1.330 (0.071)A | 30.656 | 37.970 (2.680)b | 1.265 (0.106)A | 30.016 |

*Note* The data at the table is the average data for the 5 sampling. *Lowercase letters* as carbon concentration variance test. *Uppercase letters* as nitrogen concentration variance test. Letter of the same shows no significant difference ($P > 0.05$)

**Table 5** The C, N concentration of plant roots in different successional stages of alpine Kobresia pygmae meadow (standard error)

| Soil depth (cm) | Lightly degraded meadow | | | Heavily degraded meadow | | |
|---|---|---|---|---|---|---|
| | Total carbon (%) | Total nitrogen (%) | C/N | Total carbon (%) | Total nitrogen (%) | C/N |
| 0–20 | 35.915 (1.117)a | 0.563 (0.045)A | 63.792 | 37.423 (1.132)a | 0.654 (0.074)A | 57.222 |
| 20–40 | 32.173 (1.044)b | 0.477 (0.048)A | 67.449 | 35.368 (1.518)ab | 0.637 (0.062)A | 55.523 |

*Note* The data at the table is the average data for the 6 sampling. *A, a* are same as Table 2, 3, and 4

degraded grassland, in addition to the carbon concentration of 0–20 cm and 20–40 cm plant root, the carbon concentration of 20–40 cm plant root in lightly degraded grassland showed different significantly ($P < 0.05$) compared with the severe degraded grassland, carbon and nitrogen concentration had no significant difference among the other treatments ($P > 0.05$). And in comparison for the main functional groups on carbon and total nitrogen concentrations, both lightly degraded grassland and heavily degraded grassland, the carbon and nitrogen concentrations of aboveground plant biomass were higher than those in roots. The nitrogen concentration difference is obvious, and the ground portion of the nitrogen concentration is almost two times than that of underground part.

### *Storage Characteristics of Plant Carbon and Nitrogen in Different Succession Stages*

Based on the calculation of aboveground biomass at the peak point (August 30), the carbon storage of main function groups in lightly degraded grassland and severe degraded grassland was forbs > grass > sedges (Table 6).

The carbon storage of aboveground of severe degraded grassland (59.79 g/m$^2$) reduced by 15.4 % compared with mild degraded grassland (70.67 g/m$^2$). Among them, grass carbon storage decreased by 90.4 %, sedges by 82.2 %, and forbs increased by 71.6 %. Trends of carbon storage and change trend of N storage were similar. Aboveground total nitrogen (2.04 g/m$^2$) of severe degraded grassland than that of mild degraded grassland aboveground (2.30 g/m$^2$) reduced by 11.3 %.

**Table 6** The C, N contents of main function in different successional stages of alpine Kobresia pygmae meadow (standard error) (g/m$^2$)

| | Lightly degraded meadow | | Heavily degraded meadow | |
|---|---|---|---|---|
| | Total carbon | Total nitrogen | Total carbon | Total nitrogen |
| Grass | 30.81 (5.60)a | 0.98 (0.18)A | 2.96 (1.34)b | 0.10 (0.05)C |
| Forbs | 7.51 (1.82)b | 0.25 (0.15)B | 1.34 (0.72)b | 0.05 (0.03)C |
| Sedges | 32.34 (4.64)a | 1.08 (0.15)A | 55.49 (6.55)d | 1.89 (0.22)D |
| Total | 70.67 (10.23)e | 2.30 (0.33)D | 59.79 (8.49)e | 2.04 (0.29)D |

*Note* The data at the table is the average data for the 5 sampling. *Lowercase letters* as carbon content variance test. *Uppercase letters* as nitrogen content variance test. Letter of the same show no significant difference ($P > 0.05$)

Among them, 98.8 % reduction in nitrogen reserves grasses, sedges, and forbs decreased by 80 %, and increased by 75 %, respectively.

The carbon, nitrogen storage of underground part (root) was significantly higher than those of aboveground part (stem) at alpine Kobresia meadow, which mainly concentrated in 0–20 cm soil layer. The root carbon storage of mild and severe degraded grassland in the 0–20 cm layers occupied 90.3 and 97.3 % (Table 7) to the 0–40 cm soil depth root total carbon. In the severe degraded grassland, the root carbon and nitrogen reserves (0–40 cm soil layer) decreased by 60.5 and 57.1 %, respectively, when compared with mild degraded grassland. Among them, in 0–20 cm soil carbon and nitrogen reserves were reduced by 57.4 and 53.6 %, in 20–40 cm soil carbon and nitrogen reserves were reduced by 88.8 and 88.5 %, respectively.

## The Soil Characteristics in Different Degraded Succession Stage

### *Soil C and N Characteristics of Different Degradation*

Grassland degradation caused soil carbon and nitrogen concentrations decrease, loss of soil organic carbon was far greater than the nitrogen (Table 8). Soil organic carbon concentration of the severe degraded grassland at the soil layer of 0–20, 20–40 cm decreased by 22.6, 20.4 %, respectively, and the soil total nitrogen concentrations of them were reduced by 9.85 and 5.2 %. Soil C/N ratio from severe

**Table 7** The C, N contents of plant roots in different successional stages of alpine Kobresia pygmae meadow (standard error) ($g/m^2$)

| Soil depth | Lightly degraded meadow | | Heavily degraded meadow | |
|---|---|---|---|---|
| | Total carbon | Total nitrogen | Total carbon | Total nitrogen |
| 0–20 cm | 817.93 (77.10)a | 13.76A | 348.25 (60.96)c | 6.39C |
| 20–40 cm | 87.74 (33.33)b | 1.57B | 9.84 (1.97)b | 0.18B |
| Total | 905.67 (78.14)a | 15.33A | 358.10 (62.32)c | 6.57C |

*Note* The data at the table is the average data for the 9 sampling. The means of A, b, c, A, B, C are same as Table 4

**Table 8** The TOC, total N contents (standard error) and C/N of soil in different successional stages of alpine Kobresia pygmae meadow

| Soil depth | Lightly degraded meadow | | | Heavily degraded meadow | | |
|---|---|---|---|---|---|---|
| | Total organic carbon (TOC) (%) | Total nitrogen (%) | C/N | Total organic carbon (TOC) (%) | Total nitrogen (%) | C/N |
| 0–20 cm | 4.821 (0.428)a | 1.138 (0.045)A | 4.118 | 3.733 (0.248)b | 1.027 (0.058)A | 3.635 |
| 20–40 cm | 4.180 (0.853)a | 1.053 (0.052)A | 3.973 | 3.322 (0.311)b | 0.998 (0.058)A | 3.329 |

*Note* The data at the table is the average data for the 6 sampling. The means of A, b, c, A, B, C are same as Table 4

degraded grassland to slightly degraded grassland decreased by 14.29 and 18.4 %, respectively. The difference of the carbon concentration was significantly ($P < 0.05$), but there was no significant difference in nitrogen concentration.

## The Degradation Effect on Soil Enzyme Activities and Soil Nutrients of Alpine Grassland

### *Degradation of Soil Nutrients and Structure*

Along with the alpine meadow degradation, soil bulk density increased. This is closely related with the lush plants density and rich roots of the native meadow or lightly degraded meadow and reverse characters of severe and extremely degraded meadow. Soil organic matter in the lightly degraded stage was higher, and in the other degradation stages was low. With the decrease of vegetation coverage, soil erosion was becoming more and more serious, which had a tendency to decrease the content of organic matter, which was correspondent with the different stages of grassland degradation. The content of available N in the extreme degradation stage is low, but higher in other stages (Table 9). Total and available phosphorus content

**Table 9** The soil characters in different degraded alpine meadow at 0–20 cm

| Degraded level | Lightly degraded alpine meadow | Moderately degraded alpine meadow | Heavily degraded alpine meadow | Extremely degraded alpine meadow |
|---|---|---|---|---|
| Soil bulk density (g/cm$^3$) | 1.20 | 1.32 | 1.36 | 1.57d |
| Soil moisture content % | 30.60 | 26.18 | 25.59 | 17.36 |
| Organic matter (%) | 8.44 | 5.54 | 6.85 | 6.38 |
| Available nitrogen (ppm) | 59.97 | 66.62 | 50.64 | 14.66 |
| Total Nitrogen content (%) | 0.41 | 0.26 | 0.33 | 0.34 |
| Available Phosphate (ppm) | 10.74 | 6.86 | 5.08 | 9.70 |
| Total Phosphorous content (%) | 0.071 | 0.035 | 0.048 | 0.051 |
| Available Kalium (ppm) | 200.90 | 169.82 | 245.00 | 147.93 |
| Total Kalium content (%) | 1.86 | 1.85 | 2.09 | 1.84 |
| Total salt content (%) | 0.107 | 0.081 | 0.081 | 0.068 |

between the grassland degraded stages is not obvious, which was higher in the lightly degraded stage, but lower in the other degradation stages, relatively (Table 9). Compared with the soil phosphorus content in other areas of alpine meadow, the degradation of soil phosphorus content lies at the middle level, and can meet the need of plant growth. Total potassium and available potassium of soil in the different degraded meadow have no obvious change pattern, and that at the extreme degradation was the lowest (Table 9), but which can meet the need of potassium in plant growth. With the increase of grassland degradation, the total salt content decreased. With the increase of grassland degradation, soil properties will change. In the moderate degraded stage, the pest harm was the most serious matter. With the grassland degraded to severe degradation process, soil bulk density, hardness, and other physical characteristics changed severely. With the increase of grassland degradation, total salt content, soil moisture decreased, soil bulk density increased. With the alpine meadow degradation degree increasing, the loss of organic matter in the soil surface was serious. Soil available nitrogen content was difficult to meet the need of plant growth in the extreme degradation stage. The changes of total nitrogen, total phosphorus, total potassium, available phosphorus, and available potassium content in different degraded were not obvious.

When the alpine grassland degraded from native and mildly degradation to severe and moderate degraded stage, soil water content (%), soil organic matter content, available phosphorus, and nitrate nitrogen decreased significantly, the nitrate nitrogen content in heavy degradation stage cannot meet the need of plant growth. With the alpine steppe degradation degree increasing, the content of organic matter is serious loss in the surface soil. With the degradation of alpine steppe increase, the available potassium and soil compaction decrease, the content of ammonium nitrogen in severe degradation stage was higher than that in other stages, and pH value wwas higher than 8, and did not change significantly (Table 10). With the degradation of vegetation succession in alpine grassland, soil degradation was more and more serious, and soils more and more barren. At the severe degradation stage, xeric sandy plants appeared, and which present initial desertification landscape.

So, on the whole, with the degradation succession of vegetation, soil degradation was more and more serious, and while the soil more and more barren.

*Effects of Alpine Meadow Degradation on Soil Enzyme Activity*

Urease activities in different degraded alpine meadow were compared in Table 11. The difference of urease activity of not degraded and lightly degraded stage was not significant ($P > 0.05$); the urease activity of moderate and severe degradation stage activity decreased. Variance analysis showed the urease activities in nondegradation and lightly degraded meadow were significantly higher than those in moderate and severe degraded meadow ($P < 0.01$). Mild degradation to the grassland have not too much influence on the alpine meadow vegetation composition and physicochemical properties of soil, although interfered by grazing disturbance, the soil urease activity was similar with nondegraded stage under the protection of soil humic substances

**Table 10** The soil characters in different degraded alpine steppe at 0–20 cm layer

| Degraded level | Soil depth (cm) | PH | Organic matter (%) | Total Nitrogen content (%) | Total Phosphorous content (%) | Ammonia Nitrogen (ppm) | Nitrate Nitrogen (ppm) | Available Phosphate (ppm) | Available Kalium (ppm) |
|---|---|---|---|---|---|---|---|---|---|
| Nondegraded | 0–10 | 8.5 | 1.43 | 0.06 | 0.04 | 22.99 | 25.95 | 3.38 | 258.99 |
| | 10–20 | 8.3 | 2.46 | 0.12 | 0.05 | 31.00 | 39.43 | 3.84 | 313.99 |
| Lightly degraded | 0–10 | 8.5 | 2.77 | 0.21 | 0.06 | 17.99 | 25.95 | 4.70 | 276.71 |
| | 10–20 | 8.85 | 1.59 | 0.14 | 0.06 | 16.99 | 46.99 | 2.74 | 127.67 |
| Moderately degraded | 0–10 | 8.55 | 0.71 | 0.07 | 0.05 | 18.99 | 16.09 | 4.66 | 182.21 |
| | 10–20 | 8.6 | 0.86 | 0.11 | 0.06 | 22.66 | 32.20 | 2.15 | 142.23 |
| Heavily degraded | 0–10 | 8.5 | 1.21 | 0.21 | 0.05 | 32.00 | 21.02 | 3.47 | 186.30 |
| | 10–20 | 8.7 | 0.67 | 0.20 | 0.04 | 64.69 | 16.74 | 1.69 | 105.86 |

**Table 11** The comparison between different layers of soil enzyme activity

| Enzyme activity | Soil depth (cm) | Nondegraded alpine meadow | Lightly degraded alpine meadow | Moderately degraded alpine meadow | Heavily degraded alpine meadow |
|---|---|---|---|---|---|
| Urease activity | 0–10 | 2.49 ± 0.34a | 3.04 ± 0.26a | 1.65 ± 0.25a | 1.32 ± 0.68a |
| | 10–20 | 1.58 ± 0.09b | 1.59 ± 0.09b | 0.84 ± 0.06b | 1.35 ± 0.12a |
| | 20–30 | 1.24 ± 0.07c | 1.09 ± 0.08c | 0.56 ± 0.63c | 1.10 ± 0.13a |
| Invertase activity | 0–10 | 2.25 ± 0.03a | 1.77 ± 0.02a | 2.10 ± 0.01a | 2.24 ± 0.05a |
| | 10–20 | 2.23 ± 0.03a | 1.79 ± 0.01a | 2.08 ± 0.03a | 2.20 ± 0.03a |
| | 20–30 | 2.28 ± 0.03a | 1.77 ± 0.02a | 2.10 ± 0.06a | 2.22 ± 0.04a |
| Urease activity | 0–30 | 1.77 ± 0.75A | 1.91 ± 0.13A | 1.02 ± 0.12C | 1.26 ± 0.51B |
| Invertase activity | 0–30 | 2.25 ± 0.25A | 1.77 ± 0.15B | 2.09 ± 0.14C | 2.22 ± 0.02A |

*Note* The same column with different small letters indicate significant difference ($P < 0.05$). The same for the different capital letters indicate significant difference

or clay; but in moderate and severe degradation stages, along with the vegetation degradation, vegetation coverage reduced, species changed, as well as soil water content, pH, soil aggregate structure, microbial groups, etc. correspondingly, which resulted in the soil urease activity decreased significantly. Urease activity of each degradation stage showed obvious stratification, with the deepening of the soil, urease activity decreased in turn as the upper, middle, and lower layers, which is consistent with other studies on vertical distribution of enzyme activity in soil layers. Analysis of significant difference in different degraded stages of urease activity showed that: the urease activity in the first three stages of soil degradation were significantly higher than those of middle and lower layers of the soil urease activity ($P < 0.01$), middle soil urease activity was significantly higher than that in the lower soil urea enzyme activity ($P < 0.05$); soil urease activity of severe degradation stage with the increase of soil depth did not show significant difference ($P > 0.05$). Secretion of soil enzyme is mainly from microbial and plant roots, plant roots in alpine meadow were mainly distributed in the soil of 0–20 cm depth. Soil microbes were distributed mainly in the surface soil layer, and the number of microbes reduced with the depth increase. So, we can conclude that urease vertical distribution in the soil is closely related to the distribution of plant root system and microbial activity.

Sucrose enzyme belongs to the hydrolase enzyme hydrolysis in the soil, which can participate in organic compounds, decomposed into soluble nutrients to plants and microorganisms, plays an important role in the transformation of organic matter in soil. The data in Tables 2, 3, and 11 show that sucrose enzyme activity showed no significant difference ($P > 0.05$) among the three layers of the same degradation plot, and did not show the trend of no further activity decrease with soil depth increase. The headwater region of three rivers due to the long-term stress of cold temperature, plant root, and microbial have the short growth period and weak activity, alpine meadow plant debris and dead roots did not fully resolved, the surface has obvious litter accumulation, which is probably the reason why there is little difference of soil invertase in alpine meadow activity. Comparing sucrose enzyme activity of the different degraded alpine meadows (Table 11), it had no significant difference between nondegrading meadow and severe degradation meadow ($P > 0.05$), but invertase activities of lightly degraded meadow and moderately degraded meadow were low, sucrose enzyme activities of nondegradation and lightly degraded meadow were significantly different ($P < 0.01$), sucrose enzyme activities of lightly degraded meadow and heavily degraded meadow were very significantly different ($P < 0.01$). The results as above showed that the meadow degradation influenced the sucrose enzyme activity relatively large.

## Grassland Degradation Can Affect Soil Water Conservation

Grassland vegetation coverage affects soil moisture content greatly. With the increase of vegetation coverage, soil water retention capacity increased (Fig. 5). Soil water content of lightly degraded grassland was significantly higher than that in severe degraded grassland. Among them, compared with heavily degraded grassland, 0–20 cm soil water content in lightly degraded grassland increased by 55.2 %, however, 20–40 cm soil water content increased by 21 %, and the difference was significant. With the increase of soil depth, the difference reduced gradually. There was a positive correlation between vegetation coverage and soil water content. Due to the difference of vegetation coverage, the different sun's radiation to soil, soil water evaporation is also different, which lead to soil moisture variation. Changes of vegetation coverage was according to the order of lightly degraded meadow (92.92) > moderately degraded meadow (80.67) > heavily degraded meadow (71.75) > extremely degraded meadow (55.05), soil moisture also exhibited the same trend.

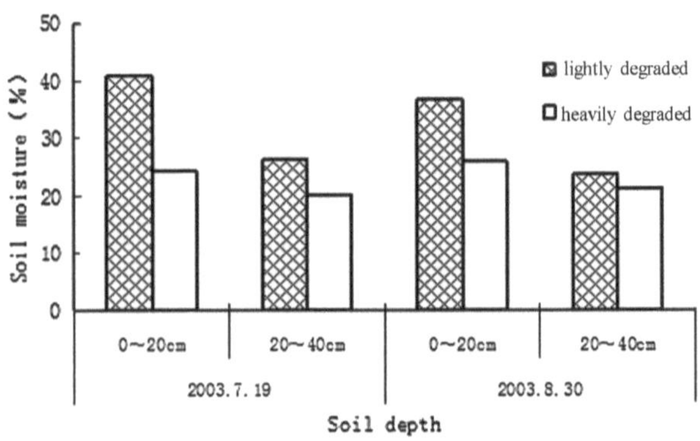


**Fig. 5** The soil moisture in different succession stages of degradation

### Close Relationship of Alpine Grassland Vegetation Degradation and Soil Degradation

Vegetation degradation is the direct cause of alpine meadow soil degradation, and soil degradation will cause the degradation of vegetation, they are reciprocal causation. Alpine meadow ecosystem in moderate grazing condition, flow and matters are basically in a state of balance, the production level is relatively stable, subsystem of the soil maintains the good structure, and have the function correctly, to provide the space nutrient and water for plant growth, and plant litter and livestock put back the excretion of soil, which can hold a good cycling and balance among the grass, livestock, and soil. Matter of serious overgrazing existed close to the settlements in general. After the excessive grazing and trampling of livestock, the normal growth of herbaceous plants inhibited, the stable material balance is destroyed, grass had become low sparse, changing structure, reducing coverage, the decreased grass yield, and the number of species, and excellent forage decreased obviously. At the same time, soil degradation is also significantly obvious, organic matter, soil moisture, and available *N* content decreased, which showed the typical characteristics of grassland degradation.

### Responses of Soil Seed Bank of Grassland Degradation

Through the study on soil seed bank in different degraded grasslands, as shown in Table 12: the soil seed bank density in moderate degraded grassland lies at maximum, with the increase of grassland degradation, soil seed bank density showed a "low-high-low" variation. There were significant differences in soil seed density between nondegradation and other degraded meadow.

In the first year, there was no significant effects of fertilizing and fencing measures to the soil seed bank, seed density and species, soil seed bank dynamics, and spatial distribution pattern. The vertical spatial structure of soil seed bank of alpine meadow vegetation: absolute advantage in the number of seed bank at the 0–10 cm soil layers, the proportion varies from 75 to 98 %, with an average of 86.5 %. The number of the seeds ratio of 10–20 cm layer was from 25 to 2 %, the average was 13.5 %.

**Table 12** The extent of alpine grassland soil seed bank density in different degradation stages

| Plot (0–10 cm layer) | Seed number (seed/m$^2$) | $P < 0.05$ |
|---|---|---|
| Nondegraded grassland | 2483.7 ± 648.2 | a |
| Lightly degraded grassland | 8438.5 ± 2207.1 | b |
| Moderately degraded grassland | 7873.6 ± 1836.7 | b |
| Heavily degraded grassland | 6242.5 ± 1040.5 | b |

*Note* Duncan multiple comparisons, each column represents the difference between the different letters samples significantly, $P < 0.05$

Temporal dynamics of soil seed bank of alpine meadow vegetation showed that the seed ratio to total seed bank density of the grass family and sedges family plants in the withering period changed from 9 to 7 % into 4.4 and 4.5 % in turning green periods, and seed ratio of broad-leaved plants changed from withering period 84 % into turning green period 91.1 %.

The soil seed bank of alpine meadow vegetation mainly consists of a large number of broadleaf weeds species seeds, sedges family and the grass family seeds were relatively small, the ratio closed to 87:7:6. With the grassland degradation, the dominance of weeds increased, grasses and sedges advantages decreased obviously. The proportions of different seed types had the same changing trends. The characteristics of soil seed bank were correlated with the different plant life forms, reproductive modes, and the external harsh environment.

## Effects of Grazing on Grassland Productivity, Community Structure

### *Changes of Plant Biomass of Different Functional Groups Under Grazing*

In order to effectively reveal the grazing disturbance on the QTP, especially the contribution and role of degraded grassland, the study was carried out in Haibei Alpine Meadow Ecosystem Research Station respect to the effects of different grazing intensities of long-term grazing on alpine *Potentilla fruticosa* shrub. With the grazing rate increase, the change of plant species diversity index was a typical single peak curve model. Long-term heavy grazing simplified the alpine shrub communities, and decreased the aboveground biomass, particularly for fine herbage. Height, total cover, and grass coverage of plant community decreased with increasing grazing intensity, the green plant coverage was the highest in the moderately grazed plot. From light grazing to heavy grazing, shrubs and grasses dominance were replaced by typical forbs. The response of natural grassland under long time grazing (more than 50 % utilization rate) was degraded, and the degradation was more and more serious with the grazing intensities. We can conclude that long-term heavy grazing plays an important role in the process of grassland degradation on the QTP. The implementation of grazing principle of "half taking with half leaving" on the QTP will benefit to prevent the degradation of grassland, improve the forage utilization and maintenance of high biodiversity.

With grazing intensity increased, the biomass of the grassland communities decreased (Fig. 6). The biomass from the heavy grazing intensity to moderate and light grazing intensity decreased by 61.72, 65.09 %. The proportion of fine herbage, litter ratio decreased with the increase of grazing intensity, the proportion of forbs, sedge ratio increased with the increase of grazing intensity. Compared with mild grazing conditions, severe grazing decreased the grass and litter proportions by 46.09 and 88.17 %, respectively, and increased forbs and sedges by 160.23 and 111.62 %, respectively. The shrubs percentage showed no difference, and under

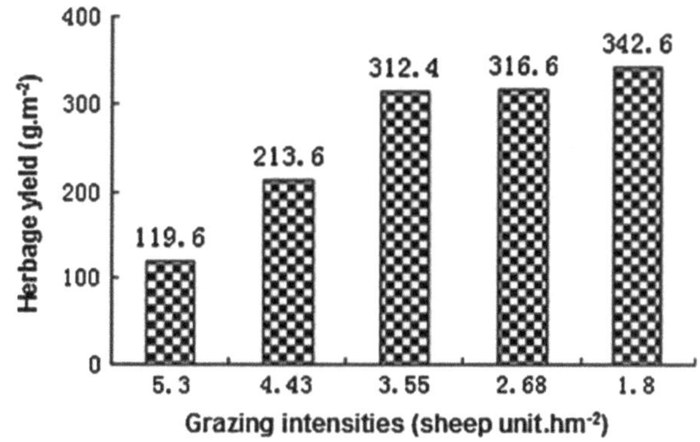


**Fig. 6** Effects of grazing on grassland aboveground biomass

moderate grazing conditions it was slightly higher than that of light and heavy grazing.

*Changes of Main Plant Populations under Grazing Conditions*

In alpine meadow, all plants can be divided into four categories with the characteristics of great fostering, sensitive, tolerant, and indifferent responses under the long-term heavy grazing disturbance. From light to heavy grazing pasture, *Prunus fruticosa* and grass gradually would be replaced by typical forbs. In the heavy grazing and the second heavy grazing plot, *Leontopodium nanum*, *Gueldenstaedtia diversifolia,* and *Podospora anserina* were significantly greater than those in light grazing and nongrazing plots, these weeds benefit from continued heavy grazing and have a large population with the increase of grazing intensity, which also belonged to the poor quality of forage and greatly reducing the forage value. Some excellent forage decreased significantly in the continuous heavy grazing disturbance, such as grasses and sedges biomass had been reduced along with treatments from exclousure, light grazing to heavy grazing. With the increasing of grazing intensity, the important values of grasses and sedges (such as *Stipa aliena*, *Festuca ovina*, *Platyraspedum tibeticum Kobresia capillifolia,* etc.) continued to decrease, and *P. fruticosa* and *Ligularia virgaurea* also had the same changes, which often had characteristics of sensitive reaction to continue heavy grazing disturbance. *Kobresia humilis* is a typical sedge on the QTP, an alpine Kobresia plant characterized by tolerance on long-term heavy grazing disturbance, with resistant to grazing and trampling showing the important value of *K. humilis* in heavy grazing. In the second heavy grazing and moderate grazing plots, the important value of *K. humilis* was higher than that of light grazing, the second light grazing, and nongrazing plot. Some other forbs, such as *Anaphalis lacteall*, *Saussurea katochaete*, *Shorea superba*, *Panchlora nivea*, *Gentiana farreri*, *Taraxacum mongolicum,* and *Aster flaccidus* did not response actively on the long-term heavy grazing disturbance in each sample, and the important value had no obvious change with the increase of grazing intensity.

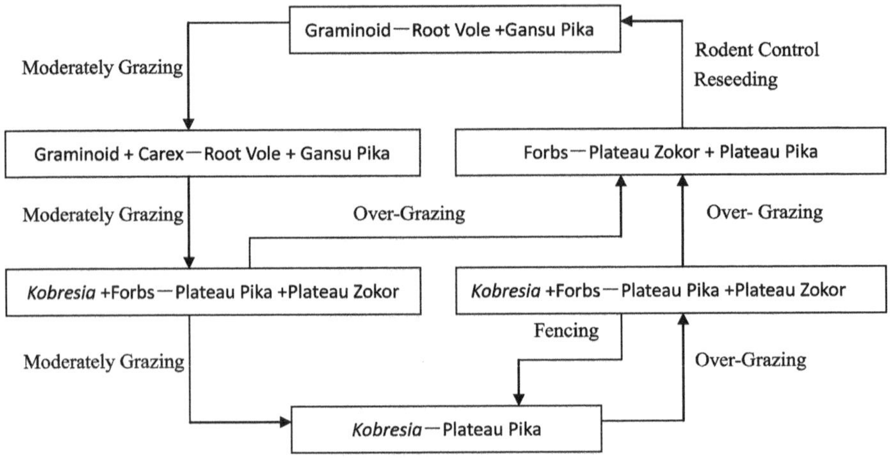


**Fig. 7** Successional process and mechanism of plant and small mammal communities of alpine grassland on the QTP

#### The Community Succession Process of Plant and Animal in Alpine Kobresia Meadow

Study on the relationship between vegetation succession and rodent population dynamics showed that zokor like eating all the juicy taproot plants, such as *P. anserina*, *Ajania tenuifolia*, *Pedicularis kansuensis,* etc., the representative plants after grassland degraded seriously. The plateau pika like to eat the Kobresia and Carex plants and grass. The trophic niche of two rodents is different largely, so they can live in the same area peacefully. However, food resource spectrum of Gansu Pika and root voles had similar with plateau pika, but the habitat difference of Gansu pika and root voles was large with plateau pika, so they rarely live in the same area. With the succession of plant communities, the community also changed with the changes of food resources and habitat (Fig. 7). The plateau pika and plateau zokor live in particular habitat of alpine meadow of the QTP, their living habitat and biology were closely linked, once its habitat change or destruct, there would be a serious threat to its survival and reproduction.

### 3.2.3 Forage Cultivation and Different Function of Artificial Vegetation Construction

#### Screening of Annual Forage Varieties

After a long-term product ratio experiments, the researchers determined that the Denmark 444, Ba Yan No. 3, and Ba Yan No. 18 as annual masters oat varieties in alpine grassland of Qinghai province from about 10 oat varieties, such as Qinghai

444, Qinghai 18, Ba Yan No. 3, Ba Yan No. 4, Ba Yan No. 5, Ba Yan No. 6, Yongjiu 108, Yongjiu 473, Yongjiu 233, and Yongjiu 001 (Zhao 2011).

Perennial Forage Breeding, Cultivation, and Domestication

*The Screening of Fine Perennial Grass*

The biological characteristics, cultivation techniques, and others were systematically studied from the introduced varieties of 13 genera 39 grass. According to the standard of the overwintering rate and coverage degree above 80 %, the beginning of the second year to complete the full reproductive growth, grass yield more than 800 g/m$^2$, we chose the top grass species as follows such as *E. sibiricus* cv. Qingmu No. 1, *E. nutans*, *Elymus sibircus* CV Tong De, and *F. Sinensis* Keng; grass yield more than 400 g/m$^2$ under the grass forage varieties such as *P. poophagorum*, *P. crymophila*, *F. kirilowii*, *F. kryloviana*, *P. tenuiflora*, *S. aliena*, and *Roegneria thoroldiana*; the overwintering rate and coverage degree above 50 %, but with the quick occupation ability was *Bromus inermis* Leyss. *Leymus secalinus* and *Agropyron cristatum*, these four species can be used as appropriate grass species at the headwater region of three rivers and other alpine meadow areas in "black soil type" degraded grassland vegetation restoration.

*Main Points of Cultivation*

*E. nutans*: sowing time for a mid May to early June, unicast uses kind of quantity 30–45 kg/hm$^2$, 2–3 cm sowing depth. No grazing is necessary for growth period of sowing year, mild grazing can be used after soil freezing period, the green period of second year is also needed for fencing. In order to maintain community stability of *E. nutans* dominated artificial grassland, *E. nutans* always can mix with other grasses ensued seed weight of 22.5–30 kg/hm$^2$.

*Elymus breviaristatus*: sowing time for a mid May to early June, unicast uses kind of quantity 30–45 kg/hm$^2$, 2–3 cm sowing depth. No grazing is necessary for growth period of sowing year, mild grazing can be used after soil freezing period, the green period of second year is also needed for fencing. Suitable cutting period is in the early August, stubble height is at 5–10 cm. In order to maintain community stability of *E. breviaristatus* dominated artificial grassland, *E. breviaristatus* always can be mixed with other grasses ensued seed weight of 22.5–30 kg/hm$^2$.

*E. sibiricus* cv. Qingmu No. 1: deep plowing and level ground are needed before sowing. It can be planted in spring, summer, and autumn. In spring, we need to prevent the harm of annual weeds in spring. Autumn planting time is 30–40 days before the first frost. Broken awn before sowing seeds enhances liquidity, the planter seeding gear clearance increased or the seed conveying pipe removed if it is necessary. Seeding process should pay attention to the seed flow to prevent clogging, and seed quality assurance. The sowing depth is 4–5 cm, seeding rate is generally 16–22 kg per ha. Lime sulfur and zineb sprayed to eliminate rust and superphosphate with 200 kg at tillering per $hm^2$ need to strength the fertilizer if necessarily.

*P. crymophila*. Cv. Qinghai: Suppression, weed control, shallow trenching, shallow overburden on the land before sowing. It can be planted in spring and autumn. Seed weight is 7.5–10 kg per ha. If sowing in line, with a row of 15–30 cm, 1–2 cm depth, and suppressed after sowing. Then treatments such as prevent livestock trampling, timely weed control, tillering, jointing irrigation, and fertilization, can increase the yield of planting grass.

*F. Sinensis* Keng cr Qinghai: Soil preparation of special fine, deep tillage, rake thin, worn flat before sowing is necessary. Input tillage and fertilizer rotted manure (15,000–20,000 kg/$hm^2$), superphosphate (30–60 kg/$hm^2$) while plowing. The time of sowing is in late April to early May during spring soil thawing. Sowing in line with a row of 15–25 cm, mixed sowing with a row of 15–30 cm, and the seed field with a row of 30 cm. Sowing rate: 10–15 kg/$hm^2$ of clean culture, 7.5–10 kg/$hm^2$ of mixture sowing. The sowing depth varies with soil types, 2–3 cm to sandy soil or light loam, 1–2 cm to clay. The grazing is absolute prohibited at the sowing year and second years. The grassland can be used to graze after 2 years with the stubble height 4–5 cm. When the seeds mature above 70 %, it can be harvested for the seed shattering. It is recommended to kill weeds with pesticides.

*P. Pratensis* L. var. anceps Gaund. cv. Qinghai: controlling weed, flatting land, and repressing soil with rake at the sowing year. The time of sowing is from late May to the early June. The seeding quantity is 15 kg/$hm^2$, row spacing with 15 cm; 7.5 kg/$hm^2$ seeding quantity for sowing seed production, row spacing with 30 cm. Weed control at the seedling time and weeding at other period. The urea need 60 kg/hm $^2$ at the late tillering stage of sowing year, then 60 kg/$hm^2$ at the late tillering stage of next year. Artificial pollination at the blossom period can help to improve the seed yield. Moderate graze was permitted at the second years after planting.

*P. tenuiflora*: Deep plow at summer and autumn, raking level land and the enough base fertilizer at the previous year before sowing. Tillage is carried at the 5–7 days after irrigation. Weed control and suppression by machinery are needed before sowing in order to control the depth of sowing to overcome the phenomenon of broken bar at the area of without irrigation. The sowing times of *P. tenuiflora* are not very strict, the latest time should not be more than mid July. 0.5–1 kg per mu of unicast amount; 0.4–0.6 kg per mu of seed field amount, a row of 15 g to 30 cm to sowing in line, 1–2 cm depth of seeding, repress after sowing. Livestock feeding and trampling are prohibited at the sowing year.

*P. Pratensis* L. Cv. Qinghai: Plow at summer and autumn and fine soil preparation at the previous year before sowing. It is required to suppress the land, keep soil moisture before and after planting. Sowing time should be from April to May. As the artificial grassland, general seeding rate is about 7.5–12 kg per hectare. A row of 15–30 cm to sowing in line, 2–3 cm sowing depth. The fertilizing time is in early autumn and early spring of vigorous growth season. The amount of nitrogen fertilizer can be in 75–150 kg per hectare while the phosphate fertilizer be in 75–120 kg per hectare.

Planting Methods of Artificial Grassland

*The Land Selection*

To establish the artificial grassland, we always choose "black soil type" degraded grassland with a relatively flat topography, soil layer above 30 cm, convenient for mechanical operation, and near winter spring pasture from the herdsman settlement. Planting area of grassland is appropriate with 200–500 μ/household, according to the household management ability after construction. Too large area cause the inconvenient management and high investment cost, and too small area cannot achieve the economic benefit.

*Plowing*

The tillage needs below 15 cm depth to increase the water storage capacity of soil, which can also be reached through the measure of two heavy harrow tillages at the loose sections of top soil.

*The Land Leveling*

The heavily degraded land is needed for flat by heavy harrow because ground rugged or some remnants of turf after the tractor plowing can create troubles for the next step of seeding and fertilization.

*Sowing*

When unicast sowing, the amount of large seeds lies in the 30–40 $kg/hm^2$, the small seed lies in the 10–15 $kg/hm^2$, and the mixture sowing is always 50–70 % the amount of unicast sowing.

*Fertilizing*

When sowing, using $(NH_4)_2HPO_4$ or sheep manure as base fertilizer, and the $(NH_4)_2HPO_4$ application is 150–300 $kg/hm^2$; with urea as topdressing when the phenology of grass is tillering and jointing period, 1–2 times topdressing with the amount of 75–150 $kg/hm^2$.

*Earthing*

Earthing measures should be carried out with light harrow after sowing and fertilization and the depth is in 2–3 cm.

*Suppressing*

The process of soil suppression is very important, which is not only closely integrated seed with the soil to make the seed germination, but also improve the soil moisture and reduce wind erosion. In addition, the absolute prohibition of returning green stage is necessary at the first years to second years after the establishment of artificial grassland.

## Classification of Artificial Grassland

Classification of three types of artificial grassland introduced below was determined by the different criterions of uses, cultivation level, forage combinations, and life form.

*Annual Artificial Grassland*

Temporary artificial grass based on annual oats planted in cow and sheep pens that free in summer or a small area of forage base built on winter pasture near pens. Its purpose is to use the micro-environment of soil and climatic conditions and the quality characteristics and high yield of oats to produce the high quality forage and solve problems of some domestic animals winter feeding. The practice proves that the grassland plays an important role in production of animal husbandry in the headwater region of three rivers, and has gradually been accepted by herdsmen.

*Unicast-Based Artificial Grassland with Perennial Grasses*

Unicast-based artificial grassland is to plant one forage species or varieties in the same land. Unicast-based grassland has simple seeding method, and is easy to cultivate and harvest with lower planting and management costs. Unicast-based grassland based on *E. nutans* is main body of perennial artificial grassland in the headwater region of three rivers, especially in the region that lack of suitable forage varieties for cultivation time.

*Mixed Artificial Grassland with Perennial Grasses*

Taking full advantage of effect of interspecific complementary and space to establish a relatively stable artificial vegetation community by planting suitable perennial gramineae grasses in the headwater region of three rivers with different types of life growth form, perennial gramineae mixed artificial grassland is not only for grazing but also can be used as pasture. Commonly used grass seeds are some hardy grass seeds, such as *E. nutans*, *E. sibiricus* L, *Poa Festuca* Linn, Puccinellia, etc.

### Optimization of Artificial Grassland Community

According to request that improve artificial grassland aboveground biomass and prolong the service life of grass, through the rational allocation among grass species and optimization vegetation spatial pattern of plant community of grassland and other means, the best artificial community group planted in mixed artificial grassland on "black soil beach" that belongs to the headwater region of three rivers of Qinghai province is "*E. nutans* + *F. Sinensis* + *P. crymophila* + *P. poophagorum*".

### The Technology Integration of Artificial Grassland Establishment

First, alpine pastoral area suitable for planting perennial artificial grassland area should be determined in the relatively high rainfall zone of the alpine meadow grassland, and secondary bare land in the area required of more than 80 %. Second, it should have a certain suitable soil, climate, and topography. Generally in the grassland of soil thickness above 30 cm, altitude below 4300 m, flat terrain is suitable for mechanical operation. Key techniques for artificial grassland establishment are below:

(i) the planting time selection: artificial grassland planting period should be mastered in early May to early June.
(ii) the selection of grass seeds: annual artificial grassland planting oats in early maturing varieties, perennial with the hay production for the purpose of artificial grassland should choose high yield varieties of *E. nutans*, *E. sibiricus* Cv. Qingmu No. 1, *E. sibircus* CV Tong De, *F. Sinensis* Keng Cv Qinghai, etc. Grazing and ecological restoration-based artificial perennial grassland should increase the grass proportion in order to increase palatability of artificial community stability and grasses, which including *P. Pratensis* L. Cv. Qinghai, *P. pratensis* L. var anceps Gaund Cv Qinghai, *P. crymophila* Cv Qinghai, and some short grasses of Festuca genus.
(iii) agronomic measures of artificial perennial grassland: rodent poisoned—tillage —rake level land—fertilization—sowing (sow or drill)—soil covering—fencing.
(iv) the artificial grassland management: artificial pasture cultivation management measures include: rodent control, weed control, fertilization, and reasonable grazing (Zhao 2011). Rodent control techniques mainly include: physical control, chemical control, and biological control. The main process of rodent control is under the step of "measures of chemical control, physical control, chemical control, biological control". Artificial grass weed control in general is always from the beginning of the third year, including chemical control and cutting machine in two ways. Methods for chemical control use the one thousand times intermixture of Metsulfuron-methyl 75 g/hm$^2$ + 2-4D butyl ester emulsion 1500 g/hm$^2$ to field weed control, cutting mechanical control is in poisonous weeds flowering by mechanical control cutting. The amount of

nitrogen fertilizer and phosphate fertilizer is 30−60 kg/hm², 60–120 kg/hm², respectively, and nitrogen and phosphorus ratio is 1:2, cattle and sheep manure is 22,500−30,000 kg/hm². Artificial vegetation demands absolute prohibition in growing season of first years after construction and the annual spring regreening period. After grassland construction, using urea as topdressing once, the total amount is 75–150 kg/hm² since the third year or every other year in forage tillering and jointing stage (late June—early July). Artificial grassland should timely weed control since fourth year under the established grassland. According to the rodent density, degree of harm in winter and spring, rodent-killing measures are carried once in every year. The headwater region of three rivers grassland grazing intensity: in warm season pasture was 2.89 yaks/hm² (14.45 sheep unit/hm²), cold season pastureland in nutritional impairment and grazing time conversion was 1.07 yaks/hm² (5.35 sheep unit/hm²), and best grazing intensity was 4.19 sheep units/hm².

## The Economic Benefits of Artificial Grassland

The investment of artificial grassland comprehensive cultivation measures includes seed, tillage, plowing fee, cost of setting up the fence, as well as the necessary management measures such as fertilizer, rodent control, and other investment specifics as shown in Tables 13, 14, and 15. The total direct investment of 6 years is 2887.5 yuan RMB, while the natural succession is only 1897.5 yuan RMB, the differences is 990 yuan RMB; the annual hay production under artificial regulation is 35,648 kg/hm², but it is only 19,189 kg/hm² under the natural succession, the production difference is 16,459 kg/hm². If we calculate by the basic market price 0.40 yuan RMB per kilogram hay, the price difference is 6583.6 yuan RMB. The cost of artificial control of herbage production fell by 0.46 yuan RMB/kg in the first

**Table 13** Investment in the establishment of artificial grassland (yuan/hm²)

| Planting ages | Seed | Tractor fee | Fence | Fertilizer | Rodent control | Weed control | Sum |
|---|---|---|---|---|---|---|---|
| First year | 615.0 | 600.0 | 300.0 | 360.0 | 22.5 | 0 | 1897.5 |
| Second year | 0 | 0 | 0 | 180.0 | 0 | 0 | 180.0 |
| Third year | 0 | 0 | 0 | 180.0 | 22.5 | 0 | 202.5 |
| Fourth year | 0 | 0 | 0 | 180.0 | 0 | 0 | 180.0 |
| Fifth year | 0 | 0 | 0 | 180.0 | 22.5 | 0 | 202.5 |
| Sixth year | 0 | 0 | 0 | 180.0 | 22.5 | 22.5 | 225.0 |
| Total | 615.0 | 600.0 | 300.0 | 1260.0 | 90.0 | 22.5 | 2887.5 |

**Table 14** Direct income of artificial grassland establishment

| Planting ages | Direct investment (yuan/hm$^2$) | Hay yield per year (kg/hm$^2$) | Cumulative production (kg/hm$^2$) | Cost of production (yuan/kg) | Annual earnings (yuan/hm$^2$) | Cumulative revenue (yuan/hm$^2$) | Input–output ratio |
|---|---|---|---|---|---|---|---|
| First year | 1897.5 | 4142 | 4142 | 0.46 | 1242.60 | 1656.80 | 1:0.87 |
| Second year | 180.0 | 7768 | 11,910 | 0.17 | 2330.40 | 4764.00 | 1:2.29 |
| Third year | 202.5 | 6340 | 18,250 | 0.13 | 1902.00 | 7300.00 | 1:3.20 |
| Fourth year | 180.0 | 6128 | 24,378 | 0.10 | 1838.40 | 9751.20 | 1:3.96 |
| Fifth year | 202.5 | 5640 | 30,018 | 0.09 | 1692.00 | 12,007.20 | 1:4.51 |
| Sixth year | 225.0 | 5630 | 35,648 | 0.08 | 1689.00 | 14,259.20 | 1:4.94 |
| Total | 2887.5 | 35,648 | | 0.08 | 10,694.4 | | 1:4.94 |

*Note* 0.40 yuan/kg hay (2006)

**Table 15** Direct income of natural succession

| Planting ages | Direct investment (yuan/hm$^2$) | Hay yield per year (kg/hm$^2$) | Cumulative production (kg/hm$^2$) | Cost of production (yuan/kg) | Annual earnings (yuan/hm$^2$) | Cumulative revenue (yuan/hm$^2$) | Input–output ratio |
|---|---|---|---|---|---|---|---|
| First year | 1897.5 | 4142 | 4142 | 0.46 | 1242.6 | 1656.8 | 1:0.87 |
| Second year | 0.0 | 5200 | 9342 | 0.206 | 1560.0 | 3736.8 | 1:1.97 |
| Third year | 0.0 | 3862 | 13204 | 0.146 | 1158.6 | 5281.6 | 1:2.78 |
| Fourth year | 0.0 | 3036 | 16240 | 0.12 | 910.8 | 6496.0 | 1:3.42 |
| Fifth year | 0.0 | 1849 | 18089 | 0.11 | 554.7 | 7235.6 | 1:3.81 |
| Sixth year | 0.0 | 1100 | 19189 | 0.10 | 330.0 | 7675.6 | 1:4.41 |
| Total | 1897.5 | 19189 | | 0.10 | 5756.7 | | 1:4.41 |

year to 0.08 yuan RMB/kg of sixth year, down nearly 82.6 %, while the cost fell by only 72.3 % under natural succession. According to Tables 13, 14, 15, both of the cumulative output or cumulative revenue, artificial control values were about two times as much as the natural succession. Thus, constructing the artificial grassland can obtain higher forage yield, when choosing flat terrain areas with better soil conditions in a vegetation type, plant high quality forage such as *E. nutans*, and using artificial regulation necessary such as fences, fertilization, weed control, rodent control measures; and can bear part or most of the carrying capacity, thus will relieve the natural grassland grazing pressure, and reverse the degradation process.

### 3.2.4 The Scientific Use of the Alpine Grassland Resource

#### The Rational Use of Natural Grassland Resource

People usually think that reasonable utilization of grassland is 50 % of aboveground biomass, namely "half taking and half leaving" grazing principles and as the rational use of natural grassland around the world. Of course, this is in terms of nondegraded meadow or the most of aboveground biomass of pasture can be utilized by herbivorous animal, and small proportion of inedible forage. In view of the condition of the QTP pasture of short growth period and poor natural conditions, the optimum utilization rate of nondegraded grassland is 45 %. Under system of two season pasture rotational grazing, the maximum grazing intensity of summer and autumn grassland without degradation is 4.30 Tibetan sheep/$hm^2$, winter spring pasture is 4.75 Tibetan sheep/$hm^2$, maximum grazing intensity does not exceed the 2.5 sheep unit/$hm^2$. Through the research of optimal grazing scheme and production structure of alpine grassland, the ratio of Tibetan sheep and yak is 3:1 in alpine meadow area. Tibetan sheep aged female proportion is 50–60 %, and yak school-age female ratio between 30 and 40 % is more reasonable (Zhao 2011).

#### The Moderate Grazing Maintain Species Diversity

Effects of different intensity of long-term grazing on species diversity and evenness can be explained by Shannon–Wiener index and Pielou evenness index. With increased grazing intensity, the two indices were showing a single peak mode. In moderate grazing intensities (C), Shannon–Wiener index and Pielou evenness index maximum (Table 16), Duncan multiple test showed that the difference in the Shannon–Wiener index of grazing after 17 years of 6 sample plots was not significant ($P > 0.05$, $n = 6$, 2-3-16), moderate grazing plots (C) of the Pielou evenness index was significantly higher than that of heavy grazing plot (A) and nongrazing plot ($P < 0.05$, $n = 6$), and no significant difference among other three grazing plots ($P > 0.05$, $n = 6$).

**Table 16** The species diversity and evenness changes of different long-term grazing intensity

| Grazing plot | A | B | C | D | E | F |
|---|---|---|---|---|---|---|
| Shannon–Wiener index | $2.47 \pm 0.38^{a}$ | $2.64 \pm 0.23^{a}$ | $2.72 \pm 0.20^{a}$ | $2.59 \pm 0.12^{a}$ | $2.51 \pm 0.11^{a}$ | $2.36 \pm 0.11^{a}$ |
| Pielou index | $0.67 \pm 0.10^{ab}$ | $0.74 \pm 0.07^{bc}$ | $0.75 \pm 0.06^{c}$ | $0.72 \pm 0.03^{abc}$ | $0.69 \pm 0.03^{abc}$ | $0.65 \pm 0.03^{a}$ |

*Note* The data at the above table showed the average value ± standard division. In the same row data, there is no significant difference in the same letter ($P > 0.05$)

**Table 17** The relationship between grazing intensity and the plant species diversity

| Pasture | Index | Regression equation | $R$ value | $P$ value |
|---|---|---|---|---|
| Cold season pasture | The number of plant species | $Y = -12.25x^2 + 57.55x - 8.75$ | 0.9940 ($R^2$) | <0.001 |
| | Evenness index | $Y = -0.0172x^2 + 0.0874x + 0.8081$ | 0.9077 ($R^2$) | <0.05 |
| | Diversity indices | $Y = -0.3164x^2 + 1.742x + 3.1763$ | 0.8876 ($R^2$) | <0.05 |
| Warm season pasture | The number of plant species | $Y = 0.1753x + 4.8633$ | 0.9920 | <0.001 |
| | Evenness index | $Y = 0.0347x + 0.8975$ | 0.9993 | <0.001 |
| | Diversity indices | $Y = -x + 46.667$ | −0.9305 | <0.05 |

There was a significant positive correlation between grazing intensity and the plant community diversity index, and evenness index, a significant negative correlation with plant community constituent species in warm season pasture (Table 17). This can be well explained by the theory of "intrinsic redundancy". Because constitute intrinsic redundancy of plant (weed) cannot be ingested by yak, some plants can be used by other animal, which plays an important role in the grassland community biodiversity and evenness. Due to the intrinsic redundancy, in the case of grazing intensity increased, compensation of edible plant community will be strengthened. The number of population and biomass will also increase, while, function of communities decrease under over high grazing intensity. But when grazing intensities are 2.3, 2.4, and 2.5 sheep unit/hm$^2$ , respectively, the number of constituent species, diversity index, and evenness index of plant community reached the maximum, and then began to decrease. This suggests that the intrinsic redundancy was conditional. In winter pasture of alpine meadow, when grazing intensity increased to some degree, adjustment and maintenance of intrinsic redundancy to diversity index, evenness index, and the number of constituent species of plant community will weaken, component redundancy strengthen, structure of plant community structure change, and stability decreased. (Zhao 2011).

#### Yak Nutrition Balance Strategy Under Different Grazing Intensities

Under the condition of light grazing and adequate supply of summer forage, yak's forage digestibility was higher, the yak adjust its forage on the high digestibility and lower intake to maintain nutritional balance at this time. Under the condition of digestive and nutritional were sufficient, the effect of grazing intensity on yaks defecation was significant, but the influence of grazing time (seasonal change) on

**Table 18** Dynamic changes of yak dung dry matter under different grazing intensities (kg/yak day)

| Grazing treatment | Grazing time (m.d–m.d) | | | | | |
|---|---|---|---|---|---|---|
| | 6.20–7.5 | 7.5–7.20 | 7.20–8.5 | 8.5–8.20 | 8.20–9.5 | 9.5–9.20 |
| Extremely lightly grazing | 2.25Aa | 2.28Aa | 1.92Aa | 2.25Aa | 2.25Aa | 2.35Aa |
| Lightly grazing | 2.17Ab | 2.01Ab | 1.73Ab | 1.98Ab | 1.97Ab | 2.21Ab |
| Moderately grazing | 1.93Ac | 2.21Ac | 2.05Ac | 2.23Ac | 1.75Ac | 2.01Ac |
| Heavily grazing | 1.84B | 1.75B | 1.63B | 1.71B | 1.48B | 1.69B |

Different uppercase letters in the same row or column show the very significant difference ($P < 0.01$). Different lowercase letters in the same row or column show the significant difference ($P < 0.05$)

**Table 19** Dynamic changes of yak dry matter digestibility under different grazing intensities (%)

| Grazing treatment | Grazing time (m.d–m.d) | | | | | |
|---|---|---|---|---|---|---|
| | 6.20–7.5 | 7.5–7.20 | 7.20–8.5 | 8.5–8.20 | 8.20–9.5 | 9.5–9.20 |
| Extremely lightly grazing | 66.61Aa | 68.41Aa | 73.13Bb | 70.13Bb | 67.41Aa | 65.11Aa |
| Lightly grazing | 63.52Aa | 68.47Aa | 73.11Bb | 70.14Bb | 67.44Aa | 65.13Aa |
| Moderately grazing | 62.11Bc | 62.83Bc | 63.84Ba | 65.41Ba | 69.43Ab | 62.41Ba |
| Heavily grazing | 55.61Cc | 59.82Cc | 62.12Aa | 63.51Aa | 66.92Aa | 58.91Cc |

Different uppercase letters in the same row or column show the very significant difference ($P < 0.01$). Different lowercase letters in the same row or column show the significant difference ($P < 0.05$)

yaks defecation was not significant (Table 18.). Yaks defecation in the very light, light, and moderate grazing area were significantly higher than those in heavy grazing area in the summer, and the differences of yak dung dry matter among very light, light, and moderate grazing area were significantly obvious.

The dynamic change of yak's dry matter digestibility was shown in Table 19. The dry matter digestibility of yak was the maximum in the very light and light grazing area in July to August, while that in the moderate and severe grazing area reached the maximum in mid August to early September. The differences between grazing intensity, time, and yak dry matter digestibility reached extremely significant level (Table 19). Under the condition of light grazing, differences between yak dry matter digestibility and grazing time were not significant in summer. Under the condition of moderate grazing, dry matter digestibility was significantly higher than that of other grazing time in summer. Under the condition of heavy grazing, the differences between yak dry matter and the digestibility of other grazing time were not significant ($P > 0.05$), but significant difference between them ($P < 0.01$) in July 20–August 5, August 5–August 20, and August 20–September 5.

The Grazing Use of Artificial Grassland

Grazing experiments in the headwater region of three river showed that, in *E. nutans* + *P. crymophila* mixed artificial grassland average annual change of excellent forage and individual yak weight varied along the grazing intensity corresponding to the two point of intersection of a straight line (9.97 yaks/hm$^2$), this can basically maintain the ratio of good quality forage and the individual yak weight increase unchanged annually. Therefore, it can be considered as about alpine grassland (grazing in grass growing season) maximum grazing intensity without degradation of the alpine grassland. In addition, on the basis of forage nutrition loss in withered season at the grazing pasture, the maximum grazing intensity that cannot cause degradation was about 4.01 yaks/hm$^2$ (Zhao 2011).

### 3.2.5 The Development Patterns of Ecological Animal Husbandry and Sustainable Management of Alpine Grassland

Low Transformation Efficiency of Traditional Animal Husbandry in Resource Productions

After a lush growth during the summer and autumn, grasses on the QTP grasslands generally had rich nutrition and good palatability, with the characteristics of "three high and one low": high contents of crude fat, crude protein and free nitrogen extractions, and low crude fiber, which congenitally provided favorable breeding conditions to cattle and sheep. However, the climate conditions on the QTP is harsh, in which seasonal variation of precipitation is obvious, the rainfall concentrated mainly from June to September, and the season with dry and cold climate was long. This caused plenty of grass withered in the winter and spring season, both of the storage and nutritive value of forage greatly reduced, the storage of pasture was only about to 43 % of the warm season, and the crude protein content of only about 5 %. All these caused the low efficiency of transformation of animal husbandry resource production, and the bad cycle of livestock as "strong in summer, fat in autumn, thin in winter and dead in spring".

*Seasonal Dynamics of Body Weigh Change of Yak*

From birth to eight months of age (from April to December), yak's weight increase was 47.2 kg, in the subsequent first winter of fat off period (from December to February in the next year) the weigh loss was 6.5 kg, 13.8 % of the accumulation weight of the birth year, and in the first spring of fat off period (from February to May of second years) the weight loss was 5.6 kg, 11.9 % of the accumulation weight of the birth year. The difference of weight losses of the first winter and spring was not statistically significant ($P > 0.05$). Therefore, the yak's weight added to 47.2 kg in the birth year, of which 25.6 % (from December to May of second year) was consumed in the first cold season, which was slightly lower than the

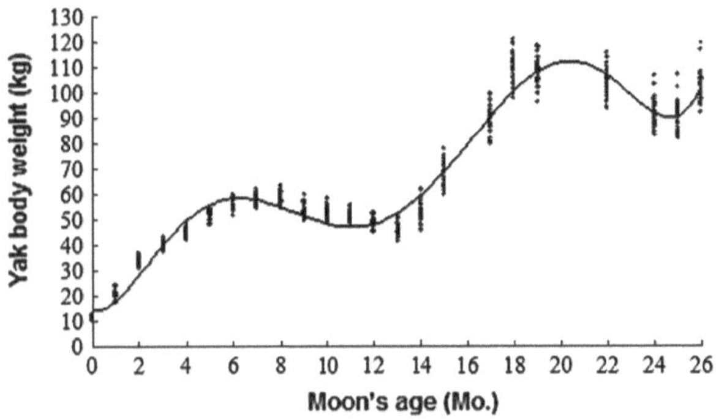


**Fig. 8** Weight change in different months of age yak under natural grazing condition

alpine fine wool sheep in the same period ($P > 0.05$). In May, the yak began weight accumulation of the second warm season with the grass germination (Fig. 8), until in October of the same year, the yak added weight was 62.9 kg, in which the summer weight addition (from May to July of the second year) was 22.2 kg, 363.9 g daily on the average; the fall weight addition (from July to October of the second year) is 40.7 kg, 442.4 g daily on the average, and the summer daily weight addition of the yak was significantly high than the autumn ($P < 0.05$). Following by the weight addition of the second warm season was weight loss of second cold season (from October of second year to May third year), when total weight loss was 18.7 kg, of which in second winter (second years in October to third years in February) was 6.0 kg, accounting for 9.5 % of the second weight accumulation, and in second spring (from February to May of third years) 12.7 kg, 20.2 % of the second weight accumulation. Therefore, in the second cold season, the yak weight losses in the spring was greater than in the winter ($P < 0.01$), and 29.7 % of the increased weight in the second warm season was consumed in the second cold season.

### *Seasonal Dynamics of Body Weight Change of Tibetan Sheep*

Sheep weight changed seasonally with the changes of forage storage and nutrient contents, and seasonal variation showed a "S" curve (Fig. 9). In most alpine grassland, Tibetan sheep was the livestock grazing on natural grassland throughout the whole year. In traditional grassland animal husbandry, there was difficult to achieve the dynamic balance of forage supply and livestock for low livestock slaughter rate, slow turnover, inconformity of seasonal herd numbers and forage yield. From the point of view of plant production, there were two critical period in grass growth. One was the period of autumn storage nourishment of seed and root, and another was the period of spring consumption of root storage nourishment by germination and sprouting. In these two periods, grass was the less resistant to grazing, so called "avoid grazing period". But using period and grazing period of cold season pasture were just overlapped in this time, causing bad conditions of forage growth and development. In the grass growth season, storage capacity of

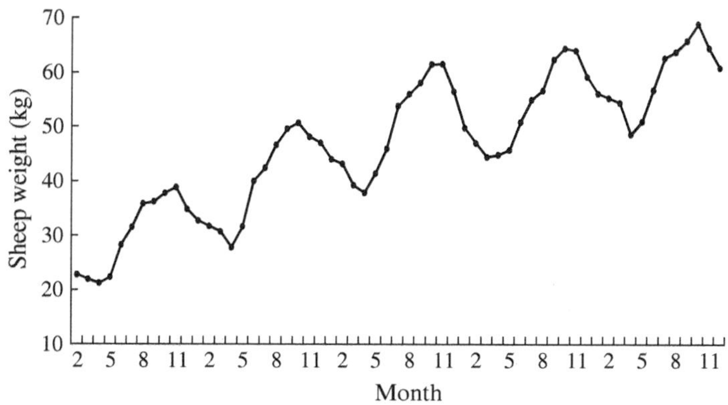


**Fig. 9** The weight change of Tibetan sheep (2–7 ages) under the natural grazing conditions

forage grass was sufficient. To the end of August, forage storage was, respectively, to 221.7 and 160 % of demand. Since then the storage of forage grass decreased, but the nutritive value of which was still high, the increasing body weight of Tibetan sheep sustained, and the storage of forage was abundant to the end of summer pasturing (from June 1 to October 31). So, the summer grazing season is the golden season of alpine grassland industry development of animal husbandry, when cattle and sheep gained weight fast for sufficient grass storage and suitable climate. In the period of November to December, the storage of forage grass was still sufficient to the sheep demands, but livestock began losing weight for decreased forage nutrients and palatability, and dropped temperature. From February to April, the contradiction between livestock and grass storage was more outstanding, and shortage forage accounted for only 85 % of the demand of livestock, 15 % deficiency, causing the nutrients forage to provide much less than that of livestock demand.

*Low Resource Use Efficiency*

Feed Conversion Ratio (FCR) refers to the weight of forage consumption when livestock or poultry increase 1 kg body weight. It is an important index for evaluating feed reward, and can directly reflect the animal husbandry forage resource conversion efficiency (Hu 2000a, b). In the single natural grazing mode, FCR of 1–7-year-old Tibetan sheep reached 130.5, meaning under the such mode every 1 kg of meat growth need consume weigh of natural herbage reaching 58.5 and 130.5 kg for the 2 and 7 age Tibetan sheep, respectively. However, in the mode of feeding fattening, the FCR increased significantly, to 8.9 for 58 days Tibetan sheep, meaning under such mode every 1 kg of meat growth need to consume herbage only 8.9 kg, so the utilization efficiency of forage resources significantly improved (Table 20).

In order to maintain the life activities, livestock had to consume the body fat deposition, protein, and carbohydrate, which cause a sharp decline in body weight

**Table 20** The ratio of feed and meat of Tibetan sheep under grazing and fattening treatment

| | Age | 1 | 2 | 3 | 4 | 5 | 6 | 7 |
|---|---|---|---|---|---|---|---|---|
| Grazing | Forage consumption (kg) | 246 | 900 | 1610 | 2020 | 2580 | 3140 | 3703.33 |
| | Ratio of feed and meat | 32.11 | 58.40 | 75.48 | 74.40 | 88.47 | 102.14 | 130.49 |
| Fattening | Fattening time (day) | 16 | 23 | 30 | 37 | 44 | 51 | 58 |
| | weight gain (kg) | 1.41 | 2.63 | 3.74 | 4.84 | 6.00 | 6.92 | 7.20 |
| | Ratio of feed and meat | 11.93 | 9.63 | 8.83 | 8.41 | 8.07 | 8.10 | 8.86 |

and even death of the elderly patient. Digestible energy from intake energy was always offset more seriously for the breeding Tibetan sheep, especially in the period from February to April. This was the greatest weakness in the production process of alpine grassland animal husbandry, and also the main causes of low economic benefit and low conversion efficiency of energy and material.

## The Development Pattern of Alpine Grassland Ecological Animal Husbandry

### *"Two Segments" Mode of the Grassland Animal Husbandry Production*

For a long time, the livestock ecosystem of the QTP was characterized by imbalance between supply and demand under the traditional livestock grazing pattern: grassland was in "surplus" period during summer and autumn, and "deficiency" period during winter and spring season. This led to increased grazing intensity in winter and spring season, and the grazing intensity could reach to 1.5–6 times of summer and autumn season. This imbalance had affected livestock production, at the same time, harmed the sustainable ability of grassland ecosystem, such as sharp declines of forage yield, water conservation, sand fixation, soil conservation, and ecosystem services.

Implementation of the concept of the "two segments" mode of the grassland animal husbandry production (Fig. 10) should make full use of natural grassland for grazing and breeding in the summer and autumn season, and fatten the lamb and ewe in the end of autumn. This can not only improve the slaughter rate, speed up the breeding cycle, increase household income, at the same time, can reduce the winter season grazing pressure to maintain the sustainable ability of grassland.

### *"Three Zones Coupling" Mode Based on the Coupling Theory of Ecological System*

Currently, the grassland animal industry of the QTP was in a condition of vicious cycle as "overgrazing of grassland-degradation grassland—intensified conflicts

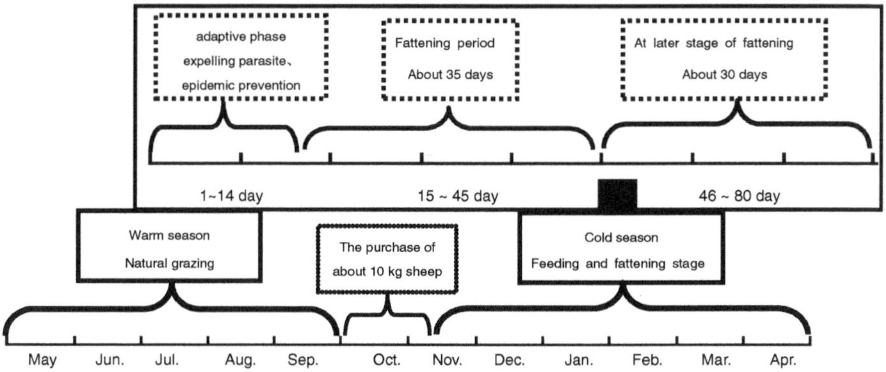


**Fig. 10** "Two segments" mode of the grassland animal husbandry production

between grassland and livestocks-declined second production capacity". In order to improve the ecological environment of the QTP, based on the theory of ecosystem coupling, interaction, it will be of great significance and prospect to generalize the animal husbandry production paradigm of "Three Zone Coupling" model that suited to local conditions of the QTP (Fig. 11).

In pastoral area of natural grassland, herd optimal management implementation was based on, the implementation of the "Seasonal Livestock Husbandry" mode, through strengthening and improvement of livestock breeding, selling a large number of livestock to the agriculture and livestock husbandry intercrossed zone and valley agricultural area before the winter, to transfer the winter spring pasture grazing pressure, make full use of forage resources in agricultural area, and realize the complementary of forage resources and livestock resources in time and space. For the large-scale farming pastoral regions, it should establish integrated construction and supporting the processing technology of forage base to provide a strong material base for the transferring grazing pressure of natural grassland, delivering part of forage productions to the source area of livestock grazing base, providing implementation feed stock for the winter animal husbandry. For the valley agricultural area, it should take advantage of breeding livestock that from pastoral areas, raise farmers' small scale cattle and sheep fattening, and part of forage could transfer to pastoral areas, causing interaction effect of plant and animal resources between agricultural areas and pastoral areas, making the resource utilization benefit beyond simple its added value, and increasing the overall business benefit.

### *"120" Forage Replacement Mode*

In the region of more severely degraded grassland, elevation below 4000 m, precipitation 400 mm, or the eastern the headwater region of three rivers of the returning farmland to forest (grassland), establishing artificial perennial vegetation with the main purpose of the restoration and update of degraded ecosystems suit its

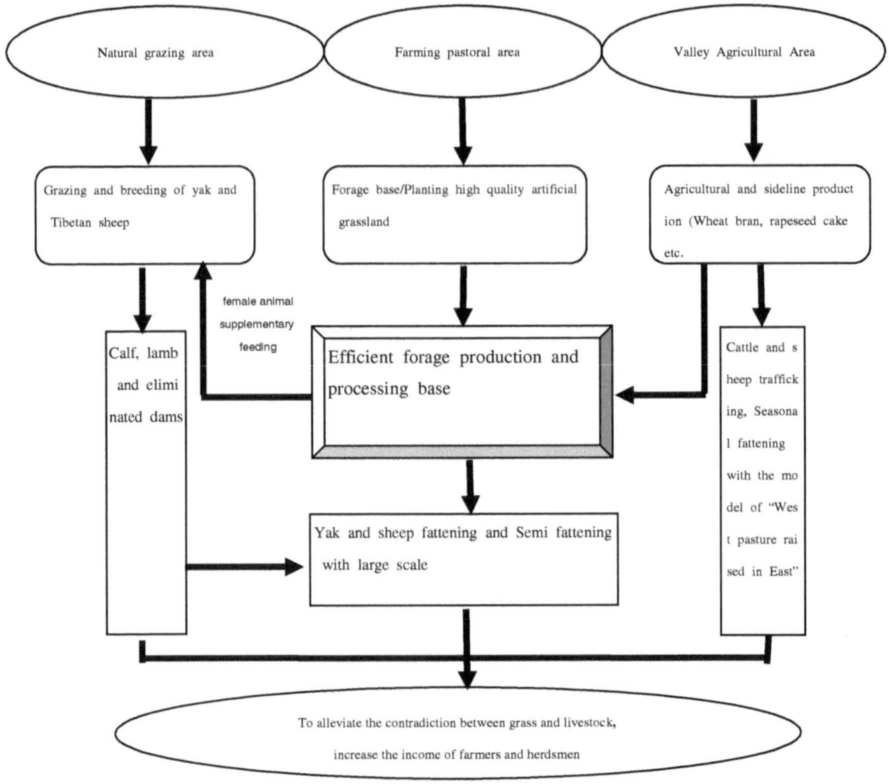


**Fig. 11** Three zone coupling mode

measures to local conditions, the grassland available for grazing livestock was equal to 27.5 times the amount of natural grassland, and the herd load capacity increased by 25.24 times in the same area (Table 21). Through the implementation of "120" forage resources replacement mode, in Qinghai Province, the headwater region of three rivers, suiting its measures to local conditions, it can provide excellent rich forage grass resources by establish artificial grassland especially in the extreme cold season lacking of natural pasture forage grass for livestock herd, alleviate contradictions of grass and livestock; at the same time, construction of artificial grassland plants every 1 acres of at least can recuperate and build up strength of natural pasture more than 20 acres.

### *"324" Lamb Production Mode*

Feed fattening was a basic method of investment in many animal husbandry developed regions of the world, but due to the affect of traditional consciousness of "depend on Heaven for livestock", which was not acceptable for the local herdsmen. In the condition of lacking in both quality and quantity of the grass, carrying

**Table 21** Comparison of before and after degradation under grassland management

| Item | Artificial grassland | Improved grassland | Fencing grassland | Native grassland |
|---|---|---|---|---|
| Grass fresh weight (kg/μ) | 2200 | 870 | 526 | 300 |
| The times of grass yield increase | 7.33 | 2.9 | 1.75 | 1.00 |
| Harvest forage ratio (%) | 75 | 80 | 40 | 20 |
| Fresh weight of harvest forage (kg/μ) | 1650 | 696 | 210 | 60 |
| The times of harvest forage yield increase | 27.5 | 11.6 | 3.50 | 1.00 |
| Load capacity (sheep unit/μ year) | 0.76 | 0.32 | 0.10 | 0.03 |
| The times of load capacity increase | 25.24 | 10.67 | 3.33 | 1.00 |

*Note* 15 μ = 1 $hm^2$

out cattle and sheep fattening, making timely slaughter cattle market, and shorten the time of livestock, which can not only reduce the grazing pressure, especially to the winter pasture, and protect natural grasslands, but improve overwintering ability of the pastoral households and increase the economic income of herdsmen. In alpine pastoral area with winter lamb production, the weight of the lamb in the winter feeding fattening 3 months will reach to the 2 years sheep (24 months) growing under traditional natural grazing condition. Through the "324" can accelerate the popularization and application of marketing mode, improve livestock slaughtering rate, and reduce natural grassland grazing pressure, meanwhile increase the income of the herdsmen.

The fattening Tibetan lamb weight reached to 8.25 kg, while Tibetan lamb weights of the herdsman A and herdsman B decreased 3 and 1.86 kg, respectively, differences reached to 11.25 and 10.11 kg, respectively (Fig. 12); according to the

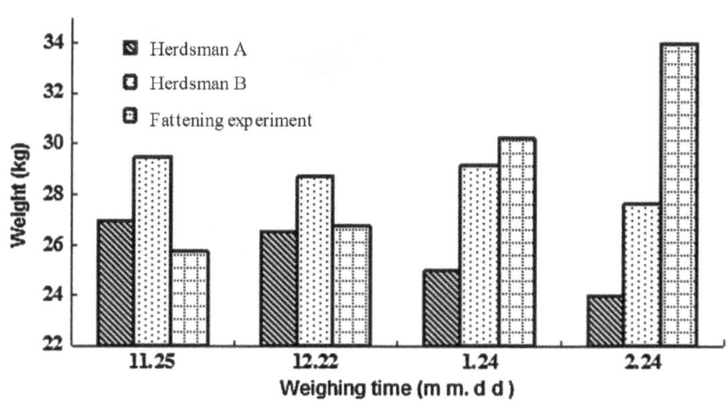


**Fig. 12** The weight gain of fattening Tibetan sheep lamb

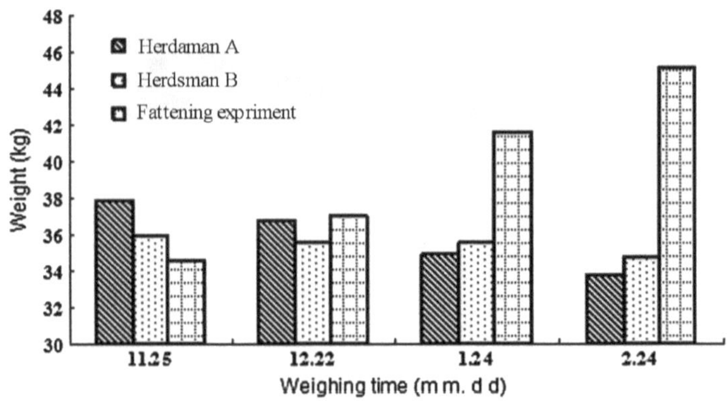


**Fig. 13** The weight gain of Tibetan sheep fattening

market prices of live weight Tibetan lamb in 2009, which is equivalent to 157.5 yuan RMB and 141.54 yuan RMB, respectively.

The growth of fattening Tibetan sheep weight reached to 10.5 kg, while the decreases of Tibetan sheep weight growth of herdsman A and herdsman B were 4.16 and 1.25 kg, respectively; weight difference reached to 14.66 and 11.3 kg, respectively (Fig. 13); according to the market prices of live weight Tibetan sheep in 2009, which was equivalent to RMB of 205.24 yuan RMB and 158.2 yuan RMB, respectively.

### *The Production of Effective Organization Mode*

In alpine pasture of the headwater region of three rivers, the introduction of leading enterprises could promote the organizational model of the herdsman (cooperatives) + base + coproduction enterprise, and enhance the comprehensive benefits of industrialization of animal husbandry production (Fig. 14). Using the characteristics in different production systems, based on the principle of ecosystem coupling system, to solve production practice problem that contrary to the systems at different levels. Mode of operation should be changed from extensive management to the intensive management to realize scale management and specialized production; feeding mode should be changed from natural grazing to the feeding and semi feeding; growth mode should be changed from a single quantity type to quality type, through the strengthening of deep processing of animal products and extending the industrial chain to improve the added value; market development should rely on leading enterprise to expand from a small local market to domestic and international market.

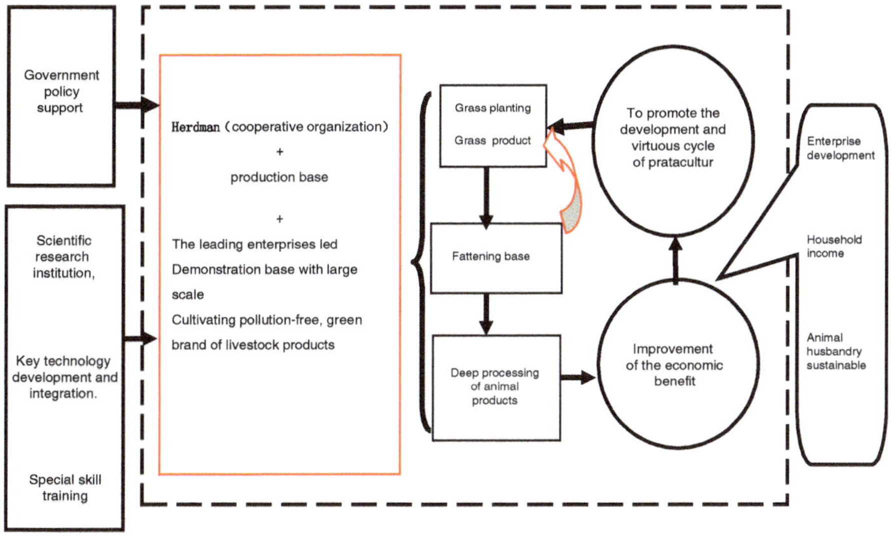


**Fig. 14** Schematic diagram of ecological animal husbandry production organization pattern in alpine region

## 3.3 *Conclusions and Prospects*

According to the present situation of the alpine area, combined with the actual needs of the protection of the ecological environment and the ecological industry development, it should focus on the plateau wetland ecosystem protection, biodiversity conservation mode and technology, different typical alpine region ecosystem service function protection and maintenance, high effective artificial grassland planting technology and forage production technology, comprehensive regulation of grassland and forestry ecosystem, construction technology of agriculture and forestry ecological system, sustainable management technology of ecological forest, ecological management mode and technology system of animal husbandry, reduction of natural grassland animal husbandry to optimize the business model and technology, etc. in the future.

In basic research, we should combine the actual needs of ecological environment protection and construction in high cold area, continue to improve the research on related theories of alpine meadow, further strengthen the alpine grassland, alpine shrub, alpine desert, and other major ecosystem types structure and its evolution, degradation mechanism and other aspects of the in-depth study, at the same time, strengthen the evaluation of biodiversity research and function of water conservation, lay a solid foundation for maintenance ecological system of follow-up and the restoration and reconstruction of degraded ecosystem. According to the survey, "black soil beach" degraded grassland in the alpine meadow in Qinghai province are located from east to west, from south to north, and only the headwater region of three rivers is of 73.6 million mu. "Ecological Protection and Construction of the Qinghai Head Water Region of Three Rivers Nature Reserve Master Plan", "Qinghai Lake Basin Ecological Environment Protection and Comprehensive Treatment Planning", and other major ecological projects are all take the black soil beach management as an important content. Among them, one phase project plan of black soil land management of the headwater region of three rivers is 5.2 million mu, but to currently only completed about half. The main reason is that governance is lake of mature technology, so the plan have not blindly implemented in large area. But through the implementation of a number of research projects in recent years, the research of "black soil beach" degraded grassland from reasonable utilization of origin, classification type, appropriate grass seed breeding, recovery mode research, grassland management after the restoration, etc. have achieved a major breakthrough. There is an urgent need to bring these new technologies and new achievements ripening, integration, and demonstration further, to promote the "black soil beach" degraded grassland in the progress of the project integrated management.

Further improvement of the relevant technology of the alpine meadow ecosystem should be conducted in the aspect of ecosystem restoration and reconstruction of degraded grassland, at the same time, of researches that aimed at reconstruction technology of alpine grassland, alpine shrub, alpine desert, and other taxa by employing different ways of introduction, integration, and innovation, to establish suitable modes for different types of alpine ecosystem function recovery and continuously improving technical support, focusing on research and development or integrated innovation ecosystem restoration and reconstruction technology of degraded grassland, land desertification prevention and control technology, comprehensive controlling of soil erosion and sustainable technology, techniques of nonpolluting ecological rodent control, comprehensive remediation technology integration, demonstration in a typical ecological function degradation area, etc.

With the application and demonstration of ecological animal husbandry production technology, it can change the traditional pattern of alpine pastoral livestock production, implementing economic development, technological progress, environmental friendly, the effective unification of herdsmen's income and living environment changes, and making useful exploration for the development of modern animal husbandry. The key promotions are livestock balance technology, artificial feed processing technology, processing configuration of multiuse matching feed technology, complete Ration Mixing Technique (TMR), grain seeds steamed flattening technology, production technology of lamb hybridizing, demonstration and popularization of animal health breeding technology. The technical level should take combination of feeding and grazing, planting grass and raising livestock, annual species and perennial grass, company and farmer, the association and the market, science and technology department and the department of agriculture and animal husbandry, ecological engineering and ecological animal husbandry, and traditional and modern, respectively. Operation pattern is market oriented, starting from the forage processing, distribution, and cattle fattening, arousing the enthusiasm of the company, cooperative, households and governments and industries to raise funds and attract private capital. Establishing professional cooperatives of different forms, content, and scale with the principle of "voluntary, private, public management, and public benefit", through the forms of "transferring, helping, and heading" and the contract and others to self-development and actively participate in market competition. Actively promoting three business modes suitable for the production of efficient animal husbandry in different regions: grass and livestock is the mode of "pasture + herdsman" production service derived by state-owned farm; fattening lamb is " cooperatives + farmers" lamb industrial production type; livestock by buying grass is the mode of the whole village promotion production particularly in immigration village implemented by purchasing grass. At the same time, should actively explore the land circulation system and mechanism in agricultural and pastoral areas, pattern of paid use of land; establish the animal husbandry cooperatives and breeding community patterns; construct the system of organic animal husbandry and operation model; establish linkage scale between leading enterprises and farmers to achieve intensive, standardized, professional production of animal husbandry, and leapfrog and low carbon developments in animal husbandry production. The construction of the service system includes ecological animal husbandry grass, grass storing, and training of scientific health breeding technology, and technical services; training of organic livestock farming technology; construction of information technology service of organic animal husbandry; and comprehensive technical training of lamb fattening. Promoting efforts of implement of scientific and technological achievements, improve the forms of technology service organization such as science and technology correspondent, farmer brokers, information platform of science and technology in project area, etc.

## 4 Sustainable Restoration Ecology: A Conceptual Framework and Application in Vegetation Restoration Southern China

### *4.1 Introduction*

The critical question in the light of rapid environmental changes facing us to realize that the theoretical and practical underpinnings of restoration have to be reconsidered (Hobbs and Cramer 2008), and the goal of restoration should be self-sustainability (Parker 1997). Sustainability was deemed as one of the criteria to judge the success of ecosystem restoration (Ewel 1987). Specifically, ecological restoration is a series of activities undertaken to return a degraded ecosystem to a healthy state, and to some form of cover that is protective, productive, aesthetically pleasing (Hobbs and Norton 1996, Palmer and Filoso 2009).

Restoration ecology and sustainability science have become increasingly important. However, the traditional theoretical foundation for restoration ecology is limited in scope and inadequate for meeting the needs of sustainable development. Thus, here we proposed a conceptual framework—"sustainable restoration ecology (SRE)"—to simultaneously achieve the goals of both ecological restoration and sustainable development.

## *4.2 The Framework of Sustainable Restoration Ecology*

This new conceptual framework is based on the integration of a number of related disciplines across scales and organizational levels. Its core concept is consisted of restoration ecology and sustainability science; its support disciplines included conservation biology, landscape ecology, ecological engineering, landscape architecture, and ecological economics; and its basic science foundation relies primarily on autecology, population ecology, community ecology, ecosystem ecology, global change biology, soil ecology, and hydrology (Fig. 15).

### 4.2.1 The Core Disciplines of SRE

The main point of SRE is the integration of restoration ecology and sustainability science. Actually, ecological sustainability is a dominant paradigm in restoration ecology. Variously defined, sustainability implies species, community, and ecosystem persistence over time, and they are often used as implicit or explicit restoration goals (Millar and Brubaker 2006). Besides, SRE emphasized the

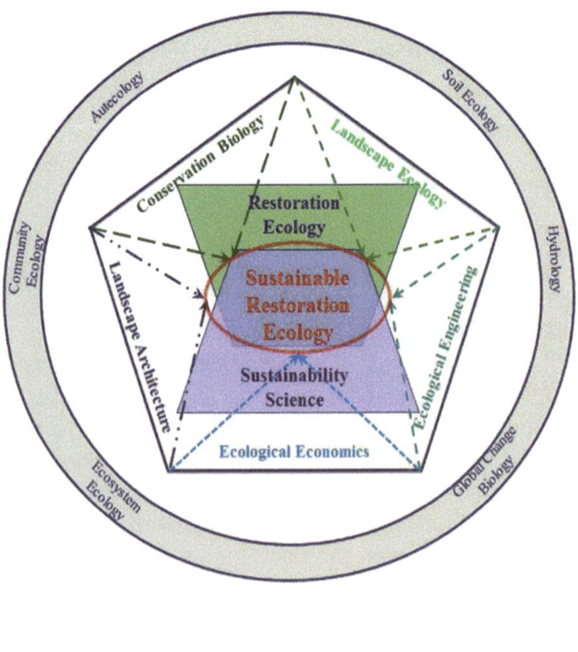


**Fig. 15** The framework of sustainable restoration ecology

sustainable development of ecology, society, and economy. Ecological restoration that satisfied socioeconomic principles can be sustainable and have ecological resilience to confront various disturbance in the future.

#### 4.2.2 The Supportive Disciplines of SRE

Among several support disciplines, integrating the principles of "conservation biology", we try to emphasize that balance the protection and application of ecosystems. Hierarchical patch dynamics in "landscape ecology" is needed because of that SRE is based on multihierarchical spatiotemporal scales, and restoration efforts can only be well documented after long series of data have been amassed (Fuhlendorf and Smeins 1997, Ehrenfeld 2000). In addition, the artistic element of restoration, and the important role played by social values have been emphasized (Davis and Slobodkin 2004). So the concepts and technology of "landscape architecture" were added to SRE to plan the landscape after ecological restoration. Specific to the restoration practice, much attention should be paid on "ecological engineering", especially in pre- or post-restoration monitoring of some projects (Ewel et al. 2001). However, to make the case for ecological restoration, we must analyze in economic terms the ecological benefits that will obtain, and also the full costs if we fail to intervene. "Ecological economics" can help us to quantify the full value of restoration, and thus bridge the gap between restoration costs and restored value (Holl and Howarth 2000).

#### 4.2.3 The Basic Disciplines of SRE

Various interactions of ecological basic disciplines are essential to SRE, and its basic science foundation relies primarily on autecology, population ecology, community ecology, ecosystem ecology, global change biology, soil ecology, and hydrology.

### *4.3 The Application of Sustainable Restoration Ecology*

#### 4.3.1 The Application Foundation of SRE

SRE integrated the change in future and practice of ecological restoration, and make it possible to restore ecosystem aimed at sustainable in the future. First, SRE tried to solve the problems of restoration ecology and ecological restoration; second, SRE achieved the sustainable development under the pressure of degraded ecosystem and increased human demand, and the ultimate success of sustainable development is likely to be tied with the ecological, economic and social sustainability. The application success of SRE in practice is not only about the science of ecology but it

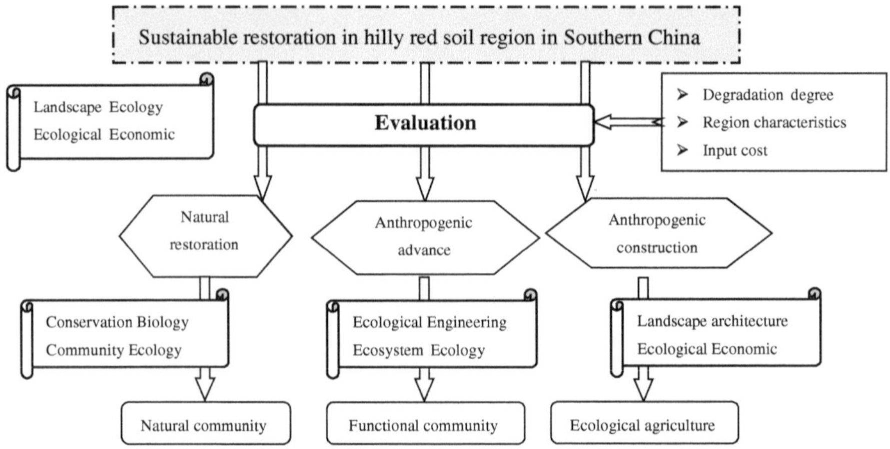


**Fig. 16** The main paths of sustainable ecological restoration in hilly red soil region of southern China

also includes societal decisions on appropriate end points for restoration, economics of restoration, policy, education, and other social and philosophical issues (Davis and Slobodkin 2004). It is sustainable to connect the ecosystem health and social benefits together.

### 4.3.2 The Application in Vegetation Restoration of Hilly Red Soil Region in Southern China

Hilly red soil region in Southern China is one of the most productive regions in China. However, the biodiversity decreased for natural disturbance and anthropogenic unreasonable utilization, and it changed to ecological fragile region (Wang et al. 2010). The concept of SRE can be used in the construction of ecological restoration as Fig. 16.

### 4.3.3 The Application Restoration of Forest–Fruit–Grass–Fish Complex Ecosystem in Heshan, Guangdong, China

The forest–fruit–grass–fish ecosystem in hilly region, Heshan, Guangdong, was changed from degraded hilly region in 1986 (Qi et al. 2007). This is a successful case of ecological restoration by integrating several compositive principles.

### 4.4 Conclusion

We proposed a conceptual framework—"sustainable restoration ecology (SRE)"—to simultaneously achieve the goals of both ecological restoration and sustainable development. This new conceptual framework was based on the integration of a number of related disciplines across scales and organizational levels. Besides, the application in vegetation restoration of hilly red soil region in Southern China, and in the development of forest–fruit–grass–fish ecosystem in hilly region, Heshan were addressed. The framework is helpful to the theory and application development of restoration ecology.

## 5 Advancement of Karst Ecosystem in Southwest China

### 5.1 Introduction

As a key component of the earth surface system, the karst ecosystem has three key characteristics: (i) Karst is widely distributed. Worldwide, carbonate rocks cover 12.34–21.09 million km$^2$, representing 9.3–15.9 % of the Earth's total land area (Wang et al. 1999; Cao et al. 2012). China is home to large areas of karst, totaling 3.34 million km$^2$ (including exposed, covered and buried karst), accounting for a third of the country's land area (Li and Duo 1983; Li 1985). The karst area in southwest China, with Guizhou as the center, has an area of 540,000 km$^2$ and is one of the three largest areas globally, in which karst is concentrated (i.e., the Mediterranean in Europe, eastern US, and southwest China) (Ford and Paul 2008). (ii) There is a close relationship between karst processes and global change. Carbonate rock dissolution consumes large amounts of atmospheric/soil $CO_2$ and results in carbon sequestration. Existing research data indicate that the carbon flux due to carbonate rock dissolution and consumption of atmospheric/soil $CO_2$ is 0.1–0.6 PgC/a worldwide, or 1/3–1/4 of the "missing carbon sink". At the same time, when carbonate deposits in caves, the climatic and environmental change information that it carried is stored in the deposits (stalagmites). (iii) Karst areas are obviously fragile eco-environments. Limited soil resources: the karst areas of southwest China feature old, hard carbonate rocks which contain small amounts of acid-insoluble material, thus these areas have limited soil resources. Double-layer karst aquifer: as carbonate rocks dissolve, a karst hydrogeological structure with both surface and underground layers is formed, and underground river systems are develop which focus underground water resources in karst areas. Harsh habitats for vegetation: karst areas are characterized by a high percentage of exposed rocks, discontinuous soil cover, thin soil, alkaline and Ca-rich soils, and harsh habitats for vegetation.

Within this context, karst ecosystems can not only contribute to studies of global climate change, but provide support for socioeconomic development and poverty alleviation efforts in karst areas.

## 5.2 Karst Ecosystem: Definition, Structure, and Function

The karst ecosystem was first proposed by Yuan (2001): The karst ecosystem can be explained as the ecosystem that is constrained by the karst environment (Fig. 17). The karst ecosystem is composed of two parts: the inorganic environment and living systems. The inorganic environment generally has two significant characteristics: (i) The dissoluble rocks contain large quantities of Ca and Mg, are alkaline and have poor soils; and (ii) The existence of underground voids makes it possible to create a double-layer structure for water and air, forming a unique karst hydroecological system. The living system has two primary components: (i) karst vegetation featuring xerophytes, chomophytes, and calciphilous vegetation; and (ii) underground biological communities living in dark, moist conditions with relatively constant temperature.

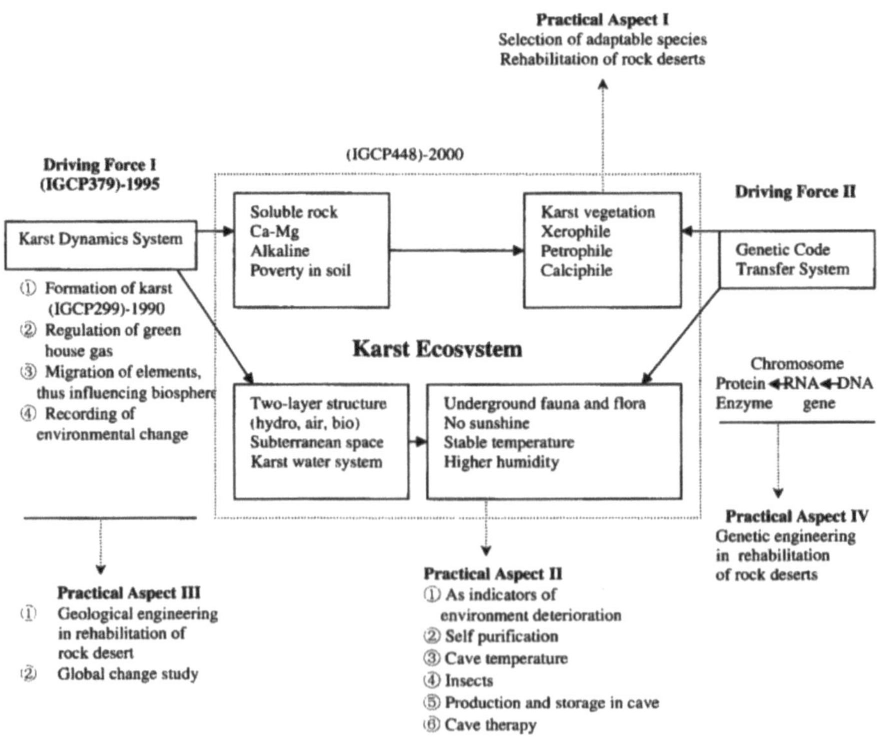


**Fig. 17** The structure, driving forces, and function of a karst ecosystem (Yuan 2001)

The functions of the karst ecosystem include the karst dynamic system (KDS) function and the Genetic Code Transfer System (GCTS) function. The KDS function is to drive karst formation and development; to contribute to the regulation of greenhouse gases in the atmosphere and mitigation of environmental acidification; to drive the migration of certain elements, and thereby influence the development of life and species selection in karst areas; and to record the processes of environmental change. These functions can be used to study global (climate) change, as well as geological engineering for the rehabilitation of karst rocky desertification.

The GCTS function is to transfer various environmental factors in karst areas, such as Ca richness, shortage of soil and water, double-layer structure, and dark, humid conditions with relatively constant temperature, into living organisms, for speciation or modification of organism habits, to help form a unifying community of producers, consumers, and decomposers in karst areas. This function provides a possible way for human beings to approach and realize sustainable development in karst areas using genetic engineering.

The KDS impacts the abiotic aspect, couples with impact of the GCTS on the biotic aspect of the operation of the karst ecosystem.

KDS involves the transfer of energy and matter within the carbon, water, and calcium (and other nutrient elements) cycles. It occurs at the interfaces of the lithosphere, hydrosphere, atmosphere, and biosphere, but is subject to the control of already formed karst features.

GCTS involves the transfer of genetic information from DNA through RNA to protein according to the Central Dogma of genetics. It constrains the formation and evolution of special producer, consumer, and decomposer communities in both the aboveground environment that has no soil, lacks water and is Ca-rich, and the underground environment that is dark, moist and at a relatively constant temperature.

## *5.3 Composition of Karst Ecosystem in Southwest China*

According to Ma (1991), founder of Ecological Society of China, ecology is defined as the science that studies the patterns and mechanisms of the relationship between the biological system and the environmental system. An ecosystem refers to a natural system that consists of a biological community and its surrounding geographical environment. It is made up of four parts: inorganic environment, producers, consumers, and decomposers.

### 5.3.1 Inorganic Environment of the Karst Ecosystem in Southwest China

(i) Ca-rich and alkaline. The calcium richness of the karst ecosystem originates from the carbonate rocks which contain large quantities of calcium, and it is

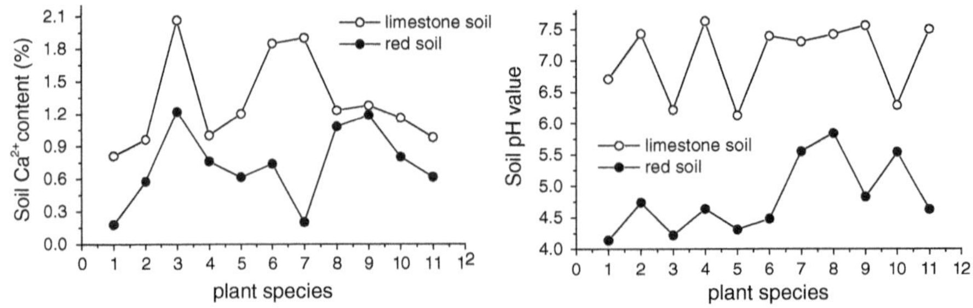


**Fig. 18** Soil pH and calcium content in plants' rhizosphere at Maocun Karst Ecological Station, Guilin, China. 1 *Pinus massoniana* 2 *Vitex negundo* 3 *Loropetalum chinense* 4 *Fractus rosae* 5 *Liquidamba formosana* 6 *Cinnamonum camphora* 7 *Agrimonia pilosa* 8 *Dendran thema* 9 *Senecio scandens* 10 *Opsmanthus fragrans* 11 *Rhus chinensis*

evident in the hydrosphere, atmosphere, and biosphere. The airborne dust (particles) in karst areas is often calcareous, so the rainwater is rich in calcium. For instance, rainwater in Guangxi and Guizhou karst areas contains 2.9–6 mg/l calcium, while that in those non-karst areas often contains less than 1 mg/l (Yuan and Cai 1988; Jiang et al. 1996; Jiang and Lu 1991). Karst underground water contains up to 50–120 mg/l bicarbonate, often decreasing with increase in latitude (Yuan 1990). The pH of limestone soil is 2.16 units greater than that of red soil, while the average calcium content in limestone soil is 3.6 times that of acid soil (Lu et al. 2006).

(ii) Large underground voids. There are numerous caves in the karst regions of China. The existing data show more than 500,000 caves can be found in bare karst areas in China, with a density of 0.8 caves per km$^2$. Among them, over 1000 caves have a measured length of over 500 m, approximately 150 caves have measured length exceeding 3000 m, and about 400 caves have been exploited as show caves (Chen et al. 2005). In karst areas, permeability ranges from 0.3 to 0.6, and some times up to 0.8. A total of 2836 subterranean streams have been identified in karst mountain areas which are continuously distributed in southwest China, with a total flow of 1482 m$^3$/s, and a total length of 13,919 km, which is equivalent to the Yellow River. They are the dominant medium for movement of water and storage of water resources (Fig. 18 and Table 22)

### 5.3.2 Producers in the Karst Ecosystem of Southwest China: Regional Karst Vegetation

Karst vegetation is regional vegetation rather than a zonal one (Li et al. 2003). Its presence is mainly subject to the impact of regional geological conditions, rather than the constraints of climate. This is particularly true in tropical and subtropical

**Table 22** Distribution of karst underground water resources in southwest China

| Province/Autonomous region | Carbonate rock outcrops | | Karst underground water resources | |
|---|---|---|---|---|
| | Area ($\times 10^4$ km$^2$) | Percentage (%) | Volume ($\times 10^8$ m$^3$/a) | Percentage (%) |
| Guizhou | 11.61 | 61.2 | 177.77 | 83.1 |
| Guangxi | 8.21 | 34.8 | 514.8 | 66 |
| Hunan | 6.36 | 30.1 | 263.4 | 57 |
| Yunnan | 10.83 | 30.0 | 325.15 | 43.2 |
| Hubei | 5.18 | 27.9 | 186.8 | 44.8 |
| Sichuan | 10.04 | 17.8 | 293.63 | 41.4 |
| Guangdong | 1.03 | 5.8 | 5.19 | 6.5 |

areas, where the karst environment has vegetation that is not zonal, but rather represents local climate.

Karst vegetation features calciphilous vegetation, chomophytes, and xerophytes. Due to shallow soil layers and high rates of bare rock, plants in karst areas have a more direct relationship with the nutrient elements of the underlying rocks. Also, due to rugged nature of karst landforms, the nutrients and water necessary for the growth of plants exhibit significant spatial differences. Different karst geochemical contexts result in the differences of local niches, and thus exert influence on the characteristics of communities.

The biodiversity of karst forest ecosystem in Maolan is much richer than that in other non-karst areas of Guizhou. Niche diversity is a major factor in biodiversity. Maolan's karst forest contains various types of plants, ranging from xerophyte to hygrophyte, from heliophile to shade-tolerant plants, from barren-resistant trees to fertilizer-absorbing trees, and from the widely distributed plants to narrowly distributed calciphilous plants, which occupy niches that fit their characteristics (Zhu and He 1993; Zhu et al. 1995).

In the 4000 m$^2$ Maolan forest area, there are a total of 344 recorded species of vascular plants which belong to 180 genera of 84 families. Plant species and genera compositions vary significantly under different karst geochemical contexts, even at higher taxonomic units like family (Hou and Jiang 2006).

The regional characteristics of karst vegetation are also demonstrated by large numbers of endemic genera and species that can only grow in karst areas. For instance, in Guangxi, China, 16 out of 38 currently known genera of Gesneriaceae are endemic to karst areas; and 91 out of 166 species of Gesneriaceae uniquely grow on karst substrates (Wei et al. 2004a, b).

The difficulty of restoring the vegetation in karst mountain areas is evidenced by the distribution status and characteristics of existing vegetation in Guangxi. In the 1960s–1980s, the forests in Guangxi suffered from several periods of large-scaled deforestation. Since the mid to late 1980s, hillsides in karst mountain areas have been closed to facilitate afforestation. So far, great changes have taken place in these areas of Guangxi. The coverage, on an average, of arbor, bush, and grass of

Guangxi is 21.18, 8.63, and 2.10 %, respectively. If karst county is defined as the county with the ratio of carbonate rock exposure over 30 %, the average coverage rate of bush in karst counties reaches 14.81 %, and the average coverage rate of arber is 12.13 %; while the average coverage rate of bush in non-karst counties is only 1.92 %, however, the coverage rate of arber averages 31.32 %. There is a good corresponding relationship between the spatial distribution of arber/bush and distribution of carbonate rocks: the coverage rate of bush is positively correlated with carbonate rock exposure ratios ($r = 0.69$), while a negative correlation exists between coverage rate of arbor and carbonate rock ratios ($r = -0.75$) (Table 23).

### 5.3.3 Consumers in the Karst Ecosystem in Southwest China: White-Headed Langur (Presbytis Luecocephalus Tan), a Rare and Endangered Animal

The white-headed langur is only found in karst fengcong areas of southwest Guangxi, China. The selection of habitats is generally a result of animals' long-term adaptation to their living environment. As an animal under the first-class state protection in China, white-headed langur is only found in some of the karst mountain areas within four counties in southern Guangxi, with a population of about 1000, covering a total habitat area of 200 km$^2$.

At night, white-headed langurs stay in high-altitude karst caves in rocky karst mountains. Most feed on the leaves of calciphilous plants endemic to karst areas. According to a survey of vegetation in the white-headed langurs' habitat, the rocky mountains feature northern tropical evergreen seasonal rain forest, including 15 families of tropical vegetation and 11 families of subtropical vegetation, of which 170 species are calciphilous plants endemic to karst areas (Huang et al. 2000).

Each year, white-headed langurs' average habitat utilization rates are: 66.45 % (foot), 21.15 % (hillside) and 12.78 % (mountain top). The dense vegetation at the foot of mountains provides hiding places, shade, and different types of food. The hillside is used for movement and rest. On mountain tops, the langurs can bathe in the sun's rays during winter, taking advantage of the bare rock on the mountain tops to maximize exposure to the sun.

### 5.3.4 Decomposers in the Karst Ecosystem in Southwest China: Active Soil Microorganisms

The number and activity of soil microorganisms are subject to the impact of various factors, including different types of vegetation structure, vegetation growth and development status, litter quantity, soil surface cover extent, and differences in soil temperature and humidity. Meanwhile, different geological conditions also have significant impact on soil microorganism communities.

Using traditional analytic techniques to detect the type and quantity of microorganisms in 0–20 cm soils in woodland, bushes, and grass within karst areas

**Table 23** Importance value (IV) of the main families of trees and shrubs in regions with different lithologies, Maolan, Guizhou, China

| Pure limestone areas | | | | Pure dolomite areas | | | | Impure carbonate rocky areas | | | |
|---|---|---|---|---|---|---|---|---|---|---|---|
| Tree layer | | Shrub layer | | Tree layer | | Shrub layer | | Tree layer | | Shrub layer | |
| Family | Importance value | Family | Importance value | Family | Importance value | Family | Importance value | Family | Importance value | Family | Importance value |
| Aceraceae | 29.02 | Poaceae | 72.00 | Corylaceae | 39.21 | Poaceae | 93.93 | Anacardiaceae | 40.59 | Poaceae | 145.62 |
| Lauraceae | 28.24 | Rosaceae | 49.32 | Fagaceae | 29.80 | Myrsinaceae | 55.25 | Fagaceae | 34.06 | Rosaceae | 31.46 |
| Juglandaceae | 25.49 | Berberidaceae | 37.14 | Aceraceae | 21.68 | Pittosporaceae | 15.36 | Lauraceae | 28.95 | Berberidaceae | 28.15 |
| Ulmaceae | 21.45 | Fagaceae | 10.82 | Cornaceae | 21.03 | Proteaceae | 14.21 | Juglandaceae | 24.38 | Smilacaceae | 22.56 |
| Fagaceae | 21.12 | Myrsinaceae | 10.77 | Ebenaceae | 19.78 | Palmae | 14.17 | Rutaceae | 23.26 | Caesalpiniaceae | 17.96 |
| Total | 125.32 | | 180.05 | | 131.50 | | 192.92 | | 151.24 | | 245.75 |

(limestone and dolomite) and clastic rock areas (sandstone and shale) in Maocun, Guilin, the data show that: the average quantities of soil microorganism in woodland, bushes, and grass in karst areas are calculated to be $39.21 \times 10^6$, $31.70 \times 10^6$, and $8.50 \times 10^6$ , respectively; while those in non-karst areas are calculated to be $2.34 \times 10^6$,$1.35 \times 10^6$ and $6.04 \times 10^6$ , respectively.

Due to the differences in the quantity of soil microorganisms in karst versus non-karst areas, indoor simulation experiments were employed to study the degradation of corn stalks with the same source, by soil microorganism in karst (limestone soil) and non-karst areas(red soil).

The results indicated that the peak degradation rate in karst was found at the 28th day, while the 42nd day for non-karst, about 15 days later. The percentage of the degradated corn stalks was to be 77 % in karst and 75 % in non-karst. These imply the corn stalks decomposed more easy and faster.

When the corn stalks were added into soils, the number of microorganism began to increase sharply, During the whole process of degradation, the number of soil microorganism in limestone soil was greater than that in red soil: the number of soil microorganism in red soil averaged $104.8 \times 10^4$/g of dry soil, compared with $220.3 \times 10^4$/g of dry soil in limestone soil. Of the three types of microorganism detected, bacteria comprised the largest share, followed by Actinomycete and fungi. In red soil, the microorganism composition was 36.0–86.3 % bacteria, 3.2–29.9 % fungi, and 6.3–60.2 % Actinomycete. In limestone soil, the microorganism composition was 16.6–93.2 % bacteria, 0.3–2.6 % fungi, and 6.3–78.1 % Actinomycete. Corn stalks were found to have different roles in promoting the growth of different microorganisms: bacteria and fungi were more sensitive, and their numbers began to increase on the third day, while the number of Actinomycete only started to increase after the 15th day (Fig. 19).

## 5.4 *Material Cycles in the Karst Ecosystem in Southwest China*

### 5.4.1 The Carbonate Rocks Dissolution/Weathering Is Actively Involved in the Global Carbon Cycle

*Carbonate rocks are the largest carbon pool in the earth system* Carbonate provides the material basis for the KDS. According to existing data, the formation of carbonate rocks is closely related to the evolution of atmospheric $CO_2$. In terms of geologic history, carbonate rocks are relatively young, mainly dating from the Phanerozoic. Over geologic time scales the geobiological forces of the processes of silicate rock weathering and carbonate rock formation, produced a huge reduction in atmospheric $CO_2$. Prior to carbonate formation, the earth's atmospheric $CO_2$ concentration was above 25 %. In contrast, the modern atmospheric $CO_2$ concentration is only 0.03–0.04 %, while $61 \times 10^{15}$ tons of carbon are stored in carbonate

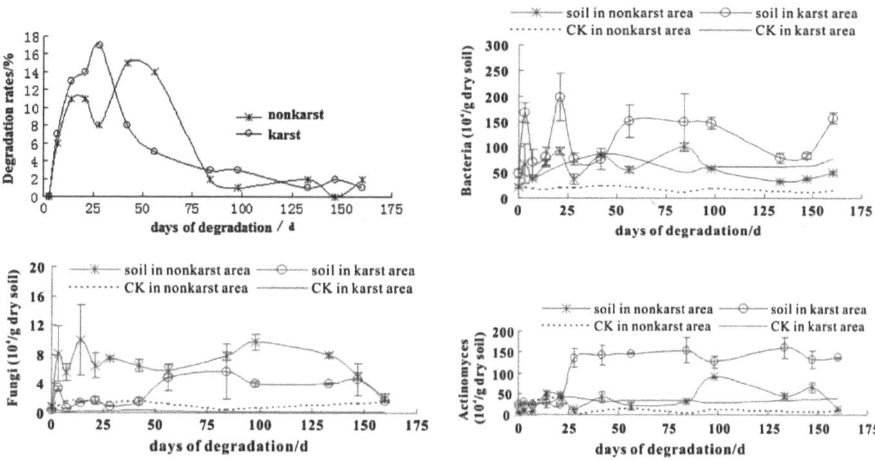


**Fig. 19** The degradation rates of corn stalks in red soil and limestone soil, and the changes of the number of bacteria, fungi, and Actinomycete at different degradation phases of corn stalks

rocks—the largest carbon reservoir in the modern earth (accounting for 99.55 % of total global carbon).

*The modern dissolution of carbonate rocks is actively involved in the global carbon cycle* In studies on the carbon sequestration process and the effects generated by terrestrial weathering, it was found that carbonate rock distribution areas, which represent about 10 % of total land area, play a bigger role in carbon sequestration. Gaillardet et al. (1999) estimated the atmospheric $CO_2$ flux consumed by global rock weathering was 0.288 PgC/a, among which 0.148 PgC/a was consumed by karstification, accounting for 51.4 %. Similarly, Munhoven (2002) estimated the atmospheric $CO_2$ flux consumed by global rock weathering as 0.221 PgC/a, among which 0.088 PgC/a was consumed by karstification, or 37 %.

According to existing estimates, the atmospheric/soil $CO_2$ flux dissolved and consumed by carbonate rocks dissolution worldwide is 0.1–0.6 PgC/a. If the median value of 0.3 PgC/a is taken, the carbon flux accounts for 17.65 % of total carbon sink fluxes of terrestrial forest, and 37.5 % of the potential for soil carbon sequestration globally. The atmospheric/soil $CO_2$ flux dissolved and consumed by carbonate rocks in China was estimated to be 0.016 PgC/a, which was equivalent to 21.3 %, 66.7–114.3 %, 2.3 times and 22.9–40 % of forest, bush, grass, and soil carbon sink fluxes, respectively (Cao et al. 2012).

*Monitoring data providing evidence for the karst carbon sink effect* Comparison of soil carbon transfer in karst and casolite areas with monitoring in Maocun karst subterranean river basins, Guilin, China (Cao et al. 2011a), showed that: (i) The rate of $CO_2$ emission from limestone soil respiration in the karst area (ranging from 23.12 to 271.26 mgC/m$^2$ h) is significantly lower than that from red soil respiration in the clasolite area (ranging from 51.60 to 326.28 mgC/m$^2$ h). The average annual $CO_2$ emitted by limestone soil respiration in the karst area was 25.12 % less than that by

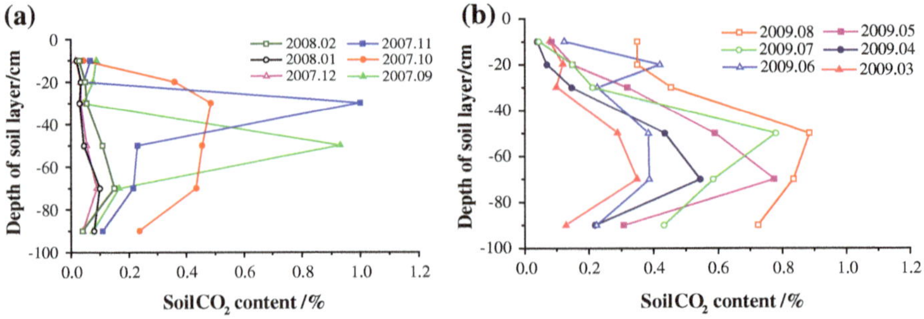


**Fig. 20** Dynamic changes in the $CO_2$ concentration in limestone soil profiles in the karst area. **a** September to February, **b** March to August

red soil respiration in the clasolite area; (ii) The $CO_2$ concentration at the limestone soil profile in the karst area exhibited a bidirectional gradient, and this phenomenon became more obvious during the seasons with good hydrothermal conditions. In contrast, the $CO_2$ concentration at the red soil profile in the clasolite areas had a unidirectional gradient, in which the soil $CO_2$ concentration increases with the depth of the red soil layer. In terms of mean values of $CO_2$ concentration in soil profiles, the $CO_2$ concentration in limestone soil ranged between 0.05 and 0.60 %, with an average annual value of 0.25 %; while that in red soil varied between 0.05 and 1.09 %, with an average annual value of 0.57 %. This indicates that the soil $CO_2$ in the lower part of the profile could be consumed and absorbed by the carbonate rock dissolution at the soil/rock interface in karst areas. In other words, the karst process in soil is a process of carbon sequestration (Figs. 20 and 21).

### 5.4.2 The Water Use Strategy of Karst Vegetation Following the Special Properties of the Water-Bearing Media in Karst Area

Carbonate rocks are dissolvable rocks. The long-term karst process forms a double-layer (i.e., surface and underground) karst hydrogeological structure, and

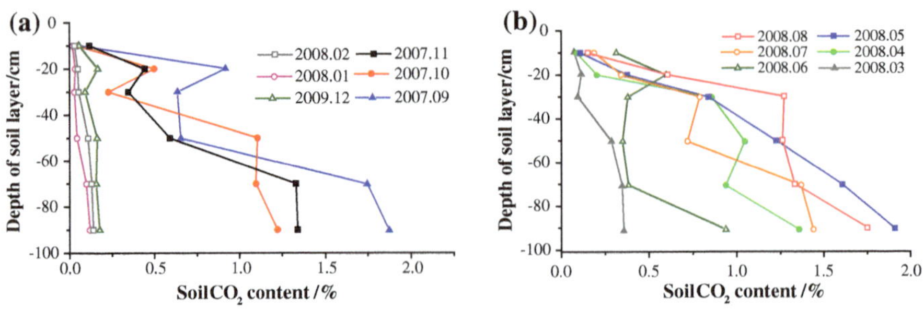


**Fig. 21** Dynamic changes in the $CO_2$ concentration in red soil profiles in the clasolite area. **a** September to February, **b** March to August

water resources are mainly stored in underground. Karst water-bearing media contain conduits, fissures, and pores. They are dominated by underground conduits, whose formation and development is closely associated with the long-term and frequent hydrodynamic processes near the underground water level. As a result, the underground water in karst is often deep, and available groundwater resources are very limited, which may bring some difficulty to the growth and development of surface vegetation, particularly in the dry season.

Liang and Wang (1998) categorized 18 drainage basins with different rocks into five types by cluster analysis, and then studies the relationship between these five drainage types and their hydrological effects in karst areas of Guizhou. The results were shown that the percentage of limestone was negative correlation with the module of low runoff; while there was a positive relationship between the percentage of dolomite and the module of low runoff, with a confidence level of 0.95, suggesting a statistically significant correlation. That is to say, in a subterranean river system, the higher the percentage of limestone, the smaller the low runoff module; while the higher the percentage of dolomite, the larger the low runoff module. No statistically significant correlation was found between the module of flood peak and percentage of lithology in a river basin (Table 24).

To adapt to the harsh and special karst hydrological conditions, karst vegetation often increases underground biomass to enhance the capacity of plants in absorbing and retaining water and using underground water. Luo et al. (2010) conducted a survey and analysis on the field plot of root biomass for karst vegetation in Maolan Nature Reserve (subtropical zone), Guizhou. The results indicated that: the aboveground biomass in the karst area was smaller than that in non-karst forest areas at the same latitude, and was only equivalent to that in non-karst temperate forest areas. However, the underground biomass in the karst area was larger than that in non-karst forest areas at the same latitude and that in non-karst both tropical and temperate forest areas. The underground biomass in the karst area reached

**Table 24** River basins' lithology and their hydrological characteristics

| Rock type | Lithological composition (average) | | | | Average low runoff value | | | Average flood runoff value | |
|---|---|---|---|---|---|---|---|---|---|
| | Limestone (%) | Dolomite (%) | Tuff slate (%) | Sandstone (%) | $\frac{\bar{M}_{10-d}}{Cv_{10-d}}$ | $\frac{\bar{M}_{10-d}}{Cv_{10-d}}$ | $\frac{\bar{M}_{m}}{Cv_{m}}$ | $\frac{\bar{M}_{max-in}}{Cv_{max-in}}$ | $\frac{\bar{M}_{max-d}}{Cv_{max-d}}$ |
| I | 72.65 | 15.22 | 0.24 | 11.87 | $\frac{1.87}{0.39}$ | $\frac{2.21}{0.33}$ | $\frac{2.59}{0.33}$ | $\frac{462}{0.37}$ | $\frac{313}{0.36}$ |
| II | 17.38 | 68.41 | 5.79 | 9.77 | $\frac{4.15}{0.27}$ | $\frac{4.61}{0.24}$ | $\frac{5.51}{0.34}$ | $\frac{630}{0.47}$ | $\frac{323}{0.40}$ |
| III | 34.32 | 41.40 | 11.61 | 12.84 | $\frac{3.16}{0.26}$ | $\frac{4.14}{0.24}$ | $\frac{4.92}{0.25}$ | $\frac{807}{0.49}$ | $\frac{391}{0.36}$ |
| IV | 50.96 | 0.00 | 0.00 | 49.04 | $\frac{2.83}{0.16}$ | $\frac{3.28}{0.10}$ | $\frac{4.21}{0.24}$ | $\frac{312}{0.42}$ | $\frac{191}{0.40}$ |
| V | 2.83 | 0.00 | 94.83 | 2.83 | $\frac{2.63}{0.24}$ | $\frac{3.04}{0.24}$ | $\frac{3.47}{0.24}$ | $\frac{781}{0.72}$ | $\frac{431}{0.66}$ |

$M_d$, $M_{10-d}$, $M_m$ modules of the minimum low runoff in a day, 10 days and a month in a year, in l s/km$^2$; $M_{max-in}$, and $M_{max-d}$: maximum instant and daily flood runoff modules in a year; $Cv_d$, $Cv_{10-d}$, $Cv_m$, $Cv_{max-in}$, $Cv_{max-d}$: variation coefficient of annual low and flood runoffs

**Table 25** Comparison of the drought resistance of *Cyclobalanopsis glauca* under different geomorphological conditions

| | Average leaf width (cm) | Average leaf length (cm) | Average leaf thickness (mm) | Water content (%) | Water potential (bar) | Proline (μg/g fresh weight) | Peroxidase ($\Delta A_{470}$/min g) |
|---|---|---|---|---|---|---|---|
| Hillside | 2.74 | 9.42 | 2.67 | 45.50 | −9.70 | 140 | 0.88 |
| top | 1.57 | 5.17 | 3.73 | 31.9 | −17.30 | 150 | 1.15 |

57.49–58.15 mg/hm$^2$, almost twice as much as that figure in non-karst tropical and temperate forest areas.

Meanwhile, karst plants reduce water transpiration and enhance water use efficiency to adapt to water stress by reducing leaf area, thickening cuticula and reducing stomatal number. In Jidanbao Hill, Nongla Village, Mashan County, Guangxi, China, Oriental white oaks (*Cyclobalanopsis glauca* (Thunb.) Oerst.) grow both at the top and side of the Hill. Although their altitude difference is only dozens of meters, their conditions in terms of solar radiation, air temperature, water, or humidity are distinctly different. Due to harsh conditions at the top the hill, the epidermal structures (i.e., cuticle, epidermal hair, epidermal cell, and stoma) of the leaf of Oriental white oaks growing there tend to be xeric (Deng et al. 2004) (Table 25).

### 5.4.3 Calcium Transfer Constrains the Geochemical Behaviors of Various Elements in the Karst Ecosystem

As the most distinct element in the karst ecosystem, calcium is a necessary nutrient element for plants' growing and a base cation in soil. Plants can be divided into calciphilous and calcifuge types, depending on their utilization strategy response to calcium in soil. Tyler and Ström (1995) used the seeds of 10 species of typical calciphilous plants and 10 species of typical calcifuge plants under dark conditions with a constant temperature of 20 °C and constant humidity to conduct germination tests. The results revealed that in the process of seed germination, the amount of small molecular organic acid produced by the seeds of calciphilous plants per unit weight was 3–4 times as much as that produced by the seeds of calcifuge plants. The amount of tribasic carboxylic acid and dicarboxylic acid of the former was twice and 3–4 times of the latter, respectively, however, the amount of the monocarboxylic acid of former was only a third of that of the latter. In the meanwhile, the biomass of evergreen–deciduous broad-leaved forest communities in subtropical karst areas was only equivalent to that of zonal coniferous forest communities in temperate areas. Insufficient intake of trace elements was considered to be one of the major obstacles affecting the growth and development of plants in karst areas. It was believed that when the pH approaches neutral, calcium will exhibit a stronger capacity to complete organic ligand, compared with other cations, thus affecting the activation of other metal ions and restricting the absorption of nutrient elements by plants.

Comparative study between limestone soil and red soil was conducted at karst ecological experimental site in Maocun Village, Guilin, Guangxi. The geochemical behaviors of Ca, Mg, Fe, Mn, Cu, Zn, Mo, B, Co, and P were analyzed. The results indicated that: (i) the total content of all the nutrient elements except B in limestone soil was larger than that in red soil. Specifically, the contents of Ca, Mg, and Zn in limestone soil were 3.68, 4.64, and 3.96 times as much as those in red soil; the figures for Mn, Cu, Co, and Fe were 1.68, 1.64, 1.39, and 1.25 times. However, the contents of P and Mo in limestone soil were only slightly higher than that in red soil, while the content of B in limestone soil is lower than that in red soil. (ii) The contents of available nutrient elements, except Ca, Mg and Cu, in limestone soil, were lower than those in red soil. Specifically, the contents of available Mn and Zn in limestone soil were 60 % of those in red soil; while the figures for Fe, P, and Mo were 30–40 %; and the figure for B was only 10 %. It suggests the nutrient elements in limestone soils are strongly impacted by karst geochemical environment with Ca-rich and alkaline (Fig. 22).

The calcium in the leaves of calciphilous plants is mainly calcium pectinate, which is found in cell walls (Cao et al. 2011b). The Banzhai karst subterranean river basin and the non-karst basin (sandstone and shale areas) within Maolan Nature Reserve were selected as sites for a comparative study. Thirteen plant species each in karst and non-karst areas were collected, including six species each endemic to karst and non-karst areas, to analyze the total content, form, and distribution parts (subcell components) of calcium in leaves. The results showed that: (i) The calcium content of plant leaves in karst area was 58.45 % more than that in non-karst area; (ii) Calcium in the leaves of calciphilous plants in karst area is mainly calcium pectinate, which accounts for 27.91–32.82 % of the total calcium content; while calcium in the leaves of calcifuge plants in non-karst areas is mainly calcium oxalate, which represents 33.69–34.34 %; (iii) Calcium in the leaves of calciphilous plants in karst area mainly occurs in cell walls, accounting for 59.05–66.54 % of the total calcium content; while calcium in the leaves of calcifuge plants in non-karst

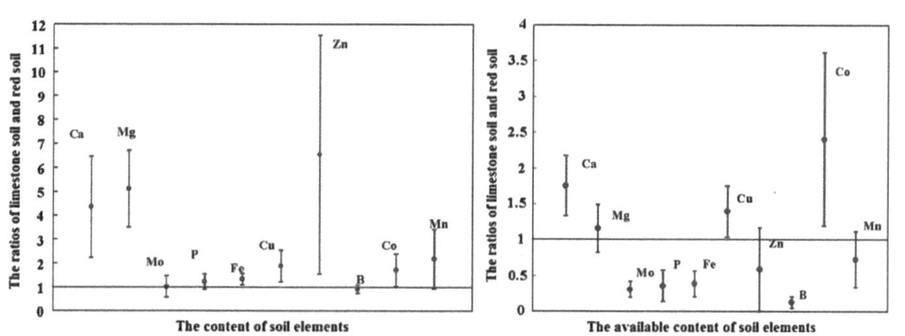


**Fig. 22** Average ratios of total content and available content of soil nutrients with a confidence level of 95 % (limestone soil versus red soil)

**Table 26** Average content and percentage (by subcell component) of calcium in plant leaves growing in karst area and non-karst area

| Group | | Cytoderm mg/kg (%) | Cytoplasm mg/kg (%) | Organelle mg/kg (%) | Total content mg/kg |
|---|---|---|---|---|---|
| Karst area | Mature leaves | 744.02 (60.29) | 339.72 (27.53) | 150.2 (12.17) | 1233.94 |
| | Immature leaves | 566.16 (47.99) | 432.63 (36.67) | 181.02 (15.34) | 1179.81 |
| | Average | 687.86 (56.53) | 369.06 (30.33) | 159.93 (13.14) | 1216.85 |
| Non-karst area | Mature leaves | 284.56 (36.52) | 340.53 (36.67) | 154.09 (15.34) | 779.18 |
| | Immature leaves | 285.14 (38.34) | 325.56 (43.77) | 132.97 (17.88) | 743.67 |
| | Average | 284.74 (37.08) | 335.8 (43.73) | 147.42 (19.20) | 767.96 |

areas mainly occurs in Cytoplasm, accounting for 36.67–43.77 % of the total calcium content (Table 26).

## 5.5 *Discussions*

Over the last 15 years, karst ecosystem study has gradually become an interdisciplinary area between karst geology and ecology. Great progress has been made in this field. In particular, the findings of "Research on Karst Ecosystems in Southwest China", a key S&T project during the period of 2001–2005 under the Ministry of Land and Resources of the People's Republic of China, played an important role in providing S&T support for the state to prepare the Guideline of Planning the Comprehensive Control of Rocky Desertification in Karst Areas in China (2006–2015). However, many areas still need to be addressed, such as the national ecological program on comprehensive control of rocky desertification in karst areas, and policy options to addressing global climate change in karst areas. Additionally, Karst ecosystem theory needs to be improved.

### 5.5.1 Karst Ecosystem in Southwest China that Needs to be Improved

Karst ecosystem is defined as an ecosystem that is restrained by karst environment, consisting of inorganic environment and living systems. Although we have had a good understanding about inorganic environment of karst ecosystem, comprehensive review and further study on living systems of karst ecosystem still need to be conducted, particularly on the study of consumers and decomposers in surface karst ecosystem, as well as underground karst cave ecosystem. Therefore, ecologists and geologists in China should work closely to make fully efforts, conduct dynamic monitoring and research on different types of karst ecosystem in China on a long-term basis, understand the operational patterns of karst ecosystem, and gradually well understand karst ecosystem in southwest China.

### 5.5.2 Vulnerability of Karst Ecosystem and Comprehensive Control of Karst Rocky Desertification

Among the factors that influence the formation and evolution of an ecosystem, climatic and hydrological factors serve as the slow and long-term driving forces shaping the ecosystem, while geological and geomorphological factors provide the material basis for the existence and development of the ecosystem. As the karst ecosystem is obviously constrained by geological conditions. Therefore, the Chinese national program to comprehensively control rocky desertification is confronted with the following three major challenges, due to the vulnerability of the karst ecosystem.

Water: comprehensive exploitation and utilization of surface and underground water with karst river basins. The accurate assessment and effective management of surface and underground water resources depend on understanding the water cycle pattern in the karst ecosystem, particularly the impact of plants' water consumption in the water cycle of karst river basins.

Soil: conservation soil resources, minimization of soil loss, and improvement of land productivity. It is important to select the appropriate plants for water and soil conservation and species of plants that can help promote dissolution of carbonate rocks and increase the rate of soil formation.

Plants: Studies on species selection and elements biogeochemical cycles are one of the fundamental approaches to promote ecological and economic development in karst areas affected by rocky desertification. The Ca-rich and alkaline biogeochemical conditions constrain the transfer of other biological elements, and contribute to the formation of a large group of plants that can adapt to karst ecological conditions and have economic value. For instance, the Chinese honeysuckle (Lonicera japonica Thunb), a medicinal plant with economic importance, has many ecological and physiological characters that can adapt to the karst ecological environment. Research findings show that genuine Chinese honeysuckle contains high levels of Ca, Sr, and Fe, but low levels of Cr and Pb, and the levels of Ca, , and Fe in limestone soil in karst areas are higher than those in red soil and brown soil in non-karst areas at the same latitude.

### 5.5.3 Material Cycle and Karst Carbon Sink Effect in Karst Ecosystem

The "missing" carbon sink has long been recognized as an important component of global carbon cycle studies, and the search for "missing" carbon sink has become a frontier scientific point. According to recent studies, carbonate rocks can act as an important carbon sink by dissolving and consuming large quantities of atmospheric/soil $CO_2$: (i) the distribution area of carbonate rocks worldwide ranged from 12.34 to 21.09 × $10^6$ $km^2$, accounting for 9.3–15.9 % of the total continent's area. Carbonate rocks dissolve and absorb atmospheric/soil $CO_2$ to generate dissolved inorganic carbon (DIC) in water bodies, serving as the major carrier of transporting DIC from terrestrial water bodies to the sea; (ii) At a short time scale (from several hundred to thousand years), the carbon sink flux of atmospheric/soil $CO_2$ dissolved and consumed by carbonate rocks varied between 0.1 and 0.6 PgC/a. When a median value of 0.3 PgC/a was used, it was equivalent to 17.65 % of carbon sink flux of terrestrial forests and 37.5 % of the potential for soil carbon sequestration; (iii) Karst water with high concentration of $HCO_3^-$ can stimulate photosynthesis of aquatic plants (e.g., algae), transforming DIC into dissolved organic carbon (DOC) and particulate organic carbon (POC). Some of the organic carbon deposits in terrestrial water bodies (rivers and lakes), while the remaining is transported to the sea. This finding will significantly increase the potential of carbon sequestration dissolved by carbonate rocks; and (iv) Organic carbon in Ca-rich and alkaline soils in karst areas tends to be more stable and have a longer cycle.

### 5.5.4 Evolution and Maintaining Mechanism of Biodiversity, and Eco-Safety of Karst Ecosystem

Compared with the study on zonal vegetation, for example, evergreen broad-leaved forests, the research on regional vegetation is still weak. Biodiversity of karst ecosystems is, first of all, subject to the heterogenicity of karst habitats, and, next, subject to the slow succession of karst populations and communities. Although some karst forest ecosystems are survived in southwest China (i.e., Longgang, Guangxi; Mulun, Guangxi; Maolan, Guizhou), most of these karst ecosystems have suffered different degrees of degradation, and the areas that are affected by rocky desertification have reached a total area of 126,000 $km^2$. Quantitative analysis on the succession stages of karst ecosystem is needed to reveal the succession process of karst ecosystem and to control the rocky desertification. Meanwhile, it is necessary to send out an early warning on the degradation process of karst ecosystem. To achieve these two objectives, it is important to develop feasible assessment indicators, approaches and models.

In addition, the total content of nutrient and heavy metal elements in Ca-rich and alkaline limestone soil is rather high. In particular, the content of some heavy metals exceeds that under Class 2 and Class 3 of the National Environmental Quality Standards for Soils in China (GB 15618-1995), this makes it difficult for those karst areas to apply to be green food production sites. Compared with red soil at the same latitude, limestone soil has a lower biologically liable nutrient element content, which constrains the normal growth and product quality of economic crops. For this reason, revealing the patterns of element transfer and loading in the rock-soil-crop system from a perspective of elements' biogeochemical cycle, and exploring methods for improving the quality of economic crops and overall eco-environmental health, will play a significant role in helping alleviate poverty and achieve prosperity, demonstrating and building green production sites for special economic crops, and achieving sustainable development in karst areas in southwest China.

# 6 Progress in Fundamental Research and Practice for Ecological Restoration of the Arid Valley Across the Hengduan Mountains of Southwest China

## *6.1 Introduction*

The Arid Valley mainly distributes in the valley of major rivers, including the Nujiang, Lancang and Jinsha River, Dadu River, Ya-lung River, Minjiang River, Bailong River, etc., and its chief tributary across the Hengduan Mountains in the southwestern China. The Arid Valley located in the lower section or the base belt of the mountain-gorge systems with the relative elevation range from 200 to 1000 m. The book, The Arid Valley across the Hengduan Mountains, reports that its area is approximately 11,230 km$^2$ (Zhang 1992). However, recent preliminary studies show that its area is at least 26,500 km$^2$, accounting for 5–6 % of the total area of the whole Hengduan Mountains (Bao et al. 2012). The Arid Valley is a distinctive geographical landscape across the Hengduan Mountains at least by three unique aspects. First, this area is characterized by arid climate under the background of the prevailing tropical, subtropical, and warm temperate climates in the regions. According to the data analysis of 23 weather observation stations within the Arid Valley, the annual precipitation is approximately 300–700 mm, whereas the annual evaporation is 1500–2200 mm, which is approximately 2–6 folds of the precipitation. More precipitation occurs in summer and autumn than in winter and spring. The heat condition indicates that the habitat of the Arid Valley is not such harsh as the global arid zone. The plants can grow as long as there is enough water in some dry-hot valleys. Second, the climate of the Arid Valley root strongly from high mountain-gorge landscape systems due to the foehn effects. It is the uniqueness of the geographical landscape formed the featured climate (Zhang 1992). Third, the vegetation in the Arid Valley, which is characterized by the dwarf shrubs, is controlled by the local arid climate. Thus, the vegetation is similar to that in global arid climatic zones rather than that in the regional tropical, subtropical, or temperate zones. Therefore, the Arid Valley ecosystem is distinctive from the arid ecosystem in north China. The Arid Valley across the Hengduan Mountains can be classified into various subclasses such as dry-heat, dry-warm, dry-temperatures, etc., based on the heat, temperature, and aridity (Zhang 1992). The dry-heat type locates in the south of 25°N across the Hengduan Mountains, while the northern part are the dry-warm and dry temperature types. Thus, the valleys along some rivers are dry-warm, such as the Minjiang River valley and the Bailong River valley. Others, however, maybe more complicated, for example, the valley along the Jinsha River being dry-heat downstream (Panzihua), being dry-warm in the middle (Benzilan and Rongxian), and being dry-warm and dry-cool upstream (Litang).

The Arid Valley characterized by closed terrain, unstable geological conditions, complicated climate, shallow soil layer, and steep slopes. Thus, the Arid Valley is of ecologically high vulnerability, low resistance, and difficulties for the restoration in case vegetation degradation. With no doubt, the Arid Valley is always core area with a high frequency of the geological hazards and disasters (Bao et al. 2007, 2012). In the past few decades, the Arid Valley also deeply disturbed by increasing human activities with the rapid population growth and economic development, such as hydropower and transportation construction, village expansion, urban development, cultivation on steep slopes, cattle and sheep grazing, stone collection, land preparation and afforestation. These disturbances make the Arid Valley became most "trouble" zone across the Hengduan Mountains, leading to the serious degradation of the local vegetation, rapid environmental deterioration, especially severe water loss and soil erosion, and frequent mountain land hazards. Not only the development of regional economy is influenced and the foundation of regional sustainable development is weakened, but the development and ecological safety of downstream are threatened.

The Arid Valley is also prominent and irreplaceable across the Hengduan Mountains in terms of the economic and social development. Among the Arid Valley and the nearby transitional area along the rivers across the Hengduan Mountains, there are 75–80 % of the villages, 55–58 % of the towns (township, district, county, and state), 80 % of the population, 55–60 % of the farmland, and more than 90 % of the enterprises. Therefore, the Arid Valley becomes the social, economic, and cultural center of the region (Bao et al. 2012). It is also the channel and key link of the regional transportation, information, resources, energy, and communication, as well as the corridor of the social and economic development among the alpine valley. Thus, the conservation, restoration, and sustainable management of the Arid Valley-mountain ecosystem is of the core and utmost importance in the regional development.

The scientific research, which focusing the Arid Valley across the Hengduan Mountains, mainly began in the 1970s based on a 100-year historical background. The ecological restoration studies of it, however, started from the early 1980s concentrated on the Arid Valley in the Minjiang River. Currently, the research and practice of ecological restoration mainly focused on it being typical of the dry-hot and dry-warm types downstream of the Jinsha River. Over the past 30 years, a lot of work has been done on the theoretical understanding, restoration mode and technology, utilization of unique plant resources, as well as the engineering practice. In conclusion, the work consists of three aspects: (i) The theories and basic knowledge of the Arid Valley ecological restoration. (ii) The schemes and techniques of the Arid Valley ecological restoration, as well as the result evaluation. (iii) The resource exploitation of the Arid Valley ecosystem, the regional ecological restoration, and the sustainable management strategies. Thus, this section will sum up the fundamental understanding and improvement into the above three aspects. The objective of this section is to propose the future direction for the ecological restoration study and practice, as well as to provide theoretical implications and guidance to the regional management.

## *6.2 Fundamental Research Related to the Ecological Restoration*

In the past 30 years, theories for the ecological restoration of the Arid Valley across the Hengduan Mountains have been further and deeply improved. The main progresses are as follows:

### 6.2.1 Ecological Degradation Evaluation on the Arid Valley

Human disturbance is the necessary prerequisite to drive the ecosystem degradation. This degradation only occurs when the disturbance pressure exceeds the system resistant threshold. The degradation itself and the degree of it are based on the historical strength, frequency, and the spatial pattern of the human disturbance. Before the planning and design of the ecological restoration, it should be scientifically evaluated that whether the Arid Valley across the Hengduan Mountains degenerate or not; if yes, which areas degenerate, and which do not? Moreover, the degree, stage, condition, and the restrain factors of the degradation need scientifically diagnosis too. The prerequisite and basis for the reconstruction of the ecological restoration is to clarify the progress, reason, and degree of the degradation. The efficient reconstruction following the natural succession will not be fulfilled until the reasonable decision for the position and progress of the degradation in the system succession is made (Bao and Chen 1999a; Sun and Bao 2005).

There is an obvious misunderstanding in the range and degree of the Arid Valley degradation. The Arid Valley is mountain ecosystem with the unique zonal climate affected by the topography and landscape. Consequently, some researchers used to mix the arid and degradation being not able to recognize the form and fact of the degradation (Tang et al. 2003; Sun et al. 2005). Most misunderstandings come from the inadequate consideration when analyze the development pattern of the Arid Valley, or from its own formation and succession patterns and the selection of non-Arid Valley vegetation like comparatively moist subalpine vegetation as criteria and reference by mistake. The afforestation designed and applied based on the misunderstanding of the Arid Valley degradation do not achieve the expected goals but cause new degradation (Tang et al. 2003). Recent studies show that the ecological degradation in the Arid Valley mainly occurs in the area partially affected by the human disturbance rather than the whole area. At least, five types of regeneration areas or zones have been clarified.

(i) Severe degradation zones along the transportation line exist broadly. The road construction is mostly based on the slope cutting in the valley, whereas the soil and rock dust are usually dumped downward along the slope or into the river. The original natural plantation or slope farmland is then covered forming naked ground exposed to the strong natural erosions. Only a small part of the steep slope obtains the preliminary treatment, but the majority does not. Thus, the 40–200 m wide transportation road line often becomes the most severe degradation zone. A 500–600 m wide degradation zone could be seen in the part of the winds around the mountain road. According to the detailed survey of the road along the Minjiang River Arid Valley, the severe regeneration area has reached to 86 $km^2$ accounting for 12 % in the total area (Bao et al. 2012).

(ii) The degradation islands surrounding the villages. The village, the living place for the human beings, distributes among the mountain lands in the Arid Valley as an island form. However, there are varied human activities (such as firewood cutting, rock sampling, manure collecting, land reclamation, house constructing, grazing, plantation, etc.) in the villages also treated as ecological degradation area. Eventually, the ecological degradation island with the core of village has been developed (Bao and Liu 1999; Bao et al. 2000). That island exhibits the degradation of the vegetation structure and productivity, the reduction of the plant sexual reproduction, the decline of the sciophilous plants, soil compaction, the reduction of the biogeochemical cycle ability, and severe water loss and soil erosion. The extent of the degradation of the island is related to the village size, the social economic situation, the vegetation resource surrounding, the topography and landscape and the traditional life style. The village degradation islands vary in the scattering distribution and the degradation area with 0.5–2.0 $km^2$ (Bao et al. 2012). In certain areas, the islands are overlapped with the barren mountain forestation and the road construction. Multi-disturbance regimes make the extent of the degradation worse.

(iii) The degradation zone in the afforestation area. Since the 1980s, the Arid Valley across the Hengduan Mountains has been considered as the key area for the afforestation with the development of series forestation. Forestation and land preparation cut off the slope, disturb the surface area violently, stir the soil, and destory the original plant cover and the soil biological crust. Because strong wind with a pretty high frequency in the valley, the forestation and land preparation would cause severe water loss and soil erosion and the air pollution by means of disturbance. The soil exposed over a long period of time would be further affected by the (heavy) rain and wind erosion, which makes the water loss and soil erosion even worse. The plant cover in many forestation areas could not recover back to the initial condition even after 6–16 years, whereas the newly planted trees become dwarf trees with a very slow growth constrained by the harsh environment (Zhu et al. 2009). Alternatively, even temporary forest establishment could not prevent the degradation of the forest environment and the weak ecological function. Hence, irrigation could lead to local landslide and collapse, which makes the water loss and soil erosion worse. Generally, the steeper the slope, the higher ratio of the naked surface area caused by forestation and land preparation, the worse the plant cover damage, the worse soil degradation, and the worse the water loss and soil erosion. Although a large number of labor power and material resources have been devoted and the provincial efforts have been paid, limited effect is obtained from the barren mountain afforestation in the Arid Valley but causes newly formed degradation. The severe disturbance by forestation and the unique weak environment in the Arid Valley are the fundamental reasons for the afforestation land degradation, besides, the forestation technique and the proper tree species are the other two reasons.

(iv) The heavy degradation zone driven by the hydropower engineering. The Arid Valley across the Hengduan Mountains is also one of the earliest developed hydropower areas in China. Certainly, the hydropower stations take place and damage the land resource, which results in the ecological degradation in certain areas. The hydropower station construction itself could cause landslide, collapse, and mudrock flow by road construction and mountain land cutoff. Consequently, the land and vegetation are destroyed further. Various spoil areas have been developed. A large number of earth and rock are dumped into the river bank or river channel led to higher riverbed, stuck river channel, and varied degradation zones.

On the other hand, the hydropower station construction has damaged extensively the shape, structure, and function of the river resulting in a devastating damage to the biomes. The hydropower scheme and the hydropower station design do not connect closely with the ecological water demand and the connectivity of the river. Moreover, the optimal monitor and sustainable management for the structure and function of the river ecosystem have not even been taken into consideration during post-construction. The river ecosystem is getting worse under the circumstances of the jams and the river division, which makes the river ecosystem even worse. Water division makes the river rock naked, the plants along the river bank wilt and even fatal (Fan et al. 2006).

The river section at a certain length above the floodgate turns into reservoir with changes in the water flow, depth, water quality and the bottom materials in the riverbed. The area between the dam and the station would be to some extent free of water resulting in a large change in the water environment. Static water, which used to be fluid, exhibits the tremendously weak self-purification ability resulting from chemical, thermal, and physical changes. The water quality deterioration and eutrophication have been developed. The domestic sewage pollution leads to severe waste accumulation and the deterioration of water quality, besides, severe reservoir and downstream pollutions.

Certainly, the river environment changed by the hydropower stations would also cause serious damages to the aquatic biome, riverbank vegetation structure, and productivity. During the change of the natural river section above the dam into the reservoir, the number of the species in Cyanophyta and Chlorophyta increases as there is transformation of the rapid flow into slow flow and the accumulation of silt and organic materials. Furthermore, the aerobic and rapid flow like Bacillarophyta disappears; the eutrophicated water like Annelid and Chironomid larvae increase gradually. However, the aerobic and rapid flow like nymphs such as ephemera, caddis, and stoneflies disappear and the aquatic algae appear in the river sections free of water. Especially, the biomass and species of the adherent algae and invertebrates decrease significantly (Fan et al. 2006).

(v) Extreme degradation sites driven by various geological hazards. The Arid Valley is of steep slope, complex topography and landscape, unsteady slope, soft and destructed rock, loose slope surface, overloaded ground, weak stability, more fissures, mudrock flow caused by water infiltration, hazard by unbalanced slope. Therefore, there are abundant varied landslide, mudrock flow, and collapse sites in the valley. The sparse vegetation, naked ground, severe water loss, and soil erosion in this area being difficult for the natural vegetation restoration makes it as the extreme Arid Valley degradation zone across the Hengduan Mountains.

(vi) The extreme degradation sites on steep slope farmland. The Arid Valley across the Hengduan Mountains, the regional farmland area, accommodates a large scale of dry slopes. Dry slope includes terrace on the hillside and slope farmland with varied gradient (15–35 °C). The main plantation mode in that area is the interplanting of Chinese red pepper, beans, corn, and tomato. It has always been one of the most important ways for the local peasants to increase their economic income. The average dry slope farmland per family is around 0.13–0.3 hectares. For a long time, traditional plantation has been applied in this farmland with naked ground from autumn to winter and spring and extensive management in summer. The severe water loss and soil erosion (with both water and wind erosion) and continuous soil fertility degradation of the farmland cause one of the most important ecological degradation problems today. Although the steep area of the dry slope farmland is the key site for the returning after 2000, most areas have not been returned completely. The plantation and severe water loss and soil erosion still exist.

The Arid Valley ecological restoration is used to concentrate on the forestation on the barren mountains and slopes. However, the most severe degradation areas and zones, such as steep slope farmland, landslide area, mudrock flow bottomland, area along the transportation, area affected by the hydropower engineering, etc., have still been overlooked. Thus, it is urgent that the ecological restoration should aim to the degradation areas. Properly evaluate the phenomenon, facts, degree, and the key restraint factors of the ecological degradation, is the prerequisite and basis in succeeding in the ecological restoration, which needs to be strengthened.

### 6.2.2 The Vegetation Origin and the Restoration Aim Selection in the Arid Valley

The species composition of the native plants is the key point for us to understand the regional vegetation diversity, and is the prerequisite and foundation for the protection of the vegetation and its diversity and restoration. The shrub dominated by the dwarf shrub and semishrub characters the current natural vegetation in the Arid Valley (Zhang 1992; Jin and Ou 2000; Yang 2007). Although the shrub with comparatively simple community structure and low productivity, it accommodates abundant native mesophytic and xerophytic shrub, semishrub and herbaceous plants with an exception of drought tolerance forest covering partially (Zhang 1992; Bao et al. 2007, 2012). From the point of view of the floral composition and features, the vegetation in the Arid Valley differs significantly from that of its nearby subalpine type or the higher altitudinal area above it (Jin and Ou 2000; Bao et al. 2012). Bao et al. (2012) indicated that the subalpine vegetation has not nothing to do with that in the Arid Valley in terms of succession, which supports the theory that shrub–grass vegetation is of original nature with local origin. Furthermore, many local unique plants distributed around the Arid Valley could not be found in the nearby subalpine area with higher altitude in terms of the floral specific species. Thus, the vegetation in the Arid Valley is of distinctiveness. The comparative studies about the vegetation transition in the Arid Valley during the past 100–150 years also show that the shrub–grass dominated vegetation type could be still found in the area without human disturbance, which in general, no significant vegetation degradation has been detected. Recently, the analysis of climate changes during the past 200 years in the Zagunao River area in Li Province, Sichuan, China dendroclimatologically also shows that the Arid Valley is always stick to the drought climate, which is of special local character determined by the local landscape (Moseley & Tang, 2006). In conclusion, xerophytic shrub and grass are the typical vegetation types in the Arid Valley as well as the highest plant in the drought (Jin and Ou 2000; Yang 2007; Bao et al. 2012). The unique mosaic xerophytic forest could be found only in some local areas, such as the cypress (*Cuppress chengjiana*) forests in the Minjiang River valley, the evergreen Quercus forests, and the pine forest in the Jinsha River dry-hot valley.

Unfortunately, most large-scale forestations for the forest restoration in the Arid Valley failed during the past 30 years (Yang et al. 2007). Contrary to the hypothesis, afforestation, especially soil preparation, destroys the original natural vegetation and its renewal ability. Furthermore, the soil structure and its function are directly damaged; the water balance gets worse, and the wind erosion is even worse (Wang et al. 2004; Yang et al. 2007; Zhu et al. 2009). It seems that we did not aim adequately to the Arid Valley restoration. Ecological restoration for it should focus on improving the physical environment including sparse vegetation, severe water loss and soil erosion, and fertile soil condition, and protecting and recovering the key resources to enhance the vegetation coverage (including the soil biological crust coverage), not on reconstructing the forest overwhelmingly.

### 6.2.3 Spatial Patterns of the Ecological Key Elements and the Strategies for the Ecological Restoration Planning Across the Arid Valley

The Arid Valley develops from the high mountain-gorge systems. There is generally an individually relatively complete Arid Valley System along each gorge with a near South-North axial across the Hengduan Mountains. Since the broad range of the spatial distribution of the Arid Valley, the natural diversity in climate, soil, and vegetation and the civilian complexity (social structure, economic basis, cultural tradition, etc.) are classified as different types. The Arid Valley is of distinct regional diversity, whereas it is grouped as different climatic types and subtypes including dry-hot, dry-warm, dry-cool, and dry-cold from south to north. Bao et al. (2012) thought that theoretically the ecological restoration and the sustainable management for the Arid Valley post-degradation across the Hengduan Mountains should be based on its regional traits. They also speculated that the most effective way was to develop a strategy compatible to the natural social economy.

Typical cases for the Arid Valley along the Minjiang River have pointed out that the natural renewal abilities of the soil, vegetation structure, diversity, population, and community in that area mainly depend on the degree of drought, i.e., precipitation. Compared to the climate of the transitional zone between the upstream and downstream and the high altitudinal zone in the Arid Valley, that of the core area and the low altitudinal zone exhibits drier, poorer soil water content and fertility, shorter plants, smaller vegetation coverage and soil seed bank, weaker vegetation natural renewal ability (Li et al. 2011). Therefore, the selection of that transitional or higher altitudinal degradation zone could fulfill easily the ecological restoration aims and get better results for its arrangement and strategy.

Practically, proper partition and layout need to be based on the rock property and soil water condition to obtain the ideal vegetation restoration results (Zhang et al. 2003). During the study in the Yuanmou dry-hot valley, Zhang and his cooperator found that the rock property is one of the key factors in determining soil water condition and vegetation type in that area. They thought that it should be considered as an important site during the vegetation restoration. Besides, low slope infiltration ability is another main reason for the dry soil in that valley. They tried to differentiate the vegetation restoration zones in the dry-hot valley based on the rock property and climate types, whereas their development is compatible to the vegetation restoration modes and proper restoration aims for different zones. Consequently, their strategies have made remarkable progresses. Furthermore, vegetation restoration modes for different stone areas and micro-water forestation technique succeeded in practice (Xiong et al. 2005), which makes a good example for studying the vegetation restoration in the dry-hot valley.

### 6.2.4 The Approaches and Methodology for Ecological Restoration

The plant growth and vegetation development in Arid Valley are jointly restricted by drought and infertile soil condition (Bao et al. 1999a; Wang et al. 2003; Wu et al. 2008; Li et al. 2008). The key point of the ecological restoration for the Arid Valley is to improve the plant growth and vegetation development. Water loss and soil erosion, such as wind erosion, would be efficiently cut down only by enhancing the vegetation coverage. Eventually, the ecological restoration achievement would be fulfilled (Bao et al. 2012).

Zhou et al. (2008, 2009) and Li et al. (2010a, b) found that the dominant shrubs in the Arid Valley are of very strong both sexual and asexual reproductive abilities. Those plants with natural renewal potentials result from the well developed seed bank. Hence, vegetation and their renewal potentials protection are the major strategies for the ecological restoration in the Arid Valley. However, the practices in the past did not think the vegetation natural renewal ability in consideration and the human disturbance actions were applied to restore the ecological conditions. Consequently, post-ecological degradation in the target area occurs making the ecological restoration aim hardly achieved (Zhu et al. 2009). Therefore, the real and best feasible way is to utilize adequately the vegetation natural renewal ability in the Arid Valley plus some artificial supplementaries (Bao et al. 2012). The method combining the native plant natural renewal improvement and the appropriate sapling replantation has been partially succeeded in some experiments (Shen et al. 2003; Li et al. 2009).

The degradation in the Arid Valley is caused by intensively strong human disturbance. The activities such as forestation, grazing, and harvesting in the barren mountains has not only destroyed the vegetation and soil structure in those areas but restricted severely the vegetation natural renewal abilities. Afforestation using a large area of the field, itself, would also result in soil degradation and vegetation disturbance. Thus, reasonable control strategies reducing the disturbance pressure caused by human being are the prerequisite and guarantee for the ecological restoration in the degenerated Arid Valley. Studies have showed that the proper use of enclosure strategy could inhibit the Arid Valley degradation as well as significantly enhance the biodiversity recovery, the productivity improvement and the efficient control for water loss and soil erosion (Shen et al. 2003; Luo and Wang 2006; Li et al. 2010a, b). In conclusion, reducing the human disturbance pressure by all means is one of the most indispensible and effective methods for the ecological restoration in the Arid Valley across the Hengduan Mountains.

## *6.3 Techniques for the Ecological Restoration in the Arid Valley*

We have sorted out some technical measures based on the practice and experimental studies of the ecological restoration in the Arid Valley during the past few decades.

### 6.3.1 Suitable Plant Selection for the Ecological Restoration in the Arid Valley

In the past few decades, more than 50 tree species were used for the ecological restoration practice in the Arid Valley. These trees did not show the ideal growth based on the survey on the Arid Valley in the northern section (Dadu River, Minjiang River, Bailong Jiang River) (Bao et al. 2007), suggesting moderate to high phanerophytes (>3 m), especially large trees were difficult to adapt to the environments in the Arid Valley. Afforestation using native tree species Cupressus chengiana made some short-term effects in their natural distribution, such as the Arid Valley along the Dadu River and the Minjiang River (Wu et al. 2006; Zheng et al. 2007), but it does not succeed in the long run (Zhu et al. 2009).

In the hot-dry valley along the Jinsha River, tests for screening the afforestating species showed that the majority of the fast-growing exotic species can be more well adapt to the environment through quickly covering the surface, but its long-term ecological effects lack systematic studies. However, Wang et al. (2004) and Li et al. (2007) have found that afforestation using exotic species reduced soil moisture and the native biodiversity. On the contrary, restoration by native shrubs or small trees, such as slope willow, *S. davidii*, *V. negundo*, *C. macyocarpa*, *A. kalkora*, *P. emblica*, and *D. obtusifolia* has showed good results in the valley (Yang et al. 2007).

Studies indicated that species with larger individuals and a great amount of transpiration per unit of time was not adapt to the Arid Valley environment. Therefore, native xeric shrubs and herbs have more advantages than the trees to restore vegetation from degraded habitats in the Arid Valley. Life forms, growth forms, and leaf traits are integrated for plants to adapt to their environment reflecting the adaptive strategies to some extent. Vascular plants in the Arid Valley were characterized by the traits of monocotyledons, leaflets, herbaceous, deciduous, etc. Natural vegetation mainly was constituted by phanerophytes with dry leaflets or deciduous microleaf or hemicryptophytes (semishrub or perennial herb) (Liu et al. 2007; Bao et al. 2012). These plants have small seeds and great vegetative propagation, by which better adapt to the Arid Valley. Thus, plant species for the ecological restoration in the Arid Valley should be primarily selected from native shrubs, semishrubs, or herbaceous plants (Bao et al. 2012).

### 6.3.2 Ineffective Model of Site Preparation for Afforestation

Over the past 30 years, afforestation and the ecological restoration have been carried out to alleviate environmental degradation in the Arid Valley across the Hengduan Mountains (Yang et al. 2007; Bao et al. 2007). Afforestation practices from the Jinsha River and the Minjiang River Basin indicated that although field preparation for afforestation formed forests temporarily and the target tree species grew well, it brought significant ecological degradation (Wang et al. 2004,

Zhu et al. 2009). Field preparation for afforestation is not successful for the ecological restoration in the Arid Valley.

Wang et al. (2004) found that the woodland exhibit "soil drying" based on the water environment monitoring on the artificial vegetation in the Jinsha River valley. Soil moisture content of native slope willow shrub was 42.68 % higher than that of the woods, whereas it was also higher in natural grass than those in the woods (34.36 %) and shrub (22.22 %). Afforestation has caused deterioration slope hydrology in the hot-dry valley, and brought far-reaching consequences for the regional ecological environment. In the Arid Valley along the Minjiang River, Cupressus chengiana still did not adapt to the environment even after 7–16 years in the measure of the contour level trench field preparation with this native tree species. Its survival rate has reduced with afforestation, and high growth or diameter growth also has shown declining trend. Most trees fruited with small size, which resulted in the phenomenon of dwarf trees. Meanwhile, the contour level trench field preparation reduced vegetation cover and soil water conservation against the exception of the afforestation. Thus, the preparations for afforestation were not effective measures for the ecological restoration in the Arid Valley along the Minjiang River. It did not reach the expected aim for the vegetation restoration, and even exacerbate the system degradation.

Soil crusts are common in the Arid Valley across the Hengduan Mountains, including soil physical and biological crusts. Physical crust generally refers to a thin layer of soil crust caused by the raindrops splashing, soil physical and chemical dispersion, or deposition of fine particles when water flowing through the soil surface. Physical crust is an important protecting mechanism for soil weathering. Biological crust is cemented by biological components including different species of mosses, lichens, algae, fungi, bacteria, etc. with their underlying thin soil particles. It is an important cover type in the Arid Valley, which covers up to 40 % of the total crusts. Soil crusts play an important role in the desert landscape processes, soil ecological processes, soil hydrological processes, soil biological processes, and geochemical cycling, and ecological restoration in the arid and semiarid regions in different bioclimatic zones (Li et al. 2009). Afforestation practices (especially field preparation) destroyed the original natural vegetation and its regeneration ability as well as the soil structure and function. Most soil crusts that formed under drought conditions are also damaged causing serious soil erosion.

Field preparation techniques in the Arid Valley also need more data. Field preparation aims to improve the soil structure, intercept rainfall and the soil moisture content, but it also damages the surface, exposes the soil, and causes water erosion or wind erosion. Field preparation can play an important role in mountain region with greater rainfall and stronger resilience by significantly reducing soil erosion and improving forest vegetation restoration efficiency. However, in the Arid Valley where there are less precipitation events, small rainfall in a single event and few event of moderate to heavy rain, the contour level trench field preparation resulted in less result in improving the soil moisture. On the contrary, the negative effect of the contour level trench field preparation was amplified in the Arid Valley characterized by the low coverage and serious erosion and further resulting in the ecological degradation (soil degradation and destruction of vegetation). Thus, natural conditions in the Arid Valley should be considered during vegetation restoration, whereas new techniques require a scientific evaluation of its feasibility before utilization. In addition, most afforestation plan does not design appropriately with proper management strategies, including monitoring, analysis, evaluation, and adjustment programs. Short-term survey and evaluation are conducted after the implementation of afforestation, and its purpose is not to assess the effect of the ecological restoration, but the survival rate of seedling and growth status, making an unsuccessful practice which may lose the opportunity for correction.

From a technical way, the human role was overemphasized during the current practice of the ecological restoration, but the attention on natural restore capacity is seriously lack in promoting replantation. Over the past 50 years, restoration aims to restore the forest primarily (including economic forest) using two approaches: seedling cultivation and seeding, which ignores the ability of the natural vegetation restoration (Bao et al. 2007). Therefore, tree planting patterns are not feasible in the Arid Valley due to the inherently contrary to the natural rule of the succession and mismatch to the local unique ecological conditions from a technical point. Meanwhile, this pattern is completely unrealistic for weak economic condition, which cannot afford the high cost of afforestation projects in a wide range of the Arid Valley.

Afforestation is characterized by "generalization" without practical focus on the degradation in the Arid Valley. The planting density is not rational, and it is over 73 hectares in most regions of the Arid Valley. High density means more soil water consumption, and this condition is exacerbated by the uneven distribution of rainfall with a few rainy days. Further, poor soil structure and severe water deficit are contributed to the soil moisture disappearance through transpiration during afforestation.

In the future, vegetation restoration in the Arid Valley should be based on the clarification of the factors and mechanisms, and minimizing human disturbance of high intensity. Restoration would achieve the goal of the vegetation and soil benign ultimately by promoting the restoration and the reproduction of the natural vegetation and improving the vegetation cover and soil quality (Yang et al. 2007; Zhu et al. 2009).

### 6.3.3 Effectiveness of Seeding and Seedling Transplanted with Native Plants

Seeding is an important means of the vegetation restoration. Native shrubs and grasses are selected and the seeds are planted in the different sections and different types of microhabitats in the Arid Valley along the Minjiang River (Li et al. 2009). The results from seeding experiments showed that the germination percentage is very low, whereas the mortality rate of the seedlings is fairly high under the arid condition. Compared with the core area in the Arid Valley, more seedlings emerge and survive for a longer time at the transition zone due to the greater precipitation and higher humidity in the air and soil. In the same habitat condition, significant differences of the germination rates are observed among species. Seed germination and seedling survival are relatively high of the *Bauhinia faberi* var. Microphylla, *Indigofera lenticellata* and *Oryzopsis munroi* indicating a stronger capacity to tolerate the environment stress in the Arid Valley. These tree species are more appropriate to adopt the seeding approach for vegetation restoration practices. *Sophora davidii* and *Miscanthus szechuanensis* emerge difficultly under natural condition. Consequently, seedling transplanting would be a better approach for the vegetation restoration. The peak of seedling appearance is within 1 month after sowing, but the survival number of seedlings decreases a month later with no seedlings surviving till August. The results suggest that the appropriate planting time is critical for the vegetation restoration with the native species. Sowing should be conducted in April, thus seedling could form woody structures and consequently show the stronger ability to resist hot and dry stresses in July and August. Sowing in the container can significantly improve the seed germination in a short period of time, but it ultimately failed to effectively improve the survival rate and the growth rate of the seedlings.

Transplanting seedlings has better effect for the vegetation restoration compared with seeding in the case of two rose species (Bao et al. 2012). Seed germination percentage was higher, but the emergence rate was very low, with most seedlings dying within one month for both rose species. Seed germination and seedling survival did not differ significantly under each microhabitat conditions. Transplanting seedlings has relatively low overall mortality (>15 %), especially for biennial seedlings with mortality less than 5 %. Further, seedlings grew well throughout the growing season after transplanting. Thus, transplanting seedlings is one of the feasible vegetation restoration measures compared with seeding. Habitat type also affects the seedling growth and the biomass accumulation, but it does not affect the seedlings survival. Seedlings grew more rapidly in bare land indicating that the plant seedling in the bare ground is more conducive to increase the plant cover. Compared with the 1-year-old seedlings, the 2-year-old seedlings have a higher survival rate.

### 6.3.4 The Appropriate Amount of N Fertilizer Can Promote the Growth of the Transplanted Seedlings

Soil nutrient supplement by fertilization is a common way to ensure nutrients requirement and to promote plant growth in the poor area. Soil available nutrients supplied by fertilizing will change all aspects of the plant adaptation strategies, whereas the adaptation of the growth and the morphological characteristics are often the most basic mechanisms. Since N is an important component for the enzyme, plants should balance the resource that allocate N to photosynthetic organs or organs to absorb nutrients. Due to the complementation between limited factors, N fertilization can promote seedling cultivation in dry conditions. We designed the experiment with the varied moisture and nitrogen fertilization gradient, which the purpose is to investigate if the nitrogen fertilizer could enhance the ability of plants to withstand drought stress. Two-year observations showed that the application of N cannot completely change the inhibitory effect by the drought stress for *S. davidii* seedlings, but moderately applied N (92 mg N/kg soil) can alleviate the limitation of drought stress on the plant growth to some extent. Moderate N application increased soil N availability, consequently affecting the seedling growth and their physiological characteristics. It improved the plant structure and the resource allocation pattern and the photosynthetic capacity, whereas the efficiency of other limited resources (such as water and P) was also increased. However, excessive applied N (184 mg N/kg soil) did not improve the plant growth under the drought stress. However, it made the damage of drought stress on plant more serious indicating a tradeoff between the effects of the moisture and the nutrient on plants.

Therefore, appropriate N application can improve the soil environment, then the plant growth and the population regeneration when a pioneer plant species of *S. davidii* used for the vegetation restoration in the Arid Valley (Wu et al. 2008).

### 6.3.5 Multifunctional Plant Hedgerow Was Effective in Preventing the Soil Erosion

Slope farmland is an important area with degradation in the Arid Valley. How to reduce the soil erosion, restore the soil fertility, and improve the production efficiency is a key issue in the Arid Valley across the Hengduan Mountains. The climate is characterized by the obvious seasonal shifts, distinct wet and dry seasons with precipitation mainly in summer (April to September). Therefore, in terms of the slope land without vegetation cover, soil erosion is extremely serious and forms a great contribution for the river sediment. Artificial Hedgerow with legumes cross slope along the contour not only can significantly abate erosion and curb the deterioration of the soil fertility, but improve the soil fertility through nitrogen fixation by legume. This creates good conditions of water and fertilizer for the crop growth. Tang and his team have developed a hedgerow mode in the mid-1990s in the Jinsha River Basin. Their studies showed that the hedgerow has great ecological efficiency in scattering stagnated surface runoff, reducing the flow rate, increasing the infiltration and intercepting the sediment. Hedgerows with crops have some economic benefits, except for the ecological benefits in improving the microclimate between the hedgerow lines and providing a good environment for interrow crops (Sun et al. 2001; Tang et al. 2002). Compared with traditional terraces, the biggest advantage of the hedgerows is low cost, simple, and practical. In the Arid Valley along the Minjiang River, the contour hedgerow with *Astragalus adsurgens* also has proved that the hedgerow with shrub and grass, legume achieved better results at the degraded shrub slope and landslide area.

### 6.3.6 The Ecological Restoration Supported by Cash Crop Cultivation and the Agroforestry

The Arid Valley across the Hengduan Mountains is densely populated with the agriculture dominated by farming, which account for more than 40–60 % of the household income. The contradiction between people and land is outstanding causing serious unreasonable land use and environmental problems. The exploitation of the unique plant species is the fundamental way for the local people to get rich, and thus promotes the social and economic development and supports the ecological restoration and protection (Liu et al. 2000; Liu and Bao 2001). In fact, intercrop, agroforestry, orchards, the construction of the special plant resource basis, and its industrial development have been the main ways for socioeconomic development in the Arid Valley across the Hengduan Mountains (Bao et al. 1999a, c). Agricultural models in this region can be roughly summarized as follows: fruit and vegetable, fruit and medicine, fruit and livestock, and fruit and grain (Xiang et al. 2007).

Three-dimensional planting mode (e.g., planting vegetable under or between the fruit trees) is of high economic and ecological benefits in the Arid Valley based on the special climatic and terrain conditions. The number of new varieties of vegetable species has increased, whereas the area of planting has expanded. Crisp plum, sweet cherries, and golden delicious apple have achieved great economic benefits with the net income per hectare from 30,000 to 120,000 CNY per year. This mode makes the farmland covered with plants all the four seasons and the vegetation cover rate is more than 99 %. It also enhances the biodiversity in the farmland, which is a preparation for the pollution free agriculture model.

Exploitation of special plants developes rapidly for their obvious ecological and economic effects. Special plants are rich in the Arid Valley, such as pepper, sugar cane, raspberry, rose, and tamarind. Pepper has the greatest cultivate area mainly in the Arid Valley along the Dadu River (Hanyuan, Xichang, Mianning, etc.) and the Minjiang River (Maowen, Jiuzhaigou, etc.) (Cui et al. 2008). Pepper industry has become a pillar industry locally. Planting pepper has also increased the forest coverage rate effectively, and thus has controlled the soil erosion and desertification. Recently, rose planted area has increased rapidly in the Arid Valley, Panxi. The latest international fashion flavor varieties Damascus has been introduced and planted. Damascus Rose III has been successfully cultivated, and the techniques for the seedling efficient production and the rose oil industrial development were applied. Currently, the average yield of oil rose buds is 600 kg, with the average net income per hectare up to 768 CNY. Rose industry has an important role in promoting the local economy and ecological effects such as preventing the soil erosion.

Special planting and agroforestry are effective models for the economic development compatible to the special environmental condition in the Arid Valley across the Hengduan Mountains. It promotes the economic development and poverty alleviation by solving the problem of the labor surplus and firewood prevention (Bao et al. 1999b, c, 2007, 2012). Furthermore, special planting and agroforestry are very effective in solving the contradiction of the economic development and the environmental protection. It also provides a useful exploration for the environmental construction in the degraded regions with high population density.

## *6.4 The Ecological Restoration and the Sustainable Management Strategy for the Arid Valley*

After 30 years research, the scientific strategy for the ecological restoration and sustainable management of the Arid Valley across the Hengduan Mountains have been preliminarily established based on the experiment and practice on ecological engineering.

### 6.4.1 Ecological Spatial Pattern Is the Scientific Basis for the Ecological Restoration and the Sustainable Management in the Arid Valley

The The Arid Valley across the Hengduan Mountains mainly consists of the Bailong River, Minjiang River, Dadu River, Yalong River, Jinsha River, Lancang River, Honghe River, and their tributaries. These rivers distributed extensively with varied natural conditions. The causes and impacts of the ecological degradation across these drainage basins are quite different. Due to the varied natural and social economic conditions, the sort management is the key point for the ecological restoration (Bao et al. 2012). The project should take the basin as a unit, types of the The Arid Valley as the basis, the ecological restoration and sustainable management as the fundamental points. Moreover, identifying the spatial pattern of the natural and social elements in the The Arid Valley is the scientific foundation to develop the strategies for the ecological restoration and sustainable management.

### 6.4.2 Land Use Planning and Classification Are Important Steps for the Ecological Restoration and Sustainable Management in the Arid Valley

Population is gathering in the Arid Valley, whereas the land is a limited resource. The main point driving the ecological degradation here is the demand for the social economic development. Whereas, the land use type and degree are quite different. Farmland and barren hill are the two main types for the land use in the Arid Valley which account for more than 90 % of the total area, among which the area of woodland is generally more than 90 %. Other land use for construction or for traffic is generally transformed from the two types. Farmland comprises the grain land, the agroforest land, the orchard, and the vegetable land. Moreover, the woodland mainly includes grass, shrubs and the farmland returned (patch forest), etc. Animal husbandry is mainly depended on the woodland but not specifically defined to the land ownership. Therefore, identification and the sort management for the land use types are necessary for the ecological restoration and sustainable management in the Arid Valley. Furthermore, comprehensive planning and sort management should be made in the unit of basin. The restoration goals, methods, and detailed techniques for each kind of the land use types must be clarified.

### 6.4.3 Clarify the Goals and Methods for the Ecological Restoration and Sustainable Management in Different Land Use Types

According to the 30-year research and practice in the Arid Valley, the goals for the ecological restoration in the barren hills are supposed to regenerate vegetation of grass, shrubs, or shrub savanna rather than the woodland (Chai and Fan 2001; Jin and Ou 2000; Bao et al. 2012). Furthermore, the reconstruction of the forest patches can be achieved within local key regions (Bao et al. 2012). Specifically, the project should aim to increase vegetation cover and decrease soil erosion. Ecological conservation is supposed to be the major method in the large area and supplemented by the necessary artificial restoration with native grass, shrub and small tree species in important local regions. For the farmland, the project should focus on monitoring the soil erosion and the degradation quality with intensive management. We try to increase the productivity, benefit and income in the farmland, and then promote the social economic sustainable development in the Arid Valley. Local characteristic plant and animal products should be promoted with great efforts. The production efficiency can be raised and the value chain can be extended by large scale based on the construction and intensive management.

## *6.5 Prospects*

The fact is that the ecosystem in the Arid Valley across the Hengduan Mountains had degraded completely in past three decades. It has been developed a great deal of theories and practices for the ecological restoration. First, the ecological restoration styles should be developed properly in the Arid Valley. Second, the strategies have been well known for the ecological restoration and sustainable management. However, this question has not yet been answered, which should be improved as following:

### 6.5.1 Deep Understanding About the Ecosystem Structure and Function and the Regularity of the Spatial and Temporal Patterns of the Arid Valley Across the Hengduan Mountains Is Required

Because of the poor understanding, there are not enough information obtained sufficiently in the vegetation succession and the natural resilience. Therefore, it limits the formation and the improvement for the vegetation restoration technology in the degradation area. In this area, natural vegetation is sparse, whereas the development of the soil biological crusts is covered. The soil biological crust could monitor the water loss and soil erosion, fix and storage the organic carbon, which is the starting point during the primary vegetation succession and is important for the ecological protection. However, the structure, the function, the spatial distribution, the ecological effect, and the methods for the restoration and reconstruction for the soil biological crust in the Arid Valley across the Hengduan Mountains are still largely unknown. Especially, the research facilitates the afforestation and vegetation replantation under the different interference intensity. All of the results from those work can provide the science and technology support directly for the natural vegetation protection and sustainable management in a large scale.

### 6.5.2 Ecological Protection for the Endemic Plant Species Fit for the Drought Conditions, i.e., Being Adapt to the Arid Valley is Needed

These species are different from those from subalpine areas. We knew well that the resource trait rather than most plants still lacks enough attention. Therefore, it is urgent to strengthen the theoretical and technological research for these plant resources including the plants for industrial raw materials, flowers resources, fruit and vegetable resources. Moreover, we should develop the industry with further processing of the resource, which can not only enhance the regional economy but protect the native plant resources. Beside, there are approximately more than 500 different kinds of native plant species with highly scientific research value. However, only a few plant species have been protected, whereas others, on the contrary, have not got enough attention and resource restoration in the Arid Valley. Thus, investigation and survey for the endemic plants in the Arid Valley should be done as soon as possible since it will provide a scientific basis for the native plant protection and the ecological restoration.

### 6.5.3 Ecological Restoration Priority to Some Key Degraded Areas

Due to the high cost of the ecological management in the Arid Valley, the investment is limited. We need to select some sensitive areas or zones for the prior rehabilitation. The environmental problems are of prominent seriously restricting the socioeconomic development and human productivity and daily life. Hence, the ecological restoration for the damaged ecosystems with sensitivity and ecological fragility should be conducted first. Because of the damage to the ecological environment, the negative effects generally not appear visibly and immediately. Studies have shown that earthquake damage caused abundant difficulties in the arid ecosystem, in which the restoration is more complex than any other kinds of damaged ecosystems. Ecosystems, especially in the fragile areas, would show the amplification and the chain expansion effect as long as they are destroyed. If the ecosystems are not recovered and management immediately, the damage will be expanded and the degree of degradation will aggravate. Further, once we miss the best restore period, postmanagement and investment will be greater with the poor results. In this case, our country has got many painful lessons. The transportation line, ground water protection sites, lakes, rural settlement sites, and the scenic spots in the Arid Valley are the urgent and important targets for the restoration and management. These areas present serious water loss and soil erosion, vegetation degradation, disasters with high frequency. The greater the threatening is, the worse the economic loss is.

### 6.5.4 Sustainable Utilization of the Animal and Plant Resources and Further Processing of the Key Technologies

Establishing the special resources production base, value-added products and resource utilization are urgent in terms of practice. In future, we should focus on the direction of the efficient use of the resources, the increase of the income and economic growth. The current resource exploitations in the Arid Valley across the Hengduan Mountains are mainly classified as two aspects:

(i) The majority of the plant-based products and raw materials are primary products with limited economy value. We should gradually shift from the production of the raw materials into the production of the further processed products. Furthermore, the development of the deep processing products, the further resources utilization, and greatly improvement of the utilization efficiency should also be taken into consideration.

(ii) Abundant plant resources lack complete utilization synthetically. For example, the byproducts of walnut oil comprise low fat food and beverages, pastries, and so on. Additionally, the utilization of walnut including its peels, leaves, shells, etc. is to produce the pigment, lignin and cellulose, activated carbon, etc. In addition, we should develop the products from the walnut peel and leaves for the chemical studies. Walnut spikes and other organs can also develop food and crafts. Besides, organic vegetable production, further processing and other key technologies are still in urgent need for promoting the vegetable base construction and product diversification, and providing the sustainable development of the regional vegetables efficiently.

# 7 Ecological Restoration of Abandoned Mine Land in China

## 7.1 Introduction

The exploitation of mineral resources not only provided the energy and raw materials for the development of the national economy, but also caused the destruction of land and ecological environment such as land subsidence, solid waste, and geological disasters; in the western ecological fragile area, coal mining also resulted in the intensifies of soil erosion and depletion of groundwater resources; the toxic gases produced by the spontaneous combustion of the coal waste plies and the harmful substance leached from the coal wastes jointly polluted the air, water, and soil in the mining area, bringing serious harm to the industrial and agricultural production and people's lives (Peng 2009).

Land reclamation and ecological restoration is an effective way to use land resources economically and achieve harmony between people and land in mining area. In our country, large-scale and organized ecological restoration in mining area began in the 1980s. So far, significant progress has been achieved in the ecological restoration of mining subsidence land, excavated land, occupied land, and other fields, providing a theoretical basis for the government to formulate and implement environment protection strategy in mining area. At the same time, also providing technical countermeasures for coal enterprises to implement the ecological restoration of abandoned mine land.

## 7.2 Ecological Restoration of Abandoned Mine Land

Abandoned mine land referred to the land that were destroyed in the mining process and could not be used without restoration or the land use function declined, including the open pit, waste dump, tailings pond, subsidence land and the land that lost economic value in use caused by heavy metal contamination.

The ecological restoration of abandoned mine land was a complex system engineering problem, which was related to many disciplines, such as ecology, geology, mining, soil, crop cultivation, forestry, environment, aesthetics, agronomy, geography, land, and so on.

### 7.2.1 Ecological Restoration of Abandoned Mine Land Caused by Surface Mining

The destruction to land caused by surface mining was most direct and obvious because surface mining needed to strip the topsoil and rock above the coal seam (Hu 1996). The land destruction types of surface coal mine included excavation, occupation and pollution, of which excavation and occupation were most direct (Hu 1995a).

The ecological restoration of the waste dump of surface coal mine included three major parts: landform reshaping, soil reconstruction, and vegetation recovery.

The landform reshaping system of the waste dump of surface coal mine included the base construct, main part construct, platform construct, and slope construct. Wei et al. (2001) described the concept of the soil reconstruction method of outer waste dump platform. Hu (1997) proposed the soil reconstruction method named "layered peeling, staggered backfill". Staggered backfill was the core principle of soil reconstruction theory (Hu 1997). The mode of vegetation recovery that reasonable configuration of grass, forestry and agriculture and the combination of short-term and long-term benefits could be used as well as the mode called "grass first" and "grass main" in the ecological restoration of the waste dump. Wei et al. (2004a, b) analyzed the effect of reducing stream and sediment of different vegetations and their configuration modes of south waste dump in Antaibu surface coal mine and proposed suitable vegetation improvement measures. Tai et al. (2002) found that sea-buckthorn was the ideal reclamation plant of the waste dump of surface coal mine in grassland area.

The ecological restoration effect of the waste dump of Pingshuo surface coal mine in Shanxi province was shown in Fig. 23.

### 7.2.2 Ecological Restoration of Abandoned Mine Land Caused by Underground Mining

Ecological restoration technologies of mining subsidence land include planning technology, reclamation engineering technology, ecological agricultural reclamation technology, and biological reclamation technology.

The improvement plan of mining subsidence land belonged to the special plan in the land use planning system, and earlier research began in the 1990s (Hu et al. 1994). Besides the traditional planning techniques, the principles, and methods of landscape ecological planning were also applied in the reclamation planning process of mining subsidence land.

After more than 20 years of research and practice, the engineering technology system of ecological restoration of mining subsidence land in China has been initially formed, including the land reclamation technology with the mud pump, towed scraper and excavator, land leveling, dredging and draining method, filling reclamation with the coal gangue, fly ash, lake mud, and so on (Hu et al. 2008).

There were many types of ecological agricultural reclamation technologies, material recycling type of amphibious exchange and complementary was the most typical one. Biological reclamation technology referred to restore the soil fertility and biological production capacity, which was also considered as the continuation of engineering reclamation and an integral part of the land reclamation process.

The ecological restoration effect of mining subsidence land of Tangshan coal mine in Hebei province was shown in Fig. 24.

**Fig. 23** Ecological restoration effect of the waste dump of Pingshuo surface coal mine in Shanxi province

### 7.2.3 Ecological Restoration of Coal Waste Piles

The ecological restoration of coal waste piles mainly included three key stages: analysis and evaluation of site conditions of coal waste piles, reshaping and soil preparation and vegetation recovery.

Numerous studies showed that the coal waste pile had coarse particle, large porosity, high permeability coefficient, barren nutrient content (Hu 1995a; Duan et al. 1999; Wang et al. 2008a, b), low field capacity, low wilting coefficient, low cumulative evaporation (Duan et al. 1999; Wang et al. 2008a, b), and other features. The main limiting factors of the ecological restoration of coal waste piles were texture, moisture, nutrients, pH, salinity, surface temperature, heavy metals in turn (Wei et al. 2009).

The selection of reshaping form of coal waste piles and design of mountain roads, drainage systems and erosion-resistant slopes were introduced in detail by Zhang et al. (1997). Based on the ecological restoration practice of coal waste piles, the cave-shaped site preparation and terracing land preparation were used more often in our country (Hu 1995c).

Due to the limiting factors for the growth of many plants, matrix improvement, selection of greening species, vegetation recovery planting and scientific tending management should be implemented. The matrix improvement technologies of coal waste piles mainly included physical, chemical and biological improvement measures. And the greening of coal waste piles should follow the principle named "Greening comes firstly, Economics comes secondly". The good varieties with the resistance to drought and barren, strong germination force, high survival rate, fast growth, especially the indigenous plants, would be the first choice of pioneer plants, as well as the plants with developed root systems (Hu 1995b).Vegetation community structure should simulate the natural vegetation structure which was stratified mixed with the trees, bushes, and grass (Li et al. 2006b).

Vegetation planting technology of coal waste piles included covering soil planting, no covering soil planting (Li et al. 2006a) and drought-resistant planting technology (Hu et al. 2006). For the poor site conditions of coal waste piles, it is

**Fig. 24** Ecological restoration effect of mining subsidence land of Tangshan coal mine in Hebei province

better to use the planting holes on the next season or interval season after excavation (Li et al. 2006b).

The forestation effect of coal waste piles of Wangzhuang coal mine in 1991 and 1994 was shown in Fig. 25.

## *7.3 Conclusion and Discussion*

The ecological restoration of abandoned mine land is a complex system engineering problem with characteristics of comprehensiveness and regional differences, which is related to many disciplines, such as ecology, geology, mining, soil, crop cultivation, forestry, environment, aesthetics, agronomy, geography, land, and so on. Therefore, intensive research should be carried out by the ecological and environmental scientists, in conjunction with the scientific and technical personnel in mining. Judging from the current situation, the following theoretical and technical issues need to be solved: (i) Basic theories of ecological restoration of abandoned mine land. (ii) Ecosystem succession process and mechanism of abandoned mine land. (iii) Structure optimization of land use of abandoned mine land. (iv) Key technologies of ecological restoration of abandoned mine land. (v) Establishment of ecological restoration demonstration bases.

# 8 Restoration of Declined Coastal Ecosystem

## *8.1 Introduction*

Coastal wetlands play an important role in nutrient cycling, sediment accretion, pollution filtration, and erosion control in the world. In addition, they are known for their distinctive flora and rich spectrum of wildlife, especially waterfowl, which

**Fig. 25** Forestation effect of coal waste piles of Wangzhuang coal mine in 1991 and 1994

makes them more valuable and more prone to human impact than other ecosystems (Costanza et al. 1997; Mitsch and Gosselink 2007). However, only a small percentage of the original wetlands remain around the world after over two centuries of intensive development and urbanization. With so many wetlands lost, it seems that there are many opportunities for wetland restoration along coastal lines, rivers, lakes, etc. In the wetland restoration process, ecological engineering is an important strategy to follow, as it is designed with natural components and strives to achieve balance between human beings and nature. Many case studies have proven that ecological engineering obviously benefits the wetland restoration and promotes sustainable development for districts, countries, even the world. For example, Spartina ecological engineering in China designed by Chung and Qin has gained much benefit and its ecological-economic value had been estimated at US$ 20,000/ha (Qin et al. 1997, 1998; Chung et al. 2004). Much saline land was deteriorated wetland for many formation reasons and favorable to halophytes and salt-tolerant plants. The research on salt-tolerant plants and halophytes is very important, because these plants could be used for saline agriculture and biomass energy on saline land, and for overcoming the worldwide problem of food shortage and energy crisis. A new series of papers reflects the research in this field and shows the potential of salt-tolerant plants and halophytes (Wang et al. 2008a, b; Li et al. 2010a, b; Zhang et al. 2011).

This study regards following issues: (i) Biological substitution and Spartina ecological engineering; (ii) Halophyte and saline agriculture.

## 8.2 Restoration and Reclamation of Declined Coastal Ecosystem

### 8.2.1 Spartina Ecological Engineering and Biological Substitution

*Spartina alterniflora* is a perennial deciduous grass which is found in intertidal wetlands, especially estuarine salt marshes. It was introduced into China in the 1979 by Dr. Chung-Hsin Chung as an efficient countermeasure to solve problems of disasters derived from storm, soil erosion, tidal pollution, even sea level rise, etc. (Chung et al. 2004).

Due to the high nutrient content in *S. alterniflora* shoot, 11.85 % crude protein, 2.26 % fat, 27.1 % cellulose, 0.235 % calcium, 0.191 % phosphorus, and various amino acid, vitamin, and trace elements, its fodder use for stock breeding has been studied in China for many years. Living on the ecotone between terrestrial and oceanic ecosystems, *S. alterniflora* contains not only rich nutrients as mentioned above, but also high content of submetabolic materials, e.g., flavonoids, vitamines, etc. which benefit human health. According to principles of ecological engineering, we invented "Biomineral Liquid" extracted from *S. alterniflora* and produced a series of green health food, such as "Weduo Beer" (with multi-trace elements), "Spartina Capsule for Lipid Falling", etc. All these formulated the important loops of Spartina ecological engineering and favored its hierarchical utilization (Qin et al. 1998).

However, Spartina is an invasive species in China, and has got many ecological problems, including competition with native species and preventing propagation of the pure native strain, impacting of aquaculture and fishing, navigation, etc. In Chongming dongtan, China, Spartina is a threat to birds in estuaries.

Considering the positive and negative effects of *S. alterniflora*, we prefer to take measures of ecological regulation based on its integrative use and biological substitution, rather than eradicate and kill the exotic invader with some extreme methods including herbicides. It is nearly impossible to eradicate a species in its favorite habitat entirely. An old Chinese saying states that "it cannot be burned to death and it will tiller when spring comes". Furthermore, when you use herbicides to kill it you will pay high prices not only economically but also environmentally.

Several means of control and eradication have been employed against *S. alterniflora* where it has become a pest. In our study, we tried an ecological engineering of substitution of Spartina by *Phragmites australis* on salt marshes in North Jiangsu, China. Phragmites is a native species in North Jiangsu and a habitat for endangered species red-crowned cranes and some other waterfowls. In our experiments, we succeeded in substitution of Spartina by *P. australis* about 50 ha in 4 years (Wang et al. 2008a, b). This situation occured via modifying the micro-geomorghic and hydrological regime of salt marsh, when gradient change of elevation, tide strength, and salinity took place in the habitat. When natural or artificial disturbances take place above variables could experience a dramatic change, and the normal vegetation succession could make a turn, even biological substitution could happen. For example, in *S. alterniflora* marsh of North Jiangsu where we constructed a rotund cofferdamed area with 55 $hm^2$, tides and sea water were obstructed, rain water was accumulated and the salinity of the soil solution decreased in the cofferdamed area, resulting in fast growth of the native Phragmites community and substitution for *S. alterniflora* to the end.

To study the involved mechanism of this substitution, we focus on the aspect of allelopathy, using water extraction and decomposing products derived from the litter of *P. australis* to investigate the effects on growth of *S. alterniflora* and a unique fungus strain isolated from the topsoil of Spartina community in North Jiangsu, China. A series of experiments proved that the fungus benefits growth of Spartina. The results showed that both water extraction and decomposing products of *P. australis* decreased the germination ratio and seed-sprouting speed, inhibited the growth of young shoots of *S. alterniflora*, and inhibited the growth of unique fungus strain significantly.

To further identify the unique fungus isolated from the topsoil of Spartaina community, we deduced the taxonomic status of the strain according to the results of plate culture observation, PCR amplification and sequencing. The results showed that this strain presents the general characteristics of Mucorales morphologically, and molecular identification showed that the strain has the nearest genetic distance with two species of soil fungi, one is Saksenaea vasiformis (Mucorales-Saksenaeaceae), and the other is *Apophysomyces elegans* (Mucorales-Mucoraceae). Based on those evidences, we concluded that the strain is a new family under the Mucorales order.

Our experimental results provide direct evidence that allelopathy involved in the substitution of Spartina by Phragmites community. Further work on the identification of allelopathic substance and characteristics of unique fungus will provide further insight on the precise mechanisms of this community succession.

### 8.2.2 Halophyte and Saline Agriculture

Reclamation is another option for the restoration of coastal ecosystem, especially for that of saline land in coastal region. Seashore mallow (*Kosteletzkya virginica*) is a perennial salt-tolerant economic plant, with oil, fodder, ornamental, and even medicine uses (Ruan et al. 2008). Over 20 years of studies of our Lab on this species has proved that the plant could be used as the preferred pioneer in the restoration of saline land, and benefited the amelioration of soil properties and structure (Zhou et al. 2010). Its flower type is beautiful, big, and much more flowers concentrates in its stem, so this is a major species for landscape prettification of coastal region. The seed from bred lines is bigger, with around 24 % of crude protein, and over 20 % of oil content, which is higher than that of soybean, is one of the important plants for biofuel (Li et al. 2010a, b).

Also, choose this salt-tolerant oil plant as a major dominant species for the construction of "Carbon Sequestration Forest" in North Jiangsu, China. The capacity of fixing $CO_2$ from seashore mallow plantation is rather high, and the $CO_2$ sequestration quantity is as follows:

$$MCO_2 = A \times NPP \times 1.63 \qquad (1)$$

In (1), $MCO_2$ is the total quantity of $CO_2$ sequestration in certain area per year (g, kg, t/year), A means the area of seashore mallow plantation ($m^2$, $\mu$*), NPP is net primary production ($g/m^2$ year), and 1.63 is the coefficient of fixing $CO_2$ from the dry matter of plant.

$$*\mu \text{ is Chinese unit of area, } 1\,\text{hm}^2 \text{ equals } 15\,\mu.$$

According to (1), 1 million mu of seashore mallow every year could fix $CO_2$ as:

$$MCO_2 = 1{,}000{,}000 \times 2.2\,\text{t}/\mu\,\text{year} \times 1.63 = 3{,}580{,}000\,\text{t/year} \tag{2}$$

Of course, NEP, that is the carbon sequestration quantity of plantation ecosystem every year as follows:

$$\text{NEP} = \text{NPP}(\text{gm} - 2\text{a} - 1) \times \text{CSE} \times \text{A}\ (\text{m}^2) \times 1.63 \tag{3}$$

In (3), CSE is the efficiency of carbon sink. Thus, 1 million of the ecosystem of seashore mallow could fix carbon 128,900 t every year.

# 9 Resources and Environment Carrying Capacity and Ecological Restoration of Stricken Areas

## *9.1 Introduction*

As the anomaly occurred events in the nature, natural disasters cause various and severe destructions on natural and human system of stricken areas. Therefore, it is of urgent and special requirements for ecological restoration and reconstruction in such areas. It is quite important for identifying the space that was destroyed during the natural disasters and also the extent of being destroyed. Evaluation of resource and environment carrying capacity (RECC for short hereafter) could reflect the special task and target of ecological restoration and reconstruction, and is of important supporting role for carrying out economic reconstruction and ecological restoration in stricken areas.

## *9.2 Theory and Practice of RECC Evaluation in Stricken Areas*

### 9.2.1 Definition and Theoretical Basis of RECC Evaluation

RECC means the ability of certain spatial unit to withstand all kinds of human social and economic activities of regional resource and environment systems (Abernethy 2001; Zhang et al. 2009). RECC is a comprehensive concept integrating various essential factors from the aspect of ecology, resources and environment (House 1974; Price 1999). The goal of regional RECC research is to find the representation of carrying capacity with qualitative or quantitative methods under certain temporal and spatial conditions (Dhondt 1988; Sleeser 1990). Up to now, relevant researches focus on carrying capacity of certain factors and comprehensive carry capacity (Liu and Hou 2009). The former mainly aims at indicating the carrying capacity of certain resources and environmental factors, such as land, water, energy, environmental capacity of water or air, etc. The later integrates the role of various factors, and gives a comprehensive result of all factors.

However, as the regional differentiation of resources and environment, and the different role of each factor on location and distribution of productive and living activities, it is of great significance for making sure the theoretical basis, technical method and objective of RECC research, especially in stricken areas. Such research could not only help understanding the changes of carrying capacity before and after the disasters, but also help defining the orientation of restoration and reconstruction.

### 9.2.2 Role and Content of Ecological Evaluation for RECC Research

Ecological research first introduces the concept of carrying capacity and regards it as a law of ecosystem (Del Monte-Luna et al. 2004). From the aspect of RECC research, the need on ecological research mainly focuses on the expansion and improvement of index system and technical method. For the index system, ecological factors, such as vegetation, NPP, biodiversity, are among the important affecting factors of RECC evaluation (Wang et al. 2000). There is upper limit for carrying capacity of natural ecosystems that are essential for the survival and development of human being (Gao 2001). It is necessary for adding the ecological factors to the index of RECC evaluation and considering the threshold of pressure that the natural ecosystem could bear. Such supplement helps reflecting the true regional carrying capacity in a more objective way (Gao 2001; Zhang 2009). For the technical aspect, the research methods of ecology, such as Logistic equation, resources supply and demand balance method, comprehensive index method, system and model analysis method, are widely adopted for RECC evaluation (Zhang et al. 2007).

As an important part of RECC research, the following aspects are major tasks of ecological evaluation. The first one is how to determine the main ecological factors that affect regional RECC. From above introduction, ecological factors are various including landform, soil, vegetation, climate, hydrology, biodiversity, etc., and every factor was of different role for calculating regional carrying capacity (Gao 2001). Therefore, the first task is to determine the chosen ecological factors through integrated analysis on regional ecosystem, social and economic development conditions and objective of RECC research. The second one is to make sure the threshold of choosed ecological factors on regional RECC. The role of different ecological factors would be found in the definition of its weight in the index system or the identification of threshold (Deng 2009). As one of the most important role of RECC research, it is necessary to make sure the pressure that the ecosystem could bear would not be exceeded. Therefore, the definition of weight and threshold are of great significance for RECC evaluation.

### 9.2.3 Practice and Application of RECC Evaluation in Stricken Areas

First used in the reconstruction planning of stricken areas in Wenchuan earthquake in the year of 2008, the RECC evaluation has been widely adopted in the restoration and reconstruction in the Zhouqu debris flow and Yushu earthquake (Fan et al. 2008; Deng 2010). They helped to formulate and implement the planning of postdisaster reconstruction in a short time and scientific approach by providing many suggestions on the population & economic carrying capacity, and the spatial location of restoration and reconstruction. However, as the natural ecosystem and social and economic base were severely destroyed during the natural disasters, the restraining factors changed greatly comparing with that before the disasters. Moreover, the goal of restoration and reconstruction also changed. Therefore, technical methods that were widely used before should be amended accordingly. In the practice of Wenchuan earthquake, consequential amendment took full consideration of the regional characteristic, destroy of earthquake and the goal of restoration and reconstruction (Fan 2009).

The stricken areas of Wenchuan earthquake included northeast Sichuan province, Southern Gansu province, and southwest Shaanxi province. It was the transition zone from west Sichuan Plateau and Qinling Mountain to the Sichuan basin from the aspect of topography, and also the transition zone from development-prioritized zone to development-restricted zone from the aspect of National Major Function Oriented Zoning of China (Fan 2009; Deng 2010). Therefore, the geographical conditions, RECC and developing goals would be specific for different areas. Based on integrated evaluation of geographical environment, geological conditions, secondary hazards, and population & economic base, the goal of RECC evaluation was defined as the following aspects: (1) carrying out zoning based on suitability of reconstruction conditions and identifying the extent of zones with high, middle and low suitability; (2) calculating the reasonable capacity of population and giving suggestions on regional economic development. The suitability evaluation of reconstruction conditions were of essential importance, and RECC evaluation including importance of ecological protection was its foundation. Importance of ecological protection included the evaluation of ecological sensitivity, ecological importance and the extent of damage on vegetation. Ecological sensitivity was reflected by rainfall erosivity, soil texture, topographic relief and land cover, and ecological importance was reflected by ecological value on maintaining biodiversity and conserving water resources. The extent of damage on vegetation was defined as an auxiliary index and was expressed by the proportion of being destroyed vegetation.

Similar methods were adopted in Zhouqu and Yushu. As the spatial differentiation was not as obvious as Wenchuan and their ecological importance was much higher, the role of ecological evaluation was given higher importance (Fan 2010).

## 9.3 Conclusion and Discussion

Based on above theoretical thinking and practical application, the flowing conclusion and discussion are provided.

(i) As the temporal and spatial differentiation, the index and threshold should be defined based on detailed analysis of objective area. Some special factors should be adopted beside for general index that were widely used in regional RECC evaluation. Moreover, the goal of RECC research should also be fully considered (Fan 2009; Zhang 2009).

(ii) Zoning based on RECC should reflect the spatial differentiation as far as possible (Fan et al. 2008; Deng 2009). Spatial units of one group should be with similar RECC, but units of two groups should with higher difference. Therefore, it is necessary for providing result of RECC evaluation with clear break point, which is quite difficult to do. Multilevel zoning is carried out accordingly (Gao 2001). The next level zoning is adopted to solve the difficulty for identifying break points that could not be identified at upper level zoning, but the function of next level zones should not disturb the development of function of upper level zone at upper level.

## References


Abernethy, V. D. (2001). Carrying capacity: The tradition and policy implications of limits. *Ethics in Science and Environmental Politics, 2001*, 9–18.

Bao, W. K., & Chen, Q. H. (1999). The processes and features of the ecosystem degradation. *Chinese Journal of Ecology, 18*, 36–42. (In Chinese with English abstract).

Bao, W. K., & Liu, Z. G. (1999). Human-induced disturbance regime in the Dagou valley in the upper reaches of the Minjiang River. *Chinese Journal of Applied and Environmental Biology, 5*, 233–239. (In Chinese with English abstract).

Bao, W. K., Chen, Q. H., & Chen, K. M. (1999a). Environment control techniques for the vegetation restoration in the dry valley in the upper reaches of the Minjiang River. *Chinese Journal of Applied Ecology, 10*, 95–98. (In Chinese with English abstract).

Bao, W. K., Chen, Q. H., & Liu, Z. G. (1999b). Productivity dynamics and optimal adjustment of nine interplanting communities in dry valley in the upper reaches of the Minjiang River. *Chinese Journal of Applied and Environmental Biology, 5*, 136–141. (In Chinese with English abstract).

Bao, W. K., Chen, J. Z., & Qiao, Y. K. (1999c). Eco-agricultural construction in the dry valley in the upper reaches of the Minjiang River. *Eco-Agricultural Research, 7*, 66–68. (In Chinese with English abstract).

Bao, W. K., Chen, Q. H., & Liu, Z. G. (2000). Changes of the structure and the species composition of the degraded Quercus community along varied disturbance gradients. *Yunnan Acta Botany, 22*, 307–316. (In Chinese with English abstract).

Bao, W. K., Pang, X. Y., Wang, C. M., & He, Q. H. (2007). The arid valley and its ecological restoration. In N. Wu (Ed.), *Restoration and reconstruction for the mountain degradation ecosystem—the theory and practice of the upper Minjiang River*. Chengdu: Sichuan Science and Technology Press. (In Chinese with English abstract).

Bao, W. K., Pang, X. Y., Li, F. L., Zhou, Z. Q., et al. (2012). *The scientific basis of ecological restoration and sustainable management for the Arid valley*. Beijing: Science Press. (In Chinese).

Cai, G., & Guo, Q. (2004). The problems and idea of ecological animal husbandry development. *Chinese Qinghai Journal of Animal and Veterinary Sciences, 34*(3), 39–40.

Cao, J. H., Zhou, L., & Yang, H. (2011a). Comparative study of soil carbon transfer between karst area and clasolite area covered forest and the karst carbon sink effect in Maocun village of Guilin. *Quaternary Sciences, 31*(3), 431–436. (in Chinese with English abstract).

Cao, J. H., Zhu, M. J., & Huang, F. (2011b). Comparison study on calcium forms in plant leaves under different geological backgrounds—a case study in Maolan, Guizhou Province. *Bulletin of Mineralogy, Petrology and Geochemistry, 30*(3), 251–260. (in Chinese with English abstract).

Cao, J. H., Yuan, D. X., Groves, C., et al. (2012). Carbon fluxes and sinks: The consumption of atmospheric and soil $CO_2$ by carbonate rock dissolution. *Acta Geologica Sinca, 86*(4), 963–972.

Chai, Z. X., & Fan, J. R. (2001). Study on the restoration of vegetation in arid-hot valleys along the Jinshajiang River. *Journal of Montain Science, 19*, 381–384. (In Chinese with English abstract).

Chambers, J. C. (1997). Restoring alpine ecosystems in the western United States: Environmental constraints, disturbance characteristics, and restoration success. In K. M. Urbanska, N. R. Webb, & P. J. Edwards (Eds.), *Restoration ecology and sustainable development* (pp. 161–187). Cambridge: Cambridge University Press.

Chen, S. (1990). *Population problem on loess plateau* (pp. 177–178). Beijing: China Economic Publishing House.

Chen, Y., & Gao, J. (1984). Time characteristic of great climatic disasters of died people over ten thousand in Chinese history. *Discovery of Nature, 2*.

Chen, W. H., Zhu, D. H., & Dai, A. D. (2005). Study of data base construction of karst caves in China. *Land and Resources Information, 1*, 14–17. (in Chinese).

Chung, C. H., Zhuo, R. Z., & Xu, G. W. (2004). Creation of Spartina plantations for reclaiming Dongtai, China, tidal flats and offshore sands. *Ecological Engineering, 23*, 135–150.

Costanza, R., et al. (1997). The value of the world's ecosystem services and natural capital. *Nature, 387*, 253–260.

Cui, J., & Li, M. L. (2008). Progress in the development and utilization of Chinese red pepper. *China Forest Science and Technology, 22*, 9–14.

Davis, M. A., & Slobodkin, L. B. (2004). The science and values of restoration ecology. *Restoration Ecology, 12*, 1–3.

Del Monte-Luna, P., Brook, B. W., Zetina-rejo, M. J., et al. (2004). The carrying capacity of ecosystems. *Global Ecology and Biogeography, 13*(6), 485–495.

Deng, W. (2009). The prospective study of reconstruction planning after "Wenchuan Earthqueke" based on the distribution of resources and environment carry capacity. *Bulletin of Chinese Academy of Sciences, 24*(1), 28–33.

Deng, W. (2010). Discussion on mountain area resource and environment carry capacity. *Geographical Research, 29*(6), 959–969.

Deng, Y., Jiang, Z. C., Cao, J. H., et al. (2004). Characteristics comparison of the leaf anatomy of *Cyclobalanopsis glauca* and its adaption to the environment of typical karst peak cluster areas in Nongla. *Guihaia, 24*(4), 317–322. (in Chinese with English abstract).

Dhondt, A. A. (1988). Carrying capacity: A confusing concept. *Acta Oecologica/Oceologia Generalis, 9*(4), 337–346.

Duan, Y. H., Bai, Z. K., & Zhao, J. K. (1999). The preliminary study on water properties of the weathered products of gangue in the shallow layers of gangue mountains in Shanxi province. *Journal of China Coal Society, 24*(5), 533–537. (in Chinese).

Ehrenfeld, J. G. (2000). Defining the limits of restoration: The need for realistic goals. *Restoration Ecology, 8*, 2–9.

Ewel, J. J. (1987). Restoration is the ultimate test of ecological theory. In W. R. Jordan, M. E. Gilpin, & J. D. Aber (Eds.), *Restoration ecology: A synthetic approach to ecological research* (pp. 31–33). Cambridge: Cambridge University Press.

Ewel, K. C., Cressa, C., Kneib, R. T., Lake, P. S., Levin, L. A., Palmer, M. A., et al. (2001). Managing critical transition zones. *Ecosystems, 4*, 452–460.

Fan, J. (2009). *Evaluation of resources and environment carrying capacity of reconstruction planning in Wenchuan Earthquake*. Beijing: Science Press.

Fan, J. (2010). *Evaluation of resources and environment carrying capacity of restoration and reconstruction in Yushu Earthquake*. Beijing: Science Press.

Fan, J. H., Liu, Q., Ma, Z. L., Zhang, J. Y., & Chen, G. W. (2006). Environmental impact of hydropower development in the upper reaches of Minjiang River. *Sichuan Environment, 25*, 23–27. (In Chinese with English abstract).

Fan, J., Tao, A., Chen, T., et al. (2008). Fundamental function in resource and environment carrying capacity evaluation in the state planning for post-Weuchuan Earthquake restoration and reconstruction. *Bulletin of Chinese Academy of Sciences, 23*(5), 387–392.

Ford, D. C., & Paul, W. (2008). *Karst geomorphology and hydrology*. West Sussex, England: Wiley.

Fuhlendorf, S. D., & Smeins, F. E. (1997). Long-term vegetation dynamics mediated by herbivores, weather and fire in a Juniperus-Quercus savanna. *Journal of Vegetation Science, 8*, 819–828.

Gaillardet, J., Dupré, B., Louvat, P., & Allegre, C. (1999). Global silicate weathering and $CO_2$ consumption rates deduced from the chemistry of large rivers. *Chemical Geology, 159*(1), 3–30.

Gao, J. (2001). *Discussion of the theory of sustainable development: ecological carrying theory, method and its application*. Beijing: China Environmental Science Press.

Guillaume, M., Berg, W. A, & Herron, J. T. (1986). Performance of native and introduced species seven years after seeding on alpine disturbance. In M. A. Shuster & R. H. Zuck (Eds.), *Proceedings: High altitude revegetation workshop no. 7, information series no. 58, fort collins* (pp. 131–141). CO: Water Resources Research Institute.

He, Y., Zhou, H., Zhao, X., Lai, D., & Zhao, J. (2008). Alpine grassland degradation and its restoration on Qinghai-Tibet Plateau. *Prataculture Animal Husbandry, 11*, 1–9.

Hobbs, R. J., & Cramer, V. A. (2008). Restoration ecology: Interventionist approaches for restoring and maintaining ecosystem function in the face of rapid environmental change. *Annual Review of Environment and Resources, 33*, 39–61.

Hobbs, R. J., & Norton, D. A. (1996). Towards a conceptual framework for restoration ecology. *Restoration Ecology, 4*, 93–110.

Holl, K. D., & Howarth, R. B. (2000). Paying for restoration. *Restoration Ecology, 8*, 260–267.

Hou, M. F., & Jiang, Z. C. (2006). Species diversity of karst original forest in different geochemical environments in Maolan. *Ecology and Environment, 15*(3), 572–576. (in Chinese with English abstract).

House, P. W. (1974). The carrying capacity of a region: a planning mode. *OMEGA, 2*(5), 667–676.

Hu, Z. Q. (1995a). Afforestation of waste pile in semi-arid areas. *Journal of China Coal Society, 20*(3), 322–327. (in Chinese).

Hu, Z. Q. (1995b). *Surface mine land reclamation research*. Beijing: China Coal Industry Publishing House. (in Chinese).

Hu, Z. Q. (1995c). Waste rock mountain greening afforestation of basic technical mode. *Coal Mine Environmental Protection, 9*(6), 35–37. (in Chinese).

Hu, Z. Q. (1996). *Land resource management and reclamation of coal mining subsidence land*. Beijing: China Coal Industry Publishing House. (in Chinese).

Hu, Z. Q. (1997). The basic principles and methods of coal mines reclaimed soil profile reconstruction. *Journal of China Coal Society, 22*(6), 617–622. (in Chinese).

Hu, Z. (2000a). The importance of artificial grassland in the development of prataculture and the control of environment in China of 21 century. *Grassland and turf, 1*, 12–15.

Hu, Z. (2000b). *The prataculture development and ecological environment on Tibetan Plateau*. Beijing: China Tibetology Press.

Hu, Z. Q., Liu, H. B., Liu, Q., & Xu, F. S. (1994). Try to discuss reclamation planning and design mining in subsidence land. *Mine Surveying, 2*, 32–34. (in Chinese).

Hu, Z. Q., Li, P. B., & Zhang, G. C. (2006). *Coal waste piles reclamation*. Beijing: China Coal Industry Publishing House. (in Chinese).

Hu, Z. Q., Bian, Z. F., Cheng, S., Zhao, Y. L., Yang, F., Zheng, L. Q., et al. (2008). *Land reclamation and ecological reconstruction*. Xuzhou: China University of Mining and Technology Press. (in Chinese).

Huang, B. (1983). *Several problems on soil conservation in the middle Yellow River*. Beijing: Science Press. (Strategy on territorial management in China).

Huang, C. M., Xue, Y. G., Wei, Y., et al. (2000). Habitat vegetation and selection of white-headed leaf monkey (*Trachypithecus leucocephalus*). *Acta Therapy Iologica Sinica, 20*(3), 180–185. (in Chinese with English abstract).

Institute of soil and water conservation, Chinese academy of sciences and Ministry of water resources. (1990). *Study of water and soil conservation type of ecological agriculture in loess hilly region*. Yangling: Tianze Publishing House.

Jiang, Y. X., & Lu, J. P. (1991). *Jianfengling tropical forest ecosystem in Hainandao*. China, Beijing: Science Press. (in Chinese).

Jiang, Z. C., Wen, J. T., Xie, Y. Q., et al. (1996). *Karst in Beijing's western mountain*. Guilin: Guangxi Normal University Publishing House. (in Chinese).

Jin, Z. Z., & Ou, X. K. (2000). *Vegetation in the dry-hot valley along the Yuanjiang River, Nujiang River, Jinshajing River, Lancang River*. Kunming: Yunnan University Press. (In Chinese).

Ke, J., et al. (1983). Preliminary study on environment and rate of soil erosion. *Geographical Research, 2*(2), 21–29.

Li, D. T. (1985). *Instruction on the types of China's soluble rock*. Beijing: Sinomaps Press. (in Chinese).

Li, W. (2008). *Study on reasonable allocation of regional agricultural resources, synthetic treatment of environment and balanced development of agricultural regions in China—agricultural ecological problems and comprehensive treatment*. Beijing: China Agriculture Press, pp. 362–365.

Li, D. T., & Luo, Y. (1983). Measurement of carbonate rocks distribution area in China. *Carsologica Sinica, 2*(2), 147–150. (in Chinese with English abstract).

Li, X. K., Su, Z. M., Lu, S. H., et al. (2003). The spatial pattern of natural vegetation in the karst regions of guangxi and the ecological signality for ecosystem rehabilitation and reconstruction. *Journal of Mountain Science, 21*(2), 129–139. (in Chinese with English abstract).

Li, P. B., Hu, Z. Q., Wu, J., Kang, J. T., & Wang, J. (2006a). Research and discussion on the hazards and greening technology of coal gangue dump. *Mining Research and Development, 26* (4), 93–96. (in Chinese).

Li, P. B., Hu, Z. Q., Wu, J., Zhao, Y. L., & Zang, Z. R. (2006b). Study on model of vegetation rehabilitation technique on coal waste pile. *Journal of Shandong Forestry Science and Technology, 4*, 13–15. (in Chinese).

Li, Q., Chen, Y. Q., Guo, X., Duan, Y., Chen, Y. L., & Xu, Z. H. (2007). Diversity of ants on the ground in different habitats in Yuanmou arid-hot valley, Yunnan. *Journal Fujian College Forest, 27*, 272–277. (In Chinese with English abstract).

Li, F. L., Bao, W. K., Wu, N., et al. (2008a). Growth, biomass partitioning, and water-use efficiency of a leguminous shrub (*Bauhinia faberi* var. microphylla) in response to various water availabilities. *New Forests, 36*, 53–65.

LI, Y., ZHAO, X., & Wang, S. (2008b). The climate change and vegetation productivity features in the source area of the Yangtze River and Yellow River. *Journal of Mountain Science, 26*(3), 678–683.

Li, F. L., Bao, W. K., Pang, X. Y., & Leng, L. (2009a). Seeding emergence, survival and growth of five endemic species in the dry valley of Minjiang River. *Acta Ecology Sinica, 29*, 2219–2230. (In Chinese with English abstract).

Li, X. R., Zhang, Y. M., & Zhao, Y. G. (2009b). A study of biological soil crusts recent development, trend and prospect. *Advances in Earth Science, 24*, 11–24. (In Chinese with English abstract).

Li, Y. J., Bao, W. K., & Wu, F. Z. (2010a). Soil seed bank and natural regeneration potential of shrub land in dry valleys of Minjiang River. *Acta Ecology Sinica, 30*, 399–407. (In Chinese with English abstract).

Li, Z., Li, G., & Qin, P. (2010b). The prediction of ecological potential for developing salt-tolerant oil plants on coastal saline land in Sheyang Saltern. *China Ecological Engineering, 36*(1), 27–35.

Li, Y. J., Bao, W. K., & Wu, N. (2011). Spatial patterns of the soil seed bank and extant vegetation across the dry Minjiang River valley in Southwest China. *Journal of Arid Environments, 75*, 1083–1089.

Liang, H., & Wang, J. (1998). A relative analysis between the lithological features and the characteristics of flood discharge and low flow in karst district-case study of the rivers, Guizhou Province. *Carsologica Sinica, 17*(1), 67–76. (in Chinese with English abstract).

Liu, Z. G., & Bao, W. K. (2001). A fundamental ideology for ecological construction. *World Science-Technology, 23*, 31–35. (In Chinese with English abstract).

Liu, Y., & Hou, H. (2009). Trend and suggestion on regional resources and environment carrying capacity research. *Natural Resource Economics of China, 9*, 19–21.

Liu, D., et al. (1985). *Loess and environment* (pp. 191–199). Beijing: Science Press.

Liu, Z. G, Bao, W. K, Wu, N., et al. (2000). Ecological reconstruction in the upper reaches of the Yangtze River: Environmental problems, degraded mechanism and rehabilitation methodology. *World Science-Technology*, Suppl, pp. 32–35. (In Chinese with English abstract).

Liu, F. Y., Zhu, H., Shi, J. P., & Chen, X. M. (2007). Characteristics of plant communities and their soil fertilities in dry-hot valley of Yuanjiang County, Yunnan, China. *Chinese Journal of Applied Environmental Biology, 13*, 782–787. (In Chinese with English abstract).

Lu, M. G., Cao, J. H., & He, Y. Y. (2006). The comparative study of biogeochemistry between limestone soil and red soil in Maocun, Guilin. *Guangxi Sciences, 13*(1), 58–64. (in Chinese with English abstract).

Luo, D. H., Xia, Q., Yuan, Q. W., et al. (2010). Root biomass of karst vegetation in a mountainous area of southwestern China. *Chinese Journal of Plant Ecology, 34*(5), 611–618. (in Chinese with English abstract).

Luo, H., & Wang, K. Q. (2006). Soil seed bank and aboveground vegetation in Jinshajing hot-dry river valley hillslope vegetation restoration. *Acta Ecologica Sinica, 26*, 2432–2442. (In Chinese with English abstract).

Ma, S. J. (1991). *Strategy on China's ecological development*. Beijing: China Economic Publishing House. (in Chinese).

Ma, Y., & Li, Q. (1999). Experiment of ruderal control on "black soil type" degraded grassland prataculturae. *Science, 16*, 46–50.

Ma, Y., Zhang, Z., Dong, Q., Shi, J., Wang, Y., & Sheng, L. (2007). Application of restoration ecology in-black soil type degraded grassland rebuilding. *Journal of Gansu Agricultural University, 42*(2), 7–16.

Millar, C. I., & Brubaker, L. B. (2006). Climate change and paleoecology: New contexts for restoration ecology. In D. A. Falk, M. Palmer, & J. Zedler (Eds.), *Foundations of restoration ecology* (pp. 315–340). Washington, D.C: Island Press.

Mitsch, W. J., & Gosselink, J. G. (2007). *Wetlands* (4th ed.). New York: Wiley.

Moseley, R. K., & Ya, Tang. (2006). Vegetation dynamics in the dry valleys of Yunnan, China, during the last 150 years: implications for the ecological restoration. *Journal of Plant Ecology, 30*, 713–722.

Munhoven, G. (2002). Glacial–interglacial changes of continental weathering: estimates of the related $CO_2$ and $HCO_3^-$ flux variations and their uncertainties. *Global and Planetary Change, 33*(1), 155–176.

Palmer, M. A., & Filoso, S. (2009). Restoration of ecosystem services for environmental markets. *Science, 325*, 575–576.

Parker, V. T. (1997). The scale of successional models and restoration objectives. *Restoration Ecology, 5*, 301–306.

Peng, S., & Ren, H. (2003). *Introduction to restoration ecology*. Beijing: Science Press.

Peng, S. P. (2009). China coal resource exploitation and environmental protection. *Science and Technology Review, 27*(17), 1. (in Chinese).

Price, D. (1999). Carrying capacity reconsidered. *Population and Environment, 21*(1), 5–26.

Qi, Y., Yu, Y., & Peng, S. (2007). Ecosystem services and ecological valuation of the forest-fruit-grass-fish ecosystem in hilly region, Heshan, Guangdong, China. *Ecology and Environment, 16*, 584–591.

Qin, P., Xie, M., Jiang, Y., & Chung, C. (1997). Estimation of the ecological-economic benefits of two *Spartina alterniflora* plantations in North Jiangsu. *China. Ecol. Eng., 8*, 5–17.

Qin, P., Xie, M., & Jiang, Y. (1998). Spartina green food ecological engineering. *Ecological Engineering, 11*, 147–156.

Ruan, C. J., Li, H., Guo, Y. Q., & Qin, P. (2008). *Kosteletzkya virginica*, an agroecoengineering halophytic species for alternative agricultural production in China's east coast: Ecological adaptation and benefits, seed yield, oil content, fatty acid and biodiesel properties. *Ecological Engineering, 32*(1), 320–328.

Shen, Y., & Hong, Q. (2003). Effective prevention and countermeasure study of soil erosion on loess plateau. *Science of Soil and Water Conservation, 1*(2), 22–27.

Shen, Y., & Yang, Q. (2006). *Study on divisions of ecological and geography and agro-ecology modes on loess plateau: Physical geography and ecological construction* (pp. 443–449). Beijing: China Meteorological Press.

Shen, Y. X., Liu, W. Y., & Zhang, Y. D. (2003). The effect of rehabilitation on vegetation species composition and soil seed bank at a degraded dry valley in Dongchuan, Yunnan Province. *Acta Ecologica Sinica, 3*, 1454–1460. (In Chinese with English abstract).

Sleeser, M. (1990). *Enhancement of carrying capacity options ECCO*. London: The Resource Use Institute.

Sun, S. C., & Bao, W. K. (2005). *Restoration ecology*. Beijing: Chemical Industry Press. (In Chinese).

Sun, H., Tang, Y., Wang, C. M., et al. (2001). Contour hedgerow intercropping for exploitation and conservation of slope cropland in mountain areas. *Journal of Mountain Science, 19*, 125–159. (In Chinese with English abstract).

Sun, H., Tang, Y., Huang, X. J., et al. (2005). Present situations and its R&D of dry valleys in the Hengduan Mountains of Southwest China. *World Science-Technology, 27*, 54–61. (In Chinese with English abstract).

Synthetical expedition team on loess plateau, Chinese academy of sciences. (1990a). *Vegetation resources and rational utilization on loess plateau*. Beijing: China Science and Technology Press.

Synthetical expedition team on loess plateau, Chinese academy of sciences. (1990b). *Hierarchical dataset on slope of cultivated land on loess plateau in China*. Beijing: China Ocean Press.

Synthetical expedition team on loess plateau, Chinese academy of sciences. (1990c). *Comprehensive development and rational distribution of agriculture, forestry and animal husbandry on loess plateau*. Beijing: Science Press.

Synthetical expedition team on loess plateau, Chinese academy of sciences. (1991). *Rational utilization of energy resource and solutions of rural energy on loess plateau*. Beijing: Science Press.

Synthetical expedition team on loess plateau, Chinese academy of sciences. (1992). *Dataset on resources, environment, society and economy on loess plateau* (pp. 529–531). Beijing: China Economic Publishing House.

Tai, P. D., Sun, T. H., Jia, H. Y., Li, P. J., & Li, W. J. (2002). Restoration for refuse dump of open-cast mine steppe region. *Journal of Soil and Water Conservation, 16*(3), 90–93. (in Chinese).

Tang, Y., Xie, J. H., Chen, K. M., et al. (2002). Role and Potential of Woody Fodder Development and Its Potential in Soil Conservation. *Research Soil Water Conser, 9*, 150–154. (In Chinese with English abstract).

Tang, Y., Sun, H., Xie, J. H., et al. (2003). Sustainable mountain development in West China: some emerging issues. *Journal of Mountain Science, 21*, 1–8.

The ministry of water resources of the people's republic of China, Chinese academy of sciences, Chinese academy of engineering. (2010). *Prevention and cure of water and soil erosion and ecological safety in China—vollume in northwest and loess plateau* (pp. 1–23). Beijing: Science Press.

Tyler, G., & Ström, L. (1995). Differing organic acid exudation pattern explains calcifuge and acidifuge behaviour of plants. *Annals of Botany, 75*(1), 75–78.

Wang, S. (1988). Natural disaster and prevention in the west of loess plateau in history. In *Comprehensive treatment and development of loess plateau*. Beijing: Science Press.

Wang, G. (2010). *Heyuan area of surface processes and environmental changes in Qinghai Tibet Plateau*. Beijing: Science Press.

Wang, G., & Cheng, G. (2001). Characteristics of grassland and ecological changes of vegetations in the source regions of Yangtze and Yellow Rivers. *Journal of Desert Research, 21*, 101–107.

Wang, S. J., JI, H. J., O'yang, Z. Y., et al. (1999). Preliminary study on the carbonate rock weathering and soil formation. *Science in China (Series D), 29*(5), 441–449. (in Chinese).

Wang, J., Yao, X., Li, J., et al. (2000). Assessment for ecological carrying capacity of Heihe river basin. *Research on Environmental Sciences, 13*(2), 44–48.

Wang, C. M., Bao, W. K., Chen, J. Z., et al. (2003). Profile characteristics and nutrients of dry cinnamon soils in dry valley of the upper Minjiang River. *Chinese Journal of Applied Environmental Biology, 9*, 230–234. (In Chinese with English abstract).

Wang, K. Q., Shen, Y. X., Chen, Q. B., et al. (2004). Soil water environment of artificial vegetation in Jinshajiang dry-hot volley. *Chinese Journal of Applied Ecology, 15*, 809–813. (In Chinese with English abstract).

Wang, W., Zhang, H. J., Zhang, C. L., Zheng, G. Q., & Li, M. S. (2008a). Influencing factors of vegetation restoration in coal waste piles—a case study of 280 coal waste piles in Yangquan city, Shanxi province. *Bulletin of Soil and Water Conservation, 28*(2), 147–152. (in Chinese).

Wang, G., Qin, P., & Wan, S. (2008b). Ecological control and integral utilization of *Spartina alterniflora*. *Ecological Engineering, 32*(1), 249–255.

Wang, Z., Zuo, C., Cao, W., Qin, W., Zhang, L., & Zhang, J. (2010). Ecological strategies and approaches of vegetation restoration in red soil hilly region. In *Ecological restoration of water and soil conservation in China Second Academic Symposium, Guiyang* (pp. 100–105).

Wei, Z. Y., & Wang, Q. B. (2009). Research on limited factors of reclaimed soil in the large coal wastes pile in fushun west opencast coal mine. *Research of Soil and Water Conservation, 16* (1), 179–182. (in Chinese).

Wei, Z. Y., Hu, Z. Q., & Bai, Z. K. (2001). The loose-heaped-ground method of soil reconstruction on the stackpiles of open-pit coal mine. *Journal of China Coal Society, 26*(1), 18–21. (in Chinese).

Wei, Y. G., Zhong, S. H., & Wen, H. Q. (2004a). Studies of the flora and ecology gesneriaceae in guangxi province. *Acta Botanica Yunnanica, 26*(2), 173–182. (in Chinese with English abstract).

Wei, Z. Y., Ma, R., Bai, Z. K., & Han, W. B. (2004b). Research on water erosion characteristics and its control measures in large dump of opencast coal mine in loess area—in the case of the south dump of ATB opencast coal mine. *Journal of Soil and Water Conservation, 18*(1), 164–167. (in Chinese).

Wu, Z. X., Liu, Q. L., Huang, Q., et al. (2006). Experiments with the Site-preparation Afforestation of Pinus radiata D. Don in Arid areas. *J Sichuan Forest Sci Technology, 27*, 11–16.

Wu, F. Z., Bao, W. K., Li, F. L., et al. (2008). Effects of the drought stress and N supply on the growth, biomass partitioning and water-use efficiency of the *Sophora davidii* seedlings. *Environmental and Experimental Botany, 63*, 248–255.

Xiang, S., Ding, J. L., Chen, Q. H., et al. (2007). The mode and technology of the ecological agriculture in the Arid Valley. In N. Wu (Eds). *Restoration and reconstruction of the mountain degradation ecosystem-the theory and practice of the upper Minjiang River*. Chengdu: Sichuan Science and Technology Press (In Chinese).

Xiong, D. H., Zhou, H. Y., Zhong, Yang, et al. (2005). Studies on revegetation in the dry-hot valley of Jinsha River. *Southwest China Journal of Agricultural Science, 18*, 337–342. (In Chinese with English abstract).

Yang, Q. (1990). Preliminary study on regional differentiation of natural environmental evolution and comprehensive treatment on loess plateau. *Journal of Arid Land Resources and Environment, 4*(1).

Yang, Q. Z. (2007). Study on the arid-valley scrubs in the upper reaches of Minjing River. *Journal of Mountain Science, 25*, 1–9. (In Chinese with English abstract).

Yang, J., Ding, Y., & Chen, R. (2007a). Assessment of eco-environmental vulnerability in the source regions of the Yangtze and Yellow Rivers. *Journal of Desert Research, 27*(6), 1012–1017.

Yang, Z. Y., Su, J. R., Luo, D., et al. (2007b). Progress and perspectives on vegetation restoration in the dry-hot valley. *Forest Res, 20*, 563–568.

Yuan, D. X. (1990). Advance on China's karst geochemistry. *Hydrogeology and Engineering Geology, 5*, 41–42. (in Chinese).

Yuan, D. X. (2001). World correlation of karst ecosystem: objectives and implementation plan. *Advance in Earth Sciences, 16*(4), 461–466. (in Chinese with English abstract).

Yuan, D. X., & Cai, G. H. (1988). *Karst environment science*. Chongqing: Chongqing Publishing House of Science and Technology. (in Chinese).

Zhang, R. Z. (1992a). *The dry valleys of the Hengduan mountains region*. Beijing: Science Press. (In Chinese).

Zhang, W. (1992b). *Study on loess plateau* (pp. 18–19). Beijing: Science Press.

Zhang, L. (2009). *Theory and methods of city ecological carrying capacity*. Beijing: China Environmental Science Press.

Zhang, X. B., Yang, Z., & Zang, J. P. (2003a). Lithologic types on hill slopes and revegetation zoning in the Yuanmou hot and dry valley. *Scientia Silvae Sinicae, 39*, 16–22. (In Chinese with English abstract).

Zhang, Y., Zhao, X., & Huang, D. (2003b). Study on the sustainable utilization of the Qinghai Tibet alpine perennial artificial grassland. *Acat Prataculturae Sinica, 12*(3), 22–27.

Zhang, L., Li, W., Liu, X., et al. (2009). Carrying capacity: Origin, development and prospective. *Acta Ecologica Sinica, 29*(2), 878–888.

Zhang, H., Wu, X., & Qin, P. (2011). Interactions between arbuscular mycorrhizal fungi and phosphate-solubilizing fungus (Mortierella sp.) and their effects on Kostelelzkya virginica growth and soil enzyme activities of rhizosphere and bulk soils at different salinities. *Biology and Fertility of Soils, 47*, 543–554.

Zhao, S. (1991). Historical development and geographic distribution of agriculture (farming) in China. *Geographical Research, 10*(1).

Zhao, X. (2011). *Grassland ecosystem restoration and sustainable management of the source region of three river*. Beijing: Science Press.

Zheng, G. L., & Bian, Z. F. (1997). Coal gangue reclamation plastic design content and method. *Coal Mine Environmental Protection, 11*(2), 33–35. (in Chinese).

Zheng, S. W., Li, Y. Q., Yue, Y. J., et al. (2007). A study of forestation experimentation in arid valley areas of the Minjiang River. *Journal of Sichuan Forest Science Technology, 28*, 57–61. (In Chinese with English abstract).

Zhou, Z. Q., Bao, W. K., Wu, F. Z., et al. (2008). Differences in growth and reproductive characters of *Rosa hugonis* in the dry valley of the upper Minjiang River, Sichuan. *Acta Ecology Sinica, 28*, 1820–1828. (In Chinese with English abstract).

Zhou, Z. Q., Bao, W. K., Wu, F. Z., et al. (2009). Capability and limitation of regeneration of Rosa hugonis and Rosa soulieana in the dry valley of the upper Minjiang River. *Acta Ecology Sinica, 29*, 132–140. (In Chinese with English abstract).

Zhou, J., Wu, B., & Qin, P. (2010). Physiological factors for tolerance of *Kosteletzkya virginica* (L.) Presl to one-instar bollworms of *Helicoverpa armigera*. *Acta Physiology Plant, 32*(3), 519–529.

Zhou, H., Zhao, X., Wen, J., Chen, Z., Yao, B., Yang, Y., Xu, W., & Duan, J. (2012). The characters of soil and vegetation of the degenerated alpine steppe in Yellow River's Source Region. *Acat Prataculturae Sinica, 21*(5), 1–11.

Zhu, S. Q., & He, J. X. (1993). The preliminary study on the law of the crown structure of masson's pine in south-subtropical region. *Journal Guizhou Agricultural College, 12*(2), 36–44. (in Chinese with English abstract).

Zhu, S. Q., Wei, L. M., Chen, Z. R., et al. (1995). Preliminary study on biomass components of karst forest in Maolan of Guizhou. *China Acta Phytoecologica Sinica, 19*(4), 358–367. (in Chinese with English abstract).

Zhu, L. H., Bao, W. K., & He, B. H. (2009). Assessment on ecological restoration effect of afforestation with cupressus chengiana seedlings in the dry Minjiang River valley, Southwestern China. *Chinese Journal of Applied Environmental Biology, 15*, 774–780. (In Chinese with English abstract).

# Chapter 12
# Ecological Protection and Establishment Projects

**Shidong Li, Xiangyang Hou, Ping Li, Yating Dai, Xiliang Li, Qingli Wang, Yong Zhang, Qi Lu, Peng Cui, Yongming Lin, Jiangchun Wei, Zhou Li and Moucheng Liu**

**Abstract** In China, the ecological and environmental conditions are inborn shortage, and there is a huge population pressure; moreover, the natural resource is overexploited. With the rapid development of economy and the accelerated pace of construction, business activities, such as improper construction projects, have brought new damages to the ecological environment, which aggravate serious eco-environmental problem. The Chinese government has always attached great importance to ecology protection. Especially after rapid growth of China's economy and ecological deterioration since reforming and opening-up policy, the government has published several policies and carried out environmental engineering projects. All efforts aim at ecological restoration and protection, and adhere to the principle of

S. Li
State Forestry Administration, Beijing 100714, China

X. Hou · P. Li · Y. Dai · X. Li · Q. Wang · Y. Zhang
Chinese Academy of Agricultural Sciences, Institute of Grassland Research, Hohhot 010010, China

Q. Lu
Chinese Academy of Forestry, Institute of Desertification Studies, Beijing 100091, China

P. Cui
Key Laboratory of Mountain Hazards and Earth Surface Process, Institute of Mountain Hazards and Environment, Chinese Academy of Sciences, Chengdu 610041, China
e-mail: pengcui@imde.ac.cn

Y. Lin
College of Forestry, Fujian Agricultural and Forestry University, Fuzhou 350002, China

J. Wei
State Key Laboratory of Mycology, Institute of Microbiology, Chinese Academy of Sciences, Beijing 100101, China

Z. Li
Chinese Academy of Social Sciences, Rural Development Institute, Beijing 100732, China

M. Liu (✉)
Institute of Geographic Sciences and Natural Resources Research, Chinese Academy of Sciences, Beijing 100101, China
e-mail: liumc@igsnrr.ac.cn

 
W. Li (ed.), *Contemporary Ecology Research in China*,
DOI 10.1007/978-3-662-48376-3_12

"conservation priorities and natural restoration based; pursue efforts to the protection and construction of ecology; reverse the trend of ecological environment deterioration from the source." The government has constructed a state safety screen, strengthened protection and restoration of ecological environment, and established ecological compensation mechanism. While, most ecological projects have been practiced for over years, it is necessary to summarize the experience and lecture of project construction theories, technologies, patterns, and managements, which is necessary for improving restoration projects construction. Therefore, this chapter introduced the ecological environment protection and construction projects and the corresponding technical means for current regional eco-environmental problems. Through comparing between China and abroad, and between regions, teasing development history, combining current typical cases, this chapter made an integrated conclusion of the leading eco-construction projects of China, including natural forest protection project, conversion of cropland to forest project, returning land for grazing to pasture project, desertification control project, and water and soil erosion and debris-flow management. All of the authors proposed their own view about principles, targets, partitions, technical systems, benefit evaluations, and institutional arrangements, and policy orientations, which is beneficial for scientific researchers and decision-makers to gain a fuller understanding of ecological protection and control projects.

**Keywords** Ecological protection · CFF · Grazing forbidden project · Desertification · Debris-flow treatment · Desert lichens · Ecologically fragile areas

# 1 Conversion of Farmland to Forests Project

## *1.1 The Important Meaning of CFF*

### 1.1.1 CFF Is the Historical Breakthrough of Chinese Ecology Environment Construction

At present, the area of water and soil erosion is more than 3.6 million km$^2$, and the desertified lands up to 1.74 million km$^2$, correspond to 18.2 of the realm area (Zhou 2001). The main reason of water and soil erosion and lands desertification in China is destruction of forests and cultivation, which results in environment aggravation and more natural calamity (Zhang 1993; Zhang 1997). At the beginning of the century, Chinese government timely carried out the policy of the conversion of farmland to forests project (CFF) in order to improve the environment. It is the historical breakthrough and great epistemological progress from cultivating to converting croplands into forests, from the devastating development at the price of ecology environment to the sustainable development of ecology and economy. At present, the CFF has been widely regarded and praised.

### 1.1.2 It is the Urgent Task of Thoroughly Combating the Yangtze and the Yellow River Water Calamity to Carry Out CFF and Improve Ecology Environment

Because of the steep lands cultivation and forest destruction in the upside and midst, the Yangtze and the Yangtze River valley have become one of the most serious regions in soil and water erosion. There are more than 2 billion tons of the earth and sand flowed into the Yangtze and the Yangtze River every year, and two-thirds of them came from steep sloping lands. The increasing water and soil erosion results in the earth and soil accumulation in rivers, lakes, and reservoirs. It not only makes the water calamity of the midst and underside of the Yangtze and the Yangtze River valley more serious, but also makes North China more droughty. The contradiction of lacking water resources brings out great danger to national economic and people's lives. Every year, Chinese government has to put a large number of people, substances, and money into defending flood calamity, droughty hazard, and relieving victims of natural calamities. The Yangtze and the Yellow River would have no security forever if the soil and water erosion were not controlled in the region. So taking resolute and decisive measures and carrying out CFF is an urgent task.

### 1.1.3 Carrying Out CFF and Improving Ecology Environment is the Underlying Measure to Put into the Great Development in the West of China

According to the statistics, the steep lands above 25° in China are 6.067 million ha, and the 70 % of which is centralized in the Western regions. The increasing aggravation environment is the underlying reason restricting economic and social development of Chinese Western regions. The most import thing is to improve ecology environment for developing and flourishing Chinese Western regions. If the ecology environment of Chinese Western regions cannot be evidently improved, the strategy of transferring the emphasis of Chinese economic construction from east to west will fail. Therefore, the strategy of carrying out CFF and ecology environment improvement of Chinese Western regions is an urgent task, and is the premise of the great development in Western China. Only if the ecology environment was obviously improved, the rich resources of Chinese Western regions can be well developed, and the investment environment can be better, and the money, technology, and talents can be attracted, and the development of Chinese Western regions can be speeded up (Jia 2001).

#### 1.1.4 CFF and the Environment Improvement is the Important Measure to Make the Countryside Economic Construction Better, Promote Local Economic Development, and Make Villagers Richer

Most regions carrying out CFF are poor and mountainous, so the development conditions of these regions directly influence the national economic construction. Carrying out CFF cannot only basically preserve water and soil and improve ecology environment, but also can efficiently increase the ability to defense drought and water calamities, and can elevate the land productivity and accelerate the economic and social development.

### *1.2 The Construction Scope and the Natural and Social Conditions*

#### 1.2.1 The Construction Scope

CFF involves 1897 counties of 25 provinces (regions and municipalities). The pilot project of CFF have been carried out in 20 provinces, and they are Hebei, Shanxi, Inner Mongolia, Liaoning, Jilin, Heilongjiang, Jiangxi, Henan, Hubei, Hunan, Guangxi, Chongqing, Sichauan, Guizhou, Yunnan, Shaanxi, Gansu, Qinghai, Ningxia and Xinjiang, respectively. But the other five provinces (regions and municipalities), Beijing, Tianjin, Hainan, Anhui, and Xizang did not carry out the pilot project.

#### 1.2.2 The Natural and Social Conditions

The scope of CFF is wide and the natural conditions have great difference. The area of CFF is 0.701 billion ha, about 70.89 % of the realm area. The population of CFF is 0.705 billion people, with 0.55 billion farmers, and the average income of farmers is 232.25 US$.

According to the investigation data of the Realm Resource Minister of People's Republic of China, the forthcoming area of cultivated lands is 0.134 7 billion ha. In them, the steep lands over 25° are 5.868 7 million ha, including 0.922 million ha of terrace, 0.547 million ha of croplands being converted into forest lands, 4.4 million ha of lands needed to be meliorated. The cultivated lands from 15 to 25° are 0.122 billion ha, including 2.73 million ha terrace, 0.133 3 million ha of croplands being converted into forestlands, 9.33 million ha of forthcoming steep lands, and 8 million ha of lands required to be rapidly combated with important ecology function. The sanded lands in CFF are 7.73 million ha, and those demanded to be rapidly combated and with important ecology function are more than 2.667 million ha. In sum, about 14.7 million ha of the steep and sanded farmlands are demanded to be rapidly combated at present.

## *1.3 The Construction Purpose, Overall Layout, and Accomplishment*

### 1.3.1 The Construction Purpose

On the basis of the pilot project in 1999 and 2000, the period of CFF is from 2001 to 2017. It is carried out in 2 phases: the construction phase from 2001 to 2010, and the stability phase from 2011 to 2017.

The latest goal is to convert 6.667 million ha of croplands into forestlands from 2001 to 2005, including converting 3.333 million ha of steep lands into forest lands, about the 75.8 % of forthcoming CFF croplands above 25°, planting forests 1.333 million ha in sanded lands, about 17.2 % of forthcoming CFF sanded lands, planting forests 8.67 million ha in mountain and wasted lands. The vegetation area of forest and grass will be increased to/by 15.3 million ha, the coverage in CFF scope will be increased to/by 2.2 %, 40.7 million ha of lands with the water and soil erosion will be controlled, and 48.7 million ha of windstorm lands will be combated. The fragile ecology environment will be basically improved through CFF.

In the long run, from 2001 to 2010, 14.7 million ha of croplands will be afforested, basically including all steep lands above 25°, planting in sanded lands 2.667 million ha, about 38.9 % of forthcoming sanded lands, and plantation 17.3 million ha in mountain and wasted lands. The vegetation of forest and grass will be increased to/by 3.2 million ha, coverage of forest and grass in CFF will be increased 4.5 %, and 8.67 million ha of soil and water erosion lands will be controlled, and 0.102 7 billion ha of windstorm lands will be controlled. After CFF, the forest ecology system will be basically recovered, and the ecology environment of the Yangtze River and the Yellow River will be evidently improved.

### 1.3.2 The Overall Layout

In the scope of CFF, it is divided into 10 subregions according to the water and soil erosion extent, sandstorm danger conditions, topography, and so on. The suitable tall trees, shrubs, or grasses are selected in every subregion (Li 2001). The 10 subregions are: High mountain and deep valley subregion in Southwest China; Yunnan–Guizhou plateau subregion; Lower mountain and hills subregion in Hainan–Guangxi; Mountain and hills subregion in Sichuan–Chongqing–Hubei–Hunan; Lower mountain and hills subregion in the Middle and Lower Reaches of the Yangtze River; Frigid grassland and meadow subregion in the source of the

Yangtze River and the Yellow River; Hilly ravine subregion in loess plateau; Semiarid subregion in Inner Mongolia–Shanxi–Hebei, Arid and deserts subregion in Xinjing; and Mountain and sand land subregion in Northeast China (Li 2002), respectively.

### 1.3.3 The Accomplishment

Initially, the experimental project of CFF began in 1999, including Sichuan, Shanxi, and Gansu province, and later, the scope of CFF was widened. 1.16 million ha of farmlands were converted to forestland, and 1 million ha of plantation in mountains and wasted lands were completed in 3 years, involving 20 provinces (autonomous regions and municipalities), 400 counties, 27 thousand villages, 5.7 thousand towns, 4.1 million farm families, and 16 million farmers (Li 2003).

CFF was officially initiated in 2002, covering 1897 counties (county-level cities and districts) in 25 provinces (autonomous regions or municipalities) of China.

## *1.4 The Investment Budget and Fund Source*

### 1.4.1 Investment Budget

From 2001 to 2010, the obligatory investment is 27.77 billion US$ according to the subsidy of foods and funds, subsidy standard, plantation area, and so on. From 2011 to 2017, the obligatory investment is 13.689 billion US$ (113.621 billion RMB yuan). The whole investment of CFF is 43.083 billion US$ including pilot project. In them, the provisions subsidy is 211.9 billion kg, about 35.82 billion US$, the cash subsidy is 4.108 US$, and seeds and seedlings subsidy is 2.899 billion US$.

### 1.4.2 Fund Source

The fund source of CFF is divided into two parts, basic construction investment and financial subsidy funds from central government. The former is used to purchase seeds and seedling, all of which is invested from 2001 to 2010. The latter is 39.929 billion US$, including provisions subsidy 211.9 billion kg, about 35.82 billion US$, and subsidy in cash 4.108 US$.

### 1.4.3 Comparison Between CFF and the World Forestry Ecological Projects

CFF and other 10 world famous forestry ecological projects (Mitsch and Jorgensen 1989; Qin 1998; Wang 1998; Xiang 2001) are contrasted and analyzed in scale, scope, investment, period, and start-up time in single factor and multifactors (using AHP method, i.e., Analytic Hierarchy Process) methods (Xu 1988). The result shows that CFF ranking first in investment and third in total scores. The 10 great forestry ecological projects in the world and their compositor are: Chinese Three-north and the Middle and Lower Reaches of the Yangtze River Shelter-belt Forest Development Project (TNYR), Chinese Natural Forest Protection Project (NFP), Chinese Conversion of Farmlands to Forestlands Project (CFF), Chinese Wildlife Conservation and Nature Reserve Development Project (WCNR), American Roosevelt Project (RS), former Soviet Stalin Rebuild Nature Plan (SRN), Canadian Green Plan (GP), Japanese Combating Mountains Plan (CMP), the Green Dam Project of the Five Countries in Northern Africa (GDFC), and Chinese Fast-growing and High-yielding Timber Forest Development Project (FHTF) (Li 2001, 2002).

## *1.5 Main Policies and Measures*

### 1.5.1 Government's Provisions Subsidy for Farmer Houses with Croplands Converted into Forestlands

The subsidy standard is that 2250 kg provisions are in subsidy for per ha per year in the Yangtze River valley and Southern China, and 1500 kg provisions in the Yellow River valley and Northern China. When the croplands are converted to grass, economic forest and ecological forest, respectively, the period of subsidy is 2, 5 and 8 years, respectively.

### 1.5.2 Government's Subsidy in Cash for Farmer Houses with Croplands Converted into Forestlands

Considering everyday necessary expenditure, such as medicine, education and so on, government offers subsidy in cash in a period of time after farmers converting croplands into forestlands, with 36.23 US$ for subsidy in cash per ha per year. The period of subsidy in cash is the same with that of provisions.

### 1.5.3 Free Seeds, Seedlings, and Subsidy for Plantation from Government

90.58 US$ are supplied in cash for seeds, seedlings, and plantation subsidy.

### 1.5.4 Free Relative Taxes and Local Finance's Part Compensation

For those croplands converted to forests lands, government should provide provisions after deducting agriculture tax, and when the provisions subsidy is terminated, the agriculture tax is free. For counties that carried out CFF, the decreased income of agriculture tax is properly compensated by central government finance.

### 1.5.5 Individual Contract System

According to the principal that people who planted trees will protect them and get benefits from them, farmers are guided and backed up to combat the mountain and sloping lands, and the tasks of plantation and protection are contracted to farmer. The plantation sites contracted by farmers can be inherited and/or transferred. The usufruct of forestlands is prolonged to 50 years and the contract period can be protracted in terms of relative rules in 50 years.

### 1.5.6 Render an Account System

The undertaking of CFF is finalized into farmer houses, and farmer houses put CFF into effect. After checkup and acceptance, farmer houses obtain provisions subsidy by cards and account system.

# 2 Ecological Effects of Grazing Forbidden Project

## *2.1 Introduction*

Grasslands in China cover nearly 400 million hm$^2$, occupied more than 47 % of total land area, which is important green ecological barrier in northern China due to grassland multiple functions including ecology, production, and life. However, the grassland ecological environment has continued to deteriorate because of the long-term reclamation and overgrazing. Over 90 % of the available natural grassland degradation had different degree degradation from the beginning of this century. China carried on Grazing Forbidden Project in Inner Mongolia, Xinjiang, Qinghai, Gansu, Sichuan, Tibet, Ningxia, Yunnan, and Xinjiang Production and Construction Group since 2003. The period of Grazing Forbidden Project is from 2002 to 2015. The project was implemented in two periods, the first period was carried from 2002 to 2010 and second period is from 2011 to 2015. Grazing Forbidden Project is an important measure in Chinese grassland utilization and management history, and has also an important action for the ecological and livelihood. To cope with the problems in the project implementation, experts have done a lot of research and discuss work, including grassland degradation, livestock balance technology, fencing and fencing grazing, rotational grazing, engineering benefit evaluation. This paper briefly described was as follows.

## 2.2 The Degradation Mechanism of Grassland

Grassland degradation is defined from two perspectives: grassland management and ecology. The study of operate perspective suggests that grassland degradation is grassland carrying capacity and livestock productivity decline because of the bad changes under the influence of various factors such as biological, soil, and society; the study of ecology perspective suggests that grassland degradation is grassland ecosystem away from the state of top due to the influence of human activities, such as grazing and reclamation. They are different between grassland degradation and retrograde succession. It is not called degradation when retrograde succession with the value increase under appropriate use. Based on the study of grassland systematic, Ren (2004) suggested that coupling relationship loss between vegetation and soil subsystems caused system structural changes and functional degradation, which is the contrary for reason of grassland degradation.

On the current, the theory is accepted widely that human activity and climate change coaction lead to grassland degradation. However, the main driving mechanism for grassland degradation is still controversy, including climate theory, human disturbance theory, dualism and comprehensive theory, etc. (Fan et al. 2007). The study of Wang and Cao (2010) thinks human disturbance is the main reason for grassland degradation in the 54 county of Inner Mongolia pastoral and semi-pastoral areas during 1980–2000 by using econometric models to quantify the effect of climate change and human actives. The study of Bian (2008) livestock increase bringing increasing pressure in part of pasture areas, population increase and frequent human activities also bring serious damage to the grassland by analyzing the situation of grassland degradation in the Northwestern Tibet, but another study showed that grassland degradation results from both climate change and human activities (Hao 2006). Although the specific factors are different for grassland degradation in different regions, overgrazing is considered as main factors. Overgrazing changes vegetation condition and soil structure due to foraging vegetation and soil trampling (Li et al. 2002; Zhang et al. 2002; Shang et al. 2009), which lead to loss coordination relationship of habitat–grass–livestock, and finally, the degradation happened for grassland system.

## 2.3 The Technology of Grazing Forbidden Project

The premise of grass–livestock balance management is grassland monitoring, whereas the key of grass–livestock balance is management. Grassland productivity and livestock production show a large seasonal fluctuations, therefore for livestock balance dynamic equilibrium strategy must be adjusted by grassland types, times, and livestock breeds (Ma 2008). For the problems in the livestock balance implementation process, the new livestock balance mode in a grassland ecosystem should be established, which will replace the old model of command and control management for livestock balance. People should establish grazing rights system, grazing rights monitoring mechanisms, and long-term grassland ecological compensation mechanism in order to develop grassland animal husbandry sustainable.

Fencing is an effective grassland protective measures, including rest grazing, non-grazing, and rotational grazing. Therefore, fence build is an important part in the Grazing Forbidden Project. As an important land use type, non-grazing strongly modifies ecosystem processes and presents on the vegetation and soil. On one hand, non-grazing can improve the vegetation biomass, coverage, diversity, richness, evenness, etc. (Zhang and Du 2006; Liu 2009; Zhou et al. 2010). On the other hand, non-grazing can reduce trampling damage to the soil, and improve the soil structure, moisture, nutrient and enzymatic activity (Sun et al. 2009). However, a large number of hay and litter is kept in fencing grassland, which is not good for grass productivity and grassland diversity community stability. Therefore, non-grazing is undesirable for grassland protection, whereas moderate grazing can help to improve grassland productivity. Except for resting grazing and non-grazing, rotational grazing is another important method for grassland protection. Rotational grazing is different grazing treatment in the dividing areas by different schedules. Rotational grazing enhances grassland primary productivity by reducing the degree of redundancy, promoting grassland plant growth compensation or overcompensation. In addition, when dominant species' biomass or coverage decreased by livestock feed, other species will grow fast with the living space, and improve the grassland ecosystem biodiversity. Rotational grazing is intensive utilization of grassland, which requires strict design and management. In the nongrow season, rotational grazing has no obvious effect with the grassland and livestock. In addition, it is also hard to plan and utilize effective grassland and livestock. However, it is of high price for building fence, water resource point, and other facilities. Therefore, rotational grazing is also not fit for the poor grassland area because building fence need a lot of money and many herdsmen cannot be burdened.

## 2.4 Study on Benefit Evaluation of Grazing Forbidden Project

With the benefit evaluation of Grazing Forbidden Project, people can grasp the effects of Grazing Forbidden Project on the grassland ecological environment and sustainable development in pastoral areas' economy, accurately understand the significance and sustainability of Grazing Forbidden Project, and ensure the project successful implementation.

The project curbs the trend of ecological deterioration, improve grassland ecological environment significant in some areas and promote the grassland ecosystem to develop in the positive direction (Yin et al. 2010; Zhao and Yu 2011).

The herdsman's production and life are affected by the project because their income is reduced in the early stage of project, but their income will increase and economics of the pasturing area will grow fast when they promote intensive management development by adjusting the animal species and structure (Bai 2010; Zhang and Wang 2010).

The ecological environment and animal husbandry are improved by the implementation of the project. The herdsmen change the traditional concept on livestock, the production and management mode is changed from extensive grazing to intensive feeding, including grassland contractual management and grassland protection system are promoted, the herdsmen's enthusiasm and creativity are transferred, the capacity about disaster prevention and mitigation of pasturing area are improved, concentration profit of the social development is obvious, economics of pasturing area is developed, national unity and social stability of China are safeguarded.

## 2.5 *The Grazing Forbidden Project Policy*

Based on fencing, reseeding, non-grazing, rest grazing, and rotational grazing, the policy of Grazing Forbidden Project has improved the grassland ecological environment and vegetation productivity, coordinated development between grassland ecology and animal husbandry.

The formation of a new and higher level of complex systems in order to highlight the combination of agriculture and animal husbandry, improve the level of industrialization and transformation of animal husbandry, increase herdsmans' income, coordinate the relationship of ecological, social, and economic benefits, and promote the sustainable development of animal husbandry economy (Hou 2005).

The project started on December 16, 2002. The Sate Council authorized the project in 11 provinces. In 2005, The Sate Council adjusted and improved the project policy (Wang 2007). In 2010 Central No. 1 document proposed to build an ecological security barrier, and strengthen the dimension of the project. In 2011, China put the policy of Grassland Subside and Reward Mechanism into implementation.

The studies focus on the policy effect to the long-term mechanism of pastoral ecological, economic, and social. The purpose of the studies is to provide a great value for Grazing Forbidden Project policy.

### 2.6 Conclusion

This paper systematically summary the Grazing Forbidden Project in China. The project changes Chinese utilization and management history, and makes the herdsmen change the traditional concept on livestock. It has improved the ecological environment, social economy and nation unity in the pasturing and semi-pasturing areas.

## 3 Desertification and Its Mitigation Strategy in China

### 3.1 Introduction

China is one of the countries most severely impacted by desertification, as 37.2 % of the country's territory—some 3.57 million $km^2$—is classified as drylands[1] (including arid, semiarid, and semihumid arid areas). Of the drylands, 2.64 million $km^2$ falls under the category of desertified land in accordance with the definition of the UNCCD, and these desertified lands are distributed in 18 provinces and account for 27.5 % of the country's landmass. China's desertified lands can be largely attributed to wind erosion, water erosion, salinization, and freezing-thawing processes. Over 400 million residents are affected by desertification, and the annual direct economic losses exceed 64 billion yuan (Ci and Wu 1997; Zhu 2006).

China's desertification mitigation efforts began in the late 1950s. Through a number of high-profile programs, such as the Three-North Shelterbelt Development Program initiated in 1978, the National Program on Combating Desertification initiated in 1990, the Sandification Control Program for Areas in the Vicinity of Beijing and Tianjin launched in 2000, and the Conversion of Croplands to Forests and Grasslands Program initiated in 2000, the Government of China has poured on average 0.024 % of the country's annual gross domestic product (GDP) into desertification mitigation efforts and, as a result, some 20 % of the desertified lands have been brought under control.

Approximately $50 \times 10^4$ $km^2$ of the existing desertified lands are considered restorable given current technology. When the potential desertification increments induced by global warming are taken into account, the overall area of desertification that is subject to restoration and mitigation in the future planning horizon is projected to range from $55 \times 10^4$ to $100 \times 10^4$ $km^2$. With the approximate restoration rate of $1.5 \times 10^4$ to $2.2 \times 10^4$ $km^2$ per annum, China's antidesertification battle is expected to last some 45–70 years. The current strategic plans set restoration targets

[1]Dryland refers to the areas of "arid, semiarid and dry subhumid areas", other than polar and subpolar regions, in which the ratio (i.e., HI = humidity index) of annual precipitation to potential evapotranspiration falls within the range from 0.05 to 0.65; this range also lays out the enabling conditions and sets the boundary of the geographic region of potential desertification.

at 22 × $10^4$ $km^2$ by 2015, with an additional 33 × $10^4$ $km^2$ by 2030, and the final 45 × $10^4$ $km^2$ of the 100 × $10^4$ $km^2$ restored by 2050. The plans also specify a number of crosscutting strategies to integrate vegetation rehabilitation and planting for the improvement of local livelihoods and promotion of economic development. The mitigation approaches are required to reflect local conditions and to combine prevention, restoration, and utilization.

On the basis of an examination of state investment in mitigation and current rehabilitation strategies, the paper provides the following suggestions on required institutional arrangements and policy-making for future antidesertification efforts: (1) expanding the previous sectoral perspective to embrace a multi-stakeholder approach; (2) setting priority zones within the restorable area, and establishing National Special Eco-Zones (e.g., forest farms, protected areas, and headwater areas); (3) restructuring the state antidesertification investment portfolio by changing the government direct investment in tree plantations to government acquisition of planted/greened areas; and (4) introducing preferential policies in favor of combating sandy desertification, such as permitting land tenures for up to 70 years and compensating for ecological services (Lu et al. 2004).

## *3.2 Desertification Status in China*

### 3.2.1 Climatic Zones and Area of Potential Desertification

Based on UNCCD's desertification definition (CCICCD 1997), China has a total of 3.32 million $km^2$ that are either arid (1.43 million $km^2$), semiarid (1.14 million $km^2$), or subhumid arid (0.75 million $km^2$), covering 34.6 % of its territory. The desertification-prone areas in the vast northwestern China cross the east corner of the Qaidam Basin and extend westward to the southwestern edge of the Qinghai–Tibet Plateau. In total, the desertification-prone area encompasses a total of 498 counties (cities, banners) in 18 provinces (autonomous regions, municipalities). In addition, island-shaped hyper arid areas—with a humidity index less than 0.05, are located in Gansu, Inner Mongolia, and Xinjiang, representing a total of 25.3 × 104 $km^2$, which is equivalent to 2.6 % of China's landmass (Ci and Wu 1997).

### 3.2.2 Types and of Hazard Desertification

Four major processes of desertification are present in China: wind erosion, water erosion, salinization, and freezing-thawing. These processes occur on some 2.64 million $km^2$ or 27.5 % of China's territory, involving 18 provinces, autonomous regions and municipalities. With widely distributed desertified and sand-encroached lands, China is one of the countries that is most severely affected by desertification. Desertification causes the degradation of ecological environment, induces natural disasters (e.g., dust and sandstorms) and takes a heavy toll on human livelihoods by sharply reducing the availability of useable land, lowering soil fertility, and aggravating poverty in the affected areas. In particular, it brings about serious damages and drastic economic losses to communication and transportation systems, water facilities, and the mining industries in the affected regions. Direct annual economic losses are estimated at more than 64 billion yuan, equivalent to about US$8.3 billion (Lu and Wu 2002). Statistics shows that approximately 400 million people, or 30.7 % of China's population, are affected by desertification, resulting in a considerable number of "environmental refugees" and "disaster migrants" in severely affected regions. Severe desertification and sandification[2] have threatened ecological security and sustainable economic development. This paper, however, focuses on examining the formation, evolution, and control of wind-erosion-type desertification.

Desertification causes ecological degradation, induces natural disasters (e.g., dust and sandstorms), and takes a heavy toll on human livelihoods by sharply reducing land availability, lowering soil fertility, and aggravating poverty in the affected areas. Especially, it brings about drastic losses to communication and transportation systems, water facilities, and the mining industries. Direct annual economic losses are estimated to exceed 64 billion CNY (Lu and Wu 2002). Statistics shows that approximately 400 million people are affected, resulting in a considerable number of "environmental refugees" and "disaster migrants" in severely affected areas. Severe desertification and sandification have threatened the ecological security and sustainable economic development. This paper primarily focuses on examining the formation, evolution, and control of wind-erosion-induced desertification.

## 3.3 *Combating Desertification Through National Programs/Projects*

### 3.3.1 National Programs

Since the late 1970s, China has launched a number of high-profile initiatives, including the Three-North Shelterbelt Development Programme (1979–2050), the National Programme on Combating Desertification (1991–2000), and the Sandification Control Programme for Areas in the Vicinity of Beijing and Tianjin (2001–2010). The Three-North Shelterbelt Development Programme is also widely known as the Great Green Wall. It has been recognized as "a great initiative to

---

[2]Sandy land refers to degraded land characterized by surface sandy materials. Sandy lands can be induced by single or multiple factors under different climatic conditions; hence, its distribution is not limited geographically.

transform the nature" and "a wonder of ecological engineering in the world" (Lu et al. 2004). The formulation and implementation of the Three-North Shelterbelt Development Programme has ushered in a new era of nationwide forestry development, and the national project itself is a landmark of forestry development involving the civil society and government agencies. It has also been viewed as a significant transition in undertaking forestry development projects using engineering and systematic and standardized approaches.

Since the beginning of the new millennium, the Chinese government has launched a number of large, ecosystem-oriented desertification prevention and control initiatives, such as the Conversion of Croplands to Forests and Grasslands Programme (2000–2010) and the National Soil and Water Conservation Programme (2000–2010).

### 3.3.2 National Action Plan to Combat Desertification – Nationwide Blueprint

In 2013, the State Council approved the National Desertification Prevention and Control Plan (2011–2020), which provides a vision and the guiding principles for combating desertification in the decade, that is, through prioritizing prevention, active rehabilitation and proper utilization, to bring under control the rehabilitatable desertified lands within 10 years. Overall, 20 million ha are targeted. Specifically, 10 million ha of relegatable sandy lands are mandated to be closed off and grazing ban areas set up to halt the advance of desertification and improve the local ecological conditions by 2015. The goal is to reverse the deteriorating trend in at least 50 % of the total restorable desertified lands, particularly in the priority areas, by 2020. The Plan also specifies the following key strategies:

(i) *Prioritized regional approach:* The plan sets key antidesertification and regional demonstration programs in five major regions and 15 subregions to gather success stories and showcase best practices for other areas in the regions (Table 1);

(ii) *Layered management goals*: The key programs should reflect layered management goals in their implementation methods—prevention (access ban, establishing nature reserves, ecological migration, etc.), comprehensive rehabilitation (integrated physical, chemical, and biological approaches) and effective utilization (husbandry, plantation, product processing, and so on);

(iii) *Coordinated restoration and development*: The implementation of the restoration programs should integrate poverty alleviation, sectoral development, and regional economic growth; program implementation should also incorporate proper water resource usage and agriculture and animal husbandry.

**Table 1** National roadmap and blueprint on combating sandification

| Rehabilitation arrangements | | Priority/pilot rehabilitation programs/projects in typical areas during 11th national five-year plan (2006–2010) |
|---|---|---|
| Five Eco-rehabilitation regions | 15 Subregions | |
| I. Desert margin and oasis in arid area | I. (1) Taklimakan Desert | 1. Phase II of the sandification control program for areas in the vicinity of Beijing and Tianjin<br>2. Phase IV of the three-north shelterbelt development program<br>3. Sandified grassland control program<br>4. Soil and water conservation program<br>5. Conversion of croplands to forests program<br>6. Conversion of pasture to grassland program<br>7. Afforestation project for areas in the vicinity of Lhasa City, Tibet<br>8. Ecological restoration project in Hetian, Xinjiang<br>9. National pilot areas of integrated efforts to combat sandification<br>10. Pilot areas of integrated efforts to combat sandification along the Old Beds of the Yellow River<br>11. Pilot areas of integrated efforts to combat sandification in Southern China<br>12. Integrated rehabilitation project in Shiyanghe/Minqin Basin, Gansu Province<br>13. Integrated rehabilitation project around Qinghai Lake, Qinghai Province<br>14. Natural conservation and rehabilitation in the headwater areas of Yangtze, Yellow and Lancangjiang Rivers |
| | I. (2) Gurbantonggut Desert | |
| | I. (3) Hexi Corridor and Alex Plateau (Badan Jilin and Tengger Deserts) | |
| | I. (4) Some humid sandy land | |
| II. Sandy land in semiarid area | II. (1) Vicinity of Beijing and Tianjin | |
| | II. (2) Korqin sandy land | |
| | II. (3) Mu Us sandy land | |
| | II. (4) Hulun Baier sandy land | |
| III. Sandified land in Qinghai–Tibet Plateau | III. (1) Qaidam Basin Desert | |
| | III. (2) Gongjhe Basin Desert | |
| | III. (3) Sandy lands of river valley in Tibet | |
| IV. Sandy land in semi-humid and dry semi-humid areas | IV. (1) Sandified lands between Yellow and Huai Rivers | |
| | IV. (2) Sandified lands between Yellow and Hai Rivers | |
| V. Humid sandy land in southern China | V. (1) Coast sandy lands | |
| | V. (2) Sandy lands along middle/lower Yangtze River | |
| | V. (3) Sandy lands in river valley of southwestern China | |

## *3.4 Mid-/Long-Term Disaster Reduction Strategy: Institutional Arrangements and Policy Recommendations*

### 3.4.1 Establishing Ecological Special Zones and Zero-Access Reserves

China's northern borderline extends some 7400 km. It is also the sand shield belt and the lifeline of nearly 50 ethnic minority groups. The Xinjiang Uygur Autonomous Region, for example, borders with Mongolia, Russia, Kazakhstan, Kyrgyzstan, Tajikistan, Afghanistan, Pakistan, and India along its 5400 km borderline. Of the 5400, 3100 km are under the impact of strong gusts and sandstorms. Inner Mongolia shares its border with Mongolia and Russia along a borderline of 4221 km. In addition to the impact from gusts and sandstorms, western China is also an important region in terms of national defense and security due to the presence of a significant number of space industry towns, military sites, satellite and missile launching bases, and border patrol stations.

State-owned forest farms, including nurseries and sand control stations, and experimental and demonstration stations in desertified areas constitute the main scientific and technical capacities in implementing initiatives for rehabilitation and combating desertification. However, many research initiatives have been constrained by insufficient funding, and key technical issues remain unresolved, such as water-saving techniques and improved seed selective breeding, and pest control. Many seed production and seedling nurseries are at stake. In some cases, employee salaries have not been paid on time, let alone technical support for program implementation.

In view of the above situations, several policy recommendations are proposed:

(i) Areas containing military bases (army bases, space industry towns, military sites, satellite and missile launch bases, and border patrol checkpoints) and scientific research and experimental stations (long-term observatories, desert control stations, new species experimental bases and new technical extension service facilities) in western China may be set aside as special complex ecological zones;

(ii) Primary and secondary Gobi, deserts and sandy lands in less populated areas, alpine and rural plateau areas may be specified as zero-access nature reserves.

### 3.4.2 Establishing Specialized Rehabilitation Forest Farm Companies

On many sites for combating desertification, some unsustainable development interventions have caused negative consequences for the expansion of croplands due to pressures that forest farms and their employees faced to reduce seedling nurseries for cereal production. The outbreaks of yellow-spotted long-horned beetle and Asian long-horned beetle in recent years have devastated massive protective forests and brought drastic losses to local agricultural production in northwestern China. In the Ningxia Plain, the pest outbreaks have forced the removal of many shelterbelts, resulting in the decline of crop yields by 20–30 %. Such originally avoidable losses are directly linked to the operational and implementation difficulties mentioned above.

In consideration of the lack of efficacy via direct—and rather limited—government investment and the need for sustained effectiveness of control activities, it is proposed to engage economic entities as contractors to assume the responsibilities for the implementation of ecological restoration projects at the ministerial, provincial, prefecture, and county levels. Existing firms, rural shareholding cooperative forest farms, desert control stations, state-owned forest stations, and laid off workers should be encouraged to participate in the bidding for state and local governments' programs for combating desertification. The ecological restoration projects should undergo proper bidding processes and be put under proper contractual arrangements. The program implementation would be able to leverage benefits from incentive-based and market-oriented efficiencies frequently seen in the private sector.

### 3.4.3 Legislation of Preferential Policies for Combating Desertification

A series of preferential policies for restoration and combating desertification have been issued in China, such as low-interest loans, zero-fee use of sandy lands, and tax-free development projects on sandy lands. However, most of these policies have become outdated. To encourage further desertification control initiatives and development of the affected lands, new preferential policies should focus on the following four aspects:

(i) Providing financial assistance: The rehabilitation and development of desertified lands demand huge investments. The budgetary allocations from the Central Government for various purposes—poverty alleviation and development in various sectors, such as agriculture, forestry, animal husbandry, water resources and energy sectors, should be combined and utilized in a coordinated manner, so as to enhance the intensity and enlarge the scale of the investment, and to ensure its effectiveness;

(ii) Issuing preferential loans: The methods of discounting interest for the fixed-target, fixed-term, and fixed-rate loans need to be improved. Repayment schedules of the loans should be tailored to match the project components. For example, the term of loans for fruit plantations should be extended to 8–15 years to reflect the longer investment period needed until economic returns can be reaped. The loan-granting conditions and procedures should be simplified and the existing mortgage requirements be relaxed;

(iii) Reforming land tenure and property rights systems: Communities, civil society groups, private individuals, and foreign enterprises are encouraged to contract rehabilitation and development of desertified lands. A new tenure system —'the contributor gets the spoils'—should be built into formal institutional settings for stable implementation (50–70 years) that allows for auctions, leasing, inheritance, and transfer of the titles;
(iv) Introducing tax breaks: Forestry is a feeble undertaking with a long production period, in comparison with agriculture. The tax-free period for the "special agriculture-forest-products taxes" should be extended and the tax rates should be reduced.

### 3.4.4 Improving Ecological Compensation Mechanisms

The essence of the various initiatives to combat desertification lies in ecological restoration, which requires sustained management and maintenance over the long term. The returns of such endeavors have been traditionally ascribed primarily or entirely as public goods. The contribution and inputs made by the contributors need to be compensated through payments. These payments will serve several purposes: replenishing the depleting pools of funds for ecological rehabilitation, balancing the welfare between bearers of the costs and beneficiaries from ecological rehabilitation, and enhancing public awareness and the civil society's sense of responsibility for environmental protection. The payment system for ecological rehabilitation in desertified areas needs to incorporate the following three components:

(i) A certain portion of the funds required for ecological rehabilitation will be charged to the enterprises, communities, and individuals who benefit from the activities; the proportion will be determined on the basis of the magnitude of the benefits;
(ii) Those entities and individuals who will use the rehabilitated desertified lands must pay fees to compensate for the costs of rehabilitation;
(iii) Those who bring detriments to the ecological environment will not only receive penalties and be held responsible for restoring the damaged environment, but will also pay the prescribed ecological compensation fees. The fees collected will be used strictly for the formulation and implementation of new projects and initiatives for combating desertification. Diversion of these funds for other purposes is strictly prohibited to ensure sound management, as well as healthy and sustainable development of the desertification restoration projects.

### 3.5 Conclusions

Climatic and other natural causes are considered to be the main factors that are accountable for 70 % of China's present desertified areas. The remaining 30 % of desertified areas are largely due to anthropogenic causes and are believed to have formed during the past 3000 years. Since the unification of China in 221 B.C., vast land conversion programs that converted grasslands into farmlands, forests into cultivated areas, and subsequently deserted farmlands into desertified areas, have significantly changed the landscape of the original northern pastures. A considerable area of these cultivated lands eventually became degraded and turned into sandy lands.

Mitigation efforts in the past five decades by the Chinese government have yielded significant achievements, in that nearly 20 % of the country's desertified lands have been brought under control using various biological and engineering approaches. However, owing to the rapid increase in desertified area induced by global climate change, the total rehabilitatable desertified lands are estimated to be 500,000–1,000,000 $km^2$. Given the magnitude of desertification and the enormous tasks for combating desertification, current levels of investment, even under a fairly optimistic outlook, would prove to be inadequate to achieve the desired goals. Combined with unsatisfactory effectiveness of existing investments, fundamental changes in the investment level and methods will be necessary to advance closer to the restoration targets.

This is a vision for China's continuing battle against desertification: given the estimated 500,000–1,000,000 $km^2$ of projected area targeted for future desertification restoration and mitigation, with the current restoration rate (approximately 10,000 $km^2$ per annum), the battle for combating desertification could easily extend over 50–100 years, the time span of the strategic blueprint. China's current strategic plan, which sets restoration targets at 220,000 $km^2$ by 2015, an additional 330,000 $km^2$ by 2030 and the final 450,000 $km^2$ by 2050, is a difficult target to reach.

A few institutional arrangements and policy recommendations may be considered for future efforts to combat desertification: (1) readjust the management system and project planning mechanisms to expand the previous sectoral perspective to an all-stakeholder approach and encourage broad social participation; (2) set priority areas amongst the entire restorable area, establish state special ecological zones (e.g., forest farms, protected areas, and headwater areas), so as to channel limited funds toward priority areas rather than spreading them out thinly; (3) restructure state investment portfolio for combating desertification by changing the government direct investment in plantations to government acquisition of planted areas; and (4) adopt policies that favor desertification control and rehabilitation, enable compensation of rehabilitation activities for ecological services, and make improvements to the institutions for long-term land tenure and ecological compensation.

# 4 Debris-Flow Treatment: Integration of Botanical and Structural Countermeasures

## *4.1 Introduction*

Debris flow brings damages to engineering constructions and ecological environment such as towns, factories, roads, farmlands, and forest. It, as one of the major global hazards, has drawn more and more attention throughout the world.

### 4.1.1 Debris-Flow Hazards in China

China, over two-thirds of the land occupied by hills and mountains, suffers lots of debris flows and has over 74 million people living in risk. According to a primary statistical research, in China, there are over $8 \times 10^4$ debris-flow gullies with intensively active areas of $1.3 \times 10^6$ $km^2$ (Kang et al. 2004). In recent decades, thousands of debris flows have occurred in China. Geohazards including debris flows and landslides took 15,649 lives from 1997 to 2010, in total 1043 a year in average. The direct economic loss is estimated to be over 5 billion RMB per year. In 2010, the death toll reached 2913 including 1765 death caused by debris flow in Zhouqu. Frequent geohazards and increased risk have raised a key issue of disaster prevention in regional development of mountainous areas. And it is indispensable to improve the techniques of disaster prevention for adapting the rising risk (Cui et al. 2007).

### 4.1.2 Advances in Debris-Flow Control

After the first project achieved by professor Guan Junwei in early 1950s, from 1961, Chinese scientists had organized a series of regional debris-flow investigations in the regions of Tibet, Hengduan Mountains, etc. Based on the results of field surveys, the map of debris-flow distribution and its hazard characters regionalization was drew up (Tang et al. 1991), and the database of debris flows and landslides was established (Zhong et al. 1998). Subsequently, the technology of disaster prevention and reduction was developed by combining prevention and control, incorporating botanical and geotechnical countermeasures, as well as accommodating disaster prevention and utilization of resources (Tang et al. 1980; Cui 2009; Cui et al. 2008).

Despite achieved advances, there is still room to improve the methodology for amalgamating geotechnical engineering with botanical methods in debris-flow mitigation projects.

## *4.2 Techniques of Debris-Flow Mitigation*

### 4.2.1 Technical System for Preventing and Controlling Debris Flow

In China, a debris flow is often treated throughout whole catchment, taking both upstream and downstream into consideration and regulating gullies and slopes simultaneously, with the methodology of incorporating botanical countermeasures and geotechnical engineering (Chen et al. 1983; Zhou et al. 1991; Tang 2000; Wu et al. 1993; Li 1997) (Fig. 1).

### 4.2.2 Botanical Countermeasures

On slopes, usually, botanical countermeasures for debris-flow control are adopted by using highly effective spatial configuration pattern with multiple vegetation structures. Besides in some locations with serious gravity erosion, the designed vegetation reduces hydrodynamic conditions during debris-flow formation process.

### 4.2.3 Geotechnical Engineering Countermeasures

Geotechnical engineering methods, including the works of water storage and diversion, slope stabilization, sediment trap, drainage, and so on, are constructed to

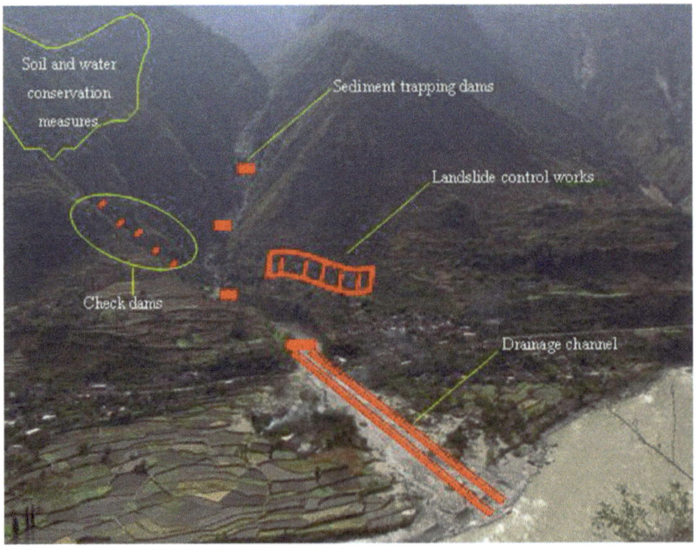


**Fig. 1** Technical system for preventing and controlling debris flow in small watershed

control the processes of debris-flow formation and to reduce damage energy in source area, transmission area, and accumulated area.

## *4.3 Integration of Botanical and Geotechnical Countermeasures*

The system of multilevel control for runoff and solid materials in "slope-gully-valley" system is set up to limit formation conditions of debris flow and regulate its movement.

### 4.3.1 Regulation for Infiltration and Runoff Process on Slope

The key point to prevent debris-flow formation is using reasonable configuration of botanical countermeasures and geotechnical structures to regulate the processes of infiltration and runoff on slope as follows:

**Regulating rainfall and runoff using multilayer structure of vegetation**: The function of multilayer vegetation can reduce rain-splash erosion and alleviate runoff. Therefore, vegetation should be planted on the slopes in the upstream of debris-flow gully to reduce soil erosion, delay runoff convergence concentration and flood formation, and limit hydrodynamic condition of debris-flow formation.

**Controlling soil erosion by contour methods**: Contour countermeasures can allay soil erosion and sediment yield by reducing the amount and velocity of surface runoff. For example, contour hedgerow, planting in twin-row belts with an interval from 4 to 8 meters, has been proved to reduce soil erosion (Sun et al. 2001) (Fig. 2).

**Mini-works of water conservancy** *on slope for regulating the process of overland flow confluence*: Mini-works of water conservancy is constructed to redistribute or trap runoff, reduce water scour and sediment discharge, store runoff and drain excessive runoff safely. The systems can regulate the processes of surface flow confluence and control the slope runoff (Fig. 3).

### 4.3.2 The Countermeasures for Controlling Retrogressive Erosion of Gully Head

Retrogressive erosion of gully head is a key point in debris-flow prevention. There are many kinds of structural works for controlling gully head erosion, such as water retaining ridge, water storing ridge, closed gully ridge, drainage ditch, waterfall, etc. In the areas with sufficient water supply, botanical methods can provide effective

**Fig. 2** Contour hedgerow used for reducing soil erosion

**Fig. 3** The diagram of mini-works of water conservancy on slope

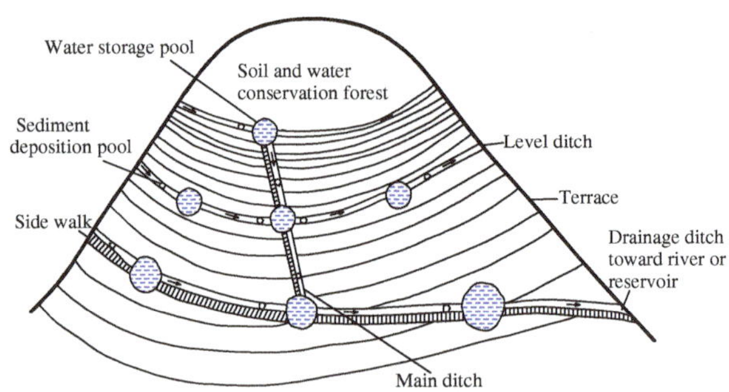


retrogressive erosion control. The combination of botanical methods and geotechnical engineering works has better effects on retrogressive erosion control than a single method.

### 4.3.3 The Techniques of Energy Dissipation in Gully

In mountainous areas, step-pool system can control river flow velocity, dissipate flow energy, regulate sedimentation in river, and also reduce the huge potential energy of water flow in steep slope to protect river bed from heavy water scour (Xu et al. 2003), which also creates favorable habitats for aquatic. Based on the properties of step-pool system, huge boulders can be used to create artificial step-pool system to prevent debris-flow initiation from the upstream of gully (Wang et al. 2009) (Fig. 4).

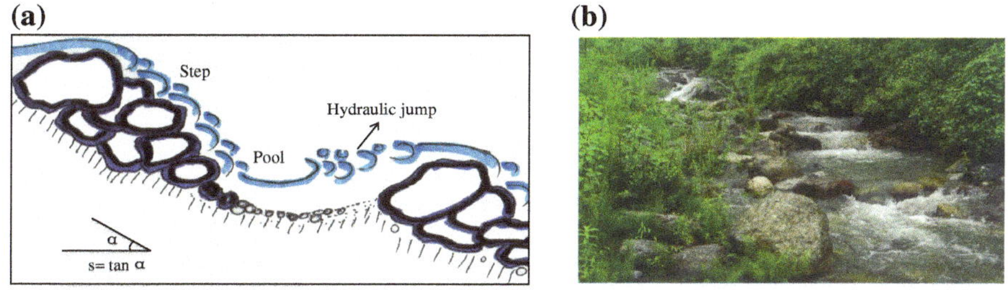


**Fig. 4** The step-pool system benefits energy dissipation, river bed stability and erosion reduction, and improves ecological status (from Wang Zhaoyin) **a** step-pool system **b** artificial step-pools and good ecological status in Shnegou gully, Dongchuan city, Yunnan Province

### 4.3.4 The Techniques for Regulating Movement and Composition of Debris Flow

For the objective of regulating the processes of a mature debris flow, step-check dams should be set up to reduce solid content, velocity, and flow rate of debris flow through blocking solid materials step by step. Sometimes, the step-check dams provide openings with various sizes reduces from upstream to downstream for trapping and sorting the gravels in debris flow. Normally, a drainage channel or a deposition barrier is constructed to control debris flow flowing over the last check dam.

### 4.3.5 Case Study: Debris Flow Treatment in Jiuzhaigou, a Site of World Heritage

In Jiuzhaigou, the above treatment, integrating botanical and geotechnical countermeasures were applied to protect the landscapes and eco-environment and to make engineering works be operated in harmony with landscapes. This methodology consists of control work in the source areas and various structures for trapping sediment and diverting flows. All works in 14 debris-flow tributaries of Jiuzhaigou have been worked well and protected landscape effectively.

## *4.4 Conclusion and Discussion*

Debris flow prevention techniques have been summarized as botanical and geotechnical engineering methods. The treatment system of multilevel runoff and unconsolidated soil in the catchment was also set up. The treatments applied in Jiuzhaigou have worked well and protected landscape effectively. However, for more effective mitigation of debris flow disaster, the following scientific issues should be approached further in future.

### 4.4.1 To Study Debris-Flow Prevention Mechanism of Botanical Countermeasures

At present, little is known about the mechanism of disaster mitigation by plants' underground parts as well as that of the "plant-soil-structure-water" system in different conditions. In future, we should highlight the study on the debris-flow mitigation mechanism of botanical methods for the planning, design, species selection, and combine with geotechnical engineering structures.

### 4.4.2 To Quantitatively Assess the Functions of Botanical Countermeasures

Up to now, little attention was paid to the function of debris-flow prevention due to plants' underground parts, especially to soil structure and soil strength. Further study should focus on soil consolidation capacity of plants' root system and critical depth of root. Furthermore, the evaluation indices of disaster reduction function of botanical countermeasures should be set up quantitatively.

### 4.4.3 To Develop Technical Specification of Botanical Countermeasure Design for Debris Flow Mitigation

The technique of botanical countermeasures should be developed for each eco-geographical zone. Afterward, the technical regulation for botanical countermeasures design in different conditions should be proposed.

### 4.4.4 To Improve the Techniques of Integrating Botanical and Geotechnical Engineering Countermeasures

So far, little is concerned about spatial configuration modes and comprehensive disaster reduction benefit due to lack of research data. The further study on optimization allocation of botanical methods and geotechnical engineering structures should provide technical support for the establishment of synthetic management system to control debris flow.

# 5 The Present Status and Prospect of the Researches on Desert Lichens and Ecological Restoration in China

This paper is dealt with the desert lichens based on the analysis of the community succession in the arid and semiarid deserts from China. The results demonstrate that the community succession of all living things in the arid and semiarid deserts depend mainly on the water balance. On the one hand the artificial vegetation provided a habitat suitable for the crust microbiota, and led to developing the crust microbiota, and on the other hand the artificial vegetation with water pump effect expended the deep soil water during the long process of ecological succession under the water balance, and led itself to decline year by year. On the contrary, the crust microbiota without water pump effect and with sand-fixating function well developed year after year. Then, the declined artificial vegetation protection system was replaced by the new protection system of the dominant carpet-like crust microbiota. This is a living example of the "Natural Selection, or the Survival of the Fittest."

The crust microbiota results from an intimate association between soil particles and cyanobacteria, algae, microfungi, lichens, and bryophytes. Soil particles are aggregated through gathering by the presence and activity of these biota, and the resultant living crust covers the surface of the ground as a coherent layer. The lichens are dominant group among the crust microbiota in the desert of the Shapotou region.

A lichen is a symbiotic association of a fungus (mycobiont) and a photosynthetic partner (photobiont), which may be an alga (phycobiont) or a cyanobacterium (cyanobiont). The association is a complicated arrangement in which the fungus produces a thallus or body within which the photobionts are housed.

Such a result provided a scientific basis for the feasibility of constructing "Bio-Carpet Engineering" on the arid desert.

The "Bio-Carpet Engineering" is to construct the crust microbiota by means of the biotechnique of isolation and inoculation of the crust microbiota, including the mycobionts and photobionts to form the lichens in the crust microbiota on the arid desert. Second, in order to improve the "Bio-Carpet Engineering," it is necessary also to study the drought-resistant transgenic sward plants using the drought-resistant genes from the desert lichens.

# 6 The Ecological Conservation and Development in the Ecologically Fragile Areas

## 6.1 Introduction

China is one of the countries in the world with the most obvious ecological fragility and the largest areas and the most types of ecological fragility. The ecological fragile areas are mainly located in northern arid and semiarid areas, southern hills, southwestern mountainous areas, Qinghai–Tibet Plateau, and the connection areas between water and land in the eastern coastal areas. Twenty one provinces (regions and/or municipalities) are involved; they are Heilongjiang, Inner Mongolia, Jilin, Liaoning, Hebei, Shanxi, Ningxia, Gansu, Qinghai, Xinjiang, Tibet,Sichuan, Yunnan, Guizhou, Guangxi, Chongqing, Hubei, Hunan, Jiangxi, and Anhui. The major types of ecological fragility include the northeastern forests and grass transfer areas, the northern agriculture and animal husbandry transfer areas, the northwestern desert and oasis transfer areas, the southern red soil and hill areas, the southwestern karst desertification areas, the Qinghai–Tibet composite erosion areas, and the coastal water and land transfer areas.

The major problems in the ecologically fragile areas are the grassland degradation and land sandification, obvious vegetation degradation and severe erosion, frequent natural disasters and difficult poverty alleviation, water shortages and drought, and wetland function decrease and biodiversity loss. Besides the original fragility, there are two other major reasons for the ecological degradation in the ecologically fragile areas, i.e., excessive disturbances of human activities, such as reclamation on the hills and in the areas with water shortages, and the government supervision departments separated in the functions, poor capacity in the coordination and inefficient in the monitoring.

## 6.2 The Conservation of Ecologically Fragile Areas

The principles for the ecological conservation in the ecologically fragile areas should be, giving the emphasis on the prevention and the priority to the protection, moving forward part by part, delivering the guidance by the types, strengthening the monitoring and moderately developing the areas, and formulating an overall plan and implementing the plan step by step. The measures to conserve the ecologically fragile areas could be divided into four categories. The first category is the adaptation measures, including to replace the growing grains by planting fruit trees and developing animal husbandry which are more adaptable to climate change and to develop nonagricultural industries safely.

With the development of the national economy, the adaptation measures for the ecologically fragile areas have expanded from agriculture to industry and services. The increase in the appropriate measures could not only reduce the pressure of production activities on the ecological system, but also promote the development of the ecologically fragile areas.

The secondary category measures to protect the ecologically fragile areas are mutually complementary. It is complementary to provide forage by the agricultural areas to pastoral areas. It should establish a mutual complementary relationship between the animal husbandry and the other industries. It should establish a complementary relationship between the ecologically fragile areas and the other areas. It also should form a complementary relationship between the production and conservation.

The third category measures are protective. The priority is to establish nature reserves, effectively protecting the fragile ecological system. The fourth category is management measures. First, it should strengthen the capacity building in the ecological monitoring and evaluation, establishing an ecological early warning system for the ecologically fragile areas. Second, it should formulate the supervision regulations for the resource development in the ecologically fragile areas, work out various technical standards and technical specifications for the resource development, ecological recovery and reconstruction in the ecologically fragile areas, and actively promote the process of ecological conservation, restoration, and reconstruction in the ecologically fragile areas. Third, it should strengthen the supervision and law enforcement of resource development and improve the environmental monitoring in the ecologically fragile areas.

## 6.3 The Development of Ecologically Fragile Areas

When the ecologically fragile areas were divided over 20 years ago, around 76 % of the counties located in the area were under the poverty line. These poor counties accounted for 73 % of the total poor counties in the provinces they were located. Around 43 % of the ecologically fragile areas were located in the poor counties, accounting for 47 % of the total area of the total poor counties within the same provinces. Around 68 % of the farmland in the ecologically fragile areas was located in the poor counties, accounting for 74 % of the total farmland in the total poor counties within the same provinces. Around 74 % of the population in the ecologically fragile areas was located in the poor counties, accounting for 81 % of the total population in the total poor counties within the same provinces.

After the over 30-year economic reform and opening up, the transformation of China's economic system has primarily completed. With the stronger and stronger market pull, the more and more laborers are employed outside the ecologically fragile areas, and the problem of absolute poverty has basically solved.

After the over 30-year rapidly economic growth, China's comprehensive national power is obviously strengthened, and the government push is stronger and stronger. With the implementation of the western development strategy, the infrastructure of water, power, and transportation in the ecologically fragile areas has been remarkably improved. And conditions for the development of nonagricultural industries are better and better. The implementation of the ecological programs and projects, including the natural forest conservation, conversion of farmland to forests, prevention and control of sandification, "Three-North" afforestation, comprehensive improvement of farmland on the hills, natural reserves, and the issue of the forest and grassland ecological compensation systems have notably improved the macro-environment of the ecological conservation and construction in the ecologically fragile areas, and pressures on the ecological environment and natural resources have continuously declined.

At the end of 1990s, China entered the stage of basic balance between the supply of and demand for the agricultural products, with surplus in the harvest years. Since then the characteristics of marketization, specialization, and regionalization are more and more remarkable. With this context, the comparative advantage has been clear to grow fruit trees and develop animal husbandry in the ecologically fragile areas.

## *6.4 Several Relationships for the Conservation and Development of the Ecologically Fragile Areas*

### 6.4.1 The Relationship Between the Ecological Improvement and the Economic Development

It is necessary to strengthen the ecological construction in the ecologically fragile areas and at the same time to promote the economic development as well. It should make the ecological system used efficiently on the one hand, while the industrial structure adjusted from unsustainable to sustainable, on the other hand. The conditions to grow and develop the organic, green and nonpollutant fruit trees, and animal husbandry are much better in the ecologically fragile areas rather than in the other areas. Therefore, the meaning to conserve the ecologically fragile areas is not to take the extreme measures such as prohibiting the logging, grazing and fishing, but to take appropriate measures which are able to keep the sustainable use of the resources.

### 6.4.2 The Relationship Between the Investments in the Ecological Improvement and in the Human Capital

Ecological conservation has a large externality. Therefore, the government should induce the residents in the ecologically fragile areas to invest in the human capital, making them have the ability to upgrade the employment structure and industrial structure and continuously reducing the pressure of economic growth on the ecological resources, while increase the investment in the improvement of the ecologically fragile areas. The government's ecological compensation should be made for both the stocks and the incremental amount of the ecological assets. If the ecological compensation is simply delivered in light of the area of the ecological assets, it would be difficult to encourage the residents in the ecologically fragile areas to protect the ecological environment, even if the standard of the ecological compensation is quite high. The residents would have no the incentives to expand the ecological assets without the establishment of the ecological compensation system in light of the incremental ecological assets.

### 6.4.3 The Relationship Between the Ecological Improvement and Reasonable Planning

First, there must be forward-looking and operational planning for the ecological improvement, ensuring the work of the ecological conservation and construction carried out smoothly. Second, it must organically unify the planning preparation and cycle management. On the one hand, it should prepare the overall planning and the implementation programs for different areas and levels, combining the short, medium, and long-term goals organically. On the other hand, it should timely (re) adjust the planning and implementation programs in light of the changes in the conditions, keeping them always suitable to the reality.

### 6.4.4 The Relationship Between the Technical and Mechanism Innovations

First, the ecological improvement must be led by technical innovation. The innovation should be conducted in the key techniques for the ecological conservation, restoration and reconstruction, really practicing the principles of priority ecological improvement. It should based on the local conditions grow fruit trees and develop the animal husbandry, eco-tourism and other nonagricultural industries, forming the industrial structures and land use structure suitable to the ecologically fragile areas, really implementing the principle of ecological and economic coordinated.

Second, it must follow the rules of gradual progress, keeping going and step by step. Third, the government should really protect the property right, maintain the fair competition, and provide public goods, completing the transformation from a management government to a service government.

## *6.5 Tasks and Measures to Maintain the Ecological Stability*

The major tasks for the ecological conservation are given below. It should improve the policy, law and regulation systems, establish a diversified comanagement mechanism for the communities, enhance the public awareness in the participation, and promote the ecological conservation and construction. It should establish the ecological compensation system, realize the balance between the demand and supply sides of the positive externalities, achieve the ecological conservation socialization, and gradually eliminate the differences between the areas, residents' income, and public services. It should strengthen the scientific and technological innovation, promoting the ecological conservation and restoration. It should establish a suitable system for the industry access, limit or reduce the human disturbance, mitigate the population pressure on the land, and effectively overcome the ecological fragility.

## References


Bai, Z. P. (2010). Study on Grazing forbidden project in wulatehou banner. *Xinjiang Agricultural Sciences, 47*(S2), 198–203.

Bian, D., Li, C., Yang, X. H., Bian, B. C., & Li, L. (2008). Analysis of the situation of grassland degradation and it's mechanism of the alpine pastoral area in northwestern tibet. *Journal of Natural Resources., 23*(2), 254–262.

CCICCD (China National Committee for the Implementation of the UNCCD). (1997). *China country paper to combat desertification*. Beijing: China Forestry Publishing House.

Chen, G. X., Wang, J. K., & Wang, L. H. (1983). *Debris-flow prevention*. Beijing: China Railway Publishing House. (in Chinses).

Ci, L. J., & Wu, B. (1997). Climatic type division and the potential extent determination of desertification in China. *Journal of Desert Research, 17*(2), 107–112. (in Chinese).

Cui, P. (2009). Advances in debris-flow prevention in China. *Science of Soil and Water Conservation, 7*(5), 7–13. (in Chinses).

Cui, P., Han, Y. S., & Ge, Y. G. (2007). China strategy of flood mitigation: From control to management. In C. Huang, C. Frey, & J. Feng (Eds.), *Proceedings of the first international conference on risk analysis and crisis response: Advances in intelligent system research* (Vol. 2, pp. 515–520). German: Springer.

Cui, P., Wang, D. J., & Zhuang, J. Q. (2008). Land exploitation and utilization on debris-flow fan. *Science of Soil and Water Conservation, 6*(s1), 6–11. (in Chinses).

Fan, J., Zhong, H. P., Chen, L. B., & Zhang, W. Y. (2007). Some scientific problems of grassland degradation in arid and semi-arid regions in northern China. *Chinese Journal of Grassland., 29* (5), 95–101.

Hao, L., Gao, J. M., & Yang, C. Y. (2006). Grey incidence analysis of factors affecting grassland degradation in Inner Mongolia. *Acta Prataculturae Sinica, 15*(6), 26–31.

Hou, X. Y. (2005). *The strategy study of ecological environment construction of grassland in China*. Beijing: China Agriculture Press.

Jia, D. (2001). Painting the blueprint of new century forestry construction. Chinese Green Times, December 31.

Kang, Z. C., Li, Z. F., Ma, A. N., et al. (2004). *Research on debris flow in China*. Beijing: Science Press. (in Chinses).

Li, D. J. (1997). *Practice and theory of debris-flow disaster reduction*. Beijing: Science Press. (in Chinses).

Li, S. (2001a). The evolvement and apocalypse of important forestry ecological projects in foreign countries. *Forestry Economy, 12*, 46–50.

Li, Y. (2001b). The review and general thought of the work on conversion croplands into forests and grass. *Forestry Economy, 9*, 3–11.

Li, S. (2002a). Study on models of conversion of cropland to forest and grassland in the Midwest China. *Scientia Silvae Sinica, 38*(3), 154–159.

Li, S. (2002b). The comparison study on forestry ecological projects in the World. *Acta Ecologica Sinica, 22*(11), 1976–1982.

Li, S. (2003). Study on the development stages of conversion of cropland to forest and grassland. *World Forestry Research, 16*(1), 36–41.

Li, S. (2005a). *Effect of mechanical disturbance to microbiotic crusts on soil hydrological process in revegetated sanddune regions. The research on the sandy desert and control for 50 Years*. Beijing: Ocean Press. (in Chinese).
Li, X. (2005b). Influence of variation of soil spatial heterogeneity on vegetation restoration. *Science in China Series D: Earth Sciences, 48*(11), 2020–2031.
Li, S. L., Chen, Y., Kang, S., & Guan, S. (2002). Relationships between soil degradation and rangeland degradation. *Journal of Arid Land Resources and Environment, 16*(1), 93–95.
Liu, D. M. (2009). Implants of fencing on 'black soil beach' sowing grassland in the Yangtze and Yellow River regions. China: Gansu Agricultural University.
Lu, Q., & Wu, B. (2002). Disaster assessment and economic loss budget of desertification in China. *Population, Resources and Environment in China, 12*(2), 29–33. (in Chinese).
Lu, Q., Yang, Y. L., Wang, S., & Liu, T. (2004). *The revelation: Combating desertification in China*. Beijing: Science Press. (in Chinese).
MA, Z. F. (2008). *Feed balance analyzer and livestock production system optimizing model building up and case study*. China: Gansu Agricultural University.
Mitsch, W. J., & Jorgensen, S. E. (1989). *Ecological engineering*. New York, USA: Wiley.
Qin, P. (1998). *Ecology engineering*. Nanjing: Nanjing University Publishing House.
Ren, J. Z. (2004). *Theory of grassland-agriculture ecosystem*. Hefei: Anhui Education Press.
Shang, Z. H., Ren, G. H., & Long, R. J. (2009). Review of soil seed bank studies: Size, pattern and impacting factors. *Acta Prataculturae Sinica, 18*(1), 144–154.
Sun, Z. J., An, S. Z., & Duan, J. J. (2009). Effect of enclosure on vegetation and soil nutrient of sagebrush desert grassland in Xinjiang. *Arid Land Geography, 26*(6), 877–881.
Sun, H., Tang, Y., & Wang, C. M. (2001). Contour hedgerow intercropping for exploitation and conservation of slope cropland in mountain areas. *Jounal of Mountain Science, 19*(2), 125–129. (in Chinses).
Tang, B. X. (2000). *Debris flow in China*. Beijing: The Commercial Press. (in Chinses).
Tang, B. X., Du, R. H., Kang, Z. C., et al. (1980). Research on debris flow in China. *Journal of Geographical Science, 35*(3), 259–265. (in Chinses).
Tang, B. X., Liu, S. Q., Liu, S. J., et al. (1991). *Distribution map of debris flow and its disaster level regionalization in China*. Chengdu: Map Press of Chengdu.
Wang, L. (1998). *Forestry ecology engineering*. Beijing: China Forestry Publishing House.
Wang, Y. C. (2007). *Study ont the restoration effects of enclosed pasture in the national project of pastureland rehabilitation areas in A-Ba County*. Ya'an: Sichuan Agricultural University.
Wang, Y., & Cao, J. (2010). Study on grassland degradation and rational utilization of Inner Mongolia pastoral and semi-pastoral areas. *Journal of Inner Mongolia Agricultural University (Social Science Edition), 12*(3), 57–59.
Wang, Z. Y., Melching, C. S., Duan, X. H., & Yu, G. A. (2009). Ecological and hydraulic studies of step-pool systems. *Journal of Hydraulic Engineering, 135*(9), 705–717.
Wu, J. S., Tian, L. Q., Kang, Z. C., et al. (1993). *Debris flow and its countermeasures*. Beijing: Science Press. (in Chinses).
Xiang, J. (2001). *Forestry ecology engineering*. Beijing: Higher Education Publishing House.
Xiao, D. (2003). Ecological reconstruction and landscape planning of ecological fragile zone. *Journal of Desert Research, 23*(1), 45–53.
Xiao, H. L., et al. (2003a). Impact of evolution of plant-soil system on the water environment during the mobile dunes stabilization. *Acta Pedologica Sinica, 40*(6), 813–826.
Xiao, H. L., et al. (2003b). Succession of plant-soil system in the process of mobile dunes stabilization. *Journal of Desert Research, 23*(6), 608–613.
Xu, S. (1988). *The elements of analytic hierarchy process methods*. Tianjin: Tianjin Publishing House.
Xu, J., & Wang, Z. Y. (2003). The formative mechanism of step-pool system and its role in bed stabilization of mountain streams. *Journal of Sediment Research, 5*, 21–27. (in Chinses).
Yin, J., Jiang, L., Xu, Z. L., Zhang, M. Y., Wang, Y. D., & Huang, A. M. (2010). Effect of "grazing withdrawal and management of grassland" project on grassland ecology and socio-economic in Diqing District of Yunnan Province. *Grassland and Livestock, 11*, 26–29.

Zhang, P. (1993). *The five important ecological projects in China in the contemporary era*. Beijing: China Forestry Publishing House.
Zhang, Z. (1997). *The ten important forestry ecological projects in China*. Beijing: China Forestry Publishing House.
Zhang, X. S. (2010). An intellectual enquiring about ecological restoration and recovery, their scientific implication and approach. *Chinese Journal of Plant Ecology, 34*(1), 112–118.
Zhang, D. J., & Du, Y. T. (2006). Effect of enclosure and grazing prohibition on the improvement of deteriorated grassland. *Grassland and Turf, 4*, 52–54.
Zhang, Y. W., Han, J. G., & Li, Z. Q. (2002). A study of the effects of different grazing intensities on soil physical properties. *Acta Agrestia Sinica., 10*(1), 74–78.
Zhang, Y., & Wang, Q-Z. (2010). Analysis of performance and sustainability of Grazing Forbidden Project in Ning Xia. *Modern animal husbandry, 1*, 53–55.
Zhao, Y., & Yu, L. X. (2011). Estimate on the effect of Grazing Forbidden Project with national debt fund in Xin Jiang. *Xin Jiang Animal Husbandry, 6*, 43–45.
Zhong, D. L., Wang, C. H., Xie, H., et al. (1998). *Catalougue, database, regional rules of debris flow and landslide in China*. Chengdu: Sichuan Publishing House of Science and Technology. (in Chinses).
Zhou, S. (2001). Make great efforts to realize the rapid progress of forestry in the new century. *China Forestry, 2B*, 3–12.
Zhou, G., Chen, G., Xu, W., Yang, L., Han, Y., & Li, J. (2010). Influences of enclosure to Achnatherum splendens steppes biomass in the Qinghai Lake Area. *Arid Land Geography, 33* (3), 434–441.
Zhou, B. F., Li, D. J., Luo, D. F., et al. (1991). *Handbook of debris-flow prevention*. Beijing: Science Press. (in Chinses).
Zhu, L. (2006). *Dynamics of desertification and sandification in China*. Beijing: China Agricultural Press Research.

# Part IV
# Global Change Ecology

# Chapter 13
# The Contribution of Ecology in International Global Change Projects

**Quansheng Ge, Bangbo Cheng, Panqin Chen, Xiuqi Fang and Xiubin Li**

**Abstract** Since 1980s, global environmental change has gradually become the hot research field of earth science and life science. Ecology, as an interdisciplinary of the two sciences, intervenes more global environmental change study. This study concerns the change of support capacity from earth system to life, which obviously accords with ecology's mission. Recognizing the importance of integral systematic concept and long-term systematic simulating observation on solving the problem of environmental change, international geological and ecological circles changes the organization formation of scientific research and continuously establish many large research projects. The research organization formation and related academic exchange activities had promoted the intercross fusion between different subjects and the meta-synthesis of studies on resource, environment, ecology, and disaster.

**Keywords** IGBP · IHDP · China's contribution · Ecological research · Academic exchange

Q. Ge · B. Cheng · X. Li (✉)
Institute of Geographic Sciences Natural Resources Research, Chinese Academy of Sciences, Beijing 100101, China
e-mail: lixb@igsnrr.ac.cn

P. Chen
Bureau of Science and Technology for Resources and Environment, Chinese Academy of Sciences, Beijing 100864, China

X. Fang
Beijing Normal University, Beijing 100875, China

 
W. Li (ed.), *Contemporary Ecology Research in China*,
DOI 10.1007/978-3-662-48376-3_13

# 1 China's Contributions to IGBP Research

## *1.1 Introduction*

The International Geosphere-Biosphere Program (IGBP) was founded by the International Council of Scientific Unions (ICSU) in 1987. In October 1988, the first meeting of Scientific Advisory Council of IGBP (SAC-IGBP) adopted the outline of IGBP, which included four main contents as follows (Chen 1989):

(i) Terrestrial biosphere–atmosphere chemical interaction
(ii) Marine biosphere–atmosphere interactions
(iii) Biosphere aspects of hydrological cycle
(iv) Effects of climate change in terrestrial ecosystem

Ecosystem is one of the few major research objects of IGBP. Many core projects of IGBP, such as GCTE, GLOBEC, iLEAPS, and IMBER are designed and implemented in the fields of ecosystem, biological processes, and biogeochemistry. Therefore, IGBP has become an important platform for the development of ecology.

As one of the initiators of IGBP, China was also one of the early global change researchers in the world and has made important contributions to the development of IGBP in many fields, including ecology.

## *1.2 Establishment and Development of China's IGBP Research Organizations and Participation of Ecological Academic Circles*

### 1.2.1 Establishment and Development of CNC-IGBP and Its Working Groups Related to Ecology

In 1988, only one year later than IGBP's foundation, the Chinese National Committee for the International Geosphere-Biosphere Program (CNC-IGBP) was established in Beijing. Under the leadership of the China Association for Science and Technology, CNC-IGBP is an academic organization that organizes and promotes Chinese scientists' participation in IGBP research.

Since the establishment of CNC-IGBP, its members' number and representativeness have both expanded. From 1988 till now, the number of CNC-IGBP's members have increased from 27 to 59, participating academic societies have increased from 13 to 16 and participating government agencies have increased from 5 to 9. Its working groups have increased from 2 to 12, with group members increasing from less than 20 to nearly 300.

Each working group of CNC-IGBP works on ecology and some of them are highly related to ecological core projects of IGBP, like working group on GCTE and working group on GLOBEC/IMBER. These groups have promoted participation of Chinese scientists in IGBP ecological research and development of China's ecological research.

### 1.2.2 Ecologists' Participation in CNC-IGBP

China's IGBP related researches are gathering many excellent ecologists and several ecological societies. Each successive session of CNC-IGBP committee has representatives from Ecological Society of China, Botanical Society of China, Chinese Society of Forestry, etc. There are many ecologists in CNC-IGBP working groups. Some of these ecologists are not only members of CNC-IGBP, but also leaders of international ecological programs like IGBP, DIVERSITAS, MAB, and IGBP ecological core projects.

## *1.3 Contribution of China's Ecological Research to IGBP*

### 1.3.1 Establishment and Development of CERN

In 1988, the Chinese Academy of Sciences established Chinese Ecosystem Research Network (CERN). In the past quarter of the century, CERN set up basic platform for China's long-term ecological experiment and data accumulation, provided firm support for deep research on ecology, and put good experience of ecosystem management into demonstration. Till now, it has made important achievements in dynamic observations of ecosystems and related researches, experiments, and demonstrations. Because of its outstanding achievements, CERN has become one of the three most important national ecosystem research networks in the world, equal to the Long Term Ecological Research (LTER) network of the United States and the Environmental Change Network (ECN) of the United Kingdom.

CERN is not only playing an important role in ecological research, but also in global change research. For example, it developed spatial climate elements database of China's terrestrial ecosystems, clarified spatial change characteristics of China's soil, carbon, and nitrogen in 1990s, and constructed a series of models as base for coupling research of carbon and hydro cycles (CERN Scientific Committee 2008; CERN Synthesis Research Center 2010; CERN website 2011).

### 1.3.2 Studies on Transects of Global Change

Terrestrial transects is an effective platform for global change studies, in order to understand how terrestrial ecosystems will respond to global change, to predict and evaluate the possible effects of global change, and to ensure the sustainable development of our life supporting system. Chinese ecologists made important contributions in this field. Northeast China Transect (NECT) and North–South Transect of Eastern China (NSTEC), which were proposed by Chinese ecologists, have been selected as IGBP standard transects and have become important bases for global change and ecosystem studies (Zhang et al. 1997; Zhou et al. 2002; Peng 2001; Li et al. 2004; Yu et al. 2010).

### 1.3.3 Development of Ecosystem Models

Accurate prediction of climate change and its possible impact to ecosystems is one of the major objectives of IGBP. In these researches, ecosystem models have been proved to be an effective tool. Chinese scientists have developed several terrestrial and marine ecosystem carbon cycle models, and made some important progresses in the development of models of typical ecosystems and coupling of ecosystem models and models of other earth subsystems. These models are playing significant roles in researches on understanding relationship between ecosystem evolution and climate change, and on estimates of characteristic and potential of carbon budget of China's terrestrial ecosystem(Zeng et al. 2005, 2008a, b; Zhou and Zhang 1995; Zheng and Zhou 2000; Gao et al. 1997; Zeng 2010; Zeng and Lin 2010; Liu et al. 2009; Mao et al. 2005).

### 1.3.4 Studies on Carbon Storage of China's Terrestrial Ecosystem and Its Spatiotemporal Distribution

Mechanisms of carbon storage change and carbon cycling process of earth system are scientific basis for analysis of climate change causes, prediction of its climate trends, and mitigation and adaptation policy making. Since 1980s, Chinese scientists have studied the carbon cycles of terrestrial ecosystem, and these studies produced important results. For example, Fang et al. (2007) estimated terrestrial vegetation carbon sinks for China's major biomes between 1981 and 2000, using China's ground observations. Tao et al. (2007) used a high-resolution climate database and an improved ecosystem process-based model to quantify spatiotemporal pattern and dynamic net ecosystem productivity (NEP) in China and its responses to climate change during 1981–2000. Ji et al. (2008) investigated the projected changes in carbon exchange between China terrestrial ecosystem and the atmosphere, vegetation, and soil carbon storage during the twenty-first century, using an atmosphere–vegetation interaction model (AVIM2).

### 1.3.5 Studies on Marine Ecosystem Dynamics

Marine ecosystem dynamics is an important part of researches on global change and marine ecosystem. Chinese scientists started studies on mega marine ecosystem from the early 1980s, and Chinese studies on marine ecosystem dynamics have been developing together with international scientific community since then. Both the Ministry of Science and Technology (MOST) of the PRC and the National Natural Science Foundation of China (NSFC) arranged key research projects on marine ecosystem dynamics, such as “Ecosystem Dynamics and Sustainable Utilization of Living Resources in the East China Sea and Yellow Sea” (1999 −2004) within national basic key research program. These research activities have made many innovative achievements (Tang et al. 2005).

### *1.4 Domestic and International Academic Exchanges*

Since Chinese scientists joined in IGBP research, there have been numerous academic exchanges in many different ways, among Chinese scientists themselves and with international scientific community. Such exchanges not only promote the understanding and cooperation among scientists, but also advanced the development of China’s global change ecology.

### *1.5 Outlook*

In future, international global change research will pay more attention to sustainable development of human society, integration of earth system factors, application of new technologies and modeling, data sharing, and international strategic cooperation (Ge et al. 2007).

As for China, ecology will play a more important role in global change research, and addressing global change will be one of the major objectives of China’s ecological research. Meanwhile, there will be need of adjustment and improvement in research contents, studying methods, basic capacity, and international cooperation (Li 2010; Li et al. 2005).

## 2 China’s Contributions to IHDP Research

The International Human Dimensions Program on global environmental change (IHDP) is an academic organization to coordinate, plan, and organize international social science researches related to global environmental change. IHDP was established in 1996, and was jointly sponsored by the International Council for Science (ICSU) and the International Social Science Council (ISSC). It is an international nongovernmental organization. The predecessor of the IHDP is Human Dimensions Program (HDP), which was founded in 1990 by the ISSC. In 2006, the United Nations University joined as an institutional sponsor. IHDP’s mission is to promote innovative interdisciplinary researches in global environmental change. These researches take the coupled human and nature system under the background of global environmental change as the research object, aiming to describe, analyze, and understand the humanistic factors of global environmental change.

In China, the basic characteristic of economic and social development in recent years is the rapid industrialization and urbanization with rapid economic growth. It has resulted in high pressure on natural environment, and it requires scholars for further improving the recognition of the earth system, especially the coupled human and nature system. Improved recognition should then be translated to relevant knowledge and techniques for solving the existing problems. In addition, human activities should be reasonably organized for sustainable development. In this context, IHDP attracted the active participation of Chinese scholars at the very beginning.

(i) The initial IHDP included four core projects, including the Global Environmental Change and Human Security Project (GECHS), the Institutional Dimensions of Global Environmental Change Project (IDGEC), the Industrial Transformation Project (IT), and the Land Use and Cover Change Project (LUCC). Of them, LUCC was the first one jointly launched by IHDP and IGBP (Turner et al. 1995). Chinese scholar Liu Yanhua participated in the formulation of the LUCC plan. The project started in 1995 and ended in 2005. During this period, another Chinese scholar Xu Jianchu, made contributions to the IGBP-LUCC project as a member of the scientific steering committee. The Urbanization and Global Environmental Change Project (UGEC) began in 2005, and Deng Xiangzheng, from the Institute of Geographical Sciences and Natural Resources Research (IGSNRR), Chinese Academy of Sciences (CAS), was elected as one of its scientific steering committee members.

After 2005, LUCC combined with another core project of the IGBP, the Global Change and Terrestrial Ecosystems Project (GCTE), and formulated a new project named the Global Land Project (GLP) in the context of the Earth System Science Partnership (ESSP) framework. Chinese scholar Liu Jiyuan participated in the formulation of the GLP plan, and was a member of the first scientific steering committee of the project. GLP established three project offices for effective execution of the project, with one hosted by IGSNRR of CAS in Beijing.

(ii) In July 2003, 26 academicians and scientists with Sun Honglie as the leader, made a suggestion to the MOST about the establishment of the Chinese National Committee for the International Human Dimensions Program on Global Environmental Change (CNC-IHDP). This suggestion was approved by the MOST. After one year's preparation, CNC-IHDP was established in Beijing on August 30th, 2004.

CNC-IHDP's mission is to organize and coordinate experts from natural sciences and humanities for cooperation in major topics on global environmental change and global social sustainable development. The first scientific steering committee of the CNC-IHDP had 69 members from 16 governmental ministries and their affiliating research institutes, 12 universities, and 2 nongovernmental organizations. Liu Yanhua was elected as the chairman of the committee; Sun Honglie was elected as the chairman of the consultant committee; Ge Quansheng was the secretary. The secretariat is hosted by IGSNRR, CAS.

(iii) A new core project named Integrated Risk Governance Project (IRG) was proposed during the IHDP scientific committee meeting in Bonn, Germany in September 2010. The IRG was the first IHDP core project led by Chinese scholars. Shi Peijun from Beijing Normal University and Carlo Jaeger from the Potsdam Institute for Climate Impact Research were elected as co-chairmen; Ye Qian from the CNC-IHDP was appointed as the executive director. International program offices (IPO) are in Beijing, Normal University and the Potsdam Institute for Climate Impact Research.

IHDP-IRG provides a platform for communications among the experts and organizations around the world who are engaged in risk research and management. It also leads the researches on future international integrated risk prevention to start from the scientific, technical, and management issues, and to carry out innovative researches on theories and methods of the integrated disaster risk prevention from multidisciplinary perspectives through case studies and comparison, aiming at promoting further development of the global integrated risk prevention practice.

Communication is a fundamental activity during the process of scientific research. It is also a catalyst for academic innovation. It is believable that Chinese scholars will make full use of the CNC-IHDP as an academic communication platform, and make further contributions to the IHDP and the sustainable development of human society.

## References


CERN Scientific Committee. (2008). Stratigic development plan of CERN (2008–2020).

CERN Synthesis Research Center. (2010). Research progress and perspectives of synthesis Research Center of Chinese Ecosystem Research Network, Key Laboratory of Ecosystem Network Observation and Modeling, CAS. *Journal of Natural Resources, 25*(9), 1458–1467.

CERN website. (2011). Introduction. http://www.cern.ac.cn/

Chen, P. (1989). Outline of global change research programme. *Advances in Earth Sciences* 1–5 (4th period of year 1989).

Fang, J. Y., Guo, Z. D., Piao, S. L., et al. (2007). Terrestrial vegetation carbon sinks in China, 1981–2000. *Science China Earth Sciences, 50*(9), 1341–1350.

Gao, Q., Yu, M., Zhang, X., et al. (1997). Dynamic modeling of Northeast China transect responses to global change-a regional vegetation model driven by remote sensing information. *Acta Botanica Sinica, 39*(9), 800–810.

Ge, Q. S., Wang, F., Chen, P. Q., et al. (2007). Review on global change research. *Advances in Earth Science, 22*(4), 417–427.

Ji, J. J., Huang, M., & Li, K. R. (2008). Prediction of carbon exchanges between China terrestrial ecosystem and atmosphere in 21st century. *Science China Earth Sciences, 51*(6), 885–898.

Li, J. L., Chen, P. Q., Ge, Q. S., et al. (2005). Global change and human activities: Priorities of the global change research in next phase in China. *Advances in earth sciences, 20*(4), 371–377.
Li, Y., Liao, S., Chi, G., et al. (2004). NPP distribution related to the terrains along the North-South transect of Eastern China. *Chinese Science Bulletin, 49*(6), 617–624.
Liu, S., Lin, Z., Jiang, X., et al. (2009). The IAP coupled atmosphere vegetation model and its numerical simulation. *Climatic and Environment al Research, 14*(3), 258–272.
Mao, J., Wang, B., Dan, L., et al. (2005). Coupling of an atmosphere-vegetation interaction model (AVIM) to a new generation grid point atmospheric general circulation model. *Chinese Journal of Atmospheric Sciences, 29*(6), 897–910.
Peng, S. (2001). Important progresses in studies on mechanisms of interactions between terrestrial agricultural ecosystem and global change in Eastern China. *China Basic Science* 18–20 (7th period of year 2001).
Tang, Q., Su, J., Song, S., et al. (2005). A study on marine ecosystem dynamics in the coastal ocean of China. *Advances in Earth Science, 20*(12), 1288–1299.
Tao, B., Cao, M. K., Li, K. R., et al. (2007). Spatial patterns of terrestrial net ecosystem productivity in China during 1981–2000. *Science China Earth Sciences, 50*(5), 745–753.
Turner, B. L., Skole, D., Sanderson, S., et al. (1995). Land-use and land-cover change science/research plan. Joint publication of the International Geosphere-Biosphere Programme (Report No. 35) and the Human Dimensions of Global Environmental Change Programme (Report No. 7). Stockholm: Royal Swedish Academy of Sciences.
Wenhua, L. (2010). Progresses and perspectives of ecological research in China. *Journal of Resources and Ecology, 1*(1), 3–14.
Yu, Z., Sun, P., & Liu, S. (2010). Phenological change of main vegetation types along a North-South transect of Eastern China. *Chinese Journal of Plant Ecology, 34*(3), 316–329.
Zeng, X. (2010). Evaluating the dependence of vegetation on climate in an improved dynamic global vegetation model. *Advances in Atmospheric Sciences, 27*(5), 977–991.
Zeng, Q. C., & Lin, Z. H. (2010). Recent progress on the earth system dynamical model and its numerical simulations. *Advances in Earth Science, 25*(1), 1–6.
Zeng, X., Wang, A., Zhao, G., et al. (2005). Ecological dynamic model of grassland and its practical verification. *Science China Life Sciences, 48*(1), 41–48.
Zeng, X., Zeng, X., Barlage, M. (2008a). Growing temperate shrubs over arid and semiarid regions in the Community Land Model–Dynamic Global Vegetation Model. *Global Biogeochemical Cycles, 22*, GB3003, doi:10.1029/2007GB003014
Zeng, Q. C., Zhou, G. Q., Pu, Y., et al. (2008b). Research on the earth system dynamic model and some related numerical simulations. *Chinese Journal of Atmospheric Sciences, 32*(4), 653–690.
Zhang, X., Gao, Q., Yang, D., et al. (1997). A gradient analysis and prediction on the northeast China transect (NECT) for global change study. *Acta Botanica Sinica, 39*(9), 785–799.
Zheng, Y. R., & Zhou, G. S. (2000). A forest vegetation NPP model based on NDVI. *Acta Phytoecologica Sinica, 24*(1), 9–12.
Zhou, G., Wang, Y., & Jiang, Y. (2002). Global change and water-driven IGBP-NECT, northeast China. *Earth Science Frontiers (China University of Geosciences, Beijing), 9*(1), 198–216.
Zhou, G., & Zhang, X. (1995). A natural vegetation NPP model. *Acta Phytoecologica Sinica, 19* (3), 193–200.

# Chapter 14
# Ecosystem Positioning Research and Models

**Guirui Yu, Nianpeng He, Guirui Yu, Biao Liang, Zhiwei Xu, Shenggong Li, Xuebing Guo, TianXiang Yue and Moucheng Liu**

**Abstract** Looking through the three tendencies of Global Change Ecology in the past several years, there are two related to this chapter: (i) insisting on a long term of systematic observation, and building a data acquisition network; and (ii) strengthening systematic simulation study, deeply understanding the mechanism and rules of the earth's coupling change and doing further prediction research. It is the scholars' consensus that in the background of global changing, a long-term and multi-scale networking observational study of ecosystem would be the most effective method to reveal its evolvement rule, and system simulation is a basic tool to build and test theoretical assumptions and to predict the ecosystem's evolvement directions. The Chinese Ecosystem Research Network (CERN) and the Chinese Terrestrial Ecosystems Flux Observation and Research Network (ChinaFLUX) take a wide participation in the International Long-Term Ecosystem Research Network (ILTER) and the Global Flux Observation and Research Programme (FLUXENT), and play an important role pushing the development of relevant academic subjects. At present, China's stations in the ecosystem observation network almost cover its main diverse ecosystem types, and with these observation data, we launched deep analysis on the main types of ecosystems, of its photosynthetic carbon fixation, respiration of carbon emissions, dynamic time and space distribution rules of ecosystem's net carbon exchange flux and the environmental and biological control mechanism, etc., with which we acquired a series of important scientific findings, and developed terrestrial ecosystem carbon-cycle process-mechanism model of forest, farmland, grassland, etc. and satellite remote sensing system of carbon balance evaluation. Readers can also get a comprehensive understanding of ecosystem modeling's theories, methods, and its practice from this chapter's massive cases and relevant essays.

**Keywords** CERN · ChinaFLUX · Multi-scale network observation · CNERN · Methods for earth surface modeling

---

G. Yu · N. He · G. Yu · B. Liang · Z. Xu · S. Li · X. Guo · T. Yue · M. Liu (✉)
Institute of Geographic Sciences and Natural Resources Research, Chinese Academy of Sciences, Beijing 100101, China
e-mail: liumc@igsnrr.ac.cn

 
W. Li (ed.), *Contemporary Ecology Research in China*,
DOI 10.1007/978-3-662-48376-3_14

# 1 Multi-scale Network Observation in Chinese Terrestrial Ecosystems: Progress and Prospects

## 1.1 Introduction

Multi-scale network observations in Chinese terrestrial ecosystems have been conducted since the 1980s. The field ecosystem station networks that were established in China in 1989 promoted their development because of the consistent observation system and monitoring parameters. The symbolic events established the Chinese ecosystem research network (CERN) by the Chinese Academy of Sciences in 1988 and formed the Chinese forest ecosystem research network (CFERN) by the Ministry of Forest of People's Republic of China. Moreover, a series of terrestrial transects, including the Northeast China temperate grassland transect (NECT in 1993), the North-South transect of eastern China (NSTEC in 1999), and the Chinese Terrestrial Ecosystem Flux Observational Research Network (ChinaFLUX in 2003), have also been important approaches in conducting the multi-scale network observations. In 2006, on the basis of the former ecosystem stations, we built the China National Ecosystem Research Network (CNERN), which was composed of 51 important field ecosystem stations. With the development of ecosystem observation networks, the multi-scale network observations in China have covered all important ecological regions (or typical ecosystems) and have a comprehensive monitoring system for the long-term monitoring of water, soil, atmosphere, and biological elements of ecosystems and important ecological processes. Here, we briefly review the progress of the multi-scale network observations in Chinese terrestrial ecosystems in the past two decades.

## 1.2 Multi-scale Network Observation in China

### 1.2.1 Ecosystem Research Networks

CERN was established in 1988. Over the past 20 years, as a major component of the Global Terrestrial Observing System and International Long-Term Ecological Research, CERN has grown to be an important research platform for the sites and partners, both at home and abroad. This has been mainly attributed to the data and measurements that were collected on a long-term and continuous basis, and to the up-to-date facilities and instruments in the field stations covering the major ecosystem types in China (Fu et al. 2010). CFERN is a large ecology research network that focuses on long-term fixed observations of forest ecosystems. It embodies 15 sites that represent diverse ecosystems and research priorities (Wang et al. 2004). CFERN uses the multi-approach ground and spatial observation technologies to study the structure, function, and feedback mechanism of Chinese forest ecosystems respond to climate change, as well as the effects of climate change on China's social and economic development. In 2006, the Ministry of Science and Technology of the People's Republic of China initially established CNERN on the basis of the former ecosystem stations, which included 51 important field ecosystem stations.

All the field stations (CERN, CFERN, and CNERN) continuously measure and record the temporal variations in hydrological, pedological, atmospheric, and biological elements of major terrestrial and aquatic ecosystems in China. Meteorological and atmospheric parameters (such as solar radiation, ultraviolet radiation, photosynthetic active radiation, net radiation, surface reflected radiation, soil heat flux, soil temperature profile, wind direction, wind speed, air temperature, air humidity, precipitation, and pan evaporation), soil physical and chemical properties, vegetation structure and function (such as plant species composition, leaf area index, and biomass), water budget (such as precipitation, evaporation, transpiration, and runoff), and nutrient budget (N, P, and K) are periodically recorded at some of the stations in accordance with standard monitoring protocols. Most stations initially implemented the routine long-term ecological and environmental monitoring and began accumulating long-term data over 20 years ago. The high-quality datasets that were collected on the long-term and continuous basis provide the foundation for the ecological research. For example, observations at CERN field stations showed that ultraviolet radiation and PAR in China increase longitudinally from east to west, and they increase latitudinally from north to south in the west and from south to north in the east (Hu et al. 2007a). Nutrient cycle experiments, particularly nitrogen cycle experiments conducted at the CERN research stations of forest, grassland, cropland, and desert ecosystems, highlight the nutrient status and dynamics.

### 1.2.2 Transect Investigation

Chinese scientists established the NECT in 1993, which was officially registered as the fifth transect of the Global Change and Terrestrial ecosystems (GCTE) by IGBP in 1994 (Chen et al. 2003; Ni 2003). This transect is located in the mid-latitude semiarid region, encompassing 42–46°N latitude and 110–132°E longitude. Later, the NSTEC, from the polar tundra to rain forests along the east coasts of the Eurasian continent, was proposed and accepted as the 14th transect of the GCTE (Hang and Zhou 2008). Moreover, other transects were established subsequently for specific scientific issues, such as the China grassland transect to study the grassland ecosystem carbon cycle and its driving mechanisms and the Euro-Asian continental grassland transect to study the carbon and water flux at the ecosystem level using eddy flux technology.

In the past decades, Chinese scientists have conducted several integrative investigations in NEST. The primary driving forces for global change are precipitation and land-use change. Research progress was made during the past decade in the following aspects: ecological database development, climate and its variability, ecophysiological response of plants to environments, vegetation and landscape changes, biodiversity patterns and their changes, plant functional types and traits relative to the climatic gradient, productivity and carbon dynamics, pollen–vegetation relationships, trace gas emissions, land-use and land-cover changes, and biogeographical and biogeochemical modeling (Chen et al. 2003; Ni and Wang 2004).

In NSTEC, scientists also conducted several investigations and collected information, including plant community structure, soil conditions, land cover, and climate. On the basis of these data, scientists have explored how primary productivity, biodiversity, and soil properties vary with changes in precipitation and temperature along the transect, and they predicted the influence of climate change and human activities on the function and structure of typical forest ecosystems in China (Yu et al. 2013). Moreover, some studies have analyzed the relationship between plant functional traits, the C:N:P stoichiometry of leaves, and non-biological factors (precipitation, temperature, and soil traits) along the transect.

### 1.2.3 Eddy-Flux Technology

ChinaFLUX is a long-term network that relies on CERN and applies eddy covariances of micrometeorology and chamber as the main research method to study fluxes in carbon dioxide, water, and heat between vegetation and soil of typical ecosystems and the atmosphere (Yu et al. 2008). ChinaFLUX is built with the support of the Knowledge Innovation Program of the Chinese Academy of Sciences "Study on Carbon Budget in Terrestrial and Marginal Sea Ecosystems of China," the National Key Basic Research and Development Program (973 Program) of the Ministry of Science and Technology of the People's Republic of China "Carbon Cycle and Driving Mechanism in Chinese Terrestrial Ecosystem," and other international cooperation programs.

Now, ChinaFLUX consists of 54 sites that apply a micrometeorological method and 17 sites that apply a chamber method. In addition to the long-term flux observations, ChinaFLUX will systematically collect related data, such as those on vegetation, soil, hydrology, and climate, and simultaneously carry out a synthesis study on the ecosystem carbon and water cycle processes.

On the basis of these observations, scientists have explored carbon and water exchange properties with seasonal and annual dynamics in these typical terrestrial ecosystems. Furthermore, the findings help us to determine the gross primary productivity, net ecosystem productivity, water-use efficiency, and (TE) at the national scale and to predict the response of terrestrial ecosystems to climate change in the future (Yu et al. 2013).

### 1.3 Future Research Outlook

Multi-scale network observations in China will further depend on the established networks of field ecosystem stations, terrestrial transects, and ChinaFLUX to explore the dynamics of ecosystem structure and function over time and the response of Chinese terrestrial ecosystems to climate change and human disturbance. The main perspectives are as follows: (i) to strengthen the integrated monitoring capacity of existing networks, particularly for the three-dimensional monitoring (traditional investigation, eddy flux observation, and remote monitoring); (ii) to provide better service for some key scientific issues, such as the response and adaptation of ecosystems to global climate change, biodiversity conservation, ecosystem restoration, and sustainability; and (iii) to establish the networks of multi-factor field control experiments and explore the underlying mechanisms of responses of terrestrial ecosystems to changes in temperature, precipitation, land-use type, and N deposition. We believe that long-term multi-scale integrative investigation will play more and more important roles in long-term ecological monitoring, research, and applications in future, and it will provide unique information services to promote the national and local socioeconomic development and environmental conditions in China.

## 2 Construction and Development of Chinese National Ecosystem Observation and Research Network

### 2.1 Introduction

According to the definition by Tansley (1935), an ecosystem is an integrated system composed of interacting biotic (Plants, animals, and microbes) and abiotic components which are closely contact with each other through material circulation and energy flow (Odum and Barrett 1971). In the 150 years since the industrial revolution, ecological problems such as vegetation degradation, land desertification, biodiversity, and ecosystem service function loss caused by climate change and economic development have become the focus of system ecology and global change. The long-term observation research station is one of the effective methods to solve these problems.

Due to the complex of ecosystem structure and function, only through the long-term ecological research network, ecosystem ecology can reveal ecosystem formation mechanism, evaluate the impact of global change on ecosystem, and predict ecosystem evolution trends in the future (Yu et al. 2009). The long-term ecological research has begun in agriculture and forestry. There already have more than 30 research stations which have observation history of more than 60 years in the world and mainly distributed in Russian, Europe, the United States, Japan, and India (Gosz 2000; Waide 2000; Parr and Lane 2000; Zhao 2004). The foreign research stations develop toward internationalization, and form different scales of monitoring, observation, and experiment research network. It put emphasis on the long-term, continuity and globalization of research, automation, and information technology of monitoring. According to the requirement of economic development, research institutes and universities have established a large number of field monitoring research stations combined with the implementation of the national research programs and projects. The Chinese Academy of Sciences already has 42 research stations including agriculture, forest, grass land, wetland and desert, five discipline branches, and one synthesized center. The stations are reasonable distributed with regional representative and ecosystem diversity. The selected stations have become the research platform and demonstration base for monitoring and experiment with advanced infrastructure (Yang et al. 2008; Fu et al. 2010).

## 2.2 *Construction and Development of CNERN*

### 2.2.1 Background and Development Course CNERN

The construction of field research station exists the following problems: (i) The spatial distribution is not reasonable with redundant construction and lack unified planning at the national level; and (ii) there are no unified technical specifications and standards of Observations with difficulty in data sharing. Therefore, it is important to resolve these problems.

In 1999, Ministry of Science and Technology of the People's Republic of China has selected 35 research stations with better infrastructure, regular staff, and higher research level to carry out the experimental work including ecology, agriculture, resources and environment, earth science, and astronomy. The stations are distributed in northeastern, north China, south China, northwest, and antarctic region which laid a solid foundation for the construction of national research network. In 2004, Ministry of Science and Technology of the People's Republic of China entrusts the panel to carry out the program the CNERN including select measures, procedures, and indicators, make national field research articles of association, management regulations, and monitoring index system. At present, CNERN is composed of 51 stations, 1 National Soil Fertility and Fertilizer Effects Long-term Monitoring Network, 1 Germplasm Resources Network, and 1 Comprehensive Research Center.

### 2.2.2 Science and Technology Resources of CNERN

All stations achieved the networking research and accumulated a lot of first-hand research data. CNERN has strong analysis ability of soil, plant, water, carbon, nitrogen, phosphorus, and morphological with field observation field of 2.80 million $m^2$, permanent plot of 1.38 million $m^2$, and 288 monitoring facilities. CNERN has saved all kinds of plants, animals, soil, water, and atmospheric composition samples of 280032 and file of 857740 and construct data management and sharing system. In 2010, CNERN has different data of 42,86,622.54 MB and construct a number of unique observation and test database, such as China agricultural meteorological information database, China 1:10,00000 grassland resources project database, China flux observation database, etc. The comprehensive research center has established specification of long-term observation data for the first time, independently developed a series of products of multi-scale elements, and published four volumes books, a total of 51 volumes of datasets, pioneering in the data sharing of ecosystem permanent observation in China (Fu et al. 2007; Yu and Yu 2013).

### 2.2.3 Development Program of CNERN

Construction of Observation and Experimental Research Platform

CNERN has implemented many special networking observation and experimental plan based on the long-term dynamic observation and scientific research which meet the requirements of national science and technology. The programs which have been carried out were ecosystem carbon and nitrogen flux and cycle, response and acclimation of ecosystem to global change, biodiversity and ecosystem function, and so on. In the future, it will put emphasis on China FLUX, China ETS, and isotope observation of terrestrial ecosystem in China (Yang et al. 2008).

Long-Term Ecology Research

The long-term ecology research is the major task of CNERN with object of understanding ecosystem structure, function, process, and pattern under the drive of global change and human activities. It provide reliable knowledge of science, and applicable technical measures for biological resources improvement and conservation, ecosystem management, food production, natural resource management, ecological environment protection combined with the long-term observation and simulation experiment, remote sensing, and other technical methods. The research mainly includes the following six aspects: (i) effects of human activities on ecosystem structure and function; (ii) carbon and nitrogen cycle processes and management; (iii) the ecosystem response and adaptation to global climate change; (iv) biodiversity conservation and biological resource utilization; (v) ecosystem restoration and regional sustainable development; and (vi) the ecological monitoring, simulation, and application technology of ecological information.

Ecosystem Network Research

Ecosystem network research is a method to solve the problem of major science and technology on regional and national scale, which making an important contribution to ecological construction, environmental protection, farming, and animal husbandry production, and to cope with climate change in recent years (Yang et al. 2008). The researches include (i) spatial and temporal pattern and mechanism of ecosystem services in China; (ii) mechanisms of C–N–$H_2O$ coupling cycle and the carbon sinks in the Chinese terrestrial ecosystem function; (iii) forest ecosystem structure, function and the ecosystem services, and its response to global change; (iv) the grassland biodiversity change mechanism and its response to global change and human activities; and (v) soil and plant C:N:P stoichiometry ecology of Chinese terrestrial ecosystem, etc.

What is more, every station should carry out ecological system optimization management demonstration work to provide technology support and demonstration model for ecological environment construction and local economic development.

### 2.3 *Conclusion and Discussion*

CNERN has reformed the infrastructure, established the specimens sample library, and realized the data integration and sharing and optimal allocation of resources, integration after continuous construction and development. The establishment of CNERN has realized the unity of all field stations across departments. In the future, CNERN should put emphasis on the construction of observation and experimental research platform, long-term ecology research, and ecosystem network research, and provide important scientific basis for the government to solve the problem of ecological environment and global change.

## 3 Chinese Ecosystem Research Network (CERN)

### 3.1 *Founding of the Chinese Ecosystem Research Network (CERN)*

Since the founding of People's Republic of China, the central and local governments have invented a lot in ecosystem monitoring and capacity building of field stations. For example, since the mid 1950s, the Chinese Academy of Sciences

(CAS) has established a lot of field stations, over 90 only in the resources and environmental science fields (Niu et al. 2006; Fu et al. 2007; Yang et al. 2008; Fu et al. 2010).

Under the auspices of the Chinese government and the World Bank Loan, the was founded in 1988. The CERN facilitates research on the China's ecosystems across large spatial and temporary scales. There are more than 2000 scientists and graduate students of various disciplines across the country to do scientific research using the CERN field stations.

There were only 29 field CERN stations in 1988, which were selected from the field stations of the CAS to represent different types of vegetation and climate. The number of the CERN field stations has increased to 42 as of 2013, encompassing diverse ecosystems in China (Table 1). Within the CERN, five disciplinary sub-centers (atmospheric, aquatic, pedological, biological, and hydrological) and one synthesis center were set up. The administrative management of the CERN is completed through the Scientific Advisory Committee, the Science Committee, and the Leading Group.

The CERN receives its funding from the CAS for its operation, but other government departments such as the China's Ministry of Science and Technology (MOST), National Natural Science Foundation of China also support a wide range

**Table 1** Distribution of the CERN field stations

| Agricultural ecosystem: 15 | | Forest ecosystem: 10 | | Aquatic ecosystem: 8 | |
|---|---|---|---|---|---|
| AKA | Akesu | MXF | Maoxian | Bay | 3 |
| ASA | Ansai | BJF | Beijing F | JZB | Jiaozhouwan |
| CSA | Changshu | ALF | Ailaoshan | SYB | Sanya |
| CWA | Changwu | BNF | Banna | DYB | Dayawan |
| FQA | Fengqiu | CBF | Changbaishan | Lake | 4 |
| HJA | Huanjiang | DHF | Dinghushan | DHL | Donghu |
| HLA | Hailun | GGF | Gonggashan | THL | Taihu |
| LCA | Luancheng | HSF | Heshan | DTL | Dongtinghu |
| LSA | Lasa | HTF | Huitong | PYL | Poyanghu |
| QYA | Qianyanzhou | SNF | Shennongjia | Marsh | 1 |
| SYA | Shengyang | | | SJM | Sanjiang |
| TYA | Taoyuan | | | | |
| YCA | Yucheng | | | | |
| YGA | Yanting | | | | |
| YTA | Yingtan | | | | |
| Desert ecosystem: 6 | | Grassland ecosystem: 2 | | Urban ecosystem: 1 | |
| CLD | Cele | HBG | Haibei | BJU | Beijing U |
| ESD | Erdos | NMG | Neimenggu | | |
| FKD | Fukang | | | | |
| LZD | Linze | | | | |
| NMD | Naiman | | | | |
| SPD | Shapotou | | | | |

of research at field station and network levels. From the onset of the CERN, many national key projects (e.g., the National Basic Research Program, i.e., the 973 program, initiated by the MOST) have been conducted at different CERN stations.

## 3.2 Missions of the CERN

The CERN's missions include (i) to carry out long-term routine observation and monitoring of ecological processes and their environmental controlling factors of China's ecosystems; (ii) to conduct experiment-based multiple disciplinary researches on dynamics of structure and functions, processes and patterns of China's ecosystems; and (iii) to make experimental demonstrations of patterns for sustainable management of ecosystems, and to develop practical techniques for national agriculture production, and ecological restoration. Further information on the missions of the CERN is available in Fu et al. (2010).

Since its establishment, CERN has collected very large amounts of long-term data with more than 200 monitoring variables across the country, and massive datasets have been built at the five disciplinary sub-centers and the synthesis center.

Enormous studies based on the CERN and its field stations have been published in Chinese and foreign journals since the CERN founding. For example, from 2001 to 2010, the CERN was published over 12,000 papers, among of which about 3500 papers were in Science Citation Index (SCI) journals. In the same period, about 200 monographs, books, and proceedings were published.

Through long-term experimental demonstration in the CERN field stations, many sustainable ecosystem management practices and eco-restoration patterns for degraded ecosystems have been tested and optimized and subsequently have been well received by the local governments and farmers.

Many scientists from the CERN field stations won the Scientific and Technological Progress Awards issued by local, provincial, and central governments because of their excellent achievements in scientific and technological development and application. In 2012, the CERN won the First Prize of the State Scientific and Technological Progress Award for the project "The Establishment, Experiment and Demonstration of CERN," leading by the Academician Prof. Sun Honglie, ranking the first among 15 award winners in the Prize.

## 3.3 Future Orientation of the CERN

The CERN has grown in size and will see a more promising future as outlined in the Strategic Plan of CERN (2008–2020) (CERN 2008; Fu et al. 2010). The new motivation is to further strengthen capacity building of the CERN, including improvement of experimental conditions (such as lab and communication facilities) and updating of monitoring instrumentation. More importantly, the long-term and network scale holistic research on ecological processes will be further emphasized. In addition, the connection with other national, continental, and global networks such as International Long-Term Ecosystem Research Network (ILTER) and its member networks will become more intensive and extensive in the future.

## 4 Methods for Earth Surface Modeling and Its Applications

Ground observation is able to obtain highly accurate data with high temporal resolution at observation points, but these observation points are too sparsely to satisfy the application requirements at regional scale. Satellite remote sensing can frequently supply spatially continuous information on earth surface, which is impossible from ground-based investigations, but remote sensing description is not able to directly obtain process parameters. In fact, in terms of fundamental theorem of surfaces, a surface is uniquely defined by the first fundamental coefficients, about the details of the surface observed when we stay on the surface, and the second fundamental coefficients, the change of the surface observed from outside the surface. A high accuracy and speed method (HASM) for surface modeling has been developed initiatively to find solutions for error problem and slow-speed problem of earth surface modeling since 1986. HASM takes global approximate information (e.g., remote sensing images or model simulation results) as its driving field and local accurate information (e.g., ground observation data and/or sampling data) as its optimum control constraints. Its output satisfies the iteration stopping criterion which is determined by application requirement for accuracy. This paper reviews problems to be solved in every development stage and applications of HASM.

### *4.1 Introduction*

Ground observation can obtain high accuracy data at observation points, but observations at fixed positions are confined within some limited dispersal points and not able to directly calculate relative parameters at regional scale. Satellite remote sensing can frequently supply surface information of geographical processes and ecological processes, but remote sensing description is not able to directly obtain process parameters. Remote-sensing data can generate information about earth surface that is impossible from ground-based studies. The timing and extent of land-cover change and the relationship between climate and phenology highlight unique information that is available only from satellite and airborne sensors. However, maps derived from satellites observations are patchy and cannot be used reliably as an independent source of information for earth surface monitoring because of the well-known limitations of satellite retrievals, such as missing data for cloud-covered pixels.

The most effective use of remote-sensing data is through its fusion with appropriate field investigation. For instances, utilizing a satellite image as secondary information decreased errors associated with yield monitor data and also allowed better prediction in areas where no reliable yield measurements were available. Gross primary production (GPP) and net ecosystem exchange (NEE) were simulated by assimilating meteorological data derived from the measurements from existing weather stations, forest volume data derived from a previous investigation, satellite data, flux tower data, and other ancillary data, which rendered the simulation more stable and accurate.

In fact, earth surface systems are controlled by a combination of global factors and local factors, which cannot be understood without accounting for both the local and global components. The system dynamic cannot be recovered from the global or local controls alone. In terms of fundamental theorem of surfaces, a surface is uniquely defined by the first fundamental coefficients and the second fundamental coefficients. The first fundamental coefficients express the information observed when we stay on the surface, about the details of the surface. The second fundamental coefficients express the change of the surface observed from outside the surface (Somasundaram 2005).

A HASM for surface modeling has been developed initiatively to efficiently assimilate remote sensing data with ground-based observation data since 1986 so that solutions could be found for error and slow-speed problems which have long-troubled earth surface modeling. This paper focuses on HASM, especially how problems, appearing in every development stage, were solved by introducing appropriate theories and methods.

## 4.2 Progress in HASM

The development process of HASM can be divided into four stages. *In the 1st stage from 1986 to 2001*, studies were based on curve theorem, dealing with a surface as a combination of its profiles. It was learnt that slope and curvature are significant variables of surface analysis. In fact, a curve is uniquely determined by its slope and its curvature in terms of curve theorem in the plane. Following this consideration, a model for modeling cirques was constructed in terms of curve theorem, which was then developed for change detection of earth surface.

*In the 2nd stage from 2001 to 2007*, studies were based on fundamental theorem of surfaces, paying attention to error problem (Yue et al. 2011). It is proven that the equation of Earth's surface can be formulated as $z = f(x, y)$, where $z$ is an attribute value of the earth's surface at location $(x, y)$. For the surface $z = f(x, y)$, an iterative formulation of HASM was developed in terms of the fundamental theorem for surfaces, which was transformed into a symmetric positive-definite and large sparse linear system.

*In the 3rd stage from 2008 to 2011*, studies were based on fundamental theorem of surfaces, paying attention to low computational-speed and large memory-requirement problems. HASM has a huge computation cost because it must use an equation set for simulating each lattice of a surface. To speed up the computation of HASM, we developed a multigrid method of HASM (HASM-MG), an adaptive method of HASM (HASM-AM), an adjustment computation of HASM (HASM-AC), and a preconditioned conjugate gradient (PCG) algorithm of HASM (HASM-PCG). Multigrid method is the fastest numerical method for solving partial differential equations, which is based on two principles that are error smoothing and coarse grid correction. The principle of the adaptive method is that grid cells where the error is large will be marked for refinement, while grid cells with a satisfied accuracy are left unchanged. The adjustment computation permits all observations, regardless of their number or type, to be entered into the adjustment and used simultaneously in the computations by means of least squares. A conjugate gradient algorithm was originally viewed as an acceleration technique for the effective solution of large linear systems by a succession of well-convergent approximations; the preconditioned conjugate gradient algorithm can be developed by introducing a preconditioner to ensure faster convergence of the conjugate gradient method.

*In the 4th stage since 2012*, Gauss–Codazii equation set was introduced into HASM. We found that accuracy of HASM was not so satisfied in a few cases of multitudinous numerical tests and real-world tests. In other words, HASM performance was not stable enough for all applications. The reason was that second-order central difference stencil, which we employed for HASM solution, had no element on the diagonal corresponding to $f_{i,j}$. This led to algebraic systems with loss of diagonal dominance.

A combination of a forward difference stencil and a backward difference stencil has produced a symmetric stencil, which has a non-zero coefficient in the diagonal and can thus restore the diagonal dominance of the corresponding matrix in the algebraic systems. This refined symmetric stencil is of second order on a uniform grid and can give a solution to the instability when solving the algebraic equations of HASM.

Let $f_{i,j}^{(0)} = \bar{f}_{i,j}$ at the sampled point $(x_i, y_j)$ in the computational domain, $(x_i, y_j) \in \Phi$, and $\Phi = \{(x_i, y_j, \bar{f}_{i,j}) | 0 \leq i \leq \mathrm{I} + 1,\ 0 \leq j \leq j+1\}$ be the set of sampling points, then the matrix formulation of HASM can be expressed as

$$\begin{bmatrix} \mathbf{A} & \mathbf{B} & \mathbf{C} & \lambda\mathbf{S} \end{bmatrix} \begin{bmatrix} \mathbf{A} \\ \mathbf{B} \\ \mathbf{C} \\ \lambda\mathbf{S} \end{bmatrix} \mathbf{z}^{(n+1)} = \begin{bmatrix} \mathbf{A} & \mathbf{B} & \mathbf{C} & \lambda\mathbf{S} \end{bmatrix} \begin{bmatrix} \mathbf{d}^{(n)} \\ \mathbf{q}^{(n)} \\ \mathbf{p}^{(n)} \\ \lambda\mathbf{k} \end{bmatrix} \tag{1}$$

where $\mathbf{z}^{(n+1)} = \left(f_{1,1}^{(n+1)}, \ldots, f_{1,J}^{(n+1)}, \ldots\ldots, f_{I,1}^{(n+1)}, \ldots, f_{I,J}^{(n+1)}\right)^T$; **A**, **B**, and **C**, respectively, represent coefficient matrixes of the first, second, and third equation of HASM master equation set; $\mathbf{d}^{(n)}$, $\mathbf{q}^{(n)}$, and $\mathbf{p}^{(n)}$ are, respectively, the right-hand side vectors of the HASM master equation set; the non-zero element of the sample matrix **S** can be expressed as $S_{p,(i-1)\times I+j} = 1$ and the non-zero element of the sample vector $k_p = \bar{f}_{i,j}$; and $\lambda$ is the weight of the sampling points and determines the contribution of the sampling points to the simulated surface.

In terms of fundamental existing theorem for surfaces, if the first and second coefficients satisfy Gauss–Codazii equations, there exists a surface uniquely determined within a Euclidean displacement. The Gauss–Codazii equations can be transformed into

$$(\varphi_{1y} - \phi_{2x} - \varphi_2 P - \phi_1 Q)^2 + (\varphi_{2x} - \phi_{1y} - \varphi_1 Q - \phi_2 P)^2 + (Q_x - P_y - \varphi_1\varphi_2 - \phi_1\phi_2)^2 = 0 \tag{2}$$

where $E$, $F$, and $G$ are the first fundamental coefficients; $L$, $M$, and $N$ represent the second fundamental coefficients; $\varphi_1 = \frac{L}{\sqrt{E}}$; $\varphi_2 = \frac{N}{\sqrt{G}}$; $P = \frac{\sqrt{E_y}}{\sqrt{G}}$; $Q = \frac{\sqrt{G_y}}{\sqrt{E}}$; $\phi_1 = \frac{M}{\sqrt{G}}$; and $\phi_2 = \frac{M}{\sqrt{E}}$.

Thus, we can design an iteration stopping criterion of the improved HASM as

$$(\varphi_{1y} - \phi_{2x} - \varphi_2 P - \phi_1 Q)^2 + (\varphi_{2x} - \phi_{1y} - \varphi_1 Q - \phi_2 P)^2 + (Q_x - P_y - \varphi_1\varphi_2 - \phi_1\phi_2)^2 < \mathrm{EI} \tag{3}$$

where EI is determined by the requirement of an application for simulation accuracy.

Although HASM performance has been considerably improved because of the introduction of the refined symmetric stencil and the Gauss–Codazii equations, as a consequence it has caused the low-speed problem once again.

## *4.3 Applications of HASM*

HASM has been successfully applied to constructing digital elevation model (Chen et al. 2013a, b), filling voids of dataset of the Shuttle Radar Topography Mission (SRTM), simulating climate change and modeling surfaces of soil properties and soil pollution. In all these applications, HASM produced the highest accurate results compared with the classical methods.

For instances, HASM was applied to simulating *elevation surfaces* of Dongzhi tableland in Loess Plateau of China. The validation results showed that HASM-AM has the highest accuracy and the fastest computation speed, compared with widely used classic methods. Dong-Zhi tableland has its area of 2724 km$^2$, consisting of 27.24 million pixels on spatial resolution of 10 m × 10 m, while area of Earth surface is 510 million km$^2$, consisting of 5.1 million pixels on spatial resolution of

10 km × 10 km. In other words, HASM has the capacity of computing more than five Earth's surfaces on spatial resolution of 10 km × 10 km.

Datasets of the SRTM, which were derived from the Space Shuttle Endeavour in February 2000, have become a useful source of elevation data and are so critical to modern imagery analysis and geospatial intelligence requirements. However, STRM data have variously sized voids, resulting in incomplete datasets. These voids account for 0.15 % of the total dataset in China. They amount to up to 30 % in rugged terrain. HASM was used to *filling voids* in China. Verification in nine regions with three different geomorphologic types of hills, plateaus, and mountains demonstrated that HASM results always had the highest accuracy compared with all the classic methods, whether auxiliary data were added or not, landform complexity was higher or lower, and void area was larger or smaller in all the nine regions of the three topographic types.

HASM has been employed to simulate *climate change trend* in China since 1961. We have found that mean annual temperature (MAT) during the period from 1961 to 2010 exhibited spatial stationarity, while mean annual precipitation (MAP) showed spatial non-stationarity. A statistical transfer function (STF) of MAT was formulated using minimized residuals output from HASM with an ordinary least squares (OLS) linear equation that used latitude and elevation as independent variables, abbreviated as HASM-OLS. The STF of MAP under a BOX-COX transformation was derived as a combination of minimized residuals output by HASM with a geographically weight regression (GWR) using latitude, longitude, elevation, impact coefficient of aspect, and sky view factor as independent variables, abbreviated as HASM-GWR-BC. Cross validation of HASM-OLS and HASM-GWR-BC indicates that mean absolute errors of MAT and MAP are −0.15 °C and 1.52 mm, respectively, which were much more accurate than the classical methods. In terms of HASM-OLS and HASM-GWR-BC, MAT had an increasing trend since 1960s in China, with an especially accelerated increasing trend since 1980. Our simulation showed that MAT has increased by 1.44 °C since the 1960s. The warming trends have spatially increased from the south to the north in China, except Qinghai-Xizang plateau. Specifically, the 2100 °C·d contour line of annual accumulated temperature (AAT) of ≥10 °C shifted northwestward 255 km in the Heilongjiang Province of northeastern China since the 1960s. MAP in Qinghai-Xizang plateau and in arid region had a continuously increasing trend. On average, China became wetter from the 1960s to the 1990s, but drier from the 1990s to 2000s. The Qinghai-Xizang Plateau and Northern China experienced more climatic extremes than Southern China since 1960s.

Temperature and precipitation data from a General Circulation Model (HadCM3) were downscaled by means of HASM-OLS and HASM-GWR-BC. Correlation coefficients of MAT and MAP between the *downscaled HadCM3 output* and the observed meteorological station data are 0.95 on average under three Special Report on Emissions Scenarios (SRES) A1FI, A2, and B2 during the baseline period T1 (1961–2010). This means that the downscaled HadCM3 output and the observed meteorological station data are comparable. Estimating MAT during the period 2010–2099 under the HadCM3A1FI, HadCM3A2, and

HadCM3B2 scenarios would increase by 4.62, 3.37, and 2 °C, respectively; MAP under the three HadCM3 scenarios would increase by 173.22, 141.55, and 79.68 mm, respectively. The decadal increasing rate (DIR) of MAT estimated between the years of 2010 and 2099 would be, respectively, 0.64, 0.51, and 0.36 °C under each scenario. MAP from 2010 to 2099 would have a DIR of 23.42, 16.48, and 11.76 mm under the three scenarios of A1FI, A2, and B2, respectively.

Taking 5374 soil profiles collected during the second national soil survey (1979–1994) as optimum control constraints of HASM and using regression relations between soil properties and their environmental factors as the driving fields, surfaces of soil properties were simulated on spatial resolution of 1 km × 1 km in China on national level. The validations indicated that HASM results were at least 17 % more accurate compared with Ordinary Kriging (OK) method. HASM was specifically once again validated in the middle part of Jiangxi Province of China, for which 150 samples were collected in different land-use types of woodlands, croplands, and grasslands. Performance of HASM on simulating soil properties, such as pH, AN, C, N, K, AI, Ca, Mg, and Zn, was evaluated by comparing with widely used methods of OK, stratified kriging (SK), and regression-kriging (RK) using a generalized linear model. The evaluation demonstrated that HASM maps of soil properties presented more details and more accurate spatial pattern.

Taking sampled biomass data in grassland as optimum control constraints and remote sensing image as a driving field, HASM was applied to simulate spatial distribution of grassland biomass in China. Output of Lund-Potsdam-Jena dynamic global vegetation model (LPJ-DGVM) was used as driving field and sampled forest biomass as optimum control constraints of HASM, and spatial distribution of forest biomass was simulated (Sun 2011). The simulation results in last 100 years showed that vegetation biomass in China first increased slowly from 1901 to 1953, and then increased faster since 1953. Total biomass under planned development scenario is 0.09 Gt higher than the one under business as usual scenario in 2030.

## 4.4 Conclusions

HASM is becoming theoretically perfect. Input of HASM is conducted by taking global approximate information (e.g., remote sensing images or simulation results) as driving field and taking local accurate information (e.g., ground observation data and/or sampling data) as optimum control constraints. Its output is the results satisfying the iteration stopping criterion which is determined by application requirement for accuracy.

HASM has been successfully applied to construct digital elevation model, filling voids of dataset of the SRTM, simulating climate change, and modeling surfaces of soil properties. In all these applications, HASM produced the highest accurate results compared with the classical methods. However, slow computational speed and large memory requirement remain the limitation of applications with huge computational work, although we have developed the adaptive method and a multi-grid method of HASM.

To meet the huge computational requirement of big data, one way is to use a faster single-processor computer, but continually pursuing the fastest computer can be very expensive and does not scale well as problem size increases. The second way is to develop a fast numerical algorithm. It has been demonstrated that PCG is the most efficient algorithm of HASM that can be transformed into a linear system with a symmetric positive-definite matrix. A third possibility is to break down the computational problem into a number of smaller problems, e.g., using adjustment method of HASM; the smaller computational problems can be solved simultaneously on less-expensive computers utilizing parallel computing methods.

Graphics Processing Units (GPUs) have become a powerful many-core processor. The massively parallel architecture offers high performance in many computing applications. Numerical algorithms can be significantly accelerated if the algorithms map well to the characteristics of the GPU. GPU-based parallel algorithm of PCG can considerably improve the efficiency and robustness of HASM. GPU implementation of PCG algorithm for HASM is up to 12 times faster compared with HASM-PCG. However, it is not faster enough for many applications, especially at high temporal and spatial resolutions on the global level.

Message Passing Interface (MPI) is the most popular choice in parallel computing environments on clusters of workstations, of which version 1 was released in 1994. MPI represents the standard adopted by most of the industries and researchers. When a parallel algorithm is implemented in a cluster of workstations using MPI, computational-speed can be greatly improved. Parallel computing with MPI on clusters of workstations is an effective way to significantly solve the slow computational-speed problem of HASM, especially for huge computation of big data.

## References


CERN (Chinese Ecosystem Research Network). (2008). The Strategic Plan of Chinese Ecosystem Research Network (CERN) (2008–2020).

Chen, X. W., Zhou, G. S., & Zhang, X. S. (2003). Spatial characteristics and change for tree species along the North East China transect (NECT). *Plant Ecology, 164*, 65–74.

Chen, Z., Yu, G. R., Ge, J. P., et al. (2013a). Temperature and precipitation control of the spatial variation of terrestrial ecosystem carbon exchange in the Asian region. *Agricultural and Forest Meteorology, 182*, 266–276.

Chen, C., Yue, T., Dai, H., & Tian, M. (2013b). The smoothness of HASM. *International Journal of Geographical Information Science, 27*(8), 1651–1667.

Chen, C., Li, Y., & Yue, T. (2013c). Surface modeling of DEMs based on a sequential adjustment method. *International Journal of Geographical Information Science, 27*(7), 1272–1291.

Somasundaram, D. ( 2005). *Differential geometry*. Harrow: Alpha Science International Ltd.

Fu, B. J., Niu, D., & Yu, G. R. (2007). The roles of ecosystem observation and research network in earth system science. *Progress in Geography, 26*(1), 1–16.

Fu, B. J., Li, S. G., Yu, X. B., et al. (2010). Chinese ecosystem research network: Progress and perspectives: Progress and perspectives. *Ecological Complexity, 7*, 225–233.
Gosz, J. R. (2000). Toward a global understanding-an introduction to the international long term ecological research network. In: *The international long term ecological research network* (pp. 4–5). Albuquerque, New Mexico: USA Academy Printers.
Hang, Y. J., & Zhou, G. S. (2008). Terrestrial transect study on driving mechanism of vegetation changes. *Science in China Series D-Earth Sciences, 51*, 984–991.
Hu, B., Wang, Y. S., & Liu, G. R. (2007). Ultraviolet radiation spatiotemporal characteristics derived from the ground-based measurements taken in China. *Atmospheric Environment, 41*, 5707–5718.
Ni, J. (2003). Plant functional types and climate along a precipitation gradient in temperate grasslands, north-east China and south-east Mongolia. *Journal of Arid Environments, 53*, 501–516.
Ni, J., & Wang, G. H. (2004). Northeast China transect (NECT): Ten-year synthesis and future challenges. *Acta Botanic Sinica, 46*, 379–391.
Niu, D., Huang, T. Q., Yang, P., & Yu, G.-R. (2006). Chinese ecosystem research network (CERN): Construction experiences, future challenge and development strategy. *Bulletin of Chinese Academy of Sciences, 21*(6), 466–471.
Odum, E. P., & Barrett, G. W. (1971). *Fundamentals of ecology*. Pacific Grove: Thomson Brooks/Cole.
Parr, T. W., & Lane, A. M. (2000). United Kingdom environmental change network. In: *The international long term ecological research network* (pp. 60–63). Albuquerque, New Mexico: USA Academy Printers.
Yue, T. X. (2011). *Surface modelling: High accuracy and high speed methods*. New York: CRC Press.
Tansley, A. G. (1935). The use and abuse of vegetational concepts and terms. *Ecology, 16*, 284–307.
Waide, B. W. (2000). The US long term ecological research network. In: *The international long term ecological research network* (pp. 89–103). Albuquerque, New Mexico: USA Academy Printers.
Wang, B., Cui, X. H., & Yang, F. W. (2004). Chinese forest ecosystem research network (CFERN) and its development. *Chinese Journal of Ecology, 23*, 84–91.
Sun, X. F. (2011). Spatial pattern simulation of terrestrial vegetation aboveground biomass (Dissertation for doctoral degree, Graduate University of Chinese Academy of Sciences, Beijing, 2011) (In Chinese).
Yang, P., Yu, X. B., Zhuang, X. L., & Niu, D. (2008). Present situation and future development of Chinese ecosystem research network (CERN) of Chinese academy of sciences. *Bulletin of Chinese Academy of Sciences, 23*(6), 555–561.
Yu, G. R., Shen, S. M., Zhao, & S. D. (2009). Carbon cycle of terrestrial ecosystem in China. In: *20 years of Chinese ecosystem research network: Comprehensive research of ecosystem*. Beijing: Science Press.
Yu, G. R., & Yu, X. B. (2013). Chinese ecosystem research network and the natural ecosystem protection. *Bulletin of Chinese Academy of Sciences, 28*(2), 275–283.
Yu, G. R., Zhang, L. M., Sun, X. M., et al. (2008). Environmental controls over carbon exchange of three forest ecosystems in eastern China. *Global Change Biology, 14*, 2555–2571.
Yu, G. R., Zhu, X. J., Fu, Y. L., et al. (2013). Spatial patterns and climate drivers of carbon fluxes in terrestrial ecosystems of China. *Global Change Biology, 19*, 798–810.
Zhao, S. D. (2004). Long-term ecological research program. *Advances in Earth Science, 19*(5), 840–844.

# Chapter 15
# Climate Change and Carbon Cycle

**Quansheng Ge, Jingyun Zheng, Haolong Liu, Guirui Yu, Huajun Fang, R.L. Wang, Jingyun Fang, Huifeng Hu, Zhaodi Guo, Haihua Shen, Yunshe Dong, Yuchun Qi, Qin Peng, Xiaoke Wang, Hong Zhao, Fei Lu, Changchun Song, Xiaomin Sun and Xuefa Wen**

**Abstract** Climate change is often called changes caused by natural factors or human activities in terms of worldwide or regional, and it usually means a giant change or lasting a long time (≥10 years) on average and statistically, which includes changes in terms of mean value and change rate. Whereas, United Nations Framework Convention on Climate Change (UNFCCC) defines it as "climate change caused directly or indirectly by human activity's changing the combination of the atmosphere after a long time of observation, apart from natural climate changes". Although the two definitions differ a lot, now people mainly concern these climate change phenomenon: global warming in climate and ocean, frequency changes of extreme climate events, polar and mountain snow and glacier melting, lining up of the global ocean surface level, global precipitation changes in terms of time and space, increasing drought and flood, and other physical and chemical factors influencing global temperature and precipitation directly. Terrestrial ecosystem's carbon cycle is the key to its material and energy cycle, and also a tie of geosphere–biosphere–atmosphere interaction. To deeply understand the carbon

Q. Ge · J. Zheng · H. Liu · G. Yu (✉) · H. Fang · R.L. Wang · Y. Dong · Y. Qi · Q. Peng
X. Sun · X. Wen
Institute of Geographic Sciences and Natural Resources Research, Chinese Academy of Sciences, Beijing 100101, China
e-mail: yugr@igsnrr.ac.cn

J. Fang · Z. Guo · H. Shen
Department of Ecology, College of Urban and Environmental Science, Peking University, Beijing 100871, China

H. Hu
State Key Laboratory of Vegetation and Environmental Change, Institute of Botany, Chinese Academy of Sciences, Beijing 100093, China

X. Wang · H. Zhao · F. Lu
State Key Laboratory of Urban and Regional Ecology, Research Center for Eco-Environmental Sciences, Chinese Academy of Sciences, Beijing 100085, China

C. Song
Northeast Institute of Geography and Agroecology, Chinese Academy of Sciences, Changchun 130102, China

 
W. Li (ed.), *Contemporary Ecology Research in China*,
DOI 10.1007/978-3-662-48376-3_15

cycle's process and mechanism is the basis to discuss ecosystem's carbon management strategy and to analyze human's progress in preventing and regulating global warming. This chapter systematically states our country's research progress on climate change's historical process in the first place, and then it makes comments on China's carbon cycle study from several perspectives as the relationship between carbon cycle and climate change, carbon budget and its cycling mechanism, human activity's influence on carbon cycle, global change's influence on carbon cycle, etc. It then summarizes research progress in studying carbon cycle process, its influencing factors and global changing response from different perspectives of forestry, grassland, farmland, wetland, etc. Finally, it makes a comprehensive comment on our country's technological progress in terrestrial ecosystem's carbon-nitrogen-water flux observation technology and applications, which provides valuable information to China's scientific research work, state-level carbon budget evaluation, and greenhouse gas management in the field of ecosystem carbon cycle.

**Keywords** Climate change · Terrestrial ecosystem carbon cycle · Forest carbon stocks · Grassland carbon cycle · Agroecosystem carbon · Wetland carbon cycle · Novel techniques on measuring carbon–nitrogen–water fluxes over terrestrial ecosystems

# 1 Historical Climate Changes of China and the Global

## *1.1 Introduction*

Historical climate change of China and the Globe is of great significance for the current ecological research, and provides a scientific basis for the actions to reduce the ecosystem vulnerability and enhance its recoverability during the process of adaptation to climate change.

The relative research in China could be traced to Zhu Kezhen, who outlined the temperature variation of eastern China during the past 5000 years using the phenological and disaster records for the first time in 1972. Since then, great progresses have been obtained in these two periods for the Holocene and the past 2000 years. Many series with high spatial and temporal resolutions were reconstructed, which formed an information network about the historical climate change all over the country.

Only until the International Geosphere–Biosphere Program (IGBP) and the World Climate Research Program (WCRP) carried out during 1990s, did significant progress of fact analysis and process reconstruction of climate change during the past thousand years achieve, especially during the nearly 1000 years in the Northern Hemisphere. Among these reconstructions, the temperature series over the past

1000 years in Northern Hemisphere with evidences of tree rings, ice cores, historical documents, and others (Mann et al. 1998, 1999) received the most attentions. The IPCC Fourth Assessment Report (2007b) conducted a comprehensive assessment on these reconstructions over the past 2000 years, and gave the associated confidence levels and uncertainties. This greatly clarifies the people's understanding of historical climate change process.

## *1.2 Climate Changes of the Global and China in Holocene*

### 1.2.1 Temperature Change

According to various global natural evidences, the Holocene climate was warm in general, interrupted by several abruptly cold events lasting for hundreds of years. Meanwhile, the Holocene climate change of global and China can both be divided into three basic stages: early, medium, and late. Among them, the earlier (about 11.5–8 ka BP) climate rapidly heated; medium-term (about 8.0–4.3 ka BP) was far warmer than the modern, known as Megathermal; The temperature decreased in late Holocene (after 4.3 ka BP) compared with that in Megathermal.

### 1.2.2 Precipitation Change

Precipitation change all over the world generally synchronized with the temperature change in Holocene. In the early term, precipitation in most parts of the world rapidly increased, and the precipitation in Chinese monsoon region reached its maximum during the period of 9.5–9.0 ka BP. In the medium term, most parts of the world were wetter than now and the highest sea level over the whole Holocene appeared about in 6 ka BP. In the late term, many lakes shrank and desert expanded significantly. After 3 ka BP, drought frequently emerged in China.

## *1.3 Global Climate Changes Over the Past 2000 Years*

### 1.3.1 Temperature Change

Currently, there exists large uncertainty of the temperature variation reconstruction for the Southern, the Northern Hemisphere, and the Globe over the past 2000 years. It could be said with a high level of confidence that global mean surface temperature was higher during the last few decades of the twentieth century than during any comparable period during the preceding four centuries (NRC 2006). The NRC Assessment Report also pointed out that less and very little confidence can be placed in large-scale surface temperature reconstructions for the two periods from

A.D. 900 to 1600 and prior to about A.D. 900, respectively. About the historical status of the twentieth century warming, the IPCC (2007a) believed that average NH temperatures during the second half of the twentieth century were warmer than in any other 50-year period over the last 500 years in a likelihood higher than 90 % and was also the warmest 50-year period in the past 1300 years in a likelihood higher than 66 %.

### 1.3.2 Precipitation Change

Existing researches showed that rainfall in Eurasia from A.D. 300 to 450 is scarce, along with serious droughts. Afterward, during the sixth to eleventh century, there were a lot of floods in Europe and the water levels of lakes in Central Asia rose. In the Medieval Warm Period (prior to about A.D. 1400), the warm and dry weather frequently appeared in most areas of the world. However, during the following Little Ice Age, the global climate trended to be wet in general.

## *1.4 Climate Changes in China Over the Past 2000 Years*

### 1.4.1 Temperature Change

The temperature variations in China since Qin dynasty have experienced seven phases, including four warm periods, i.e., 210 B.C.–A.D. 180, 541–810, 931–1320, and 1921–2000, and three cold periods, i.e., 181–540, 811–930, and 1321–1920 (Ge et al. 2010). Temperature variations over China are typically in phase with those of the Northern Hemisphere (NH) after A.D. 1000, a period which covers the Medieval Climate Anomaly, the Little Ice Age, and the Present Warm Period.

On the decadal-to-centennial timescale, the duration and amplitude of regional phases of warm/cold fluctuation were different in China. For example, the "Medieval Warm Period" characteristic of Western China (especially the Tibetan Plateau) was not as significant as that of Eastern China; and the climate of the Tibetan Plateau during the Little Ice Age was less durable and cold than that of Eastern China.

However, according to the assessment by Ge (2010), the regional temperature reconstructions for the period prior to 1470s still have large uncertainties.

### 1.4.2 Precipitation Change

There exist very obvious millennial precipitation variations in China. In the eastern monsoon region, the precipitation had two obvious trends since Qin Dynasty, i.e., climate was gradually becoming dry before the middle of the thirteenth century; and after that, it was becoming wet (Zheng et al. 2006). The eastern precipitation fluctuated around the mean status of the past 2000 years, but the wet conditions were centered at A.D. 600, 730, 820 and 900, and the dry conditions were centered at A.D. 660, 760, 800, and 850. In the West China, the dry condition was prevalent before 1500, and after that, the climate was becoming wet with a couple of short fluctuations (Shao et al. 2010).

### *1.5 Conclusion and Discussion*

At present, several key scientific issues on the historical climate researches remain unsolved: (i) uneven data quality cannot meet the needs of climate reconstruction with high spatial and temporal resolutions; (ii) few researches focused on the regional differences in climate change; (iii) the driving mechanism of climate change has not been deeply analyzed; (iv) people lacked of a comprehensive understanding to the impact of climate change and the social response.

To solve these issues, the individual researchers and groups should take the following four approaches (PAGES 2009): (i) put efforts on proxy interpretation and development, analytical innovation, and calibration refinement, with the aim to reduce uncertainty in proxy-based reconstructions; (ii) achieve a better understanding of past regional climatic and environmental dynamics through comparison of reconstructions and model simulations; (iii) focus looks at the links between regional and global-scale changes; (iv) address the long-term interactions among past climate conditions, ecological processes, and human activities.

## 2 Process of Terrestrial Ecosystem Carbon Cycle and Its Driving Factors

### *2.1 Introduction*

The increase in $CO_2$ atmospheric concentrations, greenhouse effect and atmospheric nitrogen deposition induced by human activities, are influencing the physiological and ecological characteristics of plant, animal and microbe, species competition and distribution, which may lead to the species extinction and biodiversity loss. As the important bond between the geosphere, biosphere and atmosphere, carbon cycle in terrestrial ecosystem is the core of material and energy circulation in land. Due to its diversity, terrestrial ecosystem in China is among the most powerful 'natural experiments' for global carbon circle (Yu et al. 2009). Thus, understanding the process and function of terrestrial ecosystem carbon cycle and its response mechanism to environmental changes and human disturbance in China, increasing the knowledge of terrestrial ecosystem carbon cycle and carbon balance, developing the technology to reduce the carbon emission of terrestrial ecosystem, are the urgent and important tasks for Chinese scientists.

## 2.2 *Progress in Terrestrial Ecosystem Carbon Cycle in China*

### 2.2.1 Carbon Storage in Chinese Terrestrial Ecosystem and Its Influencing Factors

The value of vegetation carbon storage in China ranged from 6.099 to 35.23 PgC, among which the forest ecosystem was the largest carbon pool, including 3.7–8.7 PgC for carbon storage and 3.2–6.2 PgC for carbon density (Li et al. 2003). Carbon storage in grassland ecosystem ranged from 1.15 to 3.32 PgC (Fan et al. 2008; Fang et al. 2007), and the average carbon density was 3.46 MgC $hm^{-2}$ in China (Fang et al. 2007). Forest vegetation carbon pools were mainly distributed in the northeast and southwest China (Fang et al. 1996) and carbon pool in grassland ecosystem was mainly in Tibet Plateau and North China (Fan et al. 2008). Temperature and precipitation were the important factors affecting the distribution pattern of vegetation ecosystem carbon pool. In addition, community composition and forest age exerted great influence on vegetation carbon storage in China (Lu and Sun 2004).

For soil organic carbon pool, carbon storage was 80–90 PgC and carbon density was 8–10 kg $m^{-2}$. Northeast area, eastern Tibet Plateau, and Yunnan–Guizhou Plateau in China had the high value of carbon density. In terms of soil inorganic carbon, carbon storage was 60 PgC and the average carbon density was 4.29 kg $m^{-2}$. The value of average carbon density in Northwest China was highest, about 36.48 kg $m^{-2}$. High precipitation and low temperature were beneficial to the formation and storage of soil organic carbon, while the area with low precipitation tended to accumulate soil inorganic carbon.

### 2.2.2 Carbon Flux in Chinese Terrestrial Ecosystem and Its Influencing Factors

Annual net primary productivity (NPP) of terrestrial vegetation in China gradually decreased from the southeast to the northwest (Piao et al. 2001), gross primary production (GPP) in grassland ecosystem was lower than that of the typical forest and farmland ecosystem (Sun et al. 2006a, b; Zhang et al. 2006a; Fu et al. 2006). Temperature, moisture, and radiation condition were the main factors to influence and control the carbon uptake of terrestrial ecosystem. Ecosystem respiration (ER) generally showed a unimodal seasonal variation pattern in natural ecosystem, consistent with the seasonal variation in temperation (Zhang et al. 2006b; Fu et al. 2006), while the pattern of ER in farmland ecosystem was different from that of natural ecology. Temperature, soil moisture, organic matter content, and microbial activity were the key factors influencing ER.

### 2.2.3 Carbon Budget in Chinese Terrestrial Ecosystem and Its Influencing Factors

Annual net ecosystem exchange (NEE) varied with latitude and longitude, which increased with decreasing latitude, indicating that the carbon absorption capacity of terrestrial ecosystems was stronger in southern China, and the potential of forest was significantly higher than that of grassland. Carbon absorption occurred mainly in the west region of Southwest China, southeastern Tibet Plateau, Northeast Plain, central and western regions of North China and Southern China region. The carbon release appeared mainly in the Sichuan Basin, Qin Ba mountain area, Zhejiang and Fujian Hilly Area, northwestern Inner Mongolia, and parts of Xinjiang (Tao et al. 2006).

NEE was closely related with temperature, precipitation, and photosynthetic active radiation.

### 2.2.4 Terrestrial Ecosystem Carbon Cycle in China Related to Anthropogenic Factors and Climate Changes

Due to the land use change, carbon pool of terrestrial biomass in China increased from 8833.04 TgC in 1991 to 12,279.33 TgC in 2006 by 0.2–0.5 PgC per year, and carbon pool of soil organic carbon increased from 25,780.72 TgC in 1991 to 29,575.26 TgC in 2006. Through a series of ecosystem managements, the carbon sequestration potential of afforestation will be 1.81 PgC $hm^{-2}$ $a^{-1}$ in the future, and for cropland, grassland and wetland in China, 160 PgC, 1.3 PgC and 0.46 PgC $hm^{-2}$ $a^{-1}$ could be sequestrated in the future 50 years, respectively.

Climate changes could have a profound impact on terrestrial ecosystem carbon fluxes. Warming reduced the carbon sequestration capacity of temperate grassland ecosystem in Inner Mongolia, while increased precipitation could ameliorate the negative impacts of climatic warming on ecosystem C fluxes (Niu et al. 2008). Nocturnal warming increased leaf respiration of dominant grass species, enhanced consumption of carbohydrates in the leaves, and consequently stimulated plant photosynthesis in the subsequent days (Wan et al. 2009). This plant photosynthetic overcompensation transferred typical grassland ecosystems from carbon source to carbon sink.

Elevated atmospheric $CO_2$ concentration increased the yield of rice and wheat by 17 and 20 %, and the soil respiration under low nitrogen and high nitrogen treatment increased by 14.8 and 15.1 %, respectively. The high $CO_2$ concentration promoted the photosynthesis of *Pinus koraiensis* seedlings, improved water use efficiency (WUE) and simulated the growth of *P. koraiensis* at the early phase, whereas showed adverse effect later.

The influence of precipitation on grassland ecosystem was more obvious and significant than on forest ecosystem. For semiarid grassland ecosystem of Inner Mongolia, increase in precipitation alleviated the soil water restriction and the negative effects of warming on ecosystem carbon and water exchange, consequently promoted gross ecosystem productivity (GEP) and ER, net ecosystem $CO_2$ exchange (NEE), and soil respiration (SR) (Niu et al. 2008). The short-term effect of precipitation change on forest was not significant (Wu et al. 2009a, b).

Atmospheric nitrogen deposition increased the nitrogen availability in terrestrial ecosystems, and then fixed a certain amount of organic carbon through increasing net primary productivity (NPP), which could slow down the increase in atmospheric $CO_2$ concentration and explained partly the distribution of missing sink. However, on the other hand, long slow nitrogen deposition may generate nitrogen saturation of terrestrial ecosystem, and thus reduce the terrestrial ecosystem productivity, improve soil nitrogen leaching, accelerate decomposition rate of litter and soil organic carbon, thereby weaken the carbon sequestration of terrestrial ecosystem.

## 2.3 Conclusion

Over the past few decades, Chinese scientists have done a lot of work and achieved remarkable results on the investigation of the carbon storage and dynamic of terrestrial ecosystem at the regional scale, control experiments on the responses of carbon, nitrogen, water cycles to global changes in typical ecosystems, the simulation of terrestrial ecosystem processes and remote sensing models. However, the interactions between carbon cycle and climate, human activities, water and nutrient cycles are extremely complex, together with the inconsistency of data sources and evaluation methods, leading to the lack of our knowledge about the temporal variation, spatial distribution of the global and regional carbon source or sink and its driving forces.

Focus of future research is to construct a new generation of coupling model on the process of carbon, nitrogen and water cycle, based on the data from the long-term observation and experiments of different regions and ecosystems. In addition, we should adopt multi-scale to observe, multi-method to verify, multi-process to reconcile and cross-scale to simulate, in order to carry out the comprehensive study to evaluate ecosystem carbon sequestration and understand the variation in the process and pattern of ecosystem carbon cycle, and regulation and management approaches of carbon sequestration of regional ecosystem driven by global change and human activities.

# 3 Forest Carbon Stocks and the Changes in China

As the largest part of terrestrial ecosystems, forests play a leading role in regional and global carbon (C) cycles. Located in the eastern margin of Eurasia, China ranks the fifth in its forest area in the world and encompasses various forest biomes, from boreal forests in the north to the subtropical/tropical evergreen broadleaf forests in the south, which provides a unique area to study the regional forest C cycle. Therefore, detailed assessment of the C stocks and dynamics of China's forests is critical to the estimation of the national C budget and can help to constitute sustainable forest management policies for climate change. Here, we summarized the C stocks and dynamics of China's forests in recent decades among the four major components, including living biomass, soil organic carbon, litter, and coarse woody detritus. Based on six periods of the national forest inventory data between 1977 and 2008, living biomass C stocks of China's forests increased from 4.7 to 6.4 PgC with an average C sequestration rate of 63.3 TgC year$^{-1}$. The area-weighted mean biomass C density also increased from 38.2 to 41.3 MgC ha$^{-1}$. Using the data from China's second national soil survey, soil organic C storage in a depth of 1 m in China's forests is 15.8–34.2 PgC. In addition, the C stocks of forest litter and coarse woody detritus in China is 0.54–1.15 PgC and 0.1–2.4 PgC, respectively. The recent study showed that the total C stock in China's forests during 2000–2007 was 24.2 PgC, including 6.5 PgC of living biomass, 16.3 PgC of soil organic carbon, 1.2 PgC of litter, and 0.1 PgC of coarse woody detritus. Therefore, China's forests have significantly contributed to the regional and global C sinks in the past several decades, mainly due to the areal expansion and forest regrowth. Furthermore, China's forests are characterized by young forest age, low C density, and a large area of planted forests, indicating China's forests have high potential to act as C sinks in the future.

# 4 Grassland Carbon Cycle and Its Response to Climate Change

## 4.1 *Introduction: Carbon Cycle and Grassland Ecosystem*

Climate changes such as global warming caused by the enrichment of atmospheric greenhouse gases, e.g., carbon dioxide ($CO_2$), nitrous oxide ($N_2O$) and methane ($CH_4$), have been the key environmental problems and received worldwide attention in recent years. Currently, the effects and the corresponding mechanism of environmental changes and human activities on the carbon (C) cycle and C budget in terrestrial ecosystems have become the hot topics in many international research projects. The responses and feedbacks of terrestrial C budget to global changes are crucial to seek the residual C sink, which is thought to come from the terrestrial ecosystems in the mid- and high latitudes of the Northern Hemisphere (IPCC 2007a).

Grassland is one of the most widespread land cover types worldwide, which covers nearly one-fifth of the natural land surface (Scurlock and Hall 1998) and accounts for approximately 34 % of the global terrestrial C stock (World Resources Institute 2000). Grassland ecosystems, ranging from the savannas of Africa to the temperate steppe of Eurasia and from the North American prairies to the pampas of South America, play a very important role in regulating global C budget through the C fixation and C turnover by plant and soil.

Grasslands comprise the largest terrestrial ecosystem in China, covering an area of approximately $4 \times 10^6$ km$^2$ or 41.7 % of the land area of China (Ren et al. 2008) and 11.8 % of the world's total grassland (Zhao et al. 2005). In addition to some small disjunct areas spreading in the tropical regions, most of the grasslands in China are distributed in temperate arid and semiarid regions of North China or in the alpine regions of West China (Hou et al. 1982; DAHV & CISNR 1994). As the major C pool of grassland ecosystem, the grassland soil C stock and its variation are very sensitive to global changes such as temperature, precipitation, and human activities because of the fragile eco-environments in the arid or alpine grassland regions. Thus, it will result in great fluctuations of grassland C budget and increases the difficulty and complexity of research on grassland C cycle (Wever et al. 2002).

Because of the wide distribution and the high sensitivity to climate change for Chinese grassland, the potential effects of global change such as multiple environmental factors (climate change, increasing atmospheric $CO_2$ concentration, nitrogen deposition, and land cover/land use change) on net C balance in grassland ecosystems of China have received much attention in recent years. These relative researches have focused on the magnitude, spatial distribution, and temporal dynamics of the grassland C emission and soil C sequestration as well as the underlying mechanisms.

The objectives of this chapter are to: (i) present an overview of the main progresses of C cycle in Chinese grassland and the main effects of environmental factors on grassland C budget, and (ii) discuss the research emphases of the related field in the future.

## 4.2 C Storage and Its Spatial Distribution in Chinese Grassland Ecosystem

The knowledge of the spatial–temporal allocation of grassland C in China is important for understanding the role of Chinese grassland in global C cycle. Studies on grassland C storage in China started from the 1990s (Ding and Peng 1991; Zhu and Jia 1991), developed rapidly in the last 10 years, and most of them were focused on the temporal dynamics of soil C storage and C emission (Wang et al.

2004; Dong et al. 2005; Qi et al. 2010), but seldom on the spatial distribution of grassland C pool. With the development of model research, more and more attentions are paid to the quantitative assessment on the grassland C storage and its distribution in plant biomass and soil at both regional and national scale. Based on lots of field experiments and model simulations, the average C density and the total C stock in biomass and soil in Chinese grasslands are estimated (Tables 1 and 2). Various estimates of total C stock in biomass ranged from 562 to 4660 TgC (Table 1) and differed by a factor of approximately eight. Soil was the largest C pool in grassland, and large differences also exist among various studies on grassland soil C stock in China, with estimates ranging from 16.7 to 41.0 PgC. Temperate grassland is the main part of the grassland in China and the estimated total soil C stock is about 16.7–28.1 PgC (Table 2). The vegetation and soil C density as well as the corresponding C storage in different Chinese grasslands have also been calculated (Ni 2001, 2002; Zhong et al. 2005). The results obtained by Ni (2002) indicated that approximately 54.5 % of the total C is stored in alpine regions and 31.6 % in temperate regions. Together, these two regions constitute more than 85 % of all the carbon stored in grasslands of China (Ni 2002). In addition, about 93 % of grassland organic C stored in the soil in China, which is about 13.5 times of biomass C (Ni 2001). In alpine grasslands, the decomposition rates of soil organic matter remain slow, as may become an important C sink and will make a great contribution to the global soil C sequestration.

**Table 1** The biomass C density and C stock in Chinese grasslands

| Study sites | Area ($10^4$ km$^2$) | Biomass C density (g/m$^2$) | Biomass C stock (TgC) | Data sources and methods |
|---|---|---|---|---|
| Chinese grasslands | 569.9 | 215.8 | 1230 | National grassland resources survey data, root to shoot ratio |
| Chinese grasslands | 405.9 | 1148.2 | 4660 | The global average biomass C density of different grassland types |
| Chinese grasslands | 299.0 | 1023.5 | 3060 | The global average biomass C density of different grassland types |
| Chinese grasslands | 331.4 | 315.3 | 1045 | National grassland resources survey data, root to shoot ratio, NDVI data |
| Chinese grasslands | 167.0 | 340.0 | 562 | CEVSA model; NDVI data |
| Chinese grasslands | 334.1 | 315.2 | 1053 | National grassland resources survey data, NDVI data |
| Chinese grasslands | 331.0 | 1002.0 | 3316 | National grassland resources survey data, field measured biomass data |

From Fang et al. (2010)

**Table 2** Soil C density and carbon pool in Chinese grasslands

| Study sites | Area ($10^4$ km$^2$) | Soil depth (cm) | Soil C density (kg/m$^2$) | Soil C pool (PgC) | Data sources and methods |
|---|---|---|---|---|---|
| Grasslands in Northern China | 168.9 | 102.7 | 16.6 | 28.1 | The first national soil survey of China and others |
| Grasslands in Northern China | 196.3 | 100 | 8.5 | 16.7 | The observational data in 2001–2005 |
| Chinese grasslands | 249.3 | 103 | 15.1 | 37.7 | The second national soil survey of China |
| Chinese grasslands | 167.0 | 100 | 10.0 | 16.7 | CEVSA model |
| Chinese grasslands | 311.8 | 100 | 13.2 | 41.0 | Global soil database |

From Fang et al. (2010)

## *4.3 The Response of Grassland C Budget to Environmental Changes and Human Activities in China*

Ecosystem C budget is decided by the differences between the fixation of atmospheric carbon dioxide ($CO_2$) through photosynthesis and the simultaneous or subsequent release of $CO_2$ through autotrophic or heterotrophic respiration. Under the background of rapid global environmental change, there is an urgent need to better understand how the above-mentioned processes response to the climate change (e.g., the variations in temperature and precipitation) and the anthropogenic disturbances (e.g., overgrazing, fire, nitrogen deposition, and land use changes). Most of the currently available researches in China have been conducted in temperate steppes and alpine meadow regions. The C budget estimation was mainly based on the following three processes: (i) C accumulation in aboveground and belowground parts of vegetation; (ii) C turnover in some standing dead grass and litter; (iii) C emission from soil to atmosphere. The results obtained from these researches are largely disputed and uncertain.

Temperature and precipitation are considered to be the primary climatic factors determining the C budget in grassland ecosystem of China. They controlled both the processes of C accumulation through net primary production (NPP) and the processes of C sequestration and C mineralization in soil. For the process of C fixation, annual precipitation was positively correlated with grassland vegetation productivity from the point of spatial view (Bai et al. 2000), and temporally, the relationship between NPP and precipitation is site specific. For example, Guo et al. (2006) gathered field data of 48 grassland sites in northern China from 31 published papers or monographs, and they found that ANPP was controlled by the precipitation of growth period. While in the temperate meadow steppe and the alpine meadow, precipitation showed its time lag effect on NPP. Precipitation has also been regarded as the most key determinant of C emissions through respiration in

Chinese grassland (as reviewed by Bao and Zhou 2010). Generally, precipitation is more important than temperature to influence the C cycle processes and C budget in most of grasslands, especially in the growing seasons of temperate grasslands. The increasing temperature plays even more complex role in driving the change of grassland vegetation biomass and the C balance which varies greatly between and within sites, and over the duration of the experiment.

Besides climate change, anthropogenic activities such as grazing, one of the important management measures in grassland ecosystem, also significantly influenced the C processes of the grasslands. It has been suggested that the impacts of grazing on vegetation growth and C emissions depended largely on the grazing intensity and duration (Wu et al. 2009a, b). Commonly, overgrazing not only reduces the plants NPP and corresponding C fixation, greatly decreasing soil C input from grassland vegetation, but also accelerates SR and stimulates C release from the soil to the atmosphere. A number of experiments have demonstrated that overgrazing can lead to a 30–50 % decrease in belowground production in Inner Mongolia temperate grasslands and reductions of 20–40 % and 30–43 % in above- and belowground biomass, respectively, in alpine grasslands (Fang et al. 2010). A recent meta-analysis suggested that grazing can lead to a large amount of C loss from soils (Huang et al. 2010). However, a few studies also indicated that grassland vegetation has certain compensation and overcompensation abilities with modest grazing intensity, and moderate grazing can reduce C emissions and increase soil organic C (SOC) storage in soils in the inner Mongolian steppe (Yang et al. 2006; Xilin et al. 2009) and alpine meadow (Gao et al. 2007). Generally, the effects of grazing on soil C are mostly within the surface layer (0–30 cm). In addition, some studies have argued that long-term heavy grazing increased SOC and total nitrogen content in surface soils compared with light grazing management (Wang et al. 2011).

Land use/cover change (LUCC) is another important global change factor in grassland ecosystem and it has been widely recognized as a key driving forces of soil C balance (Yang et al. 2009). Wang et al. (2011) synthesized 133 papers published in the last 10 years about the effects on soil C of grassland management and related land use conversions in China. The result they obtained was that the effects of conversion from native grassland to cropland on SOC were variable and even contradictory across different grassland biomes. Most of experiments showed that the cultivation of grassland would reduce the organic carbon sequestration by soil. For example, Jiao et al. (2009) found that cultivation of typical steppe worsened soil erosion, and resulted in the decrease of SOC by approximately 18 % after 20 years cultivation. But, some other studies have demonstrated that the conversion from grassland to cropland had increased soil TOC. The research results of Horqin sandy land reported by Chen et al. (2004) showed that the TOC concentration of the 0–20 cm has increased by about 5.96 % after grassland conversion to arable land for 8 years.

## 4.4 Future Perspectives

Grasslands in China, because of their extensive distribution and high C density, are a prominent part of the global C cycle. Great progresses have been made during the past several decades in the spatial–temporal dynamics of C emission and C storage, and their responses to global changes as well as the underlying mechanisms in Chinese grassland. These works have made great contributions for us to clarify the potential for grassland C sequestration in China, to provide scientific guidance on grassland management, and finally accomplish the goal of increasing global sink for atmospheric $CO_2$. Nevertheless, given the large uncertainties in the current research results, some related fields are needed to be further strengthened.

(i) The long-term and multifactor field manipulative experiments should be strengthened. The short-term researches spend less and could capture the rapid response of ecosystem to environmental change easily, but their results are often with poor stability. The large differences among various current studies may be caused by the poor stability and the different approaches used in these studies. Thus, long-term repeated observations are urgently needed to increase the stability and reliability of experimental data. Meanwhile, it is imperative to continue to initiate and support multifactor experiments to explore the interactions of multiple factors on ecosystem C cycle in different grasslands of different locations, which will be much better for the more accurate estimation in grassland C budget under the scenario of complex environmental changes.

(ii) Quantifying belowground C stocks and its sensitivity to climate change and anthropogenic disturbances are critical for ecosystem C estimation. The belowground C processes and their driving mechanisms are less well understood so far compared with relative researches on aboveground portion. Soil C pool is the main body of ecosystem C pool, and that more than 85 % of biomass occurred belowground, thus, it is important and indispensable for us to reduce uncertainties in belowground C processes and to provide more optimal parameters to improve existing dynamic models of grassland C cycle.

(iii) A uniform and standard observation method for grassland ecosystem productivity and soil C sequestration should be established to increase the data comparability and obtain more accurate estimates on grassland C stock.

(iv) Strengthen the researches about the responses of grassland C storage and C budget to the anthropogenic disturbances such as grazing, fire, and the conversion of native grassland to cropland, etc. In China, the related work initiated earlier, however, they are still sporadic and not systematical, thus become the important sources for the uncertainty in the grassland C budget estimation and cannot provide a scientific guidance for grassland C management either.

# 5 Agroecosystem Carbon Cycle and Its Response to Climate Change

## 5.1 *Introduction*

China has succeeded in feeding 22 % of the global population (1.3 billion) on 7 % of the global arable land (130 million ha). With explosive economic growth and increasing population in recent decades, the absolute growth of Chinese greenhouse gases (GHGs) (calculated in $CO_2$ eq) has been over than other countries. China has become a major country of GHGs emission, and the rapidly increasing of GHGs emission in China indicated its influence on global climate change.

As a country with a millennial history of agriculture, Chinese agricultural soils contain lower level of soil organic carbon (SOC), especially in the regions with long-term drought, and serious water and soil erosion. The soils less than 1 % level of SOC account for 31 % of upland soils. In recent decades, numbers of regions showed the accumulation of SOC, with the increasing amount of crop yield and organic carbon return. The tremendous carbon sink and potential of carbon sequestration would contribute to mitigate global warming effect.

Based on the current agricultural soil carbon pool and its dynamic change, this section focus on summarize the primary conclusions regarding of net ecosystem $CO_2$ exchange (NEE), SR, soil carbon pool and its dynamic change, and potential of carbon sequestration in China.

## 5.2 *Research Status and Primary Results*

The three methods in agroecosystem carbon cycle as follows: first, measurement of field flux on NEE and SR (Zou et al. 2004); second, long-term field experiment, SOC pool and its dynamic change were measured under long-term plots design and samples determination (Huang et al. 2006); third, model analysis, stimulate the critical factors in influence agroecosystem carbon cycle, and estimate the ability and potential of carbon sequestration in agroecosystem, by building or in adoption of available biogeochemical model (Wang et al. 2005; Yan et al. 2007).

### 5.2.1 Agroecosystem Carbon Exchange and Soil Respiration

The chamber measurement was used to measure agroecosystem NEE in early study, currently, eddy-correlation technique was the common method to determine NEE. There were significant difference in growth period NEE under different crop types, overall, the averaged NEE was in the decreasing order from maize, wheat and rice, soybean, rape to naked oats.

Since 2000, there have been reports on agricultural soil respiration in the primary agricultural regions and crop types (wheat, rice, maize, soybean, highland barley, rape, etc.). Maize, wheat, and soybean and rice cropland generally has SR rate over 20 $gCO_2/(m^2\ d)$, over 10 $gCO_2/(m^2\ d)$, and less 10 $gCO_2/(m^2\ d)$ in growing stage, respectively. The main influence factors of SR include: soil temperature and water content (Zhang et al. 2008); drought (Dai et al. 2004); precipitation triggering (Zhang et al. 2008); organic fertilizer (Dai et al. 2004) and chemical fertilizer application (Wang et al. 2002; Zhang et al. 2005); straw incorporation (Zhang et al. 2005); tillage (Meng et al. 2006); and irrigation (Yang and Cai 2004). In terms of the measurement of SR in block plot, the study obtained that root respiration account for the total soil respiration varies in crop types and regions (13–77 %, Han et al. 2008)

### 5.2.2 Agrocosystem Carbon Storage and Development

The estimation of Chinese SOC pool is mainly based on the twice general surveys of soils in the nationwide scale, however, due to in adoption of different scale of soil distribution, different amount of samples, and different estimation of soil bulk density, the results of SOC storage occurred great difference (Wang et al. 2005; Yan et al. 2007). The total carbon storage of Chinese agroecosystem is 69–180 PgC in the entire profile (1 m), with an averaged carbon density of 5.46–19.6 $kg/m^2$. The spatial pattern of Chinese agricultural topsoil SOC showed an increasing trend from West to East and from South to North. The mean SOC concentration is in the order of Northeast > Southwest, South, North, and East > Northwest. It could be attributable to the lower temperature, and higher soil moisture in Northeast, which reduce the decomposition of SOC; while due to the lower plant cover rate and drought conditions, there is lower SOC storage in Northwest region of China.

The dynamic of SOC pool primary base on the two methods: (i) general survey of soil and local soil survey data; (ii) model analysis. Based on the first and second general survey of soil, Wang et al. (2003) estimated the SOC loss 1 PgC during 1960–1980 in China. Since the second general survey of soil, there is no soil survey in the entire China. However, many local and research institutions conducted the survey work with respect to SOC since 1990s. In comparison of the second general survey of soil, the dynamic trend of Chinese agricultural SOC could be obtained. Xie et al. (2007) estimated the accumulative rate of agricultural SOC was 23.61 TgC/a in the last two decades; Yu et al. (2009) calculated the Chinese arable SOC pool increased 0.26 PgC during 1980–2000; Yan et al. (2010) obtained the averaged SOC increased 0.22 % per year from 1980 to 2008 in China; according to the literature analysis, Huang et al. (2006) found, in recent 20 years, 53–60 % of arable area's SOC concentration increased, and estimated Chinese agricultural SOC storage increased by 311.3–401.4 TgC. The increase of Chinese agricultural SOC storage could be ascribed to the adoption of fertilizer and conservational tillage in recent 20 years. There only scare studies reported the dynamic of Chinese SOC based on the model research. In terms of the DNDC model, Li (2000) estimated

Chinese agricultural SOC loss 73.8 TgC per year; Tang et al. (2006) also proved the decline in Chinese agricultural soil organic carbon.

### 5.2.3 The Rate and Potential of Carbon Sequestration in Primary Agricultural Managements

The widely accepted carbon sequestration management include minimal and no tillage, straw return, organic fertilizer and chemical fertilizer application. Jin et al. (2008) estimated the carbon sequestration rate of combination of chemical and organic fertilizer, straw return, organic fertilizer application, no tillage, and chemical fertilizer application was 0.889, 0.597, 0.545, 0.514, and 0.129 t C/(hm$^2$ a); Han et al. (2008) determined the carbon sequestration rate of chemical fertilizer application, straw return, organic fertilizer application, and no tillage was 0.38, 0.224, 0.316, and 0.284 t C/(hm$^2$ a).

There are two methods to estimate the potential of carbon sequestration: (i) based on the soil carbon saturation level, (ii) based on the carbon sequestration rate of agricultural managements. According to the result of DNDC model, Han et al. (2005) confirmed agricultural SOC saturation level was 4.8–51.4 g/kg in 1990s; by analysis the long-term experimental data, Cheng et al. (2009) indicated the potential of SOC level in upland and rice paddy could reach 17.2 and 27.7 g/kg, respectively; Jin et al. (2008) estimated the saturation of SOC in chemical fertilizer application, organic fertilizer application, straw incorporation, formula fertilizer, and no tillage was 17.4, 22.3, 25.6, 25.9, and 31.4 g/kg; Han et al. (2008) reported the potential of agricultural SOC sequestration in chemical fertilizer application, straw incorporation, organic fertilizer application, and no tillage was 94.9, 42.2, 41.4, and 3.58 TgC/a, totaling 182.1 TgC/a; Yan et al. (2007) assessed that the potential of SOC sequestration under 50–100 % of straw return, and 50–100 % of no tillage application could be reached in 23–57 TgC, and 22–43 TgC m respectively.

### 5.2.4 The Effect of Climate Change on Agroecosystem Carbon Cycle

Currently, there are rare studies reported the cycle of agroecosystem carbon in China. In the 1980s, some papers reported the effect of $CO_2$ on farmland ecosystems and its carbon cycle, for example, Wang and Lin (1999) reported the effect of elevated $CO_2$ on yield of wheat and soybeans, and Wang and Lin (1999) reported the effect of elevated $CO_2$ level on rice roots exudates. Since 2000, Zhu et al. established FACE experiments in Yangzhou, Jiangsu Province, and investigated the effect of elevated $CO_2$ level on photosynthesis of wheat, rice, crop yield, soil carbon and nitrogen, and trace element. In this FACE experiments Zheng demonstrated the effect of elevated $CO_2$ concentration on methane emission (Xu et al. 2004). The researches concerning the impact of climate warming on agroecosystem are rarely reported in China. Recently, Zhang et al. (2010) established the opening nighttime warming system to study the responses of rice and wheat to warming. The impacts of precipitation on carbon cycle of agroecosystems could primarily contribute to its changing soil moisture conditions. There are many reports on the effect of water change on agroecosystem, especially focusing on the effect of drought on crop growth and yield. But there are still rare experiments on the impact of precipitation on carbon cycle of agroecosystem. Model is the important tool to study the effect of climate change on agroecosystem carbon cycle. Currently, Li Changsheng' s DNDC, Huang Yao' s Agro-C, and Tian Hanqin's DELEM model are the primary model being applied to the study in Chinese agroecosystem carbon cycle. Additionally, further research is needed to assess the effect of future climate change scenarios on Chinese carbon cycle of agroecosystem and food security.

## 5.3 Conclusions and Discussion

Since the 1990s, researchers have estimated the carbon pool and its dynamic change, and the ability and potential of carbon sequestration in Chinese agriculture, however, there are great different and uncertainty among the different estimated results. For reduction the uncertainty in estimation the dynamic and potential of carbon sequestration, and making agroecosystem play a role in mitigation global climate change, the following research is needed:

(i) Currently, it was usually adopted soil types when estimated SOC storage. In fact, SOC density generally has great difference in the same soil type. The change of agricultural SOC pool was generally less than 1 t C/hm$^2$ a, the mean SOC pool is 100 t C/hm$^2$, which result in the difficulty to measure the change of soil carbon pool in 1–2 years. SOC pool change year by year, and it is susceptible to the carbon input and climate conditions, which would make difficulty to estimation the dynamic change of SOC pool. Consequently, it is necessary to analysis the critical factors that impact estimation accuracy.

(ii) There are considerable reports on the mechanism of SOC cycle, for example, the studies of physical, chemical, and biological fractions of SOC, SR, the decomposition of soil organic matter, and soil microorganism. However, there are no reasonable explication respecting the stability and resilience mechanism of SOC. So the above topic needs to further discussed. For example, the relationship of soil carbon structure among the physical, chemical, and biological properties, and the relationship between them and the stability of SOC; how tillage and climate change impact long-term soil respiration; and the relationship between the decomposition of soil organic matter and SR and the stability of SOC.

(iii) Recently, model analysis method had been applied in the research of Chinese terrestrial ecosystem carbon cycle, for example, the model building of CEVSA, BEPS, BIOME & BIOME-BGC, GLO-PEM, EALCO, and SIB2 is based on the interaction of energy, water, and carbon transmission. Most of the models focus on the ecosystem production and NEE, lacking of the study regarding agricultural managements, such as fertilizer application, tillage, and straw incorporation. Due to the special agricultural production in China, domestic scholars develop C-Agro model. Li (2000) have used biogeochemistry DNDC model to study the carbon cycle of Chinese agricultural soil. In the future, development of Chinese agriculture and climate change should be synthetically considered, and confirm the impact of climate change on the carbon cycle of Chinese agricultural soil.

(iv) As an important approach of mitigation greenhouse effect, soil carbon sequestration and agricultural emission reduction have received the great attention from international community, the program of cooperation in the local and international organizations is implied, and countries have legislated laws to improve the ability to combat climate change. As one of the most safe, available, economic emission reduction way, biological carbon sequestration technique attracted universal attention from international society, and become one of the hottest topics in interdisciplinary research. Currently, the primary carbon sequestration managements of agriculture include fertilizer application, straw incorporation, no tillage, and irrigation. According to the case study and the analysis of influence factors in carbon sequestration, the impact of tillage, rice paddy arable area, straw incorporation, fertilizer, cropping system, land use, and the development of society economic on Chinese agricultural carbon sequestration and emission reduction should be evaluated, and put forward the relative management for increased carbon sequestration and emission reduction.

# 6 Wetland Carbon Cycle and Its Response to Climate Change

## 6.1 Introduction

Natural wetlands, one of the significant carbon sinks in the terrestrial ecosystems, play important roles in the carbon budgets of the global carbon cycles. Wetland ecosystems hold the highest soil carbon density (Post et al. 1982), and serve as sinks or sources of greenhouse gases (GHGs), such as $CO_2$, $CH_4$, and $N_2O$ (Freeman et al. 1993; Song et al. 2009). The carbon processes in the natural wetlands could influence climate changes in the regional or global scales.

Global changes, such as water regimes, land use changes, nitrogen (N) deposition, etc., could affect the carbon cycle in wetlands. For example, water table changes can greatly influence carbon storages and carbon fluxes in wetland ecosystems (Tiiva et al. 2009; Berglund and Berglund 2011). Strom and Christensen (2007) reported that a change in hydrology may result in an increasing source of radiative forcing in the mire. N addition to wetland ecosystems alters microbial communities (Keller et al. 2005; Banger et al. 2012), and vegetation communities (Chapin et al. 2004; Keller et al. 2005), and thus the resulted changes in the three main GHGs. It is vital important to investigate the responsible mechanism of carbon cycles between soil, vegetation, and atmosphere to global changes.

In this part, the carbon cycles in responses to N deposition, water regimes, and wetland conversion were explored. Carbon sequestration in plant, soil carbon storage, the main carbon gases (i.e., $CO_2$ and $CH_4$) and litter decomposition were introduced. In the end, the long-term systematic research about wetland soil microorganisms in different environment regimes and the potential biodegradation ability of organic carbon were emphasized. More sophisticated models of carbon cycles in wetland ecosystems should be developed in the future to predict the carbon budgets from wetland ecosystems with regard to global carbon budgets.

## *6.2 Responses of Carbon Cycle in Wetlands to Global Changes*

### 6.2.1 Responses of Carbon Cycle to Water Regimes and Exogenous Nutrient Inputs

Water regimes and exogenous nutrient inputs exert large effects on carbon pools and fluxes from wetland ecosystems. Flooding condition significantly increased soil $CH_4$ emission, total amount of $CH_4$ emitted under 15 cm flooding water table was 2 and 30 times of those under 0 and −10 cm (Table 3). Water table was positively correlated to $CH_4$ emission amount. $CO_2$ emission rates increased with water table rises when the soil was inundated. The effects of N fertilization on greenhouse gases are described in Song et al. (2013). Exogenous N inputs also speeded the litter decomposition rate under submerged condition, while there were no significant differences between different N levels. The accumulated carbon released as $CO_2$–C and DOC–C was less than 26 % of the total losing carbon in litter under moist condition, however, more than 50 % of the total losing carbon released as the form of $CO_2$–C and DOC–C (Table 4), which suggested DOC is the main losing way in the decomposition process of submerged litter.

### 6.2.2 Land Use Change and Carbon Cycle in Wetland Ecosystem

Land use and management obviously affect dynamics of wetland ecosystem carbon cycle. Soil carbon storage (SOC) decreased apparently after wetland conversion in the first five years, while it begins to stabilize after 10 years of wetland conversion (Fig. 1). SOC increased significantly after cultivated soil abandonment.

**Table 3** Total emissions of $CO_2$, $CH_4$, $N_2O$ and the calculated total $CO_2$ equivalents of greenhouse under different water table levels during the whole growing season

| Water condition (cm) | $CO_2$ (g $m^{-2}$) | $CH_4$ (g $m^{-2}$) | Total C emission (g $m^{-2}$) | Total $CO_2$ equivalents (Mg $m^{-2}$) |
|---|---|---|---|---|
| 15 | 635.59 (41.09) | 130.34 (2.04) | 285.07 (12.35) | 38.94 (0.83) |
| 10 | 573.46 (22.64)[a] | 98.82 (8.64) | 241.10 (2.08) | 30.44 (1.94) |
| 5 | 552.06 (45.99) | 61.16 (13.42) | 202.98 (23.08) | 20.81 (3.75) |
| 0 | 804.01 (74.80) | 7.99 (0.63) | 226.12 (20.21) | 10.04 (0.71) |
| −5 | 1269.94 (137.89) | 8.16 (0.51) | 353.34 (37.75) | 14.74 (1.42) |
| −10 | 1257.19 (76.19) | 4.40 (1.32) | 346.64 (20.61) | 13.67 (0.78) |

*Note* This table is cited from Hou (2012)

**Table 4** DOC–C and $CO_2$–C output and output rate from litter in wetland under N addition

| Conditions | | DOC–C output (mg C/g) | DOC–C output rate (%) | $CO_2$–C flux (mg C/g) | $CO_2$–C flux rate (%) | Carbon pool loss in litter (mg C/g) |
|---|---|---|---|---|---|---|
| Moist | *K* | 9.90 | 14.00 | 4.35 | 6.15 | 70.71 |
| | 1 | 15.83 | 14.91 | 5.66 | 5.33 | 106.12 |
| | 2 | 17.14 | 19.36 | 5.29 | 5.97 | 88.54 |
| | 3 | 16.69 | 20.85 | 4.59 | 5.73 | 80.08 |
| Submergence | *K* | 16.77 | 59.46 | 3.02 | 10.70 | 28.20 |
| | 1 | 26.05 | 55.88 | 3.29 | 7.06 | 46.62 |
| | 2 | 29.99 | 63.23 | 3.16 | 6.66 | 47.43 |
| | 3 | 30.69 | 65.62 | 3.17 | 6.78 | 46.76 |

*Note* N1, 60 kgN $ha^{-1}$ $year^{-1}$; N2, 120 kgN $ha^{-1}$ $year^{-1}$; N3, 240 kgN $ha^{-1}$ $year^{-1}$. This table is cited from Liu (2009)

## 6.3 Conclusion and Discussion

Global changes, such as exogenous N input and climate warming that caused the change of wetland plant communities and productivity, may be the reason for changes in quality and quantity of litter in the soil and the underground root, resulting in the varying quantity of input carbon in the wetland and the stability in deep soil organic carbon, but the controlling factors of deep soil organic carbon stability is still unclear. Understanding of these controlling factors for the evaluation of wetland soil carbon pools in the future global change is essential. In addition, on the landscape scale, permafrost degradation process affects carbon balance in the high latitude regions. However, studies about the estimate of carbon accumulation in permafrost ecosystem are limited, and also much more studies about carbon fluxes and net carbon exchanges for these areas, especially in the context of the

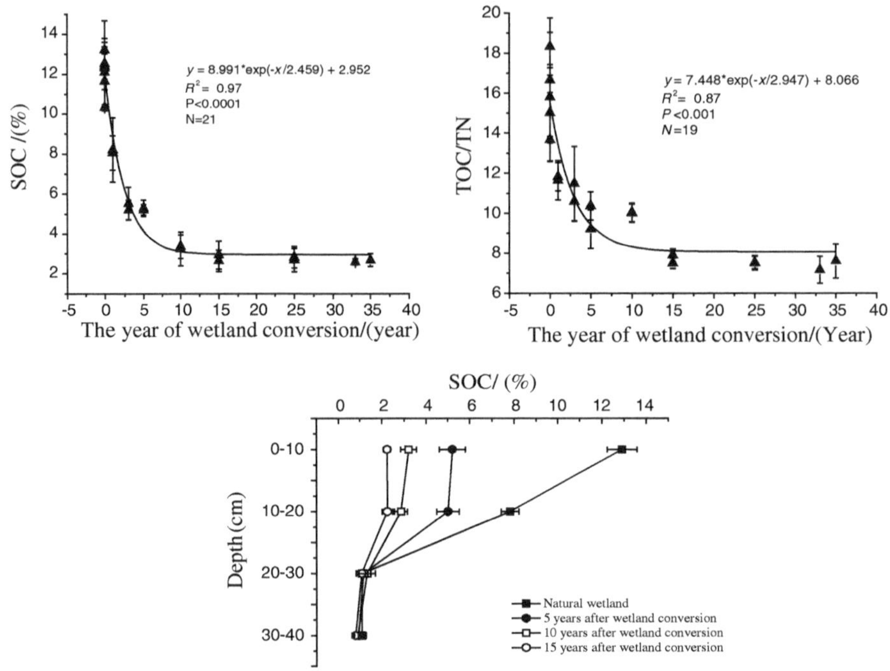


**Fig. 1** SOC changes after wetland conversion (Zhang 2006)

future global change scenarios, need to be done. The limitations also lie in insufficient understanding of some key carbon cycle processes, the quality of frozen soil organic carbon content in the soil, and the potential of carbon biodegradability within the permafrost. In conclusion, researches about the function in carbon sources/sinks in wetland ecosystems in the context of global changes still remain far from certain. Long-term field studies about the soil microorganism and carbon fluxes from wetland ecosystems are rare. Comprehensive researches in the key processes in soil microorganism and enzymatic activity regarding carbon cycle in wetlands are needed. The synthetic study with regard to soil, plant biomass, soil microorganism and enzymatic activity in the processes of wetland carbon cycle would help in comprehensively understanding the responses of carbon sequestration in wetlands to global changes. In the future, the mature dynamic carbon model for wetlands should be developed to better predict the carbon fluxes from wetland ecosystems. The combination of climate model, land use change model, and wetland carbon model might be an important direction in future wetland carbon cycle studies.

# 7 Application of Novel Techniques on Measuring the Carbon and Water Fluxes over Terrestrial Ecosystems

## 7.1 *Introduction*

The coupled cycle of carbon (C) and water ($H_2O$) in terrestrial ecosystems and their environmental and biological controlling mechanisms is one of key issues for global change ecology in the recent decades. Understanding the coupled cycles and their environmental and biological controlling mechanisms are important to explore the response and feedback of terrestrial ecosystems to global change (Yu et al. 2008b). Some novel technique, such as eddy covariance and isotope ratio infrared spectroscopy (IRIS), provides us continuous measurement of $CO_2$ and $H_2O$ fluxes and their isotopic flux ratio, which are applied in the study of the coupled cycle of carbon (C) and water ($H_2O$) in terrestrial ecosystems (Baldocchi 2008; Griffis 2013).

## 7.2 *Novel Techniques of the Carbon and Water Fluxes*

### 7.2.1 Eddy Covariance Technique

The Chinese Terrestrial Ecosystem Flux Research Network (ChinaFLUX) is a long-term national network of micrometeorological flux measurement sites that measure the net exchange of carbon dioxide, water vapor, and energy between the biosphere and atmosphere based on the eddy covariance technique. It relies on the existing Chinese Ecosystem Research Network (CERN), fills an important regional gap and increases the number of ecosystem types in FLUXNET. Expanding the scope of the FLUXNET database, ChinaFLUX offers new opportunities to quantify and compare the magnitudes and dynamics of annual ecosystem carbon and water balance and to explore the biotic and abiotic effects on ecosystem processes of carbon dioxide and water vapor exchange that are unique to ecosystems in China, such as the vegetation communities on the Qinghai–Tibet plateau. Besides, ChinaFLUX also provides more insights to help define the current status and enable future prediction of the global biogeochemical cycles of carbon, water and trace gases (Yu et al. 2006; Sun et al. 2006a, b).

Comparison of ecosystem carbon exchange among the three forests, include Changbaishan temperate mixed forest (CBS), Qianyanzhou subtropical coniferous plantation (QYZ), and Dinghushan subtropical evergreen broad-leaved forest (DHS), shows that RE was mainly determined by temperature, with the forest at CBS exhibiting the highest temperature sensitivity among the three ecosystems (Yu et al. 2008a). The RE was highly dependent on GEP across the three forests, and the ratio of RE to GEP decreased along the North–South Transect of Eastern China

(NSTEC) (i.e., from the CBS to the DHS), with an average of 0.77 ± 0.06. The slope of NEP that decreased with increasing latitude along the NSTEC was markedly different from that observed on the forest transect in the European continent.

GEP, ER, and net ecosystem productivity (NEP) of terrestrial ecosystems in China showed a significantly latitudinal pattern, declining linearly with the increase of latitude (Yu et al. 2013b). However, GEP, ER, and NEP did not present a clear longitudinal pattern. The carbon sink functional areas of terrestrial ecosystems in China were mainly located in the subtropical and temperate forests, coastal wetlands in eastern China, the temperate meadow steppe in the northeast China, and the alpine meadow in eastern edge of Qinghai–Tibetan Plateau. The forest ecosystems had stronger carbon sink than grassland ecosystems. The spatial patterns of GEP and ER in China were mainly determined by mean annual precipitation (MAP) and mean annual temperature (MAT), whereas the spatial variation in NEP was largely explained by MAT. The combined effects of MAT and MAP explained 79, 62, and 66 % of the spatial variations in GEP, ER, and NEP, respectively. The GEP, ER, and NEP in different ecosystems in China exhibited 'positive coupling correlation' in their spatial patterns. Both ER and NEP were significantly correlated with GEP, with 68 % of the per-unit GEP contributed to ER and 29 % to NEP. MAT and MAP affected the spatial patterns of ER and NEP mainly by their direct effects on the spatial pattern of GEP.

The majority of Asian terrestrial ecosystems are currently large carbon sinks based on long-term observation data from ChinaFLUX (19 sites) and published data from AsiaFlux (37 sites) and 32 other sites in Asia (Chen et al. 2013). The average net ecosystem productivity (NEP) values were 325 ± 187, 274 ± 207, 236 ± 260, 89 ± 134 g C $m^{-2}$ $year^{-1}$ in cropland, forest, wetland and grassland ecosystems, respectively. The spatial variation of gross primary production (GPP) and ecosystem respiration (Re) were mainly controlled by the mean annual temperature (MAT) and the mean annual precipitation (MAP) in the Asian region. There was a clear linear relationship between GPP and MAT, and a strong sigmoid relationship between GPP and MAP. Re was exponentially related to MAT and linearly related to MAP. Interestingly, those response modes were consistent across different ecosystem types. The different responses of GPP and Re to MAT and MAP determined the spatial variation of NEP.

### 7.2.2 In Situ Measurement of Carbon and Water Isotopic Ratio

The $\delta D$ and $\delta^{18}O$ in atmospheric water vapor provide rich information on the hydrological cycle and gaseous exchange processes between the terrestrial vegetation and the atmosphere. Isotope ratio infrared spectroscopy (IRIS) provides an in situ technology for measuring $\delta D$ and $\delta^{18}O$ in ambient conditions (Lee et al., 2005; Griffis 2013). Wen et al. (2008) have demonstrated the feasibility to simultaneously measure both δD and $\delta^{18}O$ in atmospheric water vapor using a tunable diode laser absorption spectrometer. Our laboratory tests showed that the 1-h precision (one standard deviation) was 1.1 ‰ for δD and 0.07 ‰ for $\delta^{18}O$ at the dew point temperature of 15 °C. Our atmospheric measurement captured the rapidly changing isotopic signals in both δD and $\delta^{18}O$. An intercomparison experiment was carried out with four commercial IRIS analyzers to characterize their performance and transferability of calibration methods (Wen et al. 2012b). These analyzers tracked the natural variability in ambient conditions very well and achieved an average difference between one another within 2 ‰ for δD and within 0.1 ‰ for $\delta^{18}O$ after calibration at appropriate frequencies. Two of the calibration methods (discrete liquid water injection and continuous dripping) agreed with each other within the tolerance thresholds of 2 ‰ for δD and 0.1 ‰ for $\delta^{18}O$. The concentration dependence underscores the importance of using a calibration procedure at multiple mixing ratios to bracket the range of natural variability.

Isotope ratio infrared spectroscopy (IRIS) also provides an in situ technique for measuring $^{13}C$ in atmospheric $CO_2$. A number of methods have been proposed for calibrating the IRIS measurements, but few studies have systematically evaluated their accuracy for atmospheric applications (Wen et al. 2013) carried out laboratory and ambient measurements with two commercial IRIS analyzers and compared the accuracy of four calibration strategies. Calibration based on the $^{12}C$ and $^{13}C$ mixing ratios (Bowling et al. 2003) and on linear interpolation of the measured delta using the mixing ratio of the major isotopologue (Lee et al. 2005) yielded accuracy better than 0.06 ‰. Over a 7-day atmospheric measurement in Beijing, the two analyzers agreed to within −0.02 ± 0.18 ‰ after proper calibration. The high sensitivity of the Keeling analysis to the concentration dependence underscores the challenge of IRIS for atmospheric research.

Leaf water $^{18}O$ enrichment is an important factor controlling the $H_2^{18}O$, $C^{18}OO$, and $O^{18}O$ exchanges between the biosphere and the atmosphere. At present, there is limited capacity to explain the enrichment mechanisms in field conditions. Three models of varying complexity were used to simulate the leaf water $^{18}O$ enrichment at the canopy scale. Comparisons were made among the models and with high-frequency isotopic measurements of ecosystem water pools in wheat and corn (Xiao et al. 2012). The results show that the steady state assumption was a better approximation for ecosystems with lower canopy resistance, that it is important to consider the effect of leaf water turnover in modeling the enrichment and not necessary to deal with time changes in leaf water content, and that the leaf-scale Péclet effect was incompatible with the big-leaf modeling framework for canopy–air interactions. After turbulent diffusion has been accounted for in an apparent kinetic factor parameterization, the mean $^{18}O$ composition of the canopy foliage water was a well-behaved property predictable according to the principles established by leaf-scale studies, despite substantial variations in the leaf water enrichment with leaf and canopy positions.

Dew formation has the potential to modulate the spatial and temporal variations of isotopic contents of atmospheric water vapor, oxygen and carbon dioxide (Wen et al. 2012a) demonstrate that the equilibrium fractionation played a dominant role over the kinetic fractionation in controlling the dew water isotopic compositions. A significant correlation between the isotopic compositions of leaf water and dew water suggests a large role of top-down exchange with atmospheric vapor controlling the leaf water turnover at night. According to the isotopic labeling, dew water consisted of a downward flux of water vapor from above the canopy (98 %) and upward fluxes originated from soil evaporation and transpiration of the leaves in the lower canopy (2 %).

Deuterium-excess (d) in water is a combination of the oxygen ($\delta^{18}O$) and hydrogen ($\delta D$) isotope ratios, and its variability is thought to indicate the location and environmental conditions of the marine moisture source (Welp et al. 2012) analyze d of water vapor (dv) from six sites, all between 37 and 44 N to examine patterns in the atmospheric surface layer and identify the main drivers of variability. A robust diurnal cycle was found in dv at all sites with maximum values during midday. Isotopic land surface model simulations suggest that plant transpiration is one mechanism underlying the diurnal pattern. An isotopic large-eddy simulation model shows that entrainment of the free atmosphere into the boundary layer can also produce high dv values in midday. Daily midday means of dv were negatively correlated with local midday relative humidity and positively correlated with planetary boundary layer height at the North American sites, but not the Chinese sites. The mechanism for these differences is still undetermined. These results demonstrate that within the diurnal time scale, dv of the surface air at continental locations can be significantly altered by local processes, and is therefore not a conserved tracer of humidity from the marine moisture source region as has previously been assumed.

## 7.3 Conclusion and Discussion

Carbon, nitrogen, and water cycles in terrestrial ecosystem are three critical subjects in global change science. The coupling modes of these cycles and their biological regulation mechanisms had been frontier issues in global change ecology. At present, lack of knowledge on the processes of carbon-nitrogen-water coupling cycles and the regulation mechanisms had limited assessment of terrestrial carbon sink enhancement and emission reduction (Yu et al. 2013a). This had become a major problem in predicting the effects of global change on the productivity and carbon sequestration of ecosystems. The main processes and biological regulation mechanisms of carbon-nitrogen-water coupling cycles discussed included: (i) biological processes of plant leaf canopy/root canopy and its regulation mechanism on carbon-nitrogen-water coupling cycles and the interactions among them; (ii) soil microbial functional group network and its effects on carbon, nitrogen, and water cycles; (iii) spatial and temporal variations in carbon, nitrogen, and water fluxes in terrestrial ecosystem along with the theory and practice of ecosystem stoichiometry.

## References

Bai, Y. F., Li, L. H., Wang, Q., et al. (2000). Changes in plant species diversity and productivity along gradient of precipitation and elevation in the Xilin river basin, Inner Mongolia. *Acta Phytoecologica Sinica, 6*, 667–673.

Baldocchi, D. D. (2008). Turner Review No. 15—Breathing of the terrestrial biosphere: Lessons learned from a global network of carbon dioxide flux measurement systems. *Australian Journal of Botany, 56*, 1–26.

Banger, K., Tian, H., & Lu, C. (2012). Do nitrogen fertilizers stimulate or inhibit methane emissions from rice fields? *Global Change Biology, 18*, 3259–3267.

Bao, F., & Zhou, G. S. (2010). Review of research advances in soil respiration of grassland in China. *Chinese Journal of Plant Ecology, 34*(6), 713–726.

Berglund, O., & Berglund, K. (2011). Influence of water table level and soil properties on emissions of greenhouse gases from cultivated peat soil. *Soil Biology and Biochemistry, 43*(5), 923–931.

Bowling, D. R., Pataki, D. E., & Ehleringer, J. R. (2003). Critical evaluation of micrometeorological methods for measuring ecosystem-atmosphere isotopic exchange of $CO_2$. *Agricultural and Forest Meteorology, 116*, 159–179.

Chapin, C. T., Bridgham, S. D., & Pastor, J. (2004). PH and nutrient effects on above-ground net primary production in a Minnesota USA bog and fen. *Wetlands, 24*, 186–201.

Chen, Z., Yu, G. R., Ge, J. P., et al. (2013). Temperature and precipitation control of the spatial variation of terrestrial ecosystem carbon exchange in the Asian region. *Agricultural and Forest Meteorology, 182–183*, 266–276.

Chen, F. S., Zeng, D. H., Chen, G. S., et al. (2004). Effects of reclamation on soil organic carbon of some meadow soils. *Chinese Journal of Soil Science, 35*(4), 413–419. (in Chinese).

Cheng, K., Pan, G. X., Tian, Y. G., et al. (2009). Changes in topsoil organic carbon of China's cropland evidenced from the national soil monitoring. *Journal of Agro-Envrionment Science., 28*(12), 2476–2481.

Chu, K. Z. (1972). The preliminary study of the climate change of China during the last 5000 years. *Acta Archaeol Sin, 1*, 15–38. (in Chinese).

DAHV (Department of Animal Husbandry and Veterinary, Institute of Grassland, Chinese Academy of Agricultural Sciences), CISNR (Commission for Integrated Survey of Natural Resources, Chinese Academy of Sciences). (1994). Data on grassland resources of China. Beijing: China Agricultural Science and Technology Press.

Dai, W. H., Wang, Y. Q., Huang, Y., et al. (2004). Flush of carbon dioxide caused by soil drying and rewetting. *Journal of Agro-Environment Science, 23*(2), 318–322.

Ding, S. Y., & Peng, J. (1991). A study on biomass of Castanopsis Orthacantha sprouted shrub community in Kunming region. *Journal of Southwest Forestry College, 11*(1), 41–50. (in Chinese).

Dong, Y. S., Qi, Y. C., Liu, J. Y., et al. (2005). Variation characteristics of soil respiration fluxes in four types of grassland communities under different precipitation intensity. *Chinese Science Bulletin, 50*(6), 583–591.

Fan, J. W., Zhong, H. P., Harris, W., Yu, G. R., Wang, S. Q., Hu, Z. M., & Yue, Y. Z. (2008). Carbon storage in the grasslands of China based on field measurements of above- and below-ground biomass. *Climatic Change, 86*, 375–396.

Fang, J. Y., Guo, Z. D., Piao, S. L., & Chen, A. P. (2007). Terrestrial vegetation carbon sinks in China during 1981 and 2000. *China Series D: Earth Sciences, 50*(9), 1341–1350.

Fang, J. Y., Liu, G. H., & Xu, S. L. (1996). Biomass and net production of forest vegetation in China. *Acta Ecologica Sinica, 16*(5), 497–508.

Fang, J. Y., Yang, Y. H., Ma, W. H., et al. (2010). Ecosystem carbon stocks and their changes in China's grasslands. *Science China: Life Sciences, 53*(7), 757–765.

Freeman, C., Lock, M. A., & Reynolds, B. (1993). Fluxes of $CO_2$, $CH_4$ and $N_2O$ from a Welsh peatland following simulation of water table draw-down: potential feedback to climatic change. *Biogeochemistry, 19*, 31–60.

Fu, Y. L., Yu, G. R., Wang, Y. F., et al. (2006). Effect of water stress on the photosynthesis and respiration of grassland ecosystem in Inner Mongolia. *Science in China, Series D: Earth Sciences, 36*(1), 183–193.

Gao, Y., Luo, P., Wu, N., et al. (2007). Grazing intensity impacts on carbon sequestration in an alpine meadow on the eastern Tibetan Plateau. *Research Journal of Agriculture and Biological Sciences, 3*(6), 642–647.

Ge, Q. S. (2010). *Climate change during the past Chinese dynasties*. Beijing: Science Press. (in Chinese).

Ge, Q. S., Zheng, J. Y., Hao, Z. X., et al. (2010). Temperature variation through 2000 years in China: An uncertainty analysis of reconstruction and regional difference. *Geophysical Research Letters, 37*.

Griffis, T. J. (2013). Tracing the flow of carbon dioxide and water vapor between the biosphere and atmosphere: A review of optical isotope techniques and their application. *Agricultural and Forest Meteorology, 174–175*, 85–109.

Guo, R., Wang, X. K., OuYang, Z. Y., & Li, Y. N. (2006). Spatial and temporal relationships between precipitation and ANPP of four types of grasslands in northern China. *Journal of Environmental Sciences, 18*(5), 1024–1030.

Han, B., Wang, X. K., Lu, F., et al. (2008). Soil carbon sequestration and its potential by cropland ecosystems in China. *Acta Ecologica Sinica, 28*(2), 612–619.

Han, B., Wang, X. K., & OuYang, Z. Y. (2005). Saturation levels and carbon sequestration potentials of soil carbon pools in farmland ecosystems of China. *Rural Eco-Environment, 21* (4), 6–11.

Hou, C. C. (2012). Effects of hydrological changes on soil carbon sequestration of marsh in the Sanjiang Plain. Graduate School of Chinese Academy of Sciences.

Hou, X. Y., Sun, S. Z., Zhang, J. W., et al. (1982). *Vegetation Map of the People's Republic of China*. Beijing: Map Press of China.

Huang, Y., & Sun, W. (2006). The change trend of Chinese agricultural surface soil organic carbon concentration in recent two decades. *Science of China, 51*(7), 753–763.

Huang, Y., SunWJ, Zhang W., & Yu, Y. Q. (2010). Changes in soil organic carbon of terrestrial ecosystems in China: A mini-review. *Science in China Series C, 53*(7), 766–775.

IPCC. (2007a). *Climate change 2007: The physical science basis*. Cambridge, UK: Cambridge University Press.

IPCC. (2007b). *Climate change 2007: The physical science basis*. Contribution of Working Group I to the Fourth Assessment Report of the Intergovernmental Panel on Climate Change, Cambridge University Press, Cambridge, UK, pp. 463–497.

Jiao, Y., Xu, Z., & Zhao, J. H. (2009). Effects of grassland conversion to cropland and forest on soil organic carbon and dissolved organic carbon in the farming-pastoral ecotone of Inner Mongolia. *Acta Ecologica Sinica, 29*, 150–154.

Jin, L., Li, Y. E., Gao, Q. Z., et al. (2008). Estimate of carbon sequestration under cropland management in China. *Scientia Agricultura Sinica, 41*(3), 734–743.

Keller, J. K., Bridgham, S. D., Chapin, C. T., & Iversen, C. M. (2005). Limited effects of six years of fertilization on carbon mineralization dynamics in a Minnesota fen, Soil Biol. *Biochem., 37*, 1197–1204.

Lee, X., Sargent, S., Smith, R., & Tanner, B. (2005). In-situ measurement of water vapor 18O/16O isotope ratio for atmospheric and ecological applications. *Journal of Atmospheric and Oceanic Technology, 22*, 555–565.

Li, C. S. (2000). Loss of soil carbon threatens Chinese agriculture: A comparison on agroecosysem carbon pool in China and the U.S. *Quaternary Research, 20*(4), 345–350.

Li, K. R., Wang, S. Q., & Cao, M. K. (2003). Carbon storage of vegetation and soil in China. *Science in China, Series D: Earth Sciences, 33*(1), 72–80.

Liu, D. Y. (2009). Responses of the main biogeochemical carbon processes to exogenous nitrogen of Calamagrostis angustifolia wetland ecosystem. Graduate school of Chinese Academy of Sciences.

Lu, C. Q., & Sun, S. C. (2004). A review on the distribution patterns of carbon density in terrestrial ecosystem. *Acta Phytoecologica Sinica., 28*(5), 692–703.

Mann, M. E., Bradley, R. S., & Hughes, M. K. (1998). Global-scale temperature patterns and climate forcing over the past six centuries. *Nature, 392*, 779–787.

Mann, M. E., Bradley, R. S., & Hughes, M. K. (1999). Northern hemisphere temperatures during the past millennium: inferences, uncertainties, and limitations. *Geophysical Research Letters, 26*(6), 759–762.

Meng, F. Q., Guan, G. H., Zhang, Q. Z., et al. (2006). Seasonal variation in soil respiration under different long-term cultivation practices on high yield farm land in the North China Plain. *Acta Scientiae Circumstantiate, 26*(6), 992–999.

National Research Council of the National Academy. (2006). *Surface temperature reconstructions for the last 2000 years*. Washington, DC: The National Academies Press.

Ni, J. (2001). Carbon storage in terrestrial ecosystems of China: Estimates at different spatial resolutions and their responses to climate change. *Climatic Change, 49*(3), 339–358.

Ni, J. (2002). Carbon storage in grasslands of China. *Journal of Arid Environments, 2002*(50), 205–218.

Niu, S. L., Wu, M. Y., Han, Y., et al. (2008). Water-mediated responses of ecosystem carbon fluxes to climatic change in a temperate steppe. *New Phytologist, 177*(1), 209–219.

PAGES. (2009). Science plan and implementation strategy. IGBP report no. 57, p. 67. IGBP Stockholm: Secretariat.

Piao, S. L., Fang, J. Y., & Guo, Q. H. (2001). Application of CASA model to the estimation of Chinese terrestrial net primary productivity. *Acta Phytoecologica Sinica, 25*(5), 603–608.

Post, W. M., Emanuel, W. R., Zinke, P. J., & Stangenberger, A. G. (1982). Soil carbon pool and world life zones. *Nature, 298*, 156–159.

Qi, Y. C., Dong, Y. S., Liu, L. X., et al. (2010). Spatial-temporal variation in soil respiration and its controlling factors in three steppes of Stipa L. in Inner Mongolia, China. *Science China Earth Sciences, 40*(3), 341–351.

Ren, J. Z., Hu, Z. Z., Zhao, J., et al. (2008). A grassland classification system and its application in China. *The Rangeland Journal, 30*, 199–209.

Scurlock, J. M. O., & Hall, D. O. (1998). The global carbon sink: a grassland perspective. *Global Change Biology, 4*, 229–233.

Shao, X. M., Xu, Y., Yin, Z. Y., et al. (2010). Climatic implications of a 3585-year tree-ring width chronology from the northeastern Qinghai-Tibetan Plateau. *Quat Sci Rev, 29*, 2111–2122.

Song, C. C., Wang, L. L., Tian, H. Q., et al. (2013). Effect of continued nitrogen enrichment on greenhouse gas emissions from a wetland ecosystem in the Sanjiang Plain, Northeast China: A 5-year nitrogen addition experiment. *Journal of geophysical Research: Biogeosciences, 118*.

Song, C. C., Xu, X. F., Tian, H. Q., & Wang, Y. Y. (2009). Ecosystem-atmosphere exchange of $CH_4$ and $N_2O$ and ecosystem respiration in wetlands in the Sanjiang Plain, Northeastern China. *Global Change Biology, 15*, 692–705.

Strom, L., & Christensen, T. R. (2007). Below ground carbon turnover and greenhouse gas exchanges in a sub-arctic wetland. *Soil Biology and Biochemistry, 39*, 1689–1698.

Sun, X. M., Wen, X. F., Yu, G. R., et al. (2006a). Influence of subtropical seasonal drought on carbon uptake in Qianyanzhou plantation ecosystem. *Science in China, Series D: Earth Sciences, 36*(1), 103–110.

Sun, X. M., Zhu, Z. L., Wen, X. F., et al. (2006b). The impact of averaging periods on eddy fluxes observed over ChinaFLUX sites. *Agricultural and Forest Meteorology, 137*, 188–193.

Tang, H. J., Qiu, J. J., Van Ranst, E., et al. (2006). Estimations of soil organic carbon storage in cropland of China based on DNDC model. *Geoderma, 13*(4), 200–206.

Tao, B., Cao, M. K., Li, K. R., et al. (2006). Spatial pattern and variation of terrestrial net ecosystem productivity in China between 1981 and 2000. *Science in China, Series D: Earth Sciences, 36*(12), 1131–1139.

Tiiva, P., Faubert, P., Sanna, R., et al. (2009). Contribution of vegetation and water table on isoprene emission from boreal peatland microcosms. *Atmospheric Environment, 43*(34), 5469–5475.

Wan, S., Xia, J., Liu, W., & Niu, S. (2009). Photosynthetic overcompensation under nocturnal warming enhances grassland carbon sequestration. *Ecology, 90*, 2700–2710.

Wang, X. B., Cai, D. X., Hoogmoed, W. B., et al. (2005). Scenario analysis of tillage, residue, and fertilization management effects on soil organic carbon dynamics. *Pedosphere, 15*(4), 473–483.

Wang, G. C., Du, R., Kong, Q. X., et al. (2004). Experimental study on soil respiration of temperate grassland in China. *Chinese Science Bulletin, 49*(6), 642–646.

Wang, D. L., & Lin, W. D. (1999). Effects of $CO_2$ elevation on root exudates in rice. *Acta Ecologica Sinica, 19*(4), 122–127.

Wang, L. G., Qiu, J. J., & Li, W. J. (2002). Study on the dynamics of soil respiration in the filed of summer-corn in Huanghuaihai region in China. *Soil Fetilizaer, 6*, 13–17.

Wang, S. Q., Tian, H. Q., Liu, J. Y., & Pan, S. F. (2003). Pattern and change of soil organic carbon storage in China: 1960s–1980s. *Tellus B, 55*, 416–427.

Wang, S. P., Wilkes, A., Zhang, Z. C., et al. (2011). Management and land use change effects on soil carbon in northern China's Grasslands: A synthesis. *Agriculture, Ecosystems and Environment, 142*, 329–340.

Welp, L. R., Lee, X., Griffis, T. J., et al. (2012). A meta-analysis of water vapor deuterium-excess in the midlatitude atmospheric surface layer. *Global Biogeochemical Cycles, 26*.

Wen, X. F., Lee, X., Sun, X. M., et al. (2012a). Dew water isotopic ratios and their relations to ecosystem water pools and fluxes in a cropland and a grassland in China. *Oecologia, 168*, 549–561.

Wen, X. F., Lee, X., Sun, X. M., et al. (2012b). Inter-comparison of four commercial analyzers for water vapor isotope measurement. *Journal of Atmospheric and Oceanic Technology, 29*, 235–247.

Wen, X. F., Meng, Y., Zhang, X. Y., Sun, X. M., & Lee, X. (2013). Evaluating calibration strategies for isotope ratio infrared spectroscopy for atmospheric $13CO_2/12CO_2$ measurement. *Atmospheric Measurement Techniques, 6*, 1491–1501.

Wen, X. F., Sun, X. M., Zhang, S. C., Yu, G. R., Sargent, S. D., & Lee, X. (2008). Continuous measurement of water vapor D/H and $^{18}O/^{16}O$ isotope ratios in the atmosphere. *Journal of Hydrology, 349*, 489–500.

Wever, L. A., Flanagan, L. B., & Carlson, P. J. (2002). Seasonal and interannual variation in evapotranspiration, energy balance and surface conductance in a northern temperate grassland. *Agricultural and Forest Meteorology, 112*(1), 31–49.

World Resources Institute. (2000). *World resources 2000–2001: People and ecosystems: The fraying web of life*. Canada: World Resources Institute.

Wu, G. L., Du, G. Z., Liu, Z. H., & Thirgood, S. (2009a). Effect of fencing and grazing on a Kobresia-dominated meadow in the Qinghai-Tibetan Plateau. *Plant and Soil, 319*, 115–126.

Wu, J. L., Wang, M., Lin, F., et al. (2009b). Effects of precipitation and interspecific competition on *Quercus mongolica* and *Pinus koraiensis* seedlings growth. *Chinese Journal of Applied Ecology, 20*(2), 235–240.

Xiao, W., Lee, X., Wen, X. F., Sun, X. M., & Zhang, S. C. (2012). Modeling biophysical controls on canopy foliage water $^{18}O$ enrichment in wheat and corn. *Global Change Biology, 18*, 1769–1780.

Xie, Z. B., Zhu, J. G., Liu, G., et al. (2007). Soil organic carbon stocks in China and changes from 1980s to 2000s. *Global Change Biology, 13*(9), 1989–2007.

Xilin, T. Y., Xu, Z., & Zheng, Y. (2009). Influence of different stocking rates on underground biomass and net primary productivity on Stipa krylovii Steppe in Inner Mongolia. *Chinese Journal of Grassland, 31*(3), 26–29.

Xu, Z., Zheng, X., Wang, Y., Huang, Y., Zhu, J., & Butterbach-Bahl, K. (2004). Effects of elevated $CO_2$ and N fertilization on $CH_4$ emissions from paddy rice fields. *Global Biogeochemical Cycles, 18*(3).

Yan, H., Cao, M., Liu, J., et al. (2007). Potential and sustainability for carbon sequestration with improved soil management in agricultural soils of China. *Agriculture, Ecosystems and Environment, 121*, 325–335.

Yan, X., Zhang, F. R., & Wang, J. K. (2010). Changes of soil organic carbon in different agro-ecological zones in China over 20 years. In *2010 19th World Congress of Soil Science, Soil Solutions for a Changing World, Brisbane, Australia*, August 1–6, 2010.

Yang, L. F., & Cai, Z. C. (2004). Diurnal variation of soil respiration in soil-potted with soybean (*Glycine max* L.) at various growth stages and its affecting factors. *Acta Ecologica Sinica, 24* (12), 2955–2960.

Yang, D. L., Han, G. D., Hu, Y. G., et al. (2006). Effect of grazing intensity on plant diversity and aboveground biomass of Stipa baicalensis Grassland. *Chinese Journal of Ecology, 25*(12), 1470–1475.

Yang, Y. S., Xie, J. S., Sheng, H., et al. (2009). The impact of land use/cover change on storage and quality of soil organic carbon in mid-subtropical mountainous area of southern China. *Journal of Geographical Sciences, 19*, 49–57.

Yu, G. R., Gao, Y., Wang, Q. F., Liu, S. R., & Shen, W. J. (2013a). Discussion on the key processes of carbon-nitrogen-water coupling cycles and biological regulation mechanisms in terrestrial ecosystem. *Chinese Journal of Eco-Agriculture, 21*(1), 1–13.

Yu, Y. Y., Guo, Z. T., Wu, H. B., et al. (2009). Spatial changes in soil organic carbon density and storage of cultivated soils in China from 1980 to 2000. *Global Biogeochemical Cycles, 23.*

Yu, G. R., Song, X., Wang, Q. F., et al. (2008a). Water-use efficiency of forest ecosystems in eastern China and its relations to climatic variables. *New Phytologist, 177*, 927–937.

Yu, G. R., Wen, X. F., Sun, X. M., et al. (2006). Overview of ChinaFLUX and evaluation of its eddy covariance measurement. *Agricultural and Forest Meteorology, 137*, 125–137.

Yu, G. R., Zhang, L. M., Sun, X. M., et al. (2008b). Environmental controls over carbon exchange of three forest ecosystems in eastern China. *Global Change Biology, 14*(11), 2555–2571.

Yu, G. R., Zhu, X. J., Fu, Y. L., et al. (2013b). Spatial patterns and climate drivers of carbon fluxes in terrestrial ecosystems of China. *Global Change Biology, 19*, 798–810.

Zhang, J. B. (2006). Effects of wetland cultivation and land use on soil carbon fraction in the Sanjiang Plain. Graduate school of Chinese Academy of Sciences.

Zhang, H. X., Wang, X. K., Feng, Z. W., et al. (2008). The great rainfall effect on soil respiration of wheat field in semiarid region of the Loess Plateau. *Acta Ecologica Sinica, 28*(12), 6189–6196.

Zhang, Q. Z., Wu, W. L., Wang, M. X., et al. (2005). The effects of crop residue amendment and N rate on soil respiration. *Acta Ecologica Sinica, 25*(11), 2883–2887.

Zhang, J. H., Yu, G. Y., Han, S. J., Guan, D. X., & Sun, X. M. (2006a). Seasonal and interannual variation of $CO_2$ flux in Broadleaved Korean pine mixed forest and its control mechanism in Changbai Mountain. *Science in China, Series D: Earth Sciences, 36*(1), 60–69.

Zhang, L. M., Yu, G. Y., Sun, X. M., et al. (2006b). Seasonal changes in carbon budget of the typical ecological system of Chinese Eastern forest transect. *Science in China, Series D: Earth Sciences, 36*(1), 45–59.

Zhang, B., Zheng, J. C., Tian, Y. L., et al. (2010). System design of free air temperature increased (FATI) for field nighttime warming experiment and its effects on rice-wheat cropping system. *Acta Agronomica Sinica, 36*(4), 620–628.

Zhao, H. L., Zhao, X. Y., Zhou, R. L., et al. (2005). Desertification processes due to heavy grazing in sandy grasslands, Inner Mongolia. *Journal of Arid Environments, 62*, 309–319.

Zheng, J. Y., Wang, W. C., Ge, Q. S., et al. (2006). Precipitation variability and extreme events in Eastern China during the past 1500 years. *Terrestrial Atmopheric and Oceanic Science, 17*(3), 579–592.

Zhong, H. P., Fan, J. W., Yu, G. R., et al. (2005). The research progress of carbon storage in grassland ecosystem. *Pratacultural Science, 22*(1), 4–11.

Zhu, Z. C., & Jia, D. L. (1991). A preliminary study on biomass of *Themeda japonica* community in Loess Plateau of northern Shaanxi province. *Acta Ecologica Sinica, 11*(2), 117–124. (in Chinese).

Zou, J., Huang, Y., Zong, L., et al. (2004). Carbon dioxide, methane, and nitrous oxide emissions from a rice-wheat rotation as affected by crop residue incorporation and temperature. *Advances in Atmosphere Sciences, 21*(5), 691–698.

# Chapter 16
# Impacts and Its Adaptation of Global Change

**Peili Shi, Dingpeng Xiong, Xiaoguang Yang, Zhijuan Liu, Ming Xu, Feng Ge, Yucheng Sun, Guangsheng Zhou, Qijin He, Nuyun Li and Wenhua Li**

**Abstract** It is important research fields of global change ecology to study the influence that global climate change has on ecosystem and agroforestry, environmental factor such as water resources. And it is also important scientific basis on which to cope with climate change, to manage ecosystem adaptively and to make relevant policies. If the change of the earth system's strength power is the ultimate concern of global climate change study, it will be to focus on questions to recognize the change of ecosystem's structure, process, and function motivated by climate change and human activities. The earth feeds on massive diverse natural ecosystem, and diverse semi-natural semi-artificial ecosystem, which is already a result of a long-term adaption of the earth to natural environment and human activities. In that case, the adaption research of the ecosystem to the environment change plays a key role in humans getting to know what influence global changes may have on the supporting capacity of earth system. This chapter mainly concerns about the ecosystem's adaptability. It is defined as the adaptive abilities of lowering environment changes' negative influence and making good use of favorable opportunities when environment changes but the ecosystem tries to maintain its main functions. And this balance mechanism and adaptability of ecosystem is a basis on

P. Shi · D. Xiong · W. Li (✉)
Institute of Geographic Sciences and Natural Resources Research, Chinese Academy of Sciences, Beijing 100101, China
e-mail: liwh@igsnrr.ac.cn

X. Yang · Z. Liu
China Agricultural University, Beijing 100193, China

M. Xu
Key Laboratory of Ecosystem Network Observation and Modeling, Institute of Geographic Sciences and Natural Resources, Chinese Academy of Sciences, Beijing 100101, China

F. Ge · Y. Sun
Institute of Zoology, Chinese Academy of Sciences, Beijing 100101, China

G. Zhou · Q. He
Chinese Academy of Meteorological Sciences, Beijing 100081, China

N. Li
State Forestry Administration, Beijing 100714, China

 
W. Li (ed.), *Contemporary Ecology Research in China*,
DOI 10.1007/978-3-662-48376-3_16

which to judge the influence degrees of global changes on ecosystem's structure and function. To dig into this question, we generally study from two different perspectives, one of which is global changes' influence on the ecosystem, and the other one is the ecosystem's responses to these changes, which is the principal framework of this chapter.

**Keywords** Impact · Adaptation · Global change · Agriculture · Ecosystem · Pest · Insect · Terrestrial transect

# 1 The Impacts of Global Change on the Spatial Distribution of Ecosystems

## *1.1 Introduction*

In the scenario of global warming in latter half of twentieth century (Houghton 2001), warming was also significantly in China. The global mean temperature increased by approximately 0.55 °C in recent 100 years, and mean precipitation increased by ca. 21 mm. While in China, the mean annual temperature (MAT) dramatically increased by about 1.1 °C from 1951 to the end of twentieth century. Compared with the same period over the world, the mean rate of global warming is much higher in China. Climate warming took place mainly in the 1980s, indicating a regional and seasonal variability. It was warming higher in North China than in South China, and warming is much dramatic in winter than in summer (Sha et al. 2002).

In the past 50 years, air temperature increased more in west than in East China, especially with significant warming trend in the eastern Qinghai–Tibet Plateau and northwest Xinjiang Autonomous Region (Wang et al. 2002). At the latitude of 35° N, temperature increase in the north but decrease in the south. But temperature variations were not obvious in the Yangtze River Basin and southeast area (Chen et al. 2002).

## *1.2 The Effects of Hydrothermal Condition on Primary Productivity and Soil Carbon Storage*

There are regional differences in response to the net primary productivity (NPP) of terrestrial vegetation to climate change in China. In recent two decades, NPP increased with favorable hydrothermal conditions (Zhu et al. 2007). The NPP was correlated positively with precipitation, while negatively with potential evapotranspiration in China, especially in grassland ecosystems (Xu 2004). The driving effect of precipitation on vegetation NPP seasonal variation is more significantly than that of temperature. The driving effect of hydrothermal condition on the seasonal variation of vegetation NPP is higher in north than in South China. In the interannual variation of NPP, the interactive effects of water and heat factors varied with seasons and latitude (Hou et al. 2007). In terms of relative increase in NPP, it was increased the most in Qinghai–Tibet Plateau. But NPP, in terms of absolute magnitude, increased highest in forest, least in desert (Sun and Qijiang 2001). The response of NPP and net ecosystem productivity (NEP) in Northeast China is more sensitive to elevated temperature than precipitation. With increase of 20 % of precipitation and 3 °C of temperature, the NPP and NEP of northeast forest were simulated to increase highest but the least in current scenario (Zhao et al. 2008).

Grassland is one of the important ecosystems in China. Water plays a decisive role in the distribution pattern of grasslands. Precipitation is key factor to determine NPP (Yuan et al. 2008). Interannual variation of grassland biomass in northern China is determined by precipitation from January to July, but its relationship with temperature was weak. There are differences in the relationship between grasslands and hydrothermal conditions. And therefore, the response of the NPP to climate change is different (Ma et al. 2010).

The relationship between soil carbon storage and hydrothermal factors was negative in the areas with MAT $\leq$ 10 °C, positive but more dependent on precipitation in areas with MAT higher than 10 °C and lower than 20 °C, and no trend in areas with MAT > 20 °C (Zhou et al. 2003). The soil labile carbon and soil organic carbon are dependent on precipitation positively in Northeast China Transect (NECT) (Wang et al. 2003).

## *1.3 The Spatial Distribution of Main Ecosystem in Response to Climate Change in China*

### 1.3.1 The Impact of Global Change on Zonal Vegetation Distribution

Vegetation distribution is dependent on climate and its change. In general, zonal vegetation of Northern Hemisphere will move to north in the global warming scenarios, but the responses of different species are dependent to its adaptability to climate change. In vertical distribution, the upper limits of mountain vegetation may migrate to high altitude. The response of timberline to climate change is indicative in the future climate change scenario.

In the scenario of global warming, warm temperate zone in Northeast China will expand but cold temperate zone will be reduced, and the distribution of vegetation boundaries will move northward. The area of humid area will decrease, but the area of semi-humid area and semiarid will increase, leading to shrink of forests and expansion of grasslands (Wu et al. 2003). Taking land-use constraints into consideration, deciduous broad-leaved forests are simulated to be increased and conifers, shrubs, and grasslands to be decreased in the scenario of future doubling carbon dioxide ($CO_2$).

Timberline trees are very sensitive to climate change, but its sensitivity increases with altitude (Chang et al. 2009). The response of timberline to climate change usually has time lag. Climate change influences on dynamics of timberline first from tree growth, and then to community structure and species interaction (Liu et al. 2002). And therefore vegetation ecology approach is the pathway to disentangling the relationship between ecological process, pattern, and climate factors. Growing seasonal temperature is one of the key climatic factors to determine timberline. In subalpine, the radial growths of tree rings are negatively correlated with precipitation after growing season, while positively correlated with temperature in former early autumn. But these relationships disappear in lower altitudes (Chang et al. 2009).

### 1.3.2 The Impact of Global Change on Agricultural Ecological System Structure

Global climate change will cause changes of agricultural climate resources, soil quality and as well, and thus will directly affect the agricultural planting structure, planting area, planting system, production capacity, crop variety layout (Yuan et al. 2011; Liu et al. 2010).

Thermal resources in China will increase in different extent in response to climate warming. For example, increase of accumulated temperature above 0 °C and extending growing season is beneficial to multiple cropping, consequently leading to the northern boundary of the multiple cropping moving northward. In China winter wheat, maize and double-cropping rice moving northward may enhance their yields in the cropping areas (Yang et al. 2010a, b; Zhao et al. 2010). Frost damage risk will increase in the sensitive cropping area if boundary of cropping moves northward (Li et al. 2010; Liu et al. 2010).

## *1.4 The Impacts of Global Change on Ecological Sensitive Areas*

### 1.4.1 The Change of Northern Agro-livestock Ecotones

Vegetation coverage of north agro-livestock ecotone has obvious change in response to climate change. In the last two decades of twentieth century, the area with high vegetation coverage decreased but that with low coverage increased. In the past 40 years, the climate risk increased in agro-livestock ecotones, with increase trend from southeast to northwest (Sun and Zhao 2009). Ma et al. (2011) found that climatic productivity was increasing slowly in agro-livestock ecotone of northern China, with trend higher in south than in north. In east-west direction increasing trends expand from the center to the periphery. Warmer and humid climate is most suitable for crop growth, but cold and dry climate is not profitable.

### 1.4.2 The Vegetation Response to Climate Change on the Qinghai–Tibet Plateau

Qinghai–Tibet Plateau is the sensitive area in response to climate change and thus a fragile eco-region (Sun and Zheng 1998). Climate change is earlier in the plateau than in other areas of China, and therefore the responses of ecosystems to global change are also more sensitive. Since 1980s, the vegetation coverage generally showed increasing trends with improvement in the extensive areas of, northeast, east central, while degraded in the arid and semiarid areas in the west and north plateau (Zhang et al. 2010). Earlier start of growing season and accelerated growth are the main reasons for the increase of NDVI (Yang and Piao 2006). Spring is the season for NDVI increase in maximum, but summer increases to a minimum extent. The vegetation belts will move toward higher altitudes with climate warming and horizontal vegetation belts will change from semiarid type of alpine grassland to semi humid type of alpine meadow expansion (Yu and Xu 2009).

In the past 20 years, the NPP of natural vegetation increased as result of increase of air temperature and precipitation on the Qinghai–Tibet plateau. The total amount NPP of shrubs and forest increased 1.14 and 0.88 % per year. But the NPP of grasslands increased less than forests and shrubs (Huang et al. 2008). Ye (2010) found interannual variability of NPP increased with the interannual variability of precipitation and temperature, but the effects of precipitation was the most in the past 30 years on the Qinghai–Tibet Plateau.

# 2 The Impacts of Global Change on Agriculture

## *2.1 Introduction*

During the past ten decades (1906–2005), the global surface air temperature has increased by 0.74 °C. This generally increasing temperature is especially obvious in the high latitude areas of the Northern Hemisphere (IPCC 2007a, b). Climate change in China and its impact are studied by many scholars, and a lot of stage progress has been made. The research results showed that, climate change in China was consistent with global climate change, but obvious difference still existed. Under the background of global climate change, the mean annual surface air temperature in China had significantly increased in recent 100 years, with the amplitude of 0.5–0.8 °C. In recent 50 years, MAT in northern China showed the maximum increased amplitude (Ding et al. 2006). Precipitation in the western basins of China increased the most, with the maximum amplitude of 10–15 % per decade, while precipitation in North China and the south of Northeast China decreased. The prediction trend of climate change in the twenty-first century showed that, climate in China would continue to significantly warm up, especially in the winter half year of North China, also precipitation would increase (Qin et al. 2005).

As the demands for food and energy increase due to the increases in global population, society will be pressed to increase agricultural production, especially in China. Therefore, it is important to understand the effects of climate change on cropping systems and crop productivities in China.

## 2.2 *The Effects of Climate Change on Agriculture in China*

### 2.2.1 The Effects of Climate Change on the Cropping Systems in China

#### The Effects of Climate Change on the Northern Limits of Multiple Cropping Systems in China

As the temperature rises, the accumulated temperature increases, the planting northern limits of both double-cropping system and triple-cropping systems during 1981–2007 moved northward, compared to that during the 1950s–1980. Without considering the variety change, social economy and other factors, the per unit area grain yield would increase if single-cropping system was replaced by double-cropping system or triple-cropping system. Also in the future, with the climate warming, the northern limits of double-cropping system and triple-cropping systems would move northward (Yang et al. 2011a, b).

#### The Effects of Climate Change on the Planting Areas of Winter Wheat and Double Rice

Due to climate warming, the planting northern limits of winter wheat in North China moved northward and expanded westward. The climate warming also caused the northward move of planting northern limits of double rice (Hao et al. 2001; Jin et al. 2002).

#### The Effects of Climate Change on the Northern Limits of Different Maturity-Type of Maize in China

In Northeast China with relative lower heat resources, climate warming extended the length of the potential growing period of crops. Therefore, an effective management option to counteract the negative warming effect is to plant new hybrids with a longer growing season, (Liu et al. 2013b) analyzed the effects of warming over the past five decades on the northern limits of cropping systems and the production of maize in NEC, the results indicate that climate warming would caused a northward expansion of the northern limits of maize (Fig. 1). Thus, in some places, the early-maturing maize cropping area was planted with the middle-maturing maize hybrids, and in other areas the late-maturing maize hybrids replaced the middle-maturing maize hybrids. Compared with the early- and middle-maturing maize hybrids, the yields of late-maturing maize hybrids were relatively higher because the longer growing season. Therefore, the change in maize maturity types could increase the maize yields.

### 2.2.2 The Effects of Climate Change on the Crop Phenology in China

If other factors were constant (fixed genotypes and management practices), the warming trend would result in a negative impact on maize production; the warmer climate would speed up crop development and lead to a reduction in the length of the growing season (Ma et al. 2008; Liu et al. 2010, 2012, 2013a; Xue et al. 2010; Yang et al. 2010a, b). In Northeast China, the warming trend significantly shortened the vegetative period, ranging from 1.8 to 3.1 days per decade from 1981 to 2007; the magnitude of the reproductive period trend was slightly reduced (0.5–2.4 days per decade). Thus, maize yield decreased by 4–22 % for each 1 °C increase in growing-season mean temperature (Liu et al. 2013a). In North China Plain, the warming trend shortened the growth duration of wheat if no varietal change had occurred. This was mainly attributed to a significant reduction in the length of the pre-flowering stage. The lengths of post-flowering stage remained stable. For maize, the simulated pre-flowering stage tended to decline, but not significant. The post-flowering stage remained stable (Liu et al. 2010).

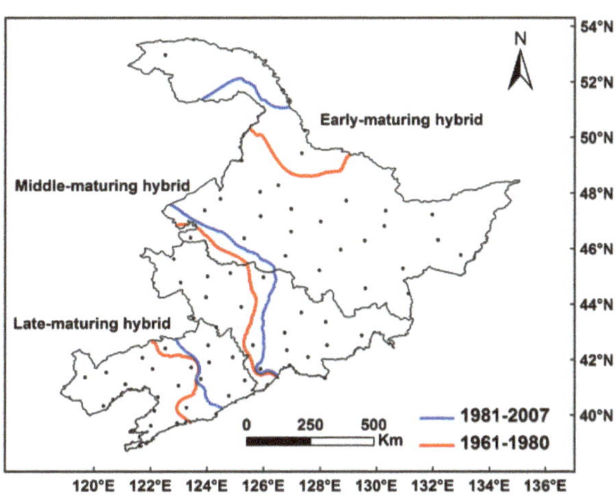


**Fig. 1** The safe planting northern limits for early-maturing, middle-maturing, and late-maturing maize hybrids during the periods 1961–1980 (*red lines*) and 1981–2007 (*blue lines*)

### 2.2.3 The Effects of Climate Change on the Crop Productivity in China

If there was no hybrid changes and management improvement, most of results indicated that warming trend would result in a negative impact on crop production in China. In the 2080s, the yield of wheat would decrease no matter whether there is irrigation under the condition with no $CO_2$ fertilizer effects (Xiong et al. 2006). The results of (Yang et al. 2010a, b) indicated that rice yield will decrease as the temperature rising in the middle and lower valley of the Yangtze River. However, it is likely that this result is also crop dependent; for instance, the results for maize in China indicates that climate change would result in the positive impacts on maize under the rain-fed condition, but the negative impacts under the irrigation condition (Xiong et al. 2005).

## *2.3 Conclusion and Discussion*

According to the above previous results, because climate warming would accelerate the crop development, hasten the maturation of crops and lead to a reduction in the length of the growing season, climate warming can lead to a negative impact on crop yield if there are no hybrid and management changes. On the other hand, climate change increased the heat resources, which can lead to the northern limits of multiple cropping systems, wheat and double rice moved northward. In the areas with relative lower heat resources, climate change can lead to the planting areas of different maturity-type varieties of maize in Northeast China moved northeastward. Thus, the effective management options to counteract the negative warming effect are to change the cropping systems and plant new hybrids with a longer growing season.

# 3 Ecosystem Adaptation to Global Change

## *3.1 Introduction*

Global change here refers to the changes, which often produce adverse impacts on ecosystems and human socioeconomics, of the environmental factors beyond the normal fluctuation range at global scale. Currently, such changes may include global climate change, atmospheric chemistry change, increasing UVB radiation due to damages of the ozone ($O_3$) layer, oceanic acidification and sea level rising, nitrogen deposition, and land-use and land cover change. More broadly, global change may also include changes in global socioeconomic conditions, such as population growth and globalization.

The ultimate impacts of global change on ecosystems depend on not only the magnitude of global change, but also the adaptation capacity of various ecosystems to the changes. Many ecosystem adaptation measures can offset the impacts of global change and, meanwhile, may mitigate the global change (Frankhauser 1996; Tol et al. 1997). For instance, improving forest management can enhance biodiversity conservation and meanwhile sequester atmospheric $CO_2$ by promoting forest growth, and thus reduces the global warming effect. Given the breadth of global change, this chapter focuses on the mechanisms of ecosystem adaptation to global climate change.

## *3.2 Mechanisms of Ecosystem Adaptation to Climate Change*

### 3.2.1 Physiological and Biochemical Acclimation/Adaptation

Thermal Acclimation/Adaptation of Photosynthesis

To many plants photosynthesis stops at either very low or very high temperatures. Photosynthesis, in general, increases with the increase of temperature and reaches the maximum photosynthetic rate at an optimal temperature. Then the photosynthetic rate declines with further increase in temperature (Berry and Björkman 1980). Previous studies found that the optimal temperature of photosynthesis (Topt) is highly related to the growing temperature (Battaglia and Beadle 1996; Cunningham and Read 2002). The change of the optimal temperature is one of the key features of plant thermal acclimation/adaptation to global warming, which is pivotal to the projection of ecosystem functions in a warming climate. However, the mechanisms of the thermal acclimation/adaptation are not fully understood with some hypotheses as follow: (i) RuBP carboxylation is a limiting factor. Under saturated $CO_2$ concentration, Rubisco activity enhances exponentially with temperature (Jordan and Ogren 1984). The maximum carboxylation rate (Vcmax), an indicator of Rubisco activity, increases with temperature through the Arrhenius model. The activation energy (Eav) of Vcmax also increases with rising temperature. The Topt increases 0.54 °C when Eav increases 1 $KJmol^{-1}$ (Hikosaka et al. 1999). (ii) RuBP regeneration causes the shift of Topt. The optimal temperature of the maximum electron transport rate (Jmax) shifts toward the high temperature under warmer growing temperature, which may cause the shift of the Topt (Yamasaki et al. 2002). Many previous studies found that the heat tolerance of the enzymes involving in RuBP regeneration varied with growing temperature. The thermal stability of photosystem II also changed with growing temperature (Yamasaki et al. 2002). (iii) The TPU may limit photosynthesis, generally under high $CO_2$ concentration. But Labate and Leegood (1988) found that the TPU limitation on photosynthesis could also happen under normal $CO_2$ concentration. The temperature effect on photosynthesis may be caused by the TPU response to temperature change. (iv) The balance between carboxylation and RuBP regeneration determines the shift of the Topt of photosynthesis. In general, the carboxylation process has a lower temperature sensitivity and lower Topt than the RuBP regeneration process. As temperature increases, photosynthesis is limited by both processes, resulting in the shift of Topt to the high temperature end. (v) stomatal limitation. Stomatal conductance determines intercellular $CO_2$ concentration (Ci) and the Topt increases with Ci (Berry and Björkman 1980). Every 1 ppm increase in Ci may lead to 0.05 °C increase in Topt.

Acclimation/Adaptation of Respiration to Warming

The temperature sensitivity of plant respiration is commonly indicated by an index, Q10. Recent studies found that the Q10 value decreased with the increase of growing temperature (Tjoelker et al. 2001). Warming has a larger impact on plants living in cold climate (high latitudes or high altitudes) than in warm temperature (tropics).

### 3.2.2 Acclimation/Adaptation at Individual Level

Adaptation Mechanisms and Strategies of Plants

In addition to the acclimation/adaptation in physiological processes, plant adaptation to climate change may also happen with morphology, growth, and reproduction. Stomatal size, density, and openness are critical to the gas exchange between plants and atmosphere (Ferris et al. 2002; Xu et al. 2009). Previous studies found that high $CO_2$ concentration significantly reduced stomatal density, stomatal index, stomatal conductance and transpiration (Luomala et al. 2005). Additionally, high $CO_2$ concentration enhances leaf and cell wall thickness, the number, width, and surface area of chloroplasts, and the size and number of starch grains in mesophyll cells, but high $CO_2$ concentration significantly decreases the number of thylakoids in chloroplasts (Teng et al. 2006). Few studies have been reported on exploring the temperature effect on plant structure and development. So far, no consistency has been made on warming effects on stomatal density and stomatal index. Furthermore, elevating $CO_2$ concentration may also affect plant flowering and reproduction. Earlier studies found that elevating $CO_2$ concentration may shorten plant flowering and growth periods and increase the production of flowers, fruits and seeds and their C/N ratio, but decrease seed nitrogen content (Ziska et al. 2004).

Adaptation Mechanisms and Strategies of Animals

With global warming in the past decades, animals in the arctic region have decreased their body size, which has relatively increased body surface area to dissipate more heat to better acclimate to the warming climate (Yoram and Jonathan 2005). Reproduction time is the most sensitive period of animals to climate change, which may determine the success rate of reproduction. Some animals may benefit from climate change and others may be adversely affected by climate change. Animals with greater dispersal ability may move faster and colonize a larger area of habitats under climate change, but animal with poor dispersal ability may face shrinking distribution ranges or even the risk of extinction. It is noted that many recent species extinctions are related to extreme climate events which are likely to become more frequent under future climate change (Parmensan 1996).

### 3.2.3 Adaptation Mechanisms at Population and Community Level

Population dynamics are determined by the competition among individuals within the population and the community dynamics are determined by the competitions among populations and the individuals. Previous studies reported that elevating $CO_2$ concentration would promote seed germination and seedling growth (Hussain et al. 2001; Steinger et al. 2000). This enhancement in seedling growth might increase competitions among seedlings for space, light, and nutrients, leading to a higher death rate and smaller number of offspring. Climate change may also alter the competition among species resulting in changes of community structure and composition. Different species may respond to climate change differently, such as variations in phenology, growth, reproduction, and dispersal. All the impacts on cellular, individual, and population levels could eventually lead to changes in communities, such as species composition and biodiversity.

### 3.2.4 Adaptation at Ecosystem and Landscape Level

The $CO_2$ fertilization effect on photosynthesis and the inhibiting effect on respiration at leaf level will eventually impact ecosystem productivity through carbon cycling. The short-term FACE experiments have shown that ecosystem productivity (NPP and NEP) has increased with the elevation of $CO_2$ concentration, but other studies showed that the $CO_2$ fertilization effect might not be as strong as expected (Long et al. 2006). Many studies have predicted that global warming induced longer growing season may enhance NPP of various ecosystems. But warming may also stimulate microbial respiration, thus the net ecosystem productivity (NEP) may not necessarily increase. Ecosystem water cycle may also be affected by elevating atmospheric $CO_2$ concentration which generally will improve water use efficiency

(WUE). In addition, ecosystem nitrogen cycle will also be affected through nitrogen deposition and the coupling of carbon and nitrogen in various ecosystem processes.

### *3.3 Facilitated Adaptation*

Adaptation of agriculture to climate change may involve many planning and managing strategies, such as adjusting sowing area for different crops, changing cropping system by enhancing the area of double-harvesting crops, adjusting sowing date, promoting the varieties with longer growing period and greater resistance to hot waves, droughts, and pest damages, and improving irrigation systems with advanced technologies. Adaptation in forest ecosystems should first protect the natural forests, which are more resistant to climate change than plantations, from logging, deforestation, and disturbances. In planting artificial forests, selecting local native tree species is very important because native species usually have greater adaptability to local climate variations through the long evolution process. In addition, southern provenances should be considered for reforestation and afforestation because of their greater thermal tolerance. Adaptation of grassland to climate change should focus on predicting and forecasting extreme climate events, such as drought and snow storms, to minimize their damages on livestock. Grassland management should also pay special attentions on biodiversity conservation which may increase the stability of grassland ecosystems to various disturbances including climate change. Ecosystem adaptation to climate change is a long-term process and many unknowns and uncertainties need to be tackled through international and multidisciplinary collaboration, large-scale multifactor experiments, and applications of new technologies.

## 4 Pest Adaptation to Global Change

### *4.1 Introduction*

Global climate change is resulted from the joint action of changes in the internal and external factors (natural and anthropogenic) of climate system. The main causes of global climate change is due to the emissions of $CO_2$ and other greenhouse gas which change the composition of the atmosphere and drive the global warming. According to the report from IPCC (2007a, b), the atmospheric $CO_2$ concentration has been increased from 280 μl/L at the year of 1700 to 379 μl/L at 2005, and is anticipant to reach 540–970 μl/L at the end of this century. The global mean surface air temperature over the past 100 years (1906–2005 years) increased by 0.74 °C, and is expected to increase 1.1–6.4 °C by the end of this century.

As the most important components of biodiversity, insect is extremely sensitive to the environmental changes, i.e., $CO_2$ concentration, drought, and temperature. Moreover, insect pest is an important factor affecting agricultural production, which is not only directly related to the structure and function of agricultural ecosystem but related to the food and ecological securities. Consequently, domestic and abroad scientists have attached great importance to the research concerning the insect responses to climate change factors including temperature, drought, and greenhouse gases ($CO_2$, $O_3$). Because these researches not only can theoretically explain the general rule in insect responses to climate change, and reveal the response mechanism of "crop-pest-natural enemy" to global climate change, but can practically predict future occurrence trend of pests and the develop new technology of biological control in order to meet the challenge of global climate change.

During recent years, China entomologists focused on how insect responses to the climate change, especially on elevated greenhouse gas ($CO_2$, $O_3$), drought and elevated temperature, and carried out vast amount of research work. This chapter will review our previous work and provide some perspectives in the future study.

## *4.2 Insects Response to Climate Change*

### 4.2.1 Insects Response to Elevated $CO_2$

Global atmospheric $CO_2$ concentrations have risen rapidly since the Industrial Revolution and are considered as a primary factor in climate change. The effects of elevated $CO_2$ on herbivore insects were found to be primarily through the $CO_2$-induced changes occurring in their host plants, which then possibly affect the intensity and frequency of pest outbreaks on crops. This section reviews several research models using primary pests of crops (cotton bollworm, whitefly, aphids) and their natural enemies (ladybeetles, parasitoids) in China to examine insect responses to elevated $CO_2$. It is generally indicated that elevated $CO_2$ prolonged the development of cotton bollworm, *Helicoverpa armigera*, a chewing insect, by decreasing the foliar nitrogen of host plants (Chen et al. 2005b; Yin et al. 2010). In contrast, the phloem sucking aphid and whitefly insects had species-specific responses to elevated $CO_2$ because of complex interactions that occur in the phloem sieve elements of plants. Some aphid species, such as cotton aphid, *Aphis gossypii* and wheat aphid, *Sitobion avenae*, were considered to represent the only feeding guild to respond positively to elevated $CO_2$ environment (Chen et al. 2005a; Sun et al. 2009). Although whitefly, *Bemisia tabaci*, a major vector of *Tomato yellow leaf curl virus*, had neutral response to elevated $CO_2$, the plants became less vulnerable to the virus infection under elevated $CO_2$ (Huang et al. 2012). The predator and parasitoid response to elevated $CO_2$ were frequently idiosyncratic (Sun et al. 2011). These documents from Chinese scientists suggested that elevated $CO_2$ initially affects the crop plant and then cascades to a higher trophic level through the food chain to encompass herbivores (pests), their natural enemies, pathogens and underground nematodes, which disrupt the natural balance observed previously in agricultural ecosystems.

### 4.2.2 Insects Response to Elevated $O_3$

The concentration of $O_3$, a major tropospheric photochemical oxidant, has risen from less than 10 nl/L a century ago to 40 nl/L today, and it is projected to continue to increase at an annual rate of 1–2 % (Vinzargan 2004; Jaffe and Ray 2007). Levels of atmospheric $O_3$ are anticipated to reach 68 nl/L by the year 2050 (Wilkinson and Davies 2010). Elevated $O_3$ can affect herbivores by altering the nutritional quality, the secondary metabolites and the resistance of plants (Percy et al. 2002; Agrell et al. 2005; Cui et al. 2011). However, the observed effects of elevated $O_3$ on herbivorous insects are often inconsistent across species (Jondrup et al. 2002). Generally, the developmental rates of chewing herbivorous insects are enhanced when these insects feed on $O_3$-exposed plants (Jackson et al. 2000; Mondor et al. 2004). Phloem-feeding insects such as aphids exhibit variable responses, including increased, decreased, or unchanged growth or oviposition rates, when fed on plants grown in conditions of increased $O_3$ (Holopainen 2002). Furthermore, elevated $O_3$ tends to reduce the nutrition of tomato plants and increases the SA content, the relative PR mRNA expression and the products of secondary metabolism in these plants, which together result in a decrease in the fitness of whiteflies on these hosts (Cui et al. 2012). Thus, it seems that the effect of elevated $O_3$ on insects via changes in host plant quality vary depending on the insect and plant species investigated, the $O_3$ level, and the fumigation time (Holopainen et al. 1995; Jondrup et al. 2002; Holopainen 2002). Moreover, the mechanism underlying fitness changes in insects that result from host plant interactions in elevated $O_3$ conditions remains unclear.

### 4.2.3 Insects Response to Elevated Temperature

Global average surface temperature has increased by around 0.74 °C during the past century and will continue to rise in the future. Understanding how these changes have affected biological systems has attracted a vast amount of research during the last two decades. As ectotherms, the growth and development of insects is affected by temperature (Deutsch et al. 2008). We here reported the existing evidence on how insects have responded to the increases in temperature (Guo et al. 2009). The fitness of insects can be predicted to change in response climate change in five ways, including changes in geographic distribution, winter survival, voltinism, dispersal/migration and phenology (Zhang et al. 2009, 2010; Stige et al. 2007; Dong and Ge 2011). However, there are still many unknowns in our understanding of the effects of climate warming on insects. Future works were needed to consider the relationships among host plants, insect herbivores and their natural enemies, and their long-term response to global warming at the population level.

### 4.2.4 Insects Response to Precipitation and Drought

The moisture factor plays a vital role in the growth, development, and fecundity of insects. Rainfall can directly kill insects through mechanical scouring. Drought can affect physiological metabolism of insect directly, and has the indirect effects through altering the host plants (Gange and Brown 1989; Masters et al. 1993). For example, drought alters the interspecific interactions among insects on the common host plants, and thereby changes the diversity and stability of the insect population and ecological community. This section summarizes the effects of rainfall and drought on the growth, development and fecundity of insects, and introduces the behavioral responses of migratory and gregarious insects, as well as the soil insects response to rainfall and drought (Dang and Chen 2011). The strategies of insects use to cope with rainfall and drought are described in detail, including behavioral adaptations, and the tactics of diapauses and migration. Artificial regulation of environmental moisture, especially soil moisture, (e.g., artificial rainfall and irrigation, etc.) is suggested as a means of controlling agricultural insect pests.

## *4.3 Conclusion and Discussion*

In the long process of evolution, plant–insect pest–natural enemy are interrelated, interact, and mutually restrict. During recent decade years, climatic factors including carbon dioxide concentration, temperature, rainfall, global warming have been changed dramatically. The tritrophic levels among plant, pest, and natural enemy have specific temporal and spatial response pattern to these environmental factors, which break the original inherent balance in tritrophic levels, and lose the biological control ability from crop resistance and natural enemies and in turn increasing the outbreak risk of pests. Meanwhile, it should be also clarified that the process of global climate change is relatively slow. IPCC (2007a, b) predicts that the atmospheric $CO_2$ concentration has been increased by 1.5 μl/L per year and temperature has been increased by 4 °C per 100 years. Since insects have relative shorter development duration, higher fecundity and polymorphism, they are able to adopt different strategies in aspects of genetic, physiological, behavioral and population levels, to maximize the survival, reproduction and population expansion and fitness in context of climate change.

Apparently, future work concerning how insects response and adapt to climate change mainly focus on resolving the following three scientific problems: (i) since insect has species-specific response to global climate change, how do these environmental factors change the spatial and temporal occurrence patterns of pests and natural enemies in China? (ii) How do pests and natural enemies adapt to these environmental factors? And (iii) how to adopt new pest control strategies to meet the challenge of pest occurrence in the context of global climate change. Furthermore, future research will focus on how the agriculture pests, regional and migrant pests, and invasive pests respond to environmental changes including $CO_2$ concentration, temperature and precipitation, from the multi-scales of genetic, molecular, species, ecosystem and landscape levels. These works should concentrate on the response characteristics, adaption mechanism and the control technology of pests through long-term monitoring, controlled experiment, and modeling establishment. Thus, there are four aspects should be emphasized as following:

(i) Occurrence and calamity law of pests affected by global climate change. The temporal and spatial pattern of occurrence and distribution of pests and their natural enemies have been changed in the context of global climate change, which need to analyze the historical data and meteorological data, and combine with the field investigation and environment controlled experiment to simulate the effects of climate change. In addition, these studies cannot only clarify new characteristics and new calamity laws of pest under climate change but establish the correlation between pest outbreak and key factor in climate change, which are the basis of monitoring, early warning, emergency prevention, and control technology of pest outbreak under global climate change.
(ii) Adaption mechanism of insect pest and natural enemy in response to climate change. Facing the challenge of climate change, pests have different adaptive strategies in genetic, physiological, behavioral, and population levels to maximize their survival, reproduction, and expansion in response to elevated temperature, $CO_2$, and drought. Moreover, study on the mechanism of adaptation can elucidate the biological effect of climate change, and enhance the stability and adaptability of ecosystem, which lays a theoretical foundation for the construction, early warning, emergency prevention, and control technology of pest outbreak.
(iii) New assessment of pest outbreak in calamity and agricultural loss in the context of climate change. With the impact of global climate change, industrial structure adjustment of agriculture and the revolution of management system of cultivation, the damage of pest to crop exhibited new pattern and rules. The virulence of insect pest in individual and population level should be quantified, and the feeding behavior, life history variation, niche determination, population growth and interspecific competition of pest should be determined. Moreover, the regional crop yield loss (i.e., assessment of economic loss, evaluation of compensation capacity, growth potential measurement and analysis of C fixation of forest), control index of major agricultural pests and invasive pest and economic threshold parameter should be considered. All information should be integrated and drafted under global climate change which can provide the basic data, evaluation index, and evaluation model parameters for plant protection.

(iv) New technologies and methods for early-warning and prevention of pest calamity. By combining the traditional pest population model (i.e., the effective accumulative temperature model, population growth model), molecular detection, pheromone monitoring, 3S technology and network technology, and integrating the remote sensing, geographic information and weather information, the prediction model for pest detriment and identification of migration and diffusion were established to monitor regional calamity rules of pest. Meanwhile, we should integrate the sustainable emergency prevention and control management system for pest damage under context of climate change, and seek for some new technologies and new methods for biological control, life history variation and crop tolerance under climate change, and finally establish national sustainable pest prevention and control system.

# 5 Terrestrial Transect Study on Global Change

## *5.1 Introduction*

The terrestrial transect has been widely used in the studies of interactive global change and terrestrial ecosystems (GCTEs), which is sponsored by the GCTE, a core project of the International Geosphere-Biosphere Programme (IGBP). The principles of transect selection are that the gradient of a particular ecological factor significantly changes along the transect, and the other ecological factors remain almost constant or their changes are quite small, so that the possible process of global change could be revealed by way of temporal–spatial substitution (Zhou et al. 2002). GCTE supported the establishment of 15 global change terrestrial transects distributed in different key regions of the Earth. Two of these are located in China: the water-driven NECT and the heat-driven North-South Transect of Eastern China (NSTEC) (Peng and Ren 2000). NECT was established in 1994 and is mainly a rainfall-driven transect centering at 43.5°N and extending 1400 km from 132°E in the east to 108°E in the west. Vegetation types along the transect are determined largely by rainfall gradient and shift from dark conifer forest, conifer–broadleaf mixed forest (*Pinus koraiensis*, *Abiesholophylla*), deciduous broadleaf forest (*Quercus mongolica*), woodlands and shrublands in the east, to meadow steppe (*Aneurolepidium chinenses*, *Stipa baicalensis*, *Filofolium sibiricum*) and cropland in the middle, and to typical steppe (*Leymus chinensis*, *Stipa grandis*), desert steppe (*Stipa krylovii*, *Stipa gobica*) in the west. The region covered by NECT has been subject to increasing human pressure, including cultivation and overgrazing. A general question guiding the NECT is how the water availability will influence the composition of plant functional types, soil organic matter, NPP, trace gas flux, and land-use distribution (Zhang et al. 1997). NSTEC is composed of two parts: 110–120°E, 15–40°N and 118–128°E, 40–57°N. It covers about 2,889,100 km$^2$, about 30.1 % of the total land area in China. The main driving forces along NSTEC are temperature and land use. Eight vegetation types from south to north have been identified: mountain tropical rain forest, tropical seasonal rain forest, low subtropical monsoon evergreen broadleaf forest, mid-latitude subtropical evergreen broadleaf forest, high latitude subtropical evergreen–deciduous broadleaf forest, warm temperate broadleaf deciduous forest, temperate coniferous and broadleaf mixed forest, and cold temperate coniferous forest.

This chapter will report on the highlights of more than 20 years researches on the impacts of global change (climate, elevated $CO_2$, and land use) on terrestrial ecosystems from these two Chinese transects, and discuss the research tasks in the future.

## *5.2 Advance in the Terrestrial Transect Studies in China*

### 5.2.1 Resources Use Efficiencies of Plant Species

Long-term WUE of plant species could be indicated by $\delta^{13}C$ values, and it is defined as the ratio of leaf photosynthetical rate to evapotranspiration rate per unit leaf area. The responses of WUE for 15 plant species along the grassland zone of the NECT showed strong differences in the responses of water use status to environmental gradients among different plant species and each species has its own adaptive strategy to environmental change. When considering restoration of degraded grasslands, it is important to consider variation between plant species in their potential to adapt to dry habitats (Su et al. 2000). The ecological plasticity of *L. chinensis* stoma density and WUE was an important mechanism for its broad ecological adaptability, and water factors were primary ecological factors influencing the stoma density of *L. chinensis*. However, the responses of the stoma density and WUE for *L. chinensis* to environmental changes were complex (Yang et al. 2007). The maximum photosynthetic rate of dominant plant species in main forests along the NSTEC was coniferous species > broad-leaved species, and deciduous species > evergreen species (Zhan et al. 2012), and the foliar WUE and nitrogen use efficiency were dominated by plant life-form, and a trade-off existed between the two resources use efficiencies.

### 5.2.2 Environmental Regulation of Plant Carbon and Nitrogen

Plant carbon (C) and nitrogen (N) metabolisms and their allocations are linked each other, both their biological processes and the abiotic environment regulation are responsible for plant net primary productivity (NPP) and nutrient status (Xu and Zhou 2007). The relationship between leaf nitrogen level and its photosynthetic capacity might depend on the chemical components of leaf nitrogen, especially the ratio of nitrogen in soluble protein (Xu and Zhou 2006). The environmental factors including temperature and precipitation are important ecophysiological factors affecting plant growth, their alone and combination changes have to influence plant biological processes from molecular, whole plant, and ecosystem levels. The effects of drought on plant nitrogen allocation might rely on different plant species and experimental conditions (Xu et al. 2004; Xu and Zhou 2006). Nocturnal warming significantly exacerbates the adverse effects of soil water stress, and their synergistic interactions might reduce the plant productivity and constrain its distribution in the region dominated by *L. chinensis* (Xu and Zhou 2005b).

Drought and high-temperature stresses have been extensively studied; however, little is known about their combined impact on plants. The ratio of root to shoot would increase under appropriate drought conditions, and not change obviously under severe drought conditions (Xu and Zhou 2005a). Drought would improve the adaptability of plant species to high temperature and keep the stronger carbon sequestration capacity (Hamerlynck et al. 2000), and severe water stress might exacerbate the adverse effects of high temperature, and their combination might reduce the plant productivity of *L. chinensis* (Xu and Zhou 2006), which depends on the two stresses' time and severity degree. High temperature, combined with severe soil drought, might reduce the function of PSII, weaken nitrogen anabolism, strengthen protein catabolism, and provoke lipid peroxidation (Xu and Zhou 2006). Usually, both elevated $CO_2$ concentration and drought would increase carbon allocation to plant root, and the combination of elevated $CO_2$ concentration with drought would promote this kind of the effect. Doubled $CO_2$ concentrations enhanced plant growth of *Carogana intermedia* under well-watered conditions but increased root growth under drought conditions resulting in an increase in root to shoot ratio (Xu et al. 2005).

### 5.2.3 Driving Mechanisms of the Transect Vegetation Changes

The vegetation distribution patterns along NECT would change significantly under future climate, and the major factors driving the vegetation changes were water and heat. However, the responses of various vegetation types to the changes in water and heat factors were obviously different. The vegetation changes were more sensitive to heat factors than to water factors (Zhang and Zhou 2008).

Land use practice is also an important driving force affecting terrestrial ecosystems. The stability of grassland community diversity was as follows: desert steppe > typical steppe > meadow steppe > saline moist meadow (Yang et al. 2010a, b), and the Shannon index was the highest in moderate grazing or heavy grazing stages, changing as follows: moderate grazing–heavy grazing > heavy grazing–moderate grazing > light grazing > over grazing. The pattern of plant diversity of the steppe region along the NECT was meadow steppe > typical steppe > desert steppe > alkaline meadow (Yang et al. 2001). Land-use practices (grassland fencing, mowing, and grazing grasslands) along the precipitation gradient result not only in changes in grassland communities but also in qualitative changes of their structure and function (Zhou et al. 2002).

### 5.3 Prospect

Global change terrestrial transects have proved to be an important and useful scientific approach to study the spatial and temporal dynamics of multiple drivers and complex responses. Although a number of studies on GCTEs have been done during the recent decade, much remains to be learned of the interactive effects of multiple drivers and their spatial and temporal dynamics. In order to improve integrative global change studies in China and develop the capability to predict the responses of terrestrial ecosystems to global change in China, the following research fields should be emphasized in the future: (i) Adaptability of ecosystems in semiarid region to climatic change; (ii) Interactive biological-physical-chemical-social processes and terrestrial ecosystem management; (iii) Global change simulation and warning system for terrestrial ecosystems in semiarid region.

## 6 Innovation and Practice of Forestry Carbon Management in China

Forests are a principal part of the terrestrial ecosystem. Accelerate the restoration of forest vegetation to increase carbon sinks, and protect forest to prevent degradation as well as reduce carbon emission have become a global consensus in response to climatic change, and also one of three voluntary commitments made by the Chinese government in controlling the greenhouse gas emission, namely: To greatly increase forest carbon sinks by increasing 40 million $hm^2$ of forest acreage and 1.3 billion $m^3$ of forest stock volume by 2020 from the level of 2005, which indicated that forestry[1] is playing a strategic role in coping with climatic change in China. In recent decades, China has made great achievements in forestry, which can be elaborated as: China now has forest with an area of 195 million $hm^2$, the forest coverage with a percentage of 20.36 %, the total growing stock volume has reached 14.913 billion $m^3$, the total carbon storage of forest vegetation has reached 7.811 billion tons, and the total forestry ecological benefit is now worth 1.001 billion CNY. China is now becoming a country of most artificial forests, with the fastest growing in forest area, and has a great increment of carbon sink in the world. In 2004, China's forests absorbed about 500 million tons of carbon dioxide, equalling to 8 % of the industrial discharge that year.

State Forestry Administration of China[2] attaches great importance to forestry carbon management and proposed the concept of Carbon Sequestration Forestry with Emphasis on Playing Forest Multiple Benefits. Five aspects included in the concept: (i)To meet to the requirement for sustainable economic and social

[1]http://cdm.unfccc.int/Projects/projsearch.html.

[2]http://www.forestry.gov.cn/main/175/content-631662.html.

development and the national strategy in responding to climatic change; (ii) Except for the carbon sink accumulation, to improve the stability, adaptability and integrated ecosystem service of forest ecosystem, promote the protection of biodiversity and ecosystem, and fully play forests' multiple benefits including community development; (iii) To establish a technical supporting system in line with international rules and China's reality; (iv) To increase the public awareness on climatic change adaptation and climate protection; (v) To develop ecological service market giving priority to carbon sinks by improving market mechanisms and legal measures. To fully play the forestry's functions in responding to climatic change, the State Forestry Administration[3] adopted forestry carbon management with the main objective of climate change adaptation from aspects of institution building, policy developing, technical standards improving and climate fund foundation establishing, and initially established a forestry carbon management system from macroscopic to microcosmic, which vigorously promoted China's climate change adaptation from forestry's perspective.

**Institution establishment and standard development** From 2002, the State Forestry Administration set up Carbon Sequestration Office, Climate Office and Energy Resource Office, released some guidance documents, and granted carbon sequestration measurement and monitoring qualification certificates to 10 units. In November, 2009, *The Forestry Action Plan to Address Climatic Change* was published, making clear the guiding ideology, basic principles, stage targets as well as key fields and main actions of China's forestry to cope with climatic changes. Over the past decade, the State Forestry Administration[4] has been conducting leading research and exploration on the development of carbon sequestration technical standard system. The methodologies and standards developed cover three aspects: the first is national-level carbon sink measurement and monitoring system, including *National Measurement and Monitoring System for Forestry Carbon Sequestration* and *National Measurement and Monitoring Guideline for Forestry Carbon Sequestration.* The second is project-level methodologies, including *Methodology on China Carbon Sequestration afforestation project, Methodology on China Bamboo Carbon Sequestration project, Methodology on China Forest Management Carbon Sequestration project* and *Methodology on China Shrub Carbon Sequestration project.* The third is market-level standards and rules, including *Examining and Verifying Guideline for Forest Carbon Sink, Forest Carbon Trading Standards, Rules on Forest Carbon Trading* and *Trading Procedure of Forest Carbon Trading.* Meanwhile, China Forest Carbon Project Registration System was set up at Academy of Forest Inventory and Planning of the State Forestry Administration for project registration.

**Set up Green Carbon Foundation** To promote China's enterprises contribute to forestation and voluntarily participate in emission reduction, in 2010, as approved

---

[3]http://www.forestry.gov.cn/main/95/content-628083.html.

[4]http://www.forestry.gov.cn/Zhuanti/content_2012tjzh/517913.html.

by the state, registered at the Ministry of Civil Affairs and supervised by the State Forestry Administration[5], China Green Carbon Foundation, the first public foundation with the main objective of coping with climatic change, was founded, with its precursor founded in July 2007. The mission of China Green Carbon Foundation is to promote activities of combating climate change including afforestation, forest management, decreasing deforestation, and other activities associated with increasing carbon sink and reducing emissions, to spread relevant knowledge so as to strengthen public capacity of combating climate change, to support and perfect the Forest Effect Compensation Mechanism of China. By June 2012, the social donation received was more than 500 million CNY and afforestation for carbon sink was more than 80,000 $hm^2$ in more than 20 provinces and autonomous regions in China. The Foundation is a public welfare platform for enterprises and the public to store carbon credits, fulfill social responsibility, improve farmer income, and improve ecological environment. Its running mode is as follows: the ownership of forests planted with the enterprise donation belongs to local farmers or those with the land use right, and the donators get carbon sink credits which are produced by each project, and measured, audited and registered by professional organizations. The purpose is to show the social responsibility of enterprises and realize low-carbon production. Many individuals also donate money to the Foundation to buy carbon sinks to balance out their daily emitted carbon dioxide. Furthermore, China Green Carbon Foundation also developed nearly 20 forest carbon neutral, including those for UN Climatic Change Tianjin Conference, Annual Summit of China Green Companies and Official Travel of International Network for Bamboo and Rattan. It also founded the "Afforest Motherland and Low-Carbon Action" Tree-planting Day to guide the public to perform the duty of tree planting, which was actively responded by the government and the public. China Green Carbon Foundation has set up 40 forestation bases accepting personal donations, which are platforms for the public to participate in "carbon compensation and carbon footmark removal." It also compiled China's first textbook *Forestry Carbon Sequestration and Climatic change* for high school students, which was studied by students in Grade 1 of the High School Affiliated to Beijing International Studies University for 32 credit hours, and thus carbon sink knowledge enters the classroom formally.

**The outlook of forestry carbon management** Key forestry ecological projects will be continually implemented and the national compulsory tree planting activities will be developed further. To carry out missions specified in the Forestry Twelfth Five Year Plan and realize the forestry "double growth" target, the afforestation is planned to reach more than 6 million $hm^2$ and the forest tending reaches more than 5.4 million $hm^2$ so as to increase forest area, promote forest quality and increase carbon storage; woodland expropriation will be strictly controlled to protect forest vegetation and soil; prevention and control of forest fire and diseases and insect pests will be strengthened to reduce carbon emission from forest; replacing plastic

[5]http://www.forestry.gov.cn/Zhuanti/content_apec/499030.html.

and steel with wood is advocated; forestry biomass energy will be developed to extend the function of forestry carbon storage and carbon emission reduction. Technical standard systems will be developed further to integrate forestry carbon sink into national carbon emission trading system. Policy advice on carbon sink/source balance at each level (country, province, city, county, and enterprise) will be given in accordance with national data on forest carbon sink and industrial discharge, including the policy of offset carbon tax with carbon sink. Maximize the function and role of forestry in coping with climate change in order to make a greater contribution to national economic development.

## References


Agrell, J., Kopper, B., McDonald, E. P., & Lindroth, R. L. (2005). $CO_2$ and $O_3$ effects on host plant preferences of the forest tent caterpillar (*Malacosoma disstria*). *Global Change Biology, 11*, 588–599.

Battaglia, M., & Beadle, C. (1996). Loughhead S. Photosynthetic temperature responses of *Eucalyptus globulus* and *Eucalyptus nitens*. *Tree Physiology, 16*, 81–89.

Berry, J., & Björkman, O. (1980). Photosynthetic response and adaptation to temperature in higher plants. *Annual Review of Plant Physiology, 31*, 491–543.

Chang, J., Wang, X., Zhang, X., et al. (2009). Alpine timberline dynamics in relation to climatic variability in the northern Daxing'an mountains. *Journal of Mountain Science, 27*(6), 703–711.

Chen, F. J., Ge, F., & Parajulee, M. N. (2005a). Impact of elevated $CO_2$ on tri-trophic interaction of *Gossypium hirsutum, Aphis gossypii*, and *Leis axyridis*. *Environmental Entomology, 34*(1), 37–46.

Chen, W., Liu, Y., & Ma, Z. (2002). The seasonal characteristics of climatic change trend in China from 1951 to 1997. *Plateau Meteorology, 21*(3), 251–257.

Chen, F. J., Wu, G., Ge, F., Parajulee, M. N., & Shrestha, R. B. (2005b). Effects of elevated $CO_2$ and transgenic Bt cotton on plant chemistry, performance and feeding of an insect herbivore, cotton bollworm Helicoverpa armigera (Hubner). *Entomologia Experimentalis et Applicata, 115*, 341–350.

Cui, H., Su, J., & Ge, F. (2011). Advances in research on the effect of elevated ozone concentration on insects. *Chinese Journal of Applied Entomology, 48*(5), 1130–1140.

Cui, H., Sun, Y., Su, J., Ren, Q., Li, C., & Ge, F. (2012). Elevated $O_3$ reduces the fitness of *Bemisia tabaci* via enhancement of the SA dependent defense of the tomato plant. *Arthropod-Plant Interactions, 6*, 425–437.

Cunningham, S. C., & Read, J. (2002). Comparison of temperate and tropical rainforest tree species: photosynthetic responses to growth temperature. *Oecologia, 133*, 112–119.

Dang, Z. H., & Chen, F. J. (2011). Responses of insects to rainfall and drought. *Chinese Journal of Applied Entomology, 48*(5), 1161–1169.

Deutsch, C. A., Tewksbury, J. J., Huey, R. B., Sheldon, K. S., & Ghalambor, C. K. (2008). Impacts of climate warming on terrestrial ectotherms across latitude. *Proceedings of the National Academy of Sciences of the United States of America, 105*(18), 6668–6672.

Ding, Y., Ren, G., Shi, G., et al. (2006). National assessment report of climate change (I): Climate change in China and its future trend. *Advances in Climate Change Research, 2*, 3–8.

Dong, Z. K., & Ge, F. (2011). The fitness of insects in response to climate warming Chinese. *Journal of Applied Entomology, 48*(5), 1141–1148.

Ferris, R., Long, L., Bunn, S. M., et al. (2002). Leaf stomatal and epidermal cell development: identification of putative quantitative trait loci in relation to elevated carbon dioxide concentration in poplar. *Tree Physiology, 22*, 633–640.

Frankhauser, S. (1996). The Potential Costs of Climate Change Adaptation. In J. B. Smith, et al. (Eds.), *Adapting to climate change: An international perspective* (pp. 80–96). New York: Springer.

Gange, A. C., & Brown, V. K. (1989). Effects of root herbivory by an insect on a foliar-feeding species, mediated through changes in the host plant. *Oecologia, 81*(1), 38–42.

Guo, K., Hao, S. G., Sun, O. J., & Kang, L. (2009). Differential responses to warming and increased precipitation among three contrasting grasshopper species. *Global Change Biology, 15*(10), 2539–2548.

Hamerlynck, E. P., Huxman, T. E., Loik, M. E., & Smith, S. D. (2000). Effects extreme high temperature, drought and elevated $CO_2$ on photosynthesis of the Mojave Desert evergreen shrub, Larreatridentata. *Plant Ecology, 148*, 185–195.

Hao, Z., Zheng, J., & Tao, X. (2001). A study on northern boundary of winter wheat during climate warming: A case study in Liaoning Province. *Progress in Geography, 20*, 254–261.

Hikosaka, K., Murakami, A., & Hirose, T. (1999). Balancing carboxylation and regeneration of ribulose-1,5-bisphosphate in leaf photosynthesis in temperature acclimation of an evergreen tree, *Quercus myrsinaefolia*. *Plant Cell & Environment, 22*, 841–849.

Holopainen, J. K. (2002). Aphid response to elevated ozone and $CO_2$. *Entomologia Experimentalis et Applicata, 104*, 137–142.

Holopainen, J. K., Kainulainen, P., & Oksanen, J. (1995). Effects of gaseous air pollutants on aphid performance on Scots pine and Norway spruce seedling. *Water Air and Soil pollution, 85*, 1431–1436.

Hou, Y., Liu, Q., Yan, H., et al. (2007). Variation trends of China terrestrial vegetation net primary productivity and its responses to climate factors in 1982–2000. *Chinese Journal of Applied Ecology, 18*(7), 1546–1553.

Houghton, J. T. (2001). *Climate change 2001: The scientific basis* (p. 2001). Cambridge, UK: Cambridge University Press.

Huang, M., Ji, J., & Peng, L. (2008). The response of vegetation net primary productivity to climate change during 1981–2000 in the Tibetan Plateau. *Climatic and Environmental Research, 13*(5), 608–616.

Huang, L., Ren, Q., Sun, Y., Ye, L., Cao, H., & Ge, F. (2012). Lower incidence and severity of tomato virus in elevated $CO_2$ is accompanied by modulated plant induced defence in tomato. *Plant Biology, 14*, 905–913.

Hussain, M., Kubiske, M. E., & Connor, K. F. (2001). Germination of $CO_2$-enriched *Pinus taeda* L. seeds and subsequent seedling growth responses to $CO_2$ enrichment. *Functional Ecology, 15*, 344–350.

Intergovernmental Panel on Climate Change (IPCC). (2001a). Third assessment report.

Intergovernmental Panel on Climate Change (IPCC). (2001b). Fourth assessment report.

Intergovernmental Panel on Climate Change (IPCC). (2007a). *Climate change 2007: The physical science basis: Summary for policy makers*. Report of Working Group I of the Intergovernmental Panel on Climate Change. http://www.ipcc.ch/pub/spm18-02.pdf

IPCC. (2007b). Climate change 2007: Synthesis Report. Intergovernmental Panel on Climate Change. Cambridge: Cambridge University Press.

Jackson, D. M., Rufty, T. W., Heagle, A. S., et al. (2000). Survival and development of tobacco hornworm larvae on tobacco plants grown under elevated levels of ozone. *Journal of Chemical Ecology, 26*, 1–19.

Jaffe, D., & Ray, J. (2007). Increase in surface ozone at rural sites in the western US. *Atmospheric Environment, 41*, 5452–5463.

Jin, Z., Ge, D., Shi, C., et al. (2002). Several strategies of food crop production in the northeast China plain for adaptation to global climate change: A modeling study. *Acta Agronomica Sinica, 28*, 24–31.

Jondrup, P. M., Barnes, J. D., & Port, G. R. (2002). The effect of ozone fumigation and different *Brassica rapa* lines on the feeding behaviour of *Pieris brassicae* larvae. *Entomologia Experimentalis et Applicata, 104*, 143–151.

Jordan, D. B., & Ogren, W. L. (1984). The $CO_2/O_2$ specificity of ribulose 1,5-bisphosphate carboxylase/oxygenase: Dependence on ribulose bisphosphate concentration, pH and temperature. *Planta, 161*, 308–313.

Labate, C. A., & Leegood, R. C. (1988). Limitation of photosynthesis by changes in temperature. *Planta, 173*, 519–527.

Li, Y., Yang, X., Wang, W., et al. (2010). The possible effects of global warming on cropping systems in China V: The possible effects of climate warming on geographical shift in safe northern limit of tropical crops and the risk analysis of cold damage in China. *Scientia Agricultura Sinica, 43*(12), 2477–2484.

Liu, H., Gu, H., Tang, Z., et al. (2002). Tree photosynthesis on alpine timberline and its relationships to environmental factors in the eastern part of temperate China. *Journal of Mountain Science, 20*(1), 32–36.

Liu, Z., Hubbard, K. G., Lin, X., et al. (2013a). Negative effects of climate warming on maize yield are reversed by the changing of sowing date and cultivar selection in Northeast China. *Global Change Biology, 19*, 3481–3492.

Liu, Y., Liu, Y., & Guo, L. (2010a). Impact of climatic change on agricultural production and response strategies in China. *Chinese Journal of Eco-agriculture, 18*(4), 905–910.

Liu, Y., Wang, E., Yang, X., et al. (2010b). Contributions of climatic and crop varietal changes to crop production in the north China plain, since 1980s. *Global Change Biology, 16*, 2287–2299.

Liu, Z., Yang, X., Chen, F., et al. (2013b). The effects of past climate change on the northern limits of maize planting in northeast China. *Climatic Change, 117*, 891–902.

Liu, Z., Yang, X., Hubbard, K. G., et al. (2012). Maize potential yields and yield gaps in the changing climate of northeast China. *Global Change Biology, 18*, 3441–3454.

Liu, Z., Yang, X., Wang, W., et al. (2010). The possible effects of global warming on cropping systems in China IV: The possible impact of future climatic warming on the northern limits of spring maize in three provinces of northeast China. *Scientia Agricultura Sinica, 43*(11), 2280–2291.

Long, S. P., Ainsworth, E. A., et al. (2006). Food for thought: Lower-than-expected crop yield stimulation with rising $CO_2$ concentrations. *Science, 312*, 1918–1921.

Luomala, E. M., Laitinen, K., Sutinen, S., et al. (2005). Stomatal density, anatomy and nutrient concentrations of scots pine needles are affected by elevated $CO_2$ and temperature. *Plant Cell & Environment, 28*, 733–749.

Ma, W., Fang, J., Yang, Y., et al. (2010). Biomass carbon stocks and their changes in northern China's grasslands during 1982–2006. *Science China Life Science, 40*(7), 632–641.

Ma, S., Wang, Q., & Luo, X. (2008). Effect of climate change on maize (*Zea mays*) growth and yield based on stage sowing. *Acta Ecologica Sinica, 28*(5), 2131–2139.

Ma, Q., Yan, J., & Du, J. (2011). Spatial and temporal variations of climate productivity in the area of pastoral transition of Datong City. *Resources & Environment, 27*(7), 641–645.

Masters, G. J., Brown, V. K., & Gange, A. C. (1993). Plant mediated interactions between above- and below-ground insect herbivores. *Oikos, 66*(1), 148–151.

Mondor, E. B., Tremblay, M. N., Awmack, C. S., & Lindroth, R. L. (2004). Divergent pheromone-mediated insect behaviour under global atmospheric change. *Global Change Biology, 10*, 1820–1824.

Parmensan, C. (1996). Climate and species' range. *Nature, 382*, 765–766.

Peng, S., & Ren, H. (2000). The north-south transect of eastern China (NSTEC) for global changes studies. *GCTE News, 2000*(16), 6.

Percy, K. E., Awmack, C. S., Lindroth, R. L., et al. (2002). Altered performance of forest pests under atmospheres enriched by $CO_2$ and $O_3$. *Nature, 420*, 403–407.

Qin, D., Ding, Y., Su, J., et al. (2005). Assessment of climate change and environment changed in China (I): Climate and environment changes in China and their projection. *Advances in Climate Change Research, 1*, 4–9.

Sha, W., Shao, X., Huang, M. (2002). Climate warming and its impact on natural geographical boundaries since 1980s. *Science in China (Series D), 32*(4), 317–326.
State Forestry Administration, P.R. China. (2009a). *China forestry resource*. Beijing: China Forestry Publishing House.
State Forestry Administration, P.R. China. (2009b). *The forestry action plan to address climatic change*. Beijing: China Forestry Publishing House.
Steinger, T., Gall, R., & Schmid, B. (2000). Maternal and direct effects of elevated $CO_2$ on seed provisioning, germination and seedling growth in *Bromus erectus*. *Oecologia, 12*, 475–480.
Stige, L. C., Chan, K. S., Zhang, Z. B., Frank, D., & Stenseth, N. C. (2007). Thousand-year-long Chinese time series reveals climatic forcing of decadal locust dynamics. In *Proceedings of the National Academy of Sciences of the United States of America*.
Su, B., Han, X., Li, L., Huang, J., Bai, Y., & Qu, C. (2000). Responses of y $\delta^{13}C$ value and water use efficiency of plant species to environmental gradients along the grassland zone of northeast China transect. *Acta Phytoecologica Sinica, 24*(6), 648–655.
Sun, Y., Chen, F. J., & Ge, F. (2009). Elevated $CO_2$ changes interspecific competition among three species of wheat aphids: *Sitobion avenae*, *Rhopalosiphum padi*, and *Schizaphis graminum*. *Environmental Entomology, 38*, 26–34.
Sun, R., & Qijiang, Z. (2001). Effect of climate change of terrestrial net primary productivity in China. *Journal of Remote Sensing, 5*(1), 58–62.
Sun, Y., Yin, J., Chen, F., Wu, G., & Ge, F. (2011). How does atmospheric elevated $CO_2$ affect crop pests and their natural enemies: The examples in China? *Insect Science, 18*(4), 393–400.
Sun, X., & Zhao, X. (2009). Assessment of ecological risk to climate change of the farming-pastoral zigzag zone in northern China. *Acta Scientiarum Naturalium Universitatis Pekinensis, 45*(4), 713–720.
Sun, H., & Zheng, D. (1998). *Tibetan Plateau's forming, evolving and development*. Guangzhou: Guangdong Science & Technology Press.
Teng, N. J., Wang, J., Chen, T., et al. (2006). Elevated $CO_2$ induces physiological, biochemical and structural changes in leaves of *Arabidopsis thaliana*. *New Phytologist, 172*, 92–103.
Tjoelker, M. G., et al. (2001). Modelling respiration of vegetation: Evidence for a general temperature-dependent $Q_{10}$. *Global Change Biology, 7*, 223–230.
Tol, R. S. J., Frankhauser, S., & Smith, J. B. (1997). *The Scope for adaptation to climate change: What can we learn from the Literature? Institute for Environmental Studies*. Amsterdam: Vrije University.
Vinzargan, R. (2004). A review of surface ozone background levels and trends. *Atmospheric Environment, 8*, 3431–3442.
Wang, S., Jingning, C., Zhu, J., et al. (2002). Stuadies on climate change in China. *Climatic and Environmental Research, 7*(2), 137–145.
Wang, S., Zhou, G., Gao, S., et al. (2003). Distribution of soil labile carbon along the northeast china transect and its response to climatic change. *Acta Phytoecologica Sinica, 27*(6), 780–785.
Wilkinson, S., & Davies, W. J. (2010). Drought, ozone, ABA and ethylene: new insights from cell to plant to community. *Plant Cell Environment, 33*, 510–525.
Wu, Z., Jin, Y., Liu, J., et al. (2003). Response of vegetation distribution to global climate change in northeast China. *Scientia Geographica Sinica, 23*(5), 564–570.
Xiong, W., Ju, H., Xu, Y., et al. (2006). Regional simulation of wheat yield in China under the climatic change conditions. *Chinese Journal of Eco-agriculture, 14*(2), 164–167.
Xiong, W., Xu, Y., Lin, E., et al. (2005). Regional simulation of maize yield under IPCC SRES A2 and B2 scenarios. *Chinese Journal of Agrometeorology, 26*(1), 11–15.
XU, X. (2004). Response of net primary productivity (NPP) of Sichuan vegetations to global climate changes. *Chinese Journal of Ecology, 23*(6), 19–24.
Xu, Z., & Zhou, G. (2005a). Effects of water stress on carbon allocation in the perennial grass *Leymus chinensis* under two nocturnal temperatures. *Physiologia Plantarum, 123*, 272–280.
Xu, Z., & Zhou, G. (2005b). Effects of water stress and high nocturnal temperature on photosynthesis and nitrogen level of a perennial grass *Leymus chinensis*. *Plant and Soil, 269*, 131–139.

Xu, Z., & Zhou, G. (2006). Combined effects of water stress and high temperature on photosynthesis, nitrogen metabolism and lipid peroxidation of a perennial grass *Leymus chinensis*. *Planta, 224*, 1080–1090.

Xu, Z., & Zhou, G. (2007). Relationship between carbon and nitrogen and environmental regulation in plants under global change—From molecule to ecosystem. *Journal of Plant Ecology, 31*(4), 738–747.

Xu, Z. Z., Zhou, G. S., & Shimizu, H. (2009). Effects of soil drought with nocturnal warming on leaf stomatal traits and mesophyll cell ultrastructure of a perennial grass. *Crop Science, 49*, 1843–1851.

Xu, Z., Zhou, G., Xiao, C., & Wang, Y. (2004). Responses of two dominated desert shrubs to soil drought under doubled $CO_2$ condition. *Acta Ecologica Sinica, 24*(10), 2186–2191.

Xu, X., Zhou, G., Xiao, C., & Wang, Y. (2005). Interactive effects of doubled atmospheric $CO_2$ concentrations and soil drought on whole plant carbon allocation in two dominant desert shrubs. *Acta Phytoecologica Sinica, 29*(2), 281–288.

Xue, C., Liu, R., & Wu, Q. (2010). Effect of climate warming on rice growing stages in Xinyang. *Chinese Journal of Agrometeorology, 31*(3), 353–357.

Yamasaki, T., Yamakawa, T., Yamane, Y., Koike, H., Satoh, K., & Katoh, S. (2002). Temperature acclimation of photosynthesis and related changes in photosystem II electron transport in winter wheat. *Plant Physiology, 128*, 1087–1097.

Yang, L., Han, M., & Li, J. (2001). Plant diversity change in grassland communities along a grazing disturbance gradient in the northeast China transect. *Acta Phytoecologica Sinica, 25*(1), 110–114.

Yang, L., Han, M., Zhou, G., & Li, J. (2007). The changes of water-use efficiency and stoma density of *Leymus chinensis* along northeast China transect. *Acta Ecologica Sinica, 27*(1), 16–24.

Yang, X., Liu, Z., Chen, F. (2010a). The possible effects of global warming on cropping systems in ChinaI: The possible effects of climate warming on northern limits of cropping systems and cop yields in China. *Scientia Agricultura Sinica, 43*(2), 329–336.

Yang, X., Liu, Z., & Chen, F. (2011a). The possible effect of climate warming on northern limits of cropping system and crop yield in China. *Agricultural Sciences in China, 10*, 585–594.

Yang, X., Liu, Z., & Chen, F. (2011b). The possible effects of global warming on cropping systems in China VI: Possible effects of future climate change on northern limits of cropping system in China. *Scientia Agricultura Sinica, 44*, 1562–1570.

Yang, Y., & Piao, S. (2006). Variations in grassland vegetation cover in relation to climatic factors on the Tibetan plateau. *Journal of Plant Ecology, 30*(1), 1–8.

Yang, S., Sheng, S., Zhao, X., et al. (2010b). Impacts of climate change and its variability on rice production in the middle and lower valley of the Yangtze River, China. *Chinese Journal of Rice Science, 36*(9), 1519–1528.

Ye, J. (2010). *Response of vegetation net primary productivity to climate change on the Tibetan Plateau.* Lanzhou: Lanzhou University.

Yin, J., Sun, Y., Wu, G., & Ge, F. (2010). Effects of elevated $CO_2$ associated with maize, a $C_4$ plant, on multiple generations of cotton bollworms *Helicoverpa armigera* Hübner (Lepidoptera: Noctuidae). *Entomologia Experimentalis et Applicata, 136*, 12–20.

Yoram, Y. T., & Jonathan, Y. T. (2005). Global warming, Bergmann's rule and body size in the masked shrew *Sorex cinereus* Kerr in Alaska. *Journal of Animal Ecology, 74*, 803–808.

Yu, H., & Xu, J. (2009). Effects of climate change on vegetations on Qinghai-Tibet Plateau : A review. *Chinese Journal o f Ecology, 28*(4), 747–754.

Yuan, B., Guo, J., Zhao, J., et al. (2011). Possible impacts of climate change on agricultural production in China and its adaptation countermeasures. *Agricultural Science & Technology, 12*(3), 420–425.

Yuan, F., Han, X., Ge, J., et al. (2008). Net primary productivity of *Leymus chinensis* steppe in Xilin River basin of Inner Mongolia and its responses to global climate change. *Chinese Journal of Applied Ecology, 19*(10), 2168–2176.

Zhan, X., Yu, G., Sheng, W., & Fang, H. (2012). Foliar water use efficiency and nitrogen use efficiency of dominant plant species in main forests along the north-south transect of east China. *Chinese Journal of Applied Ecology, 23*(3), 587–594.

Zhang, Z. B., Cazelles, B., Tian, H. D., Stige, L. C., Brauning, A., & Stenseth, N. C. (2009). Periodic temperature-associated drought/flood drives locust plagues in China. *Proceedings of the Royal Society B-Biological Sciences, 276*(1658), 823–831.

Zhang, X., Gao, Q., Yang, D., Zhou, G., Ni, J., & Wang, Q. (1997). A gradient analysis and prediction on the northeast China transect (NECT) for global change study. *Acta Botanica Sinica, 39*(9), 785–799.

Zhang, G., Ouyang, H., Zhang, X., et al. (2010a). Vegetation change and its responses to climatic variation based on eco-geographical regions of Tibetan Plateau. *Geographical Research, 29* (11), 2004–2016.

Zhang, Z. B., Tian, H. D., Cazelles, B., Kausrud, K. L., Bräuning, A., Guo, F., & Stenseth, N. C. (2010b). Periodic climate cooling enhanced natural disasters and wars in China during AD 0–1900. *Proceedings of the Royal Society B-Biological Sciences, 277*, 3745–3753.

Zhang, Y., & Zhou, G. (2008). Terrestrial transect study on driving mechanism of vegetation changes. *Science in China (Series D), 51*(7), 984–991.

Zhao, J., Yan, X., & Jia, G. (2008). Simulating the responses of forest net primary productivity and carbon budget to climate change in northeast China. *Acta Ecological Sinica, 28*(1), 92–102.

Zhao, J., Yang, X., Liu, Z., et al. (2010). The possible effects of global warming on cropping systems in China II: The characteristics of climatic variables and the possible effect on northern limits of cropping systems in South China. *Scientia Agricultura Sinica, 43*(9), 1860–1867.

Zhou, T., Shi, P., & Wang, S. (2003). Impacts of climate change and human activities on soil carbon storage in China. *Acta Geographica Sinica, 58*(5), 727–734.

Zhou, G., Wang, Y., & Wang, S. (2002). Responses of grassland ecosystems to precipitation and land use along northeast China transect. *Journal of Vegetation Science, 13*, 361–368.

Zhu, W., Pan, Y., Yang, Y., et al. (2007). The effects of climate changes on the net primary productivity of natural vegetation in the China land. *Scientific Bulletin, 52*(21), 2535–2541.

Ziska, L. H., Morris, C. F., & Goins, E. W. (2004). Quantitative and qualitative evaluation of selected wheat varieties released since 1903 to increasing atmospheric carbon dioxide: Can yield sensitivity to carbon dioxide be a factor in wheat performance? *Global Change Biology, 10*, 1810–1819.

# Part V
# Sustainable Development Ecology

# Chapter 17
# Principles and Application of Sustainable Development

**Jingzhu Zhao, Longyu Shi, Lina Tang, Lijie Gao, Gaodi Xie, Shuyan Cao, Yanying Bai, Chuanglin Fang, Chao Bao, Wenhua Li, Guangmei Yang, Moucheng Liu, Guihuan Liu, Yihui Wen, Yanmin Zhang and Huiyuan Zhang**

**Abstract** At present, sustainable development has become the consensus of governments and the people around the world, and has come from concept to practice. However, it involves the reform of population, resources, environment, production, technology, institutions, and concept. In order to achieve this goal, the road is long and full of hardships in the future. Especially for China, which is a country confronted with an enormous population, a serious shortage of resources per capita, high pressure on employment, outstanding ecological and environmental problems, and the promotion of regional sustainable development has important practical significance. To promote sustainable development, first we need to choose an index that can fully reflect regional natural, economical and social characteristics from the perspective of system, then follow certain principles to establish evaluation indicator or indicator system to evaluate the sustainable development. Single index or multi-index evaluation method can be used for the evaluation of sustainable development. Eco-compensation aims at the conservation and sustainable use of ecosystem services; it is an institutional arrangement that regulates the relationship between different stakeholders using economic method and an important means to realize sustainable development. In China, the theory and practice of eco-compensation has experienced the following stage in turn: groping spontaneously, theoretical research, and combining theory and practice. On the whole, the

J. Zhao · L. Shi · L. Tang · L. Gao
Institute of Urban Environment, Chinese Academy of Sciences, Xiamen 361021, China

G. Xie · S. Cao · Y. Bai · C. Fang · C. Bao · W. Li (✉) · M. Liu
Institute of Geographic Sciences and Natural Resources Research, Chinese Academy of Sciences, Beijing 100101, China
e-mail: liwh@igsnrr.ac.cn

G. Yang
Shanghai Hongqiao New Energy Investment Corp, Shanghai 201306, China

G. Liu · Y. Wen · Y. Zhang · H. Zhang
Academy for Environmental Planning, Beijing 100012, China

 
W. Li (ed.), *Contemporary Ecology Research in China*,
DOI 10.1007/978-3-662-48376-3_17

goal of realizing global sustainable development reflects the coordination between human and nature, and between different individuals. Coordination between human and nature provides guarantee for sustainable development, and guarantee of harmonious relations between different individuals reflect the rationality and orderly organization of sustainable development. Implementing the strategy of sustainable development in China, is not only an inevitable choice in the long term, but an inevitable conclusion to improve and promote their development ability in the contrast with countries all over the world.

**Keywords** Sustainable development · Ecological footprint · Biocapacity · Ecological deficit · Urbanization · Eco-compensation

## 1 Methods and Indicator Systems

As the global economy continues to grow rapidly, a series of crises have appeared such as overpopulation, shortage of resources, ecological destruction, and pollution. Therefore, it has become the common choice of humankind to consider changing the traditional mode of development, so as to pursue sustainable development. The grim reality and upcoming problems tell us that it is important to promote regional sustainable development for China, and it has become a key issue in current research to build a sustainable development evaluation indicator system to help answer the question, whether human activities contribute to the goal of sustainable development or not. Selection of evaluation indicators and how to determine weights, thresholds, and the overall discrimination method all present challenges for sustainable development evaluation. Regional sustainable development evaluation seeks to evaluate the sustainability of regional development based on research needs, and to choose indicators from a regional point of view reflecting natural, economic, and social characteristics, while building an evaluation indicator system using appropriate mathematical methods.

This study describes research progress toward sustainable development evaluation, both domestic and abroad; analyses several important regional sustainable development evaluation indicators; builds a sustainable development evaluation indicator system for China; notes some problems in the current evaluation of regional sustainable development; and makes proposals for the future of sustainable development evaluation.

# 2 Application of Ecological Footprint in the Evaluation of Regional Sustainable Development

## 2.1 Introduction

The biologically productive area was used in the ecological footprint method to measure the demand and impact of human activities on the natural capital (Wackernagel and Rees 1996), and further judge whether the impact is within the scope of the regional biocapacity. In this method, land use was used as a limiting factor to provide us the information of human dependence on nature. Because it has rich connotation of the concept, simple method and vivid expression of the results, it is comprehensive, comparable, and easily understood and accepted. Ecological footprint is taken as a powerful indicator for assessing the regional sustainability. In the twenty-first century, it is a big challenge facing China's sustainable development to improve biocapacity, reduce ecological footprint, and maintain an acceptable eco-environment. In this paper, we used the ecological footprint method to evaluate ecological footprint, biocapacity, and ecological deficit of China from 1980 to 2005.

## 2.2 Ecological Footprint and Biocapacity in China

### 2.2.1 Changes in Ecological Footprint

From 1980 to 2005, the consumption of ecosystem services was multiply increased in China, and the ecological footprint per capita increased from 0.98 to 2.17 $ghm^2$, with the average annual exponential growth rate of 3.2 %, which is three times the world level over the same period. With the strong growth momentum of ecological footprint, the gap between per capita ecological footprint of China and the whole world was reduced by 55 %. It reaches 0.5 $ghm^2$ in 2005. With the same trend, per capita ecological footprint of China will reach the world average in 2015 or so.

Table 1 shows the components and their changes of per capita ecological footprint of China. We can see that the land used for $CO_2$ absorption has accounted for more than half of the ecological footprint of China since 1995, and it is becoming the primary element to decide the ecosystem services of China and its growth rate.

**Table 1** Components of per capita ecological footprint of China in major years/%

| Type | 1980 | 1985 | 1990 | 1995 | 2000 | 2005 |
|---|---|---|---|---|---|---|
| Raw material consumption | 56.3 | 54.6 | 54.9 | 48.6 | 46.4 | 33.3 |
| Fossil energy consumption | 43.7 | 45.4 | 45.1 | 51.4 | 53.6 | 63.9 |
| Total | 100 | 100 | 100 | 100 | 100 | 100 |

From 1995 to 2000, the total energy consumption of China was kept almost unchanged with an increase rate of $0.6 \times 10^8$ tec, and the per capita ecological footprint level of China was also unchanged for the same period. From 2000 to 2005, the total energy consumption of China increased by 63 % and reached $22.5 \times 10^8$ tec in 2005, the per capita ecological footprint of China increased about 0.7 $ghm^2$ over the same period, of which 75 % new ecological footprint come from fossil energy consumption.

With the vast territory of China, the level of socio-economic development and consumption habits for different regions vary greatly, so there are significant regional differences in per capita ecological footprint. From 1980 to 2005, despite the reason that per capita ecological footprint is different for each province, there was an overall increasing trend and it changed significantly in the distribution pattern. At the very beginning, high value per capita ecological footprint was distributed mainly in the northeast China and three municipalities including Beijing, Tianjin, and Shanghai. With the uneven growth of per capita ecological footprint and the expand of high value areas, the per capita ecological footprint currently had the overall pattern of "higher in north and northeast China, medium in east and south China, lower in central, southwest and northwest China".

### 2.2.2 Changes in Biocapacity

From 1980 to 2005, the total biocapacity of China increased 1.5 times, varying from $5.8 \times 108$ to $15 \times 108$ $ghm^2$. The per capita biocapacity was doubled, varying from 0.59 to 1.15 $ghm^2$, however, the per capita biocapacity of the world over the same period was shrinking. China is still a country with low biological capacity, and its per capita biocapacity is only half of the world average. China supports nearly 1/4 of the world's population with only 1/11 of the world's farmland. This reality determines that the per capita biocapacity will still be relatively low in the future.

With the uneven distribution of biological resources and productivity, per capita biocapacity is greatly different for different provinces. Table 2 shows that the distribution range of per capita biocapacity for different provinces changed from 0.17–8.25 $ghm^2$ to 0.32–7.06 $ghm^2$ gradually from 1980 to 2005, with the highest value in Tibet and lowest value in Tianjin and Shanghai.

### 2.2.3 Changes in Ecological Deficit

The natural space for per capita biocapacity is relatively low in China. The contradiction between the supply and demand of ecosystem services is very prominent since it is necessary not only to protect the life of the new population, but also to meet the economic development and improve people's welfare. Table 3 showed

**Table 2** Change of per capita ecological service of China in different province/ghm2

| | 11980 | | | 11990 | | | 1995 | | | 2000 | | | 2005 | | |
|---|---|---|---|---|---|---|---|---|---|---|---|---|---|---|---|
| | TEF | BC | ED/ES | TEF | BC | ED/ES | TEF | BC | ED/ES | TEF | BC | ED/ES | TEF | BC | ED/ES |
| Beijing | 2.2 | 0.2 | −2.0 | 2.3 | 0.4 | −1.9 | 2.9 | 0.4 | −2.6 | 3.3 | 0.3 | −3.0 | 3.2 | 0.4 | −2.8 |
| Tianjin | 1.7 | 0.2 | −1.6 | 2.2 | 0.3 | −1.9 | 2.5 | 0.4 | −2.1 | 2.4 | 0.5 | −2.0 | 3.4 | 0.7 | −2.7 |
| Hebei | 0.9 | 0.4 | −0.6 | 1.2 | 0.5 | −0.7 | 1.4 | 0.7 | −0.8 | 1.5 | 0.7 | −0.8 | 2.6 | 1.0 | −1.7 |
| Shanxi | 1.5 | 0.5 | −1.0 | 1.3 | 0.6 | −0.8 | 1.7 | 0.6 | −1.2 | 1.6 | 0.5 | −1.1 | 3.1 | 0.5 | −2.6 |
| Neimeng | 1.1 | 3.1 | 2.1 | 1.6 | 2.4 | 0.8 | 1.5 | 2.4 | 0.8 | 1.8 | 2.1 | 0.4 | 3.9 | 3.3 | −0.6 |
| Liaoning | 1.9 | 0.6 | −1.3 | 2.1 | 0.8 | −1.3 | 2.3 | 1.1 | −1.2 | 2.2 | 1.3 | −0.9 | 3.3 | 1.8 | −1.5 |
| Jilin | 1.3 | 0.7 | −0.6 | 1.8 | 1.0 | −0.8 | 1.9 | 1.0 | −0.9 | 1.6 | 0.9 | −0.7 | 2.4 | 1.3 | −1.1 |
| Heilongjiang | 1.3 | 1.2 | −0.2 | 1.7 | 1.4 | −0.3 | 1.9 | 1.5 | −0.4 | 1.8 | 1.4 | −0.4 | 2.5 | 1.6 | −0.9 |
| Shanghai | 2.0 | 0.3 | −1.6 | 2.9 | 0.4 | −2.5 | 3.1 | 0.4 | −2.7 | 3.0 | 0.3 | −2.7 | 4.0 | 0.3 | −3.6 |
| Jiangsu | 1.0 | 0.3 | −0.6 | 1.2 | 0.5 | −0.6 | 1.5 | 0.7 | −0.8 | 1.5 | 0.8 | −0.7 | 2.2 | 1.0 | −1.2 |
| Zhejiang | 1.0 | 0.7 | −0.4 | 1.4 | 0.8 | −0.6 | 1.9 | 1.3 | −0.7 | 2.0 | 1.4 | −0.6 | 2.6 | 1.5 | −1.1 |
| Anhui | 0.8 | 0.3 | −0.5 | 0.9 | 0.5 | −0.4 | 1.2 | 0.6 | −0.6 | 1.2 | 0.8 | −0.4 | 1.5 | 0.9 | −0.6 |
| Fujian | 0.7 | 0.8 | 0.2 | 1.5 | 1.1 | −0.4 | 1.7 | 1.5 | −0.2 | 1.3 | 2.1 | 0.8 | 2.1 | 2.4 | 0.4 |
| Jiangxi | 0.7 | 0.6 | −0.1 | 1.2 | 0.7 | −0.5 | 1.2 | 0.8 | −0.3 | 1.0 | 0.8 | −0.2 | 1.5 | 1.0 | −0.5 |
| Shandong | 0.8 | 0.4 | −0.4 | 1.0 | 0.7 | −0.4 | 1.2 | 1.1 | −0.1 | 1.2 | 1.4 | 0.2 | 2.3 | 1.6 | −0.7 |
| Henan | 0.7 | 0.3 | −0.4 | 0.8 | 0.4 | −0.4 | 0.9 | 0.5 | −0.3 | 1.0 | 0.6 | −0.3 | 1.6 | 0.8 | −0.8 |
| Hubei | 1.0 | 0.4 | −0.6 | 1.2 | 0.6 | −0.6 | 1.3 | 0.8 | −0.5 | 1.5 | 0.9 | −0.6 | 2.0 | 1.2 | −0.8 |
| Hunan | 1.2 | 0.5 | −0.7 | 1.5 | 0.6 | −0.9 | 1.5 | 0.7 | −0.9 | 1.3 | 0.7 | −0.6 | 2.0 | 0.9 | −1.1 |
| Guangdong | 0.9 | 0.6 | −0.4 | 1.1 | 0.9 | −0.2 | 1.7 | 1.0 | −0.7 | 1.7 | 1.0 | −0.6 | 2.2 | 1.2 | −1.1 |
| Guangxi | 0.8 | 0.7 | −0.1 | 1.1 | 0.7 | −0.4 | 1.1 | 0.9 | −0.2 | 1.2 | 1.2 | 0.0 | 2.0 | 1.4 | −0.6 |
| Hainan | | | | | | | | | | 1.5 | 1.6 | 0.1 | 1.6 | 2.7 | 1.1 |
| Chongqing | | | | | | | | | | 1.6 | 0.4 | −1.2 | 2.2 | 0.6 | −1.6 |
| Sichuan | 1.0 | 0.4 | −0.5 | 1.1 | 0.6 | −0.5 | 1.3 | 0.6 | −0.6 | 1.2 | 0.7 | −0.4 | 1.6 | 0.8 | −0.8 |

(continued)

**Table 2** (continued)

| | 11980 | | | 11990 | | | 1995 | | | 2000 | | | 2005 | | |
|---|---|---|---|---|---|---|---|---|---|---|---|---|---|---|---|
| | TEF | BC | ED/ES | TEF | BC | ED/ES | TEF | BC | ED/ES | TEF | BC | ED/ES | TEF | BC | ED/ES |
| Guizhou | 0.9 | 0.5 | −0.4 | 1.4 | 0.5 | −0.9 | 1.2 | 0.5 | −0.7 | 1.6 | 0.5 | −1.1 | 2.3 | 0.5 | −1.8 |
| Yunnan | 0.9 | 1.2 | 0.2 | 1.5 | 1.1 | −0.4 | 1.5 | 1.0 | −0.5 | 1.3 | 0.9 | −0.4 | 1.8 | 1.0 | −0.7 |
| Xizang | 1.1 | 8.3 | 7.1 | 1.4 | 7.8 | 6.4 | 1.3 | 7.3 | 5.9 | 1.7 | 6.4 | 4.7 | 2.3 | 7.1 | 4.7 |
| Shaanxi | 0.8 | 0.7 | −0.1 | 1.2 | 0.7 | −0.5 | 1.2 | 0.7 | −0.5 | 1.1 | 0.6 | −0.5 | 1.5 | 0.7 | −0.8 |
| Gansu | 0.9 | 0.6 | −0.3 | 1.2 | 0.7 | −0.5 | 1.3 | 0.7 | −0.6 | 1.4 | 0.6 | −0.8 | 1.4 | 0.7 | −0.7 |
| Qinghai | 1.1 | 1.6 | 0.5 | 1.6 | 1.5 | −0.1 | 2.2 | 1.6 | −0.6 | 2.1 | 1.3 | −0.7 | 2.2 | 1.8 | −0.5 |
| Ningxia | 1.0 | 0.4 | −0.6 | 1.5 | 0.7 | −0.9 | 1.5 | 0.7 | −0.8 | 1.7 | 0.8 | −0.9 | 3.2 | 1.1 | −2.1 |
| Xinjiang | 1.7 | 0.7 | −1.0 | 2.0 | 1.2 | −0.8 | 2.0 | 1.4 | −0.6 | 2.0 | 1.4 | −0.6 | 2.1 | 1.9 | −0.2 |

*Note:* Hainan data was included into Guangdong and Chongqing data was included into Sichuan before 2000; Calculation for ecological footprint of energy used production area as standard; *TEF* refers to total ecological footprint; *BC* refers to biocapacity; *ED* refers to ecological deficit; *ES* refers to ecological surplus; Due to the limited page space, the data of 1985 was not listed in the table

**Table 3** Change of per capita ecological deficit of China/ghm$^2$

| Type | 1980 | 1985 | 1990 | 1995 | 2000 | 2005 |
|---|---|---|---|---|---|---|
| Raw material consumption | 0.09 | 0.05 | 0.02 | 0.14 | 0.27 | 0.44 |
| Fossil energy consumption | −0.30 | −0.38 | −0.54 | −0.76 | −0.79 | −1.46 |
| Total | −0.39 | −0.32 | −0.52 | −0.61 | −0.52 | −1.02 |

*Note:* Ecological deficit of fossil energy consumption = (biocapacity of woodland-ecological footprint of woodland)-ecological footprint of fossil energy consumption

that the per capita ecological deficit increased from 0.39 ghm$^2$ in 1980 to 1.02 ghm$^2$ in 2005. The natural occurrence constraint of land resources makes China to face more serious ecological deficit problems than the world average. As the world's engine of economic growth, China is in a period of rapid amplification of energy consumption, the growing trend of the ecological deficit will last for the coming period.

Every province in China has faced contradiction between supply and demand of ecosystem services. From 1980 to 2005, more than 85 % provinces has expanding ecological deficit. In 2005, only Hainan, Fujian, and Xizang have ecological surplus in China, and more than 40 % provinces need twice the biologically productive area to balance the demand of ecosystem services. A study has divided these provinces into three groups. Group A includes four municipalities (Beijing, Tianjin, Shanghai, and Chongqing). Group B includes four economically backward western provinces (Guizhou, Ningxia, Sichuan, and Shaanxi). Group C includes four populated eastern provinces (Hebei, Jiangsu, Henan, and Hunan).

The main reasons for the formation of ecological deficit differ for different groups. The reason for group A and group C is the conflict between development scale and resource constraints, while the reason for group B is both the fragile eco-environments and the deficient economic development capacity and more than 40 % provinces need twice the biologically productive area to balance the demand of ecosystem services in 2005. Compared with 1980, the ratio of biocapacity to ecological footprint of 11 provinces was declined, indicating that the gap between ecological supply and demand might widen, but the relative conflict eased.

## 2.3 *Conclusion and Discussion*

Since ecosystem service is dependent on area, the consumption of ecosystem services by regional socio-economic metabolism and its satisfaction can be measured by three indicators, including ecological footprint, biocapacity, and ecological deficit. Studies have shown that due to the increasing demand of socio-economic metabolism on ecosystem services, from 1980 to 2005, the consumption of ecosystem services is more than double in China, and the per capita ecological footprint increased from 0.98 to 2.17 $ghm^2$. At the same time, since the biological productivity was improved, the per capita biocapacity doubled from 0.59 to 1.15 $ghm^2$. However, due to the rapid growth in fossil energy consumption, per capita ecological deficit continued to expand and reached 1.02 $ghm^2$ in 2005. The ecological footprint has exceeded 89 % of the biocapacity. At the provincial level, more than 85 % provinces are in ecological deficit status, and only Hainan, Fujian, and Tibet have ecological surplus. The ecological debts in most provinces of China are soft in nature, caused mainly by the contradictions between supply and demand of ecosystem services in time, space, and structure. This kind of debt can be alleviated by way of cross-trade and advance occupation. Same as most of the world countries and regions, the main way for China to deal with the ecological deficit is also to occupy the current and future global commons, which is an inevitable result of external diseconomies of development. For a long period of time in the future, China will face more severe ecological deficit than the world average. This means that we are increasingly overdraft future ecological capital to maintain existing lifestyles and economic growth, which will cause the ecosystem degradation and collapse.

## 3 Interactive Coercing Effects Between Urbanization and Eco-environment

### *3.1 Introduction*

An extremely complex interaction exists between urbanization and eco-environment. How to harmonize their relationship has become an important issue that holds the attention of academic circles and governmental departments in China. At the same time, it has also become a global strategic issue. With the rapid urbanization throughout the world, urbanization has become an actual or potential threat to the surrounding eco-environment. All kinds of conflicts and threats have come forth between urbanization process and eco-environmental protection (Fang et al. 2008). From the view of national strategic demands, how to harmonize the relationship between urbanization and eco-environment has become an important topic for national development plan in China. It is of great importance to choose a healthy urbanization mode which can match the urbanization speed with the eco-environmental protection in China, so as to accelerate the process of regional industrialization and construct an ecological industrial structure. It can also help to build resource-saving and environment-friendly cities and eco-cities.

## *3.2 Interactive Coercing Effects Between Urbanization Process and Eco-environment*

### 3.2.1 Basic Laws and Evaluative Types of the Interactive Coercion Between Urbanization and Eco-environment

The interaction between urbanization and eco-environment is extremely complex. It manifests as three kinds of interact, interactive coercion, interactive promotion and coupling symbiosis (Fang 2004). In general, the interactive coupling and coercing system of urbanization and eco-environment should follow six basic laws, i.e., the coupling fission law, the dynamic hierarchy law, the stochastic fluctuation law, the nonlinear synergetic law, the threshold value law, and the forewarning law (Fang et al. 2006). The six basic laws have important theoretical guiding significance to reveal the interactive coercing and coupling relationship between urbanization and eco-environment.

The interactive coupling and coercing relationship between urbanization and eco-environment may be various in different areas with different eco-environmental backgrounds or urbanization modes. In general, the evolutive types of the interactive coercion between urbanization and eco-environment can be classified into nine basic coupling types, i.e., the rudimentary coordinating type, the ecological dominated type, the synchro coordinating type, the urbanization lagging type, the stepwise break-in type, the urbanization exorbitant type, the fragile ecological type, the rudimentary break-in type, and the unsustainable type (Qiao et al. 2006).

### 3.2.2 The Double-Exponential Function and Curve for the Evolutive Track of the Interaction Between Urbanization and Eco-environment

Based on the mathematic model on urbanization, economic development and eco-environmental change, it can be deduced that the coupling relationship between urbanization and eco-environment is a double-exponential function (Huang and Fang 2003) as

$$z = m - n[10^{\frac{y-b}{a}} - p]^2$$

where $z$ is the degree of eco-environmental deterioration, $y$ is the urbanization level, and $m, n, a, b, p$ are undetermined parameters.

When $y < a\lg P + b$, the eco-environment will deteriorate gradually with the increase of the urbanization level.

When $y = a\lg P + b$, the degree of eco-environmental deterioration will reach the maximum value $m$.

When $y > a\lg P + b$, the eco-environment will be improved gradually with the increase of the urbanization level.

The double-exponential function relationship indicates that as urbanization progresses the ecological pressure will raise at first and then decline. In other words, eco-environmental deterioration and pollution may appear due to urbanization at first. However, such phenomena will diminish predictably with further urbanization. This is a common rule and a general tendency to the interaction between urbanization and eco-environment.

First, a logarithmic curve and an environmental Kuznets curve (EKC) are put into the first and third quadrants of the same coordinate system, respectively. The former represents urbanization changing with economic development, while the latter represents eco-environment changing with economic development. Next, horizontal and vertical lines are drawn out from the two curves in the first and third quadrants to the second quadrant. Finally, a curve in the second quadrant is formed, which describes the coupling and coercing relationship between urbanization and eco-environment. This curve is divided into two parts by the middle inflection point. The two parts are both exponential curves. The front part (below the middle inflection point) is a monotonically increasing curve. The latter is a monotonically decreasing curve. Prior to the inflection point, the degree of eco-environmental deterioration increases with the increase of the urbanization level. After the inflection point, the degree of eco-environmental deterioration decreases. Combined with a mathematical model, it may be determined that the urbanization level at the inflection point is $y = a\lg P + b$.

### 3.2.3 Basic Stages of the Interactive Coercion Between Urbanization and Eco-environment

According to the development stages and the dynamic mechanisms of urbanization, as well as changes of the interactive coercing relationship between urbanization and eco-environment, the interactive coercing process of urbanization and eco-environment can be divided into five basic stages, namely low-grade coordinating stage, antagonistic stage, break-in stage, ameliorative stage and high-grade coordinating stage (Huang and Fang 2003; Qiao and Fang 2005; Qiao et al. 2005). When the urbanization process steps into the mature stage (urbanization level is higher than 70 %), the destroyed eco-environment will be recovered step by step, and the eco-pressure will decrease to the minimum and remain unchanged. The conflict between urbanization and eco-environment will be eliminated. The urbanization system and the eco-environmental system will coordinate with each other.

## 3.3 *Conclusions and Discussion*

(i) An extremely complex, interactive coupling, and coercive relationship exist between urbanization and eco-environment. This kind of relationship can be described by a double-exponential function or curve deduced by a power function and an exponential function. It can be divided into five basic stages, namely low-grade coordinating stage, antagonistic stage, break-in stage, ameliorative stage, and high-grade coordinating stage.

(ii) The urbanization process mainly follows the human law while the eco-environmental change mainly follows the natural law. Their relationship is extremely complex and nonlinear. Restricted by some conditions, we tentatively summarize the basic theory and method on the interactive coupling and coercing relationship between urbanization and eco-environment. However, how can we find the middle inflection point for the double-exponential curve of interactive coercing relationship between urbanization and eco-environment? How can we make the urbanization circle (or system) and the eco-environmental circle (or system) maintain the best distance and keep a dynamic balance (Wang 1993)? It is a hard question that we are querying all the time. It is also an important theoretical orientation for our further study in the future.

(iii) From the view of national strategic demands, how to practice the moderate urbanization mode, the eco-urbanization mode, the environment-friendly urbanization mode, the resource- saving mode, and the healthy urbanization mode, how to construct eco-cities, resource-saving cities, environment-friendly cities, and healthy cities (Xiao et al. 2002), how to step into a healthy urbanization road adapting to the eco-environmental capacity, are all important practical orientations for our further study in the future.

# 4 Theoretical Researches and Practices of Ecological Compensation in China

## *4.1 Theoretical Researches of Ecological Compensation in China*

Looking back to the history of ecological compensation researches in China, the features are distinctive by stages, coincided with specific historic and practical needs. In China, ecological compensation researches have been dated back to the 1980s' discussions of ecological compensation in ecological senses and exploration of ecological compensation in economical senses (Zhang 1987). After the United Nation Conference on the Human Environment, ecological compensation researches have entered into a phase of active theoretical discussions on the basis of "damager pays" principle. With the practices of ecological protection programs and the contradictions between ecological protection and economic development, ecological compensation researches have paced into a phase of theoretical and practical discussions expanding to "beneficiary pays" principle. Connotation, standard, and theoretical basis are the key scientific issues in ecological compensation researches. Connotations of ecological compensation have goes from ecological senses through economical senses to multidisciplinary senses. It is critical to clarify relationships among the ecological compensation fee, environmental and resource fee for its precise positioning in Chinese policy. Ecosystem service and externality theory is the main theoretical basis for ecological compensation. While there is disparity in benefit compensation or value compensation for ecosystem services, it is more at agreement for externality compensation of environmental economics. The standard of ecological compensation is mainly determined by monetary value of ecosystem services, cost or compromise based on analysis of supply and demand. These scientific problems are decisive for the effectiveness and efficiency to put ecological compensation into practice. So, it is in urgent need to do some researches from the angle of theoretical renovations and practical need of different regions in China.

## 4.2 *Practices in Ecological Compensation in China*

The practices of ecological compensation in China are dated back to the 1970s in the twentieth century, when 30 % ticket sales of Qingcheng Mountain were used for the conservation of forests in Chengdu of Sichuan province (Feng et al. 2009). Then, the practices in forest and mining areas became the hot spots all over the country. In Yunnan province, 0.3 yuan/ton Ming Phosphate was used as the recovery fee of local environmental damages from the year 1983. The State Council approved the energy base in Inner Mongolia Baotou and contiguous areas to levy eco-environmental fee. Then, 145 counties of 14 provinces accelerate the pilot practices. Before the year 1998, the practices of ecological compensation have focused on the environmental damage pay. Afterwards, the ecological protection and construction are given more priority due to the flood in Yangtze river, Songhua River, and Nen River in 1998, and dust storm in 2001. The central government launched six ecological projects, including natural forest protection project, three-north shelterbelt protection project, Yangtze river shelterbelt protection project, reforestation project, reverting farmland to forest and grassland project, antidesertification projects in Beijing area, wildlife conservation project, natural reserve construction project, and key areas of fast growing timber base construction project. In 1998, China's new forest law defines the legal regime of forest ecological benefit compensation fund.

Entering the twenty-first century, the exchanges and cooperation in the international area are strengthened. Governments at all levels, especially development and reform commission and environmental protection system are the key forces to promote communications internationally and summarize international experience, including "Proposal for China to promote ecological compensation practices and international cooperation."

The central government put great efforts to promote the practice of ecological compensation. Programmatic document clearly put forward the ecological compensation requirements such as the report at 17th/18th Party Congress. Ecological compensation has been the important content in environmental protection, accelerating pilot ecological compensation. The legal and policy support of ecological compensation has also been strengthened. Ecological Compensation Ordinance has been included in the legislation plan from 2010.

**Table 4** Overall frameworks of ecological compensation in China

| | Types of compensation | Contents of compensation | Ways of compensation |
|---|---|---|---|
| International compensation | Worldwide/regional/international ecological and environmental problems | Worldwide forest and biodiversity protection, pollutions transferring, greenhouse gas emission, transboundary rivers, etc | Global purchase under multilateral agreements |
| | Ecosystem compensation | Forest, grassland, wetland, ocean, farmland, etc., ecosystem services compensation | State (public)compensation by financial transfer payment, ecological compensation fund, market transaction, and participation of businesses and individuals |
| National compensation | Watershed compensation | Watershed Compensation across provincial borders, Watershed compensation in local administrative districts | Financial transfer payment, local government coordination, market transaction |
| | Regional compensation | Compensation for the eastern region of the western region | Financial transfer payment, local government coordination, market transaction |
| | Resources development compensation | Mining development, land reclamation, vegetation restoration, etc | Beneficiary payment, damage payment, development payment |

The local governments have accelerated the pilot ecological compensation, mainly in watershed areas and regional areas, including Fujian, Shandong, Zhejiang, Fujian, Henan, and Hebei provinces.

Under the active promotion of government, the market transaction of ecological compensation has important progress. By actively improving the management mechanism of market transactions, setting up carbon trading platform, promoting carbon trading pilot projects, and by other means the carbon trading market in China has been more mature. In 2003, State Forestry Administration established the office of carbon sequestration, energy offices, climate office and other agencies, and relevant policies and regulations of forest carbon management has been put forward. In 2010, China Green Carbon Foundation was incorporated, which became the first national public foundation aiming to increase carbon sequestration and countering climate change. Companies can buy carbon sequestration through donation for afforestation. In 2008, the first batch of six carbon sink afforestation projects was implemented. In 2011, China Green Carbon Foundation and the East China Forestry Exchange launched national rreen carbon trading pilot ceremony. Ten companies signed the first batch of 148,000 tons to subscribe forestry carbon sinks, the price is 18 Yuan per ton, which is China's first forestry carbon sequestration transaction under standardized operation. In 2012, Beijing, Shanghai, Chongqing, Tianjin, Guangdong, Hubei, and Shenzhen are approved to start carbon emissions trading pilot in order to gradually establish a domestic carbon emissions trading market, with a lower cost, to achieve 2020 Chinese action to control greenhouse gas emission targets (Table 4).

### 4.3 *Proposals for Improvement of the Theoretical Researches and Practices of Ecological Compensation in China*

Theoretical researches of ecological compensation in China are ecological and environmental protection-oriented to address the relationship between environmental protection and economic benefits for the target distortions. A large number of academic scholars hold for a long-time discussion on the domestic ecological compensation policy formulation and improvement of laws and regulations, which have played an important role in guiding. But theory and practice still exists disjointed. Theory is behind the practice and exploration, Ecological protection in practice, still is in the absence of structural policies, especially lack of specific ecological compensation policies and implementation guidelines. Not only the ecological protection and construction advancing toward a higher level face great difficulties, but also affect the harmony between regions and between stakeholders.

So, we propose to improve the following aspects:

(i) Unifying ecological compensation connotation and gradually improve the overall framework of ecological compensation;
(ii) Conducting partition guidance and classification implementation;
(iii) Establishing sound fiscal policies of ecological compensation system, and actively exploring financing mechanisms and multi-channel;
(iv) Handling the relationship of ecological compensation policies, including the relationship between central and local governments, the relationship between government and the market, the relationship between financial transferring and self-reliance, and the relationship between the new account and the old account;
(v) Creating ecological compensation legal environment and improving the management mechanism;
(vi) Strengthening scientific research and pilot project of ecological compensation.

## 5 River Basin Eco-compensation Progress in China

China's river basin eco-compensation has strong policy basis and rich practice progress. In this chapter, it sorts China's eco-compensation policies, laws, and regulations, and summarizes China's exploitation of typical cases and main models in river basin eco-compensation, which addressed current status of China's river basin eco-compensation in aspects of policy and regulation basis and practice basis.

### *5.1 Introduction*

Since the end of 1990s, the Chinese government began to pay close attention to research on eco-compensation in river basin; with the twenty-first century coming, the quantitative research on eco-compensation in river basin reached a peak. Now the pattern has been formed with a variety of compensation types and modes coexist after over 10 years of evolution. The eco-compensation has been really focused by the whole society of China after the "11th Five-Year Plan"; the NPC and the CPPCC members put forward many suggestions and bills every year during "two sessions", the reports on government work also highly valued it. Many local governments have started the positive exploration on eco-compensation mechanism, and carried out the pilots for river basin eco-compensation mechanism; some provinces have established the river basin eco-compensation mechanism throughout the province. The river basin eco-compensation pilots in China have initially formed a mode suitable for the regional characteristics.

## 5.2 Chinese River Basin Eco-compensation Progress

### 5.2.1 Basis of Policies and Regulations

The Related State Policies and Regulations

With worsening river basin environmental pollution and aggravating environmental management since the 1970–1980s, Chinese Government and relevant ministries have successively issued a series of laws and regulations and policy documents, which called for strengthening the river basin environmental protection and increase the related inputs. In recent years, more laws and regulations and policy documents clearly put forward the eco-compensation mechanism, especially since *Decision of the State Council on Implementing the Scientific Outlook on Development and Strengthen Environmental Protection* (State [2005] No. 39) was issued in Dec 2005; the Party Central Committee and the State Council clearly required to establish eco-compensation mechanism in the relevant files for many times, the Ministry of Environmental Protection, the Ministry of Finance (MOF), the National Development and Reform Commission, Ministry of Water Resources, etc., are also actively preparing for study and formulation of eco-compensation policy, and carry out the eco-compensation pilots in river basin.

It is stipulated in the *Water Law of the People's Republic of China,* effective since Oct 2002 that "if the livelihood and production of other units or individuals are caused losses as groundwater declined, exhausted or ground depressed and so on due to the exploited mine or underground engineer, the mining or construction entity shall take remedial measures and compensate for such losses". This kind of loss compensation actually reflects the implications of eco-compensation. In the field of eco-compensation in river basin, Chinese government has carried out some programs, such as the natural forest protection project and returning farmland to forest project, as well as the ecological protection project in three-river source, etc., since the end of last century. These policies implemented at national level, reflect the national emphasis on eco-compensation in the upstream to a certain extent.

In March 2004, the *Several Opinions of the State Council on Further Promoting Development of the Western Region* (State [2004] No. 6) was issued to specify: "to establish the compensation mechanism for ecological construction and environmental protection, and encourage all kinds of investors to participate in the ecological construction and environmental protection."

In April 2005, the *Key Works of the State Council 2005* was released (State [2005] No. 8) to stipulate: "to strengthen management of the mineral resources exploitation, rectify and standardize the order of mineral resources exploitation, improve the compensation mechanism for resources development and utilization and ecological environment restoration."

In June 2005, the *Several Opinions of the State Council on Promoting Sound Development of the Coal Industry* (State [2005] No. 18) was issued to stipulate: "to strengthen the ecological environment and water resources protection, management of wastes in the mining areas and coal-mining subsidence area according to the principle of 'developer protects, destructor recovers, beneficiary compensates, and polluter pays'; research to establish the compensation mechanism for ecological environment restoration in the mining areas; define management responsibilities of the enterprise and government; increase inputs for the ecological environment control, and gradually make the mining area environment control step into a virtuous cycle. For the historic issues during the environmental management such as mining subsidence caused by original key state-owned coal mines, the special planning shall be prepared to continuously implement the comprehensive management, the central government shall give necessary funds and policy supports, and the local governments at various levels and the coal enterprises shall arrange the supporting fund according to the regulations."

It is provided in the circular of the State Council on *Accomplishing Key Works During Construction of Conservation-Minded Society* (State [2005] No. 21) issued in June 2005: "to research how to establish and improve the resources exploitation and eco-compensation mechanism on the basis of straightening out the existing channels for charges and fund sources."

In July 2005, the Several Opinions of the State Council on *Accelerating the Circular Economy Development* (State [2005] No. 22) was issued to specify: "to positively research how to establish and improve the eco-compensation mechanism for restoration of enterprise ecological environment on the basis of straightening out the existing channels for charges and fund sources."

In December 2005, the Decision of the State Council on *Implementing the Scientific Outlook on Development and Strengthen Environmental Protection* (State [2005] No. 39) was issued to provide: "to improve the eco-compensation policies, and establish eco-compensation mechanism as soon as possible. The eco-compensation factors shall be considered in the central and local fiscal transfer payment, the State and local governments can implement some eco-compensation pilots."

In December 2005, the Several Opinions of the CPC Central Committee and the State Council on *Pushing Forward New Socialist Rural Construction* (Central [2006] No. 1) was issued to specify: "to continuously promote the ecological construction; earnestly implement key ecological projects, such as the returning farmland to forest and natural forest protection; stably improve the policies; cultivate the follow-up industries; consolidate the ecological construction results according to the requirements for construction of environment-friendly society; continuously push forward the return the grazing land to grassland and comprehensive development of mountainous area; establish and improve the eco-compensation mechanism; carry out major pest control, take effective measures to prevent the exotic biological invasion; strengthen the desertification control, and actively implement the comprehensive control project for soil and water erosion for the rocky desertification area and northeast black earth area; establish and perfect the responsibility mechanism of hydro-electric power and mining enterprises for the environmental restoration and control, extract certain funds from the revenue of hydro-electric power and mineral resources development for the restoration and management of local environment, and prevent water losses and soil erosion."

In March 2006, the *Key Works of the State Council 2006* was issued (State [2006] No. 12) to stipulate that: "the eco-compensation mechanism shall be urgently established."

In March 2006, the *Outline of 11th Five-Year Plan for the National Economic and Social Development* was issued by the National People's Congress to stipulate that: "the eco-compensation mechanism shall be established according to the principle of developer protects and beneficiary compensates."

On April 17, 2006, Premier Wen stressed at the 6th *National Environmental Protection Conference* that "the eco-compensation mechanism shall be established according to the principle of 'developer protects, destructor recovers, beneficiary compensates, and polluter pays'."

In October 2006, the Decision of the Central Committee of the CCP on Major Issues about *Constructing Harmonious Socialistic Society* (Central [2006] No. 19) was issued by the 16th CPC Central Committee at the Six Plenary Session, which provided: "to improve the industrial policy, finance and tax policy, and pricing policy in favor of environmental protection, establish an evaluation system and compensation mechanism for ecological environment, and strengthen the responsibilities of enterprise and the whole society to save resources and protect the environment."

In December 2006, the Several Opinions of the CPC Central Committee and the State Council on *Positively Developing Modern Agriculture and Steadily Pushing Forward New Socialist Rural Construction* (Central [2007] No. 1) was issued to specify: "to continuously push forward the major ecological projects, such as natural forest protection and returning farmland to forest, further improve the relevant policies and consolidate the achievements; start the comprehensive treatment engineering for rocky deserts, continuously implement the coastal shelter forest project; improve the forest eco-compensation fund system; explore to establishment of grassland eco-compensation mechanism; and speed up some projects of returning the grazing land to grassland."

In March 2007, the *Key Works of the State Council 2007* was released (State [2007] No. 8) to stipulate: "to accelerate establishment of eco-compensation mechanism."

In May 2007, the *Work Scheme on Energy Conservation and Emission Reduction* (State [2007] No. 15) was issued by the State Council to specify: "to improve the paid-use system for mineral resources, improve and perfect the eco-compensation mechanism for resources development; and conduct the pilot for trans-basin eco-compensation."

In July 2007, Opinions of the State Council on *Compiling the Planning for National Main Function Regions* (State [2007] No. 21) was released to specify that: "in order to realize the equal access to basic public service, the central and provincial financial transfer payment system shall be improved through mainly increasing the financial transfer payment for public service and eco-compensation of the development-limited and development-prohibited regions; gradually implement the investment policy according to major function regions and the relevant field, the governmental investment shall mainly support the public service facilities construction, ecological construction and environmental protection in the development-limited and development-prohibited regions; and support the infrastructure construction in the key development areas."

On October 15, 2007, General Secretary, Hu Jintao's report at the 17th National Congress definitely required: "to implement the fiscal and taxation systems conducive to the scientific development, establish and improve the paid-use systems of resources and eco-compensation mechanism."

In November 2007, the *11th Five-Year Plan for National Environmental Protection* was promulgated (State [2007] No. 37) to provide: "to implement the target responsibility system for river basin management and assessment system for water quality at the trans-provincial monitoring section; speed up establishment of eco-compensation mechanism; increase input from various channels; accelerate construction of regulation project; plan development, utilization and protection of river water resources as a whole, distribute water for life, production and ecology as a whole, so as to ensure the ecological runoff of rivers"; "define the range, leading function and development direction of key ecological function reserves, and explore establishing the evaluation index system, management mechanism, performance evaluation mechanism and eco-compensation mechanism for ecological function reserves according to the requirements for development-limited areas"; "accelerate establishment of the earnest money system for mine environment to restore; push forward mine environment management, and promote the ecological restoration for new and old mines and resource-exhausted cities"; "give priority to establish and implement the eco-compensation mechanism in the western regions"; "according to the principle of 'developer protects, destructor recovers, beneficiary compensates, and polluter pays', Three Gorges Reservoir Area, the catchments area for The South-to-North Water Diversion Project, key energy development zones and national nature reserves shall be taken as a breakthrough to expand the pilot, improve the eco-compensation policies and establish the eco-compensation mechanism"; "the finance and tax departments shall formulate the finance and tax policy favorable to environment protection, establish and improve the eco-compensation mechanism, support construction of early warning system for environmental monitoring and supervision system for environmental enforcement."

In December 2007, *Several Opinions of the State Council on Promoting the Sustainable Development of Resource-Based Cities* (Central [2007] No. 38) was issued to specify: "to improve the pricing mechanism for resource products; speed up reform of resource prices, and gradually form a pricing mechanism of resource products which can reflect the scarcity degree of resources, relation between market supply and demand, as well as cost of environmental and ecological restoration; scientifically formulate the financial accounting method for cost of resource products, the resource product cost shall be listed with such expenditures as acquisition of mining right, resources exploitation, environment management, ecological restoration, safety facilities, infrastructure construction and enterprise transformation; improve the eco-compensation system for forest benefit, prevent exteriorization of the internal cost of enterprise and socialization of private cost"; "combine with establishing the pilot of margin system for mining environmental restoration, research to establish the reserve fund system for sustainable development, the resources enterprises shall be extracted a certain proportion from the pre-tax income as the sustainable development reserve fund which is specially used for environmental restoration and eco-compensation to develop the substitution industry and solve the historic problems and rehabilitative work after the enterprises are closed, etc., the local people's governments at all levels shall strengthen supervision of the reserve fund in accordance with the principle of owned by enterprise, special fund for special purposes, deposit in the special account and regulated by the government".

In March 2008, the *Key Works of the State Council 2008* was released (State [2008] No. 15), which stipulated: "to reform resources tax and fee system, and improve the paid-use system of resources and eco-compensation mechanism."

As a milestone of eco-compensation in river basin, *Law of the People's Republic of China on Prevention and Control of Water Pollution* was amended and passed in 2008 and was implemented on Jun 1, 2008, which first put forward eco-compensation for water environmental protection in the national laws. Article 7 is stipulated that: "the State will establish and improve the eco-compensation mechanism of water environment protection for drinking water source reserves, rivers, lakes and the upstream of reservoirs through the financial transfer payment, etc." Article 4 is provided that: "The governments at or above county level shall incorporated water environment protection in the national plan for economic and social development." "The local governments at or above county level shall take countermeasures or measures to prevent and control water pollution, and be responsible for water environmental quality in their respective administrative regions," and Article 5 is stipulated that: "the State will implement the target responsibility system and evaluation system for water environmental protection, the completion situations of water environmental protection goals shall be brought as a content to evaluate the local governments and the responsible persons."

On July 22, 2008, circular of the State Council *Forwarding Opinions of National Development and Reform Commission on Deepening Reform of Economic System* (General Office [2008] No. 103) was promulgated to stipulate that: "the paid-use system of resources and eco-compensation mechanism is one of the three major mechanisms of resource conservation and environmental protection." The circular requires the MOF, MEP, and NDRC which lead to push forward the pilot work to establish the trans-provincial basin eco-compensation mechanism.

**2009 Governmental Work Report** pointed that: "to push forward the pricing reform of resource products; continuously deepen the reform of electricity price; gradually improve the pricing mechanism for feed-in electricity price, electricity price of transmission and distribution and sale, timely rationalize the pricing relation between coal and electricity; actively promote water price reform; and gradually increase water price of water conservancy projects for nonagricultural purposes; improve the management and collection system of water resources fee; accelerate establishment and improvement of the paid-use system of mineral resources and eco-compensation mechanism; and actively carry out pilot for emission trading."

Since 2005, the MEP, MOF, NDRC, and MWR have actively prepared to make eco-compensation policy and implement pilot work for eco-compensation in river basin.

In 2007, the former SEPA (State Environmental Protection Administration) promulgated the *Guidance on Implementation of Eco-Compensation Pilot* (MEP [2007] No. 130) to require the local governments to gradually establish the eco-compensation mechanism in four fields including the nature reserves, the major ecological function areas, exploitation of mineral resources, and water environmental protection; among of them, the basin eco-compensation is one of the key fields of the ecological environment compensation.

In 2008, MEP issued the *Guidance of Ministry of Environmental Protection on Prevention and Handling Trans-Provincial Water Pollution Disputes* (MEP [2008] No. 64) to require that the trans-provincial water pollution disputes shall be prevented from the beginning, and the long-term work mechanism shall be established to prevent and handle the trans-provincial water pollution disputes.

In May 2008, the MEP approved Min River Basin in Fujian province as the first batch of eco-compensation pilot areas.

Since 2008, the MOF has successively issued several transfer payment policies related to eco-compensation, mainly including the *Circular of Ministry of Finance on Issuing 2008 Transfer Payment Fund for Ecological Reserves*, such as three-river source (Financial Budget [2008] No. 495) and the *Measures for Transfer Payments to National Key Ecological Function Areas (Pilot)*, etc.; among them, the former points out that: "according to the current general transfer payment method, now the central finance hereby increases your province (autonomous region, municipality directly under the central government) the local general transfer payments in 2008 through increasing the subsidy coefficient for some counties, etc., all of these funds must be used for natural forest protection project, Qinghai Three-River Source, and Middle route Project of South-To-North Water Diversion, Danjiangkou Reservoir Area and the upstream Counties and Districts."

Since April 2010, the Chinese government launched the legislation work on eco-compensation, the *Regulations on Eco-Compensation* (hereinafter referred to as "Regulations"), which is intensively drafted with leaded by the NDRC at present; Cooperate with drafting of the regulations, NDRC also organized to prepare the *Several Opinions on Establishing and Improving the Eco-Compensation Mechanism* as the prelude of the regulations; The basin eco-compensation is the key field of the regulations, the related research and investigation and subsequent file organization are conducted by the "river basin team" of MEP before drafting the regulations.

By the end of 2010, the MOF and MEP totally allocated RMB 50 million as startup capital to Anhui province for Xin'an River as the first national trans-provincial water environment compensation pilot, which is of great significance to the cross-boundary basin eco-compensation in China.

In 2011, *Outline of 12th Five-Year Plan for National Economic and Social Development* required: "to accelerate establishment of eco-compensation mechanism according to the principle of 'developer protects, and beneficiary compensates'; strengthen the balance transfer payment for key ecological functional areas, research to establish the national special eco-compensation funds; implement the reserve fund system for sustainable development of the resource-based enterprises; encourage, guide and explore implementation of eco-compensation from the downstream to upstream, from development area to the protected area and from the ecological beneficiary area to the ecological protection area; actively explore the market-based eco-compensation mechanism; and speed up implementation of the Regulations."

In 2011, the *12th Five-Year Plan for National Environmental Protection* specified that: "the Central Finance will increase support to the western region, development-prohibited area and development-limited area, and special poverty areas, and improve the provision level for basic public services for environmental protection through general transfer payments and eco-compensation measures."

In 2011, *Opinions of the State Council on Strengthening Key Works during Environmental Protection* (State [2011] No. 35) explicitly provided to speed up establishment of eco-compensation mechanism and national special eco-compensation fund, and expand scope of eco-compensation.

In December 2011, the *7th National Environmental Protection Conference* proposed that: "it shall be adhere to integrate the development with protection, and actively explore new route for the sustainable environmental protection with lower cost, better benefit and less emission."

#### Related Regulation Documents at Local Level

On local level, the local legislation first defined that the river basin eco-compensation was an eco-compensation measure on rivers within the boundary of Changsha City, Hunan province (trial implementation), issued in February 2012. The measures have defined "river eco-compensation" as: "a public system, aimed at protecting eco-environment, promoting the harmonious development between man and nature, make an overall use of economic means and adjust the economic benefit relation between upstream and downstream of river basin and among water ecological protectors, beneficiaries and destroyers." Some departments, on provincial and municipal level have also successively launched the study and practice of trans-provincial or cross-boundary river basin eco-compensation, effectively pushing forward the establishment of two-way or one-way responsibility mechanism concerning compensation from river basin downstream to water resource and water environment protection of upstream and compensation from upstream to downstream for pollution discharge beyond the standard or environmental liability accident.

Amid the establishment of river basin eco-compensation mechanism, the launch of eco-compensation work has been promoted through river basin eco-compensation

**Table 5** Summary of agreement on river basin eco-compensation at local level

| Year | Document | Content |
|---|---|---|
| 2007 | The guideline on launching eco-compensation pilot work | Support upstream and downstream areas of river basin to reach environmental cooperation agreement based on water quantity distribution and water quality control |
| 2007 | Management measures on special fund for water environment protection of min river basin and Jiulong River Basin | "Encourage the cities (districts) in upstream and downstream areas of river basin to protect basin water environment, improve water quality through consultation, signature of agreement and other approaches, based on the assessment requirement of meeting the ecological water quantity demand, specify the compensation liabilities and control tasks of both parties, ensure the capital to yield return, gradually establish and improve eco-compensation mechanism, give reward and punishment to the cities and counties in upstream and downstream of "two river" basins according to water environment quality status of "two river" basins and regional border sections" |
| 2011 | Framework agreement on city alliance for Wei river basin environmental protection | The people's government of two cities and one district in Shaanxi and Gansu province established a basin eco-compensation mechanism, built water quality monitoring network at provincial and municipal boundaries, set leaving-the-boundary water quality objective for trans-provincial or tans-municipal boundaries, make assessment and give compensation according to water quality objective. The interim assessment factors for sections leaving the boundary include chemical oxygen demand (COD) and ammonia nitrogen. The leaving-the-boundary water quality assessment is based on monitoring result jointly recognized by Shaanxi Provincial Department of Environmental Protection and Gansu Provincial Department of Environmental Protection. Eco-compensation fund is specially used for pollution control projects, water source ecological construction projects and water quality monitoring capacity improvement projects in Wei River Basin and is not |

(continued)

**Table 5** (continued)

| Year | Document | Content |
|---|---|---|
| | | allowed to be used in balancing the financial power |
| 2011 | Management regulation on Taihu Lake Basin | Where the upstream areas have not finished key water pollutant discharge, total amount reduction and control plan and the administrative region border section water quality fails to meet the stage water quality objective, they shall give compensation to the downstream areas; where the upstream areas have finished key water pollutant discharge total amount reduction and control plan and the administrative region border section water quality meets the stage water quality objective, the downstream areas shall give compensation to the upstream areas. The compensation will be paid through financial transfer payment mode or other modes agreed by local governments through consultation. The specific approaches shall be formulated by competent departments in charge of finance and environmental protection under the State Council together with people's governments of two provinces and one city<br>The issue of the document is an important milestone of basin legislation, which will inevitably produce tremendous promotion effect and far-reaching historical influence |
| 2012 | Protection regulation on Xiangjiang River, Hunan Province | Establish and improve a handover responsibility and compensation mechanism for administrative region border section water quality in upstream and downstream water body of Xiangjiang River Basin. Where the upstream areas have not finished key water pollutant discharge total amount reduction and control plan and the administrative region border section water quality fails to meet the stage water quality objective, they shall give compensation to the downstream areas; where the upstream areas have finished key water pollutant discharge total amount reduction and control plan and the administrative region border section water quality meets the stage water |

(continued)

**Table 5** (continued)

| Year | Document | Content |
|---|---|---|
| | | quality objective, the downstream areas shall give compensation to the upstream areas. The compensation will be paid through financial transfer payment mode or other modes agreed by local governments through consultation |
| 2012 | Regulation on water pollution prevention and control of Chaohu Lake Basin (revised draft) | Where the upstream areas have not finished key water pollutant discharge total amount reduction and control plan and the administrative region border section water quality fails to meet the stage water quality objective, they shall give compensation to the downstream areas; where the upstream areas have finished key water pollutant discharge total amount reduction and control plan and the administrative region border section water quality meets the stage water quality objective, the downstream areas shall give compensation to the upstream areas |

agreement signed between local governments. Since it does not involve any change of existing administrative management system, it is less difficult. Besides, it is subject to timely adjustment according to implementation result, therefore, it has gained the support of SEPA, especially water environment eco-compensation pilot for Xin'an River basin in 2011, which has offered a good pilot and exploration specimen for award and punishment to water environment in Chinese trans-provincial major river basins and building a sharing protective mechanism (Table 5).

To sum up, two-way type eco-compensation of upstream and downstream based on river basin cross-boundary section water quality assessment, generally practiced in various regions at present is actually a comprehensive system integrating eco-compensation and pollution indemnity, which can be called as a two-way mechanism of compensation by polluters and indemnification to beneficiaries or penalty for water quality beyond the standard and reward compensation for water quality up to the standard. Under this system, downstream will give compensation to the upstream when the upstream has reached the prescribed water quality and quantity objective according to "environment liability agreement" concluded by upstream and downstream or river basin water quality objective defined in river basin environmental protection plan, where the upstream fails to meet the prescribed water quality and quantity objective, or cause water pollution accident to the downstream; the upstream shall give its compensation or indemnification to the downstream conversely. Among others, indemnification shall be compensation to downstream areas made by upstream areas for the loss caused by pollution beyond the standard. The compensation amount is related to category, concentration, and water quantity of pollutant beyond the standard and duration exceeding the standard. Compensation method by agreement shall proceed from the "fair" angle; the eco-compensation criterion shall be determined among stakeholders in a fair and reasonable way through consultation and coordination in a bid to achieve a "win-win" compensation result to both parties of interest. Therefore, it is necessary to intensify water quality liability mechanism of river basin cross-boundary sections, highlight "common but distinguishing responsibilities" assumed by river basin governments to water quality in outbound sections. As a matter of fact, there are a lot of highlights with reform of basic level, deserving our attention. Eastern region has especially condition and liability to implement and test in reform. Regulation on eco-compensation in Yangtze River delta is also being drafted.

### 5.2.2 Practice Progress

Cases of China's Eco-compensation Practices in River Basins

Presently, many local governments have begun to boldly explore river basin eco-compensation mechanism which has been already established province-wide in some provinces. So far, nearly 20 provinces have issued and implemented some river basin eco-compensation policies, of which contents cover its principles, objectives, criterion, organization, and implementation. Such policies have made certain effect in the relevant river basins, promoting management and water quality improvement in such river basins. The eco-compensation in China has initially formed its mode suitable for local characteristics, which can be divided into two types, namely economic compensation for water source protection and cross-boundary eco-compensation and pollution indemnity.

(I) **Economic Compensation Mode for Water Source Protection**
The economic compensation for water source protection is a kind of institutional arrangement with priority of economic incentives to encourage ecological protection and construction of water source, to curb ecological destruction, to adjust the distribution relations of ecological and economic benefits among the stakeholders, and to promote the regional fairness and coordinated development, so as to improve the ecological environment, sustain the ecosystem balance, and maintain ecosystem services of water source. In most cases, the economic compensation for drinking water source is a kind of compensation from the developed regions to the underdeveloped regions, in order to compensate them giving up economic development due to protection of the ecological environment. In practice, China's eco-compensation in river basin is mainly realized by means of special funds, development relocation, and the water rights transaction.

- Special fund for eco-compensation mechanism in the Min River Basin in Fujian province
  The Min River is the largest river in Fujian province with an annual runoff of 62.1 billion $m^3$. It mainly flows through Fuzhou, Nanping and Sanming, with a drainage area about half of the total area of Fujian province. As the mother river of Fujian province, Fujian provincial government set up a special fund for eco-compensation in the Min River basin. From 2005 to 2010, Fuzhou city government from the downstream portion, annually provided RMB 10 million to the upstream Sanming city and Nanping city (RMB 5 million each); and meanwhile, each of Sanming city and Nanping city annually invested RMB 5 million to the Min River basin water treatment. In addition, the provincial environmental protection bureau arranged RMB 15 million to support the special fund for eco-compensation. At the same time, the Provincial Department of Finance and Department of Environmental Protection have formulated *Administration Measures of Special Fund Water Environmental Protection of Min River Basin* to regulate forms and content of eco-compensation.
  The special fund is mainly used in implementation of the projects listed in the *Plan of Water Environmental protection in Min River Basin* approved by the Provincial Finance Department under jurisdiction of Sanming and Nanping as well as the annual treatment program. Focus is on pollution control for livestock and poultry breeding, township waste disposal, water source protection and rural non-point source pollution control, industrial pollution prevention and control, and construction of online monitoring facilities of the pollution sources.
- Special fund for eco-compensation practice in five river source areas in Jiangxi province
  As Jiangxi province is the source area of the Dongjiang River, Jiangxi Provincial Finance Department set up a special fund to establish an eco-compensation mechanism that began in 2008, aimed at rewarding the local counties and cities for their ecological environmental protection of the source areas of "five rivers and one lake" and the Dongjiang River source area with compensation range covering 40 towns. The special fund has been annually increased from RMB 50 million in 2008 to RMB 80 million in 2009 and RMB 103 million in 2010. The reward amount consists of two parts, the first is determined on the basis of area of each protection zone in the source area, accounting for 30 % of the total reward amount; the second part is determined according to effluent water quality of each protection area, accounting for the remaining 70 %.
  The special funds are arranged by each county's financial department as a whole, mainly used for expenditure related to pollution control and ecological protection. The application of funds is subject to random supervision and management by the provincial department of finance and environmental protection bureau. The provincial department of finance and

environmental protection bureau also jointly issued the *Administration Measures of Jiangxi Province for Ecological Environmental protection Reward Fund of "Five Rivers" and Dongjiang River Source Areas* for standard use and management of the funds.

- Mode of water right transaction
  Water right transaction refers to situations when the upstream adopted a series of water-saving measures to make the water amount leaving the border more than the target limit, then the downstream should pay to the upstream for utilization right of this part of water resources, even that part of which the initial water right does not completely use. Water right transaction is essentially to minimize the marginal cost of water resources utilization or maximize value-added benefits with exchange.
  The first water right transaction was initiated in the Jinhuajiang River basin, Zhejiang province. On November 24, 2001, an agreement between Yiwu city and Dongyang city in the midstream of the river basin was initially signed. Dongyang city transferred permanent water rights of an annual 49.999 million $m^3$ in the Hengjin Reservoir to Yiwu city at a price of RMB $4/m^2$ in a one-off deal with the current state of Grade-I drinking water quality guaranteed. Similar cases of water resources trading also exist between Ningxia Hui Autonomous Region and Inner Mongolia Autonomous Region, namely the upstream irrigated areas through water-saving transformation sell the redundant water to the downstream hydropower station.
- Mode of development relocation
  To avoid pollution caused by the industrial development in the upstream and compensate the loss due to limits to development rights, the industrial zone is established at the downstream city, and income tax belongs to the upstream city, which is a specific mode of development relocation.
  A case of development relocation mode was carried out in Jinhua city, Zhejiang province. Pan'an County, under jurisdiction of Jinhua city, is located in the source area of the Jinhua River, and it is an important ecological function area with a backward economy. To support economic development in Pan'an County, the "Jin-Pan Poverty Alleviation Economic and Technology Development Zone" for Pan'an County was set up within the Jinhua Industrial Park at the downstream of the Jinhua River with support in aspects of policy and infrastructure, while all income tax from the development zone belongs to Pan'an County.

(II) **Mode of Cross-Boundary Eco-compensation and Pollution Indemnity**
The cross-boundary eco-compensation and pollution indemnity means, when water resource use or pollution in river basin can be controlled within the corresponding total amount or the assessment criterion of cross-boundary section, if there is not sufficient water yield and environmental capacity to be occupied by other regions, and then the positive externality is produced, the downstream shall compensate the upstream investment for their ecological and

environmental protection costs input to provide water ecosystem services higher than the benchmark. On the contrary, if the negative externality is produced, the upstream shall undertake the downstream additional treatment cost due to pollution surpassing the standard and compensate the downstream corresponding damage, namely certain economic compensation shall be paid to the downstream.

In China, the specific practices of river basin eco-compensation and pollution indemnity is to monitor the administrative cross-boundary section water quality of a river basin, if water quality provided by the upstream meets the target requirements, the downstream areas must provide eco-compensation to the upstream; if not, the upstream must provide pollution indemnity to the downstream.

- Trans-provincial water environment compensation in the Xin'an River Basin

  The Xin'an River originates in Huangshan city, Anhui province and flows through Anhui and Zhejiang provinces with a total area of 11,674 $km^2$, 53.6 % of its total area is located in Anhui province, only behind that of Yangtze River and Huai River, ranking as the third largest water system in Anhui province. Xin'an River is also an important source of the Qiantang River, the biggest entry river of Zhejiang province. In December 2010, the Xin'an River basin was officially inaugurated for water environment compensation as the first trans-provincial pilot area in China. The MOF allocated RMB 50 million as the initial capital for pilot eco-compensation in the Xin'an River in Huangshan city, Anhui province. In October 2011, MOF and MEP jointly issued the *Implementation Plan of Water Environment Compensation in Xin'an River Basin*, which specifically provided the guiding ideology, basic principles and specific measures, etc., for the implementation of eco-compensation in the Xin'an River basin. Specific measures of the eco-compensation in the Xin'an River Basin are as follow:

  The reference of eco-compensation is the water pollution composite index of the trans-provincial monitoring site—the Jiekou site. The composite index is determined on the basis of four indices: potassium permanganate index, ammonia and nitrogen, total nitrogen, and total phosphorus confirmation. The central finance arranged RMB 300 million as a compensation fund. When Anhui province supplies water with water quality better than the basic standard, it will receive compensation of RMB 100 million from Zhejiang province, and otherwise it will lose it upon poor water quality or serious water pollution accidents in Xin'an River basin within Anhui province. The special fund for compensation is mainly used in the industrial restructuring and optimization of industrial layout, integrated river basin management, water environmental protection and water pollution control, ecological protection, etc., in the Xin'an River basin. The agreement on eco-compensation was signed by both provinces under endorsement of MEP.

- Tai Lake in Jiangsu Province
  *Measures of Jiangsu Province for Regional Compensation of Environment Resources (trial)* and *Scheme of Jiangsu Province for Regional Compensation of Environment Resources in Tai Lake Basin (trial)* were issued by Jiangsu Provincial Government, respectively, in early 2008 and early 2009, to implement regional compensation system for environment resources of major rivers in the Tai Lake basin in Jiangsu province according to the principle of "who pollutes, pays; and who destroys, compensates."
  The main idea is to determine the compensation amount of which the upstream cities at prefecture level and counties (city) under its jurisdiction receive compensation from the downstream cities at prefecture level according to the formula: "compensation capital of single factor = (sectional water quality index—sectional water quality goal) × monthly sectional water amount × compensation criterion." Jiangsu Provincial Environmental Protection Bureau and Department of Finance are responsible for accounting and consolidating the compensation amount of each quarter. The Provincial Department of Finance is responsible for collecting and allocating compensation capital, which is mainly used in water pollution control for the Tai Lake basin.
  Implementation effect: among 15 major rivers flowing into the Tai Lake Basin, 12 rivers with inferior Grade-V water quality before the pilot study was conducted sharply reduced to only one river of that grade by March 2010.
- Ziya River Basin in Hebei Province
  In 2008, *Circular on Implementing the Policy of Responsibility Assessment on Cross-Border Section Water Quality Goal for Major Rivers of Ziya River Water Body and Trying out Withholding Eco-Compensation Fund* was released by Hebei Provincial Environmental Protection Bureau. Since April 2008, this policy was implemented at five cities and 57 rivers of the Ziya River basin. Water quality of the Ziya River basin improved significantly in the first year of implementation, *Circular on Implementing Responsibility Assessment on Cross-Border Section Water Quality Goal* was issued by General Office of Hebei Province Government to specify that the eco-compensation fund policy will be fully promoted in the seven top water systems within the whole province.
  Its main idea is to determine the eco-compensation amount, according to the times of COD concentration beyond the limit at cross-border sectional water quality of each city under responsibility appraisal. If the multiples of COD concentration beyond the limit are the same, in event that the entry water quality is beyond the limit and COD concentration in the cross-border sectional water continuously increased, the compensation

amount of each city under responsibility assessment will be more than that paid when the entry water quality meets the criterion (or no entry water). The Provincial Finance Department will directly withhold the compensation amount from the annual expenditure of the city according to the total withholding amount provided by the departments of environmental protection. The eco-compensation fund must be specially used in water pollution projects of the Ziya River basin.

Implementation effect: by the end of March 2009, the total eco-compensation fund withheld from five cities within the Ziya River basin reached RMB 14.3 million, and a declining trend is presented for the overall pollution level. Compared with the previous year, the average concentration of COD reduced to 42.8 % and average concentration of ammonia nitrogen 13.7 %, which achieved the best level over many years and effectively curbed sewage discharged by the upstream to the downstream. By the end of 2009, the eco-compensation fund withheld reached a total of RMB 35.7 million. Among seven top water bodies, percentage of sections with Grade-III and superior Grade-III water quality accounts for 40.1 %, increasing 9.7 % in comparison with the same period in 2008; and percentage of sections with Grade-V and inferior Grade-V water quality reduced to 9 % in comparison with the same period in 2008.

- Shayin River Basin in Henan Province

  *Interim Measures for Water Environment Eco-Compensation in Shayin River Basin* was promulgated by the former Environmental Protection Bureau and Department of Finance of Henan province at the end of 2008, and *Interim Administrative Measures of Henan Province for Reward Fund of Water Environment Eco-Compensation in Shayin River Basin* was jointly issued in February 2009. This policy has achieved good effect. *Interim Measures of Henan Province for Water Environment Eco-Compensation* was issued jointly by Henan Provincial Environmental Protection Bureau and Department of Finance in January 2010, to fully implement the surface water environment eco-compensation mechanism within whole province.

  Its main idea is to adopt a "two-way" compensation mechanism combined with "penalty for exceeding the limit" and "reward for reaching the standard". Compensation criteria are determined according to the formula: (concentration values monitored at section water quality—target concentration value for assessment sectional water quality) × weekly assessment sectional water amount × Eco-compensation criterion. The monitoring indexes are COD and ammonia nitrogen. Provincial departments with relevant capabilities for environmental protection and water administration are responsible for verifying compensation criterion, and the provincial financial department together with the competent departments of environmental protection conduct compensation and rewards. The compensation fund will be used for river basin pollution control and compensating cities which have completed the responsibility goal of water environment well under jurisdiction within the province.

Implementation effect: the compensation fund withholding within the Shayin River basin amounted to RMB 650 million in the first half of 2009, and it reduced to RMB 180 million in the second half of 2009 due to improved water environment. In 2010, the preliminary achievement was achieved after an eco-compensation system for surface water environment and has been implemented in Henan province for only 3 months. The withheld eco-compensation fund totaled RMB 37.18 million, and surface water quality in Henan province has been improved significantly.

## Analysis on China's River Basin Eco-compensation Features

Regardless of economic compensation for water source conservation or cross-boundary eco-compensation and pollution indemnity in river basins, they are essentially government-led eco-compensation mechanisms, which have been developed as an environmental economy policy according to the current condition of China. In such mechanisms, the government promotes eco-compensation; the relationship between the upper and lower levels of the government spurs the quick and comprehensive implementation of eco-compensation in river basins. Support from their rich financial resources ensures stable fund source for eco-compensation. Moreover, the government can make supportive policies according to the characteristics of the regions which need eco-compensation such as specific financial policies for major rivers, market-based compensation policies for regions with relatively high levels of economic development, and compensation policies based on technical projects or policies that encourages development relocation for small-scale river basins. Such policy-based eco-compensation provides more lasting effects. Meanwhile, much improvement should be addressed in China's eco-compensation in river basin, mainly including the following aspects:

(I) **The Understanding of Eco-compensation is not yet Comprehensive**
Eco-compensation areas are often backward and they suffer from poverty, and eco-compensation often closely combines with poverty alleviation. Yet in fact, eco-compensation cannot be equated with poverty alleviation and poverty alleviation cannot be solved just by eco-compensation. In addition, the existing research programs, especially eco-compensation research programs of water source protection, only focus on the existence and development of the upstream contemporary issues to be met, and hardly emphasize compensation research for future generations.

(II) **Determination of Eco-compensation Criterion Lacks Scientific Rigor and Comprehensiveness**
First, COD is mainly considered during determination of eco-compensation factor in cross-border section water quality without focusing on some

characteristic pollutants according to the actual situations of the river basin pollution. It is rare to bring small tributaries into appraisal range or consider the small tributaries, but the compensation criterion differs from that of the mainstream. Second, the eco-compensation criterion in river basins was determined under the government leadership, and is mainly prepared through direct discussion among the relevant departments, so there is lack of a scientific calculation method as the basis. The eco-compensation figures set were not an agreed price despite the upstream and downstream governments repeatedly "bargaining". So, such compensation criterion are unconvincing to the wider range of stakeholders.

(III) **The Follow-up Mechanism in Research on the Eco-compensation Fund is Insufficient**

The existing research mainly focuses more on how to obtain an eco-compensation fund, and less on how to distribute, utilize, and manage the compensation fund within the compensation receiver milieu, as well as the usefulness of the compensation funds, namely the compensation monitoring and evaluation mechanism. Thus, it is a less favorable setting for the eco-compensation mechanism to play its deserved role.

(IV) **The Dual Compensation Means of Government and the Market Need to be Further Rationalized and Strengthened**

Eco-compensation policy has promoted the combined utilization of administrative command and control instrument and economic means. From the eco-compensation practice in river basins, its implementation in each region is dominated by the government with deferring promotion due to awareness of the governments at all levels to this policy. The market means are also gradually diversified, such as the mode of development relocation, mode of water right transaction and mode of emission trading, etc. The "synergy effect" of government-led means and market-based instruments is gradually emerging. However, the government is still the main buyer of eco-compensation at present, which has resulted in a strong administrative component during eco-compensation and certain instrument constraints. It is easy to cause transaction costs that are too high by market means, and thus unfavorable for further development and maturation of eco-compensation market mechanisms. In addition, the compensation form of the government is relatively singular, mainly with capital compensation. The legality issue of eco-compensation paid by the finance still exists. Other means such as policy compensation, in-kind compensation, technology compensation, and supporting compensation are less frequently adopted.

(V) **Trans-provincial Eco-compensation in River Basin is Difficult to Promote**

During establishment of trans-provincial eco-compensation mechanisms in river basins, the upstream provinces usually appeal with high enthusiasm, while the downstream provinces passively avoid it. Therefore, except for trans-provincial water environment compensation in the Xin'an River basin promoted at the national level, other domestic trans-provincial eco-compensations in river basins have not started substantive work. The reason lies in undefined stakeholder's responsibilities and unclear responsibilities between the upstream and downstream for water pollution control. Water quality protection compensation criterion need to be set further. Thus, the upstream provinces often provide more compensation to the upstream region within the province, but it is difficult to negotiate eco-compensation issues between two provinces from a whole watershed without promotion at the national level. For example, there is not much progress for the compensation issue between Guangdong province at downstream Dongjiang River and the three counties under jurisdiction of Ganzhou City, namely Anyuan, Xunwu, and Dingnan at the source of the Dongjiang River. The Eco-compensation of the Xin'an River Basin has preliminary progress just due to the policy issued by the State.

## 5.3 Conclusion and Discussion

Based on the above analysis, while giving full consideration to the specific condition of our country, we could learn from successful foreign experiences in the practice of river basin eco-compensation. Such experience includes

Promoting public involvement: The eco-compensation mechanism is a huge project, which involves a game among interested parties during its implementation. The government alone cannot solve all problems, so public participation should be fully solicited. By learning from foreign experience, making contracts based on agreement between the buyer and seller by means of negotiation among interested parties, and clearly defining the transaction amounts and expected goals of both parties, we can not only facilitate supervision and performance evaluation of compensation, but also fill the gap between the compensating party and the compensated party with various flexible compensation methods and thus helping us reach and implement the agreement. In addition, effective social participation and social supervision mechanisms should be established to construct a relatively uniform platform for the operation and coordination of policies which ensures the efficient implementation of eco-compensation in the river basins of China.

Designing transparent eco-compensation policies in river basins: The design of eco-compensation in China is mostly guided by the dominant form of government. During the actual implementation, therefore, the ecological system service provider is usually active to have equal negotiation and communication with the ecological system service beneficiary. Due to the lack of a transparent and well supervised and evaluated process, both the compensating party and the compensated party find themselves subject to weak moral and legal bonds, which seriously affect the outcomes of the implementation of compensation measures. Therefore, we should learn from foreign experience by enabling equal negotiation and talks between the ecological service provider and the ecological service beneficiary, so that both parties can timely manage and follow up the eco-compensation project, regularly sum up the effects of related policies, and thus ensure the sustainability of the policies.

## References


Fang, C. L. (2004). *The eco-economic corridor in the west of China* (pp. 78–86). Beijing: The Commercial Press.

Fang, C. L., & Yang, Y. M. (2006). Basic laws of the interactive coupling system of urbanization and ecological environment. *Arid Land Geography, 29*(1), 1–8.

Fang, C. L., Bao, Chao, & Qiao, Biao. (2008). *The urbanization process and eco-environmental effects* (pp. 1–8). Beijing: Science Press.

Feng, Y. F., Liu, Y. H., Wang, F., et al. (2009). Progress of ecological compensation practice in China. *Ecological Economy, 8*, 85–109.

Huang, J. C., & Fang, C. L. (2003). Analysis of coupling mechanism and rules between urbanization and eco-environment. *Geographical Research, 22*(2), 211–220.

Qiao, B., & Fang, C. L. (2005). The dynamic coupling model of the harmonious development between urbanization and eco-environment and its application in arid area. *Acta Ecologica Sinica, 25*(11), 3003–3009.

Qiao, B., Fang, C. L., & Huang, J. C. (2006). The coupling law and its validation of the interaction between urbanization and eco-environment in arid area. *Acta Ecologica Sinica, 26*(7), 2183–2190.

Qiao, B., Fang, C. L., & Li, M. (2005). Progress and prospect of study on interactive coercing process between urbanization and eco-environment in arid area. *Progress in Geography, 24*(6), 31–41.

Wackernagel, M., & Rees, W. (1996). *Our ecological footprint: Reducing human impact on the earth*. Philadelphia: New Society Publishers.

Wang, R. S. (1993). *Sustainable development and ecology* (pp. 121–126). Beijing: China Science and Technology Press.

Xiao, D. N., Chen, W. B., & Guo, F. L. (2002). On the basic concepts and contents of ecological security. *Chinese Journal of Applied Ecology, 13*(3), 354–358.

Yang, G. M., Li, W. H., & Min, Q. W. (2006a). Discussion on the application of evaluation of ecosystem services to ecological compensation. *Journal of Ecological Economics, 4*(1), 20–24.

Yang, G. M., Li, W. H., & Min, Q. W. (2006b). Review of foreign opinions on evaluation of ecosystem services. *Acta Ecologica Sinica, 26*(1), 205–212.

Yang, G. M., Min, Q. W., Li, W. H., et al. (2006c). Herdsmen.s willingness to accept (WTA) compensation for implement of prohibiting-grazing policy in Xilinguole steppe. *Ecology and Environment, 15*(4), 747–751.

Zhang, C. Q. (1987). Research on compensation of renewable resources. *Research of Agricultural Modernization, 5*, 22–24.

Zhang, J. H., Yu, G. Y., Han, S. J., Guan, D. X., & Sun, X. M. (2006a). Seasonal and interannual variation of $CO_2$ flux in Broadleaved Korean pine mixed forest and its control mechanism in Changbai Mountain. *Science in China Series D: Earth Sciences, 36*(supplement I), 60–69.

Zhang, L. M., Yu, G. Y., Sun, X. M., Wen, X. F., Ren, C. Y., Song, X., Liu, Y. F., Guan, D. X., Yan, J. H., & Zhang, Y. P. (2006b). Seasonal changes in carbon budget of the typical ecological system of Chinese eastern forest transect. *Science in China Series D: Earth Sciences, 36*(supplement I), 45–59.

# Chapter 18
# Industrial Ecology

**Wenliang Wu, Yifeng Zhang, Songlin Mu, Linsheng Zhong, Guofan Zhang, Huayong Que, Jianguang Fang, Xiwu Yan, Zhihua Lin, Yongyun Zheng, Yunhe Li, Kongming Wu, Yufa Peng and Gaodi Xie**

**Abstract** After a long era of fishing-and-hunting civilization, human came into agricultural civilization era about ten thousand years ago, while Industrial Revolution makes industry replace agriculture as the center industry 200 years ago, and at the same time the technical transformation of agriculture made it grow into modern agriculture. Though modern industry and agriculture have greatly promoted the development of social productivity, they caused unprecedented environmental problems. Therefore, looking for a kind of industry pattern that realize harmonious development of human and earth, can not only promote the economical development, but conserve resource and environment become the urgent affairs, ecological

---

W. Wu
College of Resources and Environmental Sciences, China Agricultural University, Beijing 100193, China

Y. Zhang · L. Zhong · G. Xie (✉)
Institute of Geographic Sciences and Natural Resources Research, CAS, Beijing 100101, China
e-mail: xiegd@igsnrr.ac.cn

S. Mu
Beijing Agriculture Information Technology Research Center, Beijing 100097, China

G. Zhang · H. Que
Institute of Oceanology, Chinese Academy of Sciences, Qingdao 266071, China

J. Fang
Yellow Sea Fisheries Research Institute, Chinese Academy of Fisheries Sciences, Qingdao 266071, China

X. Yan
Dalian Ocean University, Dalian 116023, China

Z. Lin
Zhejiang Wanli University, Ningbo 315100, China

Y. Zheng
Marine Biology Institute of Shandong Province, Jinan 250014, China

Y. Li · K. Wu · Y. Peng
State Key Laboratory for Biology of Plant Diseases and Insect Pests, Institute of Plant Protection, Chinese Academy of Agricultural Sciences, Beijing 100193, China

 
W. Li (ed.), *Contemporary Ecology Research in China*,
DOI 10.1007/978-3-662-48376-3_18

industry emerged consequently. Ecological industry is a kind of network-based, evolutionary and complex industry organized according to the principles of ecological economics that based on ecosystem carrying capacity, has a complete lifecycle, efficient metabolic process and harmonious ecological function. Compared with traditional industry, it makes material and energy be used repeatedly and output efficiently, can realize the systematic exploitation and sustainable utilization of resource and environment through the systematic coupling of two or more production systems or links. The essence of ecological industry is the application of ecological engineering to various industry, thereby ecological agriculture, ecological industry and ecological tertiary industry are formed and constitute a ecological industrial system. Both human society and nature can benefit from ecological engineering, it focuses on the ecosystem, especially the engineering technology that can enhance the sustainable development ability of social-economic-natural compound ecosystem. It promote the harmonious coexistence between human and nature, coordinated development of economy and environment, it make the simplex pursuit of economic growth or natural protection move toward compound ecosystem prosperity of affluence, health and civilization. The birth and development of ecological industry will make the mankind step into a new social form, thus a new ecological civilization will be formed.

**Keywords** Ecological agriculture · Ecological industry · Recycling economy · Ecotourism · Ecosystem carrying capacity · Cleaner production · Valley economy · Mariculture · Intertidal mudflat · Genetically modified (GM) crops · Herbicide tolerance · Insect resistance · Non-target organisms · Gene flow · Biodiversity

# 1 Ecological Agriculture

## 1.1 Introduction

### 1.1.1 Development of Ecological Agriculture

World agriculture has experienced the stages of primitive agriculture, traditional agriculture and modern agriculture (petroleum-based agriculture). Petroleum-based agriculture is characterized by high input and high energy consumption. Although oil agriculture plays a positive role in improving agricultural productivity, it also brings lots of problems, such as overpopulation, resource declination, environmental degradation, food and energy shortages. Many counties have begun to study on these issues since 1960s.

The modern ecological agriculture in China has gone through three stages including initial stage, exploration stage and stable development stage. The initial stage is from the end of 1970s to the beginning of 1980s. The symbol of this stage is the defining of ecological agriculture concept in academic community, which initially described the basic principles of ecological agriculture. The demonstration of ecological agriculture was carried out in some areas of China for trials. The period from the middle of 1980s to the beginning of 1990s is regarded as the exploration stage. The basic concepts of ecological agriculture were complementing, with a large number of research papers published and the ecological agricultural theoretical system formed with Chinese characteristics during this exploration stage. After 1990s, ecological agriculture stepped into stable development stage. The features of this stage are the beginning of pilot work of national eco-agricultural counties and deep study of its theory and methods (Chen 2004).

Ecological agriculture in China has inherited and developed the essence of traditional agriculture. Meanwhile, it integrated and used modern scientific and technological achievements, with wealth experience and valuable lessons accumulated during rural economic reform. Ecological agriculture is exploring of a stainable development model in China (Wang 2008).

### 1.1.2 Connotation of Ecological Agriculture

Ecological agriculture, guided by the principles of coordinated development of economy and environment, in line with the principles of ecology, ecological economics and biological and material recycling, summarizes and absorbs the successful experience of various agricultural production models, applies the ecosystem engineering methods and modern science and technology to create and develop reasonable arrangements to optimize agricultural production patterns and site-specific farming systems. It requires the combination of grain production and a variety of cash crop production; combination of plantation with forestry, animal husbandry, and fisheries, as well as the combination of agriculture with sector development. By combining the traditional agriculture and the essence of modern science and technology, through artificial ecological engineering design, harmonizing the relationship between resource utilization and protection and coordinating economic development and the environment, ecological agriculture can achieve a harmonious circle of ecology and economy to achieve sustainable development of agriculture (Shi 2001; Wang 2001; Li and Min 2001, 2003; Lu 2002).

Different countries have different names for ecological agriculture, but the purposes and objectives are the same, i.e., to produce clean food production, to improve human's health and promote the sustainable development of agriculture on clean land in a clean way (Ren 2004). In new era, China has entered the new development stage of industrialization of ecological agriculture.

## *1.2 Strategies for Ecological Agriculture Development*

### 1.2.1 Ecological Agriculture Industrialization

Ecological Agriculture Industrialization employs ecological agriculture as the development process of an independent industry sector. In accordance with the rules of marketing economy and effective industrial organization and operation and the inherent requirements of market economy, the economic efficiency of ecological agricultural will be improved. Laterally, intensification of capital, technology, land, labor and other factors of production will be implemented. Vertically, ecological agriculture shall be market-oriented and based on processing or economic cooperatives; farmers shall be initiated with the means of science and technology. The farming, animal husbandry, food process and storage are integrated into a holistic sector system. Ecological agriculture industrialization system will help to improve and achieve efficient harmonization of agro-ecology, economic and social benefits.

### 1.2.2 Main Types of Ecological Agriculture Industrialization

Current ecological agriculture industrialization types are diversified and consist of multi aspects. Major ecological agriculture industrialization types are as follows.

#### Comprehensive Development and Holistic Coordination

In ecosystem, the features of multi-level, multi-objective and multi-connections require the overall coordination of ecosystem during ecological agricultural industrialization process. The ultimate goal of ecological agriculture industrialization is to achieve harmonization of environmental protection and agricultural development as well as the efficient harmonization of agro-ecology, economic and social benefits.

#### Relying on the Production Base

Ecological agriculture production base is the base of ecological agriculture industry and also an important linkage between small-scale production and a large market. Hence, the development of ecological agricultural industrialization must go further to speed up reform of Rural Land Property System and guide farmers to expand the scale of operation to strengthen the construction of ecological agriculture production base.

Leading Enterprise

Leading enterprise in the industrial management system is an organizer, operations center, service center, information center, the main technological innovation and market pioneer. It plays a key role in the hub. Its drive function is the key to achieve ecological agriculture industrialization. Thus, the leading enterprises need more education and cultivation, including precise designing, focusing on support, strengthening the drive functions and so on.

Brand Cultivation and Green and Healthy Production

Under the situation of marketing economy, ecological agriculture, as the highly social and green industry, has to use brand to develop new markets, to improve product awareness, to market share and increase added value and to form of market competition and price advantage. In addition, brand cultivation can create a good brand image and corporate identity. This could help companies accurately locate and obtain legal protection.

Eco-tourism

Due to the constant deterioration of ecological environment, improvement of the economic income and increase of leisure time, people's material and cultural life of the ecological demands are also rising. Agricultural eco-tourism has become a new way of entertainment. In this way people can appreciate the pastoral scenery to enhance people's awareness of environmental protection, and also increase farmers' income and develop the rural economy.

## *1.3 Outlook of Ecological Agriculture Development in New Era in China*

### 1.3.1 Overall Principles

Modern ecological agriculture shall be guided by the scientific development perspective. To promote the coordinated development of agriculture from the perspective of sustainable development and the overall agricultural industry and consumption, a modern industrial system of ecological agriculture need to be built following the principles of "harmless, low emission, zero damage, high efficiency, sustainable and beautiful environment". In addition, it shall also save resources and adopt cleaner production technologies during the process, to recycle the agricultural wastes and promote clean consumption in rural areas; the focus shall be on saving and agricultural sector expanding; agricultural technical demonstration and organization shall be reformed. The ultimate goal is to establishing a modern ecological industrial system.

The development of modern ecological agriculture needs the change of three goals: First, the change from production function to taking into account ecological, social development. Second, change from the use of resources by the one-way loop to environmentally friendly utilization. Third, change from the extensive high resources consumption to resources efficient technology system.

### 1.3.2 Main Activities of Modern Ecological Agriculture Construction

Ecological landscape construction in rural areas; Agricultural cycle system; Relationships rebuilding of integrated innovation of agricultural diversity; Integration and renovation of ecological agriculture bio-technology system; Organization and support system construction for ecological agriculture.

### 1.3.3 Construction of Policy Support System for Modern Ecological Agriculture Development

#### To Establish Typical Comprehensive Demonstration Area of Modern Ecological Agriculture, the Technology and Compensation Policies Implementation

Based on the planning of functional areas, according to the characteristics of the regional ecological environment, fully coordinated ecology, life, the overall pattern of the production function, national typical modern ecological agriculture demonstration area should be established at provincial level to provide financial, technical and policy support and encouraged to explore successful experiences.

#### To Develop and Improve Relevant Laws and Regulations for Modern Ecological Agriculture System

"Ecological Agriculture Law of the People's Republic of China" needs to be set up timely as the basic law in accordance with the development of ecological agriculture. Based on the short and long-term development goals defined in "National Program for the Development of ecological agriculture", the stage goals, tasks and measures of ecological agriculture development shall be incorporated into the five-year plan for national economic and social development to guide the practice and implementation.

#### To Uniform Product Concepts and Standards, Improve the Modern Ecological Agriculture Market Trading Platform and Regulatory System

The regulatory and supervision system should be complemented for green food, organic products with multiple oversights of certification bodies, local authorities and farmers themselves.

#### To Establish and Improve the Modern Ecological Agriculture Compensation Mechanism and Policy Support System

For ecological agricultural products, organic agricultural production base and the scaled husbandry, effective economic compensation should be given to farmers during the transitional and stable production phase.

#### To Improve the Ecological Agriculture Model, Supporting Technology and Personnel System

According to the type of ecological resources, different types of agricultural production in the country training shall be held for the responsible persons of main agricultural sector, pilot agricultural enterprises, professional big farmer, farmer cooperatives and family farms at provincial, city and county level. From the perspective of the theory and practices, the core elements of management and action of ecological agriculture need to be understood and management and technical personnel construction of ecological agriculture in China should be improved.

#### To Establish a Special Budget Fund for Modern Ecological Agricultural Development

Special fund should be established for modern ecological agricultural demonstration area to ensure the national comprehensive demonstration of modern ecological agriculture demonstration, exploration, summarization the successful experience for the guidance of further full implementation. Based on the comprehensive exploration of various national and provincial modern ecological agriculture demonstration area, special fund for modern ecological agricultural development budget should be established and increased year by year, as the essential component of state budget in particular green budget to ensure strategic transformation of modern ecological agriculture.

## 2 Theory and Application of Valley Economic Development

Valley economy is a band new concept that is developed from the research of agriculture, farmer, rural area, production, and ecology, which combines the ecological economy and developmental characteristics of the mountainous areas, hence, it becomes a new angle of view in the research of mountain development (Zhang et al. 2009a, b).

As an important geographical type and the enforced ecological function of mountains during the past decades, there is a growing interest in mountains from both the overseas and domestic scholars. Up to now, a new research focus involved in resources, eco-environment, land and enterprise development in mountains has been formed (Li and Tian 2007; Duan et al. 2004; Balamirzoev et al. 2008; Corinne et al. 2007; Sarah et al. 2005). Most previous researches have only focused on one single factor, i.e. concept (He et al. 2010; Chen et al. 2010), characteristics (Li et al. 2011; Fan et al. 2009), development patterns (Shi et al. 2009), or spatial and organized pattern of valley economy, while the interaction between the development of valley economy and ecology was ignored. With spatial pattern as the prominent character, the propose of valley economy was much facilitated to decide the developmental trend of the mountainous area. Valley economy highlight the importance of unification of ecological conservation and economic development, which is proven to be very helpful for guiding the mountain development in a kind of macro arrangement in space, meanwhile, it also follows the principle of combining tightly with the micro examinations.

The spatially organizing process of valley economy is a process of optimizing the spatial structure of the eco-environmental elements, production elements, industries and villages within a specific area in the valley. According to the conditions of resources and the changes in rural residential areas, the spatial organizing process of valley economy can be divided into five developmental stages: (1) the stage with separating spatial structures; (2) the stage for the first concentrating spatial organization; (3) the stage of dispersing spatial organization; (4) the stage of the second concentration of spatial organization; (5) the stage of balanced spatial organization. We have built a multi-functionality evaluation index system and index analysis model used 17 indexes by taking geographic characteristics, environmental factors, resource endowment, and develop capacity into consideration from the perspective of nature and humanity. Furthermore, to optimized future development pattern of valleys' economy, we first should clarify the role of mountain according to development priority zones. Based on that, we suggest that development of deep mountain area should pay more attention in ecology and water conservation, and more front mountains could be an regions to nurture ecological industry and substitution industry, while piedmont should deal with the harmonious development of urban and rural areas and promote their function as model of the new rural construction (Fig. 1).

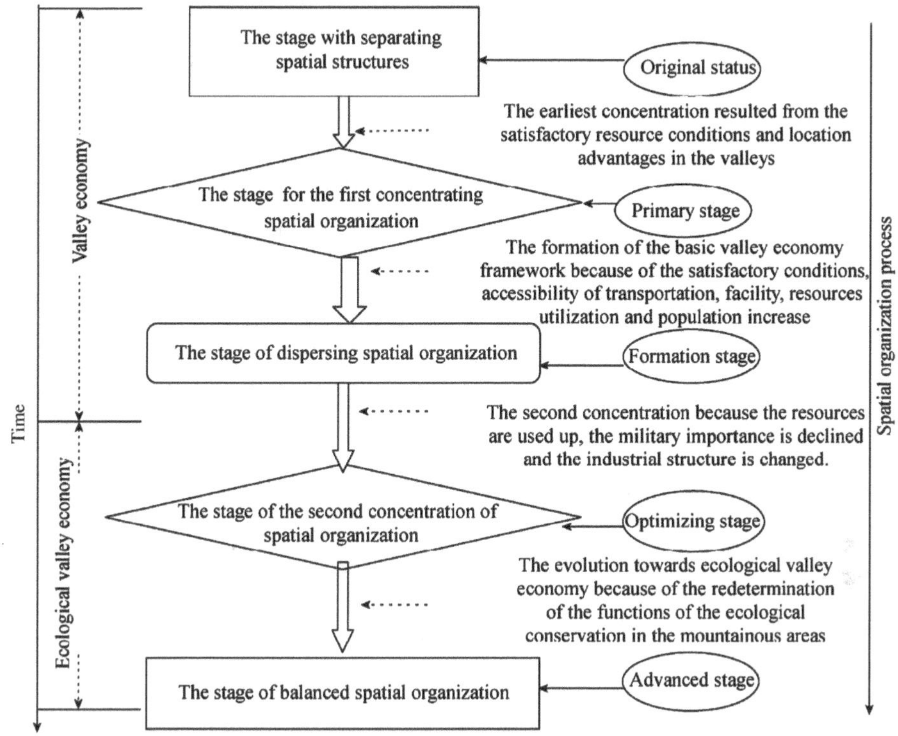


**Fig. 1** Developing stages and spatial organizing process of valley economy

Further development of valley economy should take geographical conditions, function of region system, economic development momentum, and development stages into consideration to quantify the effects and disturbance of valley economy on mountain systems and development. We have to break the administrative boundaries and make greater effort to do the research of interaction among energy flow, material flow, and human flow of valley system that include variety of provinces, counties, and towns to quantitatively depict the mutual promoting mechanism between development of valley economy and ecosystem.

# 3 Ecotourism

## *3.1 Introduction*

Due to rich resources, ecotourism experienced fasted development during past 20 years and contributed to sustainable tourism in China (Yang et al. 2010). It plays an important role in improving the structure of the tourism products, strengthening environmental education and tourism environmental protection, promoting construction of ecological civilization and energy saving in tourism industry and increasing income of community residents (Zhong et al. 2006).

### 3.2 Ecotourism Development Model

Because of China's booming development of ecotourism and the realistic requirement of developing activities, Chinese scholars has being studying the ecotourism development model (Pang et al. 2001; Cen 2003; Jiang 2004). Especially, a lot of empirical researches were carried on how to develop ecotourism in different types of destinations such as world heritage sites, man and biosphere reserves, nature reserves, forest parks, scenic spots, geological parks, the scenic spots of water, wetland parks and so on. Research contents meet the requirements of ecotourism features and tourism industry system construction (including ecotourism products, ecotourism scenic area, ecological catering, accommodation and ecotourism operators, etc.). These studies are of good guidance to promote ecotourism concept and improve the level of the regional ecological tourism management.

### 3.3 Ecotourism Planning

Ecotourism planning is involved in a relationship between tourists' activities andtheir environment. For example, based on research results from United States, Canada and other countries' experts, Rollins (1992) believed the theory of "island", "environmental capacity" and "recreation ground level theory" is suitable for China's national conditions, applied to the natural reserve tourism development planning and management. It will help with solving the conflicts between human activities and nature protection. Zhong et al. (2003) put forward the basic theories on ecotourism plan and design including ecological ethics theory, sustainable development theory, ecological economics theory, and landscape ecology. Yang et al. (2000) put forward an ecotourism development plan in protecting which was divided into 11 steps.

### 3.4 Ecotourism Management

Ecotourism management is a comprehensive system. Based on the analysis of the differences and similarities of traditional tourism and ecological tourism management, Yang et al. (2000) regarded ecotourism management includes ecotourism industry management, community management and ecological environment. In order to ensure that all the stakeholders would continue to benefit from the ecotourism activities', Ceballos-Lascurain (1996) put forward to establish a strategy system of tourism management in the protected areas.

### 3.5 Ecotourism Destination Certification

There are two main types on certification research in China. One is on the introduction of foreign certification experience (Huang et al. 2009); and the other is about the index system for evaluating ecotourism certification standards and models.

### 3.6 Ecotourism Impact

Ecotourism impact included social and cultural influence, economic effect and ecological environmental impact, usually giving the priority to the impact on destination, and with negative impact on research. Among them, introduction on related studies from western countries was paid more attention on (Liu et al. 1996). The quantitative research of tourism impact on the environment mainly focus on the scenic spot monitoring and evaluation of environmental factors and for one or two factors evaluation, but the comprehension is not enough (Jiang et al. 1996; Li et al. 1998; Ma et al. 2008). In addition, how to work out strategies for tourism environment protection is concerned about by researchers and managers. In summary, there are different ways to protect tourism environment. Of course it should be applied to the different tourist areas based on their characteristics (Lin 1999).

### 3.7 Prospect

Ecotourism research has been 20 years or so in China. It was needed to strengthen in the following aspects: (1) How to improve ecotourism residents' life in less developed areas and coordinate the relationship between ecotourism and the development of regional economy are urgent needed; (2) Chinese scholars have different understanding on the definition of ecotourism, the content of the essence, core standards and other aspects, which results in inconvenience in researching the theory and the practice of ecotourism; (3) Lack of depth on quantitative research and operability on theoretical research results; (4) Research method is relatively single; (5) Lack of research on community residents and tourists behavior.

## 4 Mariculture Ecology

### 4.1 Introduction

Mariculture has been growing rapidly in China since the 1980s. In 2010, China produced approximately 16.4 million metric tons (MMT) of seafood from aquaculture, more than two thirds of global mariculture production (FAO 2012). The dramatic increase was largely driven by rapid growing in shellfish and seaweed culture in shallow coastal waters. The common cultured species include seaweed, mollusks, shrimps, crabs, sea cucumber and fish. Farming operations take various forms ranging from inshore raft or longline culture, sea ranching, to intertidal zone cultivation and coastal ponds culture.

China is the largest producer of cultivated shellfish and seaweeds in the world with an annual production of over 10 million tons. The amount of carbon sequestered by harvesting shellfish and seaweeds is not trivial, estimated that $3.79 \pm 0.37$ and $1.20 \pm 0.11$ Mt C year$^{-1}$. Cultivation of seaweeds and shellfish therefore plays an important role in carbon coastal cycles and could improve the capacity of carbon sequestration by coastal systems.

In the past decade, unprecedented activities such as petroleum processing industry, navigation have substantially changed the environment of traditional culture sites. Meanwhile, mariculture itself has been subject to some controversy in terms of the environmental impacts. Abnormal mortality and intensive disease have hampered the industry's development and caused large economic losses. This was partly attributed to inadequate management and planning such as overstocking. Same as most farming practices, the degree of environmental impact depends on hydrography of the culture site, stock density, carrying capacity, and husbandry methods. Research on mariculure ecology has consequently gained increased attention to provide guidance and operational protocols to insure the sustainable development of marine aquaculture.

## 4.2 *Advance in Mariculture Ecology*

In recent years, the Chinese Government has paid much attention to investigate and evaluate marine living resources and their inhabiting environment for a sustainable development of mariculture and fisheries management. The project has been focused on carrying capacity in aquaculture, stock enhancement, as well as assessment of the impact of mariculture on coastal environments. Consequently, different culture models have been developed and industrialized successfully along the coast of China.

### 4.2.1 Inertial Mudflat Cultivation

At least 14 species of clams have been farmed in intertidal mudflat in China. The Manila clam farming production was about 1.8 million tons annually, accounting for about 80 % of mudflat fishery production in China and about 90 % the total global production. A new three-phase culture method has been adopted by the clam farming industry in northern China for mass production of the Manila clams. The new method involves early spawning and over-wintering in indoor greenhouse, optimized stocking size and density and substrate for mudflat grow out (Zhang and Yan 2006).

Long term intensive farming could lead to deterioration of culture mudflat environment. The mariculture industry is seeking a new efficient clean cultivation technology for clam cultivation, along with an effort to reconstructing and restoring degrading culture mudflat. Based on the on-bottom culture carrying capacity, a numerical model is established to address the relationship of seeding size, stocking density and clam production, which provide guidance and the needed information for a more efficient ecological culture model (Zhang and Yan 2010).

### 4.2.2 Coastal Pond Culture

In recent decades, pond culture has been extended to the coastal waters of China, using cages, nets and pens. Progress in shellfish pond aquaculture ecology has been made in the physiological ecology, bait biology, ecological farming model, ecological restoration, etc.

Physiological and energy metabolism of *Meretrix meretrix*, *Cycnlia sninessi*, *Tegillarca granosa* and other major shellfish were in-depth studied, which laid the foundation for the establishment of carbon income model, evaluation of shellfish growth rates, and assessment of ecological efficiency of shellfish cultivation. The effects of different microalgal diets for the growth of *M. meretrix*, *C. sninessi*, *T. granosa*, and other shellfish juveniles were estimated (Tang et al. 2006). Oriented cultivation of microalgae in ponds is under way.

Polyculture has been found to be an effective way to improve performance of cultured animals in ponds. The following combinations of multispecies culture are commonly adopted in the coastal area of southeast China: fish, shrimp and shellfish; fish, shrimp, shellfish and seaweed; shellfish and seaweed, respectively. Studies also suggested that co-cultivation of large seaweed will improve water quality and even the health of cultured animal.

### 4.2.3 Inshore Culture

Culture systems from inshore areas have been well developed along the coastal waters in China. The cage aquaculture sector has grown very rapidly and marine fish cage farming has proven to be a productive subsector within mariculture industry. Following this rapid development, marine fish cage culture has systematically been considered as a potential source of serious environmental impacts on its surrounding aquatic environment, which may cause localized hyper-eutrophication that could lead to eutrophication.

In order to reduce such effects, integrated multi-trophic aquaculture (IMTA), where “extractive” and “fed” species are grown simultaneously, has been proposed as a means of using the fed waste resource. A bioremediative approach, utilizing lower trophic levels as nutrient recyclers, could reduce waste products and sedimentation, diversify products, and provide economic gains for growers. Successful IMTA activities have been implemented in the Sungo Bay of Shandong Province, east China. A series of sustainable culture models have been set up, including polyculture of fish, shellfish, and seaweed. Other polyculture models consisted of kelp with abalone, abalone with sea cucumber and kelp, respectively, have also been performed in the bay (Zhang et al. 2009a, b). Based on the results from assessment of carrying capacity, measures were proposed and practiced to optimize culture densities in Sungo bay.

### 4.2.4 Sea Ranching

Many of the marine fisheries are no longer profitable due to over fishing and degradation of ecosystem conditions. Hatcheries allow mass production of larvae and juveniles for economically important species and facilitate the recovery of some marine stocks. The Pacific Abalone, Japanese scallop, giant cockle, sea cucumber and sea urchin have currently listed on commercial stock enhancement project in China based on comprehensive ecological assessment for designated areas. Hatchery-produced seeds are released to the bottom of various locations designated for sea ranching area, where larvae and juveniles for commercially important species will be naturally distributed and harvested. Preliminary results from this relatively new practice were promising, which in turn gained more momentum in this practice. Overall, a systematic and productive sea ranching project requires further investigation of the potential of using hatchery-produced juveniles as a means to enhance the yield of natural resources.

## *4.3 Conclusion and Discussion*

Marine culture in intertidal mudflats, seawater ponds, sheltered bays, and shallow waters has been proved successful in both economic return and social development. Due to improved understanding of ecological processes for mariculture practices and culture techniques innovation, industrial organizations have moved from primarily empirically oriented to more sophisticated ecological models. Integrated aquaculture of multi-trophic levels will be eventually dominated in future mariculture.

Culture of finfish, shellfish and seaweeds in offshore waters is now technically feasible, and has become one of the new frontiers for marine aquaculture. China is working on offshore aquaculture technology and legal regimes. Moving to the offshore areas would provide an avenue to overcome the conflict of interests for marine spatial planning and to reduce environmental impacts. Currently, most of the emphases are on offshore farming of finfish, high-value species, due to high costs and risks of practicing in offshore waters.

In general, sea ranching is a good system to enhance or to re-establish a declined population. There are likely numerous ecological factors involved in stock recruitment and overall sea ranching. Recent field experiments suggested relatively low recapture rate with large fluctuation. Therefore, there is still a long way to go.

# 5 Ecological Effects of the Commercial Cultivation of Genetically Modified Crops

## *5.1 Introduction*

Since the first planting of genetically modified (GM) crops in USA in 1996, the globally commercial growing area of GM crops has been steadily increasing, and the total area reached 170 million ha in 2012 (James 2012). To ensure the safe use of GM crops, regulatory risk assessment is necessary prior to the use of any GM crop variety. In this chapter, we summarize and analyze the current data regarding the environmental risk assessment of GM crops in the hope of providing the scientific basis for development, risk assessment and management of GM crops in China.

## *5.2 Commercial Use of GM Crops and the Related Ecological Risks*

Although a large number of GM varieties with multiple traits have been developed by genetic engineering, the primary currently grown GM crops express the traits of herbicide tolerance (HT) or insect resistance (IR) or both (James 2012). Presently *Bt* cotton and *Bt* maize are the only commercialized IRGE crops, while multiple types of HTGM crops including cotton, maize, canola, soybean and sugar beet have been widely grown in the world. Since the commercial use in 1997, *Bt* cotton has been quickly adopted in China. And in 2012, *Bt* cotton was grown by 3.9 million ha, occupying 80 % of the total cotton area (James 2012).

In spite of the great benefits with the use of GM crops (Brooks and Barfoot 2013), the potential risks to the environment associated with the cultivation of GM crops have to be critically assessed prior to commercialization (USEPA 2001; EFSA 2010). In summary, the major concerns regarding environmental risks of GM crop include: (i) effects on non-target organisms (NTOs) and the biodiversity; (ii) gene flow and the ecological consequence; (iii) GM crop-associated weediness and invasiveness; and (iv) resistance evolution of target pests (Romeis et al. 2006, 2008; Yu et al. 2011; Li et al. 2012, 2013; Chandler and Dunwell 2008; Liu et al. 2010).

### 5.2.1 Effects on Non-target Organisms

Based on the principle that risk is a function of hazard and the likelihood that this hazard will be realized, and by summarizing and analysing the previous data regarding the assessment of non-target effects of GM crops, we elaborate the basic rationale of the tiered approach that has been widely accepted for assessing the potential effects of GM crops on NTOs (Romeis et al. 2006, 2008; Garcia-Alonso et al. 2006). In addition, the potential effects of *Bt* maize and *Bt* cotton on non-target arthropods are reviewed. A general conclusion has been drawn that the current used *Bt* proteins are very specific to target pests, and have no direct toxicity on non-target arthropods (Romeis et al. 2006, 2008; Yu et al. 2011; Li et al. 2013). Moreover, the use of IRGM for pest control can apparently reduce the application of insecticides, thus provides protections for natural enemies that further enhance the control against insect pests (Lu et al. 2012).

### 5.2.2 Gene Flow and the Ecological Consequence

According to the prerequisite conditions and the principles for the occurrence of pollen-mediated gene flow of plants, we here propose a general procedure for assessing the gene flow between GM plants and their relative species (Li et al. 2012). In addition, the gene flow of GM canola and GM soybean as cases has been discussed. It is concluded that gene flow between GM and non-GM plants or their wild relatives is normal ecological phenomenon, and it is affected by multiple environmental factors (Rong et al. 2010; Li et al. 2012). By taking appropriate measures such as setting isolation belts between GM plants and non-GM plants or their wild relatives or staggering their flowering periods, gene flow can be prevented or significantly reduced (Li et al. 2012).

### 5.2.3 Effects on Biodiversity

With the introduction of GM crops, concern has been raised that genetic diversity within crop species will decrease because breeding programs will concentrate on a smaller number of high value cultivars. However, multiple studies suggest that the introduction of GM cultivars in agriculture has not significantly affected the levels of genetic diversity within crop species. Thus the development and introduction of GM crop varieties does not represent any greater risk to crop genetic diversity than the breeding programs associated with conventional agriculture (Ammann 2005; Carpenter 2011).

The evaluation of the potential impact of GM crops on farm-scale and landscape-scale diversity such as effects on soil organisms, arthropod and weed communities suggests that the current grown GM crops have reduced the impacts of agriculture on biodiversity, through enhanced adoption of conservation tillage practices, reduction of insecticide use, and use of more environmentally benign herbicides (Carpenter 2011; Bigler and Albajes 2011; Yu et al. 2011; Lu et al. 2010, 2012). Additionally increasing yields due to the use of GM crops also alleviate pressure to convert additional land into agricultural use (Carpenter 2011).

### 5.2.4 Resistance Evolution of Target Pests

One of the significant challenges associated with the use of IRGM crops is the potential for populations of target pests to develop resistance to these crops, resulting in failure of this control method (Liu et al. 2010). However, So far only a few field pest populations have developed resistance to *Bt* crops, such as the cotton pink bollworm *Pectinophora gassypiella* (Liu et al. 2010; Wan et al. 2012). This should be attributed to the successful deployment of refuge strategy. In China, although the strategy of artificial refuge is not adopted, no field populations of cotton bollworm have developed *Bt* resistance with more than 15 years of *Bt* cotton planting. This is due to the fact that cotton bollworm has many non-*Bt* host crops (e.g., maize, soybean, peanut and vegetables) that are planted around cotton fields working as the natural refuge to produce the susceptible individuals of cotton bollworm (Wu et al. 2002, 2004; Wu 2007). But natural refuge strategy does not apply to cotton pink bollworm, which feeds almost entirely on cotton in China (Wan et al. 2012).

## *5.3 Conclusion and Discussion*

GM crops have been widely grown in the world, and they have brought great benefits to the farmers and the environment. With the further development of agricultural biotechnology, they will play a larger role in improvement of agricultural productivity. However, to ensure the safe use of GM crops, regulatory risk assessments are an important part of the introduction of such products. Presently China has established a relatively well-developed regulatory system for risk assessment and management of GM plants that lays a firm basis for safe use of GM crops.

## References

Ammann, K. (2005). Effects of biotechnology on biodiversity: Herbicide–tolerant and insect-resistant GM crops. *Trends of Biotechnology, 23*, 388–394.

Balamirzoev, M. A., Mirzoev, E. M. R., & Usmanov, R. Z. (2008). Concepts of soil-agroecological zoning of mountain regions using the example of Dagestan. *Eurasian Soil Science, 41*(6), 586–594.

Bigler, F., & Albajes, R. (2011). Indirect effects of genetically modified herbicide tolerant crops on biodiversity and ecosystem services: The biological control example. *Journal of Consumer Protection and Food Safety, 6*, S79–S84.

Brookes, G., & Barfoot, P. (2013). Key environmental impacts of global genetically modified (GM) crop use 1996–2011. *GM Crops and Food Biotechnology in Agriculture and the Food Chain, 4*(2), 109–119.
Carpenter, J. E. (2011). Impact of GM crops on biodiversity. *GM Crops, 2*, 1–17.
Ceballos-Lascurain, H. (1996). *Tourism, ecotourism and protected area*. France: IUCN (The World Conservation Union).
Cen, B. (2003). The basic thinking on the development of ecotourism on Weizhou Island in Beihai city. *Tourism Tribune, 18*(2), 69–72.
Chandler, S., & Dunwell, J. M. (2008). Gene flow, risk assessment and the environmental release of transgenic plants. *Critical Reviews Plant Sciences, 27*, 25–49.
Chen, L. (2004). Background current status and outlook of Chinese ecological agriculture development. *Journal of Yunnan University (Natural Science), 26*(6A), 140–144.
Chen, J. H., Li, H., & Zhou, L. D. (2010). Gully economy development's exploration and practice in Beijing mountain area. *Ecological Economy, 1*, 57–62.
Duan, X. B., Chen, L. X., & Yu, X. X. (2004). Evaluation and analysis of sustainable development of Puwa small watershed comprehensive management in Beijing mountainous area. *Science of Soil Water Conservation, 4*, 53–57.
EFSA. (2010). Guidance on the environmental risk assessment of genetically modified plants. *The EFSA Journal, 8*, 1879.
Fan, S. Y., Lu, J., & Wei, H. W. (2009). Valley economy system coupling model research in Tibetan areas. *Journal of Southwest University for Nationalities (humanities and Social Science), 30*(1), 72–77.
FAO. (2012). *FAO statistical yearbook*. UN: Food and Agriculture Organization.
Garcia-Alonso, M., Jacobs, E., Raybould, A., Nickson, T. E., Sowig, P., Willekens, H., et al. (2006). A tiered system for assessing the risk of genetically modified plants to non-target organisms. *Environmental Biosafety Research, 5*, 57–65.
He, Z. W., Li, Y., & Wang, Y. N. (2010). The connotation and mode analysis of valley economy development in Beijing. *Agricultural Economic Problems, 9*, 105–109.
Huang, J., & Ren, X. (2009). Analysis on impetus of ecotourism accreditation and limiting factors of its implementation in China. *Tourism Forum, 2*(1), 16–19.
James, C. (2012). *Global status of commercialized biotech/GM crops*. Ithaca, NY, USA: ISAAA (ISAAA Brief 44).
Jiang, X. (2004). Development of ecotourism in minority village—taking Yao minority village in Lingui county as a example. *Economic Geography, 24*(4), 564–567.
Jiang, W., Zhu, L., Li, J., & Deng, W. (1996). The influence of tourism to ecological environment and protection countermeasures in Emei mountain. *Environmental Science, 17*(3), 48–51.
Li, W. (2003). *Ecological agriculture—The theory and practice of sustainable agriculture in China*. Beijing: Chemical Industry Press.
Li, Z., Bao, J., & Qin, C. (1998). The impact of tourist development on the vegetation cover of mount Danxia, Guangdong. *Acta Geographica Sinica, 53*(6), 554–561.
Li, P., Han, J., Ma, X., Yuan, S. Q., & Pang, C. W. (2011). Studies on current status of mountainous regions of Beijing city and development of the valley economy. *Management of Agricultural Science and Technology, 30*(1), 10–13.
Li, W., & Min, Q. (2001). *Achievements, problems and development prospects of Chinese ecological agriculture, Section of agriculture science and technology education, ministry of agriculture ecological agriculture and sustainable development* (pp. 7–12). Beijing: China Agriculture Press.
Li, Y. H., Peng, Y. F., Li, X. J., & Wu, K. M. (2012). Ecological risks of herbicide-tolerant genetically modified crops and associated management. *Chinese Bulletin of Botany, 47*, 197–208.
Li, Y. H., Romeis, J., Wu, K. M., & Peng, Y. F. (2013). Tier-1 assays for assessing the toxicity of insecticidal proteins produced by genetically engineered plants to non-target arthropods. *Insect Sciece,*. doi:10.1111/1744-7917.12044.
Li, Y. B., & Tian, Z. M. (2007). A summary on small watershed comprehensive management measures in Beijing mountains. *Environmental Science and Management, 2*, 101–103.

Lin, Y. (1999). *Tourism introduction to environmental protection*. Beijing: Tourism education press.
Liu, X., & Bao, J. (1996). Development of the researches on the environmental impacts on tourism development. *Geographical Research, 15*(4), 92–99.
Liu, C. X., Li, Y. H., Gao, Y. L., Ning, C., & Wu, K. M. (2010). Cotton bollworm resistance to *Bt* transgenic cotton: a case analysis. *Science China Life Science, 8*, 934–941.
Lu, M. (2002). *Modern ecological agriculture*. Beijing: China Agriculture Press.
Lu, Y. H., Wu, K. M., Jiang, Y. Y., Guo, Y. Y., & Desneux, N. (2012). Widespread adoption of *Bt* cotton and insecticide decrease promotes biocontrol services. *Nature, 487*(7407), 362–365.
Lundstrom, C., Kytzia, S., & Walz, A. (2007). Linking models of land use resources and economy to simulate the development of mountain regions. *Environmental Management, 40*(3), 379–393.
Ma, J., & Cheng, K. (2008). The influence of nature reserve ecotourism for wildlife. *Acta Ecologica Sinica, 28*(6), 2819–2827.
Pang, Z., & Dong, B. (2001). A study on the eco-tourism development of the urban-rural interlocking belt in Shanghai city. *Tourism Tribune, 16*(3), 76–79.
Ren, A. (2004). Comparison and reference of foreign ecological agriculture development. *Global Agriculture*, 26–27.
Rollins, R., Zhang, L. (1992). Natural conservation and recreational resorts in suburbs and countryside. *World Forestry Research, 3*, 43–49.
Romeis, J., Bartsch, D., Bigler, F., Candolfi, M. P., Gielkens, M. M., Hartley, S. E., et al. (2008). Assessment of risk of insect-resistant transgenic crops to nontarget arthropods. *Nature Biotechnology, 26*, 203–208.
Romeis, J., Meissle, M., & Bigler, F. (2006). Transgenic crops expressing *Bacillus thuringiensis* toxins and biological control. *Nature Biotechnology, 24*, 63–71.
Rong, J., Song, Z. P., De Jong, T. J., Zhang, X. S., Sun, S. G., Xu, X., et al. (2010). Modelling pollen–mediated gene flow in rice: Risk assessment and management of transgene escape. *Plant Biotechnology Journal, 8*, 452–464.
Sarah, S., Patrick, B., & Cathy, W. (2005). *Understanding the spatial heterogeneity of global environmental change in mountain regions: Global change and mountain regions, Holland* (pp. 21–30). Netherlands: Springer.
Shi, S. (2001). *Development of ecological agriculture in China*. Beijing: China Agricultural Science and Technology Press.
Shi, Y. J., Tang, H., & Huang, Y. H. (2009). Valley economy development in Beijing mountain region. *Chinese Agricultural Science Bulletin, 25*(18), 500–503.
Tang, B. J., Liu, B. Z., & Wang, G. D. (2006). Effects of various algal diets and starvation on larval growth and survival of *Meretrix meretrix*. *Aquaculture, 254*(1–4), 526–533.
United States Environmental Protection Agency (US EPA). (2001). *Biopesticide registration action document: Bacillus thuringiensis (Bt) plant-incorporated protectants*. October 15, 2001. http://www.epa.gov/oppbppd1/biopesticides/pips/bt_brad.html. Accessed September 1, 2013.
Wan, P., Huang, Y. X., Wu, H. H., Huang, M. S., Cong, S. B., Tabashnik, B. E., & Wu, K. M. (2012). Increased frequency of pink bollworm resistance to *Bt* toxin Cry1Ac in China. *PLoS ONE, 7*(1), e29975.
Wang, Z. (2001). *Chinese ecological agriculture and sustainable agricultural development*. Beijing: Beijing Publishing House.
Wang, Z. (2008). Development of Chinese ecological agriculture. *Journal of Chinese Eco-agriculture, 16*(1), 1–3.
Wu, K. M. (2007). Monitoring and management strategy for *Helicoverpa armigera* resistance to *Bt* cotton in Chinese. *Journal of Invertebrate Pathology, 95*, 220–223.
Wu, K. M., Feng, H. Q., & Guo, Y. Y. (2004). Evaluation of maize as a refuge for management of resistance to *Bt* cotton by *Helicoverpa armigera* (Hübner) in the Yellow River cotton-farming region of China. *Crop Protection, 23*, 523–530.
Wu, K. M., Guo, Y. Y., & Gao, S. S. (2002). Evaluation of the natural refuge function for *Helicoverpa armigera* (Lepidoptera: Noctuidae) within *Bacillus thuringiensis* transgenic cotton growing areas in north China. *Journal of Economic Entomology, 95*, 832–837.

Yang, G., & Wang, Y. (2000). New thinking for the protective exploitation of ecotourism. *Economic Geography, 20*(1), 87–92.
Yang, G., Zhong, L., & Ming, Q. (2010). *Ecological tourism*. Beijing: Higher Education Press.
Yu, H. L., Li, Y. H., & Wu, K. M. (2011). Risk assessment and ecological effects of transgenic *Bacillus thuringiensis* crops on non-target organisms. *Journal of Integrative Plant Biology, 53*, 520–538.
Zhang, J. H., Hansen, P. K., Fang, J. G., Wang, W., & Jiang, Z. J. (2009a). Assessment of the local environmental impact of intensive marine shellfish and seaweed farming-application of the MOM system in the Sungo Bay, China. *Aquaculture, 287*, 304–310.
Zhang, Y. F., Jia, D. M., Tan, J., Zhang, H. Y., Song, S. Y., & Sun, R. F. (2009b). The spatial structure of valley economy development in the mountainous areas of Beijing. *Acta Geographica Sinica, 64*(10), 1231–1242.
Zhang, G. F., & Yan, X. W. (2006). A new three-phase culture method for Manila clam, *Ruditapes philippinarum*, farming in northern China. *Aquaculture, 258*, 452–461.
Zhang, G. F., & Yan, X. W. (2010). *Clam aquaculture study*. Beijing: Science press.
Zhong, L., Zhao, S., & Xiang, B. (2003). *Ecological tourism planning principle and method*. Beijing: Chemical Industry Press.
Zhong, L., Zheng, Q., & Liu, M. (2006). *Ecological tourism geography in the world*. Beijing: China Forestry Publishing House.

# Chapter 19
# Eco-city Construction

**Kongjian Yu, Linbo Zhang, Zhifeng Yanng, Xiangrong Wang and Moucheng Liu**

**Abstract** City is a kind of special artificial ecosystem that is created according to the will of human. Urbanization in china have created unlimited opportunities for the theoretical development and practice of urban ecology, Chinese scholars have achieved important progress in the practice of urban ecological planning and design, especially about ecological security pattern, "anti-planning" theory and practice oriented to ecosystem service and based on ecological aesthetics. They are not only widely used in the city construction in china, but get a high evaluation of the international community. In recent decades, urbanization has become an irreversible and inevitable trend, however, the development of most city depend on high resource consumption and the damage to environment, a series of ecological problems such as declining in air quality, water environment pollution and dramatic increase in the number of waste have been brought about. In the face of growing ecological pressure, constructing eco-city has gradually become the consensus of governments all over the world. Theory of ecological carrying capacity, ecological function zoning, and construction of ecological civilization are the key theory in urban ecological planning and practice in China. In the past twenty years' theoretical research and practical exploration for eco-city construction, Chinese-featured eco-city development pattern has been formed on the basis of fully absorbing

K. Yu
College of Architecture and Landscape Architecture, Peking University,
Beijing 100871, China

L. Zhang
Chinese Research Academy of Environmental Sciences, Beijing 100012, China

Z. Yanng
School of Environment, Beijing Normal University, Beijing 100875, China

X. Wang
Department of Environmental Science and Engineering, Fudan University,
Shanghai 200433, China

M. Liu (✉)
Institute of Geographic Sciences and Natural Resources Research,
Chinese Academy of Sciences, Beijing 100101, China
e-mail: liumc@igsnrr.ac.cn

 
W. Li (ed.), *Contemporary Ecology Research in China*,
DOI 10.1007/978-3-662-48376-3_19

international experience and lessons of eco-city construction and combining with their own practical situation, it played an important role in environmental protection in the process of rapid urban development. But in practice, the related theory still have some problems and insufficiencies, in future practice of eco-city construction, the key theory and technology research need to be strengthened, the connotation of urban ecological construction need to be further improved, so as to keep the virtuous cycle of urban ecosystem, show the pattern of harmonious development of urban construction, economic construction and environmental construction.

**Keywords** Eco-city • Eco-county • Ecological infrastructure • Ecosystems services • Ecological security pattern • Regional ecological baseline • Urban ecosystem health • Regional ecological planning

## 1 Ecological Infrastructure for China Beautiful

The strategy of Ecological Infrastructure is to plan and develop land more effectively with the goal of preserving valuable ecosystem services. In China, the concept can be traced to the pre-scientific model of *Feng-shui*—the sacred landscape setting for human settlement. In the West, precedents include the nineteenth century notion of greenways as urban recreational spaces, the early twentieth century idea of greenbelts to limit sprawl, and the late twentieth century strategy of connecting ecological networks to preserve biodiversity, all of which strove to balance human habitation with natural processes (Yu 2010c; Yu et al. 2011b). From broad-scaled ecological planning for all of China, to regional-scale planning of the capital Beijing and various sized cities, to the fine-scale urban park, this article summaries the researches and practices in using the powerful tool of Ecological Infrastructure to safeguard nature's processes and to recover ecologically degraded environment and to make China beautiful, which the author and his colleagues have done in the past 15 years (Saunders 2012).

### *1.1 The Integrated Solution Across Scales: Ecological Infrastructure for Ecosystems Services*

China are now facing many ecological challenges: Floods, storm-water inundation in urban areas, draught and ground water drop at vast scale, water and soil pollution, native habitat and biodiversity loss, to name just a few. The conventional single-minded engineering solutions, such as building flood control dams and dykes, storm-water pipes and pumps, expensive sewage plants implemented as quick solutions, but usually cause more complicated problems and are not sustainable. Designing and building Ecological Infrastructure that solve the complicated problems in an integral way, and by the force of nature, becomes the practical and wise solution. Here, Ecological Infrastructure is defined as "the structural landscape network that is composed of the critical landscape elements and spatial patterns that are of strategic significance in preserving the integrity of the landscape and securing sustainable ecosystem services." (Yu 2011c, 2013a, b; Yu et al. 2011b).

The cognition of landscape in terms of structural frameworks has deep roots, both in China and the West. Among the pre-scientific models is the ancient Chinese art of geomancy, or Feng-shui, which always gave priority to the natural patterns and processes of Qi (or breath) (Lip 1979; Skinner 1982; Yu 1994).Orderly from large to small, the entire national landscape, including mountains and water courses, is considered as an interconnected dragon vein and a network of Qi movement, a sacred landscape infrastructure that any human actions have to come to terms with (Yu 1994).This model has been applied to the establishment and construction of villages and cities, roads, bridges, and even tombs, all of which are understood as connected in hierarchical patterns. From the model of *Feng-shui* or Geomancy, there are three points to be learned:

(I) the protection of landscapes of a minimum critical size that are strategic for multiple processes;
(II) the integration of natural, biological and cultural processes;
(III) the understanding of landscape in terms of hierarchical form and across scales.

Ecological Infrastructure, as a contemporary concept and methodology, is an important tool for moving built landscapes, metropolitan regions, and cities towards more sustainable conditions. But what makes the concept of Ecological Infrastructure such a powerful tool today is its marriage with the understanding of Ecosystem Services. Four categories of services have been identified: provisioning, related to production of food, water, and energy; regulating, related to the control of climate and disease and the mediation of flood and drought (i.e., the purification of water, carbon sequestration and climate regulation, waste decomposition and detoxification, crop pollination, pest and disease control); supporting, related to nutrient dispersal and cycling, seed dispersal, and habitat for wild plant and animal species; and cultural: intellectual and spiritual inspiration, recreational experiences, ecotourism and scientific discovery (Constanza and Daily 1992; Constanza et al. 1997; Daily 1997; MEA 2005).

With respect to ecosystem services, Ecological Infrastructure can be understood as the necessary assets of sustainable landscapes or ecosystems in which the output of the goods and services is maintained, and the capacity of those systems to deliver same goods and services for future generations is not undermined. The many traditions of landscape ecology and more contemporary ideas about landscape as infrastructure finally come together the understanding of natural capital and ecosystem services, and merge into the concept of Ecological Infrastructure. Although largely based in natural systems, other landscape elements such as cultural heritage corridors can also be integrated into plans for ecological infrastructure (Yu 2011c, 2013a, b; Yu et al. 2011b).

## 1.2 Defining Ecological Infrastructure Through Landscape Security Patterns

Here, ecological infrastructure is defined as the structural landscape network in which essential landscape elements are configured hierarchically. Both existing and potential spatial patterns are of strategic significance in preserving natural, biological and cultural processes, which in turn are critical in securing the integrity and identity of the natural and cultural landscapes, as well as in securing the natural capital that supports sustainable ecosystem services.

It is important to note that ecological infrastructure is process-oriented, not just a visible spatial pattern. With respect to the identification and planning of ecological infrastructure, a processes-oriented model of spatial analysis, what I term a security pattern approach, are proven to be useful (Yu 1995, 1996). Landscape, particularly ecological security patterns are composed of strategic geographic elements and spatial patterns that are critically important in safeguarding and controlling ecological processes and landscape change. Security patterns can be identified according to the properties on a general surface model of flows and processes. Potential surfaces are developed using landscape resistance to represent the dynamics of horizontal ecological processes (e.g. species movement, the spread of urban development, and water flow).

Four strategic landscape components are commonly identified on the potential surfaces: buffer zones, inter-source linkages, radiating routes, and strategic points. These components, specified by certain quantitative and qualitative parameters, together with the identified sources (e.g. native habitats), compose the landscape and ecological security patterns and can be maintained at various security levels depending on their critical significance. These security patterns can be integrated into an overall ecological infrastructure, and can be used by defenders of ecological processes as defensive frontiers and strategies of spatial bartering in landscape changes.

With these objectives in mind, the planning of Ecological Infrastructure is composed of the following steps (Fig. 1):

(I) **Process Analysis**

Processes associated with critical ecosystem functions or services are targeted to be safeguarded by ecological infrastructure, so systematic analysis is carried out using Geographical Information Systems (GIS), an efficient tool to simulate natural and cultural processes across the landscape. These processes include:

- Abiotic processes, associated with the regulation and life supporting services of ecosystems;
- Biotic processes: associated with providing habitat for wild plant and animal species, safeguarding native species and biodiversity conservation;
- Cultural processes associated with information functions including visual perception, heritage protection, and recreational activities.

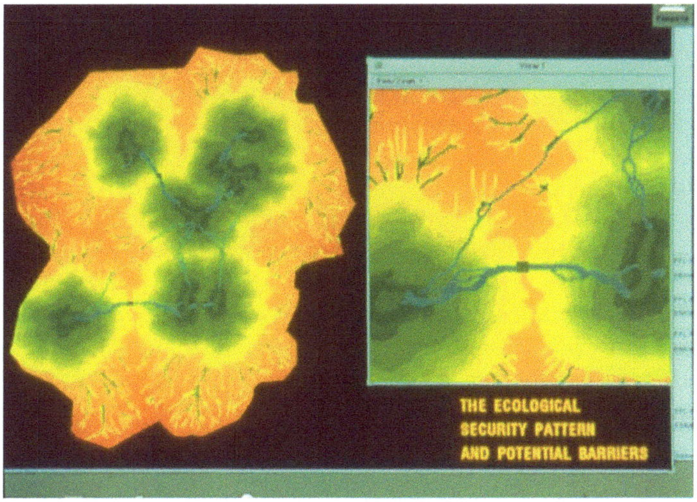


**Fig. 1** Identify ecological security pattern of bio-diversity

(II) **Defining Ecological Security Patterns**
Ecological Security Patterns are identified for the individual targeted processes. Models including suitability analysis, minimum cost distance and surface analysis are used in the identification of Security Patterns for the individual processes (Yu 1995, 1996). Alternative security levels—low, medium and high—are used to define the attributes of the security patterns in safeguarding each of the targeted processes.

(III) **Defining Ecological Infrastructure Across Scales: XL–L–M–S**
An overlay technique is used to integrate the security patterns for individual processes. Alternatives are developed at various quality levels: high, medium and low. Green lines are drawn to define and protect the ecological infrastructure.

The ecological infrastructure is planned across scales:

- National scale (X-large): The overall ecologically-protected land.
- The regional scale (hundreds to thousands of square kilometers) (large): At the regional scale, green lines are drawn to define the structural elements as corridors and restricted areas for construction.
- The intermediate scale [tens of square kilometers (medium)]: At this scale, the overall design and management guidelines are developed for ecological infrastructure, and especially for the green corridors that work as critical elements in water management, biodiversity conservation, heritage protection, and recreation.
- The small scale (less than ten square kilometers) (small): At a specific site, a green network is designed to allow ecosystem services to be delivered into the urban fabric and serve humanity.

## 1.3 Ecological Infrastructure Across Scales

### 1.3.1 Ecological Infrastructure at the X-Large Scale: The Chinese National Ecological Security Pattern

Sustainable development is crucial for China today and for the future. The population of China has grown from 541.67 million in 1949 to 1.34 billion in 2008, making it one of the most densely populated countries in the world. By 2050, the total population is expected to reach 1.41 billion and 70 % will reside in urban areas (UNFPA, State of world population 2008). Given this magnitude of development, the environment will continue to be under relentless pressure. So, ecological security in China has become a key area of scientific research for a strategy of sustainable development.

In Spring, 2007, at the request of Premier Wenjia Bao (in response to a letter from the author) and the National Bureau (now Ministry) of Environmental Protection, Peking University initiated a pilot project that aims at establishing an ecological security pattern at the national scale to protect the most sensitive ecological landscapes and to guide wise conservation and development. Critical natural processes were analyzed systematically at the national scale, including headwater conservation, prevention of soil erosion, stormwater management and flood control, combating desertification, and promoting biodiversity conservation. Individual security patterns for safeguarding each of these natural processes were identified and then integrated into an overall ecological security pattern. Three levels of National Ecological Security Pattern were defined, the lower security level, the moderate security level, and the higher security level, which encompass 35.7, 65.1 and 84.9 % of the national land respectively. This study is expected to provide a scientific basis for on-going national function zoning, incorporating ecological security patterns into land use planning at the national scale. An alternative urbanization pattern at national scale is also proposed based on the national ecological infrastructure: the foothill strategy (Yu et al. 2009, 2012; Saunders 2012; Yu 2012) (Fig. 2).

### 1.3.2 Ecological Infrastructure at the Large Scale: Defining the Regional Ecological Baseline of Urban Development Plan for Beijing

Strategic landscape elements and patterns for stormwater management and flood control, biodiversity conservation, cultural heritage protection, and recreational use are integrated using GIS models into a comprehensive ecological infrastructure, which is then used to determine future urban growth patterns. This project demonstrates how landscape planning can play a leading role in urban development through applying ecological infrastructure as a tool for smart growth and conservation (Yu et al. 2011b).

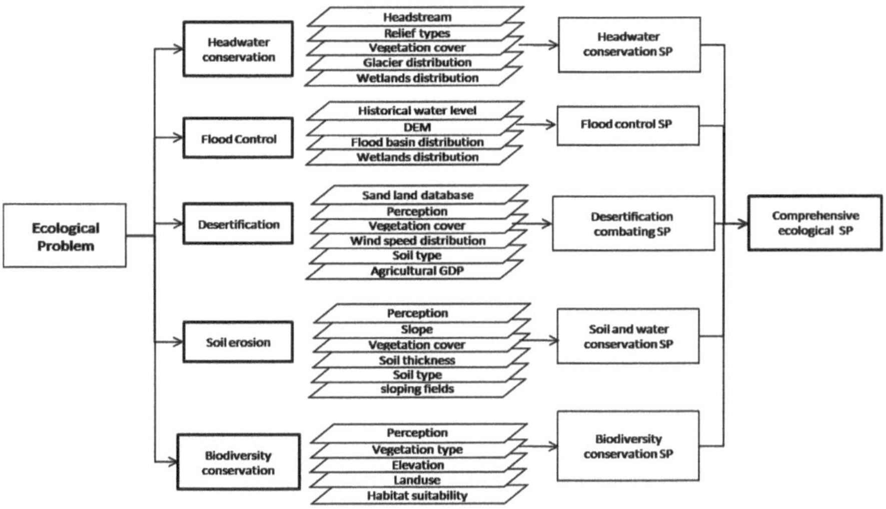


**Fig. 2** The framework for identification of national ecological security pattern and ecological infrastructure (*Source* Peking University Graduate School of Landscape Architecture)

Beijing, the capital of China, is situated in the North China Plain, with a total area of 16,410 $km^2$. Due to rapid urbanization during the last 30 years, its population has doubled from 8.7 million in 1978 to about 17 million in 2009, and its built-up area has expanded-fold and is still growing. The notorious "scrambled egg" pattern of Beijing today is the evidence of the speed and magnitude of urban sprawl and the failure of conventional planning, which tries to control urban growth using arbitrarily-located green belts. As payment for the uncontrolled spread of the growing city, local and regional natural systems and cultural heritage have all been damaged significantly. The city is now facing multiple challenges, including water shortages; increased vulnerability to geological disasters; habitat and biodiversity loss; diminished integrity and authenticity of cultural landscapes; decreased access to landscape for recreational uses; and dramatic loss of agricultural land and soil fertility.

Although greenbelts and green wedges have been planned to stop urban sprawl and maintain good landscape structure, they have largely failed. Greenbelts encircling Beijing city were planned artificially and lack an intrinsic relationship with topography and ecological systems. They also lack integration with ecosystem services, and are vulnerable to land use change—they have already been fragmented by large settlement areas. New and more effective tools have to be developed to address a wiser and sustainable development of the limited land.

Using minimum space, an ecological infrastructure for Beijing would safeguard critical ecosystem services. It would retain stormwater as much as possible to recharge the aquifer, while protecting the city from the threat of floods; minimize the risk of geological disasters; protect critical native habitats, and build an effective biological framework to maximally safeguard biodiversity; protect and regain the integrity and authenticity of cultural landscapes; increase the accessibility of the landscape for recreational uses; and maximally protect fertile land from being swallowed by urban development while not impeding urban growth. The overall objective is smart protection with smart growth (Fig. 3).

Landscape Security Patterns that safeguard individual processes are integrated into the overall comprehensive Ecological Infrastructure. Using an overlay technique to integrate the security patterns for individual processes, alternatives of regional ecological infrastructure are developed at three quality levels (Fig. 4):

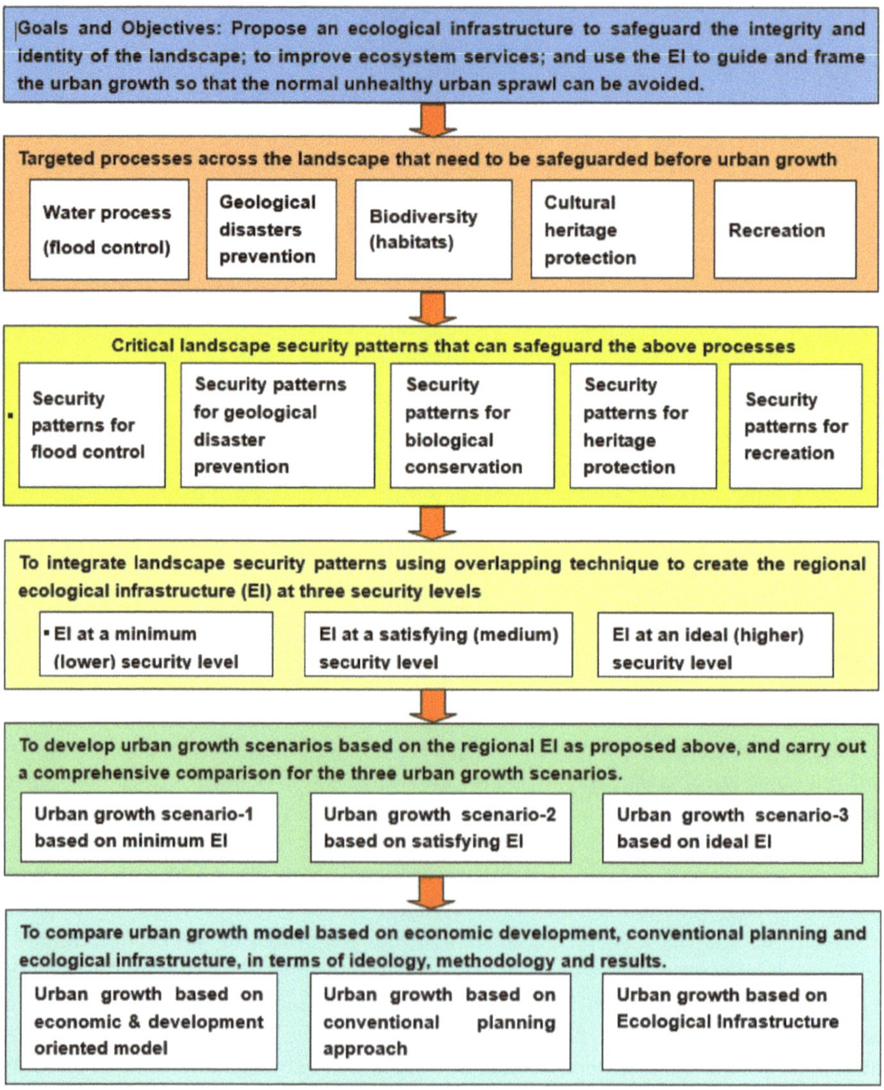


**Fig. 3** The framework of EI approach to urban growth planning of Beijing

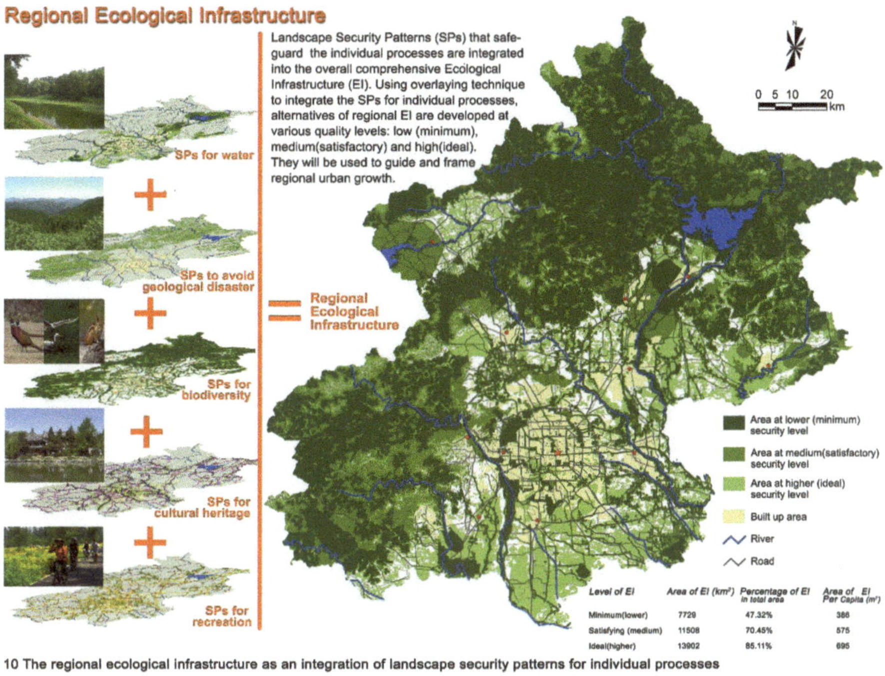


**Fig. 4** Beijing regional ecological infrastructure: the integrated ecological security patterns for various processes (*Source* Peking University Graduate School of Landscape Architecture)

(I) ecological infrastructure at low (minimum) quality: 47 % of the total land, including 24 % of the most fertile land, is protected. The integrity of the critical ecological processes will be protected at a minimum level, providing basic ecosystems services. The regional environment will be stable for the time being.

(II) ecological infrastructure at medium (satisfactory) quality. 70 % of the total land, including 45 % of the most fertile land, is protected. The integrity of the critical ecological processes will be protected at a satisfactory level, providing adequate and sustainable ecosystem services. The regional environment will be regenerated gradually.

(III) ecological infrastructure at high (ideal) quality. 85 % of the total land, including 100 % of the most fertile land, is protected; the latter for organic agriculture. The integrity of critical ecological processes will be protected at an ideal level, providing the best possible ecosystem services. The regional environment will be regenerated dramatically.

Detailed guidelines have been developed for the protection and management of ecological infrastructure components at various scales. Using the three alternative quality levels as a guide, multiple scenarios of regional urban growth patterns were simulated using GIS:

Scenario-1, The "scrambled egg": urban growth without ecological infrastructure;
Scenario-2, green infrastructure within the city: urban growth based on minimum ecological infrastructure;
Scenario-3, city on green infrastructure: urban growth based on satisfactory ecological infrastructure;
Scenario-4, urban garden: urban growth based on ideal ecological infrastructure.

Comparative evaluations are made for all four urban growth scenarios based on their impacts on ecological, cultural, and economic processes, and their capacity to address the national and regional challenges identified at the introduction of this essay. Scenario-3, city as green infrastructure, might best fulfill the goals of both conservation and development, and meet the requirement of all competing land use requirements. This is a smarter scenario, in which the limited land can be used more efficiently, through a better configuration, for both conservation and urban development (Yu et al. 2011a) (Figs. 5 and 6).

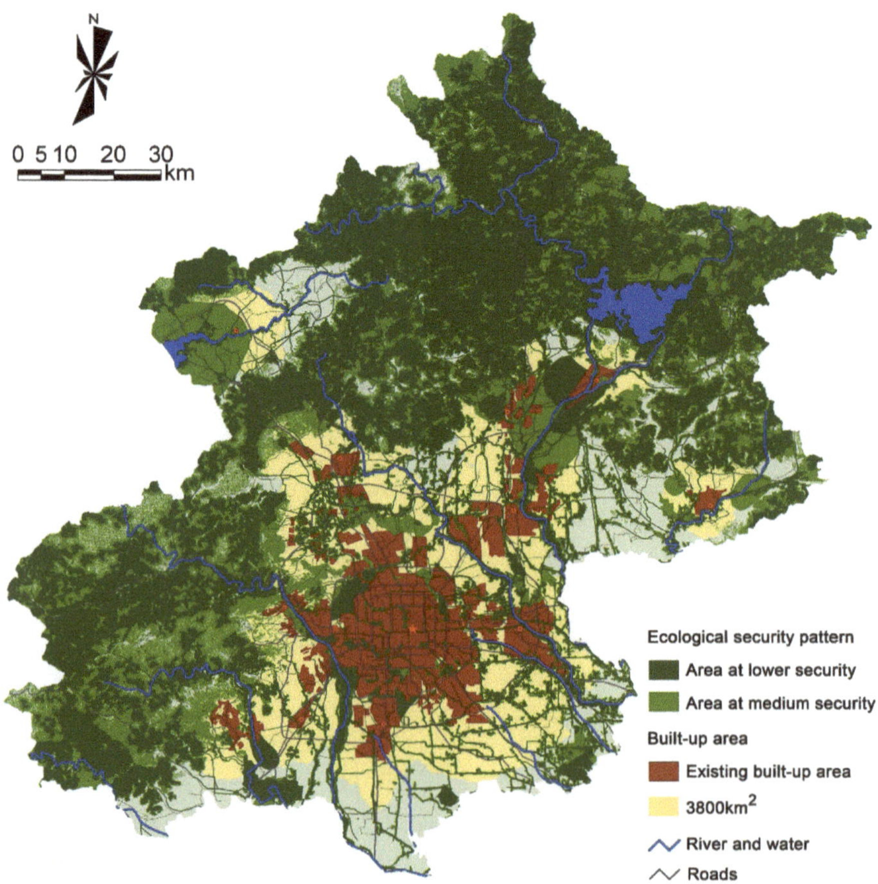


**Fig. 5** Beijing urban growth scenarios based on ecological infrastructure at medium security level: city on *green* infrastructure (*Source* Peking University Graduate School of Landscape Architecture)

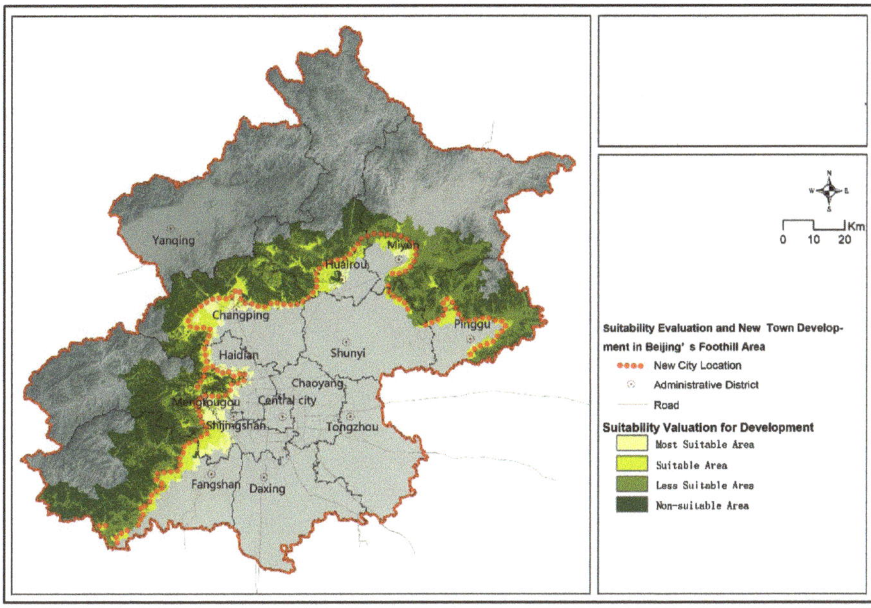


**Fig. 6** The Foothill strategy for Beijing's urbanization based on ecological Infrastructure

### 1.3.3 Ecological Infrastructure at Medium Scale: Eco-city Design Based on Ecological Infrastructure

Alternative Urbanism Based on Ecological Infrastructure

Cities are by far the largest and most complicated artificial devises that human beings have constructed, and they are considered by many to be the very testament of human civilization. From the origin of the city to its "modernized" form today, natural forces and patterns have become increasingly controlled and dependent on artificial processes. The quality of urbanity thus becomes measured by how quickly rain-water drains off our streets, how stable temperature and humidity are maintained in our rooms (or even in open spaces), how garden trees and shrubs are grown for ornamental purposes rather than for their productivity. Over time, we have drifted away from nature and become disconnected with our roots as farmers and herdsmen. This standard of civilization is built upon heavily engineered gray infrastructure. It comprises of complicated transportation systems designed for vehicles to deliver goods and services; huge pipe networks laid underground to drain excess stormwater; rivers reinforced with concrete walls to control floods; large sewage plants built to treat waste water; power lines to convey energy necessary to run all the machines and devices, etc. Built upon this gray infrastructure, are showy buildings with deformed heads and twisted bodies that that deviate from what natural forces allow.

Such model of urbanity, has unfortunately been adopted by developing countries in general and the Chinese cities in particular today. Here, landscape is largely limited to tamed gardens and parks where lawns and flowers are irrigated with tap water and stormwater is drained by underground pipes. Here, landscape is just like other components of an artificial city, it is a sink of energy and services, rather than a source. Landscape as a natural ecosystem in and around cities are largely neglected, natural processes (are) disintegrated and contaminated, and natural patterns are fragmented. The landscape therefore completely loses its capacity in providing what would have been free goods and services for the urban communities.

What would an alternative city look like if its natural forces are respectfully used and not controlled? Vegetables and food would be produced on the streets or in parks, floods would come and go to the benefits of the city, waste would be absorbed and cleansed by the natural processes, birds and native species would cohabit the city with human beings, and the beauty of nature would be appreciated in its authenticity, not tamed nor heavily maintained. This alternative practice has many names and is known as agricultural urbanism, landscape urbanism, water urbanism, new urbanism, sustainable urbanism, green urbanism, and certainly ecological urbanism. The key here is that these alternative solutions do not rely on gray infrastructure, but instead, utilize green or ecological infrastructure to deliver the goods and services the city and its urban residents need. (Yu 2010a).

## Designing Wulijie Eco-city Based on Ecological Infrastructure

This project searches for an alternative urbanism, ecologically sound, and culturally sensitive, based on the essential needs of human beings. The key of this approach to urbanism is the planning and design of ecological infrastructure that provides key services for the city, including stormwater management, food production, habitats for biodiversity, cultural and spiritual experiences, and mobility across the landscape.

Wulijie new town is 10 $km^2$ (6.2 square miles) in the eastern part of Wuhan about 30 km (19 miles) from the city center. Just to its west is a large high tech area called Optics Valley, and at its south and east is the protected, high-water-quality Liang Zhi Hu Lake, to which the water from the Wulijie is drained. The climate is hot and rainy in the summer and cold in the winter. Precipitation is around 1200 mm (47 inches) per year; the average high temperature in the summer is 38 °C (100 °F), and the average low temperature in the winter is just above freezing. The elevations are between 15 and 100 m, with a landform of rolling hills, small basins, and many ponds of different sizes that catch stormwater. Helping alleviate both floods and draughts, the ponds are important features in this previously agricultural land.

Wulijie is planned to be one of many new towns in China to accommodate rapid urbanization. It will have 100,000 residents, of which 10 % will be current locals and 90 % immigrants who will work in Optics Valley and the new town.

The main urban design concept for the town is to use landscape as ecological infrastructure to integrate various natural and cultural processes to frame the city, and to provide diverse ecosystem services for the residents. Integration and connectivity of natural, biological, and cultural processes are central to the project (Figs. 7, 8 and 9 ).

Here are the urban design strategies for the town:

(I) A water-based ecological infrastructure (EI) will organize the town. The EI is designed around the existing water systems and land forms. The key functions

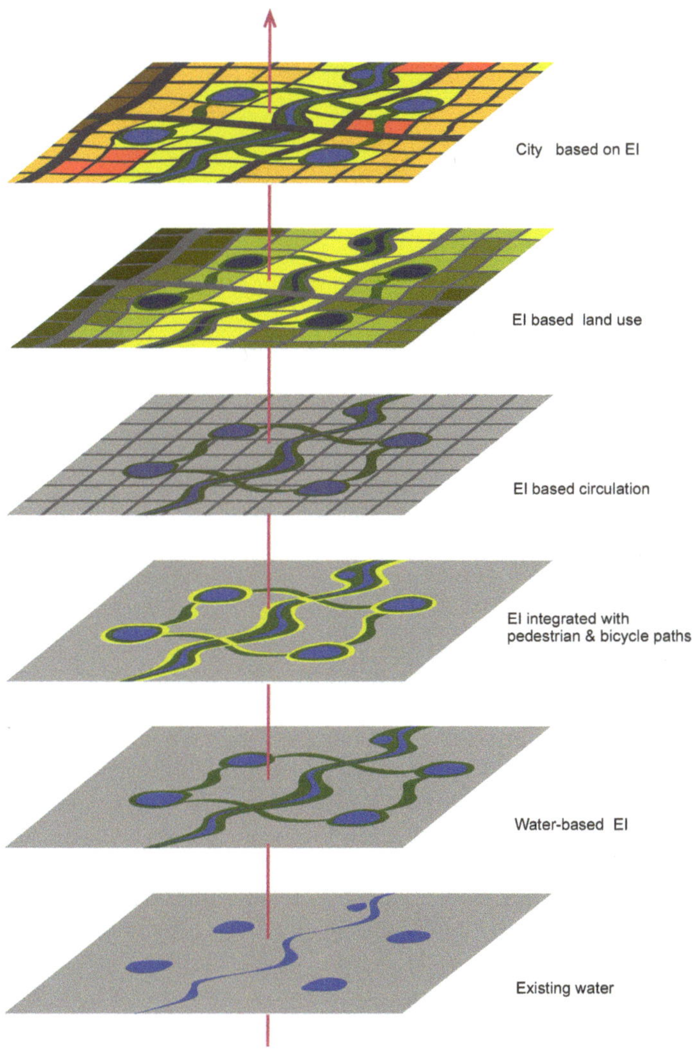


**Fig. 7** The concept model of eco-city based on ecological infrastructure that is based on the existing stormwater retention pond (*Source* Turenscape)

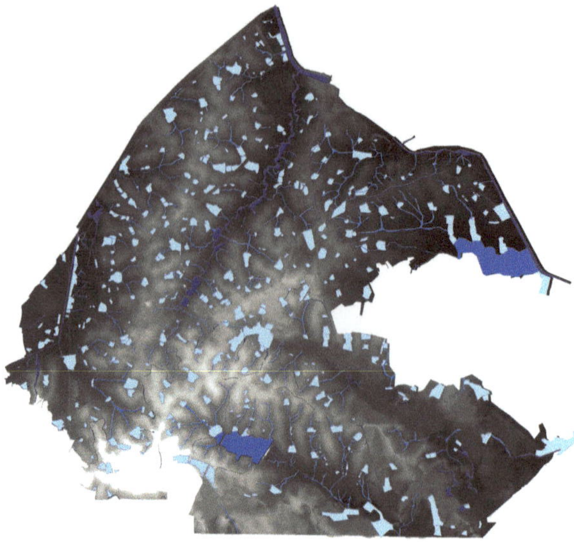

**Fig. 8** The existing water system as adaptive solution to stormwater (*Source* Turenscape)

are to retain and clean stormwater. Different rainfall amounts are simulated to determine the area and pattern of the pond/wetland system so that all stormwater will be retained on site and will have little impact on the regional water system. This will reduce the construction costs of underground drainage pipes and preserve or create habitat for native wildlife and wetland vegetation such as lotus, wild rice stem, water chestnut, water caltrop, and Chinese arrowhead. Public spaces are integrated with the EI. Three levels of green and water corridors are designed to accommodate various functions. The main corridors are 120–150 m wide and will catch runoff from the whole area during the severest storms. Secondary corridors are 60–90 m wide, and these will catch runoff from the subdivisions of the watershed during medium rainstorms. The third level corridors are 20–30 m wide and will catch the water of small storms.

(II) A network of pedestrian trails and bicycle paths will make the town completely walkable and allow residents to commute and have recreational experiences in the green space. While the regional mobility of the city is still serviced by a transit system and roads, the town will use the EI help people move around. The maximum walking distance from any corner of the town to a bus station will be 600 m, and all residents will be able to reach the green network within 5 min.

(III) Urban land is valued according to its relationship with EI. The land overlooking the ecological infrastructure will be given priority for residential development. The overall city form is defined by the EI.

(IV) The eco-friendly, environmentally sensitive, and low carbon landscape and architecture create a new aesthetic environment and a new lifestyle. Native biodiversity, low maintenance, and productivity are the main traits of the vegetation in the green space around the EI. Buildings have roof gardens and

**Fig. 9** The master plan of the Wulijie eco-city based on ecological infrastructure (*Source* Turenscape)

living green facades. Retired people can spend their time fishing in the ponds in front of their apartments or the streets; workers go along the EI to their workplaces; and children play in the productive agricultural fields where parents are growing vegetables in community gardens.

The construction of this ecological city is underway, and the executed first phase of this eco-city has proven to be successful in terms of its ecological and liveability quality.

### 1.3.4 Designed Ecologies: Four Experiments

Ecological Solution to Cleanse Contaminated Water: Landscape as a Living System and Shanghai Houtan Park

Water pollution and shortages pose bigger threats to humankind than do future oil shortages. More than 30 years of rapid urbanization and uninformed hydrological engineering for flood control in China have severely damaged China's water system. Some 75 % of surface water (lakes, streams, rivers, etc.) in China is polluted, up to 60 % of the ground water in metropolitan areas is polluted, and half of China's coastal wetlands have disappeared in the last 50 years. Some of the largest lakes in China (including Taihu, Dianchi, and Caohu) are heavily polluted. All the major rivers in China—the Yangtze, Yellow, Heilong, Pearl, and more—are equally polluted. The author proposes an ecological approach to address surface-water pollution in continuous and complete "natural" systems allowing integration with plant life and free flows between water bodies.

Houtan Park, in China's Shanghai City, became an experimental showcase of regenerative design in the middle of a densely populated urban setting. Created in conjunction with the 2010 Shanghai Expo, the park integrated site topography, flood control, and native habitat to rejuvenate ecosystem services. It reactivated valuable open space and rebuilt biodiversity while managing stormwater, cleaning contaminated river water and soil, and educating the community about the beauty of diverse native landscapes (Yu 2011a).

The 14-ha (34.6 acre) site is located on the southern boundary of the Expo grounds, locked between the east bank of the Huangpu River and Puming Road, a main city thoroughfare. Currently, the Huangpu River water is designated as Lower Grade V, the poorest quality on a scale of I–V. Water from the Huangpu River is diverted to the constructed wetland where it is biologically treated to an improved Grade III designation via a series of wetland cells that facilitate settling, aeration, and vegetative and microbial processes. The treated river water can then be used safely for landscape irrigation and other non-potable uses in adjacent parks. Full-scale pilot-testing results indicated that the Houtan Park treatment wetland has the ability to treat over 2400 $m^3$ of water per day.

In addition to its performance as a water-cleansing infrastructure, the park has proven to be a great success in other ecological aspects, and one unusual aspect of this success is that the park has become a refuge for diverse native species. Just one year after completion, some 20 species of birds have found a home in this small park. In addition to the designed plantings, many species have immigrated from the adjacent remnant riparian patch and taken root along the river corridor. It is important to note that the protection of this habitat immediately adjacent the park played an important role as one of the key sources of biodiversity, and the Huangpu River has actually played a role as a species-transporting and migratory corridor. Although it is but a node and a small-scale place, Houtan Park acts as part of landscape infrastructure at the regional and even national scale (Figs. 10 and 11).

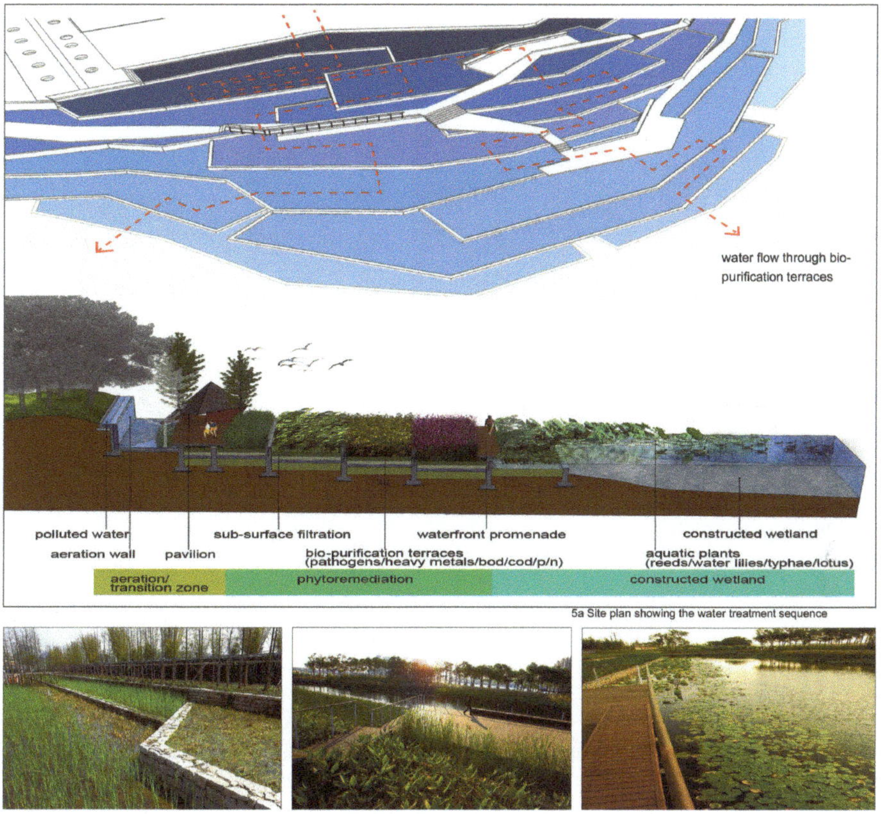


**Fig. 10** Shanghai Houtan Park: landscape as a living system that cleanse the contaminated river water and provide multiple ecosystem services (*Source* Turenscape)

## Ecological Solution to Storm-Water Management: A Green Sponge for Water-Resilient City and Qunli Storm-Water Park

Contemporary cities are not resilient when faced with inundations of surface water. Landscape architecture can play a key role in addressing this problem. This project demonstrates a Stormwater park that acts as a green sponge, cleansing, and storing urban stormwater and can be integrated with other ecosystem services including the protection of native habitats, aquifer recharge, recreational use, and aesthetic experience, in all these ways fostering urban development (Yu 2011b).

(I) **Challenges of Storm-Water and the Ecological Solution**
Beginning in 2006, a 2733 ha new urban district, Qunli New Town, was planned for the eastern outskirts of Haerbin in northern China. Thirty-two million square meters of building floor area will be constructed in the next 13–15 years. More than one-third of a million people are expected to live

**Fig. 11** The designed ecologies: the intensified constructed wetland that clean water and support wild life and sits in the middle of the metropolitan area (*Photo* Yu)

there. While about 16 % of the developable land was zoned as permeable green space, the majority of the former flat plain will be covered with impermeable concrete. The annual rainfall there is 567 mm, with the months of June, July, and August accounting for 60–70 % of annual precipitation. Floods and water logging have occurred frequently in the past, while at the same time the ground water table continues to drop due to its overuse.

In mid-2009, the landscape architect was commissioned to design a park of 34 ha right in the middle of this new town, which is listed as a protected regional wetland. The site is surrounded on four sides by roads and dense development. This wetland had thereby been severed from its water sources and was under threat. The original task given by the client was to preserve this wetland. Going beyond the original task of preserving the wetland, the landscape architect proposed to transform the area into an urban storm-water park that will provide multiple ecosystems services, and will collect, cleanse, and store stormwater and infiltrate it into the aquifer, protect, and recover the native habitats, proved a public space for recreational use and aesthetics experience, as well as foster urban development.

The challenges are obvious: How can a disappearing wetland be preserved in the middle of the city when its ecological and biological processes have been cut off by the urban context? How such an urban wetland ecosystem can be designed to provide multiple ecosystems for the city? And what is the economic way to deal with such a big landscape?

(II) **The Ecological Solutions**

Several design strategies and elements were employed (Figs. 12 and 13):

(a) The central part of the existing wetland is left along to allow the natural habitats to continue to evolve.

(b) Cut-and-fill strategy to create an outer ring of mounds and ponds. The cut-and-fill around the perimeter is a minimum earthwork strategy to transform the site. Earth is excavated and used to build up a necklace of ponds and mounds around the perimeter of the park. This ring acts as a stormwater filtrating and cleansing buffer zone for the core wetland, and a transition between nature and city. Stormwater from the newly built urban area is collected around the perimeter of the wetland and then released evenly into the wetland after having being filtered through the ponds. Native wetland grasses and meadows are grown on ponds of various depths, and natural processes are initiated. Groves of native silver Birch trees (*Betula pendula*) grow on mounds of various heights and create dense woodland. A network of paths links the ring of ponds and mounds, allowing visitors to have a walking-through-forest experience. Platforms and seats are put near the ponds to enable people to have close contact with nature.

(c) The network of paths and platforms: A network of paths links the ring of ponds and mounds, allowing visitors to have a walking-through-forest experience. Platforms and seats are put near the ponds to enable people to have close contact with nature. A skywalk links the scattered mounds allowing residents to have an above-the-wetland and in-the-canopy

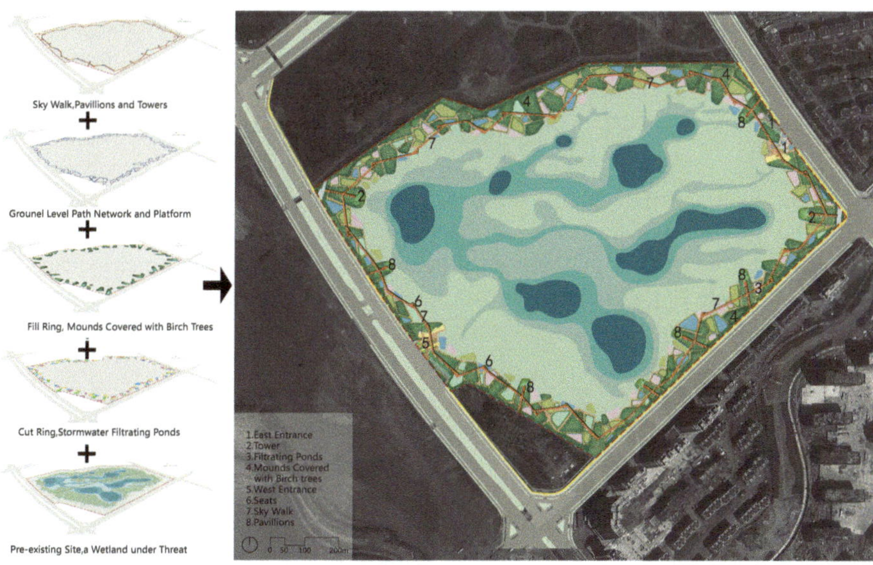


**Fig. 12** Qunli Storm-water Park: the design concept and site plan (*Source* Turenscape)

**Fig. 13** Qunli Storm-water Park: cut-and-fill to create an outer ring of mounds and ponds acting as a stormwater filtrating and cleansing buffer zone for the core wetland, and a transition between nature and city (*Photo* Yu)

experience. Platforms, five pavilions (Bamboo, Wood, Brick, Stone, and Metal), and two viewing towers (one made of steel and located at the east corner, the other one made of wood and looking like a tree at the north-west corner), are set on the mounds and connected by the skywalk, allowing visitors to have views into the distance and observation of nature in the center of the park.

The completely transformed site performs many functions, including collecting, cleansing, and storing storm-water, and recharging underground aquifers. The pre-existing wetland habitat has been restored and native biodiversity preserved. Potentially flooding stormwater now contributes to an environmental amenity in the city. The stormwater park has not only become a popular urban amenity but has also been upgraded to a National Urban Wetland Park because of its improvement to ecological and biological conditions.

## Ecological Solution to the Recovery of a Degrade Land: Adaptation Pallets and Tianjin Qiaoyuan Park

Through Regenerative Design and by changing the landform, the natural process of plant adaptation, and community evolution is introduced to transform a former deserted shooting range used as a garbage dump, into a low maintenance urban park; providing diverse nature's services for the city including containing and purifying stormwater; improving the saline-alkali soil, providing opportunities for environmental education and creating a cherished aesthetic experience (Yu 2010b).

(I) **Challenges of Degraded Land**

This is a park of 22 ha in the northern coastal city of Tianjin, China. Rapid urbanization had changed a peripheral shooting range into a garbage dump and drainage sink for urban stormwater; the site was heavy polluted, littered, deserted, and surrounded with slums and temporary rickety structures, which had been torn down before the design was commissioned. The soil is quite saline and alkaline. Densely populated at the south and east boundaries, the site is bounded on the west and north sides by a highway and an overpass.

In early 2006, in response to residents' call for environmental improvement of the site, the municipal government of Tianjin contracted the landscape architect with the difficult task of an immediate transformation of this degraded site.

The overall design goal for this project is to create a park that can provide a diversity of nature's services for the city and the surrounding urban residents, including: containing and purifying urban stormwater; improving the saline-alkali soil through natural processes; recovering the regional landscape with low maintenance native vegetation; providing opportunities for environmental education about native landscapes and natural systems, stormwater management, soil improvement, and landscape sustainability; creating a cherished aesthetic experience.

The regional landscape is flat and was once rich in wetlands and salt marshes, which had been mostly destroyed by decades of urban development and infrastructure construction. Though it is difficult to grow trees in the saline-alkali soil, the ground cover, and wetland vegetation are rich and vary in response to subtle changes in the water table and PH values.

(II) **Ecological Sotuions**

Inspired by the adaptive vegetation communities that dotted the landscape in this region, the solution for this park was developed called The Adaptation Palettes, which was designed to let the nature work. A simple landscape Regenerative design strategy was devised, one that included digging 21 pond cavities varying from 10 to 40 m in diameter, and from 1 to 5 m in depth. The garbage was handled in the earth work. Some cavities are below ground level and some above on mounds (Figs. 14 and 15).

Through the raining season and due to the shallow underground water, some cavities turn into water ponds, some into wetlands, some into seasonal pool, and some stay as dry cavities. Through seasons' rain wash and filtration, the saline-alkali soil of the dry cavities get improved, while nutrients deposit in the deeper ponds that catch stormwater runoff.

Diverse habitats were created and the natural process of plant adaptation and community evolution were initiated. Seeds of mixed plant species were sowed initially to start the vegetation, and other native species were allowed to grow

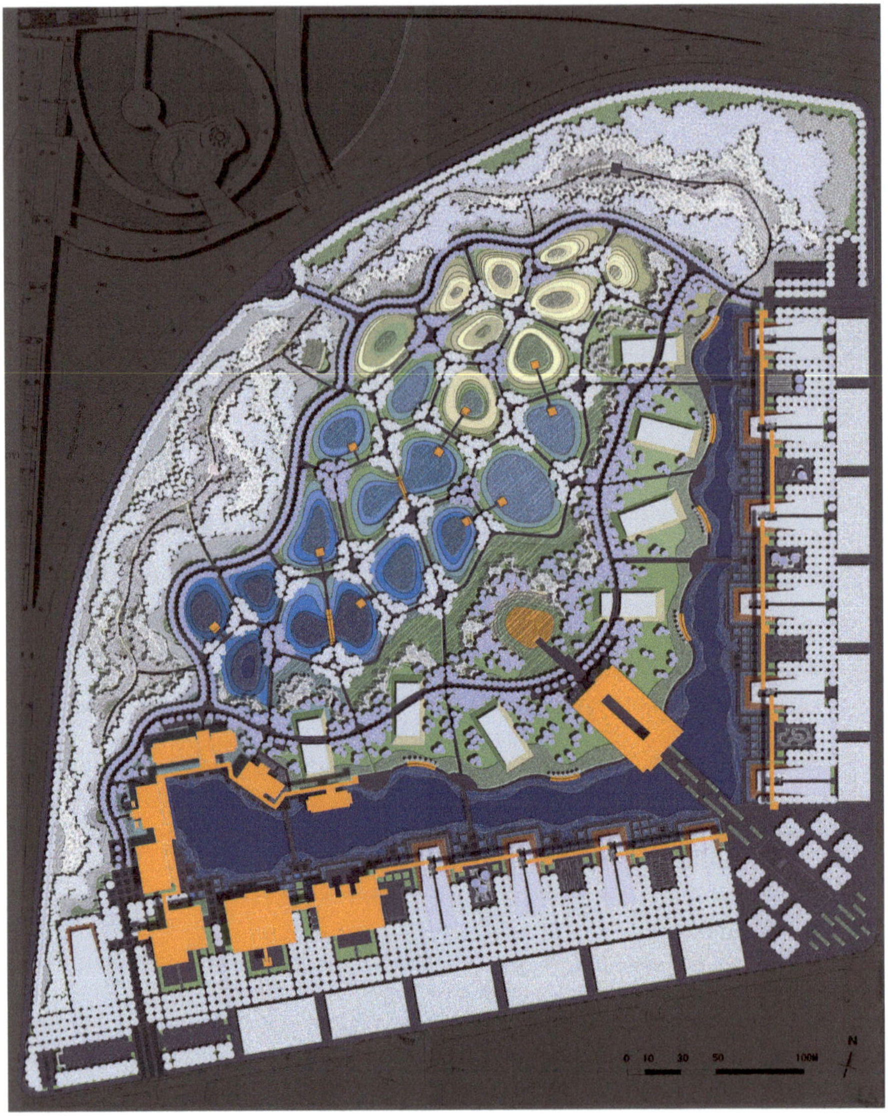


**Fig. 14** The adaptation pallets: the site plan for Tianjin Qiaoyuan Park (*Source* Turenscape)

wherever suitable. Through the seasons' evolution, patches of unique vegetation establish in correspondence to the individual wet or dry cavities, and various PH values. The allotment landscape reflects the regional water- and alkaline-sensitive vegetation.

Within some of the cavities are wood platforms that allow visitors to sit right in the middle of the vegetation patches. A network of red-colored asphalt was designed to weave through the palettes and allow visitors to stroll through the

**Fig. 15** Before: the degrade urban environment (*Photo* Yu)

patchy landscape. Along the paths is an environmental interpretation system that gives descriptions of natural patterns, processes, and native species.

The park achieved its goals in just 2 years. Stormwater is retained in the water cavities, allowing diverse water-sensitive communities to evolve. Seasonal changes in plant species occur and integrate with the beauty of the "messy" native landscape, attracting thousands of visitors every day. In the first 2 months of its opening, from the October to November of 2008, about 200,000 people visited the park. It is a successful park which changes its landscape throughout the year period, constantly visited by the community, needs very little maintenance.

This project helps to define the new aesthetics of landscape today, defined by a continuous evolving process. Untidy forms, unplanned biodiversity and nature's "messiness" keep ongoing, letting plants live and expose their genuine beauty to enrich the landscape. The ecology-driven Adaptation Palettes has become a valuable and remarkable site of the community of Tianjin (Fig. 16).

### An Integral Ecological Approach to Recover a Destroyed Mother River: An Urban Ecological Infrastructure and Sanlihe Greenway

The Qian'an Sanlihe Greenway is transformed from former garbage dump and sewage drainage. It stands as an example of how a neglected landscape can be

**Fig. 16** After: the ecologically recovered Tianjin Qiaoyuan Park (*Photo* Yu)

recovered as an ecological infrastructure and everyday landscape with restored ecosystem capacity in providing multiple services, including mediating flood and draught, providing habitats for native biodiversity, integrating pedestrian and bicycle paths for commuting and recreational uses, creating spiritual and aesthetic benefits, and catalyzing urban development (Saunders 2012; Yu 2013a, b).

(I) **Challenges of A Destroyed Urban River**

The greenway stretches 13.4 km in length and varies 100–300 m in width across the city of Qian'an. It covers approximately 135 ha and benefits a population of approximately 700,000. Qian'an City is located at the south foot of the Yanshan Mountain, at the bank of Luan River, in the northeast of Hebei Province. Although the main city lies near the Luan River to the west, one cannot see the water since Qian'an's topography is situated below the riverbed with its high embankment blocking the river view. The river is notorious for its unpredictable flooding, and has thus been kept outside of the city for decades through this high embankment. Meanwhile, as the life source of Qian'an, Sanlihe River has shouldered the long history of the city and carried the collective memory of the inhabitants. Before 1973, the Sanlihe River had crystal clear water from the groundwater recharge of Luan River, which ensured that the temperature of the city stayed cool in summer and warm enough to never ice up in winter. Although frequented by storms and heavy rain, Sanlihe River never experienced disasters of drought and flood in its history, which also provides rich water resources for nearby industries and agriculture, as legendary records witness: "reeds flourish, trees shade, and birds inhabit".

However, since the 1970s, the river has been badly polluted by sewage and waste which has resulted from the territory's continuous industrial development and urban population growth. As a consequence, with the depletion of regional water sources, the Sanlihe River became subsequently dried up and its channel blocked by solid waste. The life source of the city became festered with sores of urbanism, and the hearts and souls of local residents long for its spiritual landscape reincarnation.

(II) **Ecological Solutions to Turn A River Corridor as an Ecological Infrastructure**

The landscape architect was commissioned to recover this mother river. The scope of job included sewage management (the redesign of sewage pipes that had previously discharged directly into the river with a passive natural infiltration system), as well as ecological restoration and the urban design along the greenway (although this submission only focuses on the planning and design of the greenway itself).

The design strategies are comprehensive and developed across scales:

- Clean the site: A sewage management system was planned to separate waste water from the urban stormwater runoff. As well, organic garbage from the household was used as material to shape landforms, and industrial waste was cleaned and properly treated.
- A scenic water byway: the design for the greenway took full advantage of the existing natural elevation change between the Luan River bed and the city. A fountain was made through a pipe that goes under the high embankment, so that a constant controlled amount of water will make its way through the city before running back to the Luan River at the lower reach. This strategy turns the Sanlihe into a "scenic byway" of the larger Luan River and transforms the dangerous natural force into a pleasant amenity.
- Resilient green river strategy: the existing concrete channel of the river was removed, and a multiple water course riparian wetland system was created, including the creation of emerald-like wetland bubble chains at the edge of the main water course which regulates floods and collects and dissipates urban stormwater runoff. When the river's water level drops to its lowest point, pools of water remain in the emeralds as wetlands, creating a "Green River." Furthermore, these wetlands work as an ecological purification buffer for urban stormwater runoff from both sides. These meandering natural waterways, at various surface levels, becomes diverse habitats for wildlife.
- Tree islands: The existing trees on the site were saved and the riverbanks were transformed into a number of tree islands connected by boardwalks, creating a unique setting for daily activities of the nearby residents.
- Pedestrian and cycling paths: Along the greenway are the pedestrian and bicycle routes fully accessible to communities along the channel.

- Landscape guides Urbanism: The greenway is used as catalyst for urban development. High density residential development was envisioned at both sides of the greenway. Immediately after the greenway was built, an enormous amount of new housing development investment was attracted and completely transformed the urban morphology of the city.
- Low maintenance: The project used low-maintenance native vegetation, lush wetland species, and self-reproductive wild flowers for the undercover. The resilient green river strategy allows the water table to naturally fluctuate over seasons. The contrast between "messy" nature and the minimum design of the boardwalk and waterfront platforms turn the big greenway project into an artful everyday landscape.

Through only 3 years of design and construction, this project has transformed this seriously polluted landscape back to its previous splendour as a scenic urban ecological corridor. The mother river has been recovered and the legendary tale narrated by the grandmothers once again rings true: as a place "where reeds and lotus flourish and water abundant with fish and soft-shelled turtles." (Figs. 17, 18 and 19)

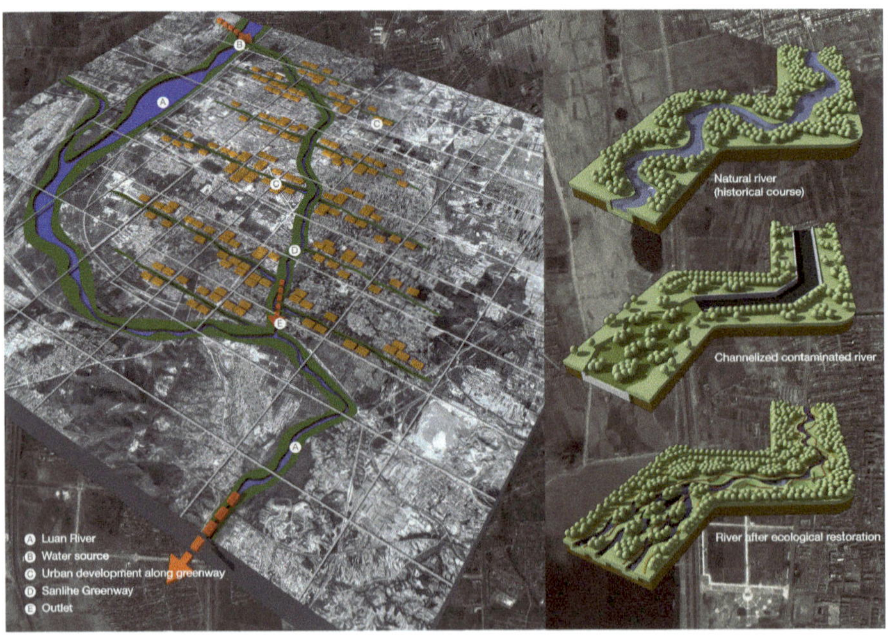


**Fig. 17** The Sanlihe Greenway: recovering the mother river of Qian'an (*Source* Turenscape)

**Fig. 18** Before: the badly degraded urban river (*Photo* Yu)

**Fig. 19** After: the ecologically recovered greenway along Sanlihe River (*Photo* Yu)

### 1.4 Conclusion

It is argued that, the current urban growth model in China is unsustainable. Recognizing this fact, the Chinese leadership is now calling for ecological civilization and new type of urbanization (Hu 2007; Xi 2013), a totally new concept proposed in Chinese language and especially worded from the top Chinese leader. It reflects an important change in the top Chinese leadership understands of development. Rather than emphasizing economic construction as the core of development as it did in the past, the Chinese leadership has come to realize that development, if sustainable, must entail a list of elements including the right relationship between man and nature. The ecological civilization concept is proposed at a time when ecological and environmental issues are at a very serious stage. Facing such a reality, the construction of ecological civilization was absolutely not rhetoric for chest thumping by officials in their speeches. It needs to be transformed into tangible measures that will change the way our economy develops and reshape the landscape that can meet the serious challenges of sustainable development.

Accordingly, it is important to recognize that the conventional approach to development planning, which is based on population projection and then built-infrastructure, is unable to meet the challenges and needs of the ecological and sustainable urban development, and certainly unable to meet the goal of ecological civilization. It is in this situation, that the negative approach is proposed (Yu et al. 2008a, b, 2011b). Using the analogy of photography in describing the film and picture, the term "negative" is used to describe the urban development model being negatively en-framed by Ecological Infrastructure, not the other way around. To say it in the other way, the EI is positively defining the urban form and growth pattern. Conventionally, landscape and green elements such as Greenbelt and Greenheart are usually negatively defined by architectural and built infrastructure. By positively defining the EI for the sake of Natural Capital and cultural integrity of the land, the urban growth pattern and urban form are negatively defined. The concept of Ecological Infrastructure builds a bridge between the disciplines of ecology and especially landscape ecology, the notion of Natural Capital and ecosystems services, and sustainable development. It is a bridge between smart growth and smart conservation.

It has been demonstrated that by implementing ecological infrastructure as a solution to urban environmental issues, such as flood control, stormwater management, soil remediation, water cleansing, and recreation, alternative, and more sustainable and pleasant urban environment and landscape can be achieved.

## 2 Theory and Methods of Urban Ecological Construction

With the development of economic globalization and continuous quickening of urbanization process, the quantity and scale of cities is continuously expanded. The increasingly serious urban problems which appear since 1960s and 1970s makes the seeking for more reasonable urban development mode and human settlement mode become an important and urgent task which the national governments, international institutions, and academic communities shall confront at present. Since 1970s, the ecological city attracts wide attention from all sectors of society, and all countries actively seek for construction of ecological city. The ecological city is a brand-new stage of urban development, and it is also a new mode for human's sustainable development.

In 1984, Ma Shijun et al. proposed a theory that the city is a complex society-economy-nature ecosystem. From perspective of geographic space, the ecological city has exceeded the concept of "city" in traditional sense; with emphasis on essential identity of inhabitation as human's living place, it is represented as a kind of new pattern of urban-rural relation, and the integrated development of city and rural area forms urban-rural network structure. From perspective of man and nature, the ecological city not only promotes human's healthy evolution and development, but also pays attention to development of the nature; therefore, the ecological city can become a new habitation environment which can make offerings to man and nature; under this environment, the man and nature realizes mutual adaptation, concerted evolution, coexistence, and co-prosperity, which shows the inseparable unity between human and nature. The ecological city not only changes the form and function of current human settlement, but also changes the humans themselves and creates a kind of new civilization and culture. The construction of ecological city can effectively improve habitation environment for urban people and enhance urban people's living quality. Currently, many cities at home and abroad are under planning and construction toward ecological city, which raises a great upsurge in construction of ecological city.

The urban ecological construction is the only road to solve current ecological environment problem, and it is also the critical link to realize sustainable development. Up to now, the idea of urban ecological construction has been developed into a construction system integrated with cultural inheritance, target integration and multiple modes, and the research fields are continuously expanded, involving philosophy, religion, ethics, aesthetics, architecture, economics, technology, etc. In China, there is deeper and deeper research on urban ecological environment problem and urban ecological construction. In the process of specific planning and practice of urban ecology, the theories (such as ecological carrying capacity, ecological function zoning, and ecological civilization construction) play the most critical role. The ecological carrying capacity is the important condition for urban development, and it influences and restricts the way and strength of urban land utilization. The urban ecological construction is restricted by ecological carrying capacity, thus all parts of ecological environment carrying capacity shall be used as reference factors for urban ecological construction; otherwise some hidden danger may be left in urban development. Therefore, it is of important significance to master the trends and features of ecological carrying capacity. The ecological function zoning is the foundation for reasonable management and continuous utilization of ecological system and natural resources, and it can provide scientific basis for construction of ecological environment and formulation of environmental management policies. On the basis of full research on features and laws of such elements as China's ecological territory, ecological service function, ecological asset, ecological sensitivity, and intimidation of human activities on ecological environment, the important content of urban ecological construction is to establish principle, method, and index system of China's ecological zoning, and then to merge and distinguish the related ecological territory. The ecological civilization pushes the urban ecological construction to a new height. The ecological civilization construction is inheritance and development of current urban ecological construction and it aims at promoting ecological civilization construction and pushing the urban ecological construction to a new height through comprehensive optimization and improvement of six systems, that is, ecological concept, ecological economy, ecological environment, ecological habitation, ecological society, and ecological system.

## 3 Urban Ecosystem Health Assessment and Management

### *3.1 Introduction*

Acting as centers for various socioeconomic activities, cities play a driving role in regional development and human civilization (Su et al. 2009, 2010b). Meanwhile, although cities have made obvious economic growth and notable wealth, we cannot ignore the negative impact of various emerging eco-environmental problems, such as air pollution, water resource scarcity and energy shortage, which further lowered the human living level and impeded the urban sustainable development (Guidotti 1995; Su et al. 2010a). Therefore, a systematic assessment of urban ecosystem health (UEH) is urgently needed for effective ecological management and sustainable development.

### *3.2 Urban Ecosystem Health Assessment Indicators*

Based on the understanding of UEH concept, i.e., (1) the ecosystem's ability for its own renewal and satisfying human demands should be integrated, and (2) various factors should be taken into account, several assessment indicators have been established (Table 1).

**Table 1** Typical urban ecosystem health indicators

| Focus | Indicators | Reference |
|---|---|---|
| External characteristics | WHO: (1) Proposed 79 indicators of a healthy urban ecosystem in 1996 from nine aspects (e.g., influence factor, progress, management, service, etc.); (2) Further developed 459 indicators in 1998 from 12 aspects | Takano and Nakamura (1998) |
| | Taking the classic framework of natural ecosystem health assessment (Mageau et al. 1995; Rapport et al. 1998), a similar framework of UEH indicators was established using 24 factors | Guo et al. (2002) |
| | Organized 30 UEH indicators using the framework of natural, economic and social subsystems in the urban ecosystem | Zhong and Peng (2003) |
| | Put forward the distance index and coordination index, for the spatial difference of UEH into account | Hu et al. (2005) |
| Internal process | Established 17 biophysical UEH indicators, by integrating energy with vigor, structure, resilience, ecosystem service maintenance and environmental impact | Su et al. (2009) |
| | Developed an energy-based UEH indicator by integrating five energy indexes | Liu et al. (2009b) |

## 3.3 *Urban Ecosystem Health Assessment Models*

Many mathematical models are developed to treat and process the indicator data to satisfy a UEH assessment. The models can be summarized into two categories (Table 2): one is based on understanding the UEH's character while another faces the problems during the UEH assessment.

**Table 2** Typical urban ecosystem health assessment models

| Emphasis | Models | Reference |
|---|---|---|
| The character of UEH | Fuzzy synthetic assessment model | Guo et al. (2002) |
| | Relative vector assessment model | Sang et al. (2006) |
| | Attribute theory model | Yan (2007) |
| | Set pair analysis model | Su et al. (2009) |
| Treating emerging problems during UEH assessment | Unascertained measure model | Shi and Yan (2007) |
| | Matter element model | Dai et al. (2007) |

## 3.4 Urban Ecosystem Health Management

### 3.4.1 Identification of Limiting Factors of Urban Development

Based on the UEH assessment results, Yang et al. (2010) identified the limiting factors of Ningbo's healthy development (Fig. 20) such as the water resources shortage, the pressure of $SO_2$ emission, insufficiency of sewage treatment, etc. Then urban ecological regulation scheme is suggested aiming at these limiting factors.

### 3.4.2 Classification of Urban Ecosystem Health Patterns

According to cluster analysis of UEH (Liu et al. 2009a), the Chinese cities are classified into four patterns (high economy development, low economy development, high economy restraint, and low economy restraint patterns in Fig. 21). Measures for different patterns are thus proposed.

### 3.4.3 Zoning Management Based on Urban Ecosystem Health

Based on the spatial distribution of UEH levels (Fig. 22), the zoning management is suggested in Guangzhou, China, where the north part (mainly including forests, grasslands, rivers, and wetlands) is defined as a conservation area, the middle and southern parts (mainly including cultivated land and general commercial area), are

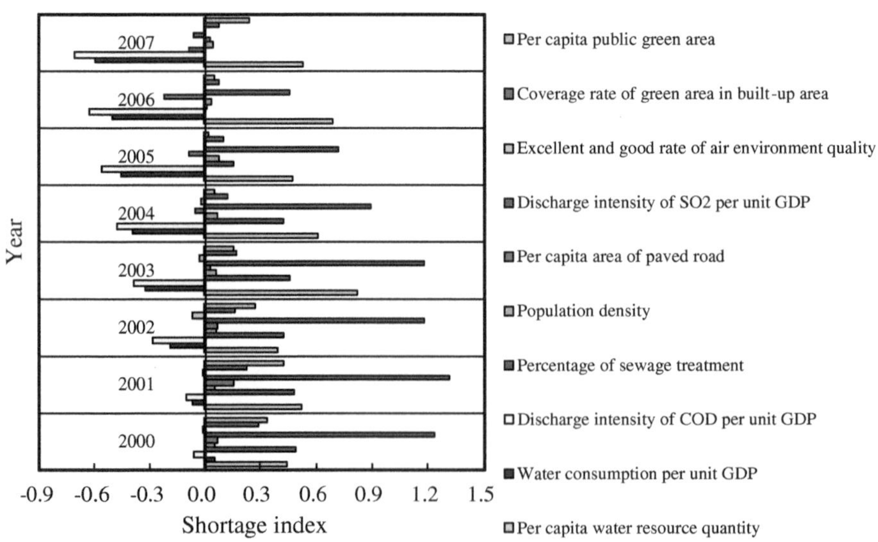


**Fig. 20** Limiting factors analysis results of Ningbo, China

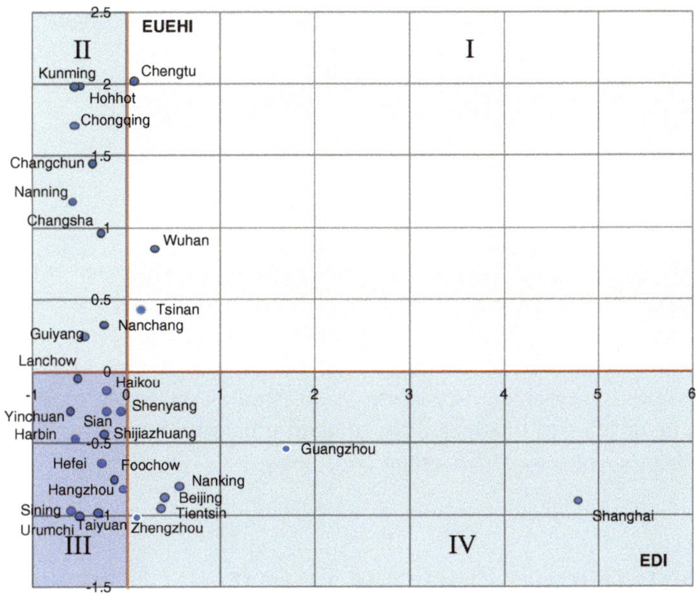


**Fig. 21** Cities classification based on urban ecosystem health patterns

**Fig. 22** Spatial distribution of urban ecosystem health levels in Guangzhou, China

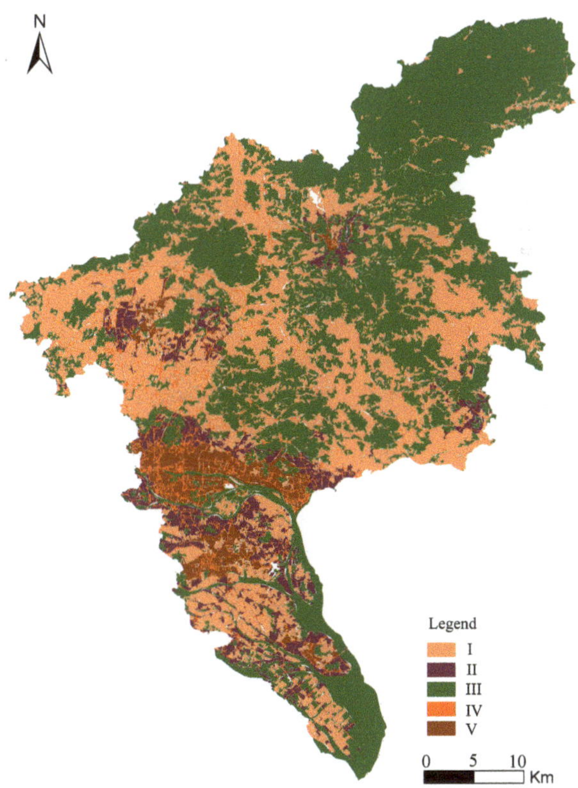

defined as maintenance areas, while the south central and southwestern parts (mainly including production and consumption centers and traffic areas) are defined as key regulation areas (Su and Fath 2012).

### 3.5 *Conclusion*

Driven by the rapid urbanization and the increasingly deteriorated urban environment, research in UEH assessment has obtained much progress in assessment indicators, assessment models, and applications to specific urban management. Future research can focus more studies on the dynamic trends of UEH and projecting possible development scenarios. Also, studies on larger scales like urban cluster can be developed to adapt to the modern urban development trend and make the research more beneficial for urban well-being.

## 4 Construction and Practice of Eco-city and Eco-county Planning

### 4.1 *Introduction*

Faced with the acceleration of worldwide urbanization trends and the severe reality of ecological environment since 1960s–1970s, the international society hoped to apply the ecology principles and methods to guide the urban construction by officially promoting the idea of 'Eco-city', whose ideological origin and theoretical basis were developed on the basis of assumptions, such as the British T. More's 'Utopia' in the sixteenth century, C. Fourier's 'Phalanstère', R. Owen's 'Village of New Harmony' and E. Howard's 'Garden city' in the eighteenth and nineteenth century, L. Corbusier's 'the Radiant City' and F. Wright's 'Broadacre City' in the 1930s–1940s.

Human settlement and urbanization work must be planned in order to avoid the adverse effects on environment, and to achieve the best interest of the society, economy, and environment. This idea was proposed from the 'Human Environment Declaration' of the Stockholm conference in 1972. Since then, the West, especially Nordic countries, has carried out a lot of research and debate on this from 1970s to 1980s. At that time, people mainly focused on making efforts to reduce human impact on the environment and to promote environmental friendly behavior. The eco-city not only involved the parks and other green spaces within the city, but also began to involve the relationship between human and nature within the urban environment in concerned with human energy flow and material flow. Urban ecology began to represent the transform of the current energy-intensive lifestyle, and was considered to be a way of reaching the ecological and social sustainability.

Urban ecological projects set out to establish environmental consciousness through the use of alternative energy sources, water supply technology, and waste disposal and drainage technologies. Eco-city construction has been widely carried out around the world since the concept of eco-city was proposed by MAB Plan (Man and the Biosphere Program) in 1972. The concept of eco-city develops in many aspects, and the understanding of the reconstruction of ecological balance also becomes more specific and ideal rather than just as common sense.

At present time, it is generally believed that 'eco-city' refers to the coordinated development of society, economy, and nature, the highly efficient utilization of material, energy and information, and the human settlements with perfect infrastructure, reasonable layout, and virtuous ecological cycles. The scientific connotation of eco-city is to advocate the social stability of civilization, the efficient economy and the harmonious ecological environment. Eco-city is not only a process of human social development, but also a desirable goal of state to achieve at the time when productive forces highly develop; social culture and human ecological environmental consciousness reach a certain level (Wang 2003).

The key to constructing an eco-city is to build an artificial complex ecosystem with a reasonable structure, functional efficiency, and coordinated relationship, which is also the current demand for modern urban people. In terms of the present overall situation of our country, whose level of urbanization only reaches about 51.3 %; its urbanization is in rapid growth stage of development. However, as the driving force of nation's economic reforms, Shanghai and the coastal economic developed area have entered the urbanization stage of 'overspeed.' Urban construction and economic development in these areas are in an unprecedented scale. The urbanization level of most of the eastern coastal cities reaches more than 75 %. At the same time, in the process of urban construction and development, there are a lot of behaviors which are limited to short-term economic benefits and ignore the long-term environmental benefits, lacking of vision. Therefore, it is a practical problem in the construction of eco-city that how to avoid building the city at the expense of urban ecological environment quality and how to avoid the phenomenon of 'first pollution, then governance' likes many foreign industrialized countries have done.

## *4.2 On the Theories of Eco-city and Eco-county Planning and Construction*

### 4.2.1 International Theory

In last two decades, European countries, the United States, Canada, Australia, and other countries have made a lot of exploration on the theory of eco-city planning and construction. Their representative work mainly includes the following aspects:

Yanistky (1981), a Russian Ecologist, considered that the process of designing and implementation of an eco-city can be divided into three kinds of knowledge levels and five action stages, namely:

Level of time and space
Level of society and function
Level of culture and history

The five action phases are basic research, applied research, design, planning, the implementation of construction and the formation of organic organizational structure.

The first level based on the urban ecological design and implementation can be understood as a natural geographic level, which is the spontaneous level of human activity and is the trend to pursue urban ecological niche and compete for balance, in order to do all it can. As human activities become more intensified, contradictions between urban and natural environment are more and more obvious in concerned with time and space. As a result, human's desire of improving the relationship between city and nature, and strengthening the function of the whole system become more and more strong. This is what the second level, namely the level of society and function, mainly discusses. The concept of eco-city is in the third level, the level of culture and consciousness. It aims to study men's ecological awareness, to change external control to internal adjustment, and to change the spontaneous behavior into conscious ones. Therefore, the ecological construction of city must develop from the first level to the second and third level. People should integrate the knowledge of sociology, economics, ecology, environment, geography, behavioral science, psychology, and other disciplines into the city planning and construction field. Changes the past simple material construction planning and construction for the social economy and the natural integration of comprehensive planning and construction, and change the past simple material construction planning into social, economic, and natural integration of comprehensive planning and construction. Yanistky's idea of ecological urban planning and construction can be summarized as 'A kind of human settlements established in according to ecology principles, where society, economy and nature develop harmoniously, material, energy and information were efficiently utilized, and ecology circles virtuously, namely efficient and harmonious human habitat.'

Gordon (1990) published a book called 'green city.' discussing the approaches of ecological construction of urban spaces. The idea of green city from an Indian scholar Dr. Rashmi Mayur is especially prominent. His idea includes: (1) Green city is the embodiment of the harmonious relationship between biological material and cultural resources, and is the condensed matter of the link between them. (2) Green city has full capacity, a balanced energy output, and even a surplus value in the nature. (3) Green city protects natural resources, eliminates or reduces waste with the minimum requirements, and recycles the waste produced inevitably. (4) Green city has a wide open space and other species coexisting with humans. (5) Green city emphasizes on human health, advocates green food, and reasonable consumption. (6) Each key elements of the city were planned according to aesthetic principles and arranged based on the relationship between creativity and nature imagination. (7) Green city provides comprehensive cultural development. (8) Green city is the final result of scientific planning of urban and human community.

Russian urban planning departments summarized the urban eco-environment planning work in 1991, and put forward the way and principle of urban eco-environment construction and protection strategies: (1) The proportion of planning layout and technology in solving the problem of urban nature conservation. (2) Urban geology, ecological border, the layout link and function link between the adjacent areas and population planning. (3) Ecological zones limiting the contamination effect and artificial load of each partition and reducing its influence. (4) The basic principle of solving the problem of land function and spatial organization against environmental damage. (5) Urban transportation, engineering, and energy infrastructure meeting the requirements of ecology. (6) Reasonable proportion of architectural space and green space, with green as the 'skeleton'. (7) Reconstruction principles of residential areas and industrial areas that meet the requirements of ecology. (8) Ecological aesthetic requirements of the spatial organization of urban construction.

The sustainable development planning of Ventura County in USA put forward the 'Eight principles of ecological construction in sustainability planning' in1991. The principles include: (1) the protection, conservation and restoration of natural environment. (2) The city should establish a net price system as a foundation for economic activity. That is to say, price should not only reflect the state of availability, but should be established from different perspectives that are long-range, recycled, and systematic. (3) Supporting local agriculture and local industrial and commercial services. (4) Developing ecological community which is scattered, functional comprehensive and has a reasonable pedestrian system. (5) Using advanced traffic, communication, and production system. (6) Protecting and developing renewable resources as far as possible. (7) Establishing cycle plan and recycled materials industry. (8) Supporting management education in popularity of to participate in.

**The Danish urban ecological committee** considers that 'Urban ecology represents a kind of environmental strategy, which takes localization and public participation as a starting point, and is engaged in solving local problems that are relevant to energy consumption, environmental strain and local natural environment'. The importance of visible localization and public participation has played a key role in this definition

**The international association of eco-city construction**, put forward the 10 principles of establishing an eco-city:

- Priority right of changing the land use management. Preferentially, developing compact, diverse, green, safe, pleasant, and energetic communities which are with mixed land uses and near the bus station and transport facilities.
- Priority right of modifying traffic construction. To place priority on walk, bikes, carriages, and public transportation than cars. Emphasis on access by proximity.

- To repair the damaged urban natural environment, especially the rivers, seashores, ridge lines, and wetlands.
- To construct economical mix-housing of decent, cheap, safe, convenient, and suitable for a variety of ethnic.
- To cultivate social impartiality and to improve the lives of women, visible minorities, and people with disabilities and social status.
- To support local agriculture, urban greening projects and to make the garden-like community come true.
- To advocate recycling, to use appropriate technology, and resource protection technology. To reduce emissions of pollutants and dangerous goods at the same time.
- To support economical activities of good ecological benefit with the business community. To control pollution, waste, and the use of dangerous toxic materials.
- To advocate consciously simple way of life and to be against the excessive consumption of resources and goods.
- To improve the local environmental and bioregion consciousness of the public by education programs and publicity campaigns aimed to raise their awareness of ecological sustainable development.

The development principles of eco-city put forward by the **Urban Ecology Australia** (UEA) are as follows:

- Restoration of degraded land.
- Coordinated development of urban and biological regions, balanced development.
- To achieve the balance of the urban development and land bearing capacity.
- To end the spread of the city.
- To optimize the energy structure, committed to the use of renewable energy, such as solar energy, wind energy, and to reduce fossil fuel consumption.
- To promote economic development and to provide community service of healthy and a sense of security.
- To encourage community participation in urban development.
- To improve social justice.
- To protect historical and cultural heritage.
- To cultivate colorful cultural landscape.
- Corrective to the destruction of the biosphere.

The ten key principles of sustainable human settlements put forward by the **European Union** are as follows:

- Budget of resource consumption.
- Energy protection and improvement of the efficiency of energy use.
- Development of renewable energy technologies.
- Building structure of long-term use.
- Residence and work place closed to each other.
- Efficient public transportation system.

- To reduce garbage output and recycle the rubbish.
- Compost with organic waste.
- Urban metabolic system following cycles.
- The main food of the local production requirements.

The relevant standards of eco-city put forward by The **United Nations** include the following six aspects:

- Guided by strategic planning and ecological theory.
- Industrial products are green products, advocating closed-loop process system.
- On the way of organic agriculture.
- Residential standards on the principle of improving the life of people.
- To protect the cultural and historical sites. Not to destroy the natural resources. Correctly handle the relationship between protection and development.
- Introduce natural into cities.

Experts majoring in ecological design believe that these are the basic concepts of eco-city. They make a very good guidance of the practice of changing the unsustainable features in the existing urban system.

The second and third international conference have passed the **Rebuilding Program of International Eco-city** and put forward the International Ecological Rebuilding Program. The program above epitomizes all kinds of the concepts of eco-city in common:

- To refactor the city and to stop the disorder spread of cities.
- To transform traditional villages, small towns, and rural areas.
- To repair the natural environment and production system of production capacity.
- To design the city according to the requirements of energy conservation and garbage recycling.
- To establish a transportation system oriented by walking, cycling, and public transportation.
- To stop all kinds of subsidy policy for car transport.
- To make efforts to provide powerful economic incentives for ecological reconstruction.
- To establish various levels of government agencies of urban, state, and national levels for ecological development.

**The fourth session of the International Eco-city Construction Forum** on August 18, 2010 was held in Chengde City of China. The Conference which centered on eco-city, ecological industry, ecological restoration, and ecological civilization focused on urban ecological civilization, reconstruction, and transformation and carried out academic exchanges and field trips from four aspects of indicators and standards, technology and engineering, science and empirical, inheritance and innovation, and so on. Famous scientists and entrepreneurs from all around the world carried out monographic academic exchanges around those topics. The forum also passed the declaration of 'revival of ecological civilization and promoting urban transformation.'

The International Eco-city Construction Forum is an international academic conference held by the international council of eco-city construction and is aimed at promoting eco-city construction. It is intended to exchange and spread new ideas, new method, and new technology of ecological urban planning construction and management at home and abroad. It also concludes experiences and lessons of industrial ecological construction in the process of urbanization and shows typical urban ecological construction, and promotes urban ecological engineering technology and capacity building. It has successfully conducted three sessions in Ningbo, Chongqing and Huaibei Since 2004, having a wide range of influence at home and abroad.

Series of seminars participated by the experts and scholars, local administrative personnel, and folk communities was held by the **Japanese Ministry of Construction** since 1992. It discusses the basic concept of eco-city construction, the implementation of the policy, and the specific steps. They think that the construction of eco-city includes at least the following three aspects of content:

**Urban system with energy conservation and circulation pattern**. To maximize pertaining to the urban system construction and the utilization efficiency of resources and energy. To maximize the waste heat, exhaust fumes, and rubbish in our city in order to reduce the environmental load of the city and to achieve coordination of the system in the city

**Water environment and water cycling**. The water system of an eco-city should meet the requirements of production and living at the same time, giving the city a moderate circulation system. It is a place to reduce the water intake and wastewater discharge from the outside, and to reduce the impact on the surrounding environment, and to provide recreation entertainment function to urban dwellers.

**Urban landscaping**. Urban greening is considered to be the foundation of eco-city construction that has the function of mitigating urban heat island effect, flood controlling, fireproofing, wind-proofing, noise-proofing, and controlling of atmospheric pollution. It can provide habitat for animals in the city, and provide a good ecological environment for urban residents. The green coverage rate of urban built-up area is suggested to be 30 % and to be up to 50 % for surrounding areas. The research of Japanese Ministry of Construction improves Japanese urban ecological research to the specific practice stage, and the reports of the seminars become guides to explore the construction of eco-city

### 4.2.2 Domestic Theories

Chinese famous ecologists, Prof. Ma Shijun and Prof. Wang Rusong proposed the theory of 'social-economic-natural complex ecosystem' in 1984, mading it clear that the city is a typical social-economic-natural complex ecosystem. Wang Rusong continued his research on urban problems and eco-city, thinking that the ecological essences of a city are those shown in Table 3.

**Table 3** Ecological essences of urban problems

| Problems | Principles | Strategies | Methodology | Goals |
|---|---|---|---|---|
| Low efficiency of resource utilization | Regeneration and competition | Technical transformation | Ecological technology | High efficiency |
| Unreasonable relationships of the system | Symbiosis/phyletic evolution | Relationship adjustment | Ecological planning | Harmonious relationships |
| Low self-regulation | Autogeny and self-learning | Act induced | Ecological management | Strong vitality |

Rusong Wang etc. also put forward the Chinese ideology of eco-city and ecological cybernetics principles, pointing out that the construction of eco-city should meet the principles of the satisfaction of human ecology, efficient principles of economical ecology, and harmonious principles of natural ecology. What's more, it should also meet the tide principles, optimum-oriented principles, promotion and restriction principles, feedback principles, compensation principles, principles of bottleneck, circulation principles, the diversity and dominant principles, principles of ecological design and principles of cleverness, etc.

Wang pointed it out in his article called 'ecological transformation theories and methods on environmental construction of urban human settlement' that at the turn of the century, the urban ecosystem research of various countries mainly focuses on various natural ecological factors, factors of technology and physics and the hierarchy, heterogeneity, and diversity of the sociocultural factor sysplex. They also pay attention to the urban metabolism process, information feedback process, the health of ecological succession process, urban economical production, social life, and the strength and vitality of natural regulating function. Ecological assets, ecological health, and ecological service function are among the current hot topics in the study of urban ecological system.

The purpose of the eco-city construction is to make the single biological link, physical link, economical link and social link assembled into a powerful ecosystem through ecological planning, ecological design, and ecological management, what's more, to regulate the structure and function of the system, to promote the coordinated development of urban society, economic, and nature and the efficient utilization of information, material and energy, the fully mix of technology and the nature from the perspective of the system reform, technological innovation, and behavior induce. The creativity and productivity of people will be limitless, system function of life supporting, and physical and mental health of residents will get maximum protection, economic, ecology, and culture will develop sustainably and healthily. What is more, it will promote the comprehensive utilization of resources, the comprehensive improvement of the environment, and comprehensive development of people.

The city of Shanghai put forward the goals of eco-city construction in the early 1990s. The scholars in Shanghai, such as Xixian Chai, Yongchang Song, Xiangrong Wang and Wu (2000), etc. in the field of urban planning, environmental protection, urban ecology, have conducted many studies of eco-city.

It is also considered by many domestic researchers that an eco-city is the sustainable subsystem to share its fair share of carrying capacity in global or regional ecosystem. It is a compound system with social justice, natural harmony, and economic efficiency, which is established on the basis of the principles of ecology. It also has its own cultural characteristics of nature and is an ideal living environment with artificial coordination and interpersonal harmony.

Guangyu Huang and his colleagues of Chongqing University cooperated with the People's Government of Leshan, Sichuan Province, making planning research and practice of 'Combining of heaven and man: new mode of eco-city structure with green heart and circular in Leshan.' Approved by the government of Sichuan Province, their achievements and paper called 'Ecopolis: Concept and Criteria' participated in the Earth Summit '1992s Global Forum' of The World Conference on Environment and Development in Brazil, achieving widespread high praises.

Academician Wenhua Li pointed it out in the Academic Conference of Chinese Ecological Society in 2011 in Changsha, Hunan Province, China that the perspective of globalization, networking, and cross research and the introduction of new technology create new conditions for the ecology research in China. China's ecology, which is at the turning point of its developing process, should seize the opportunity of global change, take advantages of scientific and technological competition, and promote the progress of ecological research development.

From the above analysis it can be recognized that the planning and construction of eco-city is actually to plan and construct the overall process of targets, programs, contents, methods, results, and implement countermeasures of the comprehensive improvement of urban ecological factors. It is also an effective way of implementing the dynamic balance of urban ecosystem and regulating the relationship between human being and the environment.

The scientific connotation of eco-city construction is embodied in the following aspects:

(I) Environmental protection system with high quality.
(II) Operation system with high efficiency.
(III) Management system with a high level.
(IV) Perfect green space system.
(V) A high degree of social civilization and eco-environmental consciousness.

## 4.3 *Practice of the Planning and Construction of Eco-cities and Counties*

Under the guidance of the eco-city theory and the initiative of the MAB (Man and Biosphere) Program by the United Nations in 1972, many cities, at home and abroad, have put ecological planning into practice. So far, there are a lot of cities carrying out the urban ecological planning and construction, trying to meet the goal of eco-cities. The Cleveland and Portland metropolitan area (U.S.A.) are among the demonstration cities, as well as Frankfurt (Germany), Bangalore(India), Curitiba and Santos City (Brazil), Whyalla and Adelaide (Australia), Whitaker City (New Zealand), Copenhagen (Denmark), Rome (Italy), Washington, D.C. (U.S.A), Tokyo and Kitakyushu (Japan), and Moscow (Russia). And in China, more than 110 cities and 500 counties, including Yichun, Ma'anshan, Hong Kong, Tianjin, Changsha, Beijing, Shenzhen, Guangzhou, Shanghai, Nanning, Xiamen, Qingdao, Rizhao, Weifang, Dalian, Nanjing, Zhenjiang, Yangzhou, Changzhou, Haining, Changshu, Zhangjiagang, Guangzhou, Shaoxing, Panjin, Mianyang, etc., have put forward the target of eco-city (or county) construction and carried out the work of eco-city (or eco-county) planning up to now. Besides, 14 provinces, such as Hainan, Fujian, Zhejiang, Jiangsu, Jilin, Heilongjiang, Shandong, Anhui, Jiangxi, Sichuan, and Shanxi Province, have carried out the work of the construction of eco-provinces.

### 4.3.1 The United States of America

The Long-Term Ecological Research (LTER) and the Eco-cities Plan (ECP)

The Long-Term Ecological Research (LTER) and the Eco-Cities Plan (ECP) devote to the construction of eco-city in the United States. The former mainly models the natural and artificial variables, which affect the natural and social factors of the ecosystem, so as to do analysis on the urban ecological system from the perspective of long-term ecological effect. The latter, started with the support of the US Forest Service and EPA, receives its funding from NSF and the US Geological Survey. With the unified leadership of the National General Committee, and the technical support of the technical committee consisting of several universities, the ECP has set up a close network among research scholars, social activists and local leaders.

Its goals are as follows:
Changing Perceptions of Nature in Cities;
Multiple Uses of Natural Spaces in Cities;
Alternative Scales of Intervention;
Institutional Flexibility and Innovation;
The Role of Science in Urban Landscape Design.

### The Eco-city Project of Berkeley

Under the influence of Professor Richard Register and the Eco-city Builder Association led by him, the practice of the construction of "eco-city" in Berkeley, a coastal city in the West Coast of the United States, has proved highly successful (Very good results have been achieved in the practice of constructing an eco-city in Berkeley, a coastal city in the West Coast of the US). Some people think that it is a model for the world's construction of eco-city, which can also be regarded as a test or an experiment of eco-city construction. In Berkeley, the overall practice is based on a series of concrete actions, that is, to build the non-motorized streets, allow the abandoned rivers to resume, plant fruit trees along the streets, build green residence using solar energy, improve energy utilization structure through energy use rules, optimize the allocation of bus lines, advocate walking instead of driving, delay and try to stop the building of the fast lane, and to hold city construction meetings with the relevant parties, etc. These actions may be insignificant to some people, but it is these small actions that make the work of eco-city construction proceeding solidly and effectively. With more than 20 years' efforts, Berkeley has walked out a successful way of building an eco-city. There is a typical special structure of urban-rural integration in Berkeley. In the residence area, with every independent house there is a piece of farmland covering an area as large as that of several houses. And the vegetables and fruits planted there is very popular among the local residents and urban residents nearby as "green food".

Since the 1960s, this kind of urban agriculture has been developed gradually in some cities in the developed countries. The "Fairview Gardens" Education Farm, located in a fertile valley in California, is completely surrounded by the development of urban areas, with the high way on one side and urban housing on the three others. The farm covers an area of 4 ha, promises not to use chemicals in the growing practices and produces about 75 different kinds of fruits and vegetables which are available for nearly 500 families to eat. It is very popular with the residents and has become an education center for a school nearby.

### The Ecological Construction Plan in Portland, Oregon

Portland is considered to be an example of a city which has successfully avoided and solved many urban issues. In the late 1960s, like many other cities, Portland was also in the trouble of problems such as traffic congestion, housing shortage, environmental degradation, and social culture decadence. Therefore, the city carried out a plan for the development of a new urban region and took comprehensive measures to stimulate the vitality of the city. The plan has changed the direction of the urban development. On one hand, improve the traffic conditions. A decision closely related to the citizens was issued—to impose restrictions on the newly bought cars entering the city, which at that time had the decisive significance. It limited the development of the urban parking lot and transformed an old one into a public square. It also converted a highway, which stretched along the river bank at the edge of the city and had separated the city and the river, into a park built along the river. Besides, a new light rail line was built to take the place of the highway. The new plan actively supported the designation and building of walking and public transport-oriented traffic system, thus, limited the use of private cars through the promotion of public transportation. On the other hand, Portland concentrated efforts on solving the housing problem and focused on the retention and provides available housing for the average families. Great attention has always been paid to the improvement of housing condition in Portland, and even if plans are in the implementation, the city has always put emphasis on housing construction. In addition, Portland also concerned about public culture and art, requiring that all new projects contribute a certain percentage of their funds to it. The protection of historical heritage was also taken into consideration and important historical buildings were protected with great efforts.

### 4.3.2 Australia

Since 1994, led by the urban ecology scholar Paul Downton, the pioneers of the ecological movement sponsored an eco-city construction project in Adelaide City in South Australia, which was called The Halifax Eco-city Project. This project was organized and implemented by UEA. And it was designed to be 1000 people on 2.4 ha in the eco-city.

The fundamental principles of this project are as follows:

- Restoring the degraded land;
- Construction projects fitting the characteristics of the local communities;
- The intensity of development shall be in harmony with the land ecological capacity, and to protect the ecological conditions of the developing areas;
- Effectively limiting the excessive extension of the city on the basis of the ecological conditions;
- Optimizing the energy structure, reducing energy consumption, using renewable energy and resources, and promoting resource recycling;
- Maintaining an appropriate level of economic development;
- Providing a health and safe living environment;
- Providing a variety of social and community service activities;
- Ensuring the fairness for social development such as gender, skin color, religion, opportunity to work, etc.;
- Respecting the past history of development and construction, protecting the historic natural landscape and human landscape;
- Promoting ecological culture construction, improving the residents' ecological consciousness;
- Improving the state of natural ecosystem, including air, water, soil, energy, biomass, food, biodiversity, biotope, ecological sensitive region, waste water recycling, etc.

Upto now, the project has been fully implemented, and attracted ecological scholars from all over the world to visit and organize a series of seminar on eco-city. The smooth implementation and success of eco-city construction lie in the first-class program plan and implementation scheme.

### 4.3.3 Denmark

The City Copenhagen, Denmark, undertook the "Danish Eco-city 1997–1999" project in the densely populated zone—Indre Norrebro. Their main experience is developing a manual to guide and promote residents to take part in. In addition, they also established the system of "green account," recording the resources consumption of a city, a school, or a family for daily activities, providing the knowledge about environmental protection. By using green account to compare the structure of resources consumption of different urban areas, they could identify the consumption of major resources and provide the basis for reducing resource consumption effectively and resource recycling.

The most famous achievement on the ecological construction in Denmark is the construction of Kalunborg Eco-industrial Park, which is also the world's most famous pioneering industrial ecological park. Located on the coast of the North Sea, about 100 km west of Copenhagen, with a population of about 20,000 people, Kalundborg is a small industrial city with a natural deep sea port. The five main industrial enterprises are: Denmark's largest thermal (coal) power plant, Denmark's biggest refinery, Denmark's largest biological engineering company (which is also one of the world's largest industrial enzymes and insulin production plants), a calcium sulfate factory and a building materials company. The distance between each of these five companies is not more than hundreds of meters. In the process of production and development, they gradually started to exchange the 'waste' spontaneously: steam, water (of different temperatures and different purity) and a variety of by-products, and gathered by using special pipe system. Among them, the power plant burnt gases exhausted from the oil refinery, the refinery share with other enterprises the cooling water, the slag of the power plants can be used as raw materials for the building materials factory and the construction company, and industrial waste heat can be used as heating for the nearby residents or enterprises, etc. Thus, a kind of "industrial symbiosis system (industrial symbiosis)" was formed. The development mode of industrial symbiosis in Kalunborg inspired people's concept and interest of plan and designs an "ecological industrial park," and provided successful experiences.

Since the 1980s, the local departments of management and development have provided support from all sides. In this small industrial city, with the pattern of interdependence and mutual utilization of steam, hot water, materials, such as gypsum, sulfuric acid, and biotechnology sludge, the early phase of an ecological industrial park formed and therefore Kalunborg has become famous of this.

### 4.3.4 Netherlands

In 1990, on the "long-term environmental policy meeting," the Netherlands put forward the five gradual stages of healthy ecological urban planning and policy making[1]:

(I) Pollution of the environment as a negative effect
(II) Pollution of the environment as a cost factor
(III) Environment as constraints
(IV) Environment as a policy guidance
(V) Environment as the target

The Eco-city Strategic Framework

The construction of eco-city is corresponding with the fourth and fiveth stage mentioned above. In 1995, the Netherlands Breda and Ankara institutions jointly studied the eco-city planning with research institutions, and put forward the strategic framework of the eco-city as below (Table 4):

(I) **A City with Sense of Responsibility**

*Saved utilization*: to use energy and material economically, for example, strongly advocate the using of bikes instead of cars, attach importance on developing energy-saving technologies, and strengthen the research on conservation measures, etc.
*Recycling*: reuse raw materials and at the same time the wastes should be classified and purified.
*Renewable using*: such as solar power, rainwater, and trees.
*Control the quality and quantity of material and energy flow*: the use of some resources should be under control even if they are renewable, taking the tropical hardwoods as an example, excessive deforestation will cause the destruction of the rainforests. To where there lack of renewable or permanent energy, try to use clean energy, such as using nature gas instead of coal.

(II) **A City Full of Vigor**

*Make use of the local natural and cultural potential*: make heavy use of the local ecological resources (both biological and no biological), and meet the needs of the residents' life.
*Combine the programming of urban space structure with the management of material/energy flow*: for example, the rainwater storage pond is more than a

[1]Meadows et al., 1996: Ecologically sound urban development, pp. 53–80.

**Table 4** Strategic framework of the eco-city

| Slogans | (A) A city with sense of responsibility | (B) A city full of vigor | (C) A city where everyone participate in |
|---|---|---|---|
| Objects | Material and energy flow | Region | The general public |
| Social goals | - Products | - Effectiveness | - Prosperity |
| | - Quality | - Attraction | - Welfare |
| | | | - Justice |
| Problems | - Wastage | - Health problems | - Alienation |
| | - Pollution | - Dysfunction | - Apathy |
| | - Interruption (stress) | - Loss of biodiversity | |
| Objectives of eco-city | - Sustainable flow management | - Sustainable development of the territory | - The commitment that the ecological relationship remain the same |
| | - Control programs | - Regional potential planning | - Organize to plan spontaneously |
| Policy themes | - Integrated flow management | - Special and regional management | - Target group policy |
| | - Management policies for the source | - Result-oriented policies | |

restriction factor of space planning, which can become the highlight of the city landscape.

*Provide a healthy and diversified human habitat*: As posed by the WHO, the city should first be a healthy place for the citizens, as well as a safe and quiet one. So the living environment should be good for the health of people and take into account people's different needs for different life style and behavior habits.

*Protect the animals and plants*: take into consideration the protection of their habitats and migration passages.

### (III) A City Where Everyone Participate in

*Create the conditions for the operation of the market*: bring environmental causes to market with/through providing some economic benefits

*Create opportunities for cooperation*: Some organizations tend not to economic interests as the goal, but a more important social impact, such as schools, government agencies, and communities.

*Popularize ecological views extensively and intensively*: and turn the thoughts into practical actions of everyone in our everyday life.

*Enforce*: through the related policies and laws.

The Guidance Model

Under the strategic framework, a series of guidance model were put forward in the Netherlands, and focus on two kinds of urban networks and the three-level structure. The two kinds of networks refer to the road network and network of rivers, and the three layers are buildings, community or city level, and regional or national level.

(I) **Development Planning for Water Resources**

The key points are:

*Saving water*: such as reduce the waste of water through the update of the technology of faucets;
*Recycling*: flush the toilet with the laundry water, bath water, or rain water.
*Storing the rainwater*: there are facilities to store rainwater in each building or block.
*Rain and sewage diversion*: the rainwater pipe and sewage pipe are under separate management.
*Preventive measures*: to get clean rainwater, use materials with no zinc as the road facilities, and take oil isolation measures at catch-basins of the parking lots. When packed into small lakes, rainwater can be purified by wetland plants such as reed or calamus.

(II) **Energy Planning Model**

The key points are:

*Energy conservation*: consider architectural lighting and leeward when deciding the building orientation, develop materials of good sealing performance, use energy-saving appliances, etc.
*Develop sustainable and renewable energy*: make full use of solar power, wind power, hydroenergy and biological gas, stop building more large power stations, and large dams and nuclear power plants are not recommended because of the high risk.
*The ultimate ways of energy utilization are electricity and hydrogen, which can minimize the pollution.*

(III) **Waste Treatment Guidance Model**

The key points are:

*Household*: waste classification; economical use of goods; ensuring that there is enough space for piling up the classified waste; collecting the wastes and getting them into the life cycle of recycling.
*Construction and destruction*: to construct sustainable buildings; recycling of buildings and materials; to destroy with caution; be vigilant and try to avoid secondary pollution in the process of production and transportation of building materials.
*Waste disposal*: Encourage individuals and enterprises to participate in the waste collection work, set up an efficient collection system; do waste disposal as regionally as possible.
*Enterprises and industry*: to save resources; does reusing plan for some of the products, and separation and recycling plan for some others.

(IV) **Traffic Guidance Model**

The key points are:

Shorten the distance between the residential areas and the companies so that the demand for transportation is reduced.

Encourage the company shuttle bus system, extend the city along the public transportation route.

Encourage the use of bicycles, build safe bikeways with good charming sights, which is in good cohesion with the public transportation.

Reduce environmental pollution: collect parking fees; promote underground parking; develop electric and hydrogen vehicles; keep cars from the city center and residential areas while retain emergency channels for fire trucks; use maglev train instead of plane for transportation more than 800 km.

Freight transportation: reduce traffic demand by centralized urbanization, raised cost of freight, and development of durable goods; increase the efficiency of traffic land use; use electricity or clean fuel; develop shipping and railroad transportation.

(V) **Urban Planning Guidance Model**

The key points are:

The design of the road made rows of buildings facing the southeast or the south, so that the buildings and green spaces can get enough sunlight;

The water supply and drainage system was built based on the circulation system of each community; therefore, there is a big pool in each community for seasonal water storage. Besides, there are a certain number of purification facilities.

Build a bicycle road system with a green belt with beautiful scenery. These roads can be access to the city center, the main station and factories within one quarter of an hour to an hour's drive.

To ensure the peace and security of the residential area, no parking is allow but the access for fire trucks, garbage trucks, the ambulance, and other emergency vehicles;

Starting from the station, transport the goods to the downtown area by electric vehicles instead of trains or trucks.

(VI) **The Suburban Area Guidance Model**

The key points are:

*Precision agriculture*: develop industries of fine products, livestock breeding, or gardening.

*Urban agriculture*: extensive management, the function of both education and conservation;

*Urban forest*: wood land growing and recuperating naturally;
*Fallow land*: quiet outdoor leisure places and lakes for swimming;
*Conservation areas*: places for the protection and rehabilitation of natural ecosystem.
*Residential areas*: building communities of low density and low rise, with some infrastructure.
*Urban parks*: community parks with ponds and in the communities or among the communities.

### 4.3.5 Brazil

Curitiba in Southern Brazil is so called the Eco-Capital City' of Brazil and is respected by the most eco-city in the world. Its main experiences focused in the following aspects: public transportation-oriented urban developing plan, paying close attention to social public welfare project, and the environmental education for the citizens. In Curitiba, urban planning and development programming make the urban design and planning and management in one system. Through the pursuit of highly systematic, progressive and thoughtful designing, the target of implementation of land use, and public transport integration were achieved in urban planning, which is a great achievement.

In the public transportation oriented urban development planning, the measures took by this city are "increasing the area of parks and improving the public transport," "the farther from bus lines, the lower the floor area ratio." The government encouraged high-density development only in the two blocks near the bus lines, and limited strictly the development of the area away from the two blocks. The integration of road system provides a high accessibility to promote the development and the utilization of land along the transport corridor. The plan also strengthened this axis measure, ensuring that there was enough space for the BRT road in the broad traffic corridor. In many built up area, similar methods were also used to make the development, which is along the existing bus lines, stretch forward along the main routes to the outer edge of the city. In addition, the planning of the high density utilization and the planning of the existing traffic corridor were fixed as an organic unity. These policies effectively guaranteed the level of bus service in the developed zone and the new developing zone of the city. At present, there is two-thirds of the citizens use the bus every day, and the bus service does not need fiscal subsidy. The city is on the road of developing an eco-city, in which the transportation is with low economic cost and environmental cost, and it is as harmonious as possible between human and nature. From this point, when decide the ecological community's location, the traffic convenience is a very important factor.

The eco-city construction of Curitiba is embodied in the harmonious development of society and the promotion of the ecological literacy education program. So far, there are hundreds of social public welfare projects, from the project of building a new library system to the project of helping homeless people. In the poorest communities, the "Line to work" project started, of which the purpose is the training of all kinds of practical skills. Curitiba also actively carried out street children rescuing projects, and organizing outdoor markets, in order to meet the requirement of the street vendors in the informal economy.

Curitiba has carried out the ecological literacy education plan in all schools. This project is rewarded as one of the world's 60 best environmental protection and civil education projects by the United Nations. Each year, schools would organize the students to participate in environmental protection activities on the World Environmental Day. Curitiba's successful experience shows that to create a green home we should not only pay attention to hardware construction, but also pay special attention to the education since people's childhood to improve the population quality.

### 4.3.6 Japan

With the Japanese industrial revolution and the rise of capitalism in the twentieth century, Kitakyushu gradually developed from a small fishing village into a large industrial city, which is mainly composed of steel industry. In the mid-1950s, the Japanese economy has entered into the stage of take-off and all kinds of environmental problems came along. From the 1960s, the governments of Kitakyushu enact a series of regulations and laws to control environmental pollution, as well as strength the supervision and management system. By the joint efforts from local governments, residents, and enterprises, they basically overcame the environmental pollution problem in the mid-1980s eventually. In 1985, the white paper of the OECD (Organization for Economic Cooperation and Development) says, "Kitakyushu has turned into a green city from the gray city." In 1990, Kitakyushu is awarded "the top 500 global cities" title by the United Nations environment organization.

Kitakyushu is in Kyushu Fukuoka County of the Western Japan, located between Kyushu and Honshu. It's a traffic artery between the western Japan and other Asian countries. The population is 1.01 million and covers an area of 482.94 $km^2$. Along with the developing of the revolution of Japanese industrial revolution and the rise of capitalism in the twentieth century, Kitakyushu gradually developed from a small fishing village into a large industrial city. The city is mainly composed of steel industry, also has the petroleum chemical industry, metal manufacturing, electronic machinery, and other relevant industries. From the mid-1950s, Japanese economy stepped into the take-off stage, with all kinds of environmental problems coming along. Kitakyushu, as one of the four big industrial districts of in Japan, experienced serious air pollution and water pollution, with dust, soot, and $SO_2$ being the main atmospheric pollutants and industrial drainage was the main sources of water pollution.

In the 1960s, the Kitakyushu government set a series of laws and regulations to control environment pollution and to strength the supervision and management system. Through the joint efforts from local governments, residents and enterprises, in the mid-1980s, they basically overcame the environment pollution, also provided many technology and experience of overcoming pollution for the developing countries. The international society thinks highly of this achievement. In 1985, the white paper of the OECD (Organization for Economic Cooperation and Development) pronounced that Kitakyushu has turned from the gray city into a green city. In 1990, Kitakyushu is awarded as "the global top 500 cites" title by the United Nations environment organization.

In the early 1990s, Kitakyushu began the program of eco-city construction, the main content of which is reducing waste and realizing the circular society. They put forward that "The wastes arising from these industries might be used for those industries and in this way the overall waste can emissions zero." This kind of eco-city construction idea was approved by Japan's trade and industry in 1997, and got much support in terms of money to accelerate the pace of eco-city construction. The Kitakyushu ecological urban program planned to construct the city in coastal landfill area. Specific planning of the program includes the following three aspects:

(I) Formation of the environmental industry. Construct a comprehensive environmental recycling industry area, including the household appliances, waste glass, waste plastics, and auto chip crushing. This plan will treat Kyushu Island and apart of the state district as the target, and the recycling industry of the waste glass and drink plastic has already begun to construct and put into production.

(II) Development of new environmental technology. Build the technology research center to develop new environmental technology, do researches about the safety of the recycling and reuse technology and the feasibility of the transition. The center also had the ability to cultivate the high new technical enterprise and the function of the international education practice. This plan began in 1997.

(III) Comprehensive development of the society. Construct the basic research and education center to foster talents in environmental policies and environmental technology. Set up the school of engineering for international environment in the University of Kitakyushu. Attract university institutes from the relevant disciplines in the area to form international environment cutting-edge technology and information center. Fukuoka University Institute of resources environment and environment suppression system has decided to construct in the place. Since 1997, Kitakyushu began to put all the plans into practice to make efforts to form large scale in all kinds of plan by 2005, and make them all fully operated.

## 4.4 *Eco-city Planning and Practice in China*

### 4.4.1 Index System of Ecological Construction [State Environmental Protection Administration (the Ministry of Environmental Protection of China), 2003 Trial Implementation, and 2005 Amendment]

Eco-city Building Indicator System (Trial Implementation)

(I) **Definition**

Eco-city (including municipal-level administrative region) is municipal administrative area with the harmonious development of social economy, ecological environment, and various fields coordinated with the requirements of the sustainable development. Eco-city is the continuation, development and the final target of ecological demonstration zone construction of municipal administrative area.

The outstanding feature of eco-city: a good ecological environment and continue to a trend of higher level balance, environmental pollution basically eliminated, natural resources to obtain the effective protection and rational utilization; basically forming stable and reliable ecological security system; environmental protection laws, regulations, and institution effectively implemented; accelerating to the development of society and economy with the characteristics of the circular economy; the human and the nature harmoniously coexistence, ecological culture have a great progress; clean and beautiful environment of the city and village; people's living standards all-round improvement.

(II) **Basic Conditions**

- Formulated the 'Eco-city construction planning', and passed the Municipal People's Congress, promulgated, and implemented.
- The more than 80 % counties in the whole city reached the index requirement of Eco-county construction; urban built-up area passed the assessment for national model city, checked, and approval of environmental protection and was named.
- The above county government (including county-level) in the whole city (including economic development zone) has independent environmental protection agency, and it is a first class administrative unit, and the township has a full-time staff of the environmental protection work. Environmental protection has brought into performance evaluation content of the county (including county-level city) party committee, government leadership, and established relevant assessment mechanism.

- The relevant environmental protection laws, regulations, and institution of the state and various environmental protection regulations and institution issued by local were effectively implemented.
- Pollution control, ecological protection, and construction are very fruitful, no major environmental pollution and ecological destruction within three years.

(III) **Construction Indicators**
There are three categories in indicators of eco-city construction, which include economic development, environmental protection, and social progress, a total of 30 indicators (Table 5).

Eco-county Building Indicator (Trial Implementation)

(I) **Definition**
Eco-county (including county-level city) is county-level administrative area with the harmonious development of social economy and ecological environment and various fields according with requirements of the sustainable development. Eco-county is the continuation, development, and the final target of county-level ecological demonstration zone construction.

(II) **Basic Conditions**

- Formulated the «ecological county construction planning», passed the County People's Congress, promulgated, and implemented.
- More than 80 % of towns reached the assessment criteria of beautiful environment towns.
- The above government has independent environmental protection agency, and it is a first-class administrative unit, and the township has a full-time staff of the environmental protection work. Environmental protection has brought into performance evaluation content of the town party committee, government leadership, and established relevant assessment mechanism.
- The relevant environmental protection laws, regulations, and institution of the state and various environmental protection regulations and institution issued by local were effectively implemented.
- Pollution control, comprehensive treatment of rural environment, ecological protection, and construction are very fruitful, no major environmental pollution and ecological destruction within 3 years.

(III) **Construction Indicators**
Index of ecological county construction has three categories, including economic development, environmental protection, and social progress, a total of 38 indicators (Table 6).

**Table 5** Index of eco-city construction

| | Serial number | Name | Unit | Index |
|---|---|---|---|---|
| Economic development | 1 | Per capita gross domestic product | RMB/capita | ≥30,000 |
| | | Economically developed areas | | ≥20,000 |
| | | Economically underdeveloped areas | | |
| | 2 | Annual per capital government receipts | RMB/capita | ≥3600 |
| | | Economically developed areas | | ≥2400 |
| | | Economically underdeveloped areas | | |
| | 3 | The annual per capita net income of farmers | RMB/capita | ≥7500 |
| | | Economically developed areas | | ≥5500 |
| | | Economically underdeveloped areas | | |
| | 4 | Average annual disposable income of urban residents | RMB/capita | ≥16,000 |
| | | Economically developed areas | | ≥13,000 |
| | | Economically underdeveloped areas | | |
| | 5 | The proportion of third industry in GDP | % | ≥50 |
| | 6 | Energy consumption per unit of GDP | Tons of standard coal/10 thousand RMB | ≤1.4 |
| | 7 | Water consumption per unit of GDP | $m^3$/10 thousand RMB | ≤150 |
| | 8 | The rate of scale enterprises through the ISO-14000 certification | % | ≥20 |
| Environmental protection | 9 | The forest coverage rate | % | ≥70 |
| | | Mountain Area | | ≥40 |
| | | Hilly area | | ≥15 |
| | | Plain area | | |
| | 10 | The ratio of protected area in the total land area | % | ≥17 |

(continued)

**Table 5** (continued)

| | Serial number | Name | Unit | Index |
|---|---|---|---|---|
| | 11 | Recovery treatment rate of degradation land | % | ≥90 |
| | 12 | City air quality | Number of days better than or equal to 2 grade standard/year | ≥330 |
| | | Southern Region | | ≥280 |
| | | Northern region | | |
| | 13 | Water qualification rate of water function area in the city | % | 100, and city without over 4 class water body |
| | | The qualification rate of water environmental quality of offshore area | | |
| | 14 | The main pollutant emission intensity | kg/10 thousand RMB (GDP) | <5.0 |
| | | sulfur dioxide | | <5.0 |
| | | COD | | Do not exceed the control index of total emission amount of major pollutants of state |
| | 15 | Water qualification rate the centralized source of drinking water | % | 100 |
| | | | | ≥70 |
| | | Centralized treatment rate of urban sewage | | ≥50 |
| | | The repetition rate of industrial water | | |
| | 16 | Coverage rate of noise standard area | % | ≥95 |
| | 17 | Harmless treatment rate of urban living garbage | | 100 |
| | | The disposal and utilization rate of industrial solid waste | % | ≥80, and no hazardous waste emission |
| | 18 | Per capita urban public green area | $m^2$/capita | ≥11 |
| | 19 | Environmental compliance rate of tourist area | % | 100 |
| | 20 | Environmental protection investment ratio in GDP | % | ≥3.5 |

(continued)

**Table 5** (continued)

| | Serial number | Name | Unit | Index |
|---|---|---|---|---|
| Social progress | 21 | The intact rate of city lifeline system | % | ≥80 |
| | 22 | City per capita area of paved roads | $m^2$/capita | ≥8 |
| | 23 | Urbanization level | % | ≥50 |
| | 24 | rate of gasification in the city | % | ≥90 |
| | 25 | The central heat-providing rate of city | % | ≥50 |
| | 26 | Engel coefficient | % | <40 |
| | 27 | Gini coefficient | | Between 0.3 and 0.4 |
| | 28 | Enrollment rate of Higher Education | % | ≥60 |
| | 29 | Technology, education funds account for the proportion of GDP | % | ≥7 |
| | 30 | Popularization rate with the environmental protection and propaganda and education | % | >85 |
| | | Public satisfaction rate with the environment | | >90 |

**Table 6** Index of eco-county construction

| | Serial number | Name | Unit | Index |
|---|---|---|---|---|
| Economic development | 1 | Per capita gross domestic product | RMB/capita | ≥25,000 |
| | | Economically developed areas | | ≥16,000 |
| | | Economically underdeveloped areas | | |
| | 2 | Annual per capital government receipts | RMB/capita | ≥3000 |
| | | Economically developed areas | | ≥1900 |
| | | Economically underdeveloped areas | | |
| | 3 | The annual per capita net income of farmers | RMB/capita | ≥7000 |
| | | Economically developed areas | | ≥5000 |
| | | Economically underdeveloped areas | | |
| | 4 | Average annual disposable income of urban residents | RMB/capita | ≥15,000 |
| | | Economically developed areas | | ≥12,000 |
| | | Economically underdeveloped areas | | |
| | 5 | Energy consumption per unit of GDP | Tons of standard coal/10 thousand RMB | ≤1.2 |
| | 6 | Water consumption per unit of GDP | $m^3$/10 thousand RMB | ≤150 |
| | 7 | Organic and green products in the proportion of mainly agricultural products | % | ≥20 |

(continued)

**Table 6** (continued)

| | Serial number | Name | Unit | Index |
|---|---|---|---|---|
| Environmental protection | 8 | The forest coverage rate | % | ≥75 |
| | | Mountain Area | | ≥45 |
| | | Hilly area | | ≥18 |
| | | Plain area | | |
| | 9 | The ratio of protected area in the total land area | % | ≥20 |
| | | Mountain Area and Hilly area | | ≥15 |
| | | Plain area | | |
| | 10 | Recovery treatment rate of degradation land | % | ≥90 |
| | 11 | Air environment quality | Up to the standard of function zone | |
| | 12 | Water environment quality Water environment quality of offshore area | | |
| | 13 | Noisy environment quality | | |
| | 14 | The emission intensity (COD) | kg/10 thousand RMB (GDP) | <4.5 |
| | | | | Do not exceed the control index of total emission amount of state |
| | 15 | Centralized treatment rate of urban sewage | % | ≥60 |
| | | The repetition rate of industrial water | | ≥40 |
| | 16 | Harmless treatment rate of urban living garbage The disposal and utilization rate of industrial solid waste | % | 100 ≥ 80, and no hazardous waste emission |
| | 17 | Per capita township public green area | $m^2$/capita | ≥12 |
| | 18 | Environmental compliance rate of tourist area | % | 100 |
| | 19 | The proportion of new energy in rural life energy consumption | % | ≥30 |
| | 20 | The rate of comprehensive utilization in straw | % | 100 |

(continued)

**Table 6** (continued)

| | Serial number | Name | Unit | Index |
|---|---|---|---|---|
| | 21 | The rate of comprehensive utilization of manure in large scale livestock and poultry breeding plant | % | ≥90 |
| | 22 | Recycling rate of agricultural plastic film | % | ≥90 |
| | 23 | Integrated Prevention and cure rate of Agriculture and Forestry Pests | % | ≥80 |
| | 24 | Fertilizer use intensity (pure) | $kg/hm^2$ | <250 |
| | 25 | Water qualification rate the centralized source of drinking water | % | 100 |
| | | The qualified rate of rural drinking water | | |
| | 26 | The popularization rate of sanitary latrines in rural areas | % | 100 |
| | 27 | Compliance rate of rural sewage irrigation | % | 100 |
| | 28 | Agricultural production system resilience (loss rate) | % | <10 |
| | 29 | Environmental protection investment ratio in GDP | % | ≥3.5 |
| Social progress | 30 | Natural population growth rate | ‰ | Conform to local standard |
| | 31 | Popularization rate of junior school education | % | ≥95 |
| | 32 | Urbanization level | % | ≥45 |
| | 33 | Engel coefficient | % | <40 |
| | 34 | The proportion of poor people | % | <0.2 |
| | | Economically developed areas | | <3 |
| | | Economically underdeveloped areas | | |
| | 35 | Gini coefficient | | Between 0.3 and 0.4 |

(continued)

**Table 6** (continued)

| | Serial number | Name | Unit | Index |
|---|---|---|---|---|
| | 36 | Technology, education funds account for the proportion of GDP | % | ≥6 |
| | 37 | Popularization rate with the environmental | % | >85 |
| | 38 | protection and propaganda and education Public satisfaction rate with the environment | % | >90 |

### 4.4.2 The Case of Eco-city Planning of Shanghai

Wang Xiangrong and his groups researched the ecological planning and construction of Shanghai, and put forward that the first-class city should have first-class ecological environment. A rational layout of the city, perfecting city infrastructure, improving environment quality, promoting sustainable development between economy, society, and ecological environment, constructing international eco-city with "The sky more blue, air cleaner, clearer water, more green, living better," are the strategic target of the development of Shanghai. According to the demand of eco-city and the context of Shanghai, Shanghai eco-city planning measurements should include index system of eco-city construction, the structure construction, and function construction of eco-city and coordination measures of ecological relationship and so on.

#### Indicator System of Construction

Eco-city should be the reasonable structure, efficient function, and relationship coordination of eco-city system and livable environment type. Reasonable structure refers to the moderate population density, reasonable land use, good quality of the environment, adequate green space system, perfect infrastructure, effective protection of nature; Efficient function refers to the optimal allocation of resources, material resources into the economy, full play of manpower, smooth and orderly physical distribution, fast and convenient information flow; Relationship coordination refers to the coordination of relationship between human and nature, the social relationship coordination, coordination of relationship between of urban and rural, coordination of resources utilization and update, coordination of environmental stress and environmental capacity. The goals of eco-city should be clean and beautiful of the environment, healthy and comfortable life, giving full scope to the talents, turning material resources to good account, useful of land, the coordinated development of man and nature, and ecological benign cycling.

The index system of Shanghai eco-city construction is a complex layer of index system including four levels structure, the highest level indicator (0 level) is the ecological comprehensive index, one class index under them which include urban ecosystem structure, function, and coordination measures index; the second class index consist of a number of similar factors which under the one class index property, the third class index consist of a number of similar factors which under the second class index property.

## The Structure Construction of Urban Ecosystem

(I) **Control the Population Size and Density and Improve the Population Quality**

At present, there are some problems in Shanghai, such as high population volume and density, the aging of the population, large floating population, although the number of people with higher education is highest in Chinese cities, there is still a gap compared with the developed international cities. The appropriate population density and the high quality of the population are the key to the construction of the eco-city, therefore we should strictly control the total volume and density of population. The urban population volume should reach the planning standard value, i.e., 3500 person/km$^2$ in 2015.

(II) **Control of High-Rising Building and Reduce the Building Density**

In recent years, the high-rising buildings in Shanghai spring up like bamboo shoots after a spring rain, improve the city visage graces, and attract foreign investment, as well as improve the living conditions. But due to the fast development, increasing impervious areas, and unreasonable layout, there are many potential ecological environment issues in Shanghai. Relevant data shows, nearly 3 years, the annual high-rising building reached more than 400 buildings. There are 4000 high-rising buildings with 16 floors, ranking no. 1 in the world in October, 2009. Because of the high building density, energy consumption, water consumption, traffic flow, and impermeable ground area increase, the heat island effect also increase. In recent years, heat island area in Shanghai expands further from mono-center to multi-center. At the same time, the most original urban waterways are blocked or filled, and affects the flood drainage capacity. Therefore, strict control of high-rising buildings, reduce building density should be imperative.

(III) **Strengthen the Greening Construction and Protect the Bio-diversity**

The goals of green construction in Shanghai city are expanding the green area, improving the coverage of green space, and improving the quality of it. Optimization of green space structure, abundant plant varieties, reasonable layout of green space system, give full play to ecological environment benefit of the green land are the targets. Green construction in Shanghai should be based on the construction of large green space, greenbelt, traffic greenbelt, and residential green space and as the focus, emphasizing on the combination of small, middle, large green, the combination of line with face mutually, and the combination of urban and rural. Green layout should further improve the existing municipal and district level parks, develop residential, and a housing estate public green space, build a series of theme parks and characteristic parks, and develop the riverside greenbelt of the Huangpu River, Suzhou River, Chuan Yang River, Taipu River and the Pudong canal, develop the green belt on both sides of radiation roads, such as Huqingping, Xinsong, Huning, Huhang, and so on. The construction of green space in the integrated structure partition of the central city improves the environmental quality.

Total 395 commonly used landscape plant species exists in Shanghai, including 210 species of arbor, 139 species of shrub, 34 species of ground cover, and 15 species of bamboo. The ratio of arbor to shrub is 1:3–6. Broad-leaved evergreen plants are 218 species, accounting for 63 % of the total, among them, only 80 are the highest frequency of use, and low bio-diversity. Greening work should be paid attention to the development of the zonal vegetation in Shanghai, at the same time, the varieties introduction and cultivating of practical and ornamental exotic plants and adapted to the ecological environment condition of Shanghai region. Vigorously develop the urban forest, ground cover plants, perennial flowers, evergreen lawn and climbing plants, in order to adapt to a variety of the needs of Shanghai as an international metropolis in the green form. At the same time, we should further strengthen the construction of urban natural reserve. In addition to the current Chongming Dongtan migratory birds National Nature Reserve and Jiuduansha National Wetland Nature reserve built so far, all or a part of place located in the Hangzhou Bay port of small and large Jinshan Island, Sheshan of Songjiang; Tianmashan, Dianshan Lake of Qingpu, Pudong, and other places can be all or a part of the nature reserve. Currently natural reservation area in Shanghai has reached about 12 %, and can make positive contribution for more effectively protection and improvement of biodiversity.

(IV) **To Strengthen Environmental Protection and Improve the Environmental Quality**

The city of Shanghai has been officially launched in early 2009 the Fourth Round Three Year Action Plan for environmental protection. This program plans to carry out 260 project construction, a total investment of over 80 billion RMB, is two times than the Third Round Three Year Environmental Protection Action of expected investment. The program cover the seven fields including the water environment management and environmental protection field, atmospheric environmental governance and protection field, noise pollution control, and solid waste utilization and disposal field and so on. Since 2000, Shanghai has continued to increase environmental protection investment. Environmental protection investment accounted for the ratio of the GDP has been more than 3 % for 8 years, and the total investment reached about 180 billion RMB. Among them, the environmental protection investment in 2007 had reached 36.6 billion RMB. After the first few round of 3 year environmental actions, Shanghai's total emissions of major pollutants and emissions intensity are declining. In 2007, the emissions of COD and $SO_2$ decreased 65 and 59 %, respectively, compared to the 2000 (Xu and Huang 2008). Recently, environmental protection still need to grasp the comprehensive environmental treatment for water, gas, and noise as the center of the regulation and the development of new energy, energy saving and emission reduction in Shanghai. In the water environment management, it mainly focus on the urban waterway pollution remediation, completely eliminates the black odor of Suzhou creek, further to make the creek clean, gradually transform it into a city sightseeing creek. However, in the suburb it mainly focus on the water quality protection of the drinking water, gradually restores the ecological function of the rural waterway, completes the expansion project of the water intake upstream and Chenhang reservoir. In-depth development of ecological construction in a water source protection areas of the Huangpu River and Yangtze River Estuary, the grass sand reservoir; In the air environment management, We should further adopt smoke abatement and desulfurization process, reduce sulfur oxide pollution, at the same time, expand the development of central heating and continuous heating, strict control of exhaust emissions of cars and pollution of CFC. The 660 $km^2$ within the outer ring exists urban dust pollution control area. Strengthen the management and disposal of the noise and solid waste pollution, and striving the standards in the pollution control.

(V) **Adjust the Structure of Land Use, Improve Infrastructure and Improve the Level of Urban Construction**

In recent years, Shanghai has made great achievements in the construction of three infrastructure including roads, housing, and hospitals. But the original foundation, especially roads and housing, compared with the advanced international city has a lot of distance. Therefore, we need to make greater efforts to completely change the traffic jam, crowded housing phenomenon. In addition, we should also complete the comprehensive planning of eco-city construction as soon as possible, pay attention to structural adjustment, strictly control the number of high-rise buildings and decrease trend of the cultivated land area, and realize the target of in 2015 initially building eco-city.

### The Function Construction of Urban Ecosystem

(I) **Improve the Treatment Rate of the Three Wastes and Strengthen the Urban Material Cycle**

Requirements of treatment rate for the three wastes reached 100 %. At the same time, We should recycle and comprehensive utilization to waste, imitate the natural ecosystem, and make the waste in the last production process into raw material of the next link, achieve material recycling and reuse. For this, we should promote the development of Industrial ecology, promote clean production, make the relevant industrial enterprise into Industrial Park, and make useful of the material and energy. At the same time, we should also practice water conservation and electronically conservation. Application of natural energy should be used in the place where the natural energy can be used. We should reduce water and fossil energy consumption, and improve the utilization efficiency of water, electricity and other natural resources.

(II) **The Construction of Fast Information Circulation System**
The development of the modern society has put forward higher requirements on information circulation. Some gaps exist in Shanghai with the international metropolis in this aspect. Therefore, we should speed up the development of the emerging modern means of communication including the Internet, global positioning satellite communication system, and so on. The government should have greater efforts to promote the hardware and software construction of information system.

(III) **To Coordinate the Relationship in Urban and Rural Ecosystems**
*The implement of the planning of urban and rural integration, optimization of the spatial structure of urban and town.* Urban and rural ecological environment is closely related because of the exchange of material, energy and information between the city and the surrounding countryside. City can provide the products to rural villages, and villages provide resources for the city, and handle the waste which the city can't disposal the waste produced by the city. The separate planning of urban and rural areas is not conducive to the mutual coordination between the various elements of the whole ecological economic system, and even produces mutual restriction. The city and countryside is a complex ecosystem, and we should break the administrative area boundary for planning of urban and rural integration, in order to clear the logistics channels, to achieve a virtuous circle of urban and rural ecological environment.
*To strengthen urban public service facilities, and improve the quality of resident life.* We should exploit concentrated plots of residential areas, improve the living standard of residents, In dwell district, we should provide a complete community service facilities, form the multilevel, multi-type of community service network, make the traffic, shopping, leisure, and education of the residents more convenient. The emphasis of environmental quality of residential area, strengthening of the residential green building, improving the design level of residential construction and outdoor environment should be made, and to meet the different needs of people. The construction of four levels of cultural infrastructure network in the city, district (county), street (township), neighborhood (Village), and the construction of three level sports

facilities network in the city, district (county), grass-roots enrich the cultural life of the people.

*Improve the ecological environment consciousness, and strengthen the construction of institutional framework.* Nowadays, many domestic and foreign city ecological environment problems are caused by the lack of ecological awareness of the decision makers, planners, managers, and the public. Therefore, we must proceed from the popularization of ecological knowledge, and popularized basic knowledge of ecology to the all publics through various educational and publicity channels. Then, they will grasp the basic principles of ecology, and understand the value of natural resources, the function of natural environment and the position, function of human in the ecological system. Promoting the masses consciously protecting the ecological environment, emphasizing the resource saving and environment friendly, actively participate in the construction of Eco-city.

The construction of eco-city need to make through leadership step by step, Therefore, Eco-city construction should have full-time leadership mechanism, and it directly responsible for the construction of eco-city in Shanghai. We should invited other administrations including economic, planning, resource management, planning and design, environmental protection, water supply, greening and city, propaganda, legal system and the District, county leaders, etc., to discuss the decision on the construction of eco-city policies, measures, major problems, and comprehensive coordination of city ecological construction. In the macroeconomic regulation and coordination decision and we should explore a new mechanism of eco-city construction and comprehensive decision according with the international metropolis.

### 4.4.3 Cases of Regional Ecological Planning in Nanning

Nanning, a capital city of Guangxi, has already a good foundation in urban ecological construction. In 1991, Nanning won the title 'One of the top 50 in comprehensive strength in China.' In 1997, it won the national title of 'garden city,' and was in the ranks of the 12 'garden city.' After then, it obtained the title of 'National Excellent Tourism City.' and won the 'Chinese Habitat Environment Prize in 2002' again. A higher goal of creating 'Chinese Green City' and 'UN Habitat Environmental Award' were pitched for new ecological construction of Nanning in the twenty-first century.

In 2003, the green coverage area was 4128.79 ha in the built-up area of Nanning, and coverage ratio reached 37.47 %. The green area of garden was 395.43 $hm^2$, of which, public green area was 851.02 $hm^2$ and the per capita public green area was 8.5 $m^2$, which were increased, respectively 4.47 % and 3.8 $m^2$ than 10 years ago; There have 13 kinds of open park, which public green area was a net increase of 504.2 $hm^2$ and an increase of 1.5 times than 10 years ago. The forest coverage ratio reached 38.38 %.

As the present achievement in Nanning, it has been into the ranks of national advanced eco-city. Environmental quality of the ambient air, water, sound reached the requirements of national environmental protection model city. The loss of soil and water in urban surrounding areas has been basically governance. Up to now, its urban green coverage ratio reached more than 45 %, and the ratio of green reached more than 40 %, the public green area of per capita reached 15 $m^2$. The planning of Nature Reserve, Scenic Area, and the Forest Park were basically completed. Totally, the whole city reached the requirements of national ecological demonstration area and established a benign ecosystem of sustainable development.

## References

Costanza, R., d'Arge, R., de Groot, R., Farberk, S., Grasso, M., Hannon, B., et al. (1997). The value of the world's ecosystem services and natural capital. *Nature, 387*, 253–259.

Costanza, R., & Daily, H. E. (1992). Natural capital and sustainable development. *Conservation Biology, 6*, 37–46.

Dai, X. L., Ji, K., Lv, F., & Yin, H. N. (2007). Urban ecosystem health assess based on matter element model. *Yunnan Geographic Environment Research, 19*(2), 58–63. (in Chinese).

Daily, G. (1997). Nature's services: Society dependence on natural ecosystems. Washington, D.C.: Island Press.

Gordon, D. (1990). *Green city: Ecologically sound approaches to urban space*. Montreal: Black Rose Books.

Guidotti, T. L. (1995). Perspective on the health of urban ecosystems. *Ecosystem Health, 1*(3), 141–149.

Guo, X. R., Yang, J. R., & Mao, X. Q. (2002). Primary studies on urban ecosystem health assessment. *China Environmental Science, 22*(6), 525–529. (in Chinese).

Hu, J. (2007). Report to the Seventeenth National Congress of the Communist Party of China on October 15, 2007.

Hu, T. L., Yang, Z. F., He, M. C., & Zhao, Y. W. (2005). An urban ecosystem health assessment method and its application. *Acta Scientiae Circumstantiae, 25*(2), 269–274. (in Chinese).

Lip, E. (1979). *Chinese geomancy*. Singapore: Times Book International.

Liu, G. Y., Yang, Z. F., Chen, B., & Ulgiati, S. (2009a). Emergy-based urban health evaluation and development pattern analysis. *Ecological Modelling, 220*(18), 2291–2301.

Liu, G. Y., Yang, Z. F., Chen, B., Zhang, L. X., Zhang, Y., Zhao, Y. W., & Jiang, M. M. (2009b). Emergy-based urban ecosystem health assessment: a case study of Baotou, China. *Communications in Nonlinear Science and Numerical Simulation, 14*(3), 972–981.

Mageau, M. T., Costanza, R., & Ulanowicz, R. E. (1995). The development and initial testing of a quantitative assessment of ecosystem health. *Ecosystem Health, 1*(4), 201–213.

MEA (Millennium Ecosystem Assessment). (2005). *Ecosystems and human well-being: synthesis* (155 pp). Washington: Island Press.

Rapport, D. J., Gaudet, C., Karr, J. R., Baron, J. S., Bohlen, C., Jackson, W., et al. (1998). Evaluating landscape health: integrating societal goals and biophysical process. *Journal of Environmental Management, 53*(1), 1–15.

Renjian, Wu, Wang, Xiangrong, et al. (2000). *The theory and pathway of eco-city construction*. Shanghai: Fudan University Press.

Sang, Y. H., Chen, X. G., Wu, R. H., & Peng, X. C. (2006). Comprehensive assessment of urban ecosystem health. *Chinese Journal of Applied Ecology, 17*(7), 1280–1285. (in Chinese).

Saunders, W. (2012). *Designed ecologies: The landscape architecture of Kongjian Yu*. Basel: Birkhauser.

Shi, F., & Yan, L. J. (2007). Application and research of unascertained measure model for urban ecosystem health assessment. *Bulletin of Science and Technology, 23*(4), 603–608. (in Chinese).

Skinner, S. (1982). *The living earth manual of Feng-shui*. London: Routledge & Kegan Paul.

Su, M. R., & Fath, B. D. (2012). Spatial distribution of urban ecosystem health in Guangzhou, China. *Ecological Indicators, 15*(1), 122–130.

Su, M. R., Fath, B. D., & Yang, Z. F. (2010a). Urban ecosystem health assessment: A review. *Science of the Total Environment, 408*(12), 2425–2434.

Su, M. R., Yang, Z. F., & Chen, B. (2010b). Relative urban ecosystem health assessment: A method integrating comprehensive evaluation and detailed analysis. *EcoHealth, 7*(4), 459–472.

Su, M. R., Yang, Z. F., Chen, B., & Ulgiati, S. (2009). Urban ecosystem health assessment based on energy and set pair analysis—a comparative study of typical Chinese cities. *Ecological Modelling, 220*(18), 2341–2348.

Takano, T., & Nakamura, K. (1998). *Indicators for healthy cities*. WHO CC HCUPR Monograph No. 2. Tokyo: World Health Organization Collaborating Center for Healthy Cities and Urban Policy Research.

UNFPA (United Nations Population Fund). (2008). State of world population.

Wang, X. (2003). *On the theory of ecological construction—a comparative analysis of the ecological construction between China and Oversea*. Nanjing: Southeast University Press.

Xi, J. (2013). Xi Jinping speaks at central urbanization work conference in Beijing, Xinhuanet (http://news.xinhuanet.com/english)

Yan, W. T. (2007). Research on urban ecosystem health attribute synthetic assessment model and application. *System Engineering, Theory and Practice, 8*, 137–145. (in Chinese).

Yang, Z. F., Su, M. R., Zhang, B., Zhang. Y., & Hu. T. L. (2010). Limiting factor analysis and regulation for urban ecosystems—a case study of Ningbo, China. *Communications in Nonlinear Science and Numerical Simulation, 15*(9), 2701–2709.

Yanitsky, O. N. (1981). *Cities and human ecology. Social problems of man's environment: Where we live and work*. Moscow: Progress Publishers.

Yu, K. (1994). Landscape into places: Feng-shui model of place making and some cross-cultural comparisn. In J. D. Clark (Ed.), *History and culture* (pp. 320–340). USA: Mississipi State University.

Yu, K. (1995). Security patterns in landscape planning: With a case in South China (Doctoral thesis, Harvard University, 1995).

Yu, K. (1996). Security patterns and surface model in landscape planning. *Landscape and Urban Planning, 36*(5), 1–17.

Yu, K. (2010a). Qiaoyuan Park, An ecosystem services-oriented regenerative design. *Topos, 70*, 28–35.

Yu, K. (2010b). Landscape as ecological infrastructure for an alternative urbanity. In M. Mostafavi (Ed.), *Implicate and explicate, Aga Khan Award for Architecture* (pp. 282–283). Switzerland: Lars Müller Publishers.

Yu, K. (2010c). Five traditions for landscape urbanism thinking. *Topos, 2010*(71), 58–63.

Yu, K. (2011a). Landscape as a living system: Shanghai 2010 Expo Houtan Park. In M. Richter & U. Weiland (Eds.), *Applied urban ecology: A global framework* (pp. 186–192).

Yu, K. (2011b). Stormwater Park for a Water Resilient City: Qunli National Urban Wetland. *TOPOS, 2011*(77), Making Space: 72–77.

Yu, K. (2011c). Ecological infrastructure leads the way: The negative approach and landscape urbanism for smart preservation and smart growth. In M. Richter & U. Weiland (Eds.), *Applied urban ecology: A global framework* (pp. 152–166).

Yu, K. (2012). Capitalizing on foothills: Restoring the relationship between people and land. *Harvard International Review, 34*(1)l, 40–45.

Yu, K. (2013a). Integration across scales: Landscape as infrastructure for the protection of biodiversity. In J. Beardsley (Ed.), *Designing wildlife habitats* (pp. 241–271). Dumbarton Oaks, Trustees for Harvard University, Washington, D.C., Printed in the United States of America.

Yu, K. (2013b). Landscape installation in the Sanlihe Greenway, Qian'an, China. *Topos,* Making Space (83), 6–7.

Yu, K., Li, H., & Li, D. (2008a). The negative approach and ecological infrastructure: The smart preservation of natural systems in the progress of urbanization. *Journal of Natural Resources, 23*, 937–958.

Yu, K., Li, H., & Li, D. (2008b) Ecological infrastructure as a tool for smart preservation and smart growth. In *Proceedings, 44th ISOCARP Congress* 2008.

Yu, K., Li, H. L., Ll, D. H., Qaio, Q., & XS, Xi. (2009a). Primary study of national scale ecological security pattern. *Acta Ecologic Sinica, 29*(10), 5163–5175.

Yu, K., Li, D.-H., LI, H.-L., & Qiao, Q. (2012). *China national ecological security patterns: A framework for reinventing the good earth.* China Building Industry Press (In Chinese).

Yu, K., Wang, S., & Li, D. (2011a). *Regional ecological security patterns: The Beijing case.* Beijing: China Building Industry Press. (In Chinese).

Yu, K., Wang, S., & Li, D. (2011b). The negative approach to urban growth planning of Beijing, China. *Journal of Environmental Planning and Management, 54*(9), 1209–1236.

Zhong, Y. X., & Peng, W. (2003). Assessment of urban ecosystem health. *Jiangxi Science, 21*(3), 253–256. (in Chinese).

# Chapter 20
# Regional Ecological Construction

**Peicheng Li, Qilei Li, Jinfeng Wang, Feimin Zheng, Jianli Yuan, Danghui Xu, Gang Wang, Xingtu Liu, Ming Wang, Zuofang Yao, Xianghao Zhong, Xiaodan Wang, Shuzhen Liu, Daming He, Shaohong Wu, Tao Pan and Qingwen Min**

**Abstract** Ecological construction is the process of using engineering or biological measures to regulate the structure and improve the ecological function of the system according to the principle of ecology and ecological economics, and making full use of modern science and technology and the nature of ecosystem. Ecological construction is aimed to make a particular ecosystem meet the growing demand of survival and development by human in that system and make the system realize sustainable development. The connotation of ecological construction widely includes urban ecological construction, rural ecological construction, the restoration and reconstruction of degraded ecosystem, ecological engineering, ecological education, and so on. Ecological construction is a key element for a region in achieving sustainable development, it can not only improve regional ecological and environmental quality, but also establish a good regional image, realize the appreciation of regional intangible assets value, and promote economic growth. At present, different regions in China have made some important progress in ecological construction according to their own characteristics. Overall, ecological construction is vital for sustainable development, our country is vast in territory, different regions

P. Li · Q. Li · J. Wang · F. Zheng
Research Institute of Water and Development, Chang'an University, Xi'an 710054, China

J. Yuan · D. Xu · G. Wang
School of Life Sciences, Lanzhou University, Lanzhou 730000, China

X. Liu · M. Wang · Z. Yao
Northeast Institute of Geography and Agroecology, Chinese Academy of Sciences, Changchun 130102, China

X. Zhong · X. Wang · S. Liu
Institute of Mountain Hazards and Environment, Chinese Academy of Sciences, Chengdu 610041, China

D. He
Asian International Rivers Center, Yunnan University, Kunming 650091, China

S. Wu · T. Pan · Q. Min (✉)
Institute of Geographic Sciences and Natural Resources Research, Chinese Academy of Sciences, Beijing 100101, China
e-mail: minqw@igsnrr.ac.cn

 
W. Li (ed.), *Contemporary Ecology Research in China*,
DOI 10.1007/978-3-662-48376-3_20

are faced with different ecological and environmental problems, for each area, it is necessary to identify the ecological security problems first, and then using principles of ecology and combing each subjects to improve the function of ecosystem and to meet the demand for survival and development by local people, thus the region could realize sustainable development.

**Keywords** Desertification · Water loss and soil erosion · Salinization · Returning farmlands to forests and grasslands · Species diversity · Seed bank · Life history strategy · Salinity tolerance · Drought resistance · Soil crust · Commodity grain base · Tibetan Plateau · Ecological security · "Corridor-barrier" functions · Longitudinal range-gorge region (LRGR) · Ecocompensation mechanism

## 1 The Ecological Environment and Reconstruction of Landscape in Northwest of China

The northwest of China is a vast area with resources superiority, but its ecological environment is vulnerable. So it is necessary to improve its condition for sustainable developing. In the last 10 years, some scientific researches on ecological environment construction and reconstruction of landscape in northwest area of China were carried by Chinese government.

### *1.1 Vast Land, Sufficient Sunlight, Rich Mineral, with Very Huge Development Potential*

The northwest of China is vast area, including Shanxi, Gansu, Ningxia, Qinghai, Xinjiang, is 3.04 million km$^2$ in total, representing 58.7 % of the total west area, occupying 31.7 % of the whole national land area. Its population is 90 million, accounts for 7 % of the whole country. In comparison, a vast territory with a sparse population, the land area per capita is 3.8 hm$^2$, which is 4.75 times the number of whole nation, including 18.53 million hm$^2$ cultivated land, per capita cultivated land is 0.203 hm$^2$, which is more than two times as much as the national per capita. Moreover, there are 65.44 million hm$^2$ grassland in northwest of China, 0.76 hm$^2$ per capita, and 14.13 million hm$^2$ forest land. Thus, the northwest area has the most abundant land resources in China (Li et al. 1999).

Besides land resources, the solar thermal resource is plentiful in northwest area also. The sunshine duration in this area is longest in China, which is benefit for developing agricultural production. Take Yulin in Shaanxi province as an example, if the efficiency for solar energy utilization increased by 2 %, the yield of sorghum will be doubled.

There are abundant mineral resources in northwest area. The coal, oil, and natural gas are top-ranked nationwide. Nonferrous metals such as nickel, copper, lead, zinc, cobalt, and molybdenum, rare metals such as palladium, iridium, manganese, beryllium, lithium, niobium, and tantalum, precious metals such as gold, silver, and platinum, chemical miners such as salt, sylvite, boron, and nitration, nonmetallic minerals such as gypsum, asbestos, mica, limestone, and silicon. They occupy very important position nationwide and provide significant material basis for large-scale developing industrial and mineral industry and energy heavy chemical industry. The Xinjiang Uygur Autonomous Region will be the Chinese oil reserve base in twenty-first century and the reserves of natural gas in north of Shaanxi is top-ranked nationwide.

This shows that the northwest area, located in the Eurasia hinterland, plays a fundamental role on the national future development and promotion on economic prosperity of the world. It is a vast area which is urgent to exploit. And it is also a huge potential market. Its accelerated development will provide broad development space and inject great vitality for national economy in west of China and the whole country.

## *1.2 The Ecological Environment is Severe and Vulnerable in Northwest of China*

The ecological environment in northwest of China is very severe and vulnerable that mainly displays in the following aspects (Li et al. 1999):

### 1.2.1 Low Precipitation, Drought, and Water Shortage

The northwest area of China locates in the Eurasia hinterland, obstructing by mountains, far away from ocean, so ocean could not affect it and the precipitation in most area is very low, the annual rainfall is lower than 400 mm, belongs to the arid and semiarid regions. The annual rainfall in loess plateau is between 300 and 500 mm, less than 200 mm in Tsaidam Basin, less than 100 mm in Hexi Corridor, only 29.5 mm in Dunhuang, less than 20 mm in Turpan, and less than 10.9 mm in Nuoqiang, where there is nearly no rain in the whole year.

Because of low rainfall, the surface water capacity in northwest is about 200 billion $m^3$/a and groundwater dynamic reserve is about 65 billion $m^3$/a. There is 265 billion $m^3$/a in total. It occupies 9.4 % of 2.81 trillion total water resources in our country. The area of northwest represents 31.7 % of national territory area, but only accounts for less than 10 % of water resources that shows the lack of water in the northwest. Moreover, the imbalance of spatial and temporal distributions and high sediment concentration in most rivers make big trouble on development and water resource utilization. In recent years, the serious pollution leads the water shortage worse and worse, hinders the economy development and people's livelihood improvement, results in vulnerable ecological environment.

It is important to emphasize that drought and water shortage are the most serious ecological environment problems. The water shortage condition of each province (or autonomous region) is showed in Table 1 on the basis of author and coauthors' research works.

Table 1 showed that if it develops as usual in the northwest area, the water deficit is 10.82 billion $m^3$/a in 2010 and will be 20.75 billion $m^3$/a in 2020. The water demand will be 100 billion $m^3$/a in 2030 and the water deficit will be 23.1–25.6 billion $m^3$/a (Li et al. 1999). We should know that the water deficit amount is smaller in fact, because the water demand amount calculation is on the basis of normal development of northwest area without considering the all demand of "large-scale development in west China". The deficit amount cannot reflect the water shortage actual condition in different regions of northwest area in different periods, so it represents average level and lessen the time and territory difference which is obvious in northwest area.

Therefore, the severe water shortage problem in northwest is the main leading factor of ecological environment vulnerable and the huge obstacle of playing superiority. According to the bottleneck theory of region development, the restrictive factors should be eliminated and the water resources sustainable supply issue on development and reconstruction in west China should be resolved.

### 1.2.2 Desert and Gobi Widespread, Strong Wind that Carries Sand and Drives Stones, Severe Habitat

The biggest deserts and gobis were spread in northwest of China. The deserts include Taklimakan desert (0.3376 million $km^2$), Gurbantunggut Desert (48,800 $km^2$), Badain Jaran Desert (44,300 $km^2$), Tengger desert (42,700 $km^2$), Mu Us Sandland (32,100 $km^2$), and so on. The total area is 0.492 million $km^2$; represents 69 % of 0.7129 million $km^2$ total desert area in China. The gobi area in northwest area is 0.4175 million $km^2$, comprises 73.3 % of the total gobi area 0.5695 million $km^2$.

Most desert and Gobi are covered by dry sand and gravel except some small area oasis. It is scorching hot in summer and bitter cold in winter. Strong wind that carries sand and drives stones and water shortage threaten all lives here. The people who work on economic activities here must make great efforts on ecological environment reconstruction.

### 1.2.3 The Area of Desertization and Desertification Extending Continuously

The desertization area in Xinjiang is up to 96,100 $km^2$ (about 9.61 million $hm^2$). The oasis lost its protective screen because of natural vegetation, such as populus euphratica forest, degenerated and died in large scale, especially the natural

**Table 1** The water shortage predication of each province (or autonomous region) in northwest area

| Provinces and regions | 2010 | | | 2020 | | |
|---|---|---|---|---|---|---|
| | Water demand (0.1 billion $m^3$) | Water deficit (0.1 billion $m^3$) | Water deficient ratio (%) | Water demand (0.1 billion $m^3$) | Water deficit (0.1 billion $m^3$) | Water deficient ratio (%) |
| Shaanxi | 144.44 | 52.74 | 36.57 | 175.10 | 83.40 | 47.63 |
| Gansu | 145.80 | 20.80 | 14.26 | 171.80 | 46.80 | 27.24 |
| Ningxia | 102.60 | 4.10 | 4.0 | 110.60 | 12.10 | 10.94 |
| Qinghai | 36.90 | 6.96 | 18.70 | 45.70 | 15.70 | 34.35 |
| Xinjiang | 468.60 | 23.60 | 5.04 | 494.50 | 49.50 | 10.01 |
| Total | 898.34 | 108.20 | 12.04 | 997.70 | 207.50 | 20.80 |

vegetation around oasis is destroyed by people. So the desertization area is increased by 400 $km^2$ every year.

The desertization area in Qinhai is up to 0.1252 million $km^2$ (about 12.52 million $hm^2$) and increased by 1300 $km^2$ (0.13 million $hm^2$) every year. The desertization problem also exists in Gansu, Ningxia, even Shaanxi province.

Grassland degeneration area is increased continuously in northwest. The meadow degeneration area in Qinghai is up to 11.73 million $hm^2$, accounting for 32.3 % of total meadow area. Yield of grass and grazing capacity declined sharply because of grassland degeneration and desertification. The grass yield per unit area in Qinghai decreased by 30–80 % in different region compared with the yield in 1950s. The average grazing capacity in Xinjiang is only one in 1.49 $hm^2$ because of grassland degeneration and desertification in large scale.

### 1.2.4 Water Loss and Soil Erosion is Still Severe and the Macroscopic Effect of Management is Still not Good Enough

Severe water loss and soil erosion is still the universal ecological environment problem through many years control and management, especially loess soil covered regions in Shaanxi, Gansu, Ningxia, and Qinghai. The distribution situation of water loss and soil erosion area is as follows: 0.118 million $km^2$ in Shaanxi, representing 57.4 % of its total land area; 0.3966 million $km^2$ in Gansu, representing 87.43 % of its total land area, including 0.105 million $km^2$ in loess plateau; 17,800 $km^2$ in Ningxia, representing 34.4 % of its total land area; and 75,000 $km^2$ in Yellow River basin region in Qinghai, representing 10.4 % of its total land area.

Moreover, the amount of soil loss in northwest is very huge, especially in Shaanxi. The sediment runoff in Shaanxi is 0.92 billion tons, constitutes one-fifth of that in the whole country, 0.8 billion tons sediment loads of the Yellow River comes from Shaanxi and 0.518 billion tons comes from Gansu. The total sediment loads of these two provinces to the Yellow River is 1.3 billion tons, make up 80 % of total amount of the Yellow River.

So water loss and soil erosion in northwest not only destroy the local production and living environment, but also threaten the life and property security in downstream of the Yellow River, even the lower reaches of Yangtze River.

### 1.2.5 Productivity of Land Decline Sharply Because of Serious Salinization Problem

The salinization area in Xinjiang is about 1.45 million $hm^2$ (21.75 million mu), takes 45 % of cultivated land area; the salinization area of cultivated land area in Ningxia is 86,700 $hm^2$, accounts for 26.6 % of cultivated land area in irrigation district. Moreover, it is possible to generate salinization in new irrigation area. The salinization problem also exists in Shaanxi, Gansu, and Qinghai.

### 1.2.6 River and Lake Shrink, Water Quality Deteriorate Seriously

The biggest river, Wei River in Guanzhong region of Shaanxi province dries up sometimes. And eight rivers of "Eight water around the Chang'an" cut off frequently from the latter half of the twentieth century. The rainfall is rare and the ground water system is sparse. In recent decades, river and lake shrink and water quality deteriorate sharply.

Tarim River in Xinjiang is the longest inland river in our country, its flow path shortens 300 km because of human activity, leading Lop Nor disappeared. In recent 100 years, the water table of famous Qinghai Lake dropped 11.12 m and the rate of descent is speeding up in recent half a century. Some rivers in Shaanxi have dried already, and some became pollution discharge channel. The charming magnificent landscape has gone and you could not find a clear river in "Eight hundreds Qin plain" (alluvial valley and plain of Wei River in the north of Qinling Mountains) now.

The water quantity is decreasing; wet land and marsh are shrinking; and ecology is deteriorating badly in Jiangheyuan, the source region of Yangtze River and Yellow River.

### 1.2.7 It is Mountainous in Northwest and the Ecology of Mountainous Area Suffered Destruction

The nationwide, even universal famous mountains, such as Qingling Mountains, Ba Mountains, Qilian Mountains, Altun Mountains, Kunlun Mountains, Tianshan Mountains, Altai Mountains, and so on, locate in the northwest. Numerous huge mountain systems make a big trouble on local traffic economy and culture development. However, there are abundant mineral, forestry, and biological resources in mountain area, where there are the repository of ice and snow and the water resource of plain and basin.

The ecology of mountain area suffered destruction by intensive human activity, especially in Qingling Mountains. According to the statistics in Ankang area, the forest vegetation coverage dropped from 36.5 % in 1949 to 27 % in 1985. Take Shangluo area as another example, it covered 0.43 million $hm^2$ (640 mu) forest in 1950s, but only 0.265 million $hm^2$ (398 mu) left in 1971. Half of the forest disappeared, which makes the timberline go up by 300–500 m, ecocatastrophe (such as water loss and soil erosion, landslide, debris flow, and so on) intensifies, water quality in river deteriorates, water quantity reduces … . It is worrying that the deteriorating trend is aggravating now.

### 1.2.8 The Survival Condition of Living Beings Is Hard, Biodiversity Suffered Threat

The survival condition of living beings is hard and biodiversity suffered threat due to natural and anthropogenic reasons. In recent years, the wildlife species which are threatened by hard habitat condition account for 22.3 % of total species in Xinjiang, some of them become extinct already. The threatened species constitute 15–20 % of total species in Qinghai, which is above the worldwide average of 10–15 %.

### 1.2.9 Flora and Fauna Diseases and Insect Pests Endanger the Ecological Environment and Life Security

The ecological environment is very vulnerable in northwest and struggle for existence is very intensive. Because of biodiversity reduction, some noxious animals with strong fertility and adaptability, such as murid, become rampant to destroy grassland and vegetation. In addition, some forest and crop pest are also kinds of ecological environment problems in the project of Reconstruction of Landscape. If we do not take it seriously, the achievement of "conversion of cropland to forest and return grazing land to grassland" will be destroyed.

### 1.2.10 Some Ecological Environment and Social Problems Caused by Local Resources Overexploitation

Pre-chairman Jiang Zemin has written instructions on the project of reconstruction of landscape. He pointed out that the severe ecological environment in northwest is because of war, natural disaster, and deforestation denudation. It shows that the ecological environment problems are caused by natural and human factors.

Take the north and central of Shaanxi as examples, where the development of history is long and human activity is intensive. In 50 years after the founding of republic, on the one hand, large-scale afforestation and small watershed integrated management were carried out to alleviate the disadvantage natural conditions restrict on agricultural production and improve the local condition of people's production and living. It did get some achievement in the short term in the face of powerful harsh nature. On the other hand, the demand of food, fuel, and forage grass increasing dramatically because of population growing, result in predatory production and operating activities, such as disorder reclamation and deforestation denudation, overgrazing, picking medicinal herbs, and so on. All these activities destroy vegetation, soil, water sources, and land, aggravate water loss and soil erosion. Though some local ecological environment was improved, it deteriorated on the whole.

In fact, many other serious ecological environment problems in northwest area are not listed above. We can see that the northwest area of China has the most serious ecological environment problem in the world. These problems will weaken resources superiority and hinder the economic development and social progress.

## *1.3 Facing Severe Ecological Environment in Northwest Area, Carrying Out Deep-Going Researches on the Project of Reconstruction of Landscape*

### 1.3.1 The Definition of Reconstruction of Landscape in Northwest Area

The "landscape" in Chinese includes high land, desert, meadow, basin, mountain, water, woods, field, road, irrigation area, town, and so on, covers first nature and second nature.

"Reconstruction of Landscape" means protecting, repairing, and remolding the first nature and the second nature with adjusting measures to local conditions to form territory social environment with everything booming, accompanying an interdependence and virtuous cycle habitat—beautiful landscape by means of advanced productivity. It is according to ecology principle, sustainable development principle, and socioeconomics principle with "civilization, glorious, rich, health and happiness" concept.

It is to create "blue sky, green land, flourish mountain and rich people" world by popular and visualizable expression.

Reconstruction of Landscape involves heaven, earth, and man and its aim is equilibrium increasing in ecology, economy, and social benefit.

It is a great and onerous undertaking and needs powerful and systematic science and technology support. Therefore, we carried out "preliminary research on reconstruction of landscape in northwest area" from 1999 to 2002, and further research project "significant science and technology issue and experiment demonstration area construction on reconstruction of landscape in different ecological region in northwest of China" from 2003 to 2008. The preliminary research was acceptance checked in 2003. We also got some significant achievement in further research, some of them are list in this section.

### 1.3.2 The Guiding Ideology, Research Objects, and Technical Route of Research on Reconstruction of Landscape in Northwest Area

The Guiding Ideology

"The science and technology action plan about reconstruction of landscape in northwest area of China" preliminary research followed the guiding ideology as follows (Li et al. 2007):

1. As mentioned previously, the nature features are much worse than before because of two powers action in long history. One is the nature force, which is the age long, evolutional, and primary one as a whole. Another is human impact. It is much shorter and secondary compared with the former, but is probably drastic and dominate in local area. Thus the nature transformation in reconstruction of landscape project should focus on two aspects. One is transforming nature following natural law, for example, how to solve water shortage problem in arid area; the other is restricting and amending human harmful activities.
2. This study is carried out for afforestation and water conservancy and desertification control in northwest area for reconstruction of landscape. Project settings require comprehensiveness, pertinence, succession, and innovativeness. It focuses on practical problem solving on the premise of integrating theory with practice to guarantee collaboration of the large-scale northwest development and reconstruction of landscape.
3. The research was carried out from three aspects: the integration and promotion of previous scientific and technological achievements, significant problem research, and experimental demonstration area construction.
4. Choosing the topics according to regionalization classification management concept considering task and target based on the actual conditions of northwest. Most task settings should consider subject category and department of ownership, particularly regional characteristic. Most researches should carry out in local area to speed up the reconstruction of landscape realization directly. It should avoid repetition and windbaggary, following realism and truth.
5. Deal with the relationship between reconstruction and exploitation, utilization and protection, damage and restoration and accomplish "combinating of exploitation and reconstruction, promoting reconstruction by exploitation and guaranteeing exploitation by reconstruction".
6. The implement of reconstruction of landscape refers to many aspects. The research subject design should deal with the relationship among areas, between the urban and the rural, between economic development and environment construction to make sure unified design, coordinate implement, division of labor responsibility and achievement sharing.
7. It should carry out joint research on some significant common problems, such as drought, water shortage, desertization, desertification, salinization, water loss and soil erosion, and so on. For interprovincial river basin management, it is better to carry out cooperative work and responsibility system.
8. Considering the big regional diversity and long process of science and technology action plan, the project designer should pay attention to joint, regionalization, and staging. The stage, territoriality, continuity, and integrality should be embodied in project design.

9. It should consider international cooperation and attract the technical and economic cooperation of some brother provinces and foreign public and private departments.
10. The government's unified command and people's positive participation should be considered in subject setting because of the large-scale, strong sociality, and practicalness of project.

Research Objects

The persistent object of this project is developing science and technology to prompt the implement of "reconstruction of landscape in northwest area" undertaking successfully. The accomplishment of "reconstruction" undertaking is a long process, so the scientific research should set stage object: near-term object (2001–2005); medium-term object (2006–2020); long-term object (2021–2030).

Technical Route

1. Integrating and matching previous achievements and promoting the application actively;
2. Applying theory to reality, studying the significant problem, and paying attention to technological innovation;
3. Strengthening experimental demonstration area construction and demonstration radiation;
4. Playing multidisciplinary superiority and strengthening comprehensive analysis and research;
5. Combining management and exploitation and prompting industrialization positively;
6. Persisting the mass route and stimulating the cadres and the masses to participate;
7. Persisting the reform and opening-up policy and striving for international cooperation;
8. Emphasizing ecology, economy, and society comprehensive benefits.

### 1.3.3 The Regionalization of Ecology–Economy–Society in Reconstruction of Landscape in Northwest Area

The ecology–economy–society factors were considered synthetically in the regionalization of reconstruction of landscape. It provides science and technology support for the reconstruction of landscape project construction and different type's experimental demonstration layout and the basis for governmental macro decision-making by carrying out and implementing scientifically reconstruction of landscape regionalization research (Li et al. 2007).

In this strategic research, the first-level regionalization research focus on ecological and economic factors, the social factor will be considered in second-level and third-level regionalization researches in the future. In the first-level regionalization, the northwest area was divided into 12 first-level regions considering ecological environment and economic activity. They are as follows:

### The Strict Conservation Area of Mountain Ecological Environment

Some famous mountains locate in northwest, such as Qingling Mountain, Ba Mountains, Qilian Mountains, Altun Mountains, Kunlun Mountains, Tian Mountains, Altai Mountains, and so on. Mountains are the natural barriers of plain and basin. The stored ice and snow and conservation moisture in mountains are the important water resources. Deforestation denudation, disordered mining and disorderly hunting lead to forest deterioration, grass coverage rate decrease, water loss and soil erosion, landslide, debris flow aggravation, river water quality deterioration, water quantity reduction and biodiversity damage … . In a word, intensive human activities result in mountain ecological environment deterioration. It should be protected by powerful measures.

### The Prevention Area of Desertification

The Taklimakan Desert locates in Tarim Basin and Gurbantunggut Desert locates in Dzungaria Basin in Xinjiang. Ulan Buh Desert, Tengger Desert, and Badain Jaran Desert surround the north boundary of Ningxia and Gansu, so it is an important ecology task to prevent deserts invading southward. The serious desertification, vegetation, and grassland degeneration influence the development of husbandry because of drought climate and sand storms hazards.

### The Ecological Active Maintenance Area in Gobi and Desert Plateau

The barren gobi is distributed widely in the east of Xinjiang and west of Gansu. The ecological environment is very atrocious there. It is a serious issue that how to change this condition. Gobi maintenance is the way to prevent its further expansion at present. If there is water diversion to gobi area in the future, it is possible to form semi-oasis and oasis gradually and develop there.

### The Reasonable Development Area of "Conversion of Cropland to Forest and Grassland" and Water Loss and Soil Erosion Management Area in Loess Plateau

Severe water loss and soil erosion is the common ecological environment problem in northwest area, especially in loess plateau in Shaanxi, Gansu, Ningxia, and Qinghai. There are thousands of furrows, loess hills, and ridges in this region because of wind erosion and water erosion. The water loss and soil erosion area is vast, and soil erosion is also very big in this region, which is the main sediments-producing area of the Yellow River. It should strengthen comprehensive treatment and carry out large-scale popularization actively for reconstruction of landscape on the basis of summarizing and applying previous achievements, like implementing in Zhuanglang County in Gansu province.

### The Ecological Intensive Governance Conservation Area in River Valley Plain

The river valley plain is the crucial economic development zone in northwest area, such as the Wei River valley plain area (the Guanzhong Plain), the Yellow River valley plains (the Yinchuan plain), and so on. The ecological environment is advantageous in this area. It is a prosperous place since ancient times. Agriculture, industry, science, education, and culture concentrated here, with developed urban traffic and prosperity economy. The high-level industrialization and dense population make a big pressure and adverse impact on ecological environment, such as environment pollution, water contamination, river dry, and so on. So it should pay attention to prevent the destroying ecological environment by human activity, the pollution prevention and control, water saving and reasonable allocation of water resources.

### The Ecological Maintenance Control Area of Intermountain Basin and Plateau Basin

The intermountain basin is rich and populous place in northwest area, such as Hami basin and Turpan basin in Xinjiang. The unique ecological environment provides condition for specialty amphisarca and economy development in basin. But the ecological environment stress is very big because of dense population. For example, more karez were excavated in Turpan basin and the water quantity is decreasing gradually. So maintenance and protection of local ecological environment should be strengthened.

Qaidam basin in Qinghai is a large-scale basin in northwest area. It is famous for its abundant mineral resources nationwide, such as sylvite, sodium salt, lithium magnesium rock (grit), mirabilite, and so on. But the strong sand wind, vegetation degradation, desertification, soil salinization, and ecological environment deteriorate caused by mine exploitation should be prevented and controlled actively.

#### The Ecological Preservation Area in Piedmont Plain (Including Oasis)

Different size of piedmont pluvial alluvial plains were distributed in submountain region of most mountains in northwest area, such as the piedmont plains in the north and south foot of Tianshan Mountains, the north foot of Kunlun Mountains in Xinjiang, the north foot of Qilian Mountains in Gansu. These plains are the important areas of industrialization and economic development with abundant land and water resources. There are different size oases in some region with good soil and water condition. Most oases locate in the middle to lower part of piedmont plain, so soil salinization is a common problem. The water quantity reduced sharply in downstream of some rivers leading desertification in natural oasis in downstream plains because of building reservoir and retaining water in upper and middle reaches. The piedmont plains in arid and semiarid area play a key role in economic development, so it is necessary to maintain ecological environment and enhance the reasonable allocation of water resources and the prevention and control of salinization.

#### The Conservation Area of Alpine Plateau Grassland, the Source Regions of Yangtze River and Yellow River and Rare Creatures

Alpine plateau in Qinghai is the headstream of Yangtze River and Yellow River. It is famous of rare animals and plants in alpine grassland. It has alpine drought, desertization, intensive desertification, wetland and greenbelt shrink, conservation water function declined, grassland degradation, severe rats and pests, biodiversity reduction, and salinization problems in this region. The ecological environment has significant meaning in this region, so it is necessary to manage and protect actively by means of effective measures.

#### The Protection Area of River, Lake, Reservoir, and Aquatic Life

The river, lake, reservoir, and groundwater are the precious water resources in northwest area and the important foundation of existence and development. The rivers and lakes shrinking, dry up, and cut off, groundwater overexploitation, water pollution and aquatic life decreasing happened because of unreasonable utilization for many years, leading water environment deterioration. Thus, water shortage phenomenon appears, such as oasis ecosystem deterioration, vegetation recession, desertification aggravation, and so on, it is urgent to protect all kinds of water bodies by means of effective measures.

### Pollution Prevention and Control Area in Urban and Peri-urban Area and New Water Loss and Soil Erosion Area

Urban beautification is the key character of urban modernization, including buildings beautification, coordination layout, gardens, greening, roads, traffic, neat typing, and civilization. It is called "garden city" for short. People who live and work in it will enjoy the beauty.

In recent years, the urban construction develops fast. Cities have completely new outlook because of paying attention to urban landscaping and water body construction. But the pollution problem in rural–urban fringe is still serious, such as black sewage, foul water, garbage, and so on. The "zip object" still exists because of repairing road and laying all kinds of pipeline, leading some environmental problems, such as water loss and soil erosion, big noise and air dust.

Urban and its periphery is a local window, so it must strengthen governance to build good urban ecological environment.

### Cropland Landscape Area and Pollution Prevention Area in Countryside, Town, and Village

The countryside, town, and village have vast area. It is necessary to change concept to build new beautiful countryside, town, and village. There is the construction model of "village beautification" in every province and region. So it should build new characteristic beautiful villages and small towns according to local condition.

Cropland construction is mainly to build protection forest and "green food" area. Meanwhile, it should pay attention to environmental health and the possible arrangement of people, livestock, and birds to create new village graceful environment and reconstruct landscape.

### Ecological Environment Construction Area of Energy Base and Township Enterprise, Prevent and Control Pollution Area and New Water Loss and Soil Erosion Area

The northwest area, with abundant resources, is the construction area of national energy resources and heavy chemical industry base. Energy base construction should conduct environmental protection exploitation, prevent and control pollution and new water loss and soil erosion and pay attention to the ecological environment restoration and landscaping construction after exploitation. Township enterprise should change the bad habits of "product regardless of ecological environment" to prevent and control pollution and new water loss and soil erosion, maintain ecological environment. These are the basic conditions of a prosperous enterprise.

## *1.4 The Main Contents of Science and Technology Action Plan Research on Landscape in Northwest Area of China*

The main content of science and technology action plan research on landscape in northwest area of China includes three sections according to three aspects guiding ideology: the integration and promotion of previous relevant achievements, significant problem research, and experimental demonstration area construction.

### 1.4.1 The Integration and Promotion of Previous Relevant Achievements

The integration and promotion of previous relevant achievements are listed as follows according to its property: territorial resources, agricultural regionalization, survey achievements integration, ecological agriculture research achievements, comprehensive treatment technological achievements of different small watershed in loess plateau, key technological achievements of soil and water conservation, wind prevention and sand fixation and desertification treatment technique achievements in wind-blown sand area, salinization prevention and control technique achievements, research achievements on tourism resource and its exploitation, dryland agricultural technical system achievements in different ecotopes, scientific research achievements on shelterbelt networks construction, meadow and grassland improvement and rodent control technique achievements, graziery production technological achievements in rural area, courtyard economy development and special cultivation technique achievements, efficient facility agriculture technological achievements, medium and low-yield field improvement comprehensive technical achievements, water-saving and water resource utilization technology achievements in irrigation area, famous quality local melon and fruit cultivation and process technology achievements, renewable energy resource exploitation and utilization technological achievements in rural area, environment protection and cropland pollution comprehensive treatment technological achievements, and so on.

### 1.4.2 Preparing to Carry Out the Significant Problems' Researches

It is preparing to carry out 16 projects, including 98 subjects. They are listed as follows:

#### The Reexamination and Supplemental Investigation of Economy–Society–Ecology Environment in Northwest Area and Information System Establishment

The northwest area is vast. Though some comprehensive investigation and regional geological survey were carried out in different regions in this area regardless of ecological environment, the information is not complete. So it is necessary to conduct comprehensive investigation on ecological environment for implementing reconstruction of landscape successfully. Moreover, it is better to set up relative integrity comprehensive information database with ecological environment, economy, and society information to adapt the accurate implementing and dynamic tracking of reconstruction of landscape science and technology action plan in northwest area ("978 plan" for short). So three researches were carried out as follows:

1. Accomplishing ecological environment background investigation and mapping of "978 plan" applying "3s" new technique (complete the investigation in 2003).
2. Establishing economy–society–ecology environment comprehensive information system on landscape in northwest area.
3. Constructing water–soil–light–heat–air–plant–human being (human activity) eco-environment system fixed-point monitoring and forecasting station in different regions (building with base construction).

#### The Research on Nature Historical Vicissitude and Heaven–Earth–Man Interaction in Northwest Area

With regard to the vulnerable ecotope in northwest area, people have the consensus that it is caused by natural deterioration and bad human activity including war. But, for the vast northwest, how does nature change? How is the intensity of process? How is future trend? For the environment deterioration, where is mainly caused by natural calamities. Where is mainly caused by man-made misfortunes. And what is percentage of the natural calamities and man-made misfortunes? How to acquire farsightedness by means of scientific research? The answers of these questions will play a very significant role in dealing with the relationship between nature and human being correctly, preventing blindness action and inaction and guiding management and "reconstruction" correctly.

The relevant subjects are as follows:

1. The geological history in northwest area and relevant loess region: Qinghai plateau, loess plateau, Tarim basin, Qaidam basin, Hexi Corridor, and Yulin-Yanchi wind-blown sand area;
2. The textual research on climate vicissitude in northwest area and adjacent regions;
3. The textual research on the loess plateau natural vicissitude, vegetation change, and human-economy-social activity vicissitude (Starting from the formation, focusing on since 5000 years);
4. Water body dynamic process and the reason analysis of shrink and dried-up in Qinghai lake and Lop Nor.

### Ecology–Economy–Society Comprehensive Regionalization and Staging Research on Reconstruction of Landscape in Northwest Area

The northwest area is vast. The nature and socioeconomic condition are distinct in different provinces and regions, even the regional disparity is prodigious in each province or region. How to act appropriately to the situation? Conducting management and exploiting by regionalization and stages, taking the experience gained at one unit and popularizing it in a whole area, from part to whole, from the recent to forward, to realize the overall objective of "reconstruction" step by step. This is the crucial decision-making thought and will be listed as important projects in this science and technology action.

The reconstruction of landscape ecology–economy first-level regionalization preliminary was finished in the earlier study by reconstruction of landscape regionalization in northwest area. The first, second, and third levels of ecology, economy, and society regionalization will be conducted in further research project. Thus, some subjects of this research are listed as follows:

1. Subarea conditions in past 50 years, including the collection of agricultural zoning information;
2. The principle of zoning and staging; the hierarchical formulation of zoning and staging;
3. Selecting content and appropriate plotting scale of regionalization map;
4. Making regionalization map;
5. Writing specification of regionalization map.

### Soil Conservation, Water Storage, Farmland Building, and Slope Revegetation Science and Technology Action in Loess Plateau

The loess plateau covers parts of Shaanxi, Gansu, Ningxia, and Qinghai. Complex geomorphic type, fragmented surface, sparse vegetation, severe water loss and soil erosion, and frequent drought are the main ecology problems in this area.

After republic foundation, central and local governments call for people to prepare soil, repair field, afforest, green slope, improve water and soil conservation, and make a great progress. Especially since 1980s, the small watershed comprehensive management has achieved a series of achievements and experiences. But the water loss and soil erosion is not controlled effectively from a macro point of view, so it is in "reconstruction of landscape" research list as follows:

1. Agriculture water conservation and conservation tillage cultivation technique in loess plateau;
2. The optimum construction of big agriculture (crop, fruit tree and grass) considering local condition and the establishment of corresponding ecological agriculture demonstration in loess area;
3. The flat parts of dryland–slope land–cheuch–watercourse water loss and soil erosion system analysis and intensified water conservation macroeffect research;
4. All kinds of land (the flat parts of dryland, cheuch, slope) efficient utilization and local fruit trees (persimmon, jujube, apricot, pear, plum, and so on) cultivation techniques;
5. Rainwater harvesting, water storage, water conservation, water-saving matching comprehensive technical standard, rained farming research;
6. Efficient terraced field and dam land water conservation system and old terraced field remolding;
7. Optimum land condition for conversion of cropland to forest and grassland, forest establishment key technology;
8. Water conservation type roads construction in loess area;
9. Water conservation type livestock selecting and the feeding model;
10. Developing chemical water conservation method and technique like surface protection coagulating agent specific to arsenic sandstone;
11. The feasibility and relevant technique of planting sea buckthorn to stabilize sands in arsenic sandstone area;
12. Water–soil–light–heat–air–plant ecology system monitoring and regulation and control mechanism in loess plateau.

### Desert Control and Soil Improvement in Tarim Basin, Qaidam Basin, and Hexi Corridor, Oasis Construction Agricultural Science Action

The ecological characteristics of above-mentioned basins are as follows: low precipitation, poor surface water, most rivers are inland rivers, in which the flow is controlled by glacier and snow; strong evaporation, most of the region is arid and semiarid region; vast desert and gobi distributed, threatening the surrounding land by desertification. And there are some severe ecological environment problems, such as rivers shrinking, lake water salinization, dry up, natural forest degeneration, land desertification, grassland degradation, soil salinization, and frequent sand–dust weather.

On the other hand, some advantages exist in this region also: abundant groundwater, some oasis with good condition, rich solar thermal resources and land resources, plentiful petroleum and natural gas, and so on. So it has a bright future of large-scale development.

For the reconstruction of landscape in this region, some research subjects should be carried out:

1. The desertification prevention and control in the downstream of Tarim River, the research and demonstration of territorial comprehensive management technology;
2. The desert area (especially in Hexi desert area) comprehensive management in Gansu province;
3. The advantages and disadvantages analysis on present development mode of inland rivers and the best development plan study;
4. The search, development, and utilization of groundwater in piedmont alluvial, pluvial regions, and desert area;
5. Oasis farming shelter forest system model;
6. The research on developing deserticulture.

### Preventing Sand, Garden Making, Forest Planting, and Seeding in Yulin-Yanchi Wind-Blown Sand Area, the Construction of Ecological Agriculture Science and Technology Action

The Yulin-Yanchi wind-blown sand area in Shaanxi locates along the Great Wall, about 50,000 km$^2$. The advantages in this region are vast land, plentiful light resource, sufficient groundwater, and abundant oil, gas, and coal. The main ecological environment problems are sandstorm striking, desertification threatening, and intensified water loss and soil erosion caused by large-scale energy development in ecologically vulnerable area.

The relevant subjects are as follows:

1. The groundwater formation condition, water and soil resource evaluation and rational development and utilization in wind-blown sand area;
2. Prevention and control of desertification comprehensive technical measures in wind-blown sand area;
3. The impact assessment of energy development on the ecological environment and relevant countermeasures;
4. The ecological environment problems caused by exploitation of energy resources, construction of cities and towns, and development of township enterprises in ecologically vulnerable area;

5. The benefits survey of the present protection forest in wind-blown sand area, the continued construction scale and density, the selection and collocation of optimum tree species;
6. The best ecological agriculture model (entity) of compound management with farming, forestry, animal husbandry, fruit trees, vegetables, and others in wind-blown sand area.

## Mountainous Area Development and Green Protection Science and Technology Action in Qingling Mountains, Qilian Mountains, Kunlun Mountains, and Tianshan Mountains

As mentioned above, the nationwide, even universal famous mountains, such as Qingling Mountains, Ba Mountains, Qilian Mountains, Altun Mountains, Kunlun Mountains, Liupan Mountains, Tianshan Mountains, Altai Mountains, and so on, locate in the northwest area. The human activity is intensive in Qingling Mountains, Qilian Mountains, Kunlun Mountains, and Kunlun Mountains. Especially in Qingling Mountains, the ecological environment caused by human activities is the most serious. The Qilian Mountains come second. The main problems are forest and vegetation destroyed, forest shrink, snow line up, water loss and soil erosion aggravation, and environment problems caused by disorderly mining. The deterioration of mountainous ecology environment directly threats the plain region, which take mountains as protection screen. Thus, "reconstruction of landscape" should give enough attention on mountainous area.

The relevant subjects are as follows:

1. Water loss and soil erosion mechanism, topsoil protection, and mountainous land and water resources rational utilization in rocky mountain;
2. The protection of forest and biodiversity in mountainous area;
3. Mountainous area clean, rational mining, the protecting of vegetation, soil and water quality;
4. The glacier dynamics and its environmental implication on rivers, forest, and vegetation;
5. Mountainous area economic development mode and its support system;
6. Mountainous area communication and transportation and corresponding machines;
7. The prevention and control of landslide and debris flow geological disaster;
8. Tourism resources development, recreation site construction, and ecological environment protection in mountainous area;
9. Man–land relationship, population policy, poverty alleviation, and poverty elimination in mountainous area.

### Large-Scale Common Water Matter Activity and Ecological Irrigation District Construction Science and Technology Action in Arid Area of Northwest

Water is the most important factor of northwest development. Besides carrying out researches in different regions as mentioned above, there are some common, large-scale, lack researches water problems which are urgent to make breakthrough in implementing "reconstruction of landscape" project.

Therefore, some subjects are selected as follows:

1. The ecological environment, water source dynamic, and the prevention deterioration countermeasures in source region of the Yangtze and the Yellow River;
2. The optimum plan of interbasin water diversion to northwest area for "reconstruction of landscape" project implementation and relevant ecological environment problems;
3. Water resources comprehensive assessment and optimum exploitation and utilization mode in mountainous area in Xinjiang and Hexi area under the glacier snow-river flow-groundwater-overflow region hydrologic cycle condition;
4. The features of inland river, the interaction between inland river and human activity and its rational exploitation;
5. Water resource efficient utilization in high-lift pumping project from the Yellow River in Shaanxi, Gansu, and Ningxia;
6. Tri-water (surface water, groundwater, and atmospheric water) transformation mechanism and regional water resources optimization;
7. Water quality protection, wastewater reclamation, and wastewater irrigation development;
8. Brackish water utilization and its development and utilization;
9. Wells and canals combination in irrigation district to regulate groundwater table and prevent and control soil salinization;
10. Irrigation district reconstruction and water-saving technique method system;
11. Ecological irrigation district construction in different regions.

### Grassland Management, Grass Industry Construction, and Animal Husbandry Development of Large-Scale Common Science and Technology

The vast grassland in northwest area is about 65.44 million $hm^2$, is a vast ecology area with its own characteristic. Carrying out "reconstruction of landscape" project here not only has significant meaning for developing animal husbandry and prataculture in northwest area, but also plays decisive role to protect ecology and realize beauty landscape.

But the grassland in northwest suffered desertification, rodent damage, over grazing, and disorder reclamation, which is the big obstacle for realizing beauty landscape. Some subjects were selected as follows:

1. The grassland presents situation investigation and analysis and ecological assessment;
2. The prevention and control of grassland degradation to improve the yield of grass;
3. The grassland protection, the determination and adjustment method of reasonable grazing capacity;
4. The rodent damage in grassland and the prevention and control of other main disasters;
5. The prevention and control of grassland desertification;
6. The development of grass industry;
7. Grassland irrigation;
8. Ecological grassland construction.

#### Trans-Provincial Shelterbelt Networks and Large-Scale Forest Network Construction Science and Technology Action

Afforestation occupies very important position in "reconstruction of landscape" project. It is one of the significant contents in almost every action mentioned above. But the large-scale forest belt and forest network may cover different province and region, occupy large area or influence other regions, involving protection, replanting, and comprehensive assessment, so it will be listed as one science and technology action in this project.

1. The construction region selection, planning, and design of future large-scale forest belt and forest network;
2. Adaptive tree species arrangement and survival rate improvement;
3. Artificial forest (including woodland) protection and ecological environment benefit observation;
4. Fast-growing tree species and the construction of fast-growing forest;
5. The long-term observation of forest environmental effect;
6. The ecological environment negative effects and countermeasure of forest tourism resource development.

#### Industrial and Agricultural Pollution and Urban and Rural Ecological Environment Co-construction Science and Technology Action

Industrial, mining, urban construction, and the township enterprises construction not only pollute local town, but also trigger ecological damage and environmental pollution in rural area and agricultural production. Moreover, the agriculture developing, agricultural activities intensifying, and plenty of pesticides and fertilizers applying will pollute cities directly or indirectly, which hinder the implement of "reconstruction of landscape" project. Thus, some subjects were selected in terms of co-construction of good ecological environment in urban and rural areas.

1. The pollution of city smoke–dust drift, sewage discharge and garbage dumping on suburban environment, and agriculture products (fruits and vegetables) and its prevention and control countermeasures;
2. The negative effect of farming activities (applying fertilizer and pesticide) on ecological environment and its countermeasures;
3. Ecological environment protection of sightseeing district in rural area;
4. The sanitation management and water and soil resources health protection of township enterprise.

### The Famous Special Rare Agriculture, Animal Husbandry, Fruits, Vegetables, Herbs, and Other Industry Products and Their Industrialization Construction in Northwest Area

Sufficient sunlight, large temperature difference between day and night, various soil, complex landform, and diverse environment in northwest area produce a great variety of famous special rare agriculture, animal husbandry, fruits, vegetables, herbs, and other industry products. Along with the economic development and social progress, it is urgent to realize product industrialization which is the constituent part of "reconstruction of landscape" project. Some science and technology problems were listed in this project. Some subjects are as follows:

1. The production place, yield, and quality evaluation of famous special rare agriculture, animal husbandry, fruits, vegetables products;
2. The industrialization of quality local agricultural products;
3. The storage, transportation, and processing of quality local fruits and vegetables products and their industrialization;
4. The industrialization of quality local animal husbandry products;
5. The rational development of quality local wild resource products and their industrialization;
6. The production modernization and merchandising of Muslim and ethnic food.

### Sinicism Rural Urbanization Process and the Optimized Model in Northwest Area

The entire world is facing urbanization process, China is no exception. But China has its own national conditions: much peasant population, less per capita cultivated land, peasant household farming style in quite a long period. Although residence dispersion, most of them are in cultivation area, the traffic is convenient, excessive concentration living results in cultivation of traffic problems. In addition, restricted by water supply, pollution discharge, and refuse disposal condition, it is demanded to develop rural urbanization in China by her own way in “reconstruction of landscape” project. Thus, some subjects are listed as follows:

1. The strategic research of urbanization overall development in northwest ecologically vulnerable area in “reconstruction of landscape” project;
2. The suitable density and scale of urbanization and regional general layout in different ecology–economy–society region in northwest;
3. The ecological small town construction model in vulnerable ecological area;
4. Save the cultivated land in various aspects of northwest urbanization process: residential building, road network, green belt, public utility, and so on.

### The Investigation, Plan, Exploitation, and Ecological Environment Protection of Tourism Resource in Northwest Area

Complex and diverse natural condition, extraordinary landscape, and colorful scenery in northwest area create many tourism resources. It should be investigated, planned, exploited, and utilized cooperating with “Development of the west regions” and “reconstruction of landscape” to develop local economy, overcome poverty and achieve prosperity, promote acceleration of development and reconstruction implementation.

But, most of these tourism resources are in vulnerable ecological environment area or its own is the ecologically vulnerable body. The unreasonable development will destroy ecological environment definitely, so it is necessary to carry out researches to prevent and control ecological environment deterioration. Some subjects were selected as follows:

1. The investigation and rational exploitation of tourism resource in different ecological landscape area, including loess land, Gobi desert, grassland pasture, mountain forest, inland water bodies, and so on.
2. A series of tourist attractions development along the Silk Road;
3. The ancient and modern tourism network composition and development in the central Shaanxi plain;
4. National customs tourism resources development;
5. Multinational tourist route choice and development;
6. The ecological environmental protection in vulnerable ecological regions tourism development.

The Sunshine Project Science and Technology Action in Reconstruction of Landscape

The sunlight is sufficient and wind is strong in northwest of China. Make full use of these plenty clean natural energy, develop, and utilize them adjusting measures to local conditions to serve production and live, alleviate villagers' pressure on energy shortage. Thereby, it is necessary to bring the enthusiasm into protecting forest and vegetation and landscaping in the meanwhile of improving their living conditions. Thus, some subjects were selected as follows:

1. The sunshine development instruments and installation in countryside of China (such as sunshine pump, solar water heater, solar cooker, solar hauling machine, and so on);
2. Pneumatic tools in sandy region of China (such as wind-driven water pump, wind mill, wind-driven generator, and so on);
3. The energy storage way and facilities of solar energy and wind energy;
4. Geothermal exploitation and utilization.

The Population, Education Policy in Northwest Area, Raising Farmers' Quality of Cultural, Science, and Technology

1. Implementation of "development of the west regions" and "reconstruction of landscape", northwest talent strategy research;
2. Talent development policy in northwest, cadre of science and technology appropriately extended length of service;
3. Expanding the scope of the compulsory education, increase national education career support and efforts appropriately;
4. The population carrying capacity of different ecological regions in northwest;
5. Carrying out population emigration policy, mode and appropriate migration region in extremely ecologically vulnerable area;
6. The possibility, migration patterns, migration region, and relevant policy on migrating to northwest from other regions;
7. The action mode of reclamation forces and engineering corps in development and "reconstruction", the co-construction of soldiers and civilians.

### 1.4.3 The Experimental Demonstration Area Construction of Reconstruction of Landscape Science and Technology Action in Northwest of China

The "reconstruction of landscape" plan is a great cause for desertification control, afforestation, water conservancy, nature transformation, overcome poverty and become prosperous in rebuilding the landscape of mountains and rivers. Its realization not only needs a long period, but also multidisciplinary joint research. According to the regional governance experience in China, select appropriate region, concentrate intellectual, financial, and material resources input, carry out science experiment and the demonstration model to popularize in large-scale area at the same time of obtaining theoretical results. The demonstration model-experiment demonstration base will play a role on personnel training and scientific propaganda.

Thus, the experimental demonstration area construction is the important part of the "reconstruction of landscape" science and technology action (Li et al. 2006).

## *1.5 Test Area Construction and Mode*

In the reconstruction of landscape scientific research in northwest area, we carried out strategic research to design 100 experimental demonstration regions which should be built in different ecological region in the future. Eight test areas of them were selected to carry out constructive research. It achieved success and formed distinctive promotion modes as follows.

### 1.5.1 Soil and Water Conservation Type Ecological Agriculture Reconstruction of Landscape Mode in Hilly-Gully Region of Loess Plateau (The Ansai County Test Area in Shaanxi Province)

1. Evaluate the land resources quality in Ansai County, divide Ansai County into three "reconstruction of landscape" construction subregions (south, middle, and north) and put forward the direction and content of soil and water conservation type ecological agriculture construction in different areas.
2. Study the key technology of soil and water conservation type ecological agriculture construction in different areas, put forward the selection and arrangement principle of afforestation tree species in suitable site, the building model and key technology of arbor–shrub–herbaceous plant in different areas.
3. Formulate whole county orchards macroscopic development planning and good quantity-high yield-increase income technical system.
4. Apply analytic hierarchy process to select drainage basin health evaluation indicators to quantize "reconstruction of landscape" scientifically.
5. Bring out the key technology for high yield by means of improving soil nutrients efficient conversion and water utilization in basic farmland. Establish "reconstruction of landscape" ecological agriculture mode and popularize it in Ansai County and Yan River Basin to acquire huge comprehensive economic benefits.

### 1.5.2 Manage Gully and Slope Reconstruction of Landscape Mode in Hilly Area of Loess Plateau (Mizhi Test Area in Shaanxi Province)

1. Put forward efficient intercropping with plastic and straw mulch and improve rainwater use efficiency by more than 20 %. The plastic mulch grain yield increased by 20–40 % of control grain, the dry land grain yield per mu is up to 850–900 kg.
2. Put forward sloping fields classification management technical specification and forest-grass model to form plastic film mulch high-yield farmland in slop land with 10°–15° gradient. In the meanwhile, put forward water detention and storage simultaneously, drought resistance and cold resistance, matching species with the site planting technology to increase farmers' economic benefits and acquire ecological benefits at the same time.
3. Introduce courtyard economy and carry out "chicken-pig-marsh gas-vegetable" cycle mode to increase farmers' income greatly.
4. Study and develop suitable local rain harvest and utilization technology, popularize water collecting, and supplying bag processed by plastic film and film mulching for water holding drip irrigation system.
5. Adjust the industrial structure and introduce selected 32 improved varieties, such as Xiangkui 9012, Qingu No. 4, Zanhuang jujube, and so on.

### 1.5.3 Terraces Efficient Development Reconstruction of Landscape Mode in Loess Hilly-Gully Region (Zhuanglang County Test Area in Gansu Province)

1. Apply 3s technology to divide experimental demonstration area. Carry out comprehensive treatment of loess hill, terrace, gully, and dam. Construction optimized ecological environment as "planting sea-buckthorn and Chinese pine on the top of hill, fruit tree in terrace, forage grass and caragana microphylla along ridge, forest and grass on the bottom of gully, construct reservoir and dam on the bottom of gully".
2. Give full play to the advantages of terrace, propel second development and carry out high standard broaden terrace construction for small pieces of terrace; popularize compound planting with economic forest, fruit tree, and grain crop; build gully dam system, sediment retaining and depositing, impounding and construct "Tri-high" agriculture with high yield, high quality, and high efficiency. Implement "conversion of cropland to forest and grassland" scientifically and improve land efficiency.
3. Study the utilization model of terrace, such as rain harvesting for irrigation mode, orchard vegetables mode, quality local product mode, and so on. Carry out researches on terrace high-efficient development technical system with

"terrace + mulch (gravel covered) + special crop + interplant + supplemental irrigation".
4. Bring in high quality, high yield, drought resisting, and excellent comprehensive characteristic crop varieties, apply water-collecting efficient agricultural technology, including winter wheat plant with water collecting on membrane sides, maize plant with double ridge and furrow and "one film two function" water-collecting planting, maize plant with mulching in autumn, spring-sown crops with plastic film mulching and injecting water seeding, potato pit planting to gather fertilizer for high yield, to improve farmland economic benefits and increase farmers' income steadily.
5. Retain rainwater with various approaches, combine rainwater harvesting agriculture with traditional dry farming techniques to form dry land rainwater harvesting high-efficiency agriculture comprehensive technical system with "terrace + water cellar + variety + mulching film".
6. Ecological energy construction technology in loess hilly and gully region: build up biogas digester, hog house, toilet, solar greenhouse, rainwater harvesting for irrigation "five in one" positive cycle production model taking solar energy as power. Moreover, popularize using solar oven.
7. Carry out the researches and development on "Dam system agriculture": build 33 key dams and warping dams in the whole county.

### 1.5.4 Water Diversion, Desertification Control, Soil Improvement, Developing Ecological Industry like Deserticulture and Grass Industry Reconstruction of Landscape Mode (Ningxia Test Area)

1. On the basis of alfalfa improved variety selection, assemble, match, and integrate former technical achievements and present technology, popularize in 80,000 $hm^2$ area in south of Ningxia to form competitive industries in Ningxia.
2. Give full play to microphylla forest on soil improvement, wind prevention, sand fixation, and good forage value; Study the plant technology of caragana microphylla, for the first time put forward stumping technique of caragana microphylla shrubwood to promote the growth and recovery of caragana microphylla.
3. The grain-economic crop–forage–livestock ecological agriculture mode popularized in south of Ningxia, is agriculture and animal husbandry combined type oriented economy, raising livestock by grass, promoting agriculture by livestock, further develop farm and pasture product intensive processing to form the positive cycle of ecology and economy.
4. Build up Chinese herbal medicine artificial planting base for ephedra and medlar by means of engineering measures like leveling sand dune for cultivation, developing irrigation and drainage, and constructing protection forest network; develop wine grape industry, combine ecological management and special deserticulture development to convert the desert into fertile land; put forward

and implement "science and technology orientation, enterprise operation, public participation, government support" operation mechanism to make manufacturer and the masses benefit at the same time, create a piece of strategic value experience for reconstruction of landscape.

### 1.5.5 Intensive Grazing Supplemented by Seeded Pasture, Protect the Source Regions of Yangtze River and Yellow River, Develop Economy in Pasturing Area Reconstruction of Landscape Mode (The Source Regions of Yangtze River and Yellow River Test Area in Qinghai)

1. Study the black soil type (in the source regions of Yangtze River and Yellow River) and grassland type (in Qinghai Lake) deteriorated grassland causes, the ecological recovery method, and corresponding matching technology: enclosure, semi-seeded pasture and seeded pasture cultivation, rodent prevention and control, and so on.
2. Construct forage seed base, study the breeding technology of forage seed, and guarantee the recovery of grassland vegetation.
3. Study dry farming forestation technology and ecological recovery and management matching technology in dry wasteland slope which converts cultivated land into forests in the upstream of the Yellow River.

The implement of project protects the ecological environment in the source regions of Yangtze River and Yellow River and promotes the ecological restoration of deteriorated grassland.

### 1.5.6 Oasis Ecological Construction and Agricultural Sustainable Development in Arid Climate Reconstruction of Landscape Mode (Xinhe County Test Area in Xinjiang)

1. Adjust industrial structure, set up technology supporting system of agricultural area and pasturing area, remold traditional grain-economic crop dual structure oasis agriculture into grain-economic crop–forage ternary structure.
2. Carry out saline–alkali low-yield land improvement experiment, besides traditional physics, chemistry, biology improvement methods, based on the local actual conditions, research and develop "break multilayer hardened clay impermeable soil layer drainage salt-leaching method" for heavy saline–alkali land to resolve the salinization problem in oasis.
3. Implement afforestation matching technology in arid region comprehensively, cultivating salt tolerance drought-resistant plant and irrigating by floodwater to recover oasis desert transition zone technology, farmland protection forest planting technology, oasis garden economy transformation technology, building

grape gallery as by-forest belt technology, and so on. Convert oasis protection forest system from ecological type into ecological economy type to realize agroforestry management in oasis.

### 1.5.7 Develop Ecological Economy, Guard Border, and Rich Soldiers in Reclamation Area of Desert and Semidesert Reconstruction of Landscape Mode (181 Regiment Test Area in Xinjiang Production and Construction Corps)

1. Change the single planting structure in desert and semidesert region to set up ecological agriculture-forestry—stockbreeding compound mode.
2. Carry out large-scale medium and low-yield field transformation, conversion of cropland to forest and grassland and cultivated land shelterbelt networks projects to increase regiment gross national product dramatically and promote continuous and stable development of local agriculture and workers' income.
3. Finish the construction of high standard seeded pasture and forage seed base, bring in some agriculture and livestock improved varieties, such as wheat, oil sunflower, Hami melon, alfalfa, beef, meat, fine wool sheep, and so on.

### 1.5.8 Oasis Agriculture Resources Efficient Utilization Mode in Xinjiang (Awat County and Yuli County Test Area in Xinjiang)

Explore High-Yield Methods of Xinjiang Cotton

On the basis of advanced research method and production technology at home and abroad, set up the experiments and demonstrations on efficient utilization of water and fertilizer resource, including cotton drip irrigation under plastic film, cotton hose irrigation under plastic film, research on relationship among water, fertilizer and salt for high-yield cotton, safe utilization of poor quality water, and so on.

In Order to Improve Oasis Agriculture Resource Efficiency, Some Key Technologies Were Applied as Follows

1. Peasant household-type cotton self-pressure hose irrigation under plastic film high-yield cultivation technology;
2. Nitrogen optimal management technology in high-yield cotton field;
3. The nitrogen testing technology on cotton field soil and plant;
4. Automatic monitoring technology on cotton bollworm dynamic condition;

5. The trap, keep, and field release of coccinella undecimpunctata linnaeus technology;
6. Saltwater utilization for develop desert industry technology.

The actual modes are not limited to the abovementioned because of the enthusiasm of local governments and people. For example, combined with conversion of cropland to forest and grassland, develop "enclosure hills and convert cropland in loess hilly region to construct ecological county reconstruction of landscape mode" (Wuqi mode in Shaanxi); "conversion of cropland to forest to manage wind-blown sand region reconstruction of landscape mode" (Yulin mode in Shaanxi), and so on (Jiang 1998).

The appearances of these mode areas, from the combination of theory with practice, propel the reconstruction of landscape cause enormously in northwest area.

## 1.6 Conclusion and Prospect

1. Reconstruction of landscape in northwest area is a great undertaking which can transform nature, promote the harmonious coexistence between man and nature, strengthen national capabilities, rich people, rich soldiers, stabilize frontier defense, and benefit us and our future generations. It should be carried out continuously generation by generation.
2. Reconstruction of landscape in northwest area involves vast scope. It needs to choose more regions according to local condition and explore new mode.
3. Conversion of cropland to forest and grassland is the one of the success measures of reconstruction of landscape in northwest area, should conduct with hill enclosure and grazing prohibition, intensive grazing and construction basic farmland measures together, prevent retrogress and relapse.
4. Deal with the relationship between management and development seriously, the relationship between ecology construction and reconstruction of landscape scientifically. Especially in Yulin-Yanchi wind-blown sand area of Shaanxi, reconstruction of landscape is a great task. Moreover, it also has the actual demand of energy development. The unreasonable development will lead to some severe environment problems inevitably, such as land subsidence, ground fracturing, river and lake shrinking, groundwater level recession, water and soil pollution, desertification aggravation, and so on. It may threat the reconstruction of landscape sustainable development significantly. So it should be deeply concerned and fulfill "combinating exploitation with reconstruction, promoting reconstruction by exploitation and guaranteeing exploitation by reconstruction".
5. It is necessary to resolve water shortage and the imbalance of land and water resources problems in northwest for long term. So we must work hard on water saving, water conserving, water shortage, water supply, water diversion, rational utilization of water and land resources and popularizing positive cycle production mode, and so on.

6. To strengthen scientific researches continuously. Reconstruction of landscape in northwest area is a great scientific undertaking referring heaven, earth, and man. It should insist and deepen scientific researches to promote multiaspect disciplinary development and intersection in regard of water, soil, light, energy, climate, earth, and creature, guided by scientific development concept.
7. To strengthen the cultivation of talents. To cultivate some management and technical elite cadres designedly, those who have strong enterprise, love northwest, and will devote themselves to reconstruction of landscape.
8. To combine reconstruction of landscape with ecological civilization construction. Reconstruction of landscape in northwest area fit the actual environment in northwest region of China and people's hope, enjoy popular support. Beauty landscape is the destination of ecological civilization construction; the ecological civilization construction is element task of beauty landscape implementation, to combine them scientifically to gain success continuously on the way forward.

## 2 Arid Zone Ecological Research

Arid areas in China include Gansu, Ningxia, Qinghai, Xinjiang, and the western area of Inner Mongolia. These areas account for approximately 25 % of the total land area and are important in Chinese economic development. Based on the characteristics of arid areas in China, arid zone ecological research combines theory with practice and has the following main aspects: understanding of the structure and function of desert ecosystems, ecophysiological research on salinity and drought resistance in desert plants, desertification control, and sustainable oasis development.

### *2.1 Understanding of the Structure and Function of Desert Ecosystems*

Understanding of the structure and function of desert ecosystems is the premise and foundation of research on ecological problems in deserts and arid areas. Geographical distribution, natural characteristic, formation and evolution of deserts, formation and evolution and the development and utilization of oasis in China were systematically analyzed in the book *Desert and Oasis in China*. The desert and oasis landscape of western China was classified with the assistance of an EU program. Therefore, all the above results extended the range of ecology research to the northeast area.

### 2.1.1 Species Diversity Investigation in Desert Ecological Systems

Species diversity investigation has important scientific significance and economic value. Lanzhou University led to an integrated scientific investigation on Anxi, an extremely arid desert national nature reserve in Gansu characterized by many types of vegetation, such as gravel desert, low-lying dampland saline meadow, swamp, sandy desertification, and forest vegetation. Three hundred and sixty-two plant species belong to 60 families and 192 geniuses; 4 of them, including *Haloxylon ammodendron,* have been placed under state protection. The reserve is home to 175 vertebrate species which belong to 26 classes and 55 families. Ten of these species are exclusive to China. Twenty-seven species are wildlife under state protection; 7 are Class I and 20 are Class II protected species in China.

### 2.1.2 Soil Seed Bank Dynamics in a Desert Ecosystem

Changes in soil seed bank dynamics precision directly determine the preservation and restoration of a desert ecosystem. Studies on soil seed bank dynamics found that 46 plant species, which belong to 17 families, appeared in soil seed banks in artificial sand-fixation vegetation in Shapotou of Ningxia Province. About 33.3 % of the total seed bank appeared in both the current species vegetation and the soil seed bank. Predicted results indicate that the succession tendency of artificial sand-fixation vegetation is dominated by shrubs, such as *Artemisia ordosica*, and grasses, such as *Bassia dasyphylla* and *Eragrostis poaeoides*. Studies analyzed the effect of different microhabitats on the soil seed bank of desert plants and the bank distribution pattern of *Stipagrostis pennata* in Gurbantunggut Desert, which provides a basis theory for vegetation restoration in an artificial sand-fixation area.

### 2.1.3 Therophyte Life History Strategy in a Desert Ecosystem

Research about germination strategy should be considered when studying therophyte life history strategy. In a desert, therophyte uses different strategies to adapt to hostile environments with unpredictable precipitation during growing season. An indoor germination experiment on *Agriophyllum squarrosum* showed that therophyte seeds have behavioral equivalence between continuous germination under suitable conditions and partial dormancy during growing season. A study on the population dynamics of therophyte grass (*E. poaeoides)* shows that the germination, growth, and dynamic state of quantity of grass depend on precipitation, which causes herbaceous plants to germinate in batches. Individual competition for water in therophyte grass results in inevitable self-thinning. Its survival curve is C type, and it has an r-type life history strategy.

#### 2.1.4 Relationship Between Species in Desert Ecosystems

Discussing relationships between interacting species and the distribution pattern of desert vegetation facilitates understanding of the structure mechanism of a desert ecosystem. Relationships between interacting species were discussed based on data about neighboring plant size and distance between plant species, such as *Salsola passerine*, *Sympegma regelii*, *Reaumuria soongorica,* and *Nitraria sphaerocarpa,* in Anxi, an extremely arid desert national nature reserve in Gansu. The results indicated that the plant size and distance of the four species have a positive interaction, that is to say, they have a competitive relationship. Intraspecific competition is more important than interspecific competition. Intraspecific competition between different combinations of the four species, except *R. soongorica* and *N. sphaerocarpa*, is more intense than interspecific competition. Therefore, the above results explained the coexisting and competitive mechanisms of dominant species in a desert ecosystem.

### 2.2 *Ecophysiological Research on Salinity Tolerance and Drought Resistance*

#### 2.2.1 Plant Ecophysiological Response to Salinity and Drought Stress

Salinity and drought stress can inhibit the height, growth of stem and root, leaf number and area, biomass, and canopy structure of trees. In *R. soongorica*, an extremely xerophytic semishrub of Tamaricaceae, the fully mature leaves die but the stem remains alive; the plant enters a dormant state under extreme drought stress. Upon rewatering, the stems are rehydrated and produce new leaves. *Caragana korshinskii* has the same characteristic as *R. soongorica.* Research indicated that salinity stress decreases sunflower leaf area, root relative activity, and relative amount of growth. With increasing salt concentration, the effect of salt on the plant intensifies.

Studies on the photo-physiological characteristics of desert plants show that photosynthesis rate decreases and the water use efficiency of the plant increases as drought stress exacerbates. As the habitat dries up, the Caragana plant saves water by decreasing transpiration rate and increasing photosynthetic rate. Under drought stress, *Ammopiptanthus mongolicus* has two protection strategies, namely, xanthophyll cycle-dependent photoprotection, which is the main mechanism for a plant to deal with excessive light energy, and reversible inactivation in optical system (PS) reaction centers, which can protect the photosynthetic system. Plants were classified into different water eco-adaption types. For example, *Panicum virgatum* has high photosynthetic rate, low transpiration rate, and high water use efficiency. *Medicago sativa* has high photosynthetic and transpiration rates but low water use efficiency. Its drought adaptation strategy is to delay dehydration by developing high water potential. Experiment shows that salt stress has a more serious effect on the photosynthetic activity of C3 non-halophytes than that of C4 non-halophytes, and salt stress level has a positive relationship with salt concentration.

*Tamarix chinensis* has a high photosynthetic rate and exhibits a single midday peak under ample soil–water content; it has a low photosynthetic rate and exhibits a bimodal curve under soil–water stress. Based on the diurnal variation of transpiration rate, xerophyte has three water ecological types, namely, nontranspiration midday depression, slight transpiration midday depression, and strong transpiration midday depression. When relative humidity is lower than 30 %, *Glycyrrhiza inflata* leaves induce stomatal oscillations with which the plant also decreases transpiration rate and increases water use efficiency; the photosynthetic rate is unaffected, which is an important adaptive strategy for desert plants under stress.

Salt stress decreases photorespiration. The degree of decreased photorespiration strengthens with increasing salt concentration. Under high salt concentration, plant leaves have high dark respiration and photorespiration at the initial stage; under low salt concentration, plant leaves have high photorespiration at the initial stage. However, dark respiration and photorespiration decrease later. Some studies have reported that NaCl treatment increases photorespiration.

### 2.2.2 Plant Ecophysiological Mechanism on Salinity Tolerance and Drought Resistance

To adapt to an arid environment, desert plants have used many strategies. Protective plants have a special morphological structure that decreases excessive water transpiration and shields leaves from strong direct sunlight. Water-conserving desert plants stay hydrated by sporting tiny leaves to minimize water transpiration and the surface area exposed to sunlight and by saving limited water before using it heavily. Tolerant plants improve water holding capacity of specified contains and storage water by mesophyll cells. Hardy plants have developed assimilation organizations (palisade tissue with lots of layers) with which the plant can decrease transpiration rate and increase photosynthetic rate. The leaves of avoidant plants fall off and the plants return to a dormant state under extreme drought stress. The stems are rehydrated and produce new leaves upon rewatering.

To alleviate salt stress, a plant decreases root $Na^+$ absorption to prevent ion transport from the roots to the higher ground parts. The following processes are involved in plant salt exclusion: root cell does not absorb $Na^+$; the plant keeps the absorbed $Na^+$ in the central vacuole of the parenchyma cell of roots, the basal part of the stem, and leaf sheath; during upward transportation, $Na^+$ is absorbed by xylem or phloem transfer cell and secreted to phloem before it is transported to roots that expel it to the environment; selective absorption of inorganic ion, especially the selective absorption of $Na^+/K^+$, uses special salt glands (salt bladder) for secreting salt that formed in the leaves and the stem, which can secrete salt absorbed by the plant.

Under drought or salt stress, a plant cell can automatically accumulate some soluble substances to reduce intracellular osmotic potential, which can ensure normal water supply when the plant is subjected to various stress factors. Plants have two osmotic adjustment mechanisms, namely, inorganic and organic.

The main physiological function of SOD is to scavenge superoxide anion-free radicals. With certain limits, SOD activity increases as salt concentration increases. SOD activity is higher in salt-tolerant plant species, and salt-induced seedling injury is less harmful in salt-tolerant plants than in salt-sensitive species. Under drought or salt stress, plant POD activity increases and then decreases as stress time extends. POD, CAT, and content MDA activity in alfalfa leaves increases as salt concentration increases under salt–alkaline mixed stress. An active oxygen scavenging system has an important function in the physiology of salt-resistant plants. Under the same drought stress, a plant has high drought resistance with rapid and high proline accumulation.

## 2.3 *Advance in Desert Biological Soil Crust Research*

### 2.3.1 Lichen Biological Soil Crust Structure

X-ray micro-CT imaging coupled with 3D image analysis based on specific procedures for porous media have been applied to the analysis of the porosity structure of the lichen crusts. The results indicated that it will provide a very accurate quantification of the internal structure of lichen crusts due to stabilization of sand dune of different age. The general trend with the age of the lichen crusts is a reduction in volume of connected pores but an increase of the size of the remaining connected ones. The lichen crusts aggregates show a general increase of average pore network tortuosity reaching the maximum value with the age, and also exhibit a significant anisotropy of the tortuosity.

### 2.3.2 Soil Respiration Characteristic of Biological Soil Crust

A moving dune was fixed in the vegetation restoration area of southeastern Tenggeli Desert. Soil respiration increased because of increasing organic matter on the soil surface and the biological soil crust. The contribution of rhizosphere respiration results in spatial heterogeneity and significantly increased soil respiration rate. Meanwhile, soil respiration rate increases with the increase in soil biological crust development.

#### 2.3.3 Wind Effect on Photosynthetic Ecophysiological Characteristics of Moss Crust

Wind has three effects on photosynthetic and respiration activities. First, wind decreases moss crust net photosynthetic activity and dark respiration rate, which decrease further as the speed of the facing wind increases. Second, wind raises the maximum value of net photosynthetic activity and dark respiration rate of moss crust during drought stress. Third, it shortens moss crust photosynthetic and respiration activity time; as the wind speed increases, the time increases. The overlay effect forms three aspects that ultimately decrease moss crust in the carbon assimilation of net photosynthesis and carbon resource consumption by dark respiration.

#### 2.3.4 Carbon Cycle Characteristic in Artificial Sand-Fixation Vegetation

Carbon exchange has largely changed after the establishment of artificial vegetation and the formation of biological soil crust. As a carbon source, the artificial sand-fixation vegetation area releases carbon dioxide into the atmosphere during the nongrowing season; as a carbon sink, it absorbs carbon dioxide from the atmosphere during the nongrowing season. Generally, as a weak carbon sink, the artificial sand-fixation vegetation areas have low annual carbon assimilation. Precipitation is the main factor that influences the carbon cycle in arid regions.

### *2.4 Research on Sustainable Oasis Development*

#### 2.4.1 Major Problems in Sustainable Development of Oasis

Water network changes and shrinkage or drying up of the lake are among the major problems involved in sustainable development of an oasis. Three inland water systems, namely, Shiyang River, Heihe River, and Shule River, originate in Qilian Mountain. Regulating reservoirs have been built where river water flows from the three inland water systems. With reservoirs, river water is controlled by the upper and middle reaches of rivers. In the lower reach of the river basin, the downstream has been cut off, and a new artificial water system pattern was formed.

A natural oasis deteriorates and forms areas characterized by soil desertification, aridization, and salinization. In the history of Hexi corridor, a natural oasis was formed by grasses, such as *Phragmites australis* and *Carex liparocarpos*, and shrubs, such as rose willow, on the flats of lakes and rivers and spring water overflowing belts because of infiltration of underground and soil surface water.

Moreover, artificial oases have degenerated, which resulted in abandonment of many lands. Water loss has turned many original artificial oases in the Hexi corridor into deserts. Soil salinization is another problem.

Ecological degradation has caused social and economic problems. The Minqin Lake area lacks drinking water for people and livestock because of the shortage of surface fresh water and deterioration of groundwater quality. This situation resulted in social problems, such as ecological refugees.

### 2.4.2 Advancement in Oasis Sustainable Development Research

#### Recognizing Oasis and Oasisization

Scientists classified Chinese oases into zones based on the geographical location of the oasis distribution area, geographical environmental differentiation, and influence of climate, oasis evolution, and human activity. These zones are oasis zones in the eastern Hetao Plain, the western inland arid area, and the Qaidam Plateau. Based on the soil and water factors of oasis formation, the following four oasis types were classified: terraces along riverbanks, submarine delta, alluvial plain, and delta oasis.

Oasis changes are closely related to the production and activities of humans. The Minqin basin oasis is used as an example. With the use of historical documents, archeological resources, maps, and remote sensing, the author identified the origins of the oasis distribution area in a typical historical period and produced an oasis distribution map. The exploitation of the Minqin oasis started during the Han Dynasty, intensified during the Wei and Jin Dynasties, reduced during the Southern and Northern Dynasties until the Yuan Dynasty, and was revived in the Ming and Qing Dynasties. Today, the oasis is thriving. In general, the spatial development of the oasis is an east-to-west migration process.

#### Changes in Oasis Vegetation

The main factors that affect vegetation coverage are ecological engineering methods, climate, and water resource. In order to study variation of vegetation coverage of Tsagan Us oasis in Dulan county in Qinghai Province of China, the grade map of vegetation coverage in Tsagan Us oasis during different period was extracted based on normalized NDVI data, then the interannual variation of the vegetation coverage was quantitatively analyzed. The results show that vegetation coverage in Tsagan Us oasis both increased and fluctuated in the past 16 years. This finding indicates that the vegetation coverage develops steadily, although some parts still degenerate.

Soil Evolution After Oasis Cultivation

With the oasis in the middle reaches of Heihe River as an example, soil physical characteristics and chemical composition in cultivating salty meadows at different years were analyzed. The results show that soil texture improved after cultivation. The clay dispersion content of salty meadow soil increased from 9.18 % before cultivation to 12.93 % after 30 years of cultivation. In general, the soil physical characteristics and chemical composition coverage improved steadily.

Countermeasures of Agricultural Sustainable Development in an Oasis

The author analyzed ecological environment characteristics, including climate, vegetation distribution, soil texture, and hydrological process in northwest desert oasis; the author also analyzed the characteristics of an oasis agricultural ecosystem and proposed countermeasures of agricultural sustainable development in an oasis. Countermeasures include maximizing superior resources, developing a special industry, proper resource allocation, and building a sustainable development model.

Water depletion is the main factor that limits agricultural sustainable development of an oasis. This problem was addressed by implementing water transfer engineering from extra-region and water conservation from intra-region. Ecological experts determined how water can be conserved, and they were able to make significant progress on this urgent issue. For example, in the Minqin irrigation area, people use "the system of completely mulched alternating narrow and wide ridges with furrow planting in dry land" as an inexpensive water-saving measure. This technique decreases irrigation amount by 2190 $m^3/hm^2$ in a cornfield and increases corn grain yield by 902 $kg/hm^2$.

## 2.5 *Prospects*

Further research can be conducted on the following:

1. The point of view of plant community construction to understand the structure and function of a desert ecosystem and to determine the formation and maintaining mechanisms of a desert ecosystem.
2. The ecological strategy used by desert plants to adapt to arid environments. This strategy can be implemented to ensure the protection of important desert plant resources.
3. Evolution of desert environment, formation of sandy desert, laws and principles of wind-blown sand movement, and disaster mechanism of wind-blown sand process under the background of global climate change.
4. Biological process of sandy desertification.

5. Integration and application of development technology and model to manage land that underwent desertification.
6. A study of water loss, water consumption, and water use efficiency of the main crop and natural plants in an oasis from different angles and at different levels, such as individual, population, and community.
7. Interaction and feedback mechanism between oasisization and processes between water, soil, air, and organisms.

## 3 Countermeasures of Agricultural-Ecological Environment Governance and Sustainable Development of Agriculture in Commodity Grain Base in Northeast China

As the most important commodity grain base, the northeast China plays an important role in ensuring national food security. Now, there is a widespread concern over how to realize the sustainable development of grain production in this area.

Ecological environment is important to the sustainable development of grain production and social economy. Although it has developed for a short time in northeast China, agricultural-ecological problems are very prominent. For example, water and soil erosion in slope cropland, black soil degeneration, soil salinization, desertification and grasslands degradation in the west, forests destruction and wetlands shrinkage, lakes pollution and groundwater recession, etc. Ecological constructions, such as comprehensive controlling of water and soil erosion in slope cropland, ameliorating saline and alkaline soil by planting rice and constructing shelterbelt network in farming areas, have been strengthened and achieved obvious benefits.

The implementation of natural forest protection and wetland conservation project in northeast forest region has supported the function of soil and water conservation, carbon sequestration, and mitigating natural disasters, and they have become the irreplaceable ecological barrier.

The lack of construction of farmland irrigation and water conservancy is the bottleneck which restricts the high and stable yields. Apparently, grain production is affected by drought and flood disasters. But in fact, the cultivated land cannot be irritated when drying and be drained when waterlogging. Therefore, it should be priority to strengthen the construction of the standardized basic farmlands, especially to expand the effective irrigation areas and to develop water-saving irrigation systems to sustain the grain production.

Expanding cultivated area of high-yield crops (i.e., rice, maize) and optimizing the planting structure play an important role in improving grain production. Grain production of Heilongjiang province is the first in China, which has a close relationship with the adjustment of planting structure.

The scientific and technological supports of sustainable increase of grain production in northeast are improving variety cultivation and application, scientific fertilization, the combination of agronomic and agricultural machinery and constructing the ecological high-yield, high-quality, and high-efficiency cultivation system, which could make the scientific and technological progress become the main power of grain production.

With the progress of science and technology and the development of social economy, the ecological construction in the core grain-producing areas of China has made great progress. However, the problems of water security and ecology security are still serious. Therefore, in the future, it should continue to implement the comprehensive management and protection of ecological environment and strengthen the construction of ecological civilization, to promote the development of grain production in northeast China.

# 4 Ecological Security and Sustainable Development of the Tibetan Plateau

## 4.1 Introduction

The Tibetan Plateau in southwest China spans an area of 2,572,400 $km^2$ along the national border (Zhang et al. 2002) and occupies 26.8 % of China. The average elevation of the plateau is above 4000 m, and approximately 960,000 $km^2$, or 80 % of the Xizang territory, is above 4500 m (Zhong 2008). The natural terrain of the plateau is unique, and its ecological status is very important. However, the eco-environment of the plateau features high elevations and a chilly and arid climate. The eco-environment is very vulnerable, primarily due to the sensitivity to exogenic action (Zhong et al. 2003). The sequence of climate change on the plateau is 10 years behind that of east China (Chang et al. 2005), and the probability ecosystem variation caused by the same exogenic action, and the various eco-environmental problems caused by it, are greater than in other areas. Grassland degradation, land desertification, soil and water loss, and other problems induced by unreasonable human activity have become increasingly serious over the last 30 years. The area of the degraded grassland in Qinghai was 9870 thousand $hm^2$ at the end of the 1990s, occupying 27.1 % of the grassland in this province (Fan 2000); the degraded grassland in Xizang was 48,228 thousand $hm^2$ in 2004, occupying 58.8 % of its grassland (Zhong 2008); and the area of land desertification caused by grassland degradation occupied 17.98 % area of Xizang by 2009 (State Forestry Administration 2011).

The ecological security and social economic development of the plateau are seriously threatened by the effects of global warming and the increasingly violent disturbances of human activity. Hence, it is quite necessary and urgent to study the ecological security and sustainable development of the plateau. The definition, theory, and evaluation method of ecological security have recently been studied primarily overseas (Mathew 1989; Michael 1989; Barnthouse 1992; Hal 1998; Malin 2002). Synthetic evaluation methods have been used multiple times to assess ecological security (Wei et al. 2002). The pressure-state-response model (PSR) was used by some domestic scholars to study ecological security in the local region of the Tibetan Plateau (Zhao et al. 2006; Li et al. 2005). Yang (2003) studied the security of the eco-environment and the sustainable development of grassland in Xizang. Cheng and Shen (2000) examined a strategy of sustainable development of the plateau, with emphasis on the aspects of population, resources, and environment. However, few synthetic studies have been conducted on the ecological security of the Tibetan Plateau, which has an important influence on the ecological security of China.

## 4.2 *Idea and Method of the Research*

### 4.2.1 Logical Relationship Between Ecological Security and Sustainable Development

When the service function of regional ecosystem enters into a benign state, not only is there a need of the ecological material products for humans to be fulfilled, but there is also the eco-environment to be protected; thus, the “natural capital” embodied by land, water, and atmosphere is to be preserved for appreciation and sustainable use in the region. The ecosystem in this area is safe, and its eco-environment and social economy are in a status of sustainable development. Ecological security is the basis of sustainable development.

### 4.2.2 Index System and Method of Synthetic Evaluation of Ecological Security

Index System for Evaluation

The service function of the ecosystem could reflect the degree of security in the regional ecosystem. By analyzing the mutual relationship between the service function of the ecosystem and environmental processing on the plateau, an index system for evaluating ecological security with a system of eco-environment as a subject of research was developed on basis of the PSR model. The choices of the indices are based on the characteristics of the data: scientific basis, simplicity, spatial distribution, and availability. The index system of the synthetic evaluation of the ecological security of the Tibetan Plateau is listed in Table 2.

**Table 2** Index system of the synthetic evaluation of the ecological security of the Tibetan Plateau

| Layer of object | Layer of item | Layer of index | |
|---|---|---|---|
| Synthetic evaluation on ecological security | Degree of ecological vulnerability ($A$) | $A_1$ Degree of ecological stability | $A_{1\text{-}1}$ Degree of stability of ecological matrix |
| | | | $A_{1\text{-}2}$ Ecological kinetic energy of nature |
| | | | $A_{1\text{-}3}$ Ecological kinetic energy of humanity |
| | | $A_2$ Degree of ecological sensitivity | $A_{2\text{-}1}$ Degree of sensitivity of soil erosion |
| | | | $A_{2\text{-}2}$ Degree of sensitivity of land desertification |
| | | | $A_{2\text{-}3}$ Degree of sensitivity of habitat |
| | | | $A_{2\text{-}4}$ Degree of sensitivity of geological hazard |
| | Service function of ecosystem ($B$) | $B_1$ Function of conservation of water resource | |
| | | $B_2$ Function of water and soil conservation | |
| | | $B_3$ Function of control of land desertification | |
| | | $B_4$ Function of biodiversity conservation | |
| | | $B_5$ Function of production | |
| | Ecological risk ($C$) | $C_1$ Natural risk | |
| | | $C_2$ Artificial risk | |

## Weight of the Index and the Model of Synthetic Evaluation

The weight of each index was determined by two completely independent means of evaluation, an expert judgment method and an analytic hierarchy process, and then the coherent assessment results were obtained, as shown in Table 3.

Based on the quantitative expression and analysis of the evaluation index of ecological security, as well as the mutual relationships of these indices and the quantitative expression of the impact on regional ecological security, a model of

**Table 3** Weight of the index of the synthetic evaluation of the ecological security of the Tibetan Plateau

| Layer of item | Layer of index |
|---|---|
| $A = 0.50$ | $A_1 = 0.70$, $A_{1\text{-}1} = 0.30$, $A_{1\text{-}2} = 0.15$, $A_{1\text{-}3} = 0.55$ |
| | $A_2 = 0.30$, $A_{2\text{-}1} = 0.30$, $A_{2\text{-}2} = 0.30$, $A_{2\text{-}3} = 0.25$, $A_{2\text{-}4} = 0.15$ |
| $B = 0.35$ | $B_1 = 0.37$, $B_2 = 0.21$, $B_3 = 0.11$, $B_4 = 0.11$, $B_5 = 0.20$ |
| $C = 0.15$ | $C_1 = 0.50$, $C_2 = 0.50$ |

synthetic evaluation of ecological security was established. The result of the quantitative expression of the ecological security status was obtained through a series of calculations.

### Formation of the Spatial Pattern of the Ecological Security

The ecological security of the Tibetan Plateau was evaluated synthetically according to the index system in Table 1, and the data were quite difficult to collect. The authors studied Xizang for many years because it is the main body of this plateau, accounting for 50 % of the plateau area; thus, the Xizang Plateau was chosen for the synthetic evaluation.

First, based on the layer structure of the aforesaid index system of evaluation on ecological security, the weight and model of the synthetic evaluation of ecological security, the layer of ecological vulnerability, the service function of the ecosystem, and the analysis of the ecological risk of the Xizang Plateau were formed. Then, the superposition of analytic layers was performed by GIS, the results were divided into five grades by natural parting, and a picture of the spatial pattern of ecological security of the plateau was produced. This picture reveals the law of regional differentiation of ecological security of the plateau.

### Functional Zonation and Localization of the Barriers of Ecological Security

The Xizang Plateau, the main body of the Tibetan Plateau, is a topographic barrier in the pattern of environmental geography in eastern China and significantly influences the stability of the climate system in this area. Additionally, the Xizang Plateau is an important region of river sources, which act as significant barriers for the security of the water resource and the eco-environment of China and the whole of East Asia (Zhong 2008). Hence, to preserve the structure of the ecosystem, the ecological processes of the plateau, the capability of the service functions of the ecosystem to satisfy the demands of modern human living and development for future generations, and the ability of the plateau to act as an important shelter for the ecological security of neighboring regions, both conservation and the construction of a barrier for ecological security are quite necessary.

Based on the spatial pattern formation of ecological security in the Xizang Plateau, the combined features of the plateau, the law of spatial distribution of the principal types of ecosystems of the plateau, the pattern of major geomorphology, and the features of the landform types, the barriers for ecological security of the Xizang Plateau were classified into 3 first-grade barrier areas and 13 sub-barrier areas. The ecological function localization and the development direction of each area have been discussed.

### 4.3 Conclusions and Discussion

Based on the results of the spatial pattern of ecological security and functional zonation of the barrier for ecological security, "Planning for conservation and construction of the barrier for ecological security on the Xizang Plateau" were compiled. The planning was rectified by the State Council of China in February 2009, with a huge investment of ¥15.5 billion from the government to be used to implement these projects of conservation and construction. A new perspective of the ecological security and sustainable development on the plateau would display the world through a full-scale implementation of the planning.

Ecological security and sustainable development are correlated but differential. Regarding the Xizang Plateau and its vulnerable ecology, ecological security is closely related to sustainable development and is the basis for sustainable development. This study focused on ecological security and the security of the ecosystem as the core research area of ecological security. Some researchers believe that the symbol of ecosystem security is its health, but a large-scale parameter of health is difficult to obtain; therefore, the evaluation of the ecosystem health was not pursued in this study. It is relatively reasonable to choose an index of evaluation of ecological security from three aspects, the status of ecological vulnerability, the ecological risk, and the service function of ecosystem, which are presented from elements of the eco-environment and the environmental processes; however, the number of layers of indices, the values of the indices, and the grade division of security must be improved.

## 5 Ecological Effects of the "Corridor-Barrier" Functions in the Longitudinal Range-Gorge Region, Southwest China

### 5.1 Introduction

The special landform with a series of longitudinal high mountains and deep gorges in southwestern China, linking southeast Asia, has been given the collective name of Longitudinal Range-Gorge Region (He et al. 2005). The LRGR covers the Yuanjiang-Red (domestic name-foreign name), Lancang-Mekong, Nujiang-Salween, Dulong-Irrawaddy rivers, plus the Yaluzangbu in the west and the upstream portion of the Yangtze in the east. It is regarded as unique landscape unit in the world because of its special topography, plentiful biodiversity, and ethnic culture diversity.

This region possesses the world unique landscape consisting from these mountains and gorges stretching from southwest China to southeast Asia in nearly north-south ways, and is the major spreading corridor and refuge for animal and plant species in Asia continent. The high mountains and deep gorges in the LRGR prevent significantly surface material movement and energy flow between the ranges and gorges in the east-west way but facilitating the movements along the ranges or gorges in the north-south direction. The LRGR landform also causes temperature inversion layers which then form vertical barriers of hydrological cycling in the valleys. These special phenomena are called the "corridor-barrier" functions of moisture–wind–energy cycling (He et al. 2005).

The LRGR landform and its "corridor-barrier" functions distinctly influence the process of forming and becoming of species, ecosystems patterns, and resettlement environments. LRGR houses every ecosystem in the northern hemisphere except desert and ocean types, and is widely acknowledged as the concentrated area for various species and global gene reserve. Different ethnic groups scatter in small basins embraced by big mountains and present vast differences in respect of social and economic development.

The corridor function also makes the region an important ecological, economic passages between China and southeast Asian countries. In China side, over the past half century, the region has served as a resource base for timber and minerals supply to support China's economic development, which led to the regional ecological degeneration. Going into the new century, the multiple proposed and ongoing economic development programs, such as China's Western Development program and its geo-cooperative programs of the Greater Mekong Subregional Economic Cooperation program (GMS), and the China–ASEAN free trade zone ("10 + 1"), further threaten unprecedented levels of regional environmental degradation and eco-risks, which have been driving the transboundary ecosystems to change dramatically and frequently. Present and emerging threats to eco-security have caught tremendous attention worldwide.

## 5.2 *Integrated Ecological Effects of the "Corridor-Barrier" Functions in LRGR*

Based on the academic field investigating, surveying, sampling, testing, modeling, and multidimensional analysis, it was found that the significant integrated ecological effects of the "corridor-barrier" functions in LRGR:

1. Wu and Pan (Wu et al. 2012; Pan et al. 2012) selected surface atmospheric water vapor content, precipitation, aridity/humidity index, and surface runoff as water indices; air temperature and accumulated temperature as temperature indices, and solar radiation as a heat index to study the hydrothermal pattern, regional differentiation of ecosystem structure and function and main influencing factors. The ANUSPLIN model, wavelet analysis, GIS spatial analysis, and landscape pattern analysis were used to reveal the effects of land surface pattern on eco-geographical regional differentiation. They found that the water, temperature, and heat have significant differences along the latitudinal direction with intermittent difference and longitudinal direction with continuous extension, which

reflects the special "corridor-barrier" effect of longitudinal range-gorge terrain, in which the vertical mountain and valleys on the surface of natural material and energy transport mainly showed significant north-south corridor and diffusion effects and the role of east-west barrier and shielding effect. The longitudinal range-gorge terrain also have important impacts on the spatial pattern of vegetation landscape diversity, ecosystem structure and function, which is the main influencing factor affecting the spatial distribution of vegetation landscape diversity and ecosystems pattern. Wavelet variance analysis reflects the spatial anisotropy of environmental factors as well as NDVI and NPP. The wavelet coherence reveals the spatial distribution of NDVI and NPP influencing factors, and also the quantitative degree of control. Conclusions believe that the longitudinal range-gorge land surface pattern is the main influencing factor of the ecogeographical elements. Under the interaction of the zonality law and the nonzonality "corridor-barrier" effects, the spatial differentiation formed in the ecogeographical system (Wu et al. 2012). They further selected surface atmospheric water vapor content, precipitation, aridity/humidity index, and surface runoff as water indices; air temperature and accumulated temperature as temperature indices, and solar radiation as a heat index to study the hydrothermal pattern, regional differentiation of ecosystem structure and function, and main influencing factors. The ANUSPLIN model, wavelet analysis, GIS spatial analysis, and landscape pattern analysis were used to reveal the effects of land surface pattern on ecogeographical regional differentiation. The longitudinal range-gorge terrain also have important impacts on the spatial pattern of vegetation landscape diversity, ecosystem structure and function, which is the main influencing factor affecting the spatial distribution of vegetation landscape diversity and ecosystems pattern. Wavelet variance analysis reflects the spatial anisotropy of environmental factors as well as NDVI and NPP. The wavelet coherence reveals the spatial distribution of NDVI and NPP influencing factors, and also the quantitative degree of control. Conclusions believe that the longitudinal range-gorge land surface pattern is the main influencing factor of the ecogeographical elements. Under the interaction of the zonality law and the nonzonality "corridor-barrier" effects, the spatial differentiation is formed in the ecogeographical system (Pan et al. 2012).

2. Based on the daily rainfall records during 1960–2001 observed at 36 meteorological stations in the Longitudinal Range-Gorge Region (LRGR), the variation and its regional differences of precipitation under the effects of the special corridor-barrier functions in LRGR are systematically analyzed (He et al. 2007b). The results indicate that there exists very closed linkage of the variation and regional differences with the special corridor-barrier function: in most areas of the northern LRGR (>26°N), the general features of precipitation intra-annual allocation (PIA) exhibit "multi-peak pattern" with "peach blossom flood period" (PBFP); the PBFP disappears gradually where the latitude is closed to 26°N, and the PIA patterns change from "multi-peak pattern" to "single-peak pattern"; In the area from 24°N to 25°N latitude belt, the "single-peak pattern" appears again; toward to the southern LRGR where the latitude is lower, the pattern

displays quasi "double-peak pattern" with the characteristic of so-called "autumn rain period", and the larger the longitude is, the more remarkable the pattern will be. In dry season, the annual variations of precipitation vary similarly because the controlling of atmospheric circulation is relative single in LRGR and the influence of the barrier function is not significant on precipitation annual variation, but in wet season, the spatial distribution of precipitation annual variations becomes more complicated.

3. He and Zhang (2007) analyzed the influences of corridor-barrier function on the spatial characteristics of temperature and precipitation in the LRGR, using the monthly observational data of air temperature and precipitation of 1960–2000 in the LRGR, also found that the annual precipitation decreases from the west to the eastern, because the longitudinal range-gorge separates southwest monsoon; there is less precipitation in the north part located in the leeward slope of longitudinal range-gorge; the action of the longitudinal range-gorge on the average air temperature distribution is smaller than that on the annual precipitation; the longitudinal range-gorge separating southwest monsoon is bigger, but separating northeast monsoon is smaller. The annual precipitation of southern and western regions is bigger than the ones of the northern and eastern regions, and the average air temperature of western part is lower than the ones of eastern part.
4. The corridor-barrier function also influences the regional water demand of paddy irrigation in the LRGR: under the influence of "corridor-barrier" function, the regional paddy irrigation water requirement was strongly linked with the latitude, the longitude, and the altitude change in LRGR; in the region of 23.5°N with altitude 1550 m, the correlation coefficient between paddy irrigation water demand and latitude is 0.857, the irrigation ration range from 6240 to 8550 $m^3/hm^2$; and mixed with the effecting of southwest and southeast monsoons in LRGR, the correlation coefficients between the paddy irrigation water demand and the longitude in the East bank and West bank of Hong He (Red river) great canyon are 0.631 and 0.913, respectively, and the irrigation rations are 7260–7440 $m^3/hm^2$ and 5970–7740 $m^3/hm^2$, respectively; the coefficient at 100°E area is 0.636, the ration is 6240–8550 $m^3/hm^2$.
5. Case studies of Wulian Mountain in the middle LRGR show that the significant effects of longititudinal ranging topography on fooding and habitating behaviors of black-crested gibbon (Jiang et al. 2006; Fan and Jiang 2008). Their field surveying on population and group distribution of the black gibbon (*Hylobates concolor*) found that 98 % groups located in the mid-montane range, and groups located in the east side and the southern region more than those in the west and the north (Jiang et al. 2006). Their further research indicated that the total home range size of the black-crested gibbon (*Nomascus nasutus*) is much bigger than that of other gibbon species; the intensity of quadrant use was significantly correlated with the distribution of food patches (Fan and Jiang 2008). The special "corridor-barrier" functions have not only the ecological effect on the behaviors of terrestrial animals but also on the aquatic animals. Ecologists

attributed the extremely high biodiversity and high degrees of endemism of fish to the diverse and spectacular geographical conditions in LRGR.

6. The study of spatial and temporal variations of runoff of Red River Basin in the east LRGR also indicates (Li et al. 2008): Under the "corridor-barrier" functions in LRGR, the spatial patterns of annual precipitation and runoff depth distribution in Red River Basin show out a NW–SE distribution, which is similar with the trend of the Red River valley and Ailao mountains; In the long temporal scale averaged over years, the most obvious effects of the "corridor-barrier" functions is on runoff variation, and the second is on the precipitation, but not obvious on the temperature; under the superposed effect of climate changes and the "corridor-barrier" functions of valleys and mountains in Red River Basin, the difference of runoff variation is obvious in the east-west direction, the runoff variation of Yuan River along the Red River Fault present an ascending trend, for example, but the Lixian River on the west side of the Fault and the Panlong River on the east present a descending trend.

## *5.3 Diverse Transboundary Eco-security Issues and Its Integrated Regulating System in LRGR*

Under the strengthening of global change and globalization of economic integrating, especially as the fast development of geo-cooperation between China and many downstream countries in Asia, the geo-issues related to transboundary ecological security are playing a very important role not only in China's Peaceful Rising but also in Asian sustainability. In LRGR, there are over 4000 km border zone between China, Burma, and Vietnam; all the great rivers, except Yangtze river, are international rivers (four great international rivers) flowing from southwest China to lower Mekong region; the transboundary ecological security is much critical to the regional sustainability.

Based on the synthesis researches of transboundary eco-security (TES), including the types, distribution pattern, characteristics, driving forces, and integrated regulating of transboundary eco-security issues in the LRGR (He et al. 2007a), the results showed: all the transboundary water and eco-security issues along the terrestrial border zones and the international watercourses in China can be found in the LRGR alone, except for transboundary sand–dust storms, such as the uncertain water-level fluctuation, path shifting of migratory fish, flooding, sediment transport change, boundary river bank erosion, and water pollution as the international watercourse change; the bioinvasion, biodiversity loss, soil erosion, landslide, and disease diffusion as the terrestrial ecosystem fragmentation along the border zones.

The TES issues show out high spacial variability and its major drivers mainly are the geo-cooperation, the large-scale hydropower development, the climate change, the huge population increasing, highway and international passage construction, international navigation development, slope land utilization, and mineral exploration. In the international rivers, the major TES issues' influence, mainly caused by cascade dams building, usually diffuses from upstream basins to downstream basins, and the contrary influence along the border zones, driven by slope land use, infrastructure construction, and mineral exploration diffuses from downstream basins to upstream basins, such as species invasion as the transbounary ecosystems fragmentation.

The integrated control systems should be established from the aspects of engineering projects, ecological threshold setting, transboundary preserve development, bioinvasion controlling, eco-security monitoring platform, ecological compensation mechanism, international laws and rules, cooperative mechanisms, and capacity building. At present, the most important task is to strengthen the participation international cooperation mechanism, establish transboundary eco-security risk funds, promote ecological compensation, develop information exchange platform and early warning system in the existing multiple international cooperation regions.

## 5.4 Perspectives

1. The longitudinal range-gorge topographic feature stretching forms the diverse environments which support the local diverse habitats; the "corridor-barrier" functions are the key ecological mechanism for the biodiversity maintaining in the LRGR. The research results offer the new scientific base for the biodiversity conservation and biological resources sustainable utilization. But the interactive mechanisms of the environments, ecosystems, climate, and land uses under the "corridor-barrier" functions in LRGR are still unclear. Further studies should analyze the interaction of special landscape and ecological process and quantitatively assess the multidimensional effects of the "corridor-barrier" functions.
2. The ecohydrological change and its transboundary impacts in LRGR have been hot topics since 1990s when the first mainstream dam (Manwan Dam in Lancang river, China) began to store water and the border trade started between the upper Mekong and the lower Mekong countries. Most of the over 50 big dams, planned in China's 12th Five Year Plan for State Energy Development (the State Council of China announced plans, 2013), will be built in LRGR, the most biologically, geographically, and ethnically diverse region in China. Linking with the "corridor-barrier" functions, the dynamics driving of interaction among dams building, land uses, and change on in the stream flow regime of large basins have much uncertainty. The quantificational assessment on environmental flow, available water, and assignable water in international watercourses, for example, could be the scientific base to make water allocation, water utilization, environmental conservation, and border management, to develop the transboundary environmental compensation mechanism among the riparian countries from whole river basin to geo-cooperative regions. It is critical

for applied ecological study to assess the accumulative effects of cascade megadams on species diversity, habitat location, migration, and ecosystem impact of both large and small hydropower projects. The linkage of the stream flow with the human activities (dams building, slope cultivation, mining, policy making, and geo-cooperation) and climate change will be identified so as to find the major driving factors; and then the change trend and the potential region-wide impact of stream flow change will be assessed. Based on the research, the suggestions for water and ecological security will be recommended for the better management of the region.

3. Under the globalization of climate change and human action influence, the transboundry issues of international rivers are showing the impacts from river basins to regions; the adaptation to these changes will address upstream–downstream relationships and needs a transboundary approach. But in actual reality, influenced by the hydropolitical situation among the riparian countries, it is very difficult to facilitate the upstream–downstream cooperation. In the LRGR, there still exist in a lot of gaps of research and knowledge for transboundary ecological security controlling: lack of models which effectively model the cumulate environmental impact and the ecological function loss of ecosystem change, especially, the aquatic ecosystem fragmentation caused by cascade dams development; no integrated model could identify the different multiple interactions between the dam building, climate change, and land use at the basin-wide level. It is still difficult to quantitatively assess diverse values of the natural flow regime for maintaining the ecosystem integrity of the river. In the international laws, the regional fair principle and the transboundary environmental compensation articles, are missed from their major principles, because the integrity of river basin was seldom emphasized as a coupling ecosystem and the great differences of biophysical environments, social and economic conditions among the co-riparian countries are usually ignored.
4. The researches of theories, methods, techniques, and index systems on the complex transboundary eco-security issues should be enhanced. The interdisciplinary knowledge of applied ecology will be much useful to help scientists and policy-makers to identify the transboundary ecological risks and jointly take the actions for transboundary eco-security. Being different with the states' traditional security issues such as the territorial sovereignty dispute, the transboundary issues of water and eco-security, as a nontraditional security, have their interdisciplinary features from an environmental or ecological point of view and sensitive features because of the ecosystems integrity crossing the border and river flow's transboundary influences. So, it is needed to further integrate the applied ecology with hydrology and water resources, policy and laws, etc.; facilitating the researches of transboundary watercourses, biodiversity, disasters (drought, flood, and pollution); developing the transboundary environmental compensation mechanism; strengthening the decision-making support for the international water laws and customary laws, water quality standards among riparian states.

## References

Barnthouse, L. W. (1992). The role of models in ecological assessment. *Environmental Toxicology and Chemistry, 11*, 1751–1760.

Chang, G., Li, F., & Li, L. (2005). Influence on eco-environment of Qinghai by climate change and its countermeasure. *Advances in Climate Change Research, 1*(4), 172.

Cheng, S., & Shen, L. (2000). Discussion on strategy of sustainable development in area of the Qinghai-Xizang Plateau. *Resource Sciences, 22*(4), 2–11.

Fan, Q. (2000). Present situation of degraded grassland in Qinghai and prevention measures. *Qinghai Prataculture, 9*(1), 22–25.

Fan, P., & Jiang, X. (2008). Effects of food and topography on ranging behavior of black crested gibbon (*Nomascus concolor jingdongensis*) inWuliang Mountain, Yunnan, China. *American Journal of Primatology, 70*, 1–8.

Hal, H. (1998, March/April). Nataral security. *Nuclear Times*, 24–26.

He, D., Li, S., & Zhang, Y. (2007a). The variation and regional differences of precipitation in the Longitudinal Range-Gorge the Region. *Chinese Science Bulletin, 52*(S), 59–73.

He, D., Liu, J., Hu, J., et al. (2007b). Transboundary eco-security and its regulation system in the Longitudinal Range-Gorge Region. *Chinese Science Bulletin, 52*(S), 1–9.

He, D., Wu, S., Peng, H., et al. (2005). A study of ecosystem changes in Longitudinal Range-Gorge region and transboundary eco-security in Southwest China. *Advances in Earth Science, 20*, 338–344.

Jiang, C. (1998). Mobilize social force to speed up ecological environment construction pace in our country. *Chinese Journal of Eco-agriculture, 6*(1), 1–6.

Jiang, X., Luo, Z., Zhao, S., et al. (2006). Status and distribution pattern of black crested gibbon (Nomascus concolor jingdon gensis) in Wuliang Mountains, Yunnan, China: implication for conservation. *Primates, 47*, 264–271.

Li, P., Cao, G., et al. (1999). Strengthen science and technology action, propel a great undertaking of reconstruction of landscape in northwest area. In *Academic Seminar Proceedings on Ecological Restoration and Coordinated Economy Development in Northwest of China*. Chengdu: Sichuan Science and Technology Press.

Li, P., Du, G., & Wei, J. (2007). *Reconstruction of landscape strategic research and demonstration test in northwest area of China*. Beijing: Science Press.

Li, Y., He, D., & Ye, C. (2008). Spatial and temporal variation of runoff of Red River Basin in Yunnan. *Journal of Geographical Sciences, 18*(03), 308–318.

Li, P., Li, Q., & Shi, G. (2006). *Ecological environment construction and reconstruction of landscape strategic research in Shaanxi*. Xi'an: Shaanxi People's Publishing House.

Li, S., Zhao, Y., & Shi, P. (2005). Method and application of evaluation on ecological security the Xizang Plateau: A case study of Qusum County, Tibet Autonomous Region. *Research of Soil and Water Conservation, 12*(6), 142–190.

Malin, F. (2002). Human livelihood security versus ecological security—an ecohydrological perspective. In *Proceedings, SIWI Seminar, Balancing Human Security Interests in Catchment* (pp. 29–36).

Mathew, S. J. T. (1989). Redefining security. *Foreign Affairs, 68*, 162–177.

Michael, R. (1989). National security: The economic environmental dimensions. World watch paper, no. 89.

Pan, T., Wu, S., He, D., Dai, E., & Liu, Y. (2012). Ecological effects of longitudinal range-gorge land surface pattern and its regional differentiation. *Acta Geographica Sinica, 67*(1), 13–26.

Shaohong, Wu, Tao, Pan, Jie, Cao, Daming, He, & Ziniu, Xiao. (2012). Barrier-corridor effect of longitudinal range-gorge terrain on monsoons in Southwest China. *Geographical Research, 31*(1), 1–13.

State Forestry Administration. (2011). *The fourth communiqué of land desertification of China*. Beijing: China Forestry Network.

Wang, Z. L., Xu, Y. Q., & Wang, G. (1998). Germination strategies of annual sandy plants under limited precipitation. *Journal of Lanzhou University (Natural Sciences), 34*(2), 98–103.
Yang, R. (2003). Study on security of eco-environment and issues of sustainable development of grassland in Tibet Autonomous Region. *Prataculturae Sinica, 12*(6), 24–29.
Yunling, H. E., & Zhang, Y. (2007). Influences of corridor-barrier function on the spatial characteristies of temperature and precipitation in the Longitudinal Range-Gorge Region (LRGR). *Journal of Mountain Science, 25*(1), 169–176.
Zhang, K., Feng, Q., YQ, L. Ü., Zhang, B., & Si, J. H. (2011). Study on spatial heterogeneity of soil water contents in oasis-desert belt of Minqin. *Journal of Desert Research, 31*(5), 1149–1155.
Zhang, J. G., Li, X. R., Wang, X. P., Wang, G., Li, J. G., & Wang, Z. L. (2001). Population dynamic of annual plant *Eragrostis poaeoides* in fixed sand dune in Shapotou Area. *Journal of Desert Research, 21*(3), 232–235.
Zhang, Y., Li, B-y, & Zheng, D. (2002). Study on scope and area of the Qinghai-Xizang Plateau. *Geographical Research, 21*(1), 1–8.
Zhao, Y., et al. (2006). Assessing the ecological security of the Tibet Plateau: Methodology and a case study for Lhaze County. *Journal of Environment Management, 80*(2), 120–131.
Zhong, X. (2008). *Ecological security of the Xizang Plateau*. Beijing: Science Press.
Zhong, X., Liu, S., Wang, X., et al. (2003). Vulnerability of eco-environment and strategy of ecological security in Tibet. *Journal of Mountain Research, S1*, 1–6.
Zuo, W., Wang, Q., Wang, W., et al. (2002). Study on index and standard of evaluation on regional ecological security. *Geography and Territory Study, 18*(1), 37–41.

# Index





W. Li (ed.), *Contemporary Ecology Research in China*,
DOI 10.1007/978-3-662-48376-3

Kolibri 360 GmbH
Ferdinand-Jühlke-Straße 7
99095 Erfurt, Deutschland
produktsicherheit@kolibri360.de